To Camilla

What good is it for a man
to gain the whole world,
yet forfeit his life?
Mc. 8, 36

B&SS – Bocconi & Springer Series

Andrea Pascucci

PDE and Martingale Methods in Option Pricing

BOCCONI
UNIVERSITY
PRESS

Springer

Andrea Pascucci
Department of Mathematics
University of Bologna
andrea.pascucci@unibo.it

B&SS – Bocconi & Springer Series
ISSN print edition: 2039-1471 ISSN electronic edition: 2039-148X

ISBN 978-88-470-1780-1 e-ISBN 978-88-470-1781-8

DOI 10.1007/978-88-470-1781-8

Library of Congress Control Number: 2010929483

Springer Milan Dordrecht Heidelberg London New York

9 8 7 6 5 4 3 2 1

Cover-Design: K Design, Heidelberg

Typesetting with LaTeX: PTP-Berlin, Protago TeX-Production GmbH, Germany (www.ptp-berlin.eu)
Printing and Binding: Grafiche Porpora, Segrate (Mi)

Printed in Italy

Springer-Verlag Italia srl – Via Decembrio 28 – 20137 Milano
Springer is a part of Springer Science+Business Media (www.springer.com)

Preface

This book gives an introduction to the mathematical, probabilistic and numerical methods used in the modern theory of option pricing. It is intended as a textbook for graduate and advanced undergraduate students, but I hope it will be useful also for researchers and professionals in the financial industry.

Stochastic calculus and its applications to the arbitrage pricing of financial derivatives form the main theme. In presenting these, by now classic, topics, the emphasis is put on the more quantitative rather than economic aspects. Being aware that the literature in this field is huge, I mention the following incomplete list of monographs whose contents overlap with those of this text: in alphabetic order, Avellaneda and Laurence [14], Benth [43], Björk [47], Dana and Jeanblanc [84], Dewynne, Howison and Wilmott [340], Dothan [100], Duffie [102], Elliott and Kopp [120], Epps [121], Follmer and Schied [134], Glasserman [158], Huang and Litzenberger [171], Ingersoll [178], Karatzas [200; 202], Lamberton and Lapeyre [226], Lipton [239], Merton [252], Musiela and Rutkowski [261], Neftci [264], Shreve [310; 311], Steele [315], Zhu, Wu and Chern [349].

What distinguishes this book from others is the attempt to present the matter by giving equal weight to the probabilistic point of view, based on the martingale theory, and the analytical one, based on partial differential equations. The present book does not claim to describe the latest developments in mathematical finance: that target would indeed be very ambitious, given the speed of progress of research in the field. Instead, I have chosen to develop some of the essential ideas of the classical pricing theory to devote space to the fundamental mathematical and numerical tools when they arise. Thus I hope to provide a sound background of basic knowledge which may facilitate the independent study of newer problems and more advanced models.

The theory of stochastic calculus, for continuous and discontinuous processes, constitutes the bulk of the book: Chapters 3 on stochastic processes, 4 on Brownian integration and 9 on stochastic differential equations may form the material for an introductory course on stochastic calculus. In these chapters, I have constantly sought to combine the theoretical concepts to the insight on the financial meaning, in order to make the presentation less abstract and more motivated: in fact many theoretical concepts naturally lend themselves to an intuitive and meaningful economic interpretation.

The origin of this book can be traced to courses on option pricing which I taught at the master program in Quantitative Finance of the University of Bologna, which I have directed with Sergio Polidoro since its beginning, in 2004. I wrote the first version as lecture notes for my courses. During these years, I substantially improved and extended the text with the inclusion of sections on numerical methods and the addition of completely new chapters on stochastic calculus for jump processes and Fourier methods. Nevertheless, during these years the original structure of the book remained essentially unchanged.

I am grateful to many people for the suggestions and helpful comments with which supported and encouraged the writing of the book: in particular I would like to thank several colleagues and PhD students for many valuable suggestions on the manuscript, including David Applebaum, Francesco Caravenna, Alessandra Cretarola, Marco Di Francesco, Piero Foscari, Paolo Foschi, Ermanno Lanconelli, Antonio Mura, Cornelis Oosterlee, Sergio Polidoro, Valentina Prezioso, Enrico Priola, Wolfgang Runggaldier, Tiziano Vargiolu, Valeria Volpe. I also express my thanks to Rossella Agliardi, co-author of Chapter 13, and to Matteo Camaggi for helping me in the translation of the book.

It is greatly appreciated if readers could forward any errors, misprints or suggested improvements to: `andrea.pascucci@unibo.it`
Corrections received after publication will be posted on the website:
`http://www.dm.unibo.it/~pascucci/`

Bologna, November 2010 *Andrea Pascucci*

Contents

General notations

- $\mathbb{N} = \{1, 2, 3, \dots\}$ is the set of natural numbers
- $\mathbb{N}_0 = \{0, 1, 2, 3, \dots\}$ is the set of non-negative integers
- $\mathbb{Q}$ is the set of rational numbers
- $\mathbb{R}$ is the set of real numbers
- $\mathbb{R}_{>0} =]0, +\infty[$
- $\mathbb{R}_{\geq 0} = [0, +\infty[$
- $\mathcal{S}_T =]0, T[\times \mathbb{R}^N$ is a strip in $\mathbb{R}^{N+1}$
- $\mathscr{B} = \mathscr{B}(\mathbb{R}^N)$ is the Borel σ-algebra in $\mathbb{R}^N$
- $|H|$ or $m(H)$ denote the Lebesgue measure of $H \in \mathscr{B}$
- $\mathbb{1}_H$ is the indicator function of H, p. 606
- $\partial_x = \frac{\partial}{\partial x}$ is the partial derivative with respect to x

For any $a, b \in \mathbb{R}$,

- $a \wedge b = \min\{a, b\}$
- $a \vee b = \max\{a, b\}$
- $a^+ = \max\{a, 0\}$
- $a^- = \max\{-a, 0\}$

For any $N \times d$-matrix $A = (a_{ij})$,

- A^* is the transpose of A
- $\mathrm{tr}A$ is the trace of A
- $\mathrm{rank}A$ is the rank of A
- $|A| = \sqrt{\sum\limits_{i=1}^{N} \sum\limits_{j=1}^{d} a_{ij}^2}$
- $\|A\| = \sup\limits_{|x|=1} |Ax|$

Note that $\|A\| \leq |A|$. The point $x \in \mathbb{R}^N$ is identified with a column vector $N \times 1$ and

$$x^* y = \langle x, y \rangle = x \cdot y = \sum_{i=1}^{N} x_i y_i$$

denotes the Euclidean scalar product in $\mathbb{R}^N$.

Depending on the context, $\mathcal{F}$ denotes the Fourier transform or the σ-algebra of a probability space. The Fourier transform of a function f is denoted by $\hat{f}$.

Shortenings

- $A := B$ means that "by definition, A equals B"
- r.v. = random variable
- s.p. = stochastic process
- a.s. = almost surely
- a.e. = almost everywhere
- i.i.d. = independent and identically distributed (referred to random variables)
- mg = martingale
- PDE = Partial Differential Equation
- SDE = Stochastic Differential Equation

Function spaces

- $m\mathscr{B}$: space of $\mathscr{B}$-measurable functions, p. 608
- $m\mathscr{B}_b$: space of bounded functions in $m\mathscr{B}$, p. 608
- BV: space of functions with bounded variation, p. 127
- Lip: space of Lipschitz continuous functions, p. 679
- $\mathrm{Lip}_{\mathrm{loc}}$: space of locally Lipschitz continuous functions, p. 679
- C^k: space of functions with continuous derivatives up to order $k \in \mathbb{N}_0$
- C_b^k: space of functions in C^k bounded together with their derivatives
- $C^{k+\alpha}$: space of functions differentiable up to order $k \in \mathbb{N}_0$ with partial derivatives that are Hölder continuous of exponent $\alpha \in]0,1[$
- $C_{\mathrm{loc}}^{k+\alpha}$: space of functions differentiable up to order $k \in \mathbb{N}_0$ with partial derivatives that are locally Hölder continuous of exponent $\alpha \in]0,1[$
- C_0^∞: space of test functions, i.e. smooth functions with compact support, p. 678
- $C^{1,2}$: space of functions $u = u(t,x)$ with continuous second order derivatives in the "spatial" variable $x \in \mathbb{R}^N$ and continuous first order derivative in the "time" variable t, p. 631
- C_P^α: space of parabolic Hölder continuous functions of exponent α, p. 258
- L^p: space of functions integrable of order p
- L_{loc}^p: space of functions locally integrable of order p
- $W^{k,p}$: Sobolev space of functions with weak derivatives up to order k in L^p, p. 679
- S^p: parabolic Sobolev space of functions with weak second order derivatives in L^p, p. 265

Spaces of processes

- $\mathbb{L}^p$: space of progressively measurable processes in $L^p([0,T]\times\Omega)$, p. 141
- $\mathbb{L}^p_{\text{loc}}$: space of progressively measurable processes X such that $X(\omega)\in L^p_{\text{loc}}([0,T])$ for almost any ω, p. 159
- $\mathcal{A}_{\text{c}}$: space of continuous processes $(X_t)_{t\in[0,T]}$, $\mathcal{F}_t$-adapted and such that

$$[\![X]\!]_T=\sqrt{E\left[\sup_{0\le t\le T}X_t^2\right]}$$

  is finite, p. 280
- $\mathscr{M}^2$: linear space of right continuous martingales $(M_t)_{t\in[0,T]}$ such that $M_0=0$ a.s. and $E\left[M_T^2\right]$ is finite, p. 115
- $\mathscr{M}^2_{\text{c}}$: linear subspace of the continuous martingales of $\mathscr{M}^2$, p. 115
- $\mathscr{M}_{\text{c,loc}}$: space of continuous local martingales M such that $M_0=0$ a.s., p. 161

1

Derivatives and arbitrage pricing

A financial derivative is a contract whose value depends on one or more securities or assets, called underlying assets. Typically the underlying asset is a stock, a bond, a currency exchange rate or the quotation of commodities such as gold, oil or wheat.

1.1 Options

An option is the simplest example of a derivative instrument. An option is a contract that gives the right (but not the obligation) to its holder to buy or sell some amount of the underlying asset at a future date, for a prespecified price. Therefore in an option contract we need to specify:

- an underlying asset;
- an exercise price K, the so-called *strike price*;
- a date T, the so-called *maturity*.

A *Call* option gives the right to buy, whilst a *Put* option gives the right to sell. An option is called *European* if the right to buy or sell can be exercised only at maturity, and it is called *American* if it can be exercised at any time before maturity.

Let us consider a European Call option with strike K, maturity T and let us denote the price of the underlying asset at maturity by S_T. At time T we have two possibilities (cf. Figure 1.1): if $S_T > K$, the payoff of the option is equal to $S_T - K$, corresponding to the profit obtained by exercising the option (i.e. by buying the underlying asset at price K and then selling it at the market price S_T). If $S_T < K$, exercising the option is not profitable and the payoff is zero. In conclusion the payoff of a European Call option is

$$(S_T - K)^+ = \max\{S_T - K, 0\}.$$

Figure 1.2 represents the graph of the payoff as a function of S_T: notice that the payoff increases with S_T and gives a potentially unlimited profit. Analo-

Pascucci A.: PDE and Martingale Methods in Option Pricing

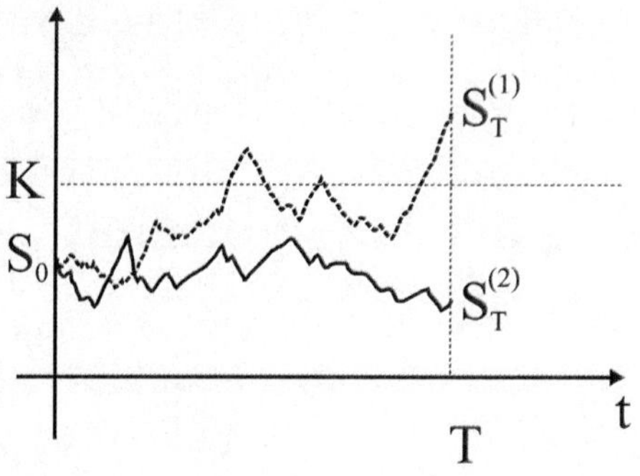

Fig. 1.1. Different scenarios for a European Call option

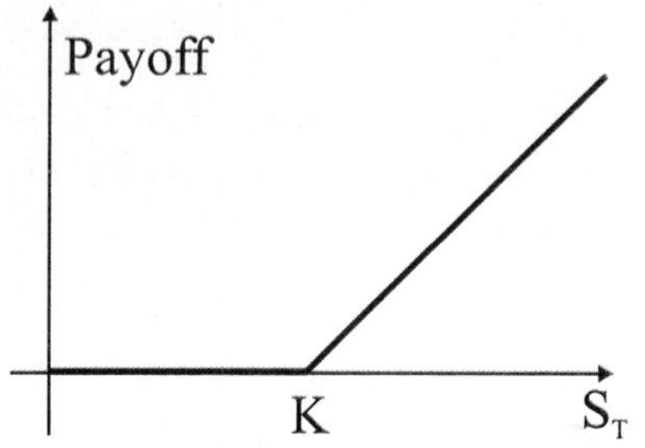

Fig. 1.2. Payoff of a European Call option

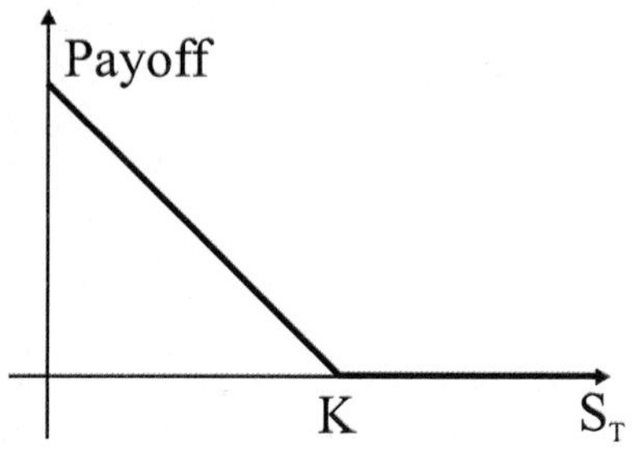

Fig. 1.3. Payoff of a European Put option

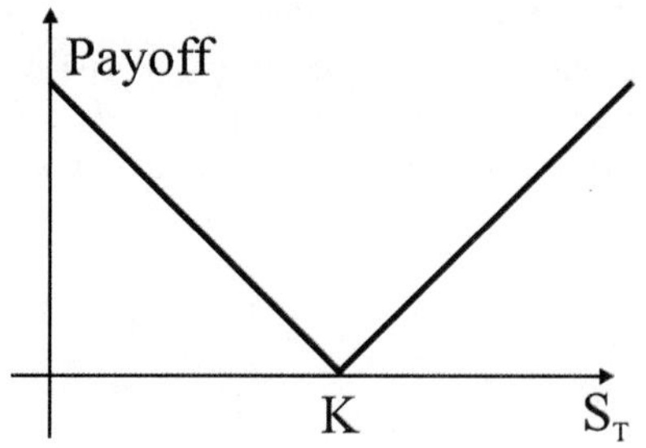

Fig. 1.4. Payoff of a Straddle

gously, we see that the payoff of a European Put option is

$$(K - S_T)^+ = \max\{K - S_T, 0\}.$$

Call and Put options are the basic derivative instruments and for this reason they are often called *plain vanilla* options. Combining such types of options it is possible to build new derivatives: for example, by buying a Call and a Put option with the same underlying asset, strike and maturity we obtain a derivative, the so-called Straddle, whose payoff increases the more S_T is far from the strike. This kind of derivative is interesting when one expects a wide movement of the price of the underlying asset without being able to foresee the direction. Evidently the pricing of this option can be reformulated in terms of the pricing of plain vanilla options. On the other hand, in the real-world markets there exists a great deal of derivatives (usually called *exotic*) having very complicated structures: the market of such derivatives is in continuous expansion and development. One can consult, for example, Zhang [344] for an encyclopedic exposition of exotic derivatives.

1.1.1 Main purposes

The use of derivatives serves mainly two purposes:

- hedging the risk;
- speculation.

For example, let us consider an investor holding the stock S: buying a Put option on S, the investor gets the right to sell S in the future at the strike price and therefore he/she hedges the risk of a crash of the price of S. Analogously, a firm using oil in its business might purchase a Call option to have the right to buy oil in the future at the fixed strike price: in this way the firm hedges the risk of a rise of the price of oil.

In recent years the use of derivatives has become widespread: not long ago a home loan was available only with fixed or variable rate, while now the offer is definitely wider. For example, it is not hard to find "protected" loans with capped variable rate: this kind of structured products contains one or more derivative instruments and pricing such objects is not really straightforward.

Derivatives can be used to speculate as well: for instance, buying Put options is the simplest way to get a profit in case of a market crash. We also remark that options have a so-called *leverage effect*: relatively minor movements in stock price can result in a huge change of the option price. For example, let us denote by S_0 the current price of the underlying asset and let us suppose that \$1 is the price of a Call option with $K = S_0 = \$10$ and maturity one year. We suppose that, at maturity, $S_T = \$13$: if we buy one unit of the underlying asset, i.e. we invest \$10, we would have a \$3 profit (i.e. 30%); if we buy a Call option, i.e. we invest only \$1, we would have a \$2 profit (i.e. 200%). On the other hand, we must also bear in mind that, if $S_T = \$10$, by investing in the Call option we would lose all our money!

1.1.2 Main problems

An option is a contract whose final value is given, this depending on the price of the underlying asset at maturity which is not known at present. Therefore the non-trivial problem of *pricing* arises, i.e. the determination of the "rational" or fair price of the option: this price is the *premium* that the buyer of the option has to pay at the initial time to get the right guaranteed by the contract.

The second problem is that of *hedging*: we have already pointed out that a Call option has a potentially unlimited payoff and consequently the institution that *sells* a Call option exposes itself to the risk of a potentially unlimited loss. A bank selling a derivative faces therefore the problem of finding an investment strategy that, by using the premium (i.e. the money received when the derivative was sold), can replicate the payoff at maturity, whatever the final value of the underlying asset will be. As we are going to see shortly, the problems of pricing and hedging are deeply connected.

1.1.3 Rules of compounding

Before going any further, it is good to recall some notions on the *time value of money* in finance: receiving \$1 today is not like receiving it after a month. We point out also that it is common practice to consider as the unit of time one year and so, for example, $T = 0.5$ corresponds to six months.

The rules of compounding express the dynamics of an investment with fixed risk-free interest rate: to put it simply, this corresponds to deposit the money on a savings account. In the financial modeling, it is always assumed that a (locally[1]) risk-free asset, the so-called *bond*, exists. If B_t denotes the value of the bond at time $t \in [0,T]$, the following rule of simple compounding with annual interest rate r

$$B_T = B_0(1 + rT),$$

states that the final value B_T is equal to the initial value B_0 plus the interest $B_0 rT$, corresponding to the interest over the period $[0,T]$ accrued on the initial wealth. Therefore, by the rule of simple compounding, the interest is only paid on the initial wealth.

Alternatively we may consider the period $[0,T]$, divide it into N sub-intervals $[t_{n-1}, t_n]$ whose common length is $\frac{T}{N}$ and assume that the simple interest is paid at the end of every sub-interval: we get

$$B_T = B_{t_{N-1}}\left(1 + r\frac{T}{N}\right) = B_{t_{N-2}}\left(1 + r\frac{T}{N}\right)^2 = \cdots = B_0\left(1 + r\frac{T}{N}\right)^N.$$

[1] This means that the official interest rate is fixed and risk-free over a brief period of time (e.g. some weeks) but in the long term it is random as well.

By taking the limit as $N \to \infty$, i.e. by assuming that the simple interest is paid more and more frequently, we obtain the formula of continuous compounding with annual interest rate r:

$$B_T = B_0 \, e^{rT}. \tag{1.1}$$

Formula (1.1) expresses the final wealth in terms of the initial investment. Conversely, since to obtain a final wealth (at time T) equal to B, it is necessary to invest the amount Be^{-rT} at the initial time, this amount is usually called *discounted value* of B.

While the rule of simple compounding is the one used in the market, the rule of continuous compounding is generally used in theoretical contexts and particularly in continuous-time models.

1.1.4 Arbitrage opportunities and Put-Call parity formula

Broadly speaking an arbitrage opportunity is the possibility of carrying out a financial operation without any investment, but leading to profit without any risk of a loss. In real-world markets arbitrage opportunities do exist, even though their life span is very brief: as soon as they arise, the market will reach a new equilibrium because of the actions of those who succeed in exploiting such opportunities. From a theoretical point of view it is evident that a sensible market model must avoid this type of profit. As a matter of fact, the no-arbitrage principle has become one of the main criteria to price financial derivatives.

The idea on which arbitrage pricing is built is that, if two financial instruments will *certainly* have the same value[2] at future date, then also in this moment they must have the same value. If this were not the case, an obvious arbitrage opportunity would arise: by selling the instrument that is more expensive and by buying the less expensive one, we would have an immediate risk-free profit since the selling position *(short position)* on the more more expensive asset is going to cancel out the buying position *(long position)* on the cheaper asset. Concisely, we can express the no-arbitrage principle in the following way:

$$X_T \le Y_T \qquad \Longrightarrow \qquad X_t \le Y_t, \quad t \le T, \tag{1.2}$$

where X_t and Y_t are the values of the two financial instruments respectively. From (1.2) in particular it follows that

$$X_T = Y_T \qquad \Longrightarrow \qquad X_t = Y_t, \quad t \le T. \tag{1.3}$$

Now let us consider a financial-market model that is free from arbitrage opportunities and consists of a bond and a stock S, that is the underlying asset

[2] We note that we need not know the future values of the two financial instruments, but only that they will certainly be equal.

of a Call option c and of a Put option p, both of European type with maturity T and strike K:

$$c_T = (S_T - K)^+, \qquad p_T = (K - S_T)^+.$$

We denote by r the risk-free interest rate and we assume that the bond follows the dynamics given by (1.1). On the basis of arbitrage arguments, we get the classical *Put-Call parity formula*, which establishes a relation between the prices c and p, and some upper and lower estimates for such prices. It is remarkable that the following formulas are "universal", i.e. independent of the market model and based only on the general no-arbitrage principle.

Corollary 1.1 (Put-Call parity) *Under the previous assumptions, we have*

$$c_t = p_t + S_t - Ke^{-r(T-t)}, \qquad t \in [0, T]. \tag{1.4}$$

Proof. It suffices to note that the investments

$$X_t = c_t + \frac{K}{B_T} B_t \quad \text{and} \quad Y_t = p_t + S_t,$$

have the same final value

$$X_T = Y_T = \max\{K, S_T\}.$$

The claim follows from (1.3). □

If the underlying asset pays a dividend D at a date between t and T, the Put-Call parity formula becomes

$$c_t = p_t + S_t - D - Ke^{-r(T-t)}.$$

Corollary 1.2 (Estimates from above and below for European options) *For every $t \in [0, T]$ we have*

$$\begin{aligned} \left(S_t - Ke^{-r(T-t)}\right)^+ &< c_t < S_t, \\ \left(Ke^{-r(T-t)} - S_t\right)^+ &< p_t < Ke^{-r(T-t)}. \end{aligned} \tag{1.5}$$

Proof. By (1.2)

$$c_t, p_t > 0. \tag{1.6}$$

Consequently by (1.4) we get

$$c_t > S_t - Ke^{-r(T-t)}.$$

Moreover, since $c_t > 0$, we get the first estimate from below. Finally $c_T < S_T$ and so by (1.2) we get the first estimate from above. The second estimate can be proved analogously and it is left as an exercise. □

1.2 Risk-neutral price and arbitrage pricing

In order to illustrate the fundamental ideas of derivative pricing by arbitrage arguments, it is useful to examine a simplified model in which we consider only two moments in time, the initial date 0 and the maturity T. As usual we assume that there exists a bond with risk-free rate r and initial value $B_0 = 1$. Further, we assume that there is a risky asset S whose final value depends on some random event: to consider the simplest possible model, we assume that the event can assume only two possible states E_1 and E_2 in which S_T takes the values S^+ and S^- respectively. To fix the ideas, let us consider the outcome of a throw of a die and let us put, for example,

$$E_1 = \{1, 2, 3, 4\}, \qquad E_2 = \{5, 6\}.$$

In this case S represents a bet on the outcome of a throw of a die: if we get a number between 1 and 4 the bet pays S^+, otherwise it pays S^-. The model can be summarized by the following table:

Time	0	T
Bond	1	e^{rT}
Risky asset	?	$S_T = \begin{cases} S^+ & \text{if } E_1, \\ S^- & \text{if } E_2. \end{cases}$

The problem is to determine the value S_0, i.e. the price of the bet.

1.2.1 Risk-neutral price

The first approach is to assign a probability to the events:

$$P(E_1) = p \quad \text{and} \quad P(E_2) = 1 - p, \tag{1.7}$$

where $p \in \,]0, 1[$. For example, if we roll a die it seems natural to set $p = \frac{4}{6}$. In this way we can have an estimate of the final average value of the bet

$$S_T = pS^+ + (1 - p)S^-.$$

By discounting that value at the present time, we get the so-called *risk-neutral price*:

$$\widetilde{S}_0 = e^{-rT} \left(pS^+ + (1 - p)S^- \right). \tag{1.8}$$

This price expresses the value that a risk-neutral investor assigns to the risky asset (i.e. the bet): indeed the current price is equal to the future discounted expected profit. On the basis of this pricing rule (that depends on the probability p of the event E_1), the investor is neither inclined nor adverse to buy the asset.

1.2.2 Risk-neutral probability

Let us suppose now that S_0 is the price given by the market and therefore it is a known quantity. The fact that S_0 is observable gives information on the random event that we are considering. Indeed by imposing that $S_0 = \widetilde{S}_0$, i.e. that the risk-neutral pricing formula holds with respect to some probability defined in terms of $q \in]0, 1[$ as in (1.7), we have

$$S_0 = e^{-rT} \left(qS^+ + (1-q)S^- \right),$$

whence we get

$$q = \frac{e^{rT} S_0 - S^-}{S^+ - S^-}, \qquad 1 - q = \frac{S^+ - e^{rT} S_0}{S^+ - S^-}. \tag{1.9}$$

Evidently $q \in]0, 1[$ if and only if

$$S^- < e^{rT} S_0 < S^+,$$

and, on the other hand, if this were not the case, obvious arbitrage opportunities would arise. The probability defined in (1.9) is called *risk-neutral probability and it represents the unique probability to be assigned to the events* E_1, E_2 *so that* S_0 *is a risk-neutral price.*

Therefore, in this simple setting there exists a bijection between prices and risk-neutral probabilities: by calculating the probabilities of the events, we determine a "rational" price for the risky asset; conversely, given a market price, there exists a unique probability of events that is consistent with the observed price.

1.2.3 Arbitrage price

Let us suppose now that there are two risky assets S and C, both depending on the same random event:

Time	0	T
Bond	1	e^{rT}
Risky asset S	S_0	$S_T = \begin{cases} S^+ & \text{if } E_1, \\ S^- & \text{if } E_2, \end{cases}$
Risky asset C	?	$C_T = \begin{cases} C^+ & \text{if } E_1, \\ C^- & \text{if } E_2. \end{cases}$

To fix the ideas, we can think of C as an option with underlying the risky asset S. If the price S_0 is quoted by the market, we can infer the corresponding risk-neutral probability q defined as in (1.9) and then find the neutral-risk price of C under the probability q:

$$\widetilde{C}_0 = e^{-rT}\left(qC^+ + (1-q)C^-\right). \tag{1.10}$$

This pricing procedure seems reasonable and consistent with the market price of the underlying asset. We emphasize the fact that *the price* $\widetilde{C}_0$ *in* (1.10) *does not depend on a subjective estimation of the probabilities of the events* E_1, E_2, *but it is implicitly contained in the quoted market value of the underlying asset.* In particular this pricing method *does not require to estimate in advance the probability of random events.* We say that $\widetilde{C}_0$ is the *risk-neutral price* of the derivative C.

An alternative approach is based upon the assumption of absence of arbitrage opportunities. We recall that the two main problems of the theory and practice of derivatives are pricing and hedging. Let us suppose to be able to determine an investment strategy on the riskless asset and on the risky asset S replicating the payoff of C. If we denote the value of this strategy by V, the replication condition is

$$V_T = C_T. \tag{1.11}$$

From the no-arbitrage condition (1.3) it follows that

$$C_0 = V_0$$

is the only price guaranteeing the absence of arbitrage opportunities. In other terms, in order to price correctly (without giving rise to arbitrage opportunities) a financial instrument, it suffices to *determine an investment strategy with the same final value* (payoff): by definition, the *arbitrage price* of the financial instrument is the current value of the replicating strategy. This price can be interpreted also as the premium that the bank receives by selling the derivative and this amount coincides with the wealth to be invested in the replicating portfolio.

Now let us show how to construct a replicating strategy for our simple model. We consider a portfolio which consists in holding a number α of shares of the risky asset and a number β of bonds. The value of such a portfolio is given by

$$V = \alpha S + \beta B.$$

By imposing the replicating condition (1.11) we have

$$\begin{cases} \alpha S^+ + \beta e^{rT} = C^+ & \text{if } E_1, \\ \alpha S^- + \beta e^{rT} = C^- & \text{if } E_2, \end{cases}$$

which is a linear system, with a unique solution under the assumption $S^+ \neq S^-$. The solution of the system is

$$\alpha = \frac{C^+ - C^-}{S^+ - S^-}, \qquad \beta = e^{-rT}\frac{S^+C^- - C^+S^-}{S^+ - S^-};$$

therefore the arbitrage price is equal to

$$
\begin{aligned}
C_0 = \alpha S_0 + \beta &= S_0 \frac{C^+ - C^-}{S^+ - S^-} + e^{-rT} \frac{S^+ C^- - C^+ S^-}{S^+ - S^-} \\
&= e^{-rT} \left(C^+ \frac{e^{rT} S_0 - S^-}{S^+ - S^-} + C^- \frac{S^+ - e^{rT} S_0}{S^+ - S^-} \right) =
\end{aligned}
$$

(recalling the expression (1.9) of the risk-neutral probability)

$$
= e^{-rT} \left(C^+ q + C^- (1 - q) \right) = \widetilde{C}_0,
$$

where $\widetilde{C}_0$ is the risk-neutral price in (1.10). The results obtained so far can be expressed in this way: *in an arbitrage-free and complete market (i.e. in which every financial instrument is replicable) the arbitrage price and the risk-neutral price coincide: they are determined by the quoted price S_0, observable on the market.*

In particular *the arbitrage price does not depend on the subjective estimation of the probability p of the event E_1*. Intuitively, the choice of p is bound to the subjective vision on the future behaviour of the risky asset: the fact of choosing p equal to 50% or 99% is due to different estimations on the events E_1, E_2. As we have seen, different choices of p determine different prices for S and C on the basis of formula (1.8) of risk-neutral valuation. Nevertheless, the only choice of p that is consistent with the market price S_0 is that corresponding to $p = q$ in (1.9). Such a choice is also the only one that avoids the introduction of arbitrage opportunities.

1.2.4 A generalization of the Put-Call parity

Let us consider again a market with two risky assets S and C, but S_0 and C_0 are not quoted:

Time	0	T
Riskless asset	1	e^{rT}
Risky asset S	?	$S_T = \begin{cases} S^+ & \text{if } E_1, \\ S^- & \text{if } E_2, \end{cases}$
Risky asset C	?	$C_T = \begin{cases} C^+ & \text{if } E_1, \\ C^- & \text{if } E_2. \end{cases}$

We consider an investment on the two risky assets

$$
V = \alpha S + \beta C
$$

and we impose that it replicates at maturity the riskless asset, $V_T = e^{rT}$:

$$\begin{cases} \alpha S^+ + \beta C^+ = e^{rT} & \text{if } E_1, \\ \alpha S^- + \beta C^- = e^{rT} & \text{if } E_2. \end{cases}$$

As we have seen earlier, we obtain a linear system that has a unique solution (provided that C and S do not coincide):

$$\bar{\alpha} = e^{rT} \frac{C^+ - C^-}{C^+ S^- - C^- S^+}, \qquad \bar{\beta} = -e^{rT} \frac{S^+ - S^-}{C^+ S^- - C^- S^+}.$$

By the no-arbitrage condition (1.3), we must have $V_0 = 1$ i.e.

$$\bar{\alpha} S_0 + \bar{\beta} C_0 = 1. \tag{1.12}$$

Condition (1.12) gives a relation between the prices of the two risky assets that must hold in order not to introduce arbitrage opportunities. For fixed S_0, the price C_0 is uniquely determined by (1.12), in line with the results of the previous section. This fact must not come as a surprise: since the two assets "depend" on the same random phenomenon, the relative prices must move consistently.

Formula (1.12) also suggests that *the pricing of a derivative does not necessarily require that the underlying asset is quoted, since we can price a derivative using the quoted price of another derivative on the same underlying asset.* A particular case of (1.12) is the Put-Call parity formula expressing the link between the price of a Call and a Put option on the same underlying asset.

1.2.5 Incomplete markets

Let us go back to the example of die rolling and suppose that the risky assets have final values according to the following table:

Time	0	T
Riskless asset	1	e^{rT}
Risky asset S	S_0	$S_T = \begin{cases} S^+ & \text{if } \{1,2,3,4\}, \\ S^- & \text{if } \{5,6\}, \end{cases}$
Riskless asset C	?	$C_T = \begin{cases} C^+ & \text{if } \{1,2\}, \\ C^- & \text{if } \{3,4,5,6\}. \end{cases}$

Now we set

$$E_1 = \{1,2\}, \qquad E_2 = \{3,4\}, \qquad E_3 = \{5,6\}.$$

If we suppose to be able to assign the probabilities to the events

$$P(E_1) = p_1, \qquad P(E_2) = p_2, \qquad P(E_3) = 1 - p_1 - p_2,$$

where $p_1, p_2 > 0$ and $p_1 + p_2 < 1$, then the risk-neutral prices are defined just as in Section 1.2.1:

$$\begin{aligned}\widetilde{S}_0 =& e^{-rT}\left(p_1 S^+ + p_2 S^+ + (1 - p_1 - p_2)S^-\right)\\ =& e^{-rT}\left((p_1 + p_2)\, S^+ + (1 - p_1 - p_2)S^-\right)\\ \widetilde{C}_0 =& e^{-rT}\left(p_1 C^+ + p_2 C^- + (1 - p_1 - p_2)C^-\right)\\ =& e^{-rT}\left(p_1 C^+ + (1 - p_1)C^-\right).\end{aligned}$$

Conversely, if S_0 is quoted on the market, by imposing $S_0 = \widetilde{S}_0$, we obtain

$$S_0 = e^{-rT}\left(q_1 S^+ + q_2 S^+ + (1 - q_1 - q_2)S^-\right)$$

and so *there exist infinitely many*[3] *risk-neutral probabilities.*

Analogously, by proceeding as in Section 1.2.3 to determine a replicating strategy for C, we obtain

$$\begin{cases} \alpha S^+ + \beta e^{rT} = C^+ & \text{if } E_1,\\ \alpha S^+ + \beta e^{rT} = C^- & \text{if } E_2,\\ \alpha S^- + \beta e^{rT} = C^- & \text{if } E_3.\end{cases} \tag{1.13}$$

In general this system is not solvable and therefore the asset C is not replicable: we say that the market model is *incomplete*. In this case it is not possible to price C on the basis of replication arguments: since we can only solve two out of three equations, we cannot build a strategy replicating C in all the possible cases and *we are able to hedge the risk only partially.*

We note that, if (α, β) solves the first and the third equation of the system (1.13), then the terminal value V_T of the corresponding strategy is equal to

$$V_T = \begin{cases} C^+ & \text{if } E_1,\\ C^+ & \text{if } E_2,\\ C^- & \text{if } E_3.\end{cases}$$

With this choice (and assuming that $C^+ > C^-$) we obtain a strategy that *super-replicates* C.

Summing up:

- *in a market model that is free from arbitrage opportunities and complete, on one hand there exists a unique the risk-neutral probability measure;*

[3] Actually, it is possible to determine a unique risk-neutral probability if we assume that both S_0 and C_0 are observable.

on the other hand, for every derivative there exists a replicating strategy. Consequently there exists a unique risk-neutral price which coincides with the arbitrage price;

- *in a market model that is free from arbitrage opportunities and incomplete, on one hand there exist infinitely many risk-neutral probabilities; on the other hand not every derivative is replicable. Consequently there exist infinitely many risk-neutral prices but it is not possible, in general, to define the arbitrage price.*

2

Discrete market models

In this chapter we describe market models in discrete time to price and hedge European and American-style derivatives. We present the classical model introduced by Cox, Ross and Rubinstein in [78] and we mention briefly the pricing problem in incomplete markets. General references on topics covered in this chapter are Dana and Jeanblanc [84], Föllmer and Schied [134], Lamberton and Lapeyre [226], Pliska [282], Shreve [310], van der Hoek and Elliott [329]: we also mention Pascucci and Runggaldier [277] where several examples and exercises can be found.

2.1 Discrete markets and arbitrage strategies

We consider a discrete-market model where, for a fixed time interval $[0, T]$, we suppose that all transactions take place only at times

$$0 = t_0 < t_1 < \cdots < t_N = T.$$

To fix the ideas, t_0 denotes today's date and t_N is the expiration date of a derivative. Let us recall that the unit of time is the year.

The market consists of one riskless asset (bond) B and d risky assets $S = (S^1, \dots, S^d)$ that are stochastic processes defined on a probability space $(\Omega, \mathcal{F}, P)$. We assume:

(H1) *Ω has a finite number of elements, $\mathcal{F} = \mathscr{P}(\Omega)$ and $P(\{\omega\}) > 0$ for any $\omega \in \Omega$.*

The dynamics of the bond is deterministic: if B_n denotes the price of the bond at time t_n, we have

$$\begin{cases} B_0 = 1, \\ B_n = B_{n-1}(1 + r_n), \qquad n = 1, \dots, N, \end{cases} \tag{2.1}$$

where r_n, such that $1 + r_n > 0$, denotes the risk-free rate in the n-th period $[t_{n-1}, t_n]$. Occasionally we also call B the *bank account.*

The risky assets have the following stochastic dynamics: if S^i_n denotes the price at time t_n of the i-th asset, then we have

$$\begin{cases} S^i_0 \in \mathbb{R}_{>0}, \\ S^i_n = S^i_{n-1}\left(1+\mu^i_n\right), \qquad n = 1,\dots,N, \end{cases} \tag{2.2}$$

where μ^i_n is a real random variable such that $1+\mu^i_n > 0$, which represents the yield rate of the i-th asset in the n-th period $[t_{n-1}, t_n]$. Then $S^i = (S^i_n)_{n=0,\dots,N}$ is a discrete stochastic process on $(\Omega, \mathcal{F}, P)$ and we say that (S, B) is a *discrete market on the probability space* $(\Omega, \mathcal{F}, P)$.

We set

$$\mu_n = (\mu^1_n, \dots, \mu^d_n), \qquad 1 \le n \le N,$$

and consider the filtration $(\mathcal{F}_n)$ defined by

$$\mathcal{F}_0 = \{\emptyset, \Omega\}, \tag{2.3}$$

$$\mathcal{F}_n = \sigma(\mu_1, \dots, \mu_n), \qquad 1 \le n \le N. \tag{2.4}$$

The σ-algebra $\mathcal{F}_n$ represents the amount of information available in the market at time t_n (cf. Appendix A.1.6): note that, by (2.2), we also have $\mathcal{F}_n = \sigma(S_0, \dots, S_n)$ for $0 \le n \le N$. Formula (2.3) is equivalent to the fact that the prices $S^1_0, \dots, S^d_0$ of the assets at the initial time are observable and so they are deterministic, i.e. positive numbers not random variables (cf. Example A.38). In the sequel we shall also assume:

(H2) $\mathcal{F}_N = \mathcal{F}$.

2.1.1 Self-financing and predictable strategies

Definition 2.1 *A strategy (or portfolio) is a stochastic process in* $\mathbb{R}^{d+1}$

$$(\alpha, \beta) = (\alpha^1_n, \dots, \alpha^d_n, \beta_n)_{n=1,\dots,N}.$$

In the preceding definition, α^i_n (resp. β_n) represents the amount of the asset S^i (resp. bond) held in the portfolio *during the* n*-th period* $[t_{n-1}, t_n]$. Therefore we denote the *value of the portfolio* (α, β) at time t_n by

$$V^{(\alpha,\beta)}_n = \alpha_n S_n + \beta_n B_n, \qquad n = 1,\dots,N, \tag{2.5}$$

and

$$V^{(\alpha,\beta)}_0 = \sum_{i=1}^d \alpha^i_1 S^i_0 + \beta_1 B_0.$$

In (2.5)

$$\alpha_n S_n = \sum_{i=1}^d \alpha^i_n S^i_n, \qquad n = 1,\dots,N,$$

denotes the scalar product in $\mathbb{R}^d$. The value $V^{(\alpha,\beta)} = (V_n^{(\alpha,\beta)})_{n=0,\dots,N}$ is a real stochastic process in discrete time: to shorten notations, sometimes we simply write V_n in place of $V_n^{(\alpha,\beta)}$. We point out that negative values of α_n^i, β_n are allowed since short selling of shares or borrowing from the bank are permitted.

Definition 2.2 *A strategy (α, β) is self-financing if the relation*

$$V_{n-1}^{(\alpha,\beta)} = \alpha_n S_{n-1} + \beta_n B_{n-1} \tag{2.6}$$

holds for every $n = 1, \dots, N$.

The self-financing property (2.6) can interpreted as follows:

> *at time t_{n-1} the wealth at our disposal is $V_{n-1}^{(\alpha,\beta)} = \alpha_{n-1}S_{n-1} + \beta_{n-1}B_{n-1}$ and we re-balance the strategy with the new quantities (α_n, β_n) in such a way that we do not modify the overall value of the portfolio.*

For example, if at $t_0 = 0$ we have at our disposal the initial wealth V_0, we construct the strategy (α_1, β_1) in such a way that its value $\alpha_1 S_0 + \beta_1 B_0$ is equal to V_0. Note that (α_n, β_n) denotes what the portfolio built at time t_{n-1} is composed of.

Example 2.3 In the case of one risky asset (i.e. $d = 1$) (2.6) is equivalent to

$$\beta_n = \beta_{n-1} - (\alpha_n - \alpha_{n-1})\frac{S_{n-1}}{B_{n-1}}.$$

The previous formula shows how β_n must vary in a self-financing portfolio if, at time t_{n-1}, we change the amount of the risky asset from α_{n-1} to α_n. Note that, in the particular case $d = 0$, a portfolio is self-financing if and only if it is constant. □

The variation, from time t_{n-1} to t_n, of the value of a self-financing strategy (α, β) is given by

$$V_n^{(\alpha,\beta)} - V_{n-1}^{(\alpha,\beta)} = \alpha_n(S_n - S_{n-1}) + \beta_n(B_n - B_{n-1}) \tag{2.7}$$

and therefore it is caused only by the variation of the prices of the assets and not by the fact that we have injected or withdrawn funds. Therefore, in a self-financing strategy we establish the wealth we want to invest at the initial time and afterwards we do not inject or withdraw funds.

In what follows we consider only investment strategies based upon the amount of information available at the moment (of course foreseeing the future is not allowed). Since in a self-financing strategy the rebalancing of the portfolio from $(\alpha_{n-1}, \beta_{n-1})$ to (α_n, β_n) occurs at time t_{n-1}, it is natural to assume that (α, β) is *predictable*:

Definition 2.4 *A strategy (α, β) is predictable if (α_n, β_n) is $\mathcal{F}_{n-1}$-measurable for every $n = 1, \dots, N$.*

Notation 2.5 *We denote by $\mathcal{A}$ the family of all self-financing and predictable strategies of the market (S, B).*

The self-financing condition establishes a relationship between the processes α and β: as a consequence, it turns out that a strategy in $\mathcal{A}$ is identified by (α, β) or equivalently by V_0 and α where $V_0 \in \mathbb{R}$ is the initial value of the strategy and α is a d-dimensional predictable process. Indeed we have the following:

Lemma 2.6 *The value of a self-financing strategy (α, β) is determined by its initial value V_0 and recursively by*

$$V_n = V_{n-1}(1 + r_n) + \sum_{i=1}^{d} \alpha_n^i S_{n-1}^i \left(\mu_n^i - r_n\right) \tag{2.8}$$

for $n = 1, \dots, N$.

Proof. By (2.7), the variation of a self-financing portfolio in the period $[t_{n-1}, t_n]$ is equal to

$$\begin{aligned} V_n - V_{n-1} &= \alpha_n \left(S_n - S_{n-1}\right) + \beta_n \left(B_n - B_{n-1}\right) \\ &= \sum_{i=1}^{d} \alpha_n^i S_{n-1}^i \mu_n^i + \beta_n B_{n-1} r_n = \end{aligned} \tag{2.9}$$

(since, by (2.6), we have $\beta_n B_{n-1} = V_{n-1} - \alpha_n S_{n-1}$)

$$= \sum_{i=1}^{d} \alpha_n^i S_{n-1}^i \left(\mu_n^i - r_n\right) + r_n V_{n-1}$$

and the claim follows. □

Proposition 2.7 *Given $V_0 \in \mathbb{R}$ and a predictable process α, there exists a unique predictable process β such that $(\alpha, \beta) \in \mathcal{A}$ and $V_0^{(\alpha,\beta)} = V_0$.*

Proof. Given $V_0 \in \mathbb{R}$ and a predictable process α, we define the process

$$\beta_n = \frac{V_{n-1} - \alpha_n S_{n-1}}{B_{n-1}}, \qquad n = 1, \dots, N,$$

where (V_n) is recursively defined by (2.8). Then by construction (β_n) is predictable and the strategy (α, β) is self-financing. □

Remark 2.8 Given $(\alpha, \beta) \in \mathcal{A}$, by summing over n in (2.9) we get

$$V_n = V_0 + g_n^{(\alpha,\beta)}, \qquad n = 1, \dots, N, \tag{2.10}$$

where

$$g_n^{(\alpha,\beta)} = \sum_{k=1}^{n} \left(\alpha_k \left(S_k - S_{k-1}\right) + \beta_k \left(B_k - B_{k-1}\right)\right)$$
$$= \sum_{k=1}^{n} \left(\sum_{i=1}^{d} \alpha_k^i S_{k-1}^i \mu_k^i + \beta_k B_{k-1} r_k \right) \tag{2.11}$$

defines the process of the *gain of the strategy*. □

2.1.2 Normalized market

For a fixed asset $Y = (Y_n)$, we define by

$$\widetilde{S}_n^i = \frac{S_n^i}{Y_n}, \qquad \widetilde{B}_n = \frac{B_n}{Y_n}, \tag{2.12}$$

the *normalized market with respect to* Y. In the normalized market we obviously have $\widetilde{Y} \equiv 1$ and the prices of the other assets are denominated in units of the asset Y: for this reason Y is usually called a *numeraire*. Often Y plays the part of the non-risky asset B corresponding to the investment in a bank account: in this case $\widetilde{S}^i$ is also called the *discounted price* of the i-th asset. In practice, by discounting one can compare quoted prices at different times.

Let us now consider the discounted market $\widetilde{S}$, that is we assume that B is the numeraire. Given a strategy (α, β), we set

$$\widetilde{V}_n^{(\alpha,\beta)} = \frac{V_n^{(\alpha,\beta)}}{B_n}.$$

Then the self-financing condition becomes

$$\widetilde{V}_{n-1}^{(\alpha,\beta)} = \alpha_n \widetilde{S}_{n-1} + \beta_n, \qquad n = 1, \dots, N,$$

or equivalently

$$\widetilde{V}_n^{(\alpha,\beta)} = \widetilde{V}_{n-1}^{(\alpha,\beta)} + \alpha_n \left(\widetilde{S}_n - \widetilde{S}_{n-1} \right), \qquad n = 1, \dots, N.$$

Therefore Lemma 2.6 has the following extension.

Lemma 2.9 *The discounted value of a self-financing strategy (α, β) is uniquely determined by its initial value V_0 and recursively by*

$$\widetilde{V}_n^{(\alpha,\beta)} = \widetilde{V}_{n-1}^{(\alpha,\beta)} + \sum_{i=1}^{d} \alpha_n^i \left(\widetilde{S}_n^i - \widetilde{S}_{n-1}^i \right)$$

for $n = 1, \dots, N$.

The following formula, analogous to (2.10), holds:

$$\widetilde{V}_n^{(\alpha,\beta)} = V_0 + G_n^{(\alpha)} \tag{2.13}$$

where

$$G_n^{(\alpha)} = \sum_{k=1}^{n} \alpha_k \left(\widetilde{S}_k - \widetilde{S}_{k-1} \right)$$

is the *normalized gain related to the predictable process* α. Note that in general $G_n^{(\alpha)}$ is different from $\frac{g_n^{(\alpha,\beta)}}{B_n}$ and that $G_n^{(\alpha)}$ does not depend on β. We also recall that $\widetilde{V}_0 = V_0$ since $B_0 = 1$ by assumption.

2.1.3 Arbitrage opportunities and admissible strategies

We recall that $\mathcal{A}$ denotes the family of self-financing and predictable strategies of the market (S, B).

Definition 2.10 *We say that* $(\alpha, \beta) \in \mathcal{A}$ *is an arbitrage strategy (or simply an arbitrage) if the value* $V = V^{(\alpha,\beta)}$ *is such that*[1]

i) $V_0 = 0$*;*

and there exists $n \geq 1$ *such that*

ii) $V_n \geq 0$ *P-a.s.;*
iii) $P(V_n > 0) > 0$.

We say that the market (S, B) *is arbitrage-free if the family* $\mathcal{A}$ *does not contain arbitrage strategies.*

An arbitrage is a strategy in $\mathcal{A}$ that does not require an initial investment, does not expose to any risk ($V_n \geq 0$ P-a.s.) and leads to a positive value with positive probability. In an arbitrage-free market it is not possible to have such a sure risk-free profit by investing in a predictable and self-financing strategy.

The absence of arbitrage opportunities is a fundamental assumption from an economic point of view and is a condition that every reasonable model must fulfill. Clearly, the fact that there is absence of arbitrage depends on the probabilistic model considered, i.e. on the space $(\Omega, \mathcal{F}, P)$ and on the kind of stochastic process (S, B) used to describe the market. In Section 2.1.4 we give a mathematical characterization of the absence of arbitrage in terms of the existence of a suitable probability measure, equivalent to P, called martingale measure. Then, in Paragraph 2.3 we examine the very easy case of the binomial model, so that we see the practical meaning of the concepts we

---

[1] By assumption (H1), the empty set is the only event with null probability: although it is superfluous to write P-a.s. after the (in-)equalities, it is convenient to do so to adapt the presentation to the continuous case in the following chapters, where the (in-)equalities hold indeed only almost surely.

have introduced. In particular we see that in the binomial model the market is arbitrage-free under very simple and intuitive assumptions.

We allowed the values of a strategy to be negative (short-selling), but it seems reasonable to require that the overall value of the portfolio does not take negative values.

Definition 2.11 *A strategy $(\alpha, \beta) \in \mathcal{A}$ is called admissible if*

$$V_n^{(\alpha,\beta)} \geq 0 \quad P\text{-}a.s.$$

for every $n \leq N$.

Implicitly the definition of arbitrage includes the admissibility condition or, more precisely, in a discrete market every arbitrage strategy can be modified to become admissible. We remark that *this result does not generalize to the continuous-time case.*

Proposition 2.12 *A discrete market is arbitrage-free if and only if there exist no admissible arbitrage strategies.*

Proof. We suppose that there exist no admissible arbitrage strategies and we have to show that no arbitrage opportunity exists. We prove the thesis by contradiction: we suppose there exists an arbitrage strategy (α, β) and we construct an admissible arbitrage strategy (α', β').

By assumption, $V_0^{(\alpha,\beta)} = \alpha_1 S_0 + \beta_1 B_0 = 0$ and there exists n (it is not restrictive to suppose $n = N$) such that $\alpha_n S_n + \beta_n B_n \geq 0$ a.s. and $P(\alpha_n S_n + \beta_n B_n > 0) > 0$. If (α, β) is not admissible there exist $k < N$ and $F \in \mathcal{F}_k$ with $P(F) > 0$ such that

$$\alpha_k S_k + \beta_k B_k < 0 \text{ on } F \quad \text{and} \quad \alpha_n S_n + \beta_n B_n \geq 0 \text{ a.s. for } k < n \leq N.$$

Then we define a new arbitrage strategy as follows: $\alpha'_n = 0$, $\beta'_n = 0$ on $\Omega \setminus F$ for every n, while on F

$$\alpha'_n = \begin{cases} 0, & n \leq k, \\ \alpha_n, & n > k, \end{cases} \qquad \beta'_n = \begin{cases} 0, & n \leq k, \\ \beta_n - (\alpha_k S_k + \beta_k B_k), & n > k. \end{cases}$$

It is straightforward to verify that (α', β') is an arbitrage strategy and it is admissible. □

2.1.4 Equivalent martingale measure

We consider a discrete market (S, B) on the space $(\Omega, \mathcal{F}, P)$ and fix a numeraire Y that is a prices process in (S, B). More generally, in the sequel we will take as numeraire the value of any strategy $(\alpha, \beta) \in \mathcal{A}$, provided that $V^{(\alpha,\beta)}$ is positive. In this section we characterize the property of absence of arbitrage in terms of the existence of a new probability measure equivalent[2] to P and with respect to which the normalized price processes are martingales. We give the following important:

Definition 2.13 *An equivalent martingale measure (in short, EMM) with numeraire Y is a probability measure Q on $(\Omega, \mathcal{F})$ such that:*

i) Q is equivalent to P;
ii) the Y-normalized prices are Q-martingales, that is

$$\frac{S_{n-1}}{Y_{n-1}} = E^Q\left[\frac{S_n}{Y_n} \mid \mathcal{F}_{n-1}\right], \qquad \frac{B_{n-1}}{Y_{n-1}} = E^Q\left[\frac{B_n}{Y_n} \mid \mathcal{F}_{n-1}\right], \tag{2.14}$$

for every $n = 1, \dots, N$.

Remark 2.14 Consider the particular case $Y = B$ and denote by $\widetilde{S}_n = \frac{S_n}{B_n}$ the discounted prices. If Q is an EMM with numeraire B, by the martingale property we also have

$$\widetilde{S}_k = E^Q\left[\widetilde{S}_n \mid \mathcal{F}_k\right], \qquad 0 \le k < n \le N,$$

and consequently

$$E^Q\left[\widetilde{S}_n\right] = E^Q\left[E^Q\left[\widetilde{S}_n \mid \mathcal{F}_0\right]\right] = \widetilde{S}_0, \qquad n \le N. \tag{2.15}$$

Formula (2.15) has an important economic interpretation: it states that the expectations of the future normalized prices are equal to the current prices. Therefore (2.15) is a *risk-neutral pricing formula* in the sense of Section 1.2.1: the mean of $\widetilde{S}_n$ with respect to the measure Q corresponds to the value given by an investor who reckons that the current market-prices of the assets are correct (and so he/she is neither disposed nor averse to buy the assets). For this reason, Q is also called a *risk-neutral probability.*

Instead, the probability measure P is usually called *objective or real-world probability* since the dynamics of the random variables μ_n is usually given under the probability P and these variables (or the parameters of the model) have to be determined a priori from observations on the market or on the basis of the historical data on the stocks. In other terms, the random variables μ_n (that are an "input" that any discrete model needs in order to be used) have to be estimated by means of observations of the real world. □

We emphasize that the notion of EMM depends on the numeraire considered. Further, we notice that, since Q is equivalent to P, the market is arbitrage-free under the measure P if and only if it is arbitrage-free under Q.

The next result, in view of its importance, is commonly known as First Fundamental Theorem of asset pricing.

---

[2] That is, having the same null and certain events, cf. Appendix A.5.1.

Theorem 2.15 (First Fundamental Theorem of asset pricing) *A discrete market is arbitrage-free if and only if there exists at least one EMM.*

We defer the proof of Theorem 2.15 to Section 2.2.3 and we analyze now some important consequences of the definition of EMM.

The following result exhibits a fundamental feature of self-financing predictable portfolios: they preserve the martingale property, that is if $(\widetilde{S}, \widetilde{B})$ is a martingale and $(\alpha, \beta) \in \mathcal{A}$ then also $\widetilde{V}^{(\alpha,\beta)}$ is a martingale.

Proposition 2.16 *Let Q be an EMM with numeraire Y and $(\alpha, \beta) \in \mathcal{A}$. Then $\widetilde{V}^{(\alpha,\beta)} = \left(\frac{V_n^{(\alpha,\beta)}}{Y_n}\right)$ is a Q-martingale:*

$$\widetilde{V}_{n-1}^{(\alpha,\beta)} = E^Q\left[\widetilde{V}_n^{(\alpha,\beta)} \mid \mathcal{F}_{n-1}\right], \qquad n = 1, \dots, N. \tag{2.16}$$

In particular the following risk-neutral pricing formula holds:

$$\widetilde{V}_0^{(\alpha,\beta)} = E^Q\left[\widetilde{V}_n^{(\alpha,\beta)}\right], \qquad n \le N. \tag{2.17}$$

Conversely, if Q is a measure equivalent to P and for every $(\alpha, \beta) \in \mathcal{A}$, equation (2.16) *holds, then Q is an EMM with numeraire Y.*

Proof. For simplicity we only consider the case $Y = B$. The result is an immediate consequence of formula (2.13) which basically expresses the fact that (α, β) *is self-financing if and only if $\widetilde{V}^{(\alpha,\beta)}$ is the transform of $\widetilde{S}$ by α* (cf. Definition A.120). Then, since α is predictable, the claim follows directly from Proposition A.121. However, for the sake of clarity, it seems to be useful to go through the proof again: by the self-financing condition (2.6), we have

$$\widetilde{V}_n^{(\alpha,\beta)} = \widetilde{V}_{n-1}^{(\alpha,\beta)} + \alpha_n(\widetilde{S}_n - \widetilde{S}_{n-1})$$

and by taking the conditional expectation given $\mathcal{F}_{n-1}$, we get

$$E^Q\left[\widetilde{V}_n^{(\alpha,\beta)} \mid \mathcal{F}_{n-1}\right] = \widetilde{V}_{n-1}^{(\alpha,\beta)} + E^Q\left[\alpha_n(\widetilde{S}_n - \widetilde{S}_{n-1}) \mid \mathcal{F}_{n-1}\right] =$$

(by the properties of conditional expectation, Proposition A.107-(7), since α is predictable)

$$= \widetilde{V}_{n-1}^{(\alpha,\beta)} + \alpha_n E^Q\left[\widetilde{S}_n - \widetilde{S}_{n-1} \mid \mathcal{F}_{n-1}\right] = \widetilde{V}_{n-1}^{(\alpha,\beta)}$$

by (2.14). The converse is trivial.

The following result expresses the main consequence, fundamental from a practical point of view, of the condition of absence of arbitrage: *if two self-financing predictable strategies have the same terminal value, then they must have the same value at all preceding times.*

Proposition 2.17 (No arbitrage principle) *In an arbitrage-free market, if* $(\alpha, \beta), (\alpha', \beta') \in \mathcal{A}$ *and*

$$V_N^{(\alpha,\beta)} = V_N^{(\alpha',\beta')} \quad P\text{-}a.s.,$$

then

$$V_n^{(\alpha,\beta)} = V_n^{(\alpha',\beta')} \quad P\text{-}a.s.$$

for every $n = 0, \dots, N$.

Proof. Since the market is arbitrage-free, there exists an EMM Q with numeraire Y. The claim follows from the fact that $\widetilde{V}^{(\alpha,\beta)}, \widetilde{V}^{(\alpha',\beta')}$ are Q-martingales with the same terminal value. Indeed, since the measures P, Q are equivalent, we have $V_N^{(\alpha,\beta)} = V_N^{(\alpha',\beta')}$ Q-a.s. and so

$$\widetilde{V}_n^{(\alpha,\beta)} = E^Q\left[\widetilde{V}_N^{(\alpha,\beta)} \mid \mathcal{F}_n\right] = E^Q\left[\widetilde{V}_N^{(\alpha',\beta')} \mid \mathcal{F}_n\right] = \widetilde{V}_n^{(\alpha',\beta')},$$

for every $n \le N$. □

Remark 2.18 Analogously, in an arbitrage-free market, if $(\alpha, \beta), (\alpha', \beta') \in \mathcal{A}$ and

$$V_N^{(\alpha,\beta)} \ge V_N^{(\alpha',\beta')} \quad P\text{-a.s.},$$

then

$$V_n^{(\alpha,\beta)} \ge V_n^{(\alpha',\beta')} \quad P\text{-a.s.}$$

for every $n = 0, \dots, N$. □

2.1.5 Change of numeraire

The choice of the numeraire is not in general unique. From a theoretical point of view, we shall see that a suitable choice of the numeraire can make computations easier (cf. Example 2.37); from a practical point of view, it is possible that different investors use different numeraires, e.g. when market prices can be expressed in different currencies (Euros, Dollars, etc.). In this section we study the relation among martingale measures relative to different numeraires: specifically, we give an explicit formula for the Radon-Nikodym derivative of a EMM with respect to another, thus showing how to switch between different numeraires. The main tool that we are going to use is Bayes' formula in Theorem A.113.

Theorem 2.19 *In a discrete market* (S, B)*, let* Q *be an EMM with numeraire* Y *and let* X *be a positive adapted process such that* $\left(\frac{X_n}{Y_n}\right)$ *is a* Q*-martingale (*X *represents the value process of another asset or strategy to be considered as the new numeraire). Then the measure* Q^X *defined by*

$$\frac{dQ^X}{dQ} = \frac{X_N}{X_0}\left(\frac{Y_N}{Y_0}\right)^{-1}, \tag{2.18}$$

is such that

$$Y_n E^Q \left[\frac{Z}{Y_N} \mid \mathcal{F}_n \right] = X_n E^{Q^X} \left[\frac{Z}{X_N} \mid \mathcal{F}_n \right], \qquad n \le N, \tag{2.19}$$

for every random variable Z. Consequently Q^X is an EMM with numeraire X.

Remark 2.20 We may rewrite formula (2.19) in the form

$$E^Q \left[D^Y(n,N) Z \mid \mathcal{F}_n \right] = E^{Q^X} \left[D^X(n,N) Z \mid \mathcal{F}_n \right], \qquad n \le N, \tag{2.20}$$

where

$$D^X(n,N) = \frac{X_n}{X_N}, \qquad D^Y(n,N) = \frac{Y_n}{Y_N}, \qquad n \le N,$$

denote the *discount factors* from N to n with respect to the numeraires X and Y, respectively.

Proof. In (2.18), $L := \frac{dQ^X}{dQ}$ denotes the Radon-Nikodym derivative of Q^X with respect to Q and therefore

$$E^{Q^X}[Z] = E^Q[ZL]$$

for any random variable Z.

From (2.18) we infer

$$E^{Q^X}[Z \mid \mathcal{F}_n] = E^Q \left[Z \frac{Y_n}{Y_N} \left(\frac{X_n}{X_N} \right)^{-1} \mid \mathcal{F}_n \right], \qquad n \le N. \tag{2.21}$$

Indeed by Bayes' formula we have

$$E^{Q^X}[Z \mid \mathcal{F}_n] = \frac{E^Q[ZL \mid \mathcal{F}_n]}{E^Q[L \mid \mathcal{F}_n]} = \frac{E^Q \left[Z \frac{X_N}{Y_N} \mid \mathcal{F}_n \right]}{E^Q \left[\frac{X_N}{Y_N} \mid \mathcal{F}_n \right]}$$

and (2.21) follows, since by assumption $\left(\frac{X_n}{Y_n} \right)$ is a Q-martingale and therefore we have

$$E^Q \left[\frac{X_N}{Y_N} \mid \mathcal{F}_n \right] = \frac{X_n}{Y_n}.$$

Now (2.19) is a simple consequence of (2.21):

$$Y_n E^Q \left[\frac{Z}{Y_N} \mid \mathcal{F}_n \right] = E^Q \left[\frac{Y_n}{Y_N} \left(\frac{X_n}{X_N} \right)^{-1} \frac{X_n Z}{X_N} \mid \mathcal{F}_n \right] =$$

(by (2.21))

$$= X_n E^{Q^X} \left[\frac{Z}{X_N} \mid \mathcal{F}_n \right].$$

Eventually from (2.19) it follows that Q^X is an EMM with numeraire Y: indeed, by definition of EMM we have

$$S_n = Y_n E^Q \left[\frac{S_N}{Y_N} \mid \mathcal{F}_n \right] =$$

(by (2.19) with $Z = S_N$)

$$= X_n E^{Q^X} \left[\frac{S_N}{X_N} \mid \mathcal{F}_n \right],$$

for $n \leq N$, and an analogous formula holds for B. □

Corollary 2.21 *Under the assumptions of Theorem 2.19, for any $n \leq N$ and $A \in \mathcal{F}_n$, we have*

$$Q^X(A) = E^Q \left[\frac{X_n}{X_0} \left(\frac{Y_n}{Y_0} \right)^{-1} \mathbb{1}_A \right], \tag{2.22}$$

that is

$$\frac{dQ^X}{dQ}|_{\mathcal{F}_n} = \frac{X_n}{X_0} \left(\frac{Y_n}{Y_0} \right)^{-1}.$$

Proof. We have

$$Q^X(A) = E^{Q^X} [\mathbb{1}_A] =$$

(by (2.18))

$$= E^Q \left[\mathbb{1}_A \frac{X_N}{X_0} \left(\frac{Y_N}{Y_0} \right)^{-1} \right] =$$

(using that $A \in \mathcal{F}_n$)

$$= E^Q \left[\mathbb{1}_A E^Q \left[\frac{X_N}{X_0} \left(\frac{Y_N}{Y_0} \right)^{-1} \mid \mathcal{F}_n \right] \right]$$

and the thesis follows from the fact that $\frac{X}{Y}$ is a Q-martingale. □

2.2 European derivatives

We consider an arbitrage-free discrete market (S, B) on the space $(\Omega, \mathcal{F}, P)$.

Definition 2.22 *A European-style derivative is an $\mathcal{F}_N$-measurable random variable X on $(\Omega, \mathcal{F}, P)$.*

To fix the ideas, X represents the terminal value (or payoff) of an option with maturity T. The $\mathcal{F}_N$-measurability condition describes the fact that X depends on the price process S, that is $S^1, \dots, S^d$ are the underlying assets of the derivative X: actually, under our assumption (H2), $\mathcal{F}_N = \mathcal{F}$ and therefore X is simply a random variable on $(\Omega, \mathcal{F}, P)$.

- A derivative X is called *path-independent* if it depends only on the *terminal* value of the underlying assets:
$$X = F(S_T), \tag{2.23}$$
where F is a given function. This is the typical case of a European Call option with strike K for which we have
$$F(x) = (x - K)^+, \qquad x > 0;$$
- a derivative X is called *path-dependent* if it depends also on the values of the underlying assets at times before maturity: for example, in the case of a Look-back option we have
$$X = S_N - \min_{0 \le n \le N} S_n.$$

The main problems in the study of a derivative X are:

1) *the pricing problem,* i.e. to determine a price for the derivative such that no arbitrage opportunities are introduced in the market;
2) *the replication problem,* i.e. to determine a strategy (if it exists) $(\alpha, \beta) \in \mathcal{A}$ that assumes the same value of the derivative at maturity:
$$V_N^{(\alpha,\beta)} = X \quad \text{a.s.}$$
If such a strategy exists, X is called *replicable* and (α, β) is called *replicating strategy.*

In an arbitrage-free market, the first problem is solvable but the solution is not necessarily unique: in other words, it is possible to find at least one value for the price of a derivative in such a way that the absence of arbitrage is preserved. Regarding the second problem, we saw in Chapter 1 that it is rather easy to construct a market model that is arbitrage-free, but in which some derivatives are not replicable. On the other hand, if a replicating strategy (α, β) for the derivative X exists, then by the no arbitrage principle (in the form of Proposition 2.17) $V_0^{(\alpha,\beta)}$ is the unique value for the initial price of X that does not introduce arbitrage opportunities.

2.2.1 Pricing in an arbitrage-free market

We introduce the families super and sub-replicating portfolios for the derivative X:
$$\mathcal{A}_X^+ = \{(\alpha, \beta) \in \mathcal{A} \mid V_N^{(\alpha,\beta)} \ge X\}, \qquad \mathcal{A}_X^- = \{(\alpha, \beta) \in \mathcal{A} \mid V_N^{(\alpha,\beta)} \le X\}.$$

Given $(\alpha,\beta)\in\mathcal{A}_X^+$, the initial value $V_0^{(\alpha,\beta)}$ represents the price at which everyone would be willing to sell the derivative: indeed $V_0^{(\alpha,\beta)}$ is an initial investment sufficient to build a strategy that super-replicates X. To fix ideas we denote by H_0 the (unknown and possibly not unique) initial price of X: it is clear that we necessarily must have

$$H_0 \le V_0^{(\alpha,\beta)}, \qquad (\alpha,\beta)\in\mathcal{A}_X^+. \tag{2.24}$$

If inequality (2.24) were not true, by introducing in the market the derivative at the price $H_0 > V_0^{(\bar\alpha,\bar\beta)}$ for a certain strategy $(\bar\alpha,\bar\beta)\in\mathcal{A}_X^+$, one could create an obvious arbitrage opportunity which consists in selling the derivative and buying the strategy $(\bar\alpha,\bar\beta)$.

Analogously we must have

$$H_0 \ge V_0^{(\alpha,\beta)}, \qquad (\alpha,\beta)\in\mathcal{A}_X^-.$$

Indeed $V_0^{(\alpha,\beta)}$, for $(\alpha,\beta)\in\mathcal{A}_X^-$, represents the price at which everyone would be willing to buy the derivative since, by selling (α,β) and buying the derivative, one could make a risk-free profit.

In conclusion any fair initial price H_0 of X must satisfy

$$\sup_{(\alpha,\beta)\in\mathcal{A}_X^-} V_0^{(\alpha,\beta)} \le H_0 \le \inf_{(\alpha,\beta)\in\mathcal{A}_X^+} V_0^{(\alpha,\beta)}. \tag{2.25}$$

Now, assuming that the market is arbitrage-free, there exists (and in general it is not unique) an EMM Q. By Theorem 2.19 it is not restrictive to assume that B is the numeraire. Then with respect to Q, the discounted prices of the assets and the discounted value of any strategy in $\mathcal{A}$ are martingales: in particular, they coincide with the conditional expectation of their terminal values. For the sake of consistency, it seems reasonable to price the derivative X in an analogous way: for a fixed EMM Q, we put

$$\widetilde{H}_n^Q = \frac{H_n^Q}{B_n} := E^Q\left[\frac{X}{B_N} \mid \mathcal{F}_n\right], \qquad n=0,\dots,N, \tag{2.26}$$

and we say that H^Q is the *risk-neutral price of X with respect to the EMM Q*.

Actually, definition (2.26) verifies the consistency assumption (2.25) for the price of X, i.e. it does not introduce arbitrage opportunities. Indeed we have the following:

Lemma 2.23 *For every EMM Q with numeraire B, we have*

$$\sup_{(\alpha,\beta)\in\mathcal{A}_X^-} \widetilde{V}_n^{(\alpha,\beta)} \le E^Q\left[\frac{X}{B_N} \mid \mathcal{F}_n\right] \le \inf_{(\alpha,\beta)\in\mathcal{A}_X^+} \widetilde{V}_n^{(\alpha,\beta)},$$

for $n=0,\dots,N$.

Proof. If $(\alpha, \beta) \in \mathcal{A}_X^-$ then, by Proposition 2.16, we have

$$\widetilde{V}_n^{(\alpha,\beta)} = E^Q \left[\widetilde{V}_N^{(\alpha,\beta)} \mid \mathcal{F}_n \right] \le E^Q \left[\frac{X}{B_N} \mid \mathcal{F}_n \right],$$

and an analogous estimate holds for $(\alpha, \beta) \in \mathcal{A}_X^+$. □

Remark 2.24 The family of EMMs is a convex set, i.e. if Q_1, Q_2 are martingale measures with numeraire B, by the linearity property of conditional expectation then also any linear combination of the form

$$\lambda Q_1 + (1 - \lambda) Q_2, \qquad \lambda \in [0, 1],$$

is an EMM. As a simple consequence we have that the set of discounted initial prices $E^Q \left[\frac{X}{B_N} \right]$ is convex and can consist of a single point only or otherwise it can be a non-trivial interval: in this last case it is an open interval (see, for example, Theorem 5.33 in [134]). □

The following theorem contains the definition of the arbitrage price of a replicable derivative.

Theorem 2.25 *Let X be a replicable derivative in an arbitrage-free market. Then for every replicating strategy $(\alpha, \beta) \in \mathcal{A}$ and for every EMM Q with numeraire B, we have*

$$E^Q \left[\frac{X}{B_N} \mid \mathcal{F}_n \right] = \frac{V_n^{(\alpha,\beta)}}{B_n}, \qquad n = 0, \dots, N. \tag{2.27}$$

The process $H := V^{(\alpha,\beta)}$ is called arbitrage price (or risk-neutral price) of X.

Proof. If $(\alpha, \beta), (\alpha', \beta') \in \mathcal{A}$ replicate X then they have the same terminal value and, by Proposition 2.17, they have the same value at all preceding times. Moreover, if $(\alpha, \beta) \in \mathcal{A}$ replicates X, then $(\alpha, \beta) \in \mathcal{A}_X^- \cap \mathcal{A}_X^+$ and by Lemma 2.23 we have

$$E^Q \left[\frac{X}{B_N} \mid \mathcal{F}_n \right] = \widetilde{V}_n^{(\alpha,\beta)},$$

for every EMM Q with numeraire B. □

The pricing formula (2.27) is extremely intuitive: for $n = 0$ it becomes

$$H_0 = E^Q \left[\frac{X}{B_N} \right];$$

then the current price of the option is given by the best estimate (expected value) of the discounted terminal value. The expectation is computed with respect to a risk-neutral measure Q, i.e. a measure that makes the mean of the prices of the assets exactly equal to the current observed price, inflated by the interest rate. This is consistent with what we had seen in the introduction, Paragraph 1.2.

Remark 2.26 The following generalization of Theorem 2.25 holds:

in an arbitrage-free market, a derivative X is replicable if and only if $E^Q\left[\frac{X}{B_N}\right]$ is independent of the particular EMM Q (with numeraire B).

For the proof of this result, based on the separation of convex sets of $\mathbb{R}^N$ (cf. Theorem A.177) we refer, for instance, to [282]. □

2.2.2 Completeness

We know that those who sell a derivative have to deal with the replication problem. For example, a bank selling a Call option takes a potentially unbounded risk of loss: therefore, from the point of view of the bank, it is important to determine an investment strategy that, by using the money obtained by selling the derivative, guarantees the replication at maturity, "hedging" the risk.

Definition 2.27 *A market is complete if every European derivative is replicable.*

In a complete market every derivative has a unique arbitrage price, defined by (2.27): moreover the price coincides with the value of any replicating strategy.

Remark 2.28 On the other hand, there exist derivatives whose underlying is not quoted and traded, that is the case for instance of a derivative on a temperature: more precisely, consider a contract that pays a certain amount of money, say 1 Euro, if at a specified date and place the temperature is below 20 degrees centigrade. Then the payoff function of the contract is

$$F(x) = \begin{cases} 1 & \text{if } x < 20, \\ 0 & \text{if } x \geq 20. \end{cases}$$

In this case it sounds more appropriate to talk about "insurance" instead of "derivative". Since the underlying of the contract is a temperature and not an asset that we can buy or sell in the market, it is not possible to build a replicating portfolio for the contract, even though we can construct a probabilistic model for the dynamics of the temperature. Clearly in this case the market is incomplete. We note that also for derivatives on quoted stocks, the completeness of the market is not always considered a desirable or realistic property.

Now we remark that *the completeness of a market model implies the uniqueness of the EMM* related to a fixed numeraire. Indeed let us first recall that, by Theorem 2.19, we may always assume B as numeraire: then if Q_1, Q_2 are EMMs with numeraire B, by (2.27) we have

$$E^{Q_1}[X] = E^{Q_2}[X]$$

for every derivative X. Since by assumption (H.2) we have $\mathcal{F}_N = \mathcal{F}$, we may consider $X = \mathbb{1}_A$, $A \in \mathcal{F}$, to conclude that $Q_1 = Q_2$.

As a matter of fact, the uniqueness of the EMM is a property that characterizes complete markets. Indeed we have the following classical result:

Theorem 2.29 (Second Fundamental Theorem of asset pricing) *An arbitrage-free market (S, B) is complete if and only if there exists a unique EMM with numeraire B.*

2.2.3 Fundamental theorems of asset pricing

We prove the First Fundamental Theorem of asset pricing which establishes the connection between the absence of arbitrage opportunities and the existence of an EMM.

Proof (of Theorem 2.15). By Theorem 2.19 it is not restrictive to consider B as the numeraire. The proof of the fact that, if there exists an EMM then (S, B) is free from arbitrage opportunities is amazingly simple. Indeed let Q be an EMM and, by contradiction, let us suppose that there exists an arbitrage portfolio $(\alpha, \beta) \in \mathcal{A}$. Then $V_0^{(\alpha,\beta)} = 0$ and there exists $n \geq 1$ such that $P(V_n^{(\alpha,\beta)} \geq 0) = 1$ and $P(V_n^{(\alpha,\beta)} > 0) > 0$. Since $Q \sim P$, we also have $Q(V_n^{(\alpha,\beta)} \geq 0) = 1$ and $Q(V_n^{(\alpha,\beta)} > 0) > 0$, and consequently $E^Q\left[\widetilde{V}_n^{(\alpha,\beta)}\right] > 0$. On the other hand, by (2.17) we obtain

$$E^Q\left[\widetilde{V}_n^{(\alpha,\beta)}\right] = \widetilde{V}_0^{(\alpha,\beta)} = 0,$$

and this is a contradiction.

Conversely, we assume that (S, B) is free from arbitrage opportunities and we prove the existence of an EMM Q with numeraire B. By using the second part of Proposition A.121 with $M = \widetilde{S}$, it is enough to prove the existence of $Q \sim P$ such that

$$E^Q\left[\sum_{n=1}^{N} \alpha_n \left(\widetilde{S}_n^i - \widetilde{S}_{n-1}^i\right)\right] = 0 \qquad (2.28)$$

for every $i = 1, \dots, d$ and for every real-valued predictable process α. Formula (2.28) expresses the fact that the expected gain is null.

Let us fix i for good; the proof of (2.28) is based upon the result of separation of convex sets (in finite dimension) of Appendix A.10. So it is useful to set the problem in the Euclidean space: we denote the cardinality of Ω by m and its elements by $\omega_1, \dots, \omega_m$. If Y is a random variable in Ω, we put $Y(\omega_j) = Y_j$ and *we identify Y with the vector in $\mathbb{R}^m$*

$$(Y_1, \dots, Y_m).$$

Therefore we have

$$E^Q[Y] = \sum_{j=1}^{m} Y_j Q(\{\omega_j\}).$$

For every *real-valued* predictable process α, we use the notation

$$G(\alpha) = \sum_{n=1}^{N} \alpha_n \left(\widetilde{S}_n^i - \widetilde{S}_{n-1}^i \right).$$

First of all we observe that the assumption of absence of arbitrage opportunities translates into the condition

$$G(\alpha) \notin \mathbb{R}_+^m := \{Y \in \mathbb{R}^m \setminus \{0\} \mid Y_j \geq 0 \text{ for } j = 1, \dots, m\}$$

for every predictable α. Indeed if there existed a real-valued predictable process α such that $G(\alpha) \in \mathbb{R}_+^m$, then, by using Proposition 2.7 and choosing $\widetilde{V}_0 = 0$, one could construct a strategy in $\mathcal{A}$ with null initial value and final value $\widetilde{V}_N = G(\alpha)$ i.e. an arbitrage strategy, violating the assumption.

Consequently

$$\mathscr{V} := \{G(\alpha) \mid \alpha \text{ predictable}\}$$

is a linear subspace of $\mathbb{R}^m$ such that

$$\mathscr{V} \cap \mathscr{K} = \emptyset,$$

with $\mathscr{K}$ defined by

$$\mathscr{K} := \{Y \in \mathbb{R}_+^m \mid Y_1 + \cdots + Y_m = 1\}.$$

Let us observe that $\mathscr{K}$ is a compact convex subset of $\mathbb{R}^m$: then the conditions to apply Corollary A.178 are fulfilled and there exists $\xi \in \mathbb{R}^m$ such that

i) $\langle \xi, Y \rangle = 0$ for every $Y \in \mathscr{V}$;
ii) $\langle \xi, Y \rangle > 0$ for every $Y \in \mathscr{K}$;

or equivalently

i) $\sum_{j=1}^{m} \xi_j G_j(\alpha) = 0$ for every predictable process α;
ii) $\sum_{j=1}^{m} \xi_j Y_j > 0$ for every $Y \in \mathscr{K}$.

In particular ii) implies that $\xi_j > 0$ for every j and so we can normalize the vector ξ to define a probability measure Q, equivalent to P, by

$$Q(\{\omega_j\}) = \xi_j \left(\sum_{i=1}^{m} \xi_i \right)^{-1}.$$

Then i) translates into

$$E^Q[G(\alpha)] = 0$$

for every predictable α, concluding the proof of (2.28). □

Next we prove the Second Fundamental Theorem of asset pricing, which establishes the connection between the completeness of the market and the uniqueness of the EMM.

Proof (of Theorem 2.29). We just have to prove that if (S, B) is free from arbitrage opportunities and the EMM Q with numeraire B is unique, then the market is complete. We proceed by contradiction: we suppose that the market is not complete and we construct an EMM with numeraire B, different from Q. We denote the linear space of normalized final values of strategies $(\alpha, \beta) \in \mathcal{A}$ by

$$\mathscr{V} = \{\widetilde{V}_N^{(\alpha,\beta)} \mid (\alpha, \beta) \in \mathcal{A}\}.$$

As in the proof of Theorem 2.15 we identify random variables with elements of $\mathbb{R}^m$. Then the fact that (S, B) is not complete translates into the condition

$$\mathscr{V} \subsetneq \mathbb{R}^m. \tag{2.29}$$

We define the scalar product in $\mathbb{R}^m$

$$\langle X, Y \rangle_Q = E^Q[XY] = \sum_{j=1}^m X_j Y_j Q(\{\omega_j\}).$$

Then, by (2.29), there exists $\xi \in \mathbb{R}^m \setminus \{0\}$ orthogonal to $\mathscr{V}$, i.e. such that

$$\langle \xi, X \rangle_Q = E^Q[\xi X] = 0, \tag{2.30}$$

for every $X = \widetilde{V}_N^{(\alpha,\beta)}$, $(\alpha, \beta) \in \mathcal{A}$. In particular, by choosing[3] $X = 1$ we infer

$$E^Q[\xi] = 0. \tag{2.31}$$

For a fixed parameter $\delta > 1$, we put

$$Q_\delta(\{\omega_j\}) = \left(1 + \frac{\xi_j}{\delta \|\xi\|_\infty}\right) Q(\{\omega_j\}), \qquad j = 1, \dots, m,$$

where

$$\|\xi\|_\infty := \max_{1 \le j \le m} |\xi_j|.$$

We prove that, for every $\delta > 1$, Q_δ defines an EMM (obviously different from Q since $\xi \neq 0$). First of all $Q_\delta(\{\omega_j\}) > 0$ for every j, since

$$1 + \frac{\xi_j}{\delta \|\xi\|_\infty} > 0.$$

[3] The constant random variable that is equal to 1 belongs to the space $\mathscr{V}$: by the representation (2.13) for $\widetilde{V}_N^{(\alpha,\beta)}$, it is enough to use Proposition 2.7 choosing $\alpha^1, \dots, \alpha^d = 0$ and $\widetilde{V}_0 = 1$.

Moreover we have

$$\begin{aligned} Q_\delta(\Omega) &= \sum_{j=1}^{m} Q_\delta(\{\omega_j\}) = \sum_{j=1}^{m} \left(1 + \frac{\xi_j}{\delta\|\xi\|_\infty}\right) Q(\{\omega_j\}) \\ &= \sum_{j=1}^{m} Q(\{\omega_j\}) + \frac{1}{\delta\|\xi\|_\infty} \sum_{j=1}^{m} \xi_j Q(\{\omega_j\}) = \\ &= Q(\Omega) + \frac{1}{\delta\|\xi\|_\infty} E^Q[\xi] = 1 \end{aligned}$$

by (2.31). Therefore Q_δ is a probability measure equivalent to Q (and to P).

Next we prove that $\widetilde{S}$ is a Q_δ-martingale. Using the second part of Proposition A.121 with $M = \widetilde{S}$, it is enough to prove that

$$E^{Q_\delta}\left[\sum_{n=1}^{N} \alpha_n \left(\widetilde{S}_n^i - \widetilde{S}_{n-1}^i\right)\right] = 0$$

for every $i = 1, \dots, d$ and for every real-valued predictable process α. For fixed i, we use the notation

$$G(\alpha) = \sum_{n=1}^{N} \alpha_n \left(\widetilde{S}_n^i - \widetilde{S}_{n-1}^i\right).$$

Then we have

$$\begin{aligned} E^{Q_\delta}[G(\alpha)] &= \sum_{j=1}^{m} \left(1 + \frac{\xi_j}{\delta\|\xi\|_\infty}\right) G_j(\alpha) Q(\{\omega_j\}) \\ &= \sum_{j=1}^{m} G_j(\alpha) Q(\{\omega_j\}) + \frac{1}{\delta\|\xi\|_\infty} \sum_{j=1}^{m} \xi_j G_j(\alpha) Q(\{\omega_j\}) \\ &= E^Q[G(\alpha)] + \frac{1}{\delta\|\xi\|_\infty} E^Q[\xi G(\alpha)] = \end{aligned}$$

(by (2.30))

$$= E^Q[G(\alpha)] = 0,$$

by Proposition A.121, since $\widetilde{S}$ is a Q-martingale and α is predictable. □

2.2.4 Markov property

Consider a discrete market (S, B) in the form (2.1)-(2.2). Under the additional assumption that *the random variables* $\mu_1, \dots, \mu_N$ *are independent*, the price process S has the Markov property: intuitively this property expresses the fact that the future expected trend of the prices depends only on the "present" and is independent of the "past". We recall the following:

Definition 2.30 *A discrete stochastic process $X = (X_n)$ on a filtered probability space $(\Omega, \mathcal{F}, P, (\mathcal{F}_n))$ has the Markov property if*

i) X is adapted to $(\mathcal{F}_n)$;
ii) for every bounded $\mathscr{B}$-measurable function φ we have

$$E[\varphi(X_n) \mid \mathcal{F}_{n-1}] = E[\varphi(X_n) \mid X_{n-1}], \qquad n \geq 1. \tag{2.32}$$

As a consequence of (2.32) and Corollary A.10, for any n there exists a measurable function g_n such that

$$E[\varphi(X_n) \mid \mathcal{F}_{n-1}] = g_n(X_{n-1}).$$

The proof of the Markov property is based upon Lemma A.108.

Theorem 2.31 *If the random variables $\mu_1, \dots, \mu_N$ are independent then the stochastic process S has the Markov property.*

Proof. We have[4]

$$E[\varphi(S_n) \mid \mathcal{F}_{n-1}] = E[\varphi(S_{n-1}(1+\mu_n)) \mid \mathcal{F}_{n-1}] =$$

(applying Lemma A.108 with $X = 1+\mu_n$, $Y = S_{n-1}$, $\mathcal{G} = \mathcal{F}_{n-1}$ and $h(X,Y) = \varphi(XY)$)

$$= g_n(S_{n-1}), \tag{2.33}$$

where

$$g_n(s) = E[\varphi(s(1+\mu_n))]$$

and the thesis follows from Remark A.109. □

2.3 Binomial model

In the binomial model, the market is composed of a non-risky asset B (bond), corresponding to the investment into a savings account in a bank, and of a risky asset S (stock), corresponding, for example, to a quoted stock in the exchange.

For the sake of simplicity, we suppose that the time intervals have the same length

$$t_n - t_{n-1} = \frac{T}{N}$$

and the interest rate is constant over the period $[0, T]$, that is $r_n = r$ for every n. Then the dynamics of the bond is given by

$$B_n = B_{n-1}(1+r), \qquad n = 1, \dots, N, \tag{2.34}$$

so that $B_n = (1+r)^n$.

[4] By assumption the empty set is the only event with null probability and so there is only one version of the conditional expectation that we denote by $E[\varphi(S_n) \mid \mathcal{F}_{n-1}]$.

For the risky asset we assume that the dynamics is stochastic: in particular we assume that when passing from time t_{n-1} to time t_n the stock can only increase or decrease its value with constant increase and decrease rates:

$$S_n = S_{n-1}(1 + \mu_n), \qquad n = 1, \dots, N, \tag{2.35}$$

where $\mu_1, \dots, \mu_N$ are independent and identically distributed (i.i.d.) random variables on a probability space $(\Omega, \mathcal{F}, P)$, whose distribution is a combination of Dirac's Deltas:

$$1 + \mu_n \sim p\delta_u + (1 - p)\delta_d, \qquad n = 1, \dots, N. \tag{2.36}$$

In (2.36) $p \in]0, 1[$, u denotes the increase rate of the stock over the period $[t_{n-1}, t_n]$ and d denotes the decrease rate[5]. We assume that

$$0 < d < u. \tag{2.37}$$

We point out that we have

$$\begin{aligned} P(S_n = uS_{n-1}) &= P(1 + \mu_n = u) = p, \\ P(S_n = dS_{n-1}) &= P(1 + \mu_n = d) = (1 - p), \end{aligned}$$

that is

$$S_n = \begin{cases} uS_{n-1}, & \text{with probability } p, \\ dS_{n-1}, & \text{with probability } 1 - p. \end{cases}$$

Hence a "trajectory" of the stock is a vector such as (for example, in the case $N = 4$)

$$(S_0, uS_0, udS_0, u^2dS_0, u^3dS_0)$$

or

$$(S_0, dS_0, d^2S_0, ud^2S_0, u^2d^2S_0)$$

which can be identified with the vectors

$$(u, d, u, u)$$

and

$$(d, d, u, u)$$

of the occurrences of the random variable $(1 + \mu_1, 1 + \mu_2, 1 + \mu_3, 1 + \mu_4)$, respectively. Therefore we can assume that the sample space Ω is the family

$$\{(e_1, \dots, e_N) \mid e_k = u \text{ or } e_k = d\}$$

containing 2^N elements and $\mathcal{F}$ is the σ-algebra of all subsets of Ω. The family of trajectories can be represented on a binomial tree as in Figure 2.1 in the case $N = 3$.

---

[5] The state u (up) corresponds to the increase of the value of the stock, whilst the state d (down) to its decrease.

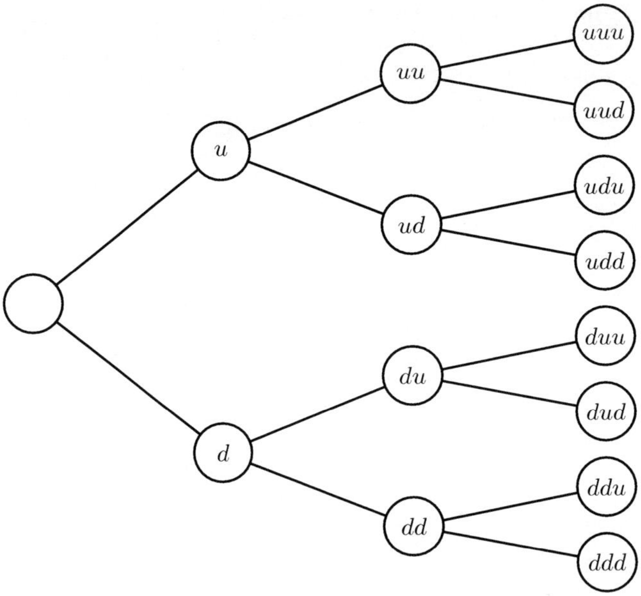

Fig. 2.1. Three-period binomial tree

Remark 2.32 The probability measure P is uniquely determined by (2.36) and the assumption of independence of the random variables $\mu_1, \dots, \mu_N$. Indeed we have

$$P(S_n = u^j d^{n-j} S_0) = \binom{n}{j} p^j (1-p)^{n-j}, \qquad j = 0, \dots, n, \tag{2.38}$$

for $n = 1, \dots, N$. Formula (2.38) corresponds to the well-known binomial distribution which represents the probability of obtaining j successes (j ups) after n trials (n time steps), when p is the probability of success of the single trial. The coefficient

$$\binom{n}{j} = \frac{n!}{j!(n-j)!}$$

represents the number of trajectories on the binomial tree that reach the price $S_n = u^j d^{n-j} S_0$.

For example, in the case $n = 2$, the probability that S_2 is equal to $u^2 S_0$ is given by

$$P(S_2 = u^2 S_0) = P((1 + \mu_1 = u) \cap (1 + \mu_2 = u)) = p^2,$$

where the last equality follows from the independence of μ_1 and μ_2. Analogously we have

$$\begin{aligned} P(S_2 = udS_0) &= P((1 + \mu_1 = u) \cap (1 + \mu_2 = d)) \\ &\quad + P((1 + \mu_1 = d) \cap (1 + \mu_2 = u)) = 2p(1-p). \end{aligned}$$

□

2.3.1 Martingale measure and arbitrage price

In this section we study existence and uniqueness of the EMM.

Theorem 2.33 *In the binomial model, the condition*

$$d < 1 + r < u, \tag{2.39}$$

is equivalent to the existence and uniqueness of the EMM Q. More precisely, if (2.39) *holds then*

$$q := \frac{1 + r - d}{u - d} \in]0, 1[, \tag{2.40}$$

and we have

$$Q(1 + \mu_n = u) = 1 - Q(1 + \mu_n = d) = q. \tag{2.41}$$

Moreover the random variables $\mu_1, \dots, \mu_N$ are Q-independent and we have

$$Q(S_n = u^k d^{n-k} S_0) = \binom{n}{k} q^k (1 - q)^{n-k}, \qquad 0 \le k \le n \le N. \tag{2.42}$$

The process S has the Markov property on the space $(\Omega, \mathcal{F}, Q, (\mathcal{F}_n))$: for every function φ we have

$$E^Q [\varphi(S_n) \mid \mathcal{F}_{n-1}] = E^Q [\varphi(S_n) \mid S_{n-1}] = q\varphi(uS_{n-1}) + (1 - q)\varphi(dS_{n-1}). \tag{2.43}$$

Proof. If an EMM Q exists, then by Definition 2.13 we have

$$\widetilde{S}_{n-1} = E^Q \left[\widetilde{S}_n \mid \mathcal{F}_{n-1} \right], \tag{2.44}$$

or equivalently

$$S_{n-1}(1 + r) = E^Q [S_{n-1} (1 + \mu_n) \mid \mathcal{F}_{n-1}] = S_{n-1} E^Q [(1 + \mu_n) \mid \mathcal{F}_{n-1}].$$

Since $S_{n-1} > 0$, we simplify the previous expression and obtain (cf. Proposition A.105)

$$\begin{aligned} r = E^Q [\mu_n \mid \mathcal{F}_{n-1}] =& (u - 1) Q (\mu_n = u - 1 \mid \mathcal{F}_{n-1}) \\ &+ (d - 1) (1 - Q (\mu_n = u - 1 \mid \mathcal{F}_{n-1})). \end{aligned}$$

Then we have

$$Q (1 + \mu_n = u \mid \mathcal{F}_{n-1}) = \frac{1 + r - d}{u - d} = q \tag{2.45}$$

and (2.39) must hold. Moreover, since the conditional probability in (2.45) is constant, by Proposition A.106 the random variables $\mu_1, \dots, \mu_N$ are Q-independent. Consequently (2.41) holds and also (2.42) can be proved as in Remark 2.32: in particular Q is uniquely determined. The Markov property of S follows from Theorem 2.31 and the fact that $\mu_1, \dots, \mu_N$ are Q-independent.

Note that $\mu_1, \dots, \mu_N$ *are Q-independent even if we do not assume that they are P-independent.* Then formula (2.43) follows from (2.33) since, in the binomial case, we have

$$g_n(s) = E^Q [\varphi(s(1 + \mu_n))] = q\varphi(us) + (1 - q)\varphi(ds).$$

Conversely, condition (2.39) is equivalent to the fact q in (2.45) belongs to the interval $]0, 1[$. Then Q, defined by (2.42), is a probability measure equivalent to P and since (2.45) is equivalent to the martingale condition (2.44), Q is an EMM. □

Condition (2.39) has a clear financial interpretation. Indeed assume that the parameters u, d, r in (2.34)-(2.36) verify (2.39), i.e. $d < 1 + r < u$: then the fact of borrowing money from the bank to invest it in the stock gives a positive probability of getting a profit, greater than leaving the money in the savings account, since $1 + r < u$. This correspond to point iii) of Definition 2.10 of arbitrage. Nevertheless, this investment strategy does not correspond to an arbitrage portfolio, since there is also exposure to the risk of loss (we have $d < 1 + r$, so there is a positive probability that the stock is worth less than the savings account) i.e. property ii) is ruled out. More generally, we have the following:

Corollary 2.34 *The binomial model is arbitrage-free and complete if and only if condition* (2.39) *holds. In this case the arbitrage price* (H_n) *of a derivative* X *is uniquely defined by the following risk-neutral pricing formula:*

$$H_n = \frac{1}{(1 + r)^{N-n}} E^Q [X \mid \mathcal{F}_n], \qquad 0 \le n \le N. \tag{2.46}$$

In particular, if $X = F(S_N)$, *we have the following explicit formula for the initial price of* X*:*

$$\begin{aligned} H_0 &= \frac{1}{(1 + r)^N} E^Q [F(S_N)] \\ &= \frac{1}{(1 + r)^N} \sum_{k=0}^{N} \binom{N}{k} q^k (1 - q)^{N-k} F(u^k d^{N-k} S_0). \end{aligned} \tag{2.47}$$

Proof. Combining Theorem 2.33 with the Fundamental Theorems of asset pricing, we prove that the binomial model is arbitrage-free and complete if and only if condition (2.39) holds. Formula (2.46) follows from (2.27). Formula (2.47) follows from (2.46) with $n = 0$ and (2.42). □

In Remark 2.14, we called P the objective or real-world probability, since it has to be determined on the basis of observations on the market, while Q is defined a posteriori. Indeed the EMM has no connection with the “real world”, but it is useful to prove theoretical results and to get simple and elegant expressions for the prices of derivatives such as formulas (2.46) and (2.47).

2.3.2 Hedging strategies

In the previous section we showed the completeness of the binomial market as a consequence of the theoretical result of Theorem 2.29: here we aim at giving a direct and constructive proof of the existence of a replicating strategy for a derivative X.

First we analyze the case of a *path-independent option* X that is $\sigma(S_N)$-measurable: in this case, by Corollary A.10, there exists a function F such that $X = F(S_N)$. If S_{N-1} denotes the price of the risky asset at time t_{N-1}, we have two possible final values

$$S_N = \begin{cases} uS_{N-1}, \\ dS_{N-1}. \end{cases}$$

For fixed $(\alpha, \beta) \in \mathcal{A}$, we set

$$V_n = \alpha_n S_n + \beta_n B_n, \qquad n = 0, \dots, N,$$

and impose the replication condition $V_N = X$: this is equivalent to

$$\begin{cases} \alpha_N u S_{N-1} + \beta_N B_N = F(uS_{N-1}), \\ \alpha_N d S_{N-1} + \beta_N B_N = F(dS_{N-1}). \end{cases} \tag{2.48}$$

Since it is necessary that both equations are satisfied, we get a linear system in the unknowns α_N and b_N, whose solution is given by

$$\bar{\alpha}_N = \frac{F(uS_{N-1}) - F(dS_{N-1})}{uS_{N-1} - dS_{N-1}}, \qquad \bar{\beta}_N = \frac{uF(dS_{N-1}) - dF(uS_{N-1})}{(1+r)^N (u-d)}. \tag{2.49}$$

Formula (2.49) expresses $\bar{\alpha}_N$ and $\bar{\beta}_N$ as functions of S_{N-1} and shows how to construct a predictable portfolio in a unique way at time t_{N-1}, replicating the derivative at time t_N for any trend of the underlying asset. We note that $\bar{\alpha}_N$ and $\bar{\beta}_N$ do not depend on the value of the parameter p (the objective probability of growth of the underlying asset). Further, $\bar{\alpha}_N$ has the form of an incremental ratio (technically called Delta).

We can now write the value of the replicating portfolio (or equivalently the arbitrage price H of the derivative) at time t_{N-1}: indeed by the self-financing condition we have

$$V_{N-1} = \bar{\alpha}_N S_{N-1} + \bar{\beta}_N B_{N-1} =$$

(by (2.49) and the definition of q in (2.40))

$$= \frac{1}{1+r} \left(qF(uS_{N-1}) + (1-q)F(dS_{N-1}) \right). \tag{2.50}$$

Recalling the Markov property (2.43) and the expression

$$q = Q(S_N = uS_{N-1}) = 1 - Q(S_N = dS_{N-1}),$$

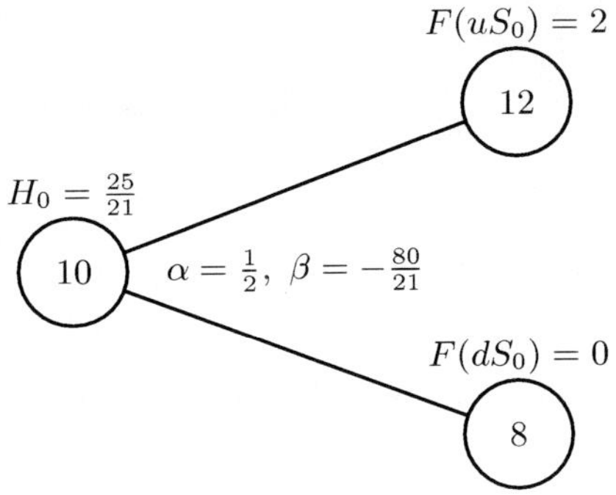

Fig. 2.2. Hedging of a Call in a one-period binomial model

we have that (2.50) is consistent with the risk-neutral pricing formula (2.46) that here reads

$$H_{N-1} = V_{N-1} = \frac{1}{1+r} E^Q \left[F(S_N) \mid S_{N-1} \right]. \tag{2.51}$$

By (2.51), H_{N-1} is a deterministic function of S_{N-1} and equals the discounted conditional expectation $F(S_N)$ given S_{N-1}.

Let us now consider a simple example to fix the ideas.

Example 2.35 We suppose that the current price of a stock is $S_0 = 10$ and that over the year the price can rise or fall within 20% of its initial value. We assume that the risk-free rate is $r = 5\%$ and we determine the hedging strategy for a Call option with maturity $T = 1$ year and strike $K = 10$. In this case $u = 1.2$ and $d = 0.8$ and the replication condition (2.48) becomes

$$\begin{cases} 12\alpha + \frac{105}{100}\beta = 2, \\ 8\alpha + \frac{105}{100}\beta = 0, \end{cases}$$

hence $\alpha = \frac{1}{2}$ and $\beta = -\frac{80}{21}$. Then the current value of the hedging portfolio (corresponding to the arbitrage price of the option) is equal to

$$V_0 = 10\alpha + \beta = \frac{25}{21}.$$

□

Let us go back to the previous argument and repeat it to compute, by a backward induction, the complete hedging strategy $(\bar{\alpha}_n, \bar{\beta}_n)$ for $n = 1, \dots, N$. More precisely, assume that the arbitrage price $H_n = H_n(S_n)$ is known. Then, since at time t_n we have two cases

$$S_n = \begin{cases} uS_{n-1}, \\ dS_{n-1}, \end{cases}$$

by imposing the replication condition $V_n = H_n$, we obtain the system

$$\begin{cases} \alpha_n u S_{n-1} + \beta_n B_n = H_n(uS_{n-1}), \\ \alpha_n d S_{n-1} + \beta_n B_n = H_n(dS_{n-1}). \end{cases} \tag{2.52}$$

The solution of (2.52) is

$$\bar{\alpha}_n = \frac{H_n(uS_{n-1}) - H_n(dS_{n-1})}{S_{n-1}(u-d)}, \qquad \bar{\beta}_n = \frac{uH_n(dS_{n-1}) - dH_n(uS_{n-1})}{(1+r)^n(u-d)}, \tag{2.53}$$

that is the hedging strategy for the n-th period $[t_{n-1}, t_n]$. By the self-financing condition we also find the arbitrage price of X at time t_{n-1}:

$$H_{n-1} := V_{n-1} = \bar{\alpha}_n S_{n-1} + \bar{\beta}_n B_{n-1}. \tag{2.54}$$

Equivalently we have

$$H_{n-1} = \frac{qH_n(uS_{n-1}) + (1-q)H_n(dS_{n-1})}{1+r} = \frac{1}{1+r} E^Q \left[H_n \mid S_{n-1} \right]. \tag{2.55}$$

More generally, we have

$$\begin{aligned} H_{N-n} &= \frac{1}{(1+r)^n} E^Q \left[F(S_N) \mid S_{N-n} \right] \\ &= \frac{1}{(1+r)^n} \sum_{k=0}^{n} \binom{n}{k} q^k (1-q)^{n-k} F(u^k d^{n-k} S_{N-n}), \end{aligned}$$

and in particular the current value of the derivative is given by

$$\begin{aligned} H_0 &= \frac{1}{(1+r)^N} E^Q \left[F(S_N) \right] \\ &= \frac{1}{(1+r)^N} \sum_{k=0}^{N} \binom{N}{k} q^k (1-q)^{N-k} F(u^k d^{N-k} S_0) \end{aligned}$$

consistently with formula (2.47).

The previous expressions can be computed explicitly as a function of the current value of the underlying asset, once F is given; nevertheless in the following section we will see that, from a practical point of view, it is easier to compute the price using a suitable iterative algorithm.

Remark 2.36 As we have already pointed out in Paragraph 1.2, *the arbitrage price of X does not depend on the probability p of growth under the real-world probability* but only on the increase and decrease rates u, d (and also on r). □

Let us consider now the general case and let X be a European derivative (possibly path-dependent). The final replication condition reads

$$\begin{cases} \alpha_N u S_{N-1} + \beta_N B_N = X^u, \\ \alpha_N d S_{N-1} + \beta_N B_N = X^d, \end{cases} \tag{2.56}$$

where X^u and X^d denote the payoffs in case of increase and decrease of the asset given the information at time t_{N-1}, respectively. The solution of the linear system (2.56) is given by

$$\bar{\alpha}_N = \frac{X^u - X^d}{(u-d)S_{N-1}}, \qquad \bar{\beta}_N = \frac{uX^d - dX^u}{(1+r)^N(u-d)}, \tag{2.57}$$

and provides the strategy for the last period that guarantees the final replication. By the self-financing condition we have that

$$H_{N-1} := V_{N-1} = \bar{\alpha}_N S_{N-1} + \bar{\beta}_N B_{N-1}$$

is the arbitrage price of X at time t_{N-1}. A direct computation shows that this result is consistent with the risk-neutral valuation formula (2.46): precisely, we have

$$\bar{\alpha}_N S_{N-1} + \bar{\beta}_N B_{N-1} = \frac{1}{1+r}\left(qX^u + (1-q)X^d\right) = \frac{1}{1+r} E^Q\left[X \mid \mathcal{F}_{N-1}\right].$$

Next we use an analogous argument in the generic n-th period. If S_{n-1} denotes the asset price at time t_{n-1}, we have

$$S_n = \begin{cases} uS_{n-1}, \\ dS_{n-1}. \end{cases}$$

We denote by H_n^u and H_n^d the arbitrage prices at time t_n, given the information at time t_{n-1}, in case of increase and decrease of the underlying asset, respectively. Imposing the replication condition $V_n = H_n$, we obtain the system

$$\begin{cases} \alpha_n u S_{n-1} + \beta_n B_n = H_n^u, \\ \alpha_n d S_{n-1} + \beta_n B_n = H_n^d, \end{cases} \tag{2.58}$$

with solution

$$\bar{\alpha}_n = \frac{H_n^u - H_n^d}{S_{n-1}(u-d)}, \qquad \bar{\beta}_n = \frac{uH_n^d - dH_n^u}{(1+r)^n(u-d)}. \tag{2.59}$$

By the self-financing condition we infer

$$H_{n-1} := V_{n-1} = \bar{\alpha}_n S_{n-1} + \bar{\beta}_n B_{n-1} \tag{2.60}$$

that is the arbitrage price of X at time t_{n-1}. Equivalently we have

$$H_{n-1} = \frac{qH_n^u + (1-q)H_n^d}{1+r} = \frac{1}{1+r} E^Q\left[H_n \mid \mathcal{F}_{n-1}\right]. \tag{2.61}$$

Example 2.37 (European Call option) We consider the payoff function of a European Call option with strike K:

$$F(S_N) = (S_N - K)^+ = \max\{S_N - K, 0\}.$$

By using formula (2.47) and recalling that $q = \frac{1+r-d}{u-d}$, the initial price C_0 of the option is given by

$$\begin{aligned} C_0 =& \frac{1}{(1+r)^N} \sum_{h=0}^{N} \binom{N}{h} q^h (1-q)^{N-h} \left(u^h d^{N-h} S_0 - K \right)^+ \\ =& S_0 \sum_{h>h_0}^{N} \binom{N}{h} \left(\frac{qu}{1+r} \right)^h \left(\frac{(1-q)d}{1+r} \right)^{N-h} \\ & - \frac{K}{(1+r)^N} \sum_{h>h_0}^{N} \binom{N}{h} q^h (1-q)^{N-h}, \end{aligned}$$

where h_0 is the smallest non-negative integer number greater than or equal to

$$\frac{\log \frac{K}{d^N S_0}}{\log \frac{u}{d}}.$$

Therefore

$$C_0 = S_0 \mathcal{N}(\widetilde{q}) - \frac{K}{(1+r)^N} \mathcal{N}(q), \tag{2.62}$$

where

$$\widetilde{q} = \frac{qu}{1+r} \tag{2.63}$$

and

$$\mathcal{N}(p) = \sum_{h>h_0}^{N} \binom{N}{h} p^h (1-p)^{N-h}, \qquad p = \widetilde{q}, q.$$

We note that $\mathcal{N}(\widetilde{q})$ and $\mathcal{N}(q)$ in formula (2.62) can be expressed in terms of probability of events with respect to suitable probability measures. Indeed, for $0 \le n \le N$, we have

$$\begin{aligned} C_n &= B_n E^Q \left[\frac{(S_N - K)^+}{B_N} \mid \mathcal{F}_n \right] \\ &= B_n E^Q \left[\frac{(S_N - K)}{B_N} \mathbb{1}_{\{S_N > K\}} \mid \mathcal{F}_n \right] \equiv I_1 - I_2, \end{aligned}$$

where

$$I_2 = \frac{B_n K}{B_N} Q \left(S_N > K \mid \mathcal{F}_n \right),$$

and

$$\begin{aligned} I_1 &= B_n E^Q \left[\frac{S_N}{B_N} \mathbb{1}_{\{S_N > K\}} \mid \mathcal{F}_n \right] \\ &= B_n \frac{S_0}{B_0} \frac{E^Q \left[\frac{S_N}{B_N} \mathbb{1}_{\{S_N > K\}} \left(\frac{S_0}{B_0} \right)^{-1} \mid \mathcal{F}_n \right]}{E^Q \left[\frac{S_N}{B_N} \left(\frac{S_0}{B_0} \right)^{-1} \mid \mathcal{F}_n \right]} E^Q \left[\frac{S_N}{B_N} \left(\frac{S_0}{B_0} \right)^{-1} \mid \mathcal{F}_n \right] = \end{aligned}$$

(by Bayes' formula and Theorem 2.19 on the change of numeraire, denoting by $\widetilde{Q}$ the EMM with numeraire S)

$$= B_n \frac{S_0}{B_0} \widetilde{Q}(S_N > K \mid \mathcal{F}_n) \frac{S_n}{B_n} \left(\frac{S_0}{B_0}\right)^{-1}.$$

In conclusion, we obtain the following formula:

$$C_n = S_n \widetilde{Q}(S_N > K \mid \mathcal{F}_n) - \frac{K}{(1+r)^{N-n}} Q\left(S_N > K \mid \mathcal{F}_n\right),$$

and in particular, with $n = 0$

$$C_0 = S_0 \widetilde{Q}(S_N > K) - \frac{K}{(1+r)^N} Q\left(S_N > K\right). \tag{2.64}$$

Comparing (2.64) to (2.62), we see that the EMM $\widetilde{Q}$ with numeraire S is the equivalent measure to P such that (cf. (2.40) and (2.63))

$$\widetilde{Q}(1 + \mu_n = u) = \widetilde{q} = \frac{qu}{1+r}.$$

It is easy to verify that $0 < \widetilde{q} < 1$ if and only if $d < 1 + r < u$.

Although formulas (2.64) and (2.62) may be more elegant from a theoretical point of view, for the numerical computation of the price of a derivative in the binomial model, it is often preferable to use a recursive algorithm as the one that we are going to present in the next section. □

2.3.3 Binomial algorithm

In this section we present an iterative scheme that is easily implementable to determine the replicating strategy and the price of a *path-independent* derivative. We discuss briefly also some particular cases of path-dependent derivatives.

Path-independent case. In this case the payoff is of the form $X = F(S_N)$. The arbitrage price H_{n-1} and the strategy (α_n, β_n) depend only on the price S_{n-1} of the underlying asset at time t_{n-1}. Since S_n is of the form

$$S_n = S_{n,k} := u^k d^{n-k} S_0, \qquad n = 0, \dots, N \text{ and } k = 0, \dots, n, \tag{2.65}$$

the value of the underlying asset is determined by the "coordinates" n (time) and k (number of movements of increase). Hence we introduce the following notation:

$$H_{n,k} = H_n(S_{n,k}), \tag{2.66}$$

for the arbitrage price of X, and analogously

$$\alpha_{n,k} = \alpha_n(S_{n-1,k}), \qquad \beta_{n,k} = \beta_n(S_{n-1,k}),$$

for the related hedging strategy. By the replication condition and the pricing formula (2.55), we get the following backward iterative formula for the price (H_n):

$$H_{N,k} = F(S_{N,k}), \qquad 0 \le k \le N, \tag{2.67}$$

$$H_{n-1,k} = \frac{1}{1+r}(qH_{n,k+1} + (1-q)H_{n,k}), \qquad 0 \le k \le n-1, \tag{2.68}$$

for $n = 1, \dots, N$ and where q defined in (2.40). Clearly the initial price of the derivative is equal to $H_{0,0}$.

Once we have determined the values $H_{n,k}$, by (2.53) the corresponding hedging strategy is given explicitly by

$$\alpha_{n,k} = \frac{H_{n,k+1} - H_{n,k}}{S_{n-1,k}(u-d)}, \qquad \beta_{n,k} = \frac{uH_{n,k} - dH_{n,k+1}}{(1+r)^n(u-d)}, \tag{2.69}$$

for $n = 1, \dots, N$ and $k = 0, \dots, n-1$. We remark explicitly that $(\alpha_{n,k}, \beta_{n,k})$ *is the strategy for the n-th period* $[t_{n-1}, t_n]$, *that is constructed at time* t_{n-1} *in the case* $S_{n-1} = S_{n-1,k}$.

Example 2.38 We consider a European Put option with strike $K = \frac{5}{2}$ and value of the underlying asset $S_0 = 1$. We set the following values for the parameters in a three-period binomial model:

$$u = 2, \qquad d = \frac{1}{2}, \qquad r = \frac{1}{2}$$

hence we obtain

$$q = \frac{1+r-d}{u-d} = \frac{2}{3}.$$

First of all we construct in Figure 2.3 the binomial tree where we put the prices of the underlying asset inside the circles and the payoff of the option at maturity outside, using notation (2.66), i.e. $H_{n,k}$ is the value of the derivative at time t_n if the underlying asset has grown k times.

Next we use the algorithm (2.67)-(2.68)

$$H_{n-1,k} = \frac{1}{1+r}(qH_{n,k+1} + (1-q)H_{n,k}) = \frac{1}{1+\frac{1}{2}}\left(\frac{2}{3}H_{n,k+1} + \frac{1}{3}H_{n,k}\right)$$

and we compute the arbitrage prices of the option, putting them outside the circles in Figure 2.4.

Eventually, using formulas (2.69), we complete the figure with the hedging strategy for the derivative in Figure 2.5. □

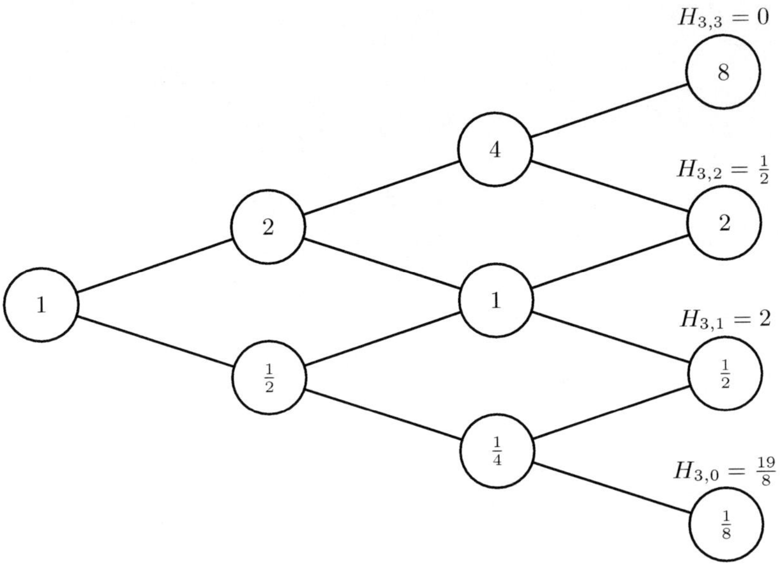

Fig. 2.3. Three-period binomial tree for a Put option with strike $K = \frac{5}{2}$ and $S_0 = 1$, with parameters $u = 2$ and $d = r = \frac{1}{2}$

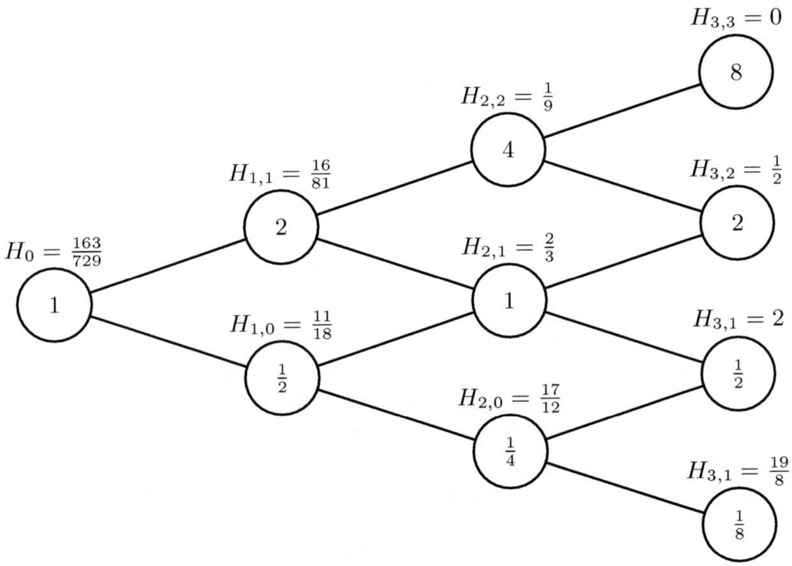

Fig. 2.4. Arbitrage prices of a Put option with strike $K = \frac{5}{2}$ and $S_0 = 1$ in a three-period binomial model with parameters $u = 2$ and $d = r = \frac{1}{2}$

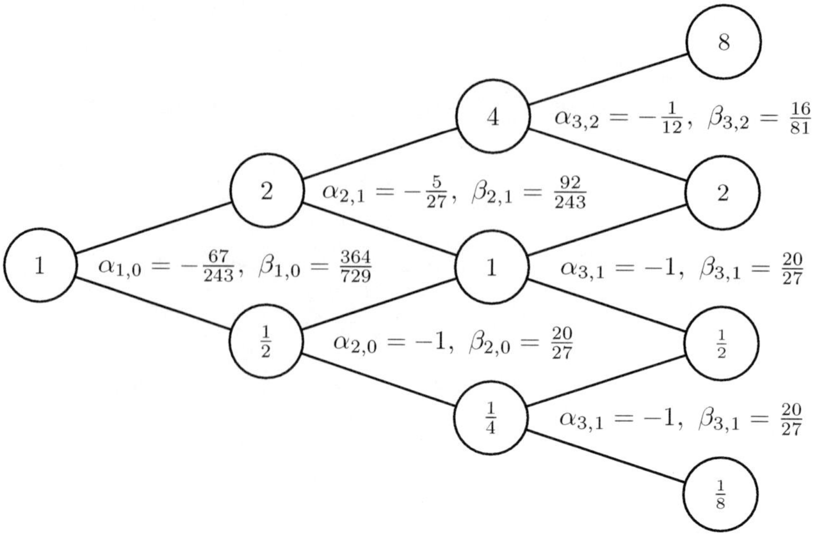

Fig. 2.5. Hedging strategy for a Put option with strike $K = \frac{5}{2}$ and $S_0 = 1$ in a three-period binomial model with parameters $u = 2$ and $d = r = \frac{1}{2}$

Path-dependent case. We examine some well-known path-dependent derivatives: Asian, look-back and barrier options. The iterative scheme (2.67)-(2.68) is based upon the fact that the price H_n of the derivative has the Markov property: so it is a function of the prices at time t_n and it does not depend on the previous prices. In particular the scheme (2.67)-(2.68) requires that at the n-th step $n+1$ equations must be solved in order to determine $(H_{n,k})_{k=0,\dots,n}$. Therefore the computational complexity grows linearly with the number of steps of the discretization.

On the contrary, we have already pointed out that, in the path-dependent case, H_n depends on the path of the underlying asset $(S_0, \dots, S_n)$ until time t_n. Since there are 2^n possible paths, the number of the equations to solve grows exponentially with the number of the steps of the discretization. For example, if we choose $N = 100$, we should solve 2^{100} equations just to compute the price at maturity and this is unfeasible.

Sometimes, by adding a state variable that incorporates the information from the past (the path-dependent variable), it is possible to make the price process Markovian: this simple idea is sometimes used also in the continuous case. We consider the following payoff:

$$F(S_N, A_N) = \begin{cases} (S_N - A_N)^+ & \text{Call option with variable strike,} \\ (A_N - K)^+ & \text{Call option with fixed strike } K, \end{cases} \qquad (2.70)$$

where A denotes the path-dependent variable: more precisely, for $n = 0, \dots, N$,

$$A_n = \begin{cases} \frac{1}{n+1}\sum_{k=0}^{n} S_k & \text{(Asian option with arithmetic average)} \\ \left(\prod_{k=0}^{n} S_k\right)^{\frac{1}{n+1}} & \text{(Asian option with geometric average)} \\ \min_{0\le k\le n} S_k & \text{(Look-back option with variable strike)} \\ \max_{0\le k\le n} S_k & \text{(Look-back option with fixed strike).} \end{cases} \tag{2.71}$$

When passing from time t_{n-1} to time t_n, we have $S_n = uS_{n-1}$ or $S_n = dS_{n-1}$ and consequently A_n takes the values A_n^u or A_n^d where

$$A_n^u = \begin{cases} \frac{nA_{n-1}+uS_{n-1}}{n+1} & \text{(Asian option with arithmetic average)} \\ ((A_{n-1})^n uS_{n-1})^{\frac{1}{n+1}} & \text{(Asian option with geometric average)} \\ \min\{A_{n-1}, uS_{n-1}\} & \text{(Look-back option with variable strike)} \\ \max\{A_{n-1}, uS_{n-1}\} & \text{(Look-back option with fixed strike)} \end{cases} \tag{2.72}$$

and A_n^d is defined analogously. The following result can be proved as Theorem 2.31.

Lemma 2.39 *The stochastic process (S, A) has the Markov property, and for every function f we have that*

$$\begin{aligned} E^Q\left[\varphi(S_{n+1}, A_{n+1}) \mid \mathcal{F}_n\right] &= E^Q\left[\varphi(S_{n+1}, A_{n+1}) \mid (S_n, A_n)\right] \\ &= q\varphi(uS_n, A_n^u) + (1-q)\varphi(dS_n, A_n^d). \end{aligned}$$

We set $S_{n,k}$ as in (2.65) and denote by $A_{n,k}(j)$ the possible values of the path-dependent variable corresponding to $S_{n,k}$, for $0 \le j \le J(n,k)$ and suitable $J(n,k) \in \mathbb{N}$.

Example 2.40 Under the assumption $ud = 1$, we have

$$S_{n,k} = \begin{cases} d^{n-2k} S_0 & \text{if } n \ge 2k, \\ u^{2k-n} S_0 & \text{if } n < 2k. \end{cases}$$

In the case of a Look-back option with fixed strike, if $n \ge 2k$ then $S_{n,k} \le S_0$ and

$$A_{n,k}(j) = u^{k-j} S_0, \qquad j = 0, \dots, n-k,$$

while, if $n \le 2k$, then $S_{n,k} \ge S_0$ and

$$A_{n,k}(j) = u^{k-j} S_0, \qquad j = 0, \dots, k.$$

Just to fix the ideas, it can be useful to construct a binomial tree with $N = 4$ and verify the previous formulas.

Also in the case of a Call option with strike K and barrier $B > K$, we can use the previous processes: here the payoff is given by

$$F(S_N, A_N) = (S_N - K)^+ \mathbb{1}_{\{A_N < B\}}.$$

□

In general we put

$$H_{n,k}(j) = H_n(S_{n,k}, A_{n,k}(j)). \tag{2.73}$$

By the previous lemma, since

$$H_n = \frac{1}{1+r} E^Q \left[H_{n+1} \mid \mathcal{F}_n \right]$$

we can price a path-dependent derivative using the following iterative scheme:

$$H_{N,k}(j) = F(S_{N,k}, A_{N,k}(j)),$$

for $0 \leq k \leq N,\ 0 \leq j \leq J(N,k)$, and

$$\begin{aligned} H_{n-1,k}(j) =& \frac{1}{1+r}\Big(qH_n \left(uS_{n-1,k}, A^u_{n-1,k}(j) \right) \\ &+ (1-q) H_n \left(dS_{n-1,k}, A^d_{n-1,k}(j) \right) \Big), \end{aligned}$$

for $0 \leq k \leq n-1,\ 0 \leq j \leq J(n-1,k)$, and $n = 1, \dots, N$. Eventually the hedging strategy is given by

$$\begin{aligned} \alpha_{n,k}(j) &= \frac{H_{n,k+1}(j) - H_{n,k}(j)}{(u-d)S_{n-1,k}}, \\ \beta_{n,k}(j) &= \frac{uH_{n,k}(j) - dH_{n,k+1}(j)}{(u-d)(1+r)^n}, \end{aligned} \tag{2.74}$$

for $n = 1, \dots, N,\ k = 0, \dots, n-1$ and $j = 0, \dots, J(n,k)$. Note that, for a Look-back option with fixed strike, $J(n,k) \leq \frac{n}{2}$ and so the computational complexity at the n-th step is of order n^2.

2.3.4 Calibration

The calibration of a model consists in determining the parameters by the observation of the current-world market. The parameters in the binomial model are the risk-free rate r over the period $[t_{n-1}, t_n]$, the increase and decrease factors u, d of the underlying asset and the objective probability p. However, we have already noticed (cf. Remark 2.36) that the arbitrage price of a derivative *does not depend on* p: therefore only r, u, d have to be determined. We point out that the parameters depend on N since obviously the increase and decrease rates depend on the amplitude of the time period $\frac{T}{N}$: nevertheless in this section N is fixed and so we do not show this dependence explicitly. In the following procedure the hypothesis of P-independence of the random variables $\mu_1, \dots, \mu_N$ plays a crucial role.

If we suppose that the *annual interest rate* $\hat{r}$ is known, then we can obtain r easily by the relation

$$1 + r = e^{\hat{r}\frac{T}{N}}. \tag{2.75}$$

Next we define the *annual rate of return* μ of the risky asset by putting

$$S_T = S_0 e^{\mu T}, \tag{2.76}$$

or equivalently

$$\mu T = \log \frac{S_T}{S_0}.$$

It is clear that μ is a random variable that plays an analogous role to the interest rate in the compounding formula. By (2.35) we have

$$\log \frac{S_T}{S_0} = \sum_{n=1}^{N} \log(1 + \mu_n),$$

and since the random variables μ_n are identically distributed by (2.36), we get the following formula that defines the *average rate of return* m:

$$mT := E[\mu T] = NE[\log(1 + \mu_1)] = N(p \log u + (1 - p) \log d). \tag{2.77}$$

Analogously the *volatility* σ is defined by the following equality:

$$\sigma^2 T := \mathrm{var}\left(\log \frac{S_T}{S_0}\right) = \mathrm{var}\left(\sum_{n=1}^{N} \log(1 + \mu_n)\right) =$$

(by the independence of the random variables μ_n)

$$= N \,\mathrm{var}\,(\log(1 + \mu_1)) =$$

(in virtue of Exercise A.36)

$$= Np(1 - p)\left(\log \frac{u}{d}\right)^2. \tag{2.78}$$

In other words, the average rate of return and the volatility are the expectation and the standard deviation of the annual rate of return, respectively. The volatility represents one of the most common and known estimators of the riskiness of the underlying asset. In principle the values of m and σ can be considered approximately observable in the current-world market. For example, one can easily get some estimates of the values of m and σ starting from a given set of historical values of S. We therefore suppose that m and σ are known and we try to draw from them the value of u and d. By equations (2.77)-(2.78) and putting $\delta = \frac{T}{N}$, we obtain the system

$$\begin{cases} m\delta = (p \log u + (1-p)\log d), \\ \sigma^2\delta = p(1-p)\left(\log \frac{u}{d}\right)^2. \end{cases} \tag{2.79}$$

Thus we have a non-linear system of two equations in the three unknowns u, d and p: in order to find a solution, we impose another condition a priori. The most common choices in the literature are the following ones: $p = \frac{1}{2}$ or

$$ud = 1. \tag{2.80}$$

Imposing $p = \frac{1}{2}$, system (2.79) becomes

$$\begin{cases} ud = e^{2\delta m}, \\ \frac{u}{d} = e^{2\sigma\sqrt{\delta}}, \end{cases}$$

and its solution is given by

$$u = e^{\sigma\sqrt{\delta}+m\delta}, \qquad d = e^{-\sigma\sqrt{\delta}+m\delta}. \tag{2.81}$$

Imposing condition (2.80), we have[6] $d < 1 < u$ and system (2.79) becomes

$$\begin{cases} 2p = 1 + \frac{m\delta}{\log u}, \\ \sigma^2\delta = 4p(p-1)(\log u)^2, \end{cases}$$

and its solution is given by

$$u = e^{\sigma\sqrt{\delta}\sqrt{1+\delta\left(\frac{m}{\sigma}\right)^2}}, \qquad d = e^{-\sigma\sqrt{\delta}\sqrt{1+\delta\left(\frac{m}{\sigma}\right)^2}}. \tag{2.82}$$

In both cases (2.81) and (2.82), we obtain[7]

$$u = e^{\sigma\sqrt{\delta}+o(\sqrt{\delta})} = 1 + \sigma\sqrt{\delta} + o(\sqrt{\delta}),$$
$$d = e^{-\sigma\sqrt{\delta}+o(\sqrt{\delta})} = 1 - \sigma\sqrt{\delta} + o(\sqrt{\delta}),$$

for $\delta \to 0$; in other terms, $\frac{u-1}{\sqrt{\delta}}$ and $\frac{1-d}{\sqrt{\delta}}$ approximate the value σ of the volatility or riskiness of the asset. For the sake of simplicity, in order to implement the binomial algorithm it is very common to choose

$$u = e^{\sigma\sqrt{\delta}}, \qquad d = e^{-\sigma\sqrt{\delta}}. \tag{2.83}$$

[6] We note that, if condition (2.80) holds, then

$$u^n d^n S_0 = S_0$$

therefore the price "moves around" its starting value.

[7] We recall that the function f is a "little-o" of the function g as $x \to x_0$ (in symbols $f(x) = o(g(x))$ as $x \to x_0$) if there exists a function w such that $f = gw$ and

$$\lim_{x \to x_0} w(x) = 0.$$

Remark 2.41 Assuming (2.83) and recalling that $\delta = \frac{T}{N}$, for the maximum and minimum values of the final price of the underlying asset, we have

$$S_N^{(\max)} = u^N S_0 = e^{\sigma\sqrt{NT}} S_0 \xrightarrow[N\to\infty]{} +\infty,$$
$$S_N^{(\min)} = d^N S_0 = e^{-\sigma\sqrt{NT}} S_0 \xrightarrow[N\to\infty]{} 0,$$

and so, when N increases, the interval of the final values of S gets larger and "covers" the whole $\mathbb{R}_{>0}$ as $x \to x_0$.

The no-arbitrage condition $d < 1 + r < u$ becomes

$$e^{-\sigma\sqrt{\delta}} < e^{\hat{r}\delta} < e^{\sigma\sqrt{\delta}}$$

or equivalently

$$-\sigma\sqrt{N} < \hat{r}\sqrt{T} < \sigma\sqrt{N}.$$

Therefore if we choose arbitrary values $\hat{r}$ and $\sigma > 0$, for the the annual risk-free rate and the volatility respectively, then the no-arbitrage condition is fulfilled provided that N is large enough: in that case, by (2.83) the EMM is defined by

$$q = \frac{1 + r - d}{u - d} = \frac{e^{\hat{r}\delta} - e^{-\sigma\sqrt{\delta}}}{e^{\sigma\sqrt{\delta}} - e^{-\sigma\sqrt{\delta}}} = \frac{1}{2} + \frac{1}{2\sigma}\left(\hat{r} - \frac{\sigma^2}{2}\right)\sqrt{\delta} + o(\sqrt{\delta})$$

as $\delta \to 0$. □

Example 2.42 We set the parameters of the market as follows: annual interest rate $\hat{r} = 5\%$ and volatility $\sigma = 30\%$. We consider a 10-period binomial model for an option with maturity in 6 months: $N = 10$ and $T = \frac{1}{2}$. By (2.75) we have

$$r = e^{\frac{5}{100}\cdot\frac{1}{2}\cdot\frac{1}{10}} - 1 \approx 0.0025.$$

Analogously, by (2.83), we have

$$u \approx e^{\frac{30}{100}\cdot\frac{1}{\sqrt{20}}} \approx 1.0693.$$ □

2.3.5 Binomial model and Black-Scholes formula

We have seen that the binomial model, with a fixed number of periods N, allows us to determine the initial arbitrage price $H_0^{(N)}$ of a given derivative X. We may wonder if the binomial model is *stable*, this meaning that, if we increase the number of steps, the price $H_0^{(N)}$ actually converges to some value, so that the situation in which the value diverges or oscillates around more than one value is avoided[8].

[8] The divergence or the oscillation around some values would cast doubts on the consistency of the model.

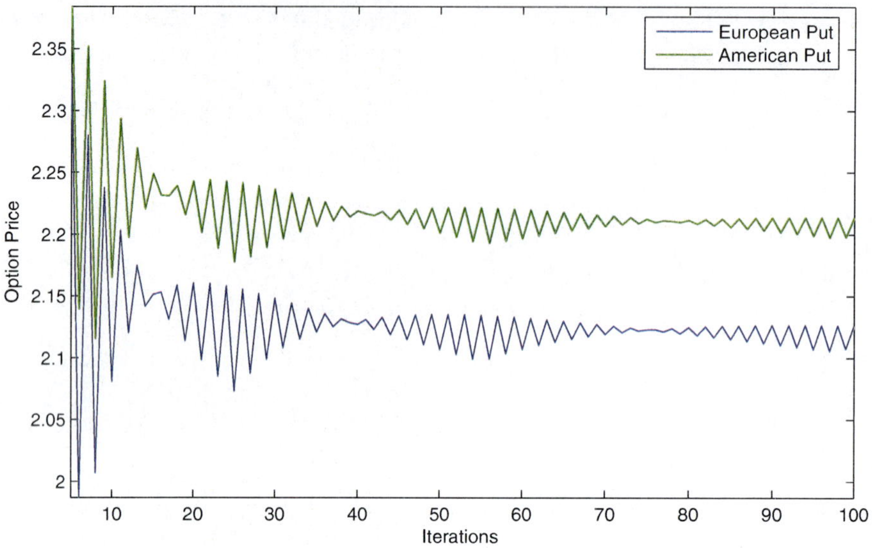

Fig. 2.6. Convergence of the price of a European and an American option in the binomial model to the corresponding Black-Scholes price when N tends to infinity

In this section we prove that the binomial model is stable and approximates the classical Black-Scholes model in a suitable way, when N tends to infinity. In what follows, the number of periods $N \in \mathbb{N}$ is variable so it is important to express the dependence on N of the parameters of the model explicitly: therefore we denote the interest rate, the increase and decrease factors by r_N, u_N, d_N respectively, the random variables in (2.35) for $k = 1, \dots, N$ by $\mu_k^{(N)}$ and the martingale probability by q_N, Q_N. Let $T > 0$ be fixed: we put

$$\delta_N = \frac{T}{N},$$

so, by (2.75), we have

$$1 + r_N = e^{r\delta_N}, \tag{2.84}$$

where we denote by r the *annual risk-free rate*. Further, we assume that u_N and d_N take the following form:

$$u_N = e^{\sigma\sqrt{\delta_N} + \alpha\delta_N}, \qquad d_N = e^{-\sigma\sqrt{\delta_N} + \beta\delta_N}, \tag{2.85}$$

where α, β are real constants. Such a choice is in line with what we saw in the previous section: indeed by imposing one of the conditions $p = \frac{1}{2}$ or $ud = 1$ for the calibration, we obtain parameters of the form (2.85). Furthermore, the simplest choice (2.83) corresponds to $\alpha = \beta = 0$.

First of all we observe that the asymptotic behaviour of the EMM is independent of α, β. Indeed we have the following:

Lemma 2.43 *If* (2.84)-(2.85) *hold, we have*

$$\lim_{N\to\infty} q_N = \frac{1}{2}. \tag{2.86}$$

Proof. By definition

$$q_N = \frac{e^{r\delta_N} - e^{-\sigma\sqrt{\delta_N}+\beta\delta_N}}{e^{\sigma\sqrt{\delta_N}+\alpha\delta_N} - e^{-\sigma\sqrt{\delta_N}+\beta\delta_N}}. \tag{2.87}$$

Then, using a Taylor expansion for the exponentials in the expression (2.87) of q_N, we get

$$\begin{aligned} 2q_N - 1 &= \frac{2e^{r\delta_N} - e^{\sigma\sqrt{\delta_N}+\alpha\delta_N} - e^{-\sigma\sqrt{\delta_N}+\beta\delta_N}}{e^{\sigma\sqrt{\delta_N}+\alpha\delta_N} - e^{-\sigma\sqrt{\delta_N}+\beta\delta_N}} \\ &= \frac{\left(r - \frac{\sigma^2}{2} - \frac{\alpha+\beta}{2}\right)\delta_N + \mathrm{o}(\delta_N)}{\sigma\sqrt{\delta_N}(1+\mathrm{o}(1))}, \qquad \text{as } N \to \infty, \end{aligned} \tag{2.88}$$

hence the claim. □

Now we consider a European Put option with strike K and maturity T: by formula (2.47), the initial price $P_0^{(N)}$ of the option in the N-period binomial model is given by

$$\begin{aligned} P_0^{(N)} &= e^{-rT} E^{Q_N}\left[\left(K - S_0 \prod_{k=1}^{N}\left(1+\mu_k^{(N)}\right)\right)^+\right] \\ &= e^{-rT} E^{Q_N}\left[\left(K - S_0 e^{X_N}\right)^+\right], \end{aligned} \tag{2.89}$$

where we put

$$X_N = \log \prod_{k=1}^{N}\left(1+\mu_k^{(N)}\right) = \sum_{k=1}^{N} Y_k^{(N)}, \tag{2.90}$$

and

$$Y_k^{(N)} := \log\left(1+\mu_k^{(N)}\right), \qquad k = 1, \dots, N,$$

are i.i.d. random variables[9]. Further, we have

$$\begin{aligned} Q_N\left(Y_k^{(N)} = \sigma\sqrt{\delta_N} + \alpha\delta_N\right) &= q_N, \\ Q_N\left(Y_k^{(N)} = -\sigma\sqrt{\delta_N} + \beta\delta_N\right) &= 1 - q_N. \end{aligned} \tag{2.91}$$

We rewrite (2.89) in the form

$$P_0^{(N)} = E^{Q_N}\left[\varphi(X_N)\right],$$

[9] By Theorem 2.33, the random variables $\mu_k^{(N)}$ are independent also under the EMM.

where

$$\varphi(x) = e^{-rT}(K - S_0 e^x)^+ \tag{2.92}$$

is a continuous bounded function on $\mathbb{R}$, $\varphi \in C_b(\mathbb{R})$. The following result provides asymptotic values for the mean and the variance of X_N in (2.90).

Lemma 2.44 *We have:*

$$\lim_{N\to\infty} E^{Q_N}[X_N] = \left(r - \frac{\sigma^2}{2}\right)T, \tag{2.93}$$

$$\lim_{N\to\infty} \mathrm{var}^{Q_N}(X_N) = \sigma^2 T. \tag{2.94}$$

Before proving the lemma, let us dwell on some remarks. By the central limit theorem[10], X_N converges in distribution to a normally distributed random variable X and so, by (2.93)-(2.94), we have

$$X \sim \mathcal{N}_{\left(r-\frac{\sigma^2}{2}\right)T, \sigma^2 T}. \tag{2.95}$$

Since the function φ is bounded and continuous, we infer[11] that

$$\lim_{N\to\infty} P_0^{(N)} = \lim_{N\to\infty} E^{Q_N}[\varphi(X_N)] = E[\varphi(X)]. \tag{2.96}$$

Since X has a normal distribution, the expectation $E[\varphi(X)]$ can be computed explicitly and, as we will see, corresponds to the classical *Black-Scholes formula*.

Proof (of Lemma 2.44). In order to prove (2.93), we compute

$$\begin{aligned} E^{Q_N}\left[Y_1^{(N)}\right] &= q_N\left(\sigma\sqrt{\delta_N} + \alpha\delta_N\right) + (1 - q_N)\left(-\sigma\sqrt{\delta_N} + \beta\delta_N\right) \\ &= (2q_N - 1)\,\sigma\sqrt{\delta_N} + \delta_N\left(\alpha q_N + \beta(1 - q_N)\right) = \end{aligned}$$

(by (2.88) and (2.86))

$$\begin{aligned} &= \frac{\left(r - \frac{\sigma^2}{2} - \frac{\alpha+\beta}{2}\right)\delta_N + \mathrm{o}(\delta_N)}{1 + \mathrm{o}(1)} + \delta_N\left(\frac{\alpha+\beta}{2} + \mathrm{o}(1)\right) \\ &= \left(r - \frac{\sigma^2}{2}\right)\delta_N + \mathrm{o}(\delta_N), \qquad \text{as } N \to \infty. \end{aligned} \tag{2.97}$$

Then we have, recalling that $\delta_N = \frac{T}{N}$,

$$E^{Q_N}[X_N] = N E^{Q_N}\left[Y_1^{(N)}\right] = \left(r - \frac{\sigma^2}{2}\right)T + \mathrm{o}(1), \qquad \text{as } N \to \infty,$$

hence (2.93).

---

[10] See Lemma 2.45 for a rigorous proof of this statement.

[11] By (A.128): this is the reason why we considered a Put option instead of a Call. The Put-Call parity formula (cf. Corollary 1.1) allows us to obtain the price of a Call option: the reader can see also Remark 2.49.

Now we prove (2.94) by using the identity

$$\mathrm{var}^{Q_N}(X_N) = N\mathrm{var}^{Q_N}(Y) = N\left(E^{Q_N}\left[Y^2\right] - E^{Q_N}\left[Y\right]^2\right) \tag{2.98}$$

where we put $Y \equiv Y_1^{(N)}$. By (A.30) in Exercise A.36, we have

$$\begin{aligned} E^{Q_N}\left[Y^2\right] &= (\log u_N + \log d_N)\, E^{Q_N}\left[Y\right] - \log u_N \log d_N \\ &= \delta_N(\alpha+\beta)E^{Q_N}\left[Y\right] - \left(\sigma\sqrt{\delta_N} + \alpha\delta_N\right)\left(-\sigma\sqrt{\delta_N} + \beta\delta_N\right) \\ &= \sigma^2\delta_N + \mathrm{o}\left(\delta_N\right), \qquad \text{as } N \to \infty, \end{aligned} \tag{2.99}$$

and so the claim follows immediately substituting the last expression into (2.98), bearing in mind also that

$$E^{Q_N}\left[Y\right]^2 = \mathrm{o}\left(\delta_N\right), \qquad \text{as } N \to \infty. \qquad \square$$

Lemma 2.45 *The sequence of random variables* (X_N) *defined in* (2.90) *converges in distribution to a random variable* X *that is normally distributed as in* (2.95).

Proof. This result is a variation of the central limit Theorem A.146: by Lévy's Theorem A.141, it is enough to verify that the sequence (φ_{X_N}) of the corresponding characteristic functions converges pointwise. We have:

$$\varphi_{X_N}(\eta) = E^{Q_N}\left[e^{i\eta X_N}\right] =$$

(since the random variables $Y_k^{(N)}$ are i.i.d. and putting $Y \equiv Y_1^{(N)}$)

$$= \left(E^{Q_N}\left[e^{i\eta Y}\right]\right)^N =$$

(by Lemma A.142, applying formula (A.129) with $\xi = \eta\sqrt{\delta_N}$ and $p = 2$)

$$= \left(1 + i\eta E^{Q_N}\left[Y\right] - \frac{\eta^2}{2}E^{Q_N}\left[Y^2\right] + \mathrm{o}\left(\delta_N\right)\right)^N \qquad \text{for } N \to \infty. \tag{2.100}$$

Now we recall formulas (2.97) and (2.99):

$$E^{Q_N}\left[Y\right] = \left(r - \frac{\sigma^2}{2}\right)\delta_N + \mathrm{o}\left(\delta_N\right), \qquad E^{Q_N}\left[Y^2\right] = \sigma^2\delta_N + \mathrm{o}\left(\delta_N\right),$$

as $N \to \infty$. Substituting those formulas into (2.100), we get

$$\varphi_{X_N}(\eta) = \left(1 + \frac{1}{N}\left(-i\eta T\left(r - \frac{\sigma^2}{2}\right) - \eta^2\frac{\sigma^2 T}{2} + \mathrm{o}(1)\right)\right)^N \qquad \text{as } N \to \infty,$$

hence

$$\lim_{N\to\infty} \varphi_{X_N}(\eta) = \exp\left(-i\eta T\left(r - \frac{\sigma^2}{2}\right) - \eta^2\frac{\sigma^2 T}{2}\right), \qquad \forall \eta \in \mathbb{R}.$$

Then, by Lévy's theorem, we have $X_N \xrightarrow{d} X$ where X is a random variable whose characteristic function is

$$\varphi_X(\eta) = \exp\left(-i\eta T\left(r - \frac{\sigma^2}{2}\right) - \eta^2\frac{\sigma^2 T}{2}\right),$$

and so, by Theorem A.89, X has normal distribution and (2.95)-(2.96) hold.

In conclusion, gathering the results of the previous lemmas, we have proved the following:

Theorem 2.46 *Let $P_0^{(N)}$ be the price of a European Put option with strike K and maturity T in an N-period binomial model with parameters*

$$u_N = e^{\sigma\sqrt{\delta_N}+\alpha\delta_N}, \qquad d_N = e^{-\sigma\sqrt{\delta_N}+\beta\delta_N}, \qquad 1 + r_N = e^{r\delta_N},$$

where α, β are real constants. Then the limit

$$\lim_{N\to\infty} P_0^{(N)} = P_0$$

exists and we have

$$P_0 = e^{-rT} E\left[\left(K - S_0 e^X\right)^+\right] \tag{2.101}$$

where X is a random variable with normal distribution

$$X \sim \mathcal{N}_{\left(r-\frac{\sigma^2}{2}\right)T,\sigma^2 T}. \tag{2.102}$$

Definition 2.47 *P_0 is called Black-Scholes price of a European Put option with strike K and maturity T.*

One of the reasons why the Black-Scholes model is renowned is the fact that the prices of European Call and Put options possess a closed-form expression.

Corollary 2.48 (Black-Scholes formula) *The following Black-Scholes formula holds:*

$$P_0 = Ke^{-rT}\Phi(-d_2) - S_0\Phi(-d_1), \tag{2.103}$$

where Φ is the standard normal distribution function

$$\Phi(x) = \frac{1}{\sqrt{2\pi}}\int_{-\infty}^{x} e^{-\frac{y^2}{2}}dy, \qquad x \in \mathbb{R}, \tag{2.104}$$

and

$$\begin{aligned} d_1 &= \frac{\log\left(\frac{S_0}{K}\right) + \left(r + \frac{\sigma^2}{2}\right)T}{\sigma\sqrt{T}}, \\ d_2 &= d_1 - \sigma\sqrt{T} = \frac{\log\left(\frac{S_0}{K}\right) + \left(r - \frac{\sigma^2}{2}\right)T}{\sigma\sqrt{T}}. \end{aligned} \tag{2.105}$$

Proof. By (2.101), we have to prove that

$$e^{-rT}E\left[\left(K - S_0e^X\right)^+\right] = Ke^{-rT}\Phi(-d_2) - S_0\Phi(-d_1), \tag{2.106}$$

where X is normally distributed as in (2.102). Now, (cf. Remark A.32)

$$X = \left(r - \frac{\sigma^2}{2}\right)T + \sigma\sqrt{T}Z$$

with $Z \sim \mathcal{N}_{0,1}$, and a simple computation shows that

$$S_T = S_0e^X < K \qquad \Longleftrightarrow \qquad Z < -d_2. \tag{2.107}$$

Then we have

$$E\left[\left(K - S_0e^X\right)^+\right] = KE\left[\mathbb{1}_{\{S_T<K\}}\right] - E\left[S_T\mathbb{1}_{\{S_T<K\}}\right] \equiv I_1 + I_2,$$

and, by (2.107),

$$I_1 = KE\left[\mathbb{1}_{\{Z<-d_2\}}\right] = K\Phi(-d_2).$$

On the other hand, we have

$$\begin{aligned} I_2 &= e^{rT}S_0E\left[e^{-\frac{\sigma^2T}{2}+\sigma\sqrt{T}Z}\mathbb{1}_{\{Z<-d_2\}}\right] \\ &= e^{rT}S_0\int_{-\infty}^{-d_2}\frac{1}{\sqrt{2\pi}}e^{-\frac{\sigma^2T}{2}+\sigma\sqrt{T}x-\frac{x^2}{2}}dx = \end{aligned}$$

(by the change of variable $y = x - \sigma\sqrt{T}$)

$$= e^{rT}S_0\int_{-\infty}^{-d_2-\sigma\sqrt{T}}\frac{e^{-\frac{y^2}{2}}}{\sqrt{2\pi}}dy,$$

and this concludes the proof of (2.106). □

Remark 2.49 (Black-Scholes formula) By the Put-Call parity formula, we have that the Black-Scholes price C_0 of a European Call option with strike K and maturity T is given by

$$C_0 = P_0 + S_0 - Ke^{-rT}.$$

Using (A.26), a simple computation shows that the following Black-Scholes formula holds:

$$C_0 = S_0\Phi(d_1) - Ke^{-rT}\Phi(d_2), \tag{2.108}$$

where d_1, d_2 are defined in (2.105) and Φ is the standard normal distribution function. □

2.3.6 Black-Scholes differential equation

In this section we continue the study of the consistency of the binomial model with the Black-Scholes model. We saw that, as N tends to infinity, the binomial price tends to the Black-Scholes price: we prove now that it is possible to interpret the iterative scheme (2.67)-(2.68) of the binomial model as a discrete version of the Cauchy problem for a parabolic differential equation, called Black-Scholes equation. By means of stochastic calculus techniques, in Chapter 7 we will present the Black-Scholes theory which gives the price of a derivative directly in terms of the solution of the Black-Scholes equation.

In the rest of the section we adopt the usual notation $\delta = \frac{T}{N}$ and assume that the parameters u, d, r_N of the binomial model with N periods (cf. (2.83)) are of the form:

$$\begin{aligned} u &= e^{\sigma\sqrt{\delta}} = 1 + \sigma\sqrt{\delta} + \frac{\sigma^2}{2}\delta + o(\delta), \\ d &= e^{-\sigma\sqrt{\delta}} = 1 - \sigma\sqrt{\delta} + \frac{\sigma^2}{2}\delta + o(\delta), \\ 1 + r_N &= e^{r\delta} = 1 + r\delta + o(\delta), \end{aligned} \tag{2.109}$$

as $\delta \to 0$. Here σ and r denote the volatility and *annual interest rate*, respectively. In this case we have

$$q = \frac{1 + r_N - d}{u - d} = \frac{1}{2} + \frac{1}{2\sigma}\left(r - \frac{\sigma^2}{2}\right)\sqrt{\delta} + o(\sqrt{\delta}) \tag{2.110}$$

as $\delta \to 0$.

Given a function $f = f(t, S)$ defined over $[0, T] \times \mathbb{R}_{>0}$ (here f plays the role of the arbitrage price of a derivative with underlying asset S), we recall the pricing formula (2.68) that, using the notation above, takes the following form:

$$f(t, S) = \frac{1}{1 + r_N}\left(qf(t + \delta, uS) + (1 - q)f(t + \delta, dS)\right). \tag{2.111}$$

If we put

$$f = f(t, S), \qquad f^u = f(t + \delta, uS), \qquad f^d = f(t + \delta, dS),$$

and if we define the discrete operator

$$J_\delta f(t, S) = -(1 + r_N)f + qf^u + (1 - q)f^d \tag{2.112}$$

(2.111) is equivalent to

$$J_\delta f(t, S) = 0.$$

Proposition 2.50 *For every $f \in C^{1,2}([0, T] \times \mathbb{R}_{>0})$ we have*

$$\lim_{\delta \to 0^+} \frac{J_\delta f(t, S)}{\delta} = L_{\text{BS}} f(t, S),$$

for every $(t, S) \in]0, T[\times \mathbb{R}_{>0}$*, where*

$$L_{\mathrm{BS}} f(t,S) := \partial_t f(t,S) + \frac{\sigma^2 S^2}{2} \partial_{SS} f(t,S) + rS\partial_S f(t,S) - rf(t,S) \quad (2.113)$$

is called Black-Scholes differential operator.

Proof. Taking a second-order Taylor expansion of f we get[12]

$$f^u - f = \partial_t f \delta + \partial_S f S(u-1) + \frac{1}{2}\partial_{SS} f S^2 (u-1)^2 + \mathrm{o}(\delta) + \mathrm{o}((u-1)^2) =$$

(by (2.109), substituting the expression for u in terms of δ and ordering the expression according to the increasing powers of $\sqrt{\delta}$)

$$= \sigma S \partial_S f \sqrt{\delta} + Lf\delta + \mathrm{o}(\delta), \qquad \delta \to 0, \quad (2.114)$$

where

$$Lf = \partial_t f + \frac{\sigma^2}{2} S\partial_S f + \frac{\sigma^2 S^2}{2} \partial_{SS} f,$$

and analogously

$$f^d - f = -\sigma S \partial_S f \sqrt{\delta} + Lf\delta + \mathrm{o}(\delta), \qquad \delta \to 0. \quad (2.115)$$

Then we have

$$\begin{aligned} J_\delta f(t,S) &= -(1+r_N)f + qf^u + (1-q)f^d \\ &= -\delta r f + q(f^u - f - (f^d - f)) + (f^d - f) + \mathrm{o}(\delta) = \end{aligned}$$

(substituting the expressions (2.114) and (2.115))

$$= -\delta r f + \delta L f + \sqrt{\delta}(2q-1)\sigma S \partial_S f + \mathrm{o}(\delta) =$$

(by (2.110))

$$\begin{aligned} &= -\delta r f + \delta L f + \sqrt{\delta}\left(\left(r - \frac{\sigma^2}{2}\right)\sqrt{\delta} + \mathrm{o}(\sqrt{\delta})\right)\sigma S \partial_S f + \mathrm{o}(\delta) \\ &= \delta L_{\mathrm{BS}} f + \mathrm{o}(\delta), \end{aligned}$$

as $\delta \to 0$ and this concludes the proof. □

By the previous proposition, the differential equation

$$L_{\mathrm{BS}} f(t,S) = 0, \qquad (t,S) \in]0,T[\times \mathbb{R}_{>0}, \quad (2.116)$$

is the asymptotic version of the pricing formula (2.68). Further, (2.67) corresponds to the final condition

$$f(T,S) = F(S), \qquad S \in \mathbb{R}_{>0}. \quad (2.117)$$

[12] In the rest of the proof we always drop the argument (t, S) of the functions.

The pair of equations (2.116)-(2.117) constitutes a *Cauchy problem* that, as we have already said, we will analyze again in Chapter 7 using the tools of stochastic calculus in continuous time.

Problem (2.116)-(2.117) is a backward problem with *final* datum such as the one examined in Appendix A.3.5. With the change of variables

$$f(t,S) = u(T-t, \log S)$$

i.e. putting $\tau = T - t$ and $x = \log S$, problem (2.116)-(2.117) is the backward version of the following *parabolic Cauchy problem with constant coefficients* (cf. Appendix A.3):

$$\begin{cases} \frac{\sigma^2}{2}\partial_{xx}u + \left(r - \frac{\sigma^2}{2}\right)\partial_x u - ru - \partial_\tau u = 0, & (\tau,x) \in]0,T[\times\mathbb{R}, \\ u(0,x) = F(e^x), & x \in \mathbb{R}. \end{cases}$$

By Theorem A.72, if the payoff $x \mapsto F(e^x)$ is a function that does not grow too rapidly, we can express the solution u in terms of the Gaussian fundamental solution Γ of the differential equation:

$$u(\tau,x) = \int_{\mathbb{R}} \Gamma(\tau, x-y) F(e^y) dy, \qquad \tau \in]0,T[, \ x \in \mathbb{R},$$

where Γ is given *explicitly* by (A.61).

The previous formula can be interpreted in terms of the expectation of the payoff that is a function of a random variable with normal distribution and density Γ. By using the expression for Γ, with a direct computation we can obtain again the Black-Scholes formulas (2.103) and (2.108) for the price of European Put and Call options.

A posteriori, the binomial algorithm can be considered as a numerical scheme for the solution of a parabolic Cauchy problem. As a matter of fact, Proposition 2.50 implicitly includes the fact that the binomial algorithm is equivalent to an explicit finite-difference scheme that will be analyzed further in Chapter 12. In their recent paper [188], Jiang and Dai extend the results for the binomial model approximating the continuous Black-Scholes case to European and American path-dependent derivatives, and they prove that the binomial model is equivalent to a finite-difference scheme for the Black-Scholes equation.

2.4 Trinomial model

In the trinomial model, the market is composed of a non-risky asset B whose dynamics is given by (2.1) with $r_n \equiv r$, and one or two risky assets whose dynamics is driven by a stochastic process $(h_n)_{n=1,\dots,N}$ with $h_1, \dots, h_N$ i.i.d.

random variables such that

$$h_n = \begin{cases} 1 & \text{with probability } p_1, \\ 2 & \text{with probability } p_2, \\ 3 & \text{with probability } p_3 = 1 - p_1 - p_2, \end{cases}$$

where $p_1, p_2 > 0$ and $p_1 + p_2 < 1$. The trinomial model with only one risky asset S^1 is called *standard trinomial model*, while in case there are two risky assets S^1, S^2, it is called *completed trinomial market*.

In general we assume that $S_0^1, S_0^2 > 0$ and

$$S_n^i = S_{n-1}^i(1 + \mu^i(h_n)), \qquad n = 1, \dots, N, \quad i = 1, 2, \tag{2.118}$$

where

$$1 + \mu^i(h) = \begin{cases} u_i & \text{if } h = 1, \\ m_i & \text{if } h = 2, \\ d_i & \text{if } h = 3, \end{cases}$$

and $0 < d_i < m_i < u_i$. In Figure 2.7 a two-period binomial tree is represented.

In the standard trinomial model S^1 typically represents the underlying asset of a derivative: the standard trinomial model is the simplest example of *incomplete* model. On the contrary, the completed trinomial model is a complete model that is typically used to price and hedge exotic options: we may think of S^1 and S^2 as an asset and a plain vanilla option on S^1 that is supposed to be quoted on the market, respectively. Then the hedging strategy of an exotic option on S^1 is constructed by using both S^1 and S^2.

We first examine the standard trinomial model and we set $S = S^1$ for convenience. In order to study the existence of an EMM Q, we proceed as in the binomial case by imposing the martingale condition (2.44): in this setting it reads

$$S_{n-1} = \frac{1}{1+r} E^Q \left[S_{n-1} (1 + \mu(h_n)) \mid \mathcal{F}_{n-1} \right], \tag{2.119}$$

where $\mu(h) = \mu^1(h)$. Then, putting

$$q_j^n = Q(h_n = j \mid \mathcal{F}_{n-1}), \qquad j = 1, 2, 3, \quad n = 1, \dots, N, \tag{2.120}$$

we obtain the following system

$$\begin{cases} u q_1^n + m q_2^n + d q_3^n = 1 + r, \\ q_1^n + q_2^n + q_3^n = 1, \end{cases} \tag{2.121}$$

that does not admit a unique solution q_1, q_2, q_3. Therefore the EMM is not unique and consequently, by Theorem 2.29, the market is incomplete. Note also that the random variables h_n are not necessarily independent under a generic EMM.

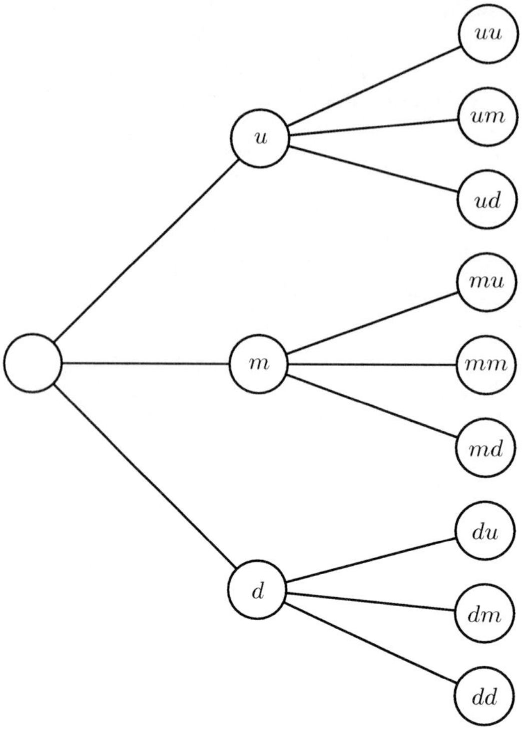

Fig. 2.7. Two-period trinomial tree

The incompleteness of the market can also be deduced more directly by examining the construction of a replicating strategy. For the sake of simplicity, we consider the one-period case $N = 1$ and we let $S_0^1 = 1$ and $r = 0$. Given a derivative $X = F(S_1)$, the replication condition $V_1 = X$ becomes

$$\alpha_1 S_1 + \beta_1 = F(S_1),$$

that is equivalent to the following linear system in the unknowns α_1, β_1:

$$\begin{cases} \alpha_1 u + \beta_1 = F(u) \\ \alpha_1 m + \beta_1 = F(m) \\ \alpha_1 d + \beta_1 = F(d). \end{cases} \tag{2.122}$$

It is interesting to note that the matrix associated to system (2.122)

$$\begin{pmatrix} u & 1 \\ m & 1 \\ d & 1 \end{pmatrix}$$

is the transpose of the matrix associated to system (2.121): this points to the duality relation between the problem of completeness and absence of arbitrage opportunities. In the binomial model an analogous part is played by the matrix

$$\begin{pmatrix} u & 1 \\ d & 1 \end{pmatrix}$$

that is a square matrix with maximum rank, so guaranteeing completeness.

It is well known from linear algebra that system (2.122) admits a solution if and only if the complete matrix

$$\begin{pmatrix} u & 1 & F(u) \\ m & 1 & F(m) \\ d & 1 & F(d) \end{pmatrix}$$

does not have maximum rank. Imposing, for example, that the second row is a linear combination (with coefficients λ, μ) of the first and of the third rows, we obtain

$$\begin{cases} m = \lambda u + \mu d \\ 1 = \lambda + \mu \\ F(m) = \lambda F(u) + \mu F(d), \end{cases}$$

hence

$$\mu = \frac{u-m}{u-d}, \qquad \lambda = \frac{m-d}{u-d},$$

and we can eventually write the condition a derivative must verify in order to be replicated:

$$F(m) = \frac{m-d}{u-d} F(u) + \frac{u-m}{u-d} F(d). \tag{2.123}$$

Condition (2.123) is tantamount to saying that the second equation of the system (2.122) is superfluous and can be dropped. In that case the system can be solved and we see that it is equivalent to the analogous system in the binomial model, whose solution is given by (2.49): in this particular case we get

$$\alpha_1 = \frac{F(u) - F(d)}{u-d}, \qquad \beta_1 = \frac{uF(d) - dF(u)}{u-d}. \tag{2.124}$$

By the self-financing condition, the arbitrage price of the derivative is defined by

$$H_0 = \alpha_1 S_0 + \beta_1 B_0$$

and it does not depend on the fixed EMM. The derivatives that do not satisfy condition (2.123) cannot be replicated and this substantiates the fact that the standard trinomial market is incomplete.

Next we consider the completed trinomial model: imposing the martingale condition (2.119) with $S = S^i$ and $\mu = \mu^i$ for $i = 1, 2$, and setting $q_j^n = Q(h_n = j \mid \mathcal{F}_{n-1})$, $j = 1, 2, 3$, we obtain the linear system

$$\begin{cases} u_1 q_1^n + m_1 q_2^n + d_1 q_3^n = 1 + r, \\ u_2 q_1^n + m_2 q_2^n + d_2 q_3^n = 1 + r, \\ q_1^n + q_2^n + q_3^n = 1, \end{cases} \tag{2.125}$$

which has a solution $q_j^n = q_j$, $j = 1, 2, 3$, independent on n. Under suitable conditions on the parameters of the model, we have that $q_j \in]0, 1[$ and therefore the EMM Q is uniquely determined. In this case the completed trinomial model is arbitrage-free and complete. Furthermore, since q_j^n are constants independent on n and $\omega \in \Omega$, we conclude that the random variables h_n are i.i.d. under the probability Q. As a consequence, S^1 *and* S^2 *are Markov processes on* $(\Omega, \mathcal{F}, Q, (\mathcal{F}_n))$ by Theorem 2.31.

On the other hand, the replication strategy of a derivative X with arbitrage price H can be determined as in the binomial case: to construct the hedging strategy $(\alpha_n^1, \alpha_n^2, \beta_n)$ for the n-th period $[t_{n-1}, t_n]$ given $\mathcal{F}_{n-1}$, we solve the linear system

$$\begin{cases} \alpha_n^1 u_1 S_{n-1}^1 + \alpha_n^2 u_2 S_{n-1}^2 + \beta_n (1+r)^n = H_n^u, \\ \alpha_n^1 m_1 S_{n-1}^1 + \alpha_n^2 m_2 S_{n-1}^2 + \beta_n (1+r)^n = H_n^m, \\ \alpha_n^1 d_1 S_{n-1}^1 + \alpha_n^2 d_2 S_{n-1}^2 + \beta_n (1+r)^n = H_n^d, \end{cases} \tag{2.126}$$

where H_n^u, H_n^m and H_n^d denote the arbitrage prices of the derivative at time t_n in the three possible scenarios. The solution to system (2.126) is

$$\alpha_n^1 = \frac{d_2 (H_n^m - H_n^u) + H_n^u m_2 - H_n^m u_2 + H_n^d (-m_2 + u_2)}{S_{n-1}^1 (d_2 (m_1 - u_1) + m_2 u_1 - m_1 u_2 + d_1 (u_2 - m_2))},$$

$$\alpha_n^2 = \frac{d_1 (H_n^m - H_n^u) + H_n^u m_1 - H_n^m u_1 + H_n^d (u_1 - m_1)}{S_{n-1}^2 (-m_2 u_1 + d_2 (u_1 - m_1) + d_1 (m_2 - u_2) + m_1 u_2)},$$

$$\beta_n = \frac{d_2 (H_n^u m_1 - H_n^m u_1) + d_1 (-H_n^u m_2 + H_n^m u_2) + H_n^d (m_2 u_1 - m_1 u_2)}{(1+r)^n (d_2 (m_1 - u_1) + m_2 u_1 - m_1 u_2 + d_1 (-m_2 + u_2))}.$$

2.4.1 Pricing and hedging in an incomplete market

In this section we briefly discuss the pricing and hedging problems in incomplete markets. We first recall the following definition given in Section 2.2.1.

Definition 2.51 *In an arbitrage-free market* (S, B)*, let* Q *be an EMM with numeraire* B*. The risk-neutral price relative to* Q *of a (not necessarily replicable) derivative* X *is defined by*

$$H_n^Q = E^Q [D(n, N) X \mid \mathcal{F}_n], \qquad 0 \le n \le N, \tag{2.127}$$

where $D(n, N) = \frac{B_n}{B_N}$ *is the discount factor.*

By Lemma 2.23, the pricing formula (2.127) does not introduce arbitrage opportunities in the sense that the augmented market (S, B, H^Q) is still arbitrage-free. Further, in the case X is replicable, by Theorem 2.25 the price H^Q does not depend on the fixed EMM and is equal to the arbitrage price.

Given a self-financing predictable strategy $(\alpha, \beta) \in \mathcal{A}$, the quantity

$$X - V_N^{(\alpha, \beta)}$$

represents the replication error, also called Profit and Loss (P&L), at maturity of the strategy. Now, if Q denotes the selected EMM and $\widetilde{V}$ is the discounted value process, we may consider the expected (discounted and squared) P&L

$$R^Q(\alpha,\beta) := E^Q\left[\left(\frac{X}{B_N} - \widetilde{V}_N^{(\alpha,\beta)}\right)^2\right] \tag{2.128}$$

as a measure of the hedging risk under the EMM Q.

We now remark that *any strategy that minimizes the risk R^Q requires an initial investment equal to the risk-neutral price $E^Q\left[\frac{X}{B_N}\right]$*. Indeed, using the fact that $\widetilde{V}^{(\alpha,\beta)}$ is a Q-martingale and the identity

$$E\left[Y^2\right] = E\left[Y\right]^2 + E\left[(Y - E(Y))^2\right],$$

with $Y = \frac{X}{B_N} - \widetilde{V}_N^{(\alpha,\beta)}$, we may rewrite (2.128) as follows:

$$R^Q(\alpha,\beta) = \left(E^Q\left[\frac{X}{B_N}\right] - \widetilde{V}_0^{(\alpha,\beta)}\right)^2 + E^Q\left[\left(\frac{X}{B_N} - E^Q\left[\frac{X}{B_N}\right] - \left(\widetilde{V}_N^{(\alpha,\beta)} - \widetilde{V}_0^{(\alpha,\beta)}\right)\right)^2\right].$$

Now, recalling that the gain $\widetilde{V}_N^{(\alpha,\beta)} - \widetilde{V}_0^{(\alpha,\beta)}$ does not depend on $\widetilde{V}_0^{(\alpha,\beta)}$ (cf. formula (2.13) and Proposition 2.7), we conclude that in order to minimize the risk R^Q it is necessary to put

$$\widetilde{V}_0^{(\alpha,\beta)} = E^Q\left[\frac{X}{B_N}\right]. \tag{2.129}$$

This motivates Definition 2.51 even if it poses some questions on the very foundations of the classical theory of arbitrage pricing. In particular this theory has two cornerstones:

i) the *uniqueness of the price* of the derivative: the arbitrage price should be objective, dependent only on the quoted prices of the underlying assets and not on the subjective estimate of the probability P;
ii) the *hedging* procedure, i.e. the neutralization of the risk that we take on the derivative by the investment in a replicating strategy.

The risk-neutral price in Definition 2.51 *is not unique* since it depends on the choice of the EMM. Furthermore, in an incomplete market, a derivative *is not generally replicable* and so it is necessary to study possible hedging strategies that limit the risks (super-hedging, risk minimization within the set of EMMs, etc.). Such a choice can be made by following the preferences of the traders or on the grounds of some objective criterion (calibration to market data). The study of these problems goes beyond the scope of this book and is treated thoroughly in monographs such as [134] by Follmer and Schied, to which we refer the interested reader.

Here we confine ourselves to the following example that shows how the hedging problem can be tackled in the standard (hence, incomplete) trinomial model. The approach is based on a classical optimization technique called Dynamic Programming. The main idea is to find a strategy minimizing the expected replication error of the payoff *under the real-world probability measure*+P.

Example 2.52 We consider a two-period standard trinomial market model where the dynamics of the risky asset is given by

$$S_0 = 1, \qquad S_n = S_{n-1}(1 + \mu_n), \qquad n = 1, 2$$

and μ_n, $n = 1, 2$, are i.i.d. random variables defined on a probability space $(\Omega, \mathcal{F}, P)$ and such that

$$1 + \mu_n = \begin{cases} \frac{1}{2} & \text{with probability } p_1 = \frac{1}{3}, \\ 1 & \text{with probability } p_2 = \frac{1}{3}, \\ 2 & \text{with probability } p_3 = \frac{1}{3}. \end{cases}$$

We assume that the short rate is null, $r = 0$.

We consider the problem of pricing and hedging a European Call option with payoff

$$F(S_2) = (S_2 - 1)^+,$$

by minimization of the "shortfall" risk criterion. More precisely, by means of the Dynamic Programming (DP) algorithm, we aim at determining a self-financing strategy with non-negative value V (that is, such that $V_n \geq 0$ for any n) that minimizes

$$E^P\left[\mathcal{U}(V_2, S_2)\right],$$

where

$$\mathcal{U}(V, S) = (F(S) - V)^+$$

is the shortfall risk function.

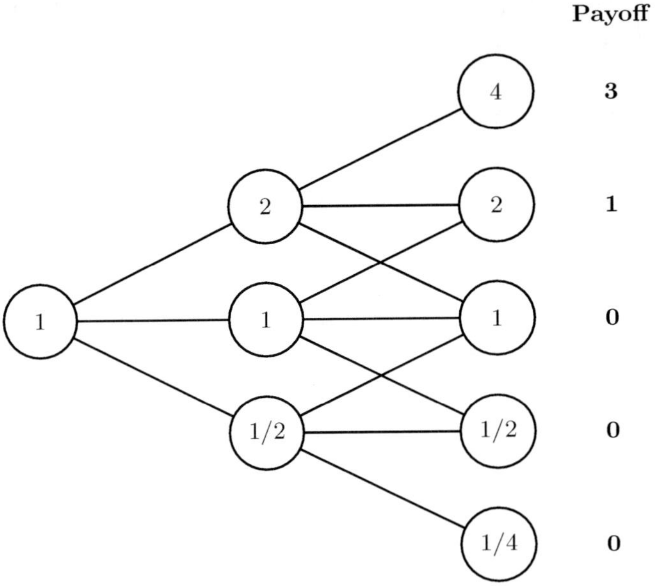

Fig. 2.8. Two-period trinomial tree: price of the underlying and payoff of a European Call with strike 1

We first represent the binomial tree with the prices of the underlying asset. By (2.8), the value V of a self-financing strategy (α, β) satisfies

$$V_n = V_{n-1} + \alpha_n S_{n-1}\mu_n = V_{n-1} + \begin{cases} \alpha_n S_{n-1}, \\ 0, \\ -\frac{\alpha_n S_{n-1}}{2}. \end{cases} \tag{2.130}$$

Then $V_n \geq 0$ for any n if and only if $V_0 \geq 0$ and

$$-\frac{V_{n-1}}{S_{n-1}} \leq \alpha_n \leq \frac{2V_{n-1}}{S_{n-1}}, \qquad n = 1, 2.$$

In the general framework of a model with N periods, the DP algorithm consists of two steps:

i) we compute

$$R_{N-1}(V, S) := \min_{\alpha \in \left[-\frac{V}{S}, \frac{2V}{S}\right]} E^P \left[\mathcal{U}\left(V + S\alpha\mu_N, S\left(1 + \mu_N\right)\right)\right]$$

for S varying among the possible values of S_{N-1}. Recalling that we are considering predictable strategies, we denote by $\alpha_N = \alpha_N(V)$ the minimum point for V varying among the possible values of V_{N-1};

ii) for $n \in \{N-1, N-2, \dots, 1\}$, we compute

$$R_{n-1}(V, S) := \min_{\alpha \in \left[-\frac{V}{S}, \frac{2V}{S}\right]} E^P \left[R_n\left(V + S\alpha\mu_n, S\left(1 + \mu_n\right)\right)\right]$$

for S varying among the possible values of S_{n-1}. We denote by $\alpha_n = \alpha_n(V)$ the minimum point for V varying among the possible values of V_{n-1}.

In our setting, as a first step of the DP algorithm we compute $R_1(V,S)$ for $S \in \{2, 1, \frac{1}{2}\}$. We have

$$\begin{aligned} R_1(V,2) &= \min_{\alpha\in[-V/2,V]} E^P\left[\mathcal{U}(V + 2\alpha\mu_2, 2(1+\mu_2))\right] \\ &= \min_{\alpha\in[-V/2,V]} E^P\left[\left((2(1+\mu_2)-1)^+ - (V+2\alpha\mu_2)\right)^+\right] \\ &= \min_{\alpha\in[-V/2,V]} \frac{1}{3}\left((3-V-2\alpha)^+ + (1-V)^+\right) = \frac{4}{3}(1-V)^+, \end{aligned}$$

and the minimum is attained at

$$\alpha_2 = V. \tag{2.131}$$

Next we have

$$\begin{aligned} R_1(V,1) &= \min_{\alpha\in[-V,2V]} E^P\left[\mathcal{U}(V+\alpha\mu_2, 1+\mu_2)\right] \\ &= \min_{\alpha\in[-V,2V]} E^P\left[\left(\mu_2^+ - (V+\alpha\mu_2)\right)^+\right] \\ &= \min_{\alpha\in[-V,2V]} \frac{1}{3}(1-V-\alpha)^+ = \frac{1}{3}(1-3V)^+, \end{aligned}$$

and the minimum is attained at

$$\alpha_2 = 2V. \tag{2.132}$$

Moreover we have

$$\begin{aligned} R_1\left(V,\frac{1}{2}\right) &= \min_{\alpha\in[-2V,4V]} E^P\left[\mathcal{U}\left(V+\frac{\alpha\mu_2}{2}, \frac{1+\mu_2}{2}\right)\right] \\ &= \min_{\alpha\in[-2V,4V]} E^P\left[\left(\underbrace{\left(\frac{1+\mu_2}{2}-1\right)^+}_{=0} - \underbrace{\left(V+\frac{\alpha\mu_2}{2}\right)}_{\geq 0}\right)^+\right] = 0, \end{aligned}$$

and the minimum is attained at any

$$\alpha_2 \in [-2V, 4V]. \tag{2.133}$$

The second step consists in computing the risk at the initial time:

$$\begin{aligned}
R_0(V,1) &= \min_{\alpha\in[-V,2V]} E^P\left[R_1\left(V+\alpha\mu_1, 1+\mu_1\right)\right] \\
&= \frac{1}{3}\min_{\alpha\in[-V,2V]}\left(R_1(V,1)+R_1(V+\alpha,2)\right) \\
&= \frac{1}{3}\min_{\alpha\in[-V,2V]}\left(\frac{1}{3}(1-3V)^+ + \frac{4}{3}(1-(V+\alpha))^+\right) \\
&= \frac{5}{9}(1-3V)^+, \qquad (2.134)
\end{aligned}$$

and the minimum is attained at

$$\alpha_1 = 2V. \qquad (2.135)$$

By formula (2.134) for $R_0(V,1)$, it is clear that an initial wealth $V \geq \frac{1}{3}$ is sufficient to make the shortfall risk null or, in more explicit terms, *to super-replicate the payoff*.

Next we determine the shortfall strategy, that is the self-financing strategy that minimizes the shortfall risk. Let us denote by V_0 the initial wealth: by (2.135) we have

$$\alpha_1 = 2V_0.$$

Consequently, by (2.130) we get

$$V_1 = V_0 + \begin{cases} 2V_0, & \text{for } \mu_1 = 1, \\ 0, & \text{for } \mu_1 = 0, \\ -V_0, & \text{for } \mu_1 = -\frac{1}{2}. \end{cases}$$

Then by (2.131)-(2.132)-(2.133) we have

$$\alpha_2 = \begin{cases} 3V_0, & \text{if } S_1 = 2, \\ 2V_0, & \text{if } S_1 = 1, \\ 0, & \text{if } S_1 = \frac{1}{2}, \end{cases}$$

and we can easily compute the final value V_2 by means of (2.130). We represent in Figure 2.9 the trinomial tree with the prices of the underlying asset and the values of the shortfall strategy inside the circles. On the right side we also indicate the final values of the option and of the shortfall strategy corresponding to $V_0 = \frac{1}{3}$. We remark that we have perfect replication in all scenarios except for the trajectory $S_0 = S_1 = S_2 = 1$ for which we have super-replication: the terminal value of the shortfall strategy $V_2 = \frac{1}{3}$ is strictly greater than the payoff of the Call option that in this case is null. □

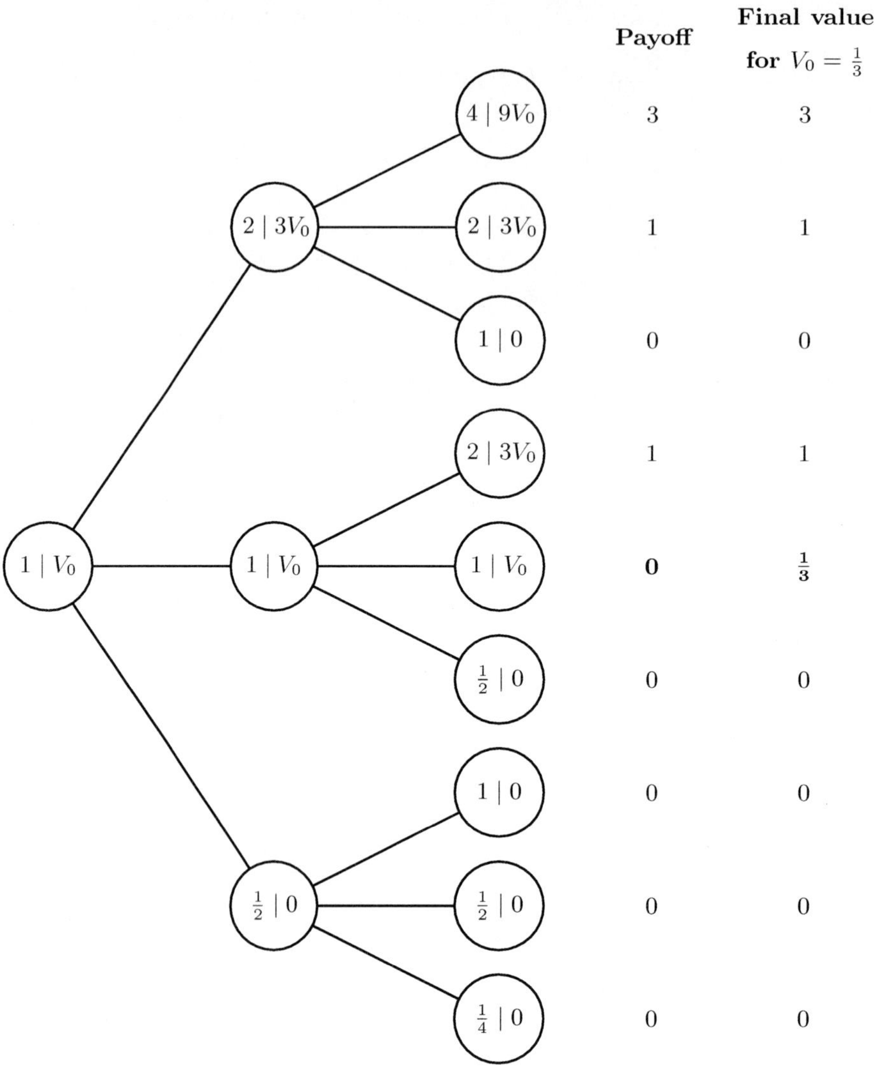

Fig. 2.9. Two-period trinomial tree: prices of the underlying asset and values of the shortfall strategy with initial wealth V_0 are inside the circles

2.5 American derivatives

In this section we examine pricing and hedging of American-style derivatives. We consider a generic discrete market (S, B) defined on the space $(\Omega, \mathcal{F}, P, (\mathcal{F}_n))$. American derivatives are characterized by the possibility of early exercise at every time t_n, $0 \le n \le N$, during the life span of the contract. To describe an American derivative it is therefore necessary to specify the premium (or payoff) that has to be paid to the owner in case he/she exercises the option at time t_n with $n \le N$. For example, in the case of an American Call option with underlying asset S and strike K, the payoff at time t_n is $X_n = (S_n - K)^+$.

Definition 2.53 *An American derivative is a non-negative discrete stochastic process $X = (X_n)$ adapted to the filtration $(\mathcal{F}_n)$.*

By definition, X_n is a non-negative $\mathcal{F}_n$-measurable random variable: the measurability condition describes the fact that the payoff X_n is known only at time t_n. We say that X is path-independent if X_n is $\sigma(S_n)$-measurable, for every n, so that there exist measurable functions φ_n such that $X_n = \varphi_n(S_n)$.

Since the choice of the best time to exercise an American option must depend only on the information available at that moment, the following definition of *exercise strategy* seems natural.

Definition 2.54 *A stopping time*

$$\nu : \Omega \longrightarrow \{0, 1, \dots, N\},$$

i.e. a random variable such that

$$\{\nu = n\} \in \mathcal{F}_n, \qquad n = 0, \dots, N, \tag{2.136}$$

is called exercise strategy (or exercise time). We denote by $\mathcal{T}_0$ the set of all exercise strategies.

Intuitively, given a path $\omega \in \Omega$ of the underlying market, the number $\nu(\omega)$ represents the moment when one decides to exercise the American derivative. Condition (2.136) merely means that the decision to exercise at time t_n depends on $\mathcal{F}_n$, i.e. on the information available at t_n.

In the rest of the paragraph we assume that the market (S, B) is arbitrage-free and so there exists at least one EMM Q equivalent to P, with numeraire B. Hereafter

$$\widetilde{Y}_n = \frac{Y_n}{B_n}$$

denotes the discounted price of any asset Y.

Definition 2.55 *Given an American derivative X and an exercise strategy $\nu \in \mathcal{T}_0$, the random variable X_ν defined by*

$$(X_\nu)(\omega) = X_{\nu(\omega)}(\omega), \qquad \omega \in \Omega,$$

is called payoff of X relative to the strategy ν. An exercise strategy ν_0 is called optimal under Q if

$$E^Q\left[\widetilde{X}_{\nu_0}\right] = \sup_{\nu \in \mathcal{T}_0} E^Q\left[\widetilde{X}_\nu\right]. \tag{2.137}$$

We observe that the random variable

$$\widetilde{X}_\nu = \frac{X_\nu}{B_\nu}$$

can be interpreted as the discounted payoff of a *European option*: in particular $E^Q\left[\widetilde{X}_\nu\right]$ gives the risk-neutral price relative to Q of the option (cf. Definition 2.51) when the option is exercised following the strategy ν. The greatest price among all exercise strategies corresponds to the optimal exercise strategy: that price is equal to the greatest expected payoff with respect to the fixed EMM.

2.5.1 Arbitrage price

In an arbitrage-free complete market, the price of a *European* option with payoff X_N is by definition equal to the value of a replicating strategy: in particular, the discounted price is a martingale with respect to the risk-neutral measure Q. Pricing an American option $X = (X_n)$ is a slightly more delicate matter since it is not possible to determine a self-financing predictable strategy (α, β) that replicates the option in the sense that $V_n^{(\alpha,\beta)} = X_n$ for every $n = 0, \dots, N$: this is simply due to the fact that $\widetilde{V}^{(\alpha,\beta)}$ is a Q-martingale while X is a generic adapted process. On the other hand, it is possible to develop a theory of arbitrage pricing for American options, essentially analogous to the European case, by using the results on stopping times, martingales and Doob's theorems collected in Appendix A.6.

Let us begin by observing that, by arbitrage arguments, it is possible to determine upper and lower bounds to the price of X: to fix ideas, as in the European case we denote by H_0 the (unknown and possibly not unique) initial price of X. Recalling that $\mathcal{A}$ denotes the family of self-financing predictable strategies, we define

$$\mathcal{A}_X^+ = \{(\alpha, \beta) \in \mathcal{A} \mid V_n^{(\alpha,\beta)} \geq X_n, \ n = 0, \dots, N\},$$

the family of those strategies in $\mathcal{A}$ that super-replicate X. By Remark 2.18, to avoid introducing arbitrage opportunities, the price H_0 must be less or equal to the initial value $V_0^{(\alpha,\beta)}$ for every $(\alpha, \beta) \in \mathcal{A}_X^+$ and so

$$H_0 \leq \inf_{(\alpha,\beta)\in\mathcal{A}_X^+} V_0^{(\alpha,\beta)}.$$

On the other hand we put

$$\mathcal{A}_X^- = \{(\alpha, \beta) \in \mathcal{A} \mid \text{ there exists } \nu \in \mathcal{T}_0 \text{ s.t. } X_\nu \geq V_\nu^{(\alpha,\beta)}\}.$$

Intuitively, an element (α, β) of $\mathcal{A}_X^-$ represents a strategy in which a short position is taken, to get money to invest in the American option. In other words, $V_0^{(\alpha,\beta)}$ represents the amount of money that one can initially borrow to buy the option X, knowing that there exists an exercise strategy ν yielding a payoff X_ν greater or equal to $V_\nu^{(\alpha,\beta)}$, corresponding to the amount necessary to close the short position in the strategy (α, β). The initial price H_0 of X must necessarily be greater or equal to $V_0^{(\alpha,\beta)}$ for every $(\alpha,\beta) \in \mathcal{A}_X^-$: if this were not true, one could easily build an arbitrage strategy. Then we have

$$\sup_{(\alpha,\beta)\in\mathcal{A}_X^-} V_0^{(\alpha,\beta)} \leq H_0.$$

Therefore we determined an interval to which the initial price H_0 must belong, in order to avoid introducing arbitrage opportunities. Let us show now that risk-neutral pricing relative to an optimal exercise strategy respects such conditions.

Proposition 2.56 *For every EMM Q, we have*

$$\sup_{(\alpha,\beta)\in\mathcal{A}_X^-} \widetilde{V}_0^{(\alpha,\beta)} \leq \sup_{\nu\in\mathcal{T}_0} E^Q\left[\widetilde{X}_\nu\right] \leq \inf_{(\alpha,\beta)\in\mathcal{A}_X^+} \widetilde{V}_0^{(\alpha,\beta)}. \tag{2.138}$$

Proof. If $(\alpha,\beta) \in \mathcal{A}_X^-$, there exists $\nu_0 \in \mathcal{T}_0$ such that $V_{\nu_0}(\alpha) \leq X_{\nu_0}$. Further, $\widetilde{V}^{(\alpha,\beta)}$ is a Q-martingale and so by the Optional sampling Theorem A.129 we have

$$\widetilde{V}_0^{(\alpha,\beta)} = E^Q\left[V_{\nu_0}^{(\alpha,\beta)}\right] \leq E^Q\left[\widetilde{X}_{\nu_0}\right] \leq \sup_{\nu\in\mathcal{T}_0} E^Q\left[\widetilde{X}_\nu\right],$$

hence we obtain the first inequality in (2.138), by the arbitrariness of $(\alpha,\beta) \in \mathcal{A}_X^-$.

On the other hand, if $(\alpha,\beta) \in \mathcal{A}_X^+$ then, again by Theorem A.129, for every $\nu \in \mathcal{T}_0$ we have

$$\widetilde{V}_0^{(\alpha,\beta)} = E^Q\left[\widetilde{V}_\nu^{(\alpha,\beta)}\right] \geq E^Q\left[\widetilde{X}_\nu\right],$$

hence we get the second inequality in (2.138), by the arbitrariness of $(\alpha,\beta) \in \mathcal{A}_X^+$ and $\nu \in \mathcal{T}_0$. □

Under the assumption that the market is arbitrage-free and complete[13], the following theorem shows how to define the initial arbitrage price of an American derivative X in a unique way.

Theorem 2.57 *Let X be an American derivative in an arbitrage-free and complete market. Then there exists $(\alpha,\beta) \in \mathcal{A}_X^+ \cap \mathcal{A}_X^-$ and so we have:*

i) $V_n^{(\alpha,\beta)} \geq X_n$, $n = 0, \dots, N$;
ii) there exists $\nu_0 \in \mathcal{T}_0$ *such that* $V_{\nu_0}^{(\alpha,\beta)} = X_{\nu_0}$.

[13] According to Definition 2.27, this means that every *European* derivative is replicable.

Consequently[14]

$$V_0^{(\alpha,\beta)} = \sup_{\nu \in \mathcal{T}_0} E^Q\left[\widetilde{X}_\nu\right] = E^Q\left[\widetilde{X}_{\nu_0}\right], \tag{2.139}$$

defines the initial arbitrage price of X.

Proof. The proof is constructive and is made up of three main steps:

1) we construct the smallest super-martingale $\widetilde{H}$ greater than $\widetilde{X}$, usually called *Snell envelope of the process* $\widetilde{X}$;
2) we use Doob's decomposition theorem to find the martingale part of the process $\widetilde{H}$ and by this we determine the strategy $(\alpha, \beta) \in \mathcal{A}_X^+ \cap \mathcal{A}_X^-$;
3) we conclude by proving that $\widetilde{H}_0 = V_0^{(\alpha,\beta)}$ and (2.139) holds.

First step. We define iteratively the stochastic process $\widetilde{H}$ by putting

$$\widetilde{H}_n = \begin{cases} \widetilde{X}_N, & n = N, \\ \max\left\{\widetilde{X}_n, E^Q\left[\widetilde{H}_{n+1} \mid \mathcal{F}_n\right]\right\}, & n = 0, \dots, N-1. \end{cases} \tag{2.140}$$

Below we will see that the process $\widetilde{H}$ defines the discounted arbitrage price process of X (cf. Definition 2.60). It is indeed an intuitive notion of price that gives rise to the definition above: indeed the derivative is worth X_N at maturity and at time t_{N-1} it is worth

- X_{N-1} if one decides to exercise it;
- the price of a European derivative with payoff X_N and maturity t_N, if one decides not to exercise it.

Consistently with the arbitrage price of a European option (2.27), it seems reasonable to define

$$H_{N-1} = \max\left\{X_{N-1}, \frac{1}{1+r} E^Q\left[H_N \mid \mathcal{F}_{N-1}\right]\right\}.$$

By repeating this argument backwards and setting $\widetilde{H}_n = \frac{H_n}{B_n}$, we get definition (2.140).

Next we show that $\widetilde{H}$ is the smallest super-martingale greater than $\widetilde{X}$. Evidently, $\widetilde{H}$ is an adapted non-negative stochastic process. Further, for every n, we have

$$\widetilde{H}_n \geq E^Q\left[\widetilde{H}_{n+1} \mid \mathcal{F}_n\right], \tag{2.141}$$

i.e. $\widetilde{H}$ is a *Q-super-martingale*. This means that $\widetilde{H}$ "decreases in mean" (cf. Section A.6): intuitively this corresponds to the fact that, moving forward in time, the advantage of the possibility of early exercise decreases. More generally, from (2.141) it follows also that

$$\widetilde{H}_k \geq E^Q\left[\widetilde{H}_n \mid \mathcal{F}_k\right], \qquad 0 \leq k \leq n \leq N.$$

---

[14] Recall that, by assumption, $B_0 = 1$ and therefore $V_0 = \widetilde{V}_0$.

We conclude by showing that $\widetilde{H}$ is *the smallest* super-martingale that dominates $\widetilde{X}$: if Y is a Q-super-martingale such that $Y_n \geq \widetilde{X}_n$, then we have

$$\widetilde{H}_N = \widetilde{X}_N \leq Y_N.$$

Then the thesis follows by induction: indeed, assuming $\widetilde{H}_n \leq Y_n$, we have

$$\begin{aligned}\widetilde{H}_{n-1} &= \max\left\{\widetilde{X}_{n-1}, E^Q\left[\widetilde{H}_n \mid \mathcal{F}_{n-1}\right]\right\} \\ &\leq \max\left\{\widetilde{X}_{n-1}, E^Q\left[Y_n \mid \mathcal{F}_{n-1}\right]\right\} \\ &\leq \max\left\{\widetilde{X}_{n-1}, Y_{n-1}\right\} = Y_{n-1}.\end{aligned}$$

Second step. We prove that there exists $(\alpha, \beta) \in \mathcal{A}_X^+ \cap \mathcal{A}_X^-$. Since $\widetilde{H}$ is a Q-super-martingale, we can apply Doob's decomposition Theorem A.119 to get

$$\widetilde{H} = M + A$$

where M is a Q-martingale such that $M_0 = \widetilde{H}_0$ and A is a predictable decreasing process with null initial value.

By assumption the market is complete, and so there exists a strategy $(\alpha, \beta) \in \mathcal{A}$ that replicates the European derivative M_N. Further, since $\widetilde{V}^{(\alpha,\beta)}$ and M are martingales with the same terminal value, they are equal:

$$\widetilde{V}_n^{(\alpha,\beta)} = E^Q\left[\widetilde{V}_N^{(\alpha,\beta)} \mid \mathcal{F}_n\right] = E^Q\left[M_N \mid \mathcal{F}_n\right] = M_n, \qquad (2.142)$$

for $0 \leq n \leq N$. Consequently, $(\alpha, \beta) \in \mathcal{A}_X^+$: indeed, since $A_n \leq 0$, we have

$$\widetilde{V}_n^{(\alpha,\beta)} = M_n \geq \widetilde{H}_n \geq \widetilde{X}_n, \qquad 0 \leq n \leq N.$$

Moreover, since $A_0 = 0$, we have

$$V_0^{(\alpha,\beta)} = M_0 = \widetilde{H}_0.$$

Then (α, β) is a hedging strategy for X that has an initial cost equal to the price of the option.

In order to verify that $(\alpha, \beta) \in \mathcal{A}_X^-$, we put:

$$\nu_0(\omega) = \min\{n \mid \widetilde{H}_n(\omega) = \widetilde{X}_n(\omega)\}, \qquad \omega \in \Omega. \qquad (2.143)$$

Since

$$\{\nu_0 = n\} = \{\widetilde{H}_0 > \widetilde{X}_0\} \cap \cdots \cap \{\widetilde{H}_{n-1} > \widetilde{X}_{n-1}\} \cap \{\widetilde{H}_n = \widetilde{X}_n\} \in \mathcal{F}_n$$

for every n, then ν_0 is a stopping time, i.e. an exercise strategy. Further, ν_0 is the first time that $\widetilde{X}_n \geq E^Q\left[\widetilde{H}_{n+1} \mid \mathcal{F}_n\right]$ and so intuitively it represents the first time that it is profitable to exercise the option.

According to Doob's decomposition theorem and in particular by (A.107), for $n = 1, \dots, N$, we have

$$M_n = \widetilde{H}_n + \sum_{k=0}^{n-1} \left(\widetilde{H}_k - E^Q \left[\widetilde{H}_{k+1} \mid \mathcal{F}_k \right] \right), \tag{2.144}$$

and consequently

$$M_{\nu_0} = \widetilde{H}_{\nu_0} \tag{2.145}$$

since

$$\widetilde{H}_k = E^Q \left[\widetilde{H}_{k+1} \mid \mathcal{F}_k \right] \quad \text{over} \quad \{k < \nu_0\}.$$

Then, by (2.142), we have

$$\widetilde{V}_{\nu_0}^{(\alpha,\beta)} = M_{\nu_0} =$$

(by (2.145))

$$= \widetilde{H}_{\nu_0} =$$

(by the definition of ν_0)

$$= \widetilde{X}_{\nu_0}, \tag{2.146}$$

and this proves that $(\alpha, \beta) \in \mathcal{A}_X^-$.

Third step. We show that ν_0 is an optimal exercise time. Since $(\alpha, \beta) \in \mathcal{A}_X^+ \cap \mathcal{A}_X^-$, by (2.138) in Proposition 2.56 we get

$$V_0^{(\alpha,\beta)} = \sup_{\nu \in \mathcal{T}_0} E^Q \left[\widetilde{X}_\nu \right].$$

On the other hand, by (2.146) and the optional sampling Theorem A.129, we have

$$V_0^{(\alpha,\beta)} = E^Q \left[\widetilde{X}_{\nu_0} \right]$$

and this concludes the proof. □

Remark 2.58 The preceding theorem is significant from both a theoretical and practical point of view: on one hand it proves that there exists a unique initial price of X that does not give rise to arbitrage opportunities. On the other hand it shows a constructive way to determine the main features of X:

i) the initial price $\widetilde{H}_0 = \sup_{\nu \in \mathcal{T}_0} E^Q \left[\widetilde{X}_\nu \right]$ that can be computed by the iterative formula (2.140) (see also (2.148) below);
ii) an optimal exercise strategy ν_0 for which we have

$$E^Q \left[\widetilde{X}_{\nu_0} \right] = \sup_{\nu \in \mathcal{T}_0} E^Q \left[\widetilde{X}_\nu \right] = \widetilde{H}_0;$$

iii) a hedging strategy $(\alpha, \beta) \in \mathcal{A}_X^+ \cap \mathcal{A}_X^-$ such that $V_n^{(\alpha,\beta)} \geq X_n$ for any n and whose initial cost equals the initial arbitrage price $\widetilde{H}_0$. More precisely, (α, β) is the replicating strategy for the European option M_N: in Section 2.5.3, we will analyze more in details how to compute (α, β). □

Remark 2.59 For fixed $n \leq N$, we denote by

$$\mathcal{T}_n = \{\nu \in \mathcal{T}_0 \mid \nu \geq n\}$$

the family of exercise strategies of an American derivative *bought at time* t_n. A strategy $\nu_n \in \mathcal{T}_n$ is optimal if

$$E^Q\left[\widetilde{X}_{\nu_n} \mid \mathcal{F}_n\right] = \sup_{\nu \in \mathcal{T}_n} E^Q\left[\widetilde{X}_\nu \mid \mathcal{F}_n\right].$$

If $\widetilde{H}$ is the process in (2.140), we denote the first time it is profitable to exercise the American derivative bought at time t_n by

$$\nu_n(\omega) = \min\{k \geq n \mid \widetilde{H}_k(\omega) = \widetilde{X}_k(\omega)\}, \qquad \omega \in \Omega.$$

We can easily extend Theorem 2.57 and prove that ν_n is the first optimal exercise time following n. To be more precise we have

$$\widetilde{H}_n = E^Q\left[\widetilde{X}_{\nu_n} \mid \mathcal{F}_n\right] = \sup_{\nu \in \mathcal{T}_n} E^Q\left[\widetilde{X}_\nu \mid \mathcal{F}_n\right]. \tag{2.147}$$

□

Definition 2.60 *The process H defined by $H_n = B_n \widetilde{H}_n$ with $\widetilde{H}$ as in* (2.140), *is called arbitrage price of X. More explicitly we have*

$$H_n = \begin{cases} X_N, & n = N, \\ \max\left\{X_n, \frac{1}{1+r} E^Q\left[H_{n+1} \mid \mathcal{F}_n\right]\right\}, & n = 0, \dots, N-1. \end{cases} \tag{2.148}$$

Remark 2.61 In the proof of Theorem 2.57 we saw that hedging X is equivalent to replicating the (European) derivative M_N. Let us point out that, by (2.144), we have

$$M_n = \widetilde{H}_n + \sum_{k=0}^{n-1} \left(\widetilde{X}_k - E^Q\left[\widetilde{H}_{k+1} \mid \mathcal{F}_k\right]\right)^+ =: \widetilde{H}_n + I_n, \qquad 1 \leq n \leq N,$$

and so M_n can be decomposed as the sum of the discounted price $\widetilde{H}_n$ and the term I_n that can be interpreted as the value of early exercises: as a matter of fact, the terms of the sum that defines I_n are positive when $\widetilde{X}_k > E^Q\left[\widetilde{H}_{k+1} \mid \mathcal{F}_k\right]$, i.e. at times that early exercise is profitable. To fix the ideas, if $n = 1$, we have

$$M_1 = \widetilde{H}_1 + \left(\widetilde{X}_0 - E^Q\left[\widetilde{H}_1\right]\right)^+. \tag{2.149}$$

□

2.5.2 Optimal exercise strategies

The optimal exercise strategy of an American derivative X is not necessarily unique. In this section we aim at giving some general characterization of optimal exercise strategies and to determine the first and the last ones of these strategies.

Hereafter we assume that the market is arbitrage-free but not necessarily complete. For a fixed EMM Q, we denote by $\widetilde{H}$ the Snell envelope of $\widetilde{X}$, with respect to Q, defined in (2.140). We recall that, by (2.137), an exercise strategy $\bar{\nu} \in \mathcal{T}_0$ is optimal for X under Q if we have

$$E^Q\left[\widetilde{X}_{\bar{\nu}}\right] = \max_{\nu \in \mathcal{T}_0} E^Q\left[\widetilde{X}_\nu\right].$$

Moreover, given a process Y and a stopping time ν, we denote by $Y^\nu = (Y_n^\nu)$ the stopped process defined as

$$Y_n^\nu(\omega) = Y_{n\wedge\nu(\omega)}(\omega), \qquad \omega \in \Omega.$$

By Lemma A.125, if Y is adapted then Y^ν is adapted; if Y is a martingale (resp. super/sub-martingale) then Y^ν is a martingale (resp. super/sub-martingale) as well.

Lemma 2.62 *For any $\nu \in \mathcal{T}_0$ we have*

$$E^Q\left[\widetilde{X}_\nu\right] \le H_0. \tag{2.150}$$

Moreover $\nu \in \mathcal{T}_0$ is optimal for X under Q if and only if

$$E^Q\left[\widetilde{X}_\nu\right] = H_0. \tag{2.151}$$

Proof. We have

$$E^Q\left[\widetilde{X}_\nu\right] \overset{(1)}{\le} E^Q\left[\widetilde{H}_\nu\right] = E^Q\left[\widetilde{H}_N^\nu\right] \overset{(2)}{\le} H_0 \tag{2.152}$$

where inequality (1) is a consequence of the fact that $X_n \le H_n$ for any n and (2) follows from the Q-super-martingale property of $\widetilde{H}$ and Doob's optional sampling Theorem A.129.

By (2.150), it is clear that (2.151) is a sufficient condition for the optimality of ν. In order to prove that (2.151) is also a necessary condition, we have to show the existence of at least one strategy for which (2.151) holds: actually, two of these strategies will be explicitly constructed in Proposition 2.64 below. The reader can check that the proof of Proposition 2.64 is independent on our thesis so that no circular argument is used. We also remark that, under the assumption of completeness of the market, an exercise strategy verifying (2.151) was already introduced in the proof of Theorem 2.57. □

Corollary 2.63 *If $\nu \in \mathcal{T}_0$ is such that*

i) $\widetilde{X}_\nu = \widetilde{H}_\nu$;

ii) $\widetilde{H}^\nu$ *is a* Q*-martingale;*

then ν is an optimal exercise strategy for X under Q.

Proof. Conditions $i)$ and $ii)$ imply that (1) and (2) in formula (2.152) are equalities. Consequently $E^Q\left[\widetilde{X}_\nu\right] = H_0$ and therefore, by Lemma 2.62, ν is optimal for X under Q. □

Next, for greater convenience, we introduce the process

$$E_n = \frac{1}{1+r} E^Q\left[H_{n+1} \mid \mathcal{F}_n\right], \qquad n \le N-1, \tag{2.153}$$

and we also set $E_N = -1$. Then by (2.148) we have

$$H_n = \max\{X_n, E_n\}, \qquad n \le N,$$

and the sets $\{n \mid X_n \ge E_n\}$ and $\{n \mid X_n > E_n\}$ are nonempty since $X_N \ge 0$ by assumption. Consequently the following definition of exercise strategies is well-posed:

$$\nu_{\min} = \min\left\{n \mid X_n \ge E_n\right\}, \tag{2.154}$$

$$\nu_{\max} = \min\left\{n \mid X_n > E_n\right\}. \tag{2.155}$$

Proposition 2.64 *The exercise strategies $\nu_{\min}$ and $\nu_{\max}$ are optimal for X under Q.*

Proof. We show that $\nu_{\min}$ and $\nu_{\max}$ are optimal by verifying the conditions $i)$ and $ii)$ of Corollary 2.63. By definition (2.154)-(2.155) we have that

$$H_{\nu_{\min}} = \max\left\{X_{\nu_{\min}}, E_{\nu_{\min}}\right\} = X_{\nu_{\min}},$$
$$H_{\nu_{\max}} = \max\left\{X_{\nu_{\max}}, E_{\nu_{\max}}\right\} = X_{\nu_{\max}},$$

and this proves $i)$. Next we recall that by Doob's decomposition theorem we have

$$\widetilde{H}_n = M_n + A_n, \qquad n \le N,$$

where M is a Q-martingale such that $M_0 = H_0$ and A is a predictable and decreasing process such that $A_0 = 0$. More precisely we have (cf. (A.108))

$$A_n = -\sum_{k=0}^{n-1}\left(\widetilde{H}_k - \widetilde{E}_k\right), \qquad n = 1, \dots, N.$$

By definition (2.154)-(2.155), we have

$$H_n = E_n \quad \text{in} \quad \{n \le \nu_{\max} - 1\},$$

so that

$$A_n = 0 \quad \text{in} \quad \{n \le \nu_{\max}\}, \tag{2.156}$$

and

$$A_n < 0 \quad \text{in} \quad \{n \ge \nu_{\max} + 1\}. \tag{2.157}$$

Thus we get

$$\widetilde{H}_n = M_n \quad \text{in} \quad \{n \le \nu_{\max}\}, \tag{2.158}$$

and since clearly $\nu_{\min} \le \nu_{\max}$, we have

$$\widetilde{H}^{\nu_{\min}} = M^{\nu_{\min}}, \qquad \widetilde{H}^{\nu_{\max}} = M^{\nu_{\max}}.$$

Consequently, by Lemma A.125, the processes $\widetilde{H}^{\nu_{\min}}$ and $\widetilde{H}^{\nu_{\max}}$ are Q-martingales: this proves *ii)* of Corollary 2.63 and concludes the proof. □

We close this section by proving that $\nu_{\min}$ and $\nu_{\max}$ are the *first* and *last* optimal exercise strategies for X under Q, respectively.

Proposition 2.65 *If $\nu \in \mathcal{T}_0$ is optimal for X under Q then*

$$\nu_{\min} \le \nu \le \nu_{\max}.$$

Proof. Let us suppose that

$$P(\nu < \nu_{\min}) > 0. \tag{2.159}$$

We aim at proving that ν cannot be optimal because (1) in (2.152) is a strict inequality. Indeed, since P and Q are equivalent, from (2.159) it follows that

$$Q\left(\widetilde{X}_\nu < \widetilde{H}_\nu\right) > 0,$$

and therefore, since $\widetilde{X}_\nu \le \widetilde{H}_\nu$, we get

$$E^Q\left[\widetilde{X}_\nu\right] < E^Q\left[\widetilde{H}_\nu\right].$$

On the other hand, let us suppose that

$$P(\nu > \nu_{\max}) > 0. \tag{2.160}$$

In this case we prove that ν cannot be optimal because (2) in (2.152) is a strict inequality. Indeed, since P, Q are equivalent and A is a decreasing and non-positive process, from (2.157) it follows that

$$E^Q[A_\nu] < 0.$$

Consequently we have

$$E^Q\left[\widetilde{H}_\nu\right] = E^Q[M_\nu] + E^Q[A_\nu] < M_0 = H_0. \qquad \square$$

2.5.3 Pricing and hedging algorithms

We consider an American derivative X in a complete market (S, B) where Q is the EMM with numeraire B. By the results of the previous sections, the arbitrage price of an American derivative X is defined by the recursive formula

$$H_n = \begin{cases} X_N, & n = N, \\ \max\{X_n, E_n\}, & n = 0, \dots, N-1, \end{cases} \tag{2.161}$$

where E is the process defined by $E_N = -1$ and

$$E_n = \frac{1}{1+r} E^Q [H_{n+1} \mid \mathcal{F}_n], \qquad n \le N-1. \tag{2.162}$$

A remarkable case is when the underlying assets are modeled by Markov processes (as in the binomial and trinomial models by Theorem 2.31) and the American derivative is path-independent, that is $X = (\varphi_n(S_n))$ where φ_n is the payoff function at time t_n. In this case, by the Markov property of the price process S, the arbitrage price is given by

$$H_n = \begin{cases} \varphi_N(S_N), & n = N, \\ \max\left\{\varphi_n(S_n), \frac{1}{1+r} E^Q [H_{n+1} \mid S_n]\right\}, & n = 0, \dots, N-1, \end{cases} \tag{2.163}$$

and therefore H_n can be expressed as a function of S_n.

Once we have determined the process E in (2.162), the minimal and maximal among optimal exercise strategies are given by

$$\nu_{\min} = \min\{n \mid X_n \ge E_n\}, \qquad \nu_{\max} = \min\{n \mid X_n > E_n\}. \tag{2.164}$$

Concerning the hedging strategy, at least from a theoretical point of view, this problem was solved in Theorem 2.57: indeed a super- and sub-replicating strategy (α, β) (i.e. a strategy $(\alpha, \beta) \in \mathcal{A}_X^+ \cap \mathcal{A}_X^-$) was defined as the replicating strategy for the European derivative M_N. We recall that M denotes the martingale part of the Doob's decomposition of $\widetilde{H}$, that is the Snell envelope of $\widetilde{X}$, and once $\widetilde{H}$ has been determined by (2.140), then the process M can be computed by the forward recursive formula (cf. (A.105))

$$M_0 = H_0, \qquad M_{n+1} = M_n + \widetilde{H}_{n+1} - E\left[\widetilde{H}_{n+1} \mid \mathcal{F}_n\right]; \tag{2.165}$$

consequently the hedging strategy can be determined proceeding as in the European case. However M_N is given by formula (2.165) in terms of a conditional expectation and therefore M_N *is a path-dependent derivative even if* X *is path-independent.* So the computation of the hedging strategy can be burdensome, since M_N is a function of the entire path of the underlying assets and not just of the final values. As a matter of fact, this approach is not used in practice.

Instead, it is worthwhile noting that the process M_n depends on the path of the underlying assets just because it has to keep track of the possible early exercises: but at the moment the derivative is exercised, hedging is no longer necessary and the problem gets definitely easier. Indeed we recall that (cf. (2.158))

$$\widetilde{H}_n = M_n \quad \text{for} \quad n \le \nu_{\max}, \tag{2.166}$$

where $\nu_{\max}$ is the *last* optimal exercise time by Proposition 2.65. In particular, before $\nu_{\max}$ *the hedging strategy can be determined by using directly the process* H *instead of* M*:* this is convenient since if X is Markovian, i.e. $X_n = \varphi_n(S_n)$, then H is Markovian as well by (2.163).

Next we consider the special case of the binomial model. We use notation (2.65) and, for a path-independent derivative with payoff $X_n = \varphi_n(S_n)$ at time t_n, by the Markovian property of the arbitrage price in (2.163), we set

$$H_{n,k} = H_n(S_{n,k}), \qquad 0 \le k \le n \le N.$$

Then the binomial algorithm that we presented in Section 2.3.3 can be easily modified to handle the possibility of early exercise. More precisely we have the following iterative pricing algorithm:

$$\begin{cases} H_{N,k} = \varphi_N(S_{N,k}), & k \le N, \\ H_{n-1,k} = \max\left\{\varphi_{n-1}(S_{n-1,k}), \frac{1}{1+r}(qH_{n,k+1} + (1-q)H_{n,k})\right\}, & k \le n-1, \end{cases} \tag{2.167}$$

with $n = 1, \dots, N$ and $q = \frac{1+r-d}{u-d}$ where u, d, r are the binomial parameters.

Concerning the hedging problem, by using identity (2.166) we have that the hedging strategy for the n-th period, $n \le \nu_{\max}$, is simply given by

$$\alpha_{n,k} = \frac{H_{n,k+1} - H_{n,k}}{(u-d)S_{n-1,k}}, \qquad \beta_{n,k} = \frac{uH_{n,k} - dH_{n,k+1}}{(1+r)^n(u-d)}, \qquad k = 0, \dots, n-1, \tag{2.168}$$

exactly as in the European case. We recall that $(\alpha_{n,k}, \beta_{n,k})$ *is the strategy for the n-th period* $[t_{n-1}, t_n]$*, that is constructed at time* t_{n-1} *in the case* $S_{n-1} = S_{n-1,k}$.

At time $t_{\nu_{\max}}$ it is not necessary to compute the strategy $(\alpha_{\nu_{\max}+1}, \beta_{\nu_{\max}+1})$ (for the $(\nu_{\max}+1)$-th period) since $t_{\nu_{\max}}$ is the last time at which it is profitable to exercise the American option. If the holder of the option erroneously does not exercise at a time preceding or equal to $t_{\nu_{\max}}$, then he/she gives rise to an arbitrage opportunity for the writer: indeed, since the value of the hedging strategy is equal to $M_{\nu_{\max}}$, for the writer it suffices to adopt the strategy (2.168) with $n = \nu_{\max} + 1$ at time $t_{\nu_{\max}}$, to get at time $t_{\nu_{\max}+1}$

$$M_{\nu_{\max}+1} > \widetilde{H}_{\nu_{\max}+1} \ge \widetilde{X}_{\nu_{\max}+1},$$

that is strictly greater than the payoff.

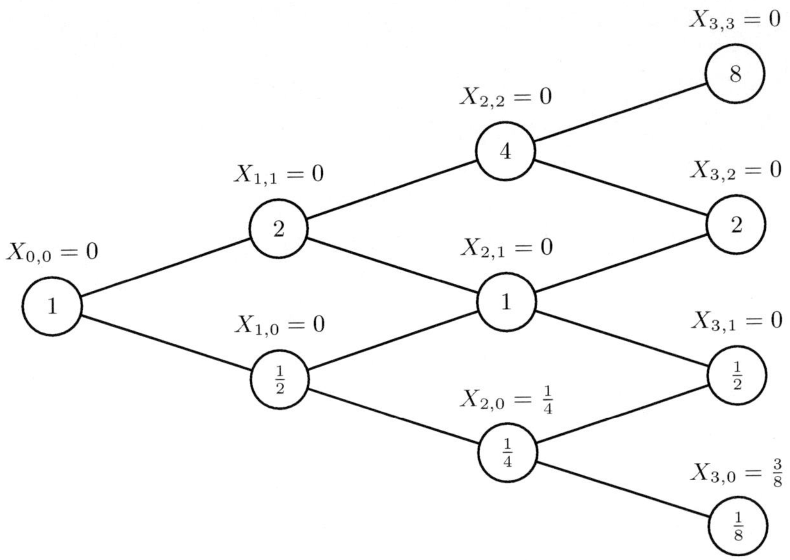

Fig. 2.10. Binomial tree with asset prices (inside the circles) and payoff X of the American Put with strike $K = \frac{1}{2}$

Example 2.66 In a three-period binomial model, we consider an American Put option with payoff $X_n = (\frac{1}{2} - S_n)^+$, $n = 0, 1, 2, 3$. We assume that $u = 2, d = r = \frac{1}{2}$ and the initial price of the underlying asset is $S_0 = 1$. In Figure 2.10 we represent the asset prices and the values of the payoff of the American Put on the binomial tree.

We first compute the arbitrage price process H and the minimal and maximal optimal exercise strategies. By (2.167) we have

$$H_n = \begin{cases} (\frac{1}{2} - S_3)^+, & n = 3, \\ \max\left\{(\frac{1}{2} - S_n)^+, E_n\right\}, & n = 0, 1, 2, \end{cases} \tag{2.169}$$

where E is the process in (2.162), that is $E_3 = -1$ and

$$E_n = \frac{1}{1+r} E^Q \left[H_{n+1} \mid \mathcal{F}_n\right], \qquad n = 0, 1, 2.$$

At maturity we have

$$\begin{cases} H_{3,3} = X_{3,3} = \left(\frac{1}{2} - 8\right)^+ = 0, \\ H_{3,2} = X_{3,2} = \left(\frac{1}{2} - 2\right)^+ = 0, \\ H_{3,1} = X_{3,1} = \left(\frac{1}{2} - \frac{1}{2}\right)^+ = 0, \\ H_{3,0} = X_{3,0} = \left(\frac{1}{2} - \frac{1}{8}\right)^+ = \frac{3}{8}. \end{cases}$$

Subsequently, by (2.169), we have

$$X_{2,2} = X_{2,1} = E_{2,2} = E_{2,1} = 0,$$

so that $H_{2,2} = H_{2,1} = 0$. Moreover

$$X_{2,0} = \left(\frac{1}{2} - \frac{1}{4}\right)^+ = \frac{1}{4},$$

and

$$E_{2,0} = \frac{1}{1+r}\left(qH_{3,1} + (1-q)H_{3,0}\right) = \frac{2}{3}\left(q \cdot 0 + (1-q)\frac{3}{8}\right) = \frac{1}{12}$$

since $q = \frac{1+r-d}{u-d} = \frac{2}{3}$. Then we have

$$H_{2,0} = \max\left\{X_{2,0}, E_{2,0}\right\} = X_{2,0} = \frac{1}{4}.$$

At the previous time we have $X_{1,1} = E_{1,1} = H_{1,1} = 0$ and

$$E_{1,0} = \frac{1}{1+r}\left(qH_{2,1} + (1-q)H_{2,0}\right) = \frac{1}{4}\left(\frac{1-q}{1+r}\right) = \frac{1}{18},$$

so that, since $X_{1,0} = 0$, we have $H_{1,0} = E_{1,0} = \frac{1}{18}$. Lastly, we have $X_{0,0} = 0$ and therefore

$$H_{0,0} = E_{0,0} = \frac{1}{1+r}\left(qH_{1,1} + (1-q)H_{1,0}\right) = \frac{1}{81}.$$

To make the following computations easier, in Figure 2.11 we represent the values of the processes X (inside the circles) and E (outside the circles), writing in bold the greater of the two values that is equal to the arbitrage price H of the American option.

Examining Figure 2.11 we can easily determine the minimal and maximal optimal exercise strategies: indeed, by definition (2.164) we have

$$\nu_{\min} = \min\{n \mid X_n \geq E_n\} = \begin{cases} 1 & \text{on } \{S_1 = S_{1,1}\} \\ 2 & \text{on } \{S_1 = S_{1,0}\}. \end{cases}$$

Analogously we have

$$\nu_{\max} = \min\{n \mid X_n > E_n\} = \begin{cases} 2 & \text{on } \{S_2 = S_{2,0}\} \\ 3 & \text{otherwise.} \end{cases}$$

These extreme optimal strategies are represented in Figure 2.12.

Next we compute the hedging strategy (α, β). As we already explained, even if (α, β) is the replicating strategy of the European derivative M_N in (2.165), it is not necessary to determine M_N explicitly: instead, we may use the usual formulas (2.168) for $n \leq \nu_{\max}$. Thus, in the first period we have

$$\alpha_{1,0} = \frac{H_{1,1} - H_{1,0}}{(u-d)S_0} = \frac{0 - \frac{1}{18}}{\frac{3}{2}} = -\frac{1}{27}, \quad \beta_{1,0} = \frac{uH_{1,0} - dH_{1,1}}{(1+r)(u-d)} = \frac{2\frac{1}{18}}{\frac{9}{4}} = \frac{4}{81}.$$

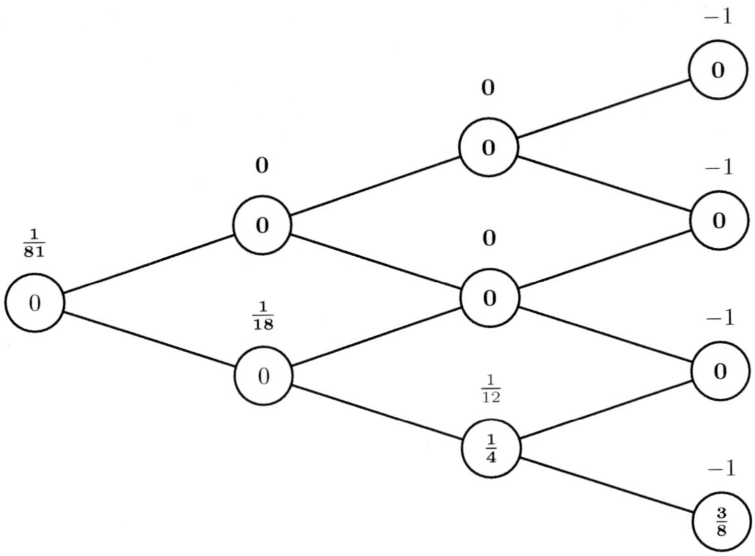

Fig. 2.11. Values of the processes X (inside the circles) and E (outside the circles)

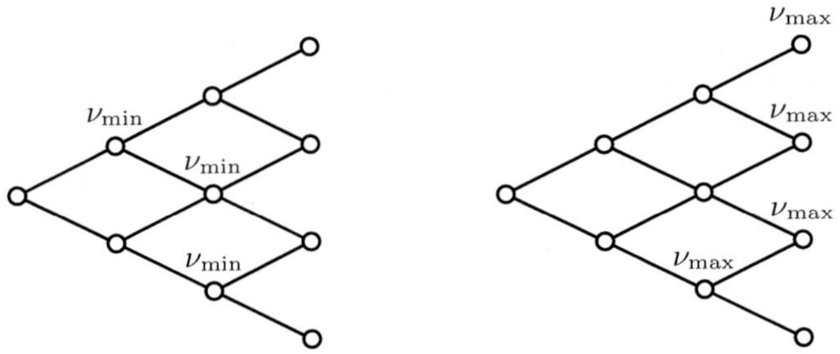

Fig. 2.12. Minimal (on the left) and maximal (on the right) optimal exercise strategies

In the second period, the strategy is the following:

$$\alpha_{2,1} = \frac{H_{2,2} - H_{2,1}}{(u-d)S_{1,1}} = 0, \qquad \beta_{2,1} = \frac{uH_{2,1} - dH_{2,2}}{(1+r)^2(u-d)} = 0,$$

$$\alpha_{2,0} = \frac{H_{2,1} - H_{2,0}}{(u-d)S_{1,0}} = \frac{-\frac{1}{4}}{\frac{3}{2}\cdot\frac{1}{2}} = -\frac{1}{3}, \qquad \beta_{2,0} = \frac{uH_{2,0} - dH_{2,1}}{(1+r)^2(u-d)} = \frac{4}{27}.$$

In the last period we have to compute the strategy only for $S_2 = S_{2,2}$ and $S_2 = S_{2,1}$ since in $S_2 = S_{2,0}$ there is the last opportunity of optimal early

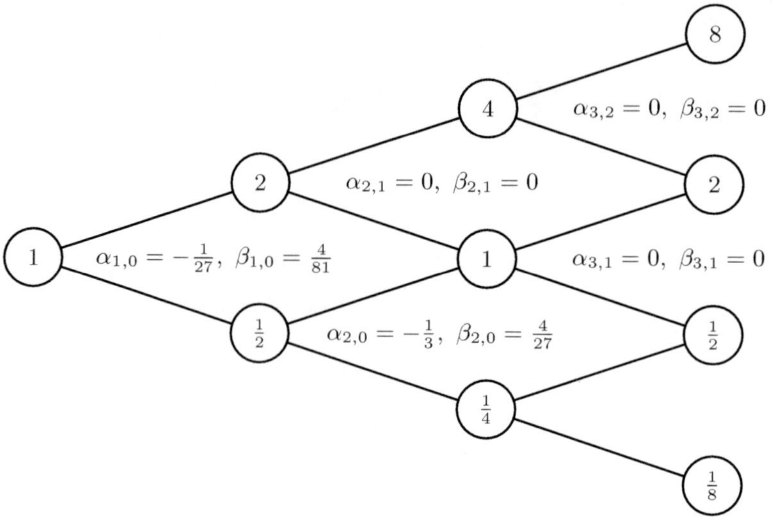

Fig. 2.13. Hedging strategy for $0 < r < 1$

exercise (see also Figure 2.12). Then we have

$$\alpha_{3,2} = \frac{H_{3,3} - H_{3,2}}{(u-d)S_{2,2}} = 0, \qquad \beta_{3,2} = \frac{uH_{3,2} - dH_{3,3}}{(1+r)^3(u-d)} = 0,$$

$$\alpha_{3,1} = \frac{H_{3,2} - H_{3,1}}{(u-d)S_{2,1}} = 0, \qquad \beta_{3,1} = \frac{uH_{3,1} - dH_{3,2}}{(1+r)^3(u-d)} = 0.$$

The hedging strategy is represented in Figure 2.13. □

2.5.4 Relations with European options

In an arbitrage-free and complete market, we denote by (H_n^A) the arbitrage price of the American derivative X and by (H_n^E) the arbitrage price of the related European derivative with payoff X_N. We recall that

$$\widetilde{H}_n^A = \max_{\nu \in \mathcal{T}_n} E^Q \left[\widetilde{X}_\nu \mid \mathcal{F}_n \right], \qquad \widetilde{H}_n^E = E^Q \left[\widetilde{X}_N \mid \mathcal{F}_n \right], \qquad n = 0, \dots, N,$$

where Q is the EMM.

The next result establishes some relations between the prices of American-style and European-style derivatives. In particular we prove that an American Call option (on a stock that does not pay dividends and assuming that $r \geq 0$) is worth just as the corresponding European option.

Proposition 2.67 *We have*

i) $H_n^A \geq H_n^E$ *for* $0 \leq n \leq N$;
ii) if $H_n^E \geq X_n$ *for every* n, *then*

$$H_n^A = H_n^E, \qquad n = 0, \dots, N,$$

and $\nu \equiv N$ *is an optimal exercise strategy.*

Proof. i) Since $\widetilde{H}^A$ is a Q-super-martingale, we have

$$\widetilde{H}_n^A \geq E^Q\left[\widetilde{H}_N^A \mid \mathcal{F}_n\right] = E^Q\left[\widetilde{X}_N \mid \mathcal{F}_n\right] = \widetilde{H}_n^E,$$

hence the claim, since $B_n > 0$. This first part states that, in general, an American derivative is worth more than its corresponding European counterpart: this fact is intuitive because an American derivative gives more rights to the holder who is free to exercise it even before maturity.

ii) By hypothesis, $\widetilde{H}^E$ is a martingale (and thus also a super-martingale) greater than $\widetilde{X}$. However $\widetilde{H}^A$ is the smallest super-martingale greater than $\widetilde{X}$ (cf. first step in the proof of Theorem 2.57): therefore we have $\widetilde{H}^A = \widetilde{H}^E$ and also $H^A = H^E$. □

Remark 2.68 Assume $r \geq 0$. We have

$$\widetilde{H}_n^E = \frac{1}{B_N} E^Q\left[(S_N - K)^+ \mid \mathcal{F}_n\right] \geq \frac{1}{B_N} E^Q\left[S_N - K \mid \mathcal{F}_n\right] = \widetilde{S}_n - \frac{K}{B_N}.$$

Since $r \geq 0$, we get

$$H_n^E \geq S_n - K\frac{B_n}{B_N} \geq S_n - K,$$

and since $H_n^E \geq 0$, we also have

$$H_n^E \geq (S_n - K)^+.$$

As a consequence of the second part of Proposition 2.67, *an American Call option is worth as the corresponding European option.*

We can also give an intuitive meaning to the preceding result: it is known that instead of exercising an American Call option before maturity it is more profitable to sell the underlying asset. Indeed, if the owner of an American Call option decided to exercise it early at time $n < N$, he/she would get a $S_n - K$ profit, that becomes $(1+r)^{N-n}(S_n - K)$ at maturity. Conversely, by selling one unit of the underlying asset at time t_n and keeping the option, at maturity he/she would get

$$(1+r)^{N-n}S_n - S_N + (S_N - K)^+ = \begin{cases} (1+r)^{N-n}S_n - K, & \text{if } S_N > K, \\ (1+r)^{N-n}S_n - S_N, & \text{if } S_N \leq K. \end{cases}$$

Therefore in all cases, if $r \geq 0$, the second strategy is worth more than the first. □

Example 2.69 The equivalence between the prices of American and European derivatives does not hold for Call options that pay dividends and for Put options. As a simple example, let us consider an American Put option in a one-period binomial model ($N = 1$) with $r > 0$ and, for the sake of simplicity,

$$q = \frac{1 + r - d}{u - d} = \frac{1}{2}.$$

Then $u+d = 2(1+r)$ and the price of the corresponding European Put option is

$$p_0 = \frac{1}{2(1+r)}((K - uS_0)^+ + (K - dS_0)^+) =$$

(if, for example, $K > uS_0$)

$$= \frac{1}{2(1+r)}(K - uS_0 + K - dS_0) = \frac{K}{1+r} - S_0.$$

For the American Put option we have

$$P_0 = \max\{K - S_0, p_0\} = K - S_0$$

and so in this case it is profitable to exercise the option immediately. □

2.5.5 Free-boundary problem for American options

In this section we study the asymptotic behaviour of the binomial model for an American option $X = \varphi(t, S)$ as N goes to infinity and we prove a consistency result for American-style derivatives, analogous to the one presented in Section 2.3.5. As we are going to see, the Black-Scholes price of an American option is the solution of a so-called "free-boundary problem" that is in general more difficult to handle than the classical Cauchy problem for European options. In this case pricing by the binomial algorithm becomes an effective alternative to the solution of the problem in continuous time.

We use the notations of Section 2.3.6: in particular we denote the arbitrage price of the derivative by $f = f(t, S)$, $(t, S) \in [0, T] \times \mathbb{R}_{>0}$, and we put $\delta = \frac{T}{N}$; the recursive pricing formula (2.167) becomes

$$\begin{cases} f(T, S) = \varphi(T, S), \\ f(t, S) = \max\left\{\frac{1}{1+r_N}\left(qf(t+\delta, uS) + (1-q)f(t+\delta, dS)\right), \varphi(t, S)\right\}. \end{cases} \tag{2.170}$$

The second equation in (2.170) is equivalent to

$$\max\left\{\frac{J_\delta f(t, S)}{\delta}, \varphi(t, S) - f(t, S)\right\} = 0$$

where J_δ is the discrete operator in (2.112). By using the consistency result of Proposition 2.50, we get the asymptotic version of the discrete problem

(2.170) as δ tends to zero:

$$\begin{cases} \max\{L_{\mathrm{BS}} f, \varphi - f\} = 0, & \text{in }]0,T[\times\mathbb{R}_{>0}, \\ f(T,S) = \varphi(T,S), & S \in \mathbb{R}_{>0}, \end{cases} \tag{2.171}$$

where

$$L_{\mathrm{BS}} f(t,S) = \partial_t f(t,S) + \frac{\sigma^2 S^2}{2} \partial_{SS} f(t,S) + rS\partial_S f(t,S) - rf(t,S)$$

is the Black-Scholes differential operator. Problem (2.171) contains a *differential inequality* and is theoretically more difficult to study than the usual parabolic Cauchy problem: we will prove the existence and the uniqueness of the solution in Paragraph 8.2. On the other hand, from a numerical point of view, the classical finite-difference methods can be adapted without difficulties to such problems.

The domain of the solution f of problem (2.171) can be divided in two regions:

$$[0,T[\times\mathbb{R}_{>0} = R_e \cup R_c,$$

where[15]

$$R_e = \{(t,S) \in [0,T[\times\mathbb{R}_{>0} \mid L_{\mathrm{BS}} f(t,S) \le 0 \ \text{ and } \ f(t,S) = \varphi(t,S)\}$$

is called *early-exercise region*, where $f = \varphi$, and

$$R_c = \{(t,S) \in [0,T[\times\mathbb{R}_{>0} \mid L_{\mathrm{BS}} f(t,S) = 0 \ \text{ and } \ f(t,S) > \varphi(t,S)\}$$

is called *continuation region*, where $f > \varphi$ (i.e. it is not profitable to exercise the option) and the price satisfies the Black-Scholes equation, as in the European case.

The boundary that separates the sets R_e, R_c depends on the solution f and is not assigned a priori in the problem: if this were the case, then problem (2.171) could be reduced to a classical Cauchy-Dirichlet problem for L_{BS} over R_c with boundary value φ. On the contrary, (2.171) is usually called a *free boundary problem* because finding the boundary is an essential part of the problem. Indeed, from a financial point of view, the free boundary *determines the optimal exercise price and time.*

Example 2.70 In the particular case of an American Put option, $\varphi(S) = (K - S)^+$ with maturity T, some properties of the free boundary can be proved by resorting solely to arbitrage arguments. Let us put

$$R_e(t) = \{S \mid (t,S) \in R_e\}.$$

[15] Since

$$\{\max\{F(x), G(x)\} = 0\} = \{F(x) = 0, \ G(x) \le 0\} \cup \{F(x) < 0, \ G(x) = 0\}.$$

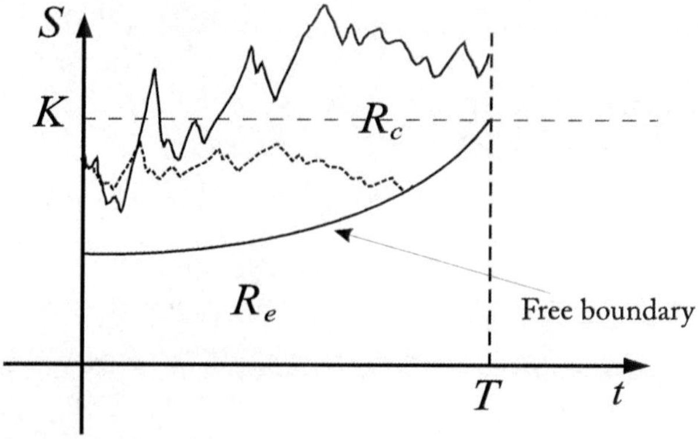

Fig. 2.14. Exercise and continuation regions of an American Put option

Then, under the assumption that the risk-free rate r is positive, for every $t \in [0, T[$ there exists $\beta(t) \in]0, K[$ such that

$$R_e(t) =]0, \beta(t)].$$

Indeed let $f(t, S)$ be the price of the option. Then $f(t, S)$ is strictly positive for every $t \in [0, T[$: on the other hand, since $\varphi(S) = 0$ for $S \geq K$, we have

$$R_e(t) \subseteq \{S < K\}, \qquad t \in [0, T[. \tag{2.172}$$

Further, by definition, $R_e(t)$ is relatively closed in $\mathbb{R}_{>0}$. $R_e(t)$ is an interval because of the convexity with respect to S of the price: if $S_1, S_2 \in R_e(t)$, then for every $\varrho \in [0, 1]$ we have

$$\varphi(\varrho S_1 + (1 - \varrho)S_2) \leq f(t, \varrho S_1 + (1 - \varrho)S_2) \leq \varrho f(t, S_1) + (1 - \varrho)f(t, S_2) =$$

(since $S_1, S_2 \in R_e(t)$ and by (2.172))

$$= \varrho(K - S_1) + (1 - \varrho)(K - S_2) = \varphi(\varrho S_1 + (1 - \varrho)S_2),$$

and so $\varrho S_1 + (1 - \varrho)S_2 \in R_e(t)$. The fact that the price function is convex can be proved by using the no-arbitrage principle.

Finally we have

$$]0, K - Ke^{-r(T-t)}] \subseteq R_e(t).$$

Indeed, if $S \leq K(1 - e^{-r(T-t)})$, then it is profitable to exercise the option, since at time t one receives the amount

$$K - S \geq Ke^{-r(T-t)},$$

that, at maturity, yields

$$(K - S)e^{r(T-t)} \geq K \geq f(T, S).$$

By arbitrage arguments it is also possible to prove that β is a continuous and monotone increasing function. Figure 2.14 depicts the exercise and continuation regions of an American Put option. □

Going back to the general case, we point out that, by definition, we have

$$R_e \subseteq \{(t,S) \in [0,T[\times\mathbb{R}_{>0} \mid L_{\mathrm{BS}}\varphi(t,S) \le 0\}, \tag{2.173}$$

and this raises the question about the regularity assumptions we have to impose on φ, and also about what kind of regularity we might expect for the solution f of (2.171). Indeed, even in the simplest case of a Put option, the payoff function $\varphi = \varphi(S)$ is not differentiable at $S = K$ and $L_{\mathrm{BS}}\varphi$ is not defined everywhere in the classical sense. However in this case, by using the theory of distributions (cf. Appendix A.9.3), we get

$$L_{\mathrm{BS}}(K-S)^+ = \frac{\sigma^2 K^2}{2}\delta_K(S) - rK\mathbb{1}_{]0,K[}(S),$$

where δ_K denotes Dirac's delta distribution, concentrated at K. Therefore, if $r \ge 0$, at least formally we have

$$L_{\mathrm{BS}}(K-S)^+ \begin{cases} \le 0, & S < K, \\ \ge 0, & S \ge K, \end{cases}$$

and (2.173) is verified, recalling (2.172). Concerning the regularity of the solution, problem (2.171) does not admit in general a classical solution: in Paragraph 8.2 we will prove the existence of a solution in a suitable Sobolev space.

We conclude the section by stating a result analogous to Theorem 2.46 on the approximation of the continuous case by the binomial model: for the proof we refer to Kushner [220] or Lamberton and Pagès [228].

Theorem 2.71 *Let $P_N^A(0,S)$ be the price at the initial time of an American Put option with strike K and maturity T in the N-period binomial model with parameters*

$$u_N = e^{\sigma\sqrt{\delta_N}+\alpha\delta_N}, \qquad d_N = e^{-\sigma\sqrt{\delta_N}+\beta\delta_N},$$

where α, β are real constants. Then the limit

$$\lim_{N\to\infty} P_N^A(0,S) = f(0,S), \qquad S > 0$$

exists and f is the solution of the free-boundary problem (2.171).

2.5.6 American and European options in the binomial model

In this section, by using the arbitrage pricing formulas in the N-period binomial model, we present a qualitative study of the graph of the price of a Put option, as a function of the price of the underlying asset, and we compare the American and European versions.

Let P^E and P^A be the prices of a European and an American Put option, respectively, with strike K on the underlying asset S: using the notations of Paragraph 2.3, and denoting the initial price of the underlying asset by $S_0 = x$, we have

$$S_n = x\psi_n, \qquad \psi_n = \prod_{k=1}^{n}(1+\mu_k)$$

and the arbitrage prices at the initial time have the following expressions:

$$P^E(x) = E^Q\left[\frac{(K - x\psi_N)^+}{(1+r)^N}\right], \tag{2.174}$$

$$P^A(x) = \sup_{\nu\in\mathcal{T}_0} E^Q\left[\frac{(K - x\psi_\nu)^+}{(1+r)^\nu}\right]. \tag{2.175}$$

Proposition 2.72 *Assume that the parameter d in the binomial model is smaller than 1. The function $x \mapsto P^E(x)$ is continuous, convex and decreasing for $x \in \mathbb{R}_{\geq 0}$. Further,*

$$P^E(0) = \frac{K}{(1+r)^N}, \qquad P^E(x) = 0, \; x \in [Kd^{-N}, +\infty[,$$

and there exists $\bar{x} \in]0, K[$ such that

$$P^E(x) < (K - x)^+, \; x \in [0, \bar{x}], \qquad P^E(x) > (K - x)^+, \; x \in [\bar{x}, Kd^{-N}]. \tag{2.176}$$

The function $x \mapsto P^A(x)$ is continuous, convex and decreasing for $x \in \mathbb{R}_{\geq 0}$. Further,

$$P^A(0) = K, \qquad P^A(x) = 0, \; x \in [Kd^{-N}, +\infty[,$$

and there exists $x^* \in]0, K[$ such that

$$P^A(x) = (K - x)^+, \; x \in [0, x^*], \qquad P^A(x) > (K - x)^+, \; x \in [x^*, Kd^{-N}].$$

Proof. We can write (2.174) more explicitly as

$$P^E(x) = \frac{1}{(1+r)^N}\sum_{h=0}^{N} c_h (K - u^h d^{N-h} x)^+,$$

where $c_h = \binom{N}{h} q^h(1-q)^{N-h}$ are positive constants. Hence we infer directly the properties of continuity, convexity and the facts that the price function is monotone decreasing and that $P^E(x) = 0$ if and only if $(K - u^h d^{N-h}x)^+ = 0$ for every h or, equivalently, if $u^h d^{N-h} x \geq K$ for every h i.e.[16] $d^N x \geq K$.

[16] Since $d < 1$.

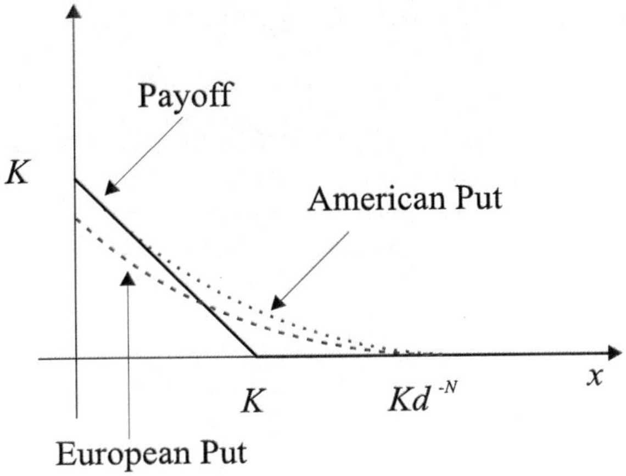

Fig. 2.15. Graph of the binomial price (dotted line) of an American Put option as a function of the price x of the underlying asset. The broken line represents the graph of the corresponding European Put option

Further, by (2.174) it is obvious that $P^E(0) = \frac{K}{(1+r)^N}$. To prove (2.176), let us consider the continuous convex function[17]

$$g(x) = P^E(x) - (K - x), \qquad x \in [0, K].$$

Since $g(0) < 0$ and $g(K) > 0$, by continuity g is null in at least one point: it remains to be seen whether such a point is unique. We put

$$x_0 = \inf\{x \mid g(x) > 0\}, \qquad x_1 = \sup\{x \mid g(x) < 0\}.$$

By continuity $g(x_0) = g(x_1) = 0$ and $x_0 \le x_1$: we want to prove that $x_0 = x_1$. If this were not the case, i.e. $x_0 < x_1$, by the convexity of g we would have

$$0 = g(x_0) \le tg(0) + (1-t)g(x_1) = tg(0) < 0$$

for some $t \in]0, 1[$ and this is a contradiction. This concludes the proof of the first part of the proposition.

The continuity of the function P^A follows from (2.140) which recursively defines P^A as the composition of continuous functions. The facts that the price function is convex and monotone decreasing follow from (2.175) since the functions

$$x \mapsto E^Q\left[\frac{(K - x\psi_\nu)^+}{(1+r)^\nu}\right]$$

are convex and decreasing and their least upper bound, when ν varies, preserves such properties.

[17] The sum of convex functions is convex.

Now, by (2.175), $P^A(x) = 0$ if and only if

$$E^Q\left[\frac{(K - x\psi_\nu)^+}{(1+r)^\nu}\right] = 0 \tag{2.177}$$

for every $\nu \in \mathcal{T}_0$. The expectation in (2.177) is a sum of terms of the form $c_{nh}(K - u^h d^{n-h}x)^+$, with c_{nh} positive constants. So $P^A(x) = 0$ if and only if $u^h d^{n-h}x \geq K$ for every[18] n, k i.e. if $d^N x \geq K$.

Finally let us consider the function

$$f(x) = P^A(x) - (K - x)^+.$$

By (2.140) $f \geq 0$ and since $\nu \geq 0$, we have

$$f(0) = K \sup_{\nu \in \mathcal{T}_0} E^Q\left[(1+r)^{-\nu}\right] - K = 0,$$

that is $P^A(0) = K$. Further

$$f(K) = K \sup_{\nu \in \mathcal{T}_0} E^Q\left[\frac{(1 - \psi_\nu)^+}{(1+r)^\nu}\right] \geq$$

(for $\nu = 1$)

$$\geq KE^Q\left[\frac{(-\mu_1)^+}{(1+r)}\right] > 0.$$

For $x \geq K$ we obviously have $f(x) = P^A(x) \geq (K - x)^+ = 0$. We put

$$x^* = \inf\{x \in [0, K] \mid f(x) > 0\}.$$

On the grounds of what we have already proved, we have $0 < x^* < K$ and, by definition, $f = 0$ over $[0, x^*]$. Finally we have that $f > 0$ over $]x^*, K]$; we prove this last fact by contradiction. Let us suppose that $f(x_1) = 0$ for some $x_1 \in]x^*, K[$. By the definition of x^*, there exists $x_0 < x_1$ such that $f(x_0) > 0$. Now we note that, over the interval $[0, K]$, the function f is convex and so

$$0 < f(x_0) \leq tf(x) + (1-t)f(x_1) = (1-t)f(x_1)$$

if $x_0 = tx + (1-t)x_1$, $t \in]0, 1[$. This concludes the proof. □

[18] Such that $0 \leq k \leq n \leq N$.

3

Continuous-time stochastic processes

In this chapter we introduce the elements of the theory of stochastic processes that we will use in continuous-time financial models. After a general presentation, we define the one-dimensional Brownian motion and we discuss some equivalence notions among stochastic processes. The most substantial part of the chapter is devoted to the study of the first and the second variation of a process: such a concept is introduced at first in the framework of the classical function theory and for Riemann-Stieltjes integration. Afterwards, we extend our analysis to the Brownian motion by determining its quadratic-variation process.

3.1 Stochastic processes and real Brownian motion

Let $(\Omega, \mathcal{F}, P)$ be a probability space and I a real interval of the form $[0, T]$ or $\mathbb{R}_{\geq 0}$.

Definition 3.1 *A measurable stochastic process (in what follows, simply a stochastic process) on $\mathbb{R}^N$ is a collection $(X_t)_{t \in I}$ of random variables with values in $\mathbb{R}^N$ such that the map*

$$X : I \times \Omega \longrightarrow \mathbb{R}^N, \qquad X(t, \omega) = X_t(\omega),$$

is measurable with respect to the product σ-algebra $\mathscr{B}(I) \otimes \mathcal{F}$. We say that X is integrable if $X_t \in L^1(\Omega, P)$ for every $t \in I$.

The concept of stochastic process extends that of deterministic function

$$f : I \longrightarrow \mathbb{R}^N.$$

Just as f associates t to the variable (the number) $f(t)$ in $\mathbb{R}^N$, similarly the stochastic process associates t to the random variable X_t in $\mathbb{R}^N$. A stochastic process can be used to describe a random phenomenon that evolves in time: for example, we can interpret a positive random variable X_t as the price of

Pascucci A.: PDE and Martingale Methods in Option Pricing

a risky asset at time t, or a random variable X_t in $\mathbb{R}^3$ as the position of a particle in space at time t.

To further help intuition, it is useful to think of a function $f : I \longrightarrow \mathbb{R}^N$ as a *curve or trajectory in* $\mathbb{R}^N$: the range of the curve f is

$$\gamma = \{f(t) \mid t \in I\}$$

and as the parameter t varies, $f(t)$ represents a point of the range γ. The idea can be extended to stochastic processes and, in this case, a different path (and therefore a possible trajectory of the price of an asset or a possible motion of a particle in space) corresponds to any $\omega \in \Omega$:

$$\gamma_\omega = \{X_t(\omega) \mid t \in I\}, \qquad \omega \in \Omega.$$

Definition 3.2 *A stochastic process X is continuous (a.s.) if the paths*

$$t \longmapsto X_t(\omega)$$

are continuous functions for every $\omega \in \Omega$ (for almost all $\omega \in \Omega$).

Analogously X is right continuous (a.s.-right continuous) if

$$X_t(\omega) = X_{t+}(\omega) := \lim_{s \to t^+} X_s(\omega)$$

for every t and for every $\omega \in \Omega$ (for almost all $\omega \in \Omega$).

The family of right-continuous processes is extremely important since, by using the density of $\mathbb{Q}$ in $\mathbb{R}$, many properties of discrete-time processes can be extended to this collection. This quite general fact will be used repeatedly later on.

Now we extend the concepts of filtration and adapted stochastic process to the continuous case. As in the discrete case, a filtration represents *a flow of information* and saying that a price is described by an adapted process means that it depends on the information available up to that moment.

Definition 3.3 *A filtration $(\mathcal{F}_t)_{t \geq 0}$ in $(\Omega, \mathcal{F}, P)$ is an increasing family of sub-σ-algebras of $\mathcal{F}$.*

We say beforehand that later on we will assume suitable hypotheses on the filtration (cf. Section 3.3.3).

Definition 3.4 *Given a stochastic process $X = (X_t)_{t \in I}$, the natural filtration for X is defined by*

$$\widetilde{\mathcal{F}}_t^X = \sigma(X_s \mid 0 \leq s \leq t) := \sigma(\{X_s^{-1}(H) \mid 0 \leq s \leq t,\, H \in \mathscr{B}\}), \qquad t \in I. \tag{3.1}$$

Definition 3.5 *A stochastic process X is adapted to a filtration $(\mathcal{F}_t)$ (or, simply, $\mathcal{F}_t$-adapted) if $\widetilde{\mathcal{F}}_t^X \subseteq \mathcal{F}_t$ for every t, or, in other terms, if X_t is $\mathcal{F}_t$-measurable for every t.*

Clearly $\widetilde{\mathcal{F}}^X$ is the smallest filtration with respect to which X is adapted.

Definition 3.6 (Real Brownian motion) *Let $(\Omega, \mathcal{F}, P, (\mathcal{F}_t))$ be a filtered probability space. A real Brownian motion is a stochastic process $W = (W_t)_{t \geq 0}$ in $\mathbb{R}$ such that*

i) $W_0 = 0$ a.s.;
ii) W is $\mathcal{F}_t$-adapted and continuous;
iii) for $t > s \geq 0$, the random variable $W_t - W_s$ has normal distribution $\mathcal{N}_{0,t-s}$ and is independent of $\mathcal{F}_s$.

It is not straightforward to prove the existence of a Brownian motion: some proofs can be found, for example, in Karatzas-Shreve [201]. A remarkable case is when the filtration is the natural one for W, i.e. $\mathcal{F}_t = \widetilde{\mathcal{F}}_t^W$.

In view of properties i) and ii) of Definition 3.6, the paths of a Brownian motion start (at $t = 0$) from the origin a.s. and they are continuous. Further, as a consequence of i) and iii), for every t we have

$$W_t \sim \mathcal{N}_{0,t} \tag{3.2}$$

since $W_t = W_t - W_0$ a.s.

Remark 3.7 (Brownian motion as random motion) Brownian motion was originally created as a probabilistic model for the motion of a particle. The following properties of Brownian motion are obvious consequences of (3.2):

a) $E[W_t] = 0$ for every $t \geq 0$, i.e. at any time the expected position of the particle is the initial one;
b) recalling the expression of the density of the normal distribution $\Gamma(t, \cdot)$ in (A.7), we have that, for every fixed $t > 0$, the probability that W_t belongs to a Borel set H decreases by translating H far from the origin. Intuitively the probability that the particle reaches H decreases moving H away from the starting point;
c) for every fixed $H \in \mathscr{B}$,
$$\lim_{t \to 0+} P(W_t \in H) = \delta_0(H).$$
Intuitively, when time decreases also the probability that the particle has moved away from the initial position decreases;
d) $E\left[W_t^2\right] = \mathrm{var}(W_t) = t$, i.e. the estimate of the distance, at time t, from the starting point of a particle moving randomly is $\sqrt{t}$: this fact is less intuitive but it corresponds to Einstein's observations [118].

Example 3.8 (Brownian motion as a model for a risky asset) A first continuous time model for the price of a risky asset S is the following:

$$S_t = S_0(1 + \mu t) + \sigma W_t, \qquad t \geq 0. \tag{3.3}$$

In (3.3), S_0 is the initial price of the asset, μ is the expected rate of return and σ denotes the riskiness of the asset or volatility. If $\sigma = 0$, the dynamics in (3.3) is *deterministic* and correspond to *simple compounding of interest* with risk-free rate μ. If $\sigma > 0$, the dynamics in (3.3) is stochastic and $S = (S_t)_{t \geq 0}$ is a Gaussian (or normal) stochastic process, i.e.

$$S_t \sim \mathcal{N}_{S_0(1+\mu t), \sigma^2 t} \tag{3.4}$$

for $t \geq 0$. From (3.4) it follows that

$$E[S_t] = S_0(1 + \mu t)$$

so that the expectation of S corresponds to a risk-free deterministic dynamics. Then a Brownian motion introduces "noise" but it does not modify the process in mean. Further, σ is directly proportional to the variance and so to the riskiness of the asset.

In practice this model is not used for two reasons: on one hand it is preferable to use a continuously compounded rate; on the other hand (3.4) implies that $P(S_t < 0) > 0$ if t is positive and this is does not make sense from an economic point of view. Nevertheless, (3.3) is sometimes used as a model for the debts/credits of a firm. □

3.1.1 Markov property

We have already commented on the meaning of the Markov property from a financial point of view: a stochastic process X, representing the price of an asset, has the Markov property if the expectation at time t of the future price X_T, $T > t$, depends only on the current price X_t and not on the past prices. While there are several ways to express this property, perhaps the following is the simplest one.

Definition 3.9 *In a filtered probability space $(\Omega, \mathcal{F}, P, (\mathcal{F}_t))$, an adapted stochastic process X has the Markov property if:*

(M) *for every bounded $\mathscr{B}$-measurable function φ we have*

$$E[\varphi(X_T) \mid \mathcal{F}_t] = E[\varphi(X_T) \mid X_t], \qquad T \geq t.$$

Remark 3.10 By Dynkin's Theorem A.9, property (M) is equivalent to the following condition that in general is easier to verify:

(M1) *for every Borel set H, we have*

$$E[X_T \in H \mid \mathcal{F}_t] = E[X_T \in H \mid X_t], \qquad T \geq t.$$ □

Note that the Markov property depends on the given filtration. The first remarkable example of Markov process is the Brownian motion: in order to illustrate more clearly this fact, we introduce some notations.

Definition 3.11 *Let W be a Brownian motion on the space $(\Omega, \mathcal{F}, P, (\mathcal{F}_t))$. For fixed $x \in \mathbb{R}$ and $t \geq 0$, the stochastic process $W^{t,x}$ defined by*

$$W_T^{t,x} = x + W_T - W_t, \qquad T \geq t,$$

is called Brownian motion starting at time t from x.

Clearly we have

i) $W_t^{t,x} = x$;
ii) $W^{t,x}$ is an adapted and continuous stochastic process;
iii) for $t \leq T \leq T + h$, the random variable $W_{T+h}^{t,x} - W_T^{t,x}$ has normal distribution $\mathcal{N}_{0,h}$ and is independent of $\mathcal{F}_T$.

Remark 3.12 As a consequence of the previous properties we have

$$W_T^{t,x} \sim \mathcal{N}_{x,T-t}, \qquad T \geq t. \tag{3.5}$$

Therefore, for fixed $x \in \mathbb{R}$ and $T > t$, the density of $W_T^{t,x}$ is

$$y \mapsto \Gamma^*(t, x; T, y),$$

where

$$\Gamma^*(t, x; T, y) = \frac{1}{\sqrt{2\pi(T-t)}} \exp\left(-\frac{(x-y)^2}{2(T-t)}\right) \tag{3.6}$$

is the fundamental solution of the *adjoint* heat equation (cf. Section 3.1.2 below). □

This justifies the following:

Definition 3.13 *The function $\Gamma^* = \Gamma^*(t, x; T, \cdot)$ is called transition density of the Brownian motion from the "initial" point (t, x) to the "final" time T.*

We prove now that a Brownian motion has the Markov property. Let φ be a bounded $\mathscr{B}$-measurable function: in view of Lemma A.108, we get

$$E\left[\varphi(W_T) \mid \mathcal{F}_t\right] = E\left[\varphi(W_T - W_t + W_t) \mid \mathcal{F}_t\right] = u(t, W_t), \qquad T \geq t, \tag{3.7}$$

where u is the $\mathscr{B}$-measurable function defined by

$$u(t, x) = E\left[\varphi(W_T - W_t + x)\right] = E\left[\varphi(W_T^{t,x})\right]. \tag{3.8}$$

Therefore we have proved the following:

Theorem 3.14 *A Brownian motion W on the space $(\Omega, \mathcal{F}, P, (\mathcal{F}_t))$ has the Markov property with respect to $(\mathcal{F}_t)$ and, in particular, formulas (3.7)-(3.8) hold: in a more compact form, we have*

$$E\left[\varphi(W_T) \mid \mathcal{F}_t\right] = E\left[\varphi(W_T^{t,x})\right]_{x=W_t}. \tag{3.9}$$

We note that (3.9) implies in particular (cf. Remark A.109) that

$$E\left[\varphi(W_T) \mid \mathcal{F}_t\right] = E\left[\varphi(W_T) \mid W_t\right], \qquad T \geq t,$$

i.e. W is a Markov stochastic process, according to Definition 3.9.

Using the expression for the transition density of Brownian motion, we can also write (3.9) more explicitly:

$$E\left[\varphi(W_T) \mid \mathcal{F}_t\right] = \int_{\mathbb{R}} \frac{\varphi(y)}{\sqrt{2\pi(T-t)}} \exp\left(-\frac{(y-W_t)^2}{2(T-t)}\right) dy.$$

Note that both sides of the equality are actually random variables.

Exercise 3.15 Given $\alpha, \beta \in \mathbb{R}$, $\alpha \neq 0$, prove that $(\alpha^{-1} W_{\alpha^2 t} + \beta)$ is a Brownian motion starting from β.

3.1.2 Brownian motion and the heat equation

We consider the adjoint heat operator in two variables:

$$L^* = \frac{1}{2}\partial_{xx} + \partial_t, \qquad (t,x) \in \mathbb{R}^2. \tag{3.10}$$

In Appendix A.3 we show that the function Γ^* in (3.6) is the fundamental solution of L^* and consequently, for every *final* datum $\varphi \in C_b(\mathbb{R})$, the Cauchy problem

$$\begin{cases} L^* u(t,x) = 0, & (t,x) \in]0,T[\, \times \mathbb{R}, \\ u(T,x) = \varphi(x) & x \in \mathbb{R}, \end{cases} \tag{3.11}$$

has a classical solution

$$u(t,x) = \int_{\mathbb{R}} \Gamma^*(t,x;T,y)\varphi(y)dy, \qquad t < T, \ x \in \mathbb{R}. \tag{3.12}$$

Since Γ^* is the transition density of Brownian motion, there exists a strong connection between Brownian motion and the heat equation; this is summed up in the following statements:

i) *the solution u in* (3.12) *of problem* (3.11) *has the following probabilistic representation:*

$$u(t,x) = E\left[\varphi\left(W_T^{t,x}\right)\right], \qquad x \in \mathbb{R}, \ t \in [0,T]; \tag{3.13}$$

ii) *by Theorem 3.14, the following formula for the conditional expectation of a Brownian motion holds:*

$$E\left[\varphi(W_T) \mid \mathcal{F}_t\right] = u(t, W_t), \qquad T \geq t, \tag{3.14}$$

where u is the solution in (3.12) *of the Cauchy problem* (3.11)*:* (3.14) *expresses the Markov property of Brownian motion.*

We remark that Monte Carlo[1] numerical methods for PDEs are based on representation formulas like (3.13).

[1] See Paragraph 12.4.

3.2 Uniqueness

3.2.1 Law of a continuous process

The law (or distribution) of a discrete stochastic process

$$(X_0, \dots, X_n)$$

is the joint distribution of the random variables $X_0, \dots, X_n$. Now we examine how to extend this notion to the case of a *continuous* stochastic process

$$X : [0,T] \times \Omega \longrightarrow \mathbb{R}^N$$

defined on a probability space $(\Omega, \mathcal{F}, P)$.

We denote by $C([0,T]) := C([0,T], \mathbb{R}^N)$ the linear space of the continuous functions on $[0,T]$ with values in $\mathbb{R}^N$ and we consider the function

$$\hat{X} : \Omega \longrightarrow C([0,T])$$

that maps the elementary event ω into the continuous trajectory $t \rightarrow X_t(\omega)$. Endowing $C([0,T])$ with a structure of probability space, we aim at showing that $\hat{X}$ *is a random variable.* First of all we recall that $C([0,T])$, with the usual maximum norm

$$\|w\|_\infty = \max_{t \in [0,T]} |w(t)|, \qquad w \in C([0,T]),$$

is a complete[2] normed space: in particular the norm defines the collection of open sets in $C([0,T])$ and consequently the Borel σ-algebra of $C([0,T])$, denoted by $\mathscr{B}(C([0,T]))$. The simplest example of Borel set is the ball with radius $r > 0$ and center w_0:

$$D(w_0, r) := \{w \in C([0,T]) \mid |w(t) - w_0(t)| < r, \ t \in [0,T]\}. \tag{3.15}$$

We recall also that $C([0,T])$ is a separable space and $\mathscr{B}(C([0,T]))$ is generated by a *countable* collection of balls of the form (3.15): for the proof of this statement, see Example A.157 in Appendix A.8.

Lemma 3.16 *For every $H \in \mathscr{B}(C([0,T]))$, we have*

$$\hat{X}^{-1}(H) = \{\omega \in \Omega \mid X(\omega) \in H\} \in \mathcal{F},$$

and therefore

$$\hat{X} : (\Omega, \mathcal{F}) \longrightarrow (C([0,T]), \mathscr{B}(C([0,T]))$$

is a random variable.

[2] Every Cauchy sequence is convergent.

Proof. Since $\mathscr{B}(C([0,T]))$ is generated by a countable collection of balls, it suffices to observe that, being X continuous, we have

$$\{X \in \overline{D}(w_0, r)\} = \bigcap_{t\in[0,T]\cap\mathbb{Q}} \underbrace{\{\omega \in \Omega \mid |X_t(\omega) - w_0(t)| \le r\}}_{\in\mathcal{F}},$$

hence the claim, in view of Remark A.20. □

Definition 3.17 *The probability measure P^X, defined by*

$$P^X(H) = P(X \in H), \qquad H \in \mathscr{B}(C([0,T])),$$

is called law of the process X.

Next we consider the map

$$\mathbb{X} : [0,T] \times C([0,T]) \longrightarrow \mathbb{R}^N, \qquad \mathbb{X}_t(w) := w(t), \tag{3.16}$$

that defines a stochastic process on $(C([0,T]), \mathscr{B}(C([0,T])))$: indeed

$$(t, w) \mapsto \mathbb{X}_t(w)$$

is a continuous function and therefore measurable with respect to the product σ-algebra $\mathscr{B}([0,T]) \otimes \mathscr{B}(C([0,T]))$. Since the process $\mathbb{X}$, defined on the probability space

$$(C([0,T]), \mathscr{B}(C([0,T])), P^X),$$

has the same law of X, we give the following:

Definition 3.18 *The process $\mathbb{X}$ on $(C([0,T]), \mathscr{B}(C([0,T])), P^X)$ is called canonical realization of X.*

Lemma 3.19 *We have that*

$$\sigma(\mathbb{X}_t, t \in [0,T]) = \mathscr{B}(C([0,T])).$$

Proof. Given a set $\tau = \{t_1, \dots, t_n\}$ consisting of a finite number of points in $[0,T]$ and $K = K_1 \times \cdots \times K_n$ with $K_i \in \mathscr{B}(\mathbb{R}^N)$, $i = 1, \dots, n$, a "cylinder" in $\mathscr{B}(C([0,T]))$ is a set of the form

$$\begin{aligned}\mathcal{H}(\tau, K) &= \{w \in C([0,T]) \mid w(t_i) \in K_i, \ i = 1, \dots, n\} \\ &= \bigcap_{i=1}^{n} \{\mathbb{X}_{t_i} \in K_i\}.\end{aligned}$$

Since $\mathbb{X}$ is a stochastic process, by Fubini's Theorem, we have

$$\sigma(\mathbb{X}_t, t \in [0,T]) \subseteq \mathscr{B}(C([0,T])).$$

To prove the reverse inclusion, we have to verify that the collection of cylinders $\mathcal{H}(\tau, K)$, when τ and K vary, generates $\mathscr{B}(C([0,T]))$. To this end, we recall that $\mathscr{B}(C([0,T]))$ is generated by a countable collection of balls. Then it suffices to prove that every closed ball $\overline{D}(w_0, r)$ is a countable intersection of cylinders: let (τ_j) be a sequence of sets of points in $[0,T]$ as above, such that

$$\bigcup_{j\geq 1} \tau_j = [0,T] \cap \mathbb{Q}.$$

Then, using the notation $\tau_j = \{t_1^j, \dots, t_{n_j}^j\}$, we have

$$\overline{D}(w_0, r) = \bigcap_{j\geq 1} \{w \in C([0,T]) \mid |w(t_i^j) - w_0(t_i^j)| \leq r,\ i = 1, \dots, n_j\}.$$

□

Notation 3.20 *We denote the natural filtration for* $\mathbb{X}$ *by*

$$\mathscr{B}_t(C([0,T])) := \sigma(\mathbb{X}_s, s \in [0,t]), \qquad 0 \leq t \leq T. \tag{3.17}$$

The previous results can be easily extended to the case $T = +\infty$. Indeed,

$$C(\mathbb{R}_{\geq 0}) := C(\mathbb{R}_{\geq 0}, \mathbb{R}^N)$$

endowed with the norm[3]

$$\|w\|_\infty = \sum_{n=1}^{\infty} \frac{1}{2^n} \max_{0\leq t\leq n} (|w(t)| \wedge 1),$$

is a normed, complete and separable space in which the Borel σ-algebra is defined in a natural way. Further, Lemmas 3.16 and 3.19 can be easily generalized and, as before, we can define the canonical realization of a continuous process X.

In particular, if X is a Brownian motion, then the process $\mathbb{X}$ in (3.16) on the space $(C(\mathbb{R}_{\geq 0}), \mathscr{B}(C(\mathbb{R}_{\geq 0})), P^X, (\mathscr{B}_t(C(\mathbb{R}_{\geq 0}))))$ is called *canonical Brownian motion* (or canonical realization of the Brownian motion).

3.2.2 Equivalence of processes

Given a finite set of points $\tau = \{t_1, \dots, t_n\}$, we say that the joint distribution of $(X_{t_1}, \dots, X_{t_n})$ is a *finite-dimensional distribution* of the process X.

Definition 3.21 *Two processes* X, Y *defined on the spaces* $(\Omega, \mathcal{F}, P)$ *and* $(\Omega', \mathcal{F}', P')$*, respectively, are called equivalent if they have the same finite-dimensional distributions for every* $\tau = \{t_1, \dots, t_n\}$ *as above.*

Proposition 3.22 *Two processes are equivalent if and only if they have the same law.*

[3] This norm induces uniform convergence on compact sets.

Proof. Let X, Y be two equivalent stochastic processes defined on $(\Omega, \mathcal{F}, P)$ and $(\Omega', \mathcal{F}', P')$ respectively. The claim follows from Proposition A.6, observing that, by assumption,

$$P(X \in \mathcal{H}(\tau, K)) = P'(Y \in \mathcal{H}(\tau, K)),$$

for every cylinder $\mathcal{H}(\tau, K)$ and the collection of cylinders is $\cap$-stable and, as we have seen in the proof of Lemma 3.19, generates the Borel σ-algebra. The reverse implication is left as an exercise. □

According to Definition 3.6 *any* stochastic process verifying properties i), ii) and iii) is a Brownian motion. Therefore, in principle, there exist different Brownian motions, possibly defined on different probability spaces. Now we show that Definition 3.6 characterizes the finite-dimensional distributions of Brownian motion uniquely and so its law as well. In particular, by Proposition 3.22, the canonical realization of a Brownian motion is unique.

The following proposition contains some useful characterizations of a Brownian motion: in particular it gives explicitly the *finite-dimensional distributions* of a Brownian motion, i.e. the joint distributions of the random variables $W_{t_1}, \dots, W_{t_N}$ for every set of points $0 \le t_1 < \dots < t_N$.

Proposition 3.23 *A Brownian motion W on the filtered probability space $(\Omega, \mathcal{F}, P, (\mathcal{F}_t))$ verifies the following properties:*

1) *W has independent and stationary increments, i.e. for $0 \le t \le T$ the random variable $W_T - W_t$ has normal distribution $\mathcal{N}_{0,T-t}$ and the random variables*
$$W_{t_2} - W_{t_1}, \dots, W_{t_N} - W_{t_{N-1}}$$
are independent for every set of points $t_1, t_2, \dots, t_N$ with $0 \le t_1 < t_2 < \dots < t_N$;
2) *for $0 \le t_1 < \dots < t_N$, the joint distribution of $W_{t_1}, \dots, W_{t_N}$ is given by*

$$\begin{aligned} P((W_{t_1}, \dots, W_{t_N}) \in & H_1 \times \dots \times H_N) = \\ &= \int_{H_1} \cdots \int_{H_N} \Gamma^*(0,0;t_1,y_1)\Gamma^*(t_1,y_1;t_2,y_2) \cdots \\ &\quad \cdots \Gamma^*(t_{N-1}, y_{N-1}; t_N, y_N) dy_1 dy_2 \dots dy_N \end{aligned} \tag{3.18}$$

where $H_1, \dots, H_N \in \mathscr{B}$ and Γ^* is defined in (3.6).

Conversely, if W is a continuous stochastic process on a probability space $(\Omega, \mathcal{F}, P)$ such that $P(W_0 = 0) = 1$ and it satisfies 1) or 2), then W is a Brownian motion with respect to the natural filtration $\widetilde{\mathcal{F}}^W$.

Sketch of the proof. It is easy to prove that, if W is a Brownian motion, then it verifies 1). First of all it suffices to prove the independence of the increments: if $N = 3$, since

$$\{(W_{t_2} - W_{t_1}) \in H\} \in \mathcal{F}_{t_2},$$

the claim is an immediate consequence of the fact that $W_{t_3} - W_{t_2}$ and $\mathcal{F}_{t_2}$ are independent. For $N > 3$, we iterate the previous argument.

To prove that a Brownian motion W verifies 2), we consider only the case $N = 2$: firstly, for $0 \le t \le T$ and $H, K \in \mathscr{B}$, we have

$$\{W_t \in H\} \cap \{W_T \in K\} = \{W_t \in H\} \cap \{(W_T - W_t) \in (K - H)\},$$

where $K - H = \{x - y \mid x \in K,\ y \in H\}$. Then, by the independence of the increments we get

$$P(W_t \in H,\ W_T \in K) = P(W_t \in H)P((W_T - W_t) \in (K - H)) =$$

(by property iii))

$$= \int_H \Gamma^*(0,0;t,x_1)dx_1 \int_{K-H} \Gamma^*(t,0;T,x_2)dx_2 =$$

(by the change of variables $x_1 = y_1$ and $x_2 = y_2 - y_1$)

$$= \int_H \int_K \Gamma^*(0,0;t,y_1)\Gamma^*(t,y_1;T,y_2)dy_1dy_2.$$

To prove the other implication, we have to verify that, if W is a continuous stochastic process on a probability space $(\Omega, \mathcal{F}, P)$ such that $P(W_0 = 0) = 1$ and it satisfies 1), then the random variable $W_T - W_t$ is independent of $\widetilde{\mathcal{F}}_t^W$, for $t \le T$. In this case we can use Dynkin's Theorem A.5: in general, if X is a stochastic process such that, for every set of points $t_1, \dots, t_N$ with $0 \le t_1 < t_2 < \dots < t_N$, the random variables

$$X_{t_1}, X_{t_2} - X_{t_1}, \dots, X_{t_N} - X_{t_{N-1}}$$

are independent, then $X_T - X_t$ is independent of $\widetilde{\mathcal{F}}_t^X$ for $0 \le t < T$.

Finally, we leave it as an exercise to prove that, for every stochastic process W such that $P(W_0 = 0) = 1$, the properties 1) and 2) are equivalent. □

3.2.3 Modifications and indistinguishable processes

We introduce some other notions of equivalence of stochastic processes.

Definition 3.24 *Let X, Y be stochastic processes defined on the same probability space $(\Omega, \mathcal{F}, P)$. We say that X is a modification of Y if $X_t = Y_t$ a.s. for every $t \ge 0$. We say that X and Y are indistinguishable if, for almost all $\omega \in \Omega$, we have*

$$X_t(\omega) = Y_t(\omega) \quad \textit{for any} \quad t \ge 0.$$

We set

$$N_t = \{\omega \in \Omega \mid X_t(\omega) \ne Y_t(\omega)\}, \qquad N = \bigcup_{t \ge 0} N_t.$$

Then X is a modification of Y if N_t is negligible, $N_t \in \mathcal{N}$, for every $t \geq 0$. As already said, since t varies in the set of real numbers that is not countable: even if $N_t \in \mathcal{N}$ for any t, we may have $N \notin \mathcal{F}$ or even $N = \Omega$. On the other hand, the processes X and Y are indistinguishable if $N \in \mathcal{N}$, that is *almost all the paths of X and Y coincide.*

In general it is clear that, if X, Y are indistinguishable, then they are also modifications, but the converse may not hold: nevertheless, if the stochastic processes are continuous, it is possible to employ the density of the set of the rational numbers in $\mathbb{R}$ to prove that the two notions coincide.

Proposition 3.25 *Let X, Y be a.s. right-continuous stochastic processes. If X is a modification of Y, then X, Y are indistinguishable. In particular we can equivalently write*

$$X_t = Y_t \text{ a.s. for every } t \qquad \text{or} \qquad X_t = Y_t \text{ for every } t \text{ a.s.}$$

Remark 3.26 The result of Proposition 3.25 still holds true if we assume that X, Y are a.s. *left-continuous* (instead of right-continuous) stochastic processes such that $X_0 = Y_0$. □

Proof (of Proposition 3.25). It suffices to consider the case $Y = 0$. Let $F \in \mathcal{N}$ be the set in which the paths of X are not right continuous. We set

$$N = \bigcup_{t \in \mathbb{R}_{\geq 0} \cap \mathbb{Q}} N_t \cup F$$

where $N_t = \{\omega \in \Omega \mid X_t \neq 0\}$ is a negligible event by assumption. Then we get $N \in \mathcal{N}$ and $X_t(\omega) = 0$ for every $\omega \in \Omega \setminus N$ and $t \in \mathbb{R}_{\geq 0} \cap \mathbb{Q}$. Further, if $t \in \mathbb{R}_{\geq 0} \setminus \mathbb{Q}$, we take a sequence $(t_n) \in \mathbb{Q}$ converging to t from the right. Then for every $\omega \in \Omega \setminus N$ we have

$$X_t(\omega) = \lim_{n \to \infty} X_{t_n}(\omega) = 0,$$

and this concludes the proof. □

Summing up: *two continuous processes are indistinguishable if and only if they are modifications; in this case they are also equivalent and have the same canonical realization.*

Example 3.27 Let $u, v \in L^1_{\mathrm{loc}}(\mathbb{R})$ such that $u = v$ almost everywhere (with respect to Lebesgue measure) and $u(0) = v(0)$. If W is a real Brownian motion, then the processes $u(W_t)$ and $v(W_t)$ are modifications: indeed, assuming that $H = \{u \neq v\}$ has null Lebesgue measure, we have

$$P(u(W_t) \neq v(W_t)) = P(W_t \in H) = \int_H \Gamma^*(0, 0; t, x)dx = 0, \qquad t \geq 0,$$

where Γ^* is defined in (3.6). Note that for $t = 0$ we use the fact that $0 \notin H$ by assumption.

However in general the processes $u(W), v(W)$ are not indistinguishable. Indeed, to fix ideas, let assume that $1 \in H$, that is $u(1) \neq v(1)$: then, for any fixed $T > 0$, we have $P(W_T > 1) > 0$ and this means that the family of trajectories ω such that $u(W_t(\omega)) = u(1) \neq v(1) = v(W_t(\omega))$ for some $t \in [0, T]$ has a positive probability. In other terms

$$P(\{\omega \in \Omega \mid u(W_t(\omega)) = v(W_t(\omega)) \text{ for any } t \in [0, T]\}) < 1,$$

and therefore $u(W), v(W)$ are not indistinguishable.

A second simple example of processes that are modifications but are not indistinguishable is the following: on the probability space $([0, 1], \mathscr{B}, m)$, where m is Lebesgue measure, the processes

$$X_t = 0, \qquad \text{and} \qquad Y_t(\omega) = \mathbb{1}_{\{\omega\}}(t), \qquad t \in [0, 1],$$

are modifications, but

$$\{\omega \mid X_t(\omega) = Y_t(\omega),\ t \in [0, 1]\}$$

is empty and therefore X, Y are not indistinguishable. □

We now introduce another weaker notion of equivalence of processes. As usual, let m denote the Lebesgue measure.

Definition 3.28 *We say that the stochastic processes X, Y are $(m \otimes P)$-equivalent if*

$$(m \otimes P)(\{(t, \omega) \mid X_t(\omega) \neq Y_t(\omega)\}) = 0 \tag{3.19}$$

that is $X = Y$ $(m \otimes P)$-almost everywhere.

The processes X, Y are $(m \otimes P)$-equivalent if $X_t = Y_t$ a.s. for almost every t or equivalently if $X_t = Y_t$ for almost every t, a.s. In particular if X, Y are modifications then they are $(m \otimes P)$-equivalent. On the other hand, the process

$$X_t = \begin{cases} 1 & t = 0, \\ 0 & t > 0, \end{cases}$$

is $(m \otimes P)$-equivalent to the null process, even if X is not a modification of the null process since $N_0 = \{\omega \mid X_0(\omega) \neq 0\}$ is not negligible. However, in the case of continuous processes, also (3.19) is equivalent to the indistinguishability property.

Proposition 3.29 *Let X, Y be $(m \otimes P)$-equivalent, a.s. right-continuous stochastic processes. Then X, Y are indistinguishable.*

Proof. It suffices to consider the case $Y = 0$. We set

$$N_t = \{\omega \in \Omega \mid X_t(\omega) \neq 0\}, \qquad I = \{t \geq 0 \mid P(N_t) > 0\}.$$

We aim at showing that X is a modification of the null process or equivalently that $I = \emptyset$: then the thesis will follow from Proposition 3.25.

We consider a countable subset J of $[0, +\infty[\setminus I$, that is dense in $[0, +\infty[\setminus I$ and we put

$$\hat{N} = \bigcup_{t \in J} N_t$$

so that $P(\hat{N}) = 0$. We fix $t \geq 0$ and consider a decreasing sequence (t_n) in J converging to t. Then for any $\omega \in \Omega \setminus \hat{N}$ we have

$$X_t(\omega) = \lim_{n \to \infty} X_{t_n}(\omega) = 0.$$

This proves that $X_t = 0$ P-a.s., that is $t \notin I$. Therefore I is empty and this concludes the proof. □

3.2.4 Adapted and progressively measurable processes

The definition of stochastic process X requires not only that, for every t, X_t is a random variable, but also the stronger condition of measurability in the pair of variables (t, ω). We shall soon see[4] that the property of being adapted must be strengthened in an analogous way.

Definition 3.30 *A stochastic process X is called progressively measurable with respect to the filtration $(\mathcal{F}_t)$ if, for every t, $X|_{[0,t]\times\Omega}$ is $\mathscr{B}([0,t]) \otimes \mathcal{F}_t$-measurable, i.e.*

$$\{(s, \omega) \in [0, t] \times \Omega \mid X_s(\omega) \in H\} \in \mathscr{B}([0, t]) \otimes \mathcal{F}_t, \qquad H \in \mathscr{B}.$$

Clearly every progressively measurable process is also measurable and, by the Fubini-Tonelli Theorem A.50, adapted. Conversely, it is not a trivial result[5] to prove that, if X is measurable and adapted, then it admits a progressively measurable modification. Nevertheless, if the processes are continuous, the situation is much simpler:

Lemma 3.31 *Every right-continuous and adapted process is progressively measurable.*

Proof. Let X be right continuous and adapted. For fixed t and $n \in \mathbb{N}$, we set $X_t^{(n)} = X_t$ and

$$X_s^{(n)} = X_{\frac{k+1}{2^n}t}, \qquad \text{for } s \in \left[\frac{k}{2^n}t, \frac{k+1}{2^n}t\right[, \qquad k + 1 \leq 2^n.$$

Since X is right continuous, $X^{(n)}$ converges pointwise to X on $[0, t] \times \Omega$ for $n \to \infty$. The claim follows from the fact that $X^{(n)}$ is progressively measurable

[4] See, for example, Theorem 3.58.

[5] See, for example, Meyer [253].

because, for every $H \in \mathscr{B}$, we have

$$\{(s,\omega) \in [0,t] \times \Omega \mid X_s^{(n)}(\omega) \in H\}$$
$$= \bigcup_{k<2^n} \left(\left[\frac{k}{2^n}t, \frac{k+1}{2^n}t \right[\times \left(X_{\frac{k+1}{2^n}t} \in H \right) \right) \cup (\{t\} \times (X_t \in H))$$

which belongs to $\mathscr{B}([0,t]) \otimes \mathcal{F}_t$. □

3.3 Martingales

We present some fundamental results on continuous-time martingales: many of these results are simple extensions of their counterparts in Appendix A.6 in the discrete-time setting.

Definition 3.32 *Let M be an integrable adapted stochastic process on the filtered probability space $(\Omega, \mathcal{F}, P, (\mathcal{F}_t))$. We say that M is*

- *a martingale with respect to $(\mathcal{F}_t)$ and to the measure P if*
$$M_s = E\left[M_t \mid \mathcal{F}_s\right], \qquad \text{for every } 0 \le s \le t;$$
- *a super-martingale if*
$$M_s \ge E\left[M_t \mid \mathcal{F}_s\right], \qquad \text{for every } 0 \le s \le t;$$
- *a sub-martingale if*
$$M_s \le E\left[M_t \mid \mathcal{F}_s\right], \qquad \text{for every } 0 \le s \le t.$$

As in the discrete case, the mean of a martingale M is constant in time: indeed

$$E\left[M_t\right] = E\left[E\left[M_t \mid \mathcal{F}_0\right]\right] = E\left[M_0\right], \qquad t \ge 0. \tag{3.20}$$

Example 3.33 Given an integrable random variable Z in $(\Omega, \mathcal{F}, P, (\mathcal{F}_t))$, the stochastic process, defined by

$$M_t = E\left[Z \mid \mathcal{F}_t\right], \qquad t \ge 0,$$

is a $\mathcal{F}_t$-martingale. □

Example 3.34 In a filtered probability space $(\Omega, \mathcal{F}, P, (\mathcal{F}_t))$, let $Q \ll_{\mathcal{F}} P$ be another probability measure on $\mathcal{F}$. Then we have

$$Q \ll_{\mathcal{F}_t} P, \quad t \in [0,T],$$

and, by the Radon-Nikodym Theorem A.96, we define the process

$$L_t = \frac{dQ}{dP} \mid_{\mathcal{F}_t} .$$

It is easy to verify that L is a P-martingale: indeed

i) $L_t \geq 0$ and $E[L_t] = Q(\Omega) = 1$, for every $t \geq 0$;
ii) by (A.101), we have $L_s = E[L_t \mid \mathcal{F}_s]$ for every $s \leq t$.

Prove as an exercise that M is a Q-martingale if and only if ML is a P-martingale. □

Example 3.35 Let M be a martingale and X an adapted and bounded process. Then we have

$$E[M_T X_t] = E[M_t X_t], \qquad t \leq T.$$

Indeed

$$E[M_t X_t] = E[E[M_T \mid \mathcal{F}_t] X_t] = E[E[M_T X_t \mid \mathcal{F}_t]] = E[M_T X_t].$$

□

Remark 3.36 If M is a martingale and φ is a convex function on $\mathbb{R}$ such that $\varphi(M)$ is integrable, then $\varphi(M)$ is a sub-martingale. Indeed

$$E[\varphi(M_t) \mid \mathcal{F}_s] \geq$$

(by Jensen's inequality in Proposition A.107)

$$\geq \varphi(E[M_t \mid \mathcal{F}_s]) = \varphi(M_s).$$

Further, if M is a sub-martingale and φ is a convex and *increasing* function on $\mathbb{R}$ such that $\varphi(M)$ is integrable, then $\varphi(M)$ is a sub-martingale. As remarkable cases, if M is a martingale, then $|M|$ and M^2 are sub-martingales. □

The next result shows some remarkable examples of martingales that can be constructed using Brownian motion.

Proposition 3.37 *If W is a Brownian motion on $(\Omega, \mathcal{F}, P, (\mathcal{F}_t))$ and $\sigma \in \mathbb{R}$, then*

i) W_t;
ii) $W_t^2 - t$;
iii) $\exp\left(\sigma W_t - \frac{\sigma^2}{2} t\right)$

are continuous $\mathcal{F}_t$-martingales.

Proof. i) By Hölder's inequality,

$$E[|W_t|]^2 \leq E\left[W_t^2\right] = t,$$

and so W is integrable. Further, for $0 \leq s \leq t$ we have

$$E[W_t \mid \mathcal{F}_s] = E[W_t - W_s \mid \mathcal{F}_s] + E[W_s \mid \mathcal{F}_s] =$$

(since $W_t - W_s$ is independent of $\mathcal{F}_s$ and W_s is $\mathcal{F}_s$-measurable)

$$= E\left[W_t - W_s\right] + W_s = W_s.$$

ii) This is left as an exercise. We shall see later on that the fact that $W_t^2 - t$ is a martingale essentially characterizes Brownian motion (cf. Theorem 5.34).

iii) Recalling Exercise A.34, clearly $\exp\left(\sigma W_t - \frac{\sigma^2}{2}t\right)$ is integrable; further, for $s < t$ we have

$$\begin{aligned}
&E\left[\exp\left(\sigma W_t - \frac{\sigma^2}{2}t\right) \mid \mathcal{F}_s\right] \\
&= \exp\left(\sigma W_s - \frac{\sigma^2}{2}t\right) E\left[\exp(\sigma(W_t - W_s)) \mid \mathcal{F}_s\right] =
\end{aligned}$$

(since $W_t - W_s$ is independent of $\mathcal{F}_s$)

$$= \exp\left(\sigma W_s - \frac{\sigma^2}{2}t\right) E\left[\exp(\sigma Z\sqrt{t-s})\right],$$

with $Z = \frac{W_t - W_s}{\sqrt{t-s}} \sim \mathcal{N}_{0,1}$. The claim follows from Exercise A.34. □

3.3.1 Doob's inequality

We extend to the continuous case Doob's inequality, Theorem A.130, by using a simple density argument.

Theorem 3.38 (Doob's inequality) *Let M be a right continuous martingale*[6] *and $p > 1$. Then for every T*

$$E\left[\sup_{t\in[0,T]} |M_t|^p\right] \le q^p E\left[|M_T|^p\right], \tag{3.21}$$

where $q = \frac{p}{p-1}$ is the conjugate exponent to p.

Proof. We denote by $(t_n)_{n\ge 0}$ an enumeration of the rational numbers in the interval $[0,T[$ with $t_0 = 0$, i.e.

$$\mathbb{Q} \cap [0,T[= \{t_0, t_1, \dots\}.$$

Let us consider the increasing sequence (ς_n) of partitions[7] of $[0,T]$

$$\varsigma_n = \{t_0, t_1, \dots, t_n, T\},$$

[6] The result holds also for every a.s. right-continuous martingale.

[7] For every n we re-assign the indexes to the points $t_0, \dots, t_n$ in such a way that $t_0 < t_1 < \dots < t_n$.

such that

$$\bigcup_{n\geq 1} \varsigma_n = [0,T[\cap \mathbb{Q} \cup \{T\}.$$

For every n, the discrete process $M^{(n)}$ defined by

$$M^{(n)} = (M_{t_0}, M_{t_1}, \dots, M_{t_n}, M_T)$$

is a martingale with respect to the filtration

$$(\mathcal{F}_{t_0}, \mathcal{F}_{t_1}, \dots, \mathcal{F}_{t_n}, \mathcal{F}_T).$$

So, by Theorem A.130, setting

$$f_n(\omega) = \max\{|M_{t_0}(\omega)|, |M_{t_1}(\omega)|, \dots, |M_{t_n}(\omega)|, |M_T(\omega)|\}, \qquad \omega \in \Omega,$$

we have

$$E\left[f_n^p\right] \leq q^p E\left[|M_T|^p\right] \tag{3.22}$$

for every $n \in \mathbb{N}$ and $p > 1$. Further, (f_n) is an increasing non-negative sequence an so, by Beppo Levi's theorem, taking the limit as n goes to infinity in (3.22), we get

$$E\left[\sup_{t\in[0,T[\cap\mathbb{Q}\cup\{T\}} |M_t|^p\right] \leq q^p E\left[|M_T|^p\right].$$

The claim follows from the fact that, being M right-continuous, we have

$$\sup_{t\in[0,T[\cap\mathbb{Q}\cup\{T\}} |M_t| = \sup_{t\in[0,T]} |M_t|.$$

□

Example 3.39 Let $\Omega = [0,1]$, let P be Lebesgue measure and

$$X_t(\omega) = \mathbb{1}_{[t,t+\varepsilon]}(\omega), \qquad \omega \in \Omega,$$

with fixed $\varepsilon \in]0,1[$. Then X has non-negative values and is such that

$$\sup_{t\in[0,T]} E\left[X_t\right] < E\left[\sup_{t\in[0,T]} X_t\right].$$ □

Example 3.40 If $M_t = E\left[Z \mid \mathcal{F}_t\right]$ is the martingale in Example 3.33 with $Z \in L^2(\Omega, P)$, then using Doob's and Jensen's inequalities we get

$$E\left[\sup_{t\in[0,T]} |M_t|^2\right] \leq 4E\left[|M_T|^2\right] \leq 4E\left[|Z|^2\right].$$ □

3.3.2 Martingale spaces: $\mathscr{M}^2$ and $\mathscr{M}^2_c$

Even though we often deal with martingales whose continuity property is known, we state the following classical result[8].

[8] For the proof see, for example, Karatzas-Shreve [201], p. 16.

Theorem 3.41 *Let M be a super-martingale on a filtered probability space $(\Omega, \mathcal{F}, P, (\mathcal{F}_t))$ satisfying the usual hypotheses on the filtration (cf. Section 3.3.3). Then M has a càdlàg*[9] *modification $\widetilde{M}$ if and only if the function $t \mapsto E[M_t]$ is right continuous. In this case, we can choose $\widetilde{M}$ in such a way that it is $\mathcal{F}_t$-adapted and so it is a $\mathcal{F}_t$-super-martingale.*

In view of the previous theorem and of (3.20), every martingale admits a right-continuous modification that is unique, up to indistiguishability, by Proposition 3.25. So the assumption of right-continuity, that will be conventionally made later on, is not actually restrictive: in other words, *we will always assume that we are taking the right continuous version of every martingale.*

Further, we note that, if M is a martingale, then $(M_t - M_0)$ is a martingale with null initial value. So every martingale can be "normalized" in such a way that $M_0 = 0$.

Notation 3.42 *For fixed $T > 0$, we denote by*

- *$\mathscr{M}^2$ the linear space of right-continuous $\mathcal{F}_t$-martingales $(M_t)_{t\in[0,T]}$ such that $M_0 = 0$ a.s. and*

$$[\![M]\!]_T := \sqrt{E\left[\sup_{0\le t\le T} |M_t|^2\right]} \tag{3.23}$$

  is finite;
- *$\mathscr{M}^2_c$ the subspace of continuous martingales in $\mathscr{M}^2$.*

The importance of the class $\mathscr{M}^2_c$ will be clear in Paragraphs 4.2 and 4.3, where we shall see that, under suitable assumptions, the stochastic integral is an element of $\mathscr{M}^2_c$.

Equation (3.23) defines a semi-norm in $\mathscr{M}^2$: we note that $[\![M]\!]_T = 0$ if and only if M is indistinguishable from (but not necessarily equal to) the null stochastic process. Further, by Doob's inequality, we have[10]

$$\|M_T\|_2 = E\left[\left|M^T\right|^2\right]^{\frac{1}{2}} \le [\![M]\!]_T \le 2\|M_T\|_2, \tag{3.24}$$

and therefore $[\![M]\!]_T$ and $\|M_T\|_2$ are equivalent semi-norms in $\mathscr{M}^2$. Next we prove that the spaces $\mathscr{M}^2$ and $\mathscr{M}^2_c$ are *complete*.

Lemma 3.43 *The space $(\mathscr{M}^2, [\![\cdot]\!]_T)$ is complete, i.e. for every Cauchy sequence (M^n) there exists $M \in \mathscr{M}^2$ such that*

$$\lim_{n\to\infty} [\![M^n - M]\!]_T = 0.$$

Further, if the sequence (M^n) is in $\mathscr{M}^2_c$, then $M \in \mathscr{M}^2_c$: in other terms, $\mathscr{M}^2_c$ is a closed subspace of $\mathscr{M}^2$.

---

9 Càdlàg is the French shortening for "right-continuous with finite left limits at all t" (continue à droite et limité à gauche).

10 We recall that $\|M_T\|_2 = \sqrt{E\left[|M_T|^2\right]}$.

Proof. The proof is similar to that of the completeness of the L^2 space. First of all, given a Cauchy sequence (M^n) in $\mathscr{M}^2$, it suffices to prove that it admits a convergent subsequence to conclude that also (M^n) converges.

Let (M^{k_n}) be a subsequence of (M^n) such that

$$[\![M^{k_n} - M^{k_{n+1}}]\!] \leq \frac{1}{2^n}, \qquad n \geq 1.$$

For the sake of simplicity, we put $v_n = M^{k_n}$ and define

$$w_N(\omega) = \sum_{n=1}^{N} \sup_{t\in[0,T]} |v_{n+1}(t,\omega) - v_n(t,\omega)|, \qquad N \geq 1.$$

Then (w_N) is a non-negative, monotone increasing sequence such that

$$E\left[w_N^2\right] \leq 2\sum_{n=1}^{N} [\![v_{n+1} - v_n]\!]_T^2 \leq 2.$$

Therefore the limit

$$w(\omega) := \lim_{N\to\infty} w_N(\omega), \qquad \omega \in \Omega,$$

exists and, by Beppo Levi's Theorem, $E\left[w^2\right] \leq 2$: in particular there exists $F \in \mathcal{N}$ such that $w(\omega) < \infty$ for every $\omega \in \Omega \setminus F$. Further, for $n \geq m \geq 2$ we have

$$\sup_{t\in[0,T]} |v_n(t,\omega) - v_m(t,\omega)| \leq w(\omega) - w_{m-1}(\omega), \tag{3.25}$$

and so $(v_n(t,\omega))$ is a Cauchy sequence in $\mathbb{R}$ for $t \in [0,T]$ and $\omega \in \Omega \setminus F$ and it converges, uniformly with respect to t for every $\omega \in \Omega \setminus F$, to a limit that we denote by $M(t,\omega)$. Since the convergence of (v_n) is uniform in t, we have that the path $M(\cdot,\omega)$ is right-continuous (continuous if $M^n \in \mathscr{M}_c^2$) for every $\omega \in \Omega \setminus F$: in particular, M is indistinguishable from a right-continuous stochastic process. We denote such a process again by M. From (3.25) it follows that

$$\sup_{t\in[0,T]} |M(t,\omega) - v_n(t,\omega)| \leq w(\omega), \qquad \omega \in \Omega \setminus F, \tag{3.26}$$

hence we infer that $[\![M]\!]_T < \infty$. Finally we can use the estimate (3.26) and Lebesgue's dominated convergence theorem to prove that

$$\lim_{n\to\infty} [\![X - v_n]\!]_T = 0.$$

Eventually we observe that M is adapted since it is a pointwise limit of adapted processes: further, for $0 \leq s < t \leq T$ and $A \in \mathcal{F}_s$, we have, by Hölder's inequality,

$$0 = \lim_{n\to\infty} E\left[(M_t^n - M_t)\mathbb{1}_A\right] = \lim_{n\to\infty} E\left[(M_s^n - M_s)\mathbb{1}_A\right],$$

and so the equality $E\left[M_t^n \mathbb{1}_A\right] = E\left[M_s^n \mathbb{1}_A\right]$ implies $E\left[M_t \mathbb{1}_A\right] = E\left[M_s \mathbb{1}_A\right]$ and in view of this we conclude that M is a martingale. □

3.3.3 The usual hypotheses

Given a probability space $(\Omega, \mathcal{F}, P)$, we recall the notation

$$\mathcal{N} = \{F \in \mathcal{F} \mid P(F) = 0\},$$

for the family of P-negligible events.

Definition 3.44 *We say that $(\mathcal{F}_t)$ satisfies the so-called "usual hypotheses" with respect to P if:*

i) $\mathcal{F}_0$ (and so also $\mathcal{F}_t$ for every $t > 0$) contains $\mathcal{N}$;
ii) the filtration is right-continuous, i.e. for every $t \geq 0$

$$\mathcal{F}_t = \bigcap_{\varepsilon>0} \mathcal{F}_{t+\varepsilon}. \tag{3.27}$$

The idea of considering only filtrations containing the collection of negligible events stems from the need of avoiding the unpleasant situation in which $X = Y$ a.s., X is $\mathcal{F}_t$-measurable but Y fails to be so. Analogously, for purely technical reasons, it is useful to know that, if a random variable X is $\mathcal{F}_s$-measurable for every $s > t$, then X is also $\mathcal{F}_t$-measurable: this is ensured by (3.27). We will shortly be using these properties, for example in the proof of Proposition 3.50 and in Remark 4.3.

The rest of the section can be skipped on first reading: we prove how to complete a filtration so that we make it satisfy the usual hypotheses. At a first glance this problem might seem technical, but actually it is of great importance in the development of stochastic calculus.

Recalling Definition 3.4, we point out that, in general, *even if X is a continuous stochastic process,* its natural filtration $\widetilde{\mathcal{F}}^X$ may not satisfy the usual hypotheses and, in particular, it may not be right-continuous. This motivates the following:

Definition 3.45 *Given a stochastic process X on the space $(\Omega, \mathcal{F}, P)$, we set, for $t \geq 0$,*

$$\mathcal{F}_t^X := \bigcap_{\varepsilon>0} \hat{\mathcal{F}}_{t+\varepsilon}^X, \qquad \textit{where} \quad \hat{\mathcal{F}}_t^X := \sigma\left(\widetilde{\mathcal{F}}_t^X \cup \mathcal{N}\right). \tag{3.28}$$

It can be easily verified that $\mathcal{F}^X := (\mathcal{F}_t^X)$ is a filtration satisfying the usual hypotheses: it is called standard filtration of X.

Remark 3.46 In what follows, unless otherwise stated, given a filtration $(\mathcal{F}_t)$, we implicitly assume that it verifies the usual hypotheses in Definition 3.44. In particular, given a stochastic process X, we will usually employ the standard filtration $\mathcal{F}^X$ instead of the natural filtration $\widetilde{\mathcal{F}}^X$. □

Now we consider the particular case of a Brownian motion W on a probability space $(\Omega, \mathcal{F}, P)$ endowed with the natural filtration $\widetilde{\mathcal{F}}^W$. We prove that, in order to make the filtration $\widetilde{\mathcal{F}}^W$ standard, it suffices to complete it with the negligible events, without having to further enrich it as in (3.28). More precisely, we define the natural filtration completed by the negligible events by setting

$$\mathcal{F}_t^W = \sigma\left(\widetilde{\mathcal{F}}_t^W \cup \mathcal{N}\right),$$

and we call $\mathcal{F}^W = \left(\mathcal{F}_t^W\right)$ *Brownian filtration.*

Theorem 3.47 *The filtration $\mathcal{F}^W$ verifies the usual hypotheses and it coincides with the standard filtration of W. Further, W is a Brownian motion on the space $(\Omega, \mathcal{F}, P, \mathcal{F}^W)$ and is called standard Brownian motion.*

Proof. The proof makes use of Dynkin's Theorems A.5 and A.9. We set

$$\mathcal{F}_{t-} := \sigma\left(\bigcup_{s<t} \mathcal{F}_s^W\right), \qquad \mathcal{F}_{t+} := \bigcap_{s>t} \mathcal{F}_s^W.$$

We point out that, in general, $\bigcup_{s<t} \mathcal{F}_s^W$ may not be a σ-algebra and this justifies the definition of $\mathcal{F}_{t-}$. Clearly we have

$$\mathcal{F}_{t-} \subseteq \mathcal{F}_t^W \subseteq \mathcal{F}_{t+}.$$

We want to prove that

$$\mathcal{F}_{t+} \subseteq \mathcal{F}_{t-}, \tag{3.29}$$

for every t. To this end, it suffices to prove that

$$E\left[X \mid \mathcal{F}_{t+}\right] = E\left[X \mid \mathcal{F}_{t-}\right] \tag{3.30}$$

for every bounded $\mathcal{F}_s^W$-measurable random variable X, with $s > t$: indeed, if this holds true in particular for every bounded $\mathcal{F}_{t+}$-measurable random variable X, then we will infer that X is also $\mathcal{F}_{t-}$-measurable, hence (3.29) holds.

We denote the imaginary unit by i. For every $\alpha \in \mathbb{R}$ and $u < t \le s$ we have

$$E\left[e^{i\alpha W_s} \mid \mathcal{F}_u^W\right] = e^{i\alpha W_u} E\left[e^{i\alpha(W_s - W_u)} \mid \mathcal{F}_u^W\right] =$$

(since $W_s - W_u$ is independent on $\mathcal{F}_u^W$)

$$= e^{i\alpha W_u} E\left[e^{i\alpha(W_s - W_u)}\right] =$$

(by Example A.34)

$$= e^{i\alpha W_u - \frac{\alpha^2}{2}(s-u)}. \tag{3.31}$$

Taking the limit as $u \to t^-$, we get

$$Z := e^{i\alpha W_t - \frac{\alpha^2}{2}(s-t)} = \lim_{u \to t^-} E\left[e^{i\alpha W_s} \mid \mathcal{F}_u^W\right].$$

Now we verify that $Z = E\left[e^{i\alpha W_s} \mid \mathcal{F}_{t-}\right]$: first of all we observe that Z is $\mathcal{F}_{t-}$-measurable, being the pointwise limit of $\mathcal{F}_{t-}$-measurable random variables. It remains to be seen that

$$E\left[Z \mathbb{1}_G\right] = E\left[e^{i\alpha W_s} \mathbb{1}_G\right], \tag{3.32}$$

for every $G \in \mathcal{F}_{t-}$. This follow from Dynkin's Theorem, in the version of Exercise A.104: indeed, if $G \in \mathcal{F}_u^W$, $u < t$, we have

$$E\left[Z \mathbb{1}_G\right] = \lim_{v \to t^-} E\left[E\left[e^{i\alpha W_s} \mid \mathcal{F}_v^W\right] \mathbb{1}_G\right] =$$

(since $\mathbb{1}_G E\left[e^{i\alpha W_s} \mid \mathcal{F}_v^W\right] = E\left[e^{i\alpha W_s} \mathbb{1}_G \mid \mathcal{F}_v^W\right]$ if $v \geq u$)

$$= \lim_{v \to t^-} E\left[E\left[e^{i\alpha W_s} \mathbb{1}_G \mid \mathcal{F}_v^W\right]\right] = E\left[e^{i\alpha W_s} \mathbb{1}_G\right].$$

So (3.32) holds for $G \in \bigcup_{u<t} \mathcal{F}_u^W$ which is an $\cap$-stable collection, containing Ω and generating $\mathcal{F}_{t-}$: consequently, (3.32) holds also for $G \in \mathcal{F}_{t-}$. In conclusion we have proved that

$$E\left[e^{i\alpha W_s} \mid \mathcal{F}_{t-}\right] = e^{i\alpha W_t - \frac{\alpha^2}{2}(s-t)} = E\left[e^{i\alpha W_s} \mid \mathcal{F}_t\right].$$

In an analogous way we can prove that

$$E\left[e^{i\alpha W_s} \mid \mathcal{F}_{t+}\right] = e^{i\alpha W_t - \frac{\alpha^2}{2}(s-t)} = E\left[e^{i\alpha W_s} \mid \mathcal{F}_t\right],$$

and so, for every $s \geq 0$ (for $s < t$ it is obvious),

$$E\left[e^{i\alpha W_s} \mid \mathcal{F}_{t-}\right] = E\left[e^{i\alpha W_s} \mid \mathcal{F}_{t+}\right].$$

More generally, proceeding as above, we can prove that

$$E\left[e^{i\left(\alpha_1 W_{s_1} + \cdots + \alpha_k W_{s_k}\right)} \mid \mathcal{F}_{t-}\right] = E\left[e^{i\left(\alpha_1 W_{s_1} + \cdots + \alpha_k W_{s_k}\right)} \mid \mathcal{F}_{t+}\right] \tag{3.33}$$

for every $\alpha_1, \dots, \alpha_k \in \mathbb{R}$ and $0 \leq s_1 < \cdots < s_k$, $k \in \mathbb{N}$. It suffices to observe that, if $k = 2$, we can prove a relation analogous to (3.31) in the following way: for $u < t \leq s_1 < s_2$ we have

$$E\left[e^{i\left(\alpha_1 W_{s_1} + \alpha_2 W_{s_2}\right)} \mid \mathcal{F}_u^W\right]$$
$$= e^{i(\alpha_1+\alpha_2)W_u} E\left[e^{i(\alpha_1+\alpha_2)(W_{s_1}-W_u)} e^{i\alpha_2(W_{s_2}-W_{s_1})} \mid \mathcal{F}_u^W\right] =$$

(since $W_{s_1} - W_u$ and $W_{s_2} - W_{s_1}$ are independent on $\mathcal{F}_u^W$)

$$= e^{i(\alpha_1+\alpha_2)W_u} E\left[e^{i(\alpha_1+\alpha_2)(W_{s_1}-W_u)} e^{i\alpha_2(W_{s_2}-W_{s_1})}\right]$$
$$= e^{i(\alpha_1+\alpha_2)W_u} e^{-\frac{(\alpha_1+\alpha_2)^2}{2}(s_1-u)} e^{-\frac{\alpha_2^2}{2}(s_2-s_1)}.$$

Now let us call $\mathcal{H}$ the collection of bounded random variables Z such that

$$E[Z \mid \mathcal{F}_{t-}] = E[Z \mid \mathcal{F}_{t+}].$$

Then $\mathcal{H}$ is a monotone family of functions (cf. Definition A.8), containing the a.s. null random variables (since $\mathcal{F}_{t-}$ and $\mathcal{F}_{t+}$ contain the negligible events) and the linear combinations and the products of $\cos(\alpha W_s) = \mathrm{Re}\left(e^{i\alpha W_s}\right)$ and $\sin(\alpha W_s) = \mathrm{Im}\left(e^{i\alpha W_s}\right)$ for $\alpha \in \mathbb{R}$ and $s \geq 0$, in view of (3.33). For fixed $s > 0$ and setting

$$\mathcal{A}_s = \{(W_{s_1} \in H_1) \cap \cdots \cap (W_{s_k} \in H_k) \mid 0 \leq s_j \leq s,\ H_j \in \mathscr{B},\ 1 \leq j \leq k \in \mathbb{N}\},$$

by density[11] $\mathcal{H}$ contains also the characteristic functions of the elements of $\mathcal{A}$ and $\mathcal{N}$. On the other hand, $\mathcal{A}$ and $\mathcal{N}$ are $\cap$-stable and $\sigma(\mathcal{A} \cup \mathcal{N}) = \mathcal{F}_s^W$: so, by Theorem A.9 $\mathcal{H}$ contains also every bounded $\mathcal{F}_s^W$-measurable function (for every $s > 0$). This concludes the proof of (3.30) and of the theorem. □

Remark 3.48 A result, analogous to that of the previous theorem, holds in general for the processes having the strong Markov property: for the details we refer to Karatzas-Shreve [201], Chapter 2.7, or Breiman [60]. □

3.3.4 Stopping times and martingales

Definition 3.49 *A random variable*

$$\tau : \Omega \longrightarrow \mathbb{R}_{\geq 0} \cup \{+\infty\}$$

is a stopping time with respect to the filtration $(\mathcal{F}_t)$ *if*

$$\{\tau \leq t\} \in \mathcal{F}_t, \tag{3.34}$$

for every $t \geq 0$.

Clearly a (constant) deterministic time $\tau \equiv t$ is a stopping time. Note also that a stopping time τ can take the value $+\infty$. The next significant result is based on the usual hypotheses on the filtration.

Proposition 3.50 *The random variable* τ *is a stopping time if and only if*

$$\{\tau < t\} \in \mathcal{F}_t, \tag{3.35}$$

for every $t > 0$. *Consequently we also have* $\{\tau = t\}, \{\tau \geq t\}, \{\tau > t\} \in \mathcal{F}_t$.

Proof. If τ is a stopping time, then

$$\{\tau < t\} = \bigcup_{n \in \mathbb{N}} \left\{\tau \leq t - \frac{1}{n}\right\}$$

[11] It is well known that the indicator function of a Borel set can be approximated by trigonometric polynomials.

with $\{\tau \le t - \frac{1}{n}\} \in \mathcal{F}_{t-\frac{1}{n}} \subseteq \mathcal{F}_t$. Conversely, for every $\varepsilon > 0$ we have

$$\{\tau \le t\} = \bigcap_{0<\delta<\varepsilon} \{\tau < t + \delta\},$$

and so $\{\tau \le t\} \in \mathcal{F}_{t+\varepsilon}$. Consequently, in view of the usual hypotheses,

$$\{\tau \le t\} \in \mathcal{F}_t = \bigcap_{\varepsilon>0} \mathcal{F}_{t+\varepsilon}. \qquad \square$$

Proposition 3.51 *Let τ, τ_1 be stopping times. Then also*

$$\tau \wedge \tau_1 = \min\{\tau, \tau_1\} \qquad \text{and} \qquad \tau \vee \tau_1 = \max\{\tau, \tau_1\}$$

are stopping times.

Proof. It suffices to observe that

$$\{\min\{\tau, \tau_1\} \le t\} = \{\tau \le t\} \cup \{\tau_1 \le t\},$$
$$\{\max\{\tau, \tau_1\} \le t\} = \{\tau \le t\} \cap \{\tau_1 \le t\}. \qquad \square$$

In mathematical finance, the typical example of stopping time is the exercise time of an American option (cf. Paragraph 2.5). Another remarkable example having a particularly intuitive geometrical interpretation is the so-called hitting time for a stochastic process of an open or closed set in $\mathbb{R}^N$.

Theorem 3.52 (Hitting time) *Let $X = (X_t)_{t\in\mathbb{R}_{\ge 0}}$ be a stochastic process in $\mathbb{R}^N$, right-continuous and $\mathcal{F}_t$-adapted and let H be an open set in $\mathbb{R}^N$. We put*

$$I(\omega) = \{t \ge 0 \mid X_t(\omega) \in H\}, \qquad \omega \in \Omega,$$

and

$$\tau(\omega) = \begin{cases} \inf I(\omega), & \text{if } I(\omega) \ne \emptyset, \\ +\infty, & \text{if } I(\omega) = \emptyset. \end{cases}$$

Then τ is an $\mathcal{F}_t$-stopping time called "hitting time" of H for X.

Proof. In view of Proposition 3.50, it suffices to verify that $\{\tau < t\} \in \mathcal{F}_t$ for every t. Since H is open and X is right-continuous, we have

$$\{\tau < t\} = \bigcup_{s\in\mathbb{Q}\cap[0,t[} \{X_s \in H\},$$

and the claim follows from the fact that, being X adapted, we have

$$\{X_s \in H\} \in \mathcal{F}_t, \qquad s \le t. \qquad \square$$

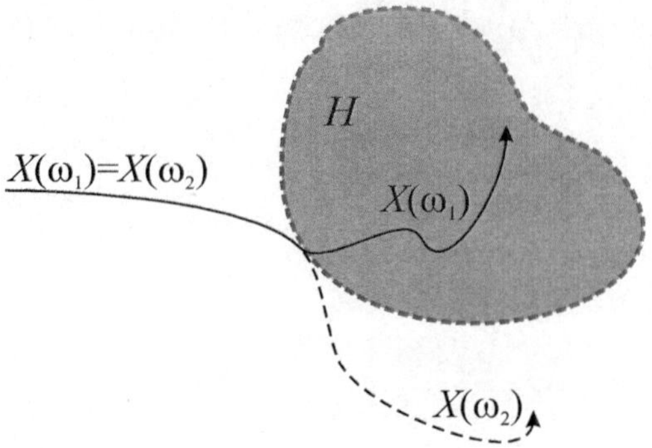

Fig. 3.1. Hitting time for a process of an open set H

Corollary 3.53 *Let $X = (X_t)_{t\in\mathbb{R}_{\geq 0}}$ be a stochastic process in $\mathbb{R}^N$, right-continuous and $\mathcal{F}_t$-adapted and let H be a closed set in $\mathbb{R}^N$. We put*

$$I(\omega) = \{t \geq 0 \mid X_t(\omega) \in H \text{ or } X_{t-}(\omega) \in H\}, \qquad \omega \in \Omega,$$

and

$$\tau(\omega) = \begin{cases} \inf I(\omega), & \text{if } I(\omega) \neq \emptyset, \\ +\infty, & \text{if } I(\omega) = \emptyset. \end{cases}$$

Then τ is an $\mathcal{F}_t$-stopping time.

Proof. We consider the sequence of open sets in $\mathbb{R}^N$

$$H_n = \left\{x \in \mathbb{R}^N \mid \operatorname{dist}(x, H) < \frac{1}{n}\right\}, \qquad n \in \mathbb{N},$$

where $\operatorname{dist}(\cdot, H)$ is the Euclidean distance from H. The claim follows from the equality[12]

$$\{\tau \leq t\} = \{X_t \in H \text{ or } X_{t-} \in H\} \cup \Big(\bigcap_{n\in\mathbb{N}} \bigcup_{s\in\mathbb{Q}\cap[0,t[} \{X_s \in H_n\}\Big). \qquad \square$$

The condition $\{\tau \leq t\} \in \mathcal{F}_t$ expresses the fact that, to know if X reaches H by time t, it suffices to observe the paths of the process until time t. Looking at Figure 3.1, the subtlety of Proposition 3.50 is apparent: intuitively, the information in $\mathcal{F}_t$ allows us to establish if X enters the open set H at time t. However the paths of $X(\omega_1), X(\omega_2)$ in the figure coincide until time t in which $X_t(\omega_1) = X_t(\omega_2) \notin H$ and afterwards the path $X(\omega_1)$ enters H (and so $\tau(\omega_1) = t$) while the path $X(\omega_2)$ does not enter H (and so $\tau(\omega_2) > t$).

[12] Since $\tau(\omega) \leq t$ if and only if $X_t(\omega) \in H$ or, for every $n \in \mathbb{N}$, there exists $s \in \mathbb{Q} \cap [0, t[$ such that $X_s(\omega) \in H_n$.

We explicitly remark that *the last exit time*

$$\widetilde{\tau} = \sup\{t \mid X_t \in H\}$$

is not in general a stopping time. Intuitively, in order to know if X leaves H at time t *for the last time,* it is necessary to observe the entire path of X.

Notation 3.54 *Let τ be a stopping time which is finite on $\Omega \setminus N$, where N is a negligible event, and let X be a stochastic process. We set*

$$X_\tau(\omega) = X_{\tau(\omega)}(\omega), \qquad \omega \in \Omega. \tag{3.36}$$

Further, we define the σ-algebra

$$\mathcal{F}_\tau = \{F \in \mathcal{F} \mid F \cap \{\tau \le t\} \in \mathcal{F}_t \text{ for every } t\} \tag{3.37}$$

that is called σ-algebra associated to the stopping time τ.

We observe that, if τ_1, τ_2 are stopping times such that $\tau_1 \le \tau_2$ a.s., then $\mathcal{F}_{\tau_1} \subseteq \mathcal{F}_{\tau_2}$. Indeed, for fixed t, by assumption

$$\{\tau_1 \le t\} \supseteq \{\tau_2 \le t\};$$

so, if $F \in \mathcal{F}_{\tau_1}$ we have

$$F \cap \{\tau_2 \le t\} = (F \cap \{\tau_1 \le t\}) \cap \{\tau_2 \le t\} \in \mathcal{F}_t.$$

Remark 3.55 For every stopping time τ and $n \in \mathbb{N}$, the equation

$$\tau_n(\omega) = \begin{cases} \frac{k+1}{2^n} & \text{if } \frac{k}{2^n} \le \tau(\omega) < \frac{k+1}{2^n}, \\ +\infty & \text{if } \tau(\omega) = +\infty, \end{cases}$$

defines a decreasing sequence (τ_n) of discrete-valued stopping times, such that

$$\tau = \lim_{n\to\infty} \tau_n. \qquad \square$$

Now we prove the continuous version of Theorem A.129: the proof is based on an approximation procedure and on the analogous result in discrete time.

Theorem 3.56 (Doob's optional sampling Theorem) *Let M be a right-continuous martingale and let τ_1, τ_2 be stopping times such that $\tau_1 \le \tau_2 \le T$ a.s., with $T > 0$. Then*

$$M_{\tau_1} = E\left[M_{\tau_2} \mid \mathcal{F}_{\tau_1}\right].$$

In particular, for every a.s. bounded stopping time τ we have

$$E\left[M_\tau\right] = E\left[M_0\right].$$

Proof. Let $(\tau_{1,n}), (\tau_{2,n})$ be sequences of discrete stopping times, constructed as in Remark 3.55, approximating τ_1 and τ_2 respectively. By the continuity assumption,

$$\lim_{n\to\infty} M_{\tau_{i,n}} = M_{\tau_i}, \qquad i = 1, 2, \text{ a.s.}$$

Further, by Theorem A.129 we have

$$M_{\tau_{2,n}} = E\left[M_T \mid \mathcal{F}_{\tau_{2,n}}\right]$$

and so, by Corollary A.151, the sequence $\left(M_{\tau_{2,n}}\right)$ is uniformly integrable. Finally, by Theorem A.129

$$M_{\tau_{1,n}} = E\left[M_{\tau_{2,n}} \mid \mathcal{F}_{\tau_{1,n}}\right]$$

and the claim follows by taking the limit in n. □

Remark 3.57 In an analogous way we prove that, if M is a right-continuous super-martingale and $\tau_1 \le \tau_2 \le T$ a.s., then

$$M_{\tau_1} \ge E\left[M_{\tau_2} \mid \mathcal{F}_{\tau_1}\right]. \tag{3.38}$$

We refer to [201] for all the details.

The boundedness assumption on the stopping times can be replaced by a boundedness assumption on the process: (3.38) is still valid if M is a super-martingale such that

$$M_t \ge E\left[M \mid \mathcal{F}_t\right], \qquad t \ge 0,$$

with $M \in L^1(\Omega, P)$, and $\tau_1 \le \tau_2$ are a.s. finite stopping times. □

Theorem 3.58 *Let X be a stochastic process on the space $(\Omega, \mathcal{F}, P, (\mathcal{F}_t))$ and let τ be an a.s. bounded stopping time. We consider the stopped process X^τ defined by*

$$X_t^\tau(\omega) = X_{t\wedge\tau(\omega)}(\omega), \qquad t \ge 0, \ \omega \in \Omega. \tag{3.39}$$

We have:

i) if X is progressively measurable, then also Y is progressively measurable;
ii) if X is progressively measurable, then the random variable X_τ is $\mathcal{F}_\tau$-measurable;
iii) if X is a right-continuous $\mathcal{F}_t$-martingale, then

$$X_{t\wedge\tau} = E\left[X_\tau \mid \mathcal{F}_t\right] \tag{3.40}$$

and consequently also X^τ is a right-continuous $\mathcal{F}_t$-martingale.

Proof. i) The function

$$\varphi : [0,t] \times \Omega \longrightarrow [0,t] \times \Omega, \qquad \varphi(s,\omega) = (s \wedge \tau(\omega), \omega),$$

is measurable with respect to the product σ-algebra $\mathscr{B}([0,t]) \otimes \mathcal{F}_t$. Since $X_{t\wedge\tau}$ is equal to the composition of X with φ

$$X \circ \varphi : ([0,t] \times \Omega, \mathscr{B}([0,t]) \otimes \mathcal{F}_t) \longrightarrow \mathbb{R}^N$$

the first part of the claim follows by the assumption of progressive measurability on X.

ii) To prove that X_τ is $\mathcal{F}_\tau$-measurable, we have to show that, for every $H \in \mathscr{B}$ and $t \geq 0$, we have $F := \{X_\tau \in H\} \cap \{\tau \leq t\} \in \mathcal{F}_t$. Since we have $F = \{X_{t\wedge\tau} \in H\} \cap \{\tau \leq t\}$, the claim follows from the fact that $(X_{t\wedge\tau})$ is progressively measurable.

iii) We apply Theorem 3.56 to get

$$\begin{aligned} X_{t\wedge\tau} &= E\left[X_\tau \mid \mathcal{F}_{t\wedge\tau}\right] \\ &= E\left[X_\tau \mathbb{1}_{\{\tau<t\}} + X_\tau \mathbb{1}_{\{\tau\geq t\}} \mid \mathcal{F}_{t\wedge\tau}\right] \\ &= X_\tau \mathbb{1}_{\{\tau<t\}} + E\left[X_\tau \mathbb{1}_{\{\tau\geq t\}} \mid \mathcal{F}_{t\wedge\tau}\right] = \end{aligned}$$

(since $A\mathbb{1}_{\{\tau\geq t\}} \in \mathcal{F}_\tau$ if $A \in \mathcal{F}_t$)

$$= X_\tau \mathbb{1}_{\{\tau<t\}} + E\left[X_\tau \mid \mathcal{F}_t\right] \mathbb{1}_{\{\tau\geq t\}} =$$

(since $X_\tau \mathbb{1}_{\{\tau<t\}}$ is $\mathcal{F}_t$-measurable)

$$= E\left[X_\tau \mid \mathcal{F}_t\right],$$

and this proves (3.40).

For fixed $t < s$, applying (3.40) with the stopping time $s \wedge \tau$ instead of τ, we get

$$X_{t\wedge\tau} = E\left[X_{s\wedge\tau} \mid \mathcal{F}_t\right],$$

and therefore X^τ is a $\mathcal{F}_t$-martingale. □

3.4 Riemann-Stieltjes integral

Let us go back to the model for a risky asset in Example 3.8 where, assuming $S_0 = 1$, we have

$$S_t = 1 + \mu t + \sigma W_t, \qquad t \in [0,T],$$

and W is a real Brownian motion starting from the origin. We consider a partition

$$\varsigma = \{t_0, t_1, \dots, t_N\}$$

of $[0,T]$ with $0 = t_0 < t_1 < \cdots < t_N = T$. Now, let $V = uS$ be the value of a *self-financing portfolio* (cf. Definition 2.2) that is composed only by the asset S. Then, for every $k = 1, \dots, N$, we have

$$\begin{aligned} V_{t_k} - V_{t_{k-1}} &= u_{t_{k-1}}(S_{t_k} - S_{t_{k-1}}) \\ &= \mu u_{t_{k-1}}(t_k - t_{k-1}) + \sigma u_{t_{k-1}}(W_{t_k} - W_{t_{k-1}}). \end{aligned}$$

Summing over k from 1 to N, we get

$$V_T = V_0 + \underbrace{\mu \sum_{k=1}^{N} u_{t_{k-1}}(t_k - t_{k-1})}_{=:I_{1,\varsigma}} + \underbrace{\sigma \sum_{k=1}^{N} u_{t_{k-1}}(W_{t_k} - W_{t_{k-1}})}_{=:I_{2,\varsigma}}. \tag{3.41}$$

To move on to continuous time, it is necessary to verify the existence of the limits $I_{1,\varsigma}$ and $I_{2,\varsigma}$ as the refinement parameter $|\varsigma|$ of the partition tends to zero: we recall that

$$|\varsigma| = \max_{1 \le k \le N} |t_k - t_{k-1}|. \tag{3.42}$$

The first term $I_{1,\varsigma}$ is a Riemann sum and so, supposing that the function $t \mapsto u_t(\omega)$ is Riemann integrable[13] in $[0,T]$ for every $\omega \in \Omega$, we simply have

$$\lim_{|\varsigma| \to 0^+} I_{1,\varsigma}(\omega) = \int_0^T u_t(\omega)dt,$$

for every $\omega \in \Omega$.

The second term $I_{2,\varsigma}$ is the transform of u by W (cf. Definition A.120). The existence of the second limit is not trivial: if we suppose that the limit

$$\lim_{|\varsigma| \to 0^+} I_{2,\varsigma} = I \tag{3.43}$$

exists and is finite, by analogy we may use the notation

$$I = \int_0^T u_t \, dW_t. \tag{3.44}$$

Therefore we get, at least formally, the following formula

$$V_T = V_0 + \mu \int_0^T u_t dt + \sigma \int_0^T u_t \, dW_t.$$

Actually the limit in (3.43) *does not exist in general, unless further assumptions on the stochastic process u are made.* In order to justify this claim,

[13] As a matter of fact, in a self-financing portfolio with one asset only, the function $t \mapsto u_t$ is necessarily constant. The situation is not trivial anymore in the case of a portfolio with at least two assets.

we ought to digress and give some mathematical details showing that the trajectories of a Brownian motion are almost surely "irregular" in a sense that will be specified later on.

Indeed, let us consider a regular path

$$t \longmapsto W_t(\bar{\omega}),$$

and assume that it belongs to $C^1([0,T])$: in this case we can easily prove that there exists the limit

$$\lim_{|\varsigma|\to 0+} I_{2,\varsigma}(\bar{\omega}) = \int_0^T u_t(\bar{\omega}) W_t'(\bar{\omega})\, dt, \tag{3.45}$$

where the integral is understood in the usual Riemann sense and $W_t'(\bar{\omega})$ denotes the derivative $\frac{d}{dt} W_t(\bar{\omega})$. Indeed, by Lagrange mean value theorem there exists $t_k^* \in [t_{k-1}, t_k]$ such that

$$I_{2,\varsigma}(\bar{\omega}) = \sum_{k=1}^{N} u_{t_{k-1}}(\bar{\omega}) W_{t_k^*}'(\bar{\omega})(t_k - t_{k-1});$$

so $I_{2,\varsigma}(\bar{\omega})$ is a Riemann sum and (3.45) follows easily.

As a matter of fact it is not difficult to prove the existence of the limit in (3.43) under the weaker assumption that $t \mapsto W_t(\bar{\omega})$ is a bounded variation function (cf. Section 3.4.1):

$$\lim_{|\varsigma|\to 0+} I_{2,\varsigma}(\bar{\omega}) = l \in \mathbb{R}.$$

The number l is usually called *Riemann-Stieltjes integral of* $u_t(\bar{\omega})$ *with respect to* $W_t(\bar{\omega})$ *over* $[0,T]$ (cf. Section 3.4.2) and the notation $l = \left(\int_0^T u_t\, dW_t \right)(\bar{\omega})$ is used.

Unfortunately, in Section 3.4.3 we show that the paths of a Brownian motion do not have bounded variation almost surely and so the integral in (3.44) cannot be defined in the Riemann-Stieltjes sense. Chapter 4 will be entirely devoted to an introduction to stochastic integration theory.

3.4.1 Bounded-variation functions

The material in this section is not essential for the rest of the treatment and can be skipped on first reading, although some concepts may help the reader understand more clearly the next chapter.

Given a real interval $[a,b]$, we consider a function

$$g : [a,b] \to \mathbb{R}^n$$

and a partition $\varsigma = \{t_0, \dots, t_N\}$ of $[a,b]$. The variation of g relative to ς is defined by

$$V_{[a,b]}(g, \varsigma) = \sum_{k=1}^{N} |g(t_k) - g(t_{k-1})|.$$

Definition 3.59 *The function g has bounded variation on $[a,b]$ (we write $g \in \mathrm{BV}([a,b])$) if the supremum of $V_{[a,b]}(g,\varsigma)$, taken over all partitions ς of $[a,b]$, is finite:*

$$V_{[a,b]}(g) := \sup_{\varsigma} V_{[a,b]}(g,\varsigma) < +\infty.$$

$V_{[a,b]}(g)$ is called (first) variation of g over $[a,b]$.

Example 3.60 i) If

$$g : [a,b] \to \mathbb{R}$$

is monotone, then $g \in \mathrm{BV}([a,b])$. For example, if g is increasing, we have

$$V_{[a,b]}(g,\varsigma) = \sum_{k=1}^{N} (g(t_k) - g(t_{k-1})) = g(b) - g(a),$$

and so

$$V_{[a,b]}(g) = g(b) - g(a);$$

ii) if g is Lipschitz continuous, i.e. there exists a constant C such that

$$|g(t) - g(s)| \le C|t-s|, \qquad t,s \in [a,b],$$

then $g \in \mathrm{BV}([a,b])$. Indeed

$$V_{[a,b]}(g,\varsigma) = \sum_{k=1}^{N} |g(t_k) - g(t_{k-1})| \le C \sum_{k=1}^{N} (t_k - t_{k-1}) = C(b-a),$$

and so

$$V_{[a,b]}(g) \le C(b-a);$$

iii) if $u \in L^1([a,b])$, then the integral function

$$g(t) := \int_a^t u(s)ds$$

has bounded variation over $[a,b]$, indeed

$$\begin{aligned} V_{[a,b]}(g,\varsigma) &= \sum_{k=1}^{N} |g(t_k) - g(t_{k-1})| = \sum_{k=1}^{N} \left| \int_{t_{k-1}}^{t_k} u(s)ds \right| \\ &\le \sum_{k=1}^{N} \int_{t_{k-1}}^{t_k} |u(s)|\, ds = \int_a^b |u(s)|\, ds, \end{aligned}$$

and so

$$V_{[a,b]}(g) \le \|u\|_{L^1};$$

iv) the function

$$g(t) = \begin{cases} 0 & \text{per } t = 0, \\ t \sin\left(\frac{1}{t}\right) & \text{per } t \in]0,1], \end{cases}$$

is continuous on $[0,1]$ but does not have bounded variation. As an exercise, prove this statement by using partitions with points of the form $t_n = \left(\frac{\pi}{2} + n\pi\right)^{-1}$.

□

Lemma 3.61 *If* $g \in \mathrm{BV} \cap C([a,b])$, *then*

$$V_{[a,b]}(g) = \lim_{|\varsigma| \to 0} V_{[a,b]}(g, \varsigma). \tag{3.46}$$

Proof. Let us first recall that $|\varsigma|$ denotes the refinement parameter of the partition ς, defined in (3.42). By contradiction, if (3.46) were not true, then there would exist a partition $\varsigma = \{t_0, \dots, t_N\}$ of $[a,b]$, a sequence of partitions (ς_n) and a positive number ε such that

$$V_{[a,b]}(g, \varsigma_n) \le V_{[a,b]}(g, \varsigma) - \varepsilon, \qquad \lim_{n \to \infty} |\varsigma_n| = 0. \tag{3.47}$$

Now we have

$$V_{[a,b]}(g, \varsigma) \le V_{[a,b]}(g, \varsigma_n) + \sum_{k=1}^{N} |g(t_k) - g(t^n_{k_n})|$$

where $t^n_{k_n}$ are points of the partition ς_n such that $|t_k - t^n_{k_n}| \le |\varsigma_n|$. On the other hand, since the function g is uniformly continuous on $[a,b]$ and $\lim_{n\to\infty} |\varsigma_n| = 0$, we can choose n large enough such that

$$\sum_{k=1}^{N} |g(t_k) - g(t^n_{k_n})| \le \frac{\varepsilon}{2},$$

contradicting (3.47). □

Example 3.62 The function $g : [0,2] \to \mathbb{R}$, identically zero except for $t = 1$ where $g(1) = 1$, is such that $V_{[0,2]}(g) = 2$. On the other hand,

$$V_{[0,2]}(g, \varsigma) = 0$$

for every partition ς not containing 1. Therefore (3.46) is not true for a generic $g \in \mathrm{BV}([a,b])$. □

Remark 3.63 Geometrically, the variation $V_{[a,b]}(g, \varsigma)$ of a function

$$g : [a,b] \to \mathbb{R}^n$$

represents the length of the broken line in $\mathbb{R}^n$ given by the union of the line segments $\overline{g(t_0)g(t_1)}, \dots, \overline{g(t_{N-1})g(t_N)}$. Intuitively if g is a *continuous* curve, then by Lemma 3.61, $V_{[a,b]}(g, \varsigma)$ approximates the length of g as $|\varsigma|$ tends to zero: therefore the curve g is of bounded variation (or rectifiable) if it has finite length, approximated by broken lines inscribed to g. □

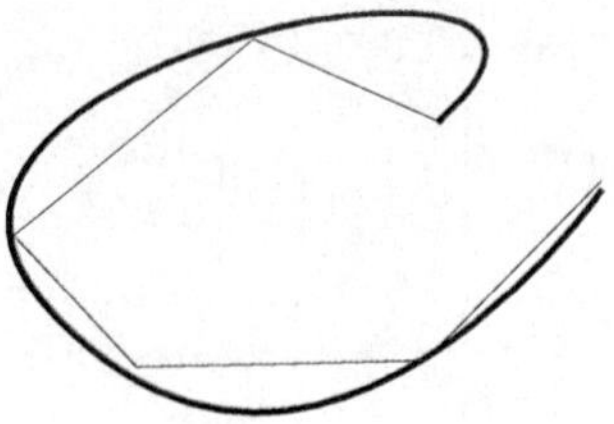

Fig. 3.2. Approximation of a continuous curve by a broken line

The following result characterizes real-valued functions of bounded variation.

Theorem 3.64 *A real function has bounded variation if and only if it is a difference of two monotone increasing functions.*

Proof. As a consequence of the triangular inequality we have

$$V_{[a,b]}(g_1 + g_2) \leq V_{[a,b]}(g_1) + V_{[a,b]}(g_2),$$

and so from Example 3.60-i) it follows that the difference of two monotone increasing function has bounded variation.

The converse is a consequence of the following property[14] of the variation: for every $t \in]a, b[$ we have

$$V_{[a,b]}(g) = V_{[a,t]}(g) + V_{[t,b]}(g). \tag{3.48}$$

First of all the function

$$\varphi(t) := V_{[a,t]}(g), \qquad t \in [a, b],$$

is monotone increasing, in view of (3.48). Further, setting $\psi = \varphi - g$, we get $\psi(t + h) \geq \psi(t)$ for $h \geq 0$, since equivalently we have

$$\varphi(t + h) \geq \varphi(t) + g(t + h) - g(t)$$

as a consequence of (3.48). □

Remark 3.65 As a simple consequence of the previous result, if $g \in \mathrm{BV}([a, b])$ then the limits

$$g(t+) = \lim_{s \to t+} g(s), \qquad \text{and} \qquad g(t-) = \lim_{s \to t^-} g(s), \tag{3.49}$$

exist and are finite, for every t. Further, the set of the discontinuity points of g is *at most countable*[15]. Consequently it is always possible to modify a

---

[14] For its simple proof see, for instance, Rudin [293].

[15] It suffices to consider g monotone increasing. The jumps of g in t are

$$\Delta g(t) = g(t+) - g(t-).$$

For every $n \in \mathbb{N}$, the set $A_n = \{t \in]a, b[\mid \Delta g(t) \geq 1/n\}$ is finite since $g(a) \leq g(t) \leq g(b)$. The claim follows from the fact that the set of discontinuity points of g is given by the countable union of the A_n.

function $g \in \mathrm{BV}([a,b])$ over a countable set (having null Lebesgue measure) in such a way that g becomes right-continuous , i.e.

$$g(t+) = g(t), \qquad t \in [a,b],$$

or left-continuous. □

3.4.2 Riemann-Stieltjes integral and Itô formula

Let us introduce some notations: given a real interval $[a,b]$, we denote by

$$\mathcal{P}_{[a,b]} = \{\varsigma = (t_0, \dots, t_N) \mid a = t_0 < t_1 < \dots < t_N = b\},$$
$$\mathcal{T}_\varsigma = \{\tau = (\tau_1, \dots, \tau_N) \mid \tau_k \in [t_{k-1}, t_k],\ k = 1, \dots, N\},$$

the collection of partitions of $[a,b]$ and the collection of the "choices of points" relative to the partition ς, respectively. Given two real functions u, g defined on $[a,b]$, we define by

$$S(u, g, \varsigma, \tau) = \sum_{k=1}^{N} u(\tau_k)(g(t_k) - g(t_{k-1}))$$

the Riemann-Stieltjes sum of u relative to g, to the partition ς and to the choice of points $\tau \in \mathcal{T}_\varsigma$. We have the following classical result:

Theorem 3.66 *If $u \in C([a,b])$ and $g \in \mathrm{BV}([a,b])$, then there exists the limit*

$$\lim_{|\varsigma| \to 0} S(u, g, \varsigma, \tau) =: \int_a^b u(t)dg(t), \tag{3.50}$$

i.e. for every $\varepsilon > 0$ there exists $\delta > 0$ such that

$$\left| \int_a^b u(t)dg(t) - S(u, g, \varsigma, \tau) \right| < \varepsilon,$$

for every $\varsigma \in \mathcal{P}_{[a,b]}$ such that $|\varsigma| < \delta$ and for every $\tau \in \mathcal{T}_\varsigma$. Formula (3.50) defines the Riemann-Stieltjes integral of u with respect to g. Further, if $g \in C^1([a,b])$ then we simply have

$$\int_a^b u(t)dg(t) = \int_a^b u(t)g'(t)dt. \tag{3.51}$$

Example 3.67 Consider the interval $[0,1]$ and the BV function $g(t) = \mathbb{1}_{\{T\}}(t)$, $t \in [0,1]$, where $T \in]0,1[$. If ς is a partition of $[0,1]$, we have two cases: if $t_0 \notin \varsigma$ then $S(u, g, \varsigma, \tau) = 0$ for any function u and $\tau \in \mathcal{T}_\varsigma$. In particular, if the Riemann-Stieltjes integral exists, we necessarily have

$$\int_0^1 u(t)dg(t) = 0.$$

On the other hand, if $T \in \varsigma = (t_0, \dots, t_N)$, say $T = t_n$, then

$$\begin{aligned} S(u, g, \varsigma, \tau) &= u(\tau_n)(g(t_n) - g(t_{n-1})) + u(\tau_{n+1})(g(t_{n+1}) - g(t_n)) \\ &= u(\tau_n) - u(\tau_{n+1}) \end{aligned}$$

for every $\tau_n \in [t_{n-1}, T]$ and $\tau_{n+1} \in [T, t_{n+1}]$. Letting $|\varsigma|$ go to zero, we see that $S(u, g, \varsigma, \tau)$ converges to 0 if and only if u is continuous at T. Thus *the continuity of the integrand is necessary to guarantee the convergence of the Riemann-Stieltjes sum for any BV function* g. The class of integrands can be extended by considering Lebesgue-Stieltjes integration (cf. Section 14.1).

Conversely, *if the Riemann-Stieltjes sum* $S(u, g, \varsigma, \tau)$ *converges to a limit for every continuous function* u *then* g *has bounded variation* (see, for instance, Theorem I-55 in Protter [287]). □

Example 3.68 Let $T \in]0, 1[$ and consider a BV function g such that $g(t) = 0$ for $t \in [0, T[$ and $g(t) = 1$ for $t \in]T, 1]$. For any $u \in C([0, 1])$, we have

$$\int_0^1 u(t)dg(t) = u(T)\Delta g(T) := u(T)\left(g(T+) - g(T-)\right).$$

Note that the value of the integral is independent of $g(T)$. On the other hand, we have

$$\int_0^T u(t)dg(t) = u(T)\left(g(T) - g(T-)\right), \tag{3.52}$$

and therefore the value of the Riemann-Stieltjes in (14.25) may vary by modifying g at the point T. It is left as an exercise to prove that, if $g = \mathbb{1}_{]0,1]}$, then

$$\int_0^T u(t)dg(t) = u(0)$$

for any $T \in]0, 1]$. □

Proof (of Theorem 3.66). We give a sketch of the proof and leave the details to the reader. To prove (3.50) using the Cauchy criterion, it suffices to verify that, for every $\varepsilon > 0$, there exists $\delta > 0$ such that

$$|S(u, g, \varsigma', \tau') - S(u, g, \varsigma'', \tau'')| < \varepsilon,$$

for every $\varsigma', \varsigma'' \in \mathcal{P}_{[a,b]}$ such that $|\varsigma'|, |\varsigma''| < \delta$ and for every $\tau' \in \mathcal{T}_{\varsigma'}$ and $\tau'' \in \mathcal{T}_{\varsigma''}$.

We put $\varsigma = \varsigma' \cup \varsigma'' = \{t_0, \dots, t_N\}$. For fixed $\varepsilon > 0$, since f is uniformly continuous on $[a, b]$, it is enough to choose $|\varsigma'|$ and $|\varsigma''|$ small enough to get

$$|S(u, g, \varsigma', \tau') - S(u, g, \varsigma'', \tau'')| \le \varepsilon \sum_{k=1}^{N} |g(t_k) - g(t_{k-1})| \le \varepsilon V_{[a,b]}(g),$$

so we can conclude.

If $g \in C^1([a,b])$, by the mean value theorem, given $\varsigma \in \mathcal{P}_{[a,b]}$ there exists $\tau \in \mathcal{T}_\varsigma$ such that

$$S(u,g,\varsigma,\tau) = \sum_{k=1}^{N} u(\tau_k)g'(\tau_k)(t_k - t_{k-1}) = S(ug', \mathrm{id}, \varsigma, \tau)$$

and (3.51) follows by taking the limit as $|\varsigma|$ tends to zero. □

Now we list some simple properties of the Riemann-Stieltjes integral; their proof is left as an exercise.

Proposition 3.69 *Let $u, v \in C([a,b])$, $f, g \in \mathrm{BV}([a,b])$ and $\lambda, \mu \in \mathbb{R}$. Then we have:*

i)

$$\int_a^b (\lambda u + v) d(f + \mu g) = \lambda \int_a^b u df + \lambda\mu \int_a^b u dg + \int_a^b v df + \mu \int_a^b v dg;$$

ii) if $u \le v$ and g is monotone increasing, then

$$\int_a^b u dg \le \int_a^b v dg;$$

iii)

$$\left| \int_a^b u dg \right| \le \max |u| V_{[a,b]}(g);$$

iv) for $c \in]a,b[$ we have

$$\int_a^b u dg = \int_a^c u dg + \int_c^b u dg.$$

Now we prove a theorem that extends the classical results concerning the notion of primitive and its role in the computation of the Riemann integral. The next theorem is the "deterministic version" of the Itô formula, the fundamental result in stochastic calculus that will be proved in Chapter 5.1.1.

Theorem 3.70 (Itô formula) *Let $F \in C^1([a,b] \times \mathbb{R})$ and $g \in \mathrm{BV} \cap C([a,b])$. Then we have*

$$F(b, g(b)) - F(a, g(a)) = \int_a^b (\partial_t F)(t, g(t)) dt + \int_a^b (\partial_g F)(t, g(t)) dg(t). \quad (3.53)$$

Before proving the theorem, we consider some examples: in the particular case $F(t,g) = g$, (3.53) becomes

$$g(b) - g(a) = \int_a^b dg(t).$$

Further, if $g \in C^1$ we have

$$g(b) - g(a) = \int_a^b g'(t)dt.$$

For $F(t, g) = f(t)g$ we get

$$f(b)g(b) - f(a)g(a) = \int_a^b f'(t)g(t)dt + \int_a^b f(t)dg(t),$$

and this extends the integration by parts formula to the case $g \in \mathrm{BV} \cap C([a, b])$. Formula (3.53) allows also to compute explicitly some integrals: for example, if $F(t, g) = g^2$ we get

$$\int_a^b g(t)dg(t) = \frac{1}{2}\left(g^2(b) - g^2(a)\right).$$

Proof (of Theorem 3.70). For every $\varsigma \in \mathcal{P}_{[a,b]}$, we have

$$F(b, g(b)) - F(a, g(a)) = \sum_{k=1}^{N} \left(F(t_k, g(t_k)) - F(t_{k-1}, g(t_{k-1}))\right) =$$

(by the mean value theorem and the continuity of g, with $t_k', t_k'' \in [t_{k-1}, t_k]$)

$$= \sum_{k=1}^{N} \left(\partial_t F(t_k', g(t_k''))(t_k - t_{k-1}) + \partial_g F(t_k', g(t_k''))(g(t_k) - g(t_{k-1}))\right)$$

and the claim follows by taking the limit as $|\varsigma| \to 0$. □

Exercise 3.71 Proceeding as in the proof of the Itô formula, prove the following integration-by-parts formula

$$f(b)g(b) - f(a)g(a) = \int_a^b f(t)dg(t) + \int_a^b g(t)df(t),$$

valid for $f, g \in \mathrm{BV} \cap C([a, b])$.

3.4.3 Regularity of the paths of a Brownian motion

In this section we prove that the family of trajectories with bounded variation of a Brownian motion W is negligible. In other words W has almost all paths that are irregular, non-rectifiable: in every time interval $[0, t]$ with $t > 0$, W covers almost surely a path of infinite length. Consequently, for almost all the paths of W it is not possible to define the integral

$$\int_0^T u_t \, dW_t$$

in the Riemann-Stieltjes sense. In order to study the regularity of the paths of Brownian motion we introduce the fundamental concept of *quadratic variation*.

Definition 3.72 *Given a function*

$$g : [0, t] \longrightarrow \mathbb{R}^n$$

and a partition $\varsigma = \{t_0, \dots, t_N\} \in \mathcal{P}_{[0,t]}$, *the quadratic variation of* g *relative to* ς *is defined by*

$$V_t^{(2)}(g, \varsigma) = \sum_{k=1}^{N} |g(t_k) - g(t_{k-1})|^2.$$

The case of continuous functions with bounded (first) variation is of particular interest.

Proposition 3.73 *If* $g \in \mathrm{BV} \cap C([0, t])$ *then*

$$\lim_{|\varsigma| \to 0} V_t^{(2)}(g, \varsigma) = 0.$$

Proof. The function g is uniformly continuous on $[0, t]$, consequently for every $\varepsilon > 0$ there exists $\delta > 0$ such that

$$|g(t_k) - g(t_{k-1})| \leq \varepsilon$$

for every $\varsigma = \{t_0, t_1, \dots, t_N\} \in \mathcal{P}_{[0,t]}$ such that $|\varsigma| < \delta$. The claim follows from the fact that

$$0 \leq V_t^{(2)}(g, \varsigma) = \sum_{k=1}^{N} |g(t_k) - g(t_{k-1})|^2 \leq \varepsilon \sum_{k=1}^{N} |g(t_k) - g(t_{k-1})| \leq \varepsilon\, V_{[0,t]}(g)$$

where the first variation $V_{[0,t]}(g)$ is finite by assumption. □

Theorem 3.74 *If* W *is a Brownian motion, then we have*

$$\lim_{|\varsigma| \to 0} V_t^{(2)}(W, \varsigma) = t \qquad \text{in } L^2(\Omega, P). \tag{3.54}$$

Consequently, for any $t > 0$, *almost all the trajectories of* W *do not have bounded variation on* $[0, t]$. *We say that the function* $f(t) = t$ *is the quadratic variation of the Brownian motion and we write* $\langle W \rangle_t = t$ *for* $t \geq 0$.

Proof. To unburden notations, for fixed $t > 0$ and the partition

$$\varsigma = \{t_0, \dots, t_N\} \in \mathcal{P}_{[0,t]},$$

we set

$$\Delta_k = W_{t_k} - W_{t_{k-1}} \quad \text{and} \quad \delta_k = t_k - t_{k-1}$$

for $k = 1, \dots, N$. We recall that $E\left[\Delta_k^2\right] = \delta_k$. Further, it is not difficult to prove[16] that

$$E\left[\Delta_k^4\right] = 3\delta_k^2. \tag{3.55}$$

Then we have

$$E\left[\left(V_t^{(2)}(W,\varsigma) - t\right)^2\right] = E\left[\left(\sum_{k=1}^N \Delta_k^2 - t\right)^2\right] = E\left[\left(\sum_{k=1}^N \left(\Delta_k^2 - \delta_k\right)\right)^2\right]$$

$$= \sum_{k=1}^N E\left[\left(\Delta_k^2 - \delta_k\right)^2\right] + 2\sum_{h<k} E\left[\left(\Delta_k^2 - \delta_k\right)\left(\Delta_h^2 - \delta_h\right)\right].$$

Now we observe that, by (3.55), we have

$$E\left[\left(\Delta_k^2 - \delta_k\right)^2\right] = E\left[\Delta_k^4\right] - 2\delta_k E\left[\Delta_k^2\right] + \delta_k^2 = 2\delta_k^2.$$

On the other hand

$$E\left[\left(\Delta_k^2 - \delta_k\right)\left(\Delta_h^2 - \delta_h\right)\right] =$$

(by the independence of the increments of Brownian motion for $h < k$)

$$= E\left[\Delta_k^2 - \delta_k\right] E\left[\Delta_h^2 - \delta_h\right] = 0.$$

In conclusion we get

$$E\left[\left(V_t^{(2)}(W,\varsigma) - t\right)^2\right] = 2\sum_{k=1}^N \delta_k^2 \le 2t|\varsigma|$$

that proves (3.54).

Now by Theorem A.136, for any sequence of partitions (ς_n) with mesh converging to zero, there exists a subsequence (ς_{k_n}) such that

$$\lim_{n\to\infty} V_t^{(2)}(W, \varsigma_{k_n}) = t \qquad \text{a.s.}$$

Thus, by Proposition 3.73, almost all the trajectories of W cannot have bounded variation on $[0,t]$. □

Remark 3.75 Since the L^2-convergence implies convergence in probability (cf. Theorem A.136), by (3.54) we also have that $V_t^{(2)}(W,\varsigma)$ converges *in probability* to t, that is

$$\lim_{|\varsigma|\to 0} P\left(\left|V_t^{(2)}(W,\varsigma) - t\right| > \varepsilon\right) = 0$$

---

[16] We have

$$E\left[\Delta_k^4\right] = \int_{\mathbb{R}} y^4 \Gamma(y, \delta_k) dy,$$

with Γ as in (A.7). Then (3.55) can be obtained integrating by parts. See also Example 5.6-(3).

for every $\varepsilon > 0$. Note however that it is not true that

$$\lim_{|\varsigma|\to 0} V_t^{(2)}(W,\varsigma) = t \qquad \text{a.s.}$$

Indeed, one can find (cf. Exercise 1.15 in [259]) a sequence (ς_n) of partitions in $\mathcal{P}_{[0,t]}$ such that $|\varsigma_n| \to 0$ as $n \to \infty$ and

$$\limsup_{n\to\infty} V_t^{(2)}(W,\varsigma_{k_n}) = +\infty \qquad \text{a.s.}$$

On the other hand,

$$\lim_{n\to\infty} V_t^{(2)}(W,\varsigma_n) = t \qquad \text{a.s.}$$

for any refining[17] sequence (ς_n) of partitions such that $\lim_{n\to\infty} |\varsigma_n| = 0$ (see, for instance, Theorem I-28 in [287]). □

Exercise 3.76 Let $f \in C([0,T])$. Prove that

$$\lim_{|\varsigma|\to 0} \sum_{k=1}^{N} f(t_{k-1})(W_{t_k} - W_{t_{k-1}})^2 = \int_0^T f(t)dt, \qquad \text{in } L^2(\Omega),$$

where, as usual, $\varsigma = \{t_0, \dots, t_N\} \in \mathcal{P}_{[0,T]}$.

Exercise 3.77 Given $g \in C([0,t])$, $p \geq 1$ and $\varsigma = \{t_0, \dots, t_N\} \in \mathcal{P}_{[0,t]}$, we define

$$V_t^{(p)}(g,\varsigma) = \sum_{k=1}^{N} |g(t_k) - g(t_{k-1})|^p$$

the p-th order variation of g over $[0,t]$ relative to the partition ς. Prove that, if

$$\lim_{|\varsigma|\to 0} V_t^{(p_0)}(g,\varsigma) \in]0,+\infty[,$$

for some p_0, then

$$\lim_{|\varsigma|\to 0} V_t^{(p)}(g,\varsigma) = \begin{cases} +\infty & p < p_0 \\ 0 & p > p_0. \end{cases}$$

The case $p > p_0$ can be proved exactly as in Proposition 3.73; the case $p < p_0$ can be proved by contradiction.

Definition 3.78 *Given two functions $f, g : [0,t] \to \mathbb{R}^n$, the co-variation of f, g over $[0,t]$ is defined by the limit (if it exists)*

$$\langle f, g\rangle_t = \lim_{\substack{|\varsigma|\to 0\\ \varsigma\in\mathcal{P}_{[0,t]}}} \sum_{k=1}^{N} \langle f(t_k) - f(t_{k-1}), g(t_k) - g(t_{k-1})\rangle.$$

The following result can be proved as Proposition 3.73.

Proposition 3.79 *If $f \in C([0,t])$ and $g \in \mathrm{BV}([0,t])$, then $\langle f, g\rangle_t = 0$.*

---

[17] A sequence (ς_n) is refining if $\varsigma_n \supset \varsigma_{n+1}$ for any n.

4

Brownian integration

In this chapter we introduce the elements of stochastic integration theory that are necessary to treat some financial models in continuous time. In Paragraph 3.4 we gave grounds for the interest in the study of the limit of a Riemann-Stieltjes sum of the form

$$\sum_{k=1}^{N} u_{t_{k-1}}(W_{t_k} - W_{t_{k-1}}) \tag{4.1}$$

as the refinement parameter of the partition $\{t_0, \dots, t_N\}$ tends to zero. In (4.1) W is a real Brownian motion that represents a risky asset and u is an adapted process that represents an investment strategy: if the strategy is self-financing, the limit of the sum in (4.1) is equal to the value of the investment.

However the paths of W do not have bounded variation a.s. and this fact prevents us to define pathwise the integral

$$\int_0^T u_t dW_t$$

in the Riemann-Stieltjes sense. On the other hand W *has finite quadratic variation* and this property makes it possible to construct the stochastic integral for suitable classes of integrands u: generally speaking, we require that u is progressively measurable and satisfies some integrability conditions.

The concept of Brownian integral was introduced by Paley, Wiener and Zygmund [275] for deterministic integrand functions. The general construction is due to Itô [179]-[180] in the case of Brownian motion, and to Kunita and Watanabe [219] in $\mathscr{M}^2$. This theory lays the foundations for a rigorous study of stochastic differential equations that describe the diffusion processes introduced by Kolmogorov [213], on which the modern stochastic models for finance are based. In this chapter we confine ourselves to the Brownian case.

The aim of this chapter is to construct the Brownian integral gradually, first considering the integration of "simple" processes, i.e. processes that are

Pascucci A.: PDE and Martingale Methods in Option Pricing

piecewise constant with respect to the time variable, then extending the definition to a sufficiently general class of progressively measurable and square-integrable processes. Among the main consequences of the definition, we have that the stochastic integral has null expectation, it is a continuous martingale in $\mathscr{M}_c^2$ and it satisfies Itô isometry. By further extending the class of integrands, some of those properties are lost and it is necessary to introduce the more general notion of local martingale.

4.1 Stochastic integral of deterministic functions

As an introductory example, useful to see in advance some of the main results we are going to prove, we consider Paley, Wiener and Zygmund's construction [275] of the stochastic integral for deterministic functions.

Let $u \in C^1([0,1])$ be a real-valued function such that $u(0) = u(1) = 0$. Given a real Brownian motion W, we define

$$\int_0^1 u(t)dW_t = -\int_0^1 u'(t)W_t dt. \tag{4.2}$$

This integral is a random variable that verifies the following properties:

i) $E\left[\int_0^1 u(t)dW_t\right] = 0$;

ii) $E\left[\left(\int_0^1 u(t)dW_t\right)^2\right] = \int_0^1 u^2(t)dt$.

Indeed

$$E\left[\int_0^1 u'(t)W_t dt\right] = \int_0^1 u'(t)E\left[W_t\right]dt = 0.$$

Further,

$$E\left[\int_0^1 u'(t)W_t dt \int_0^1 u'(s)W_s ds\right] = \int_0^1\int_0^1 u'(t)u'(s)E\left[W_tW_s\right]dtds =$$

(since $E\left[W_tW_s\right] = t \wedge s$)

$$\begin{aligned}
&= \int_0^1 u'(t)\left(\int_0^t su'(s)ds + t\int_t^1 u'(s)ds\right)dt\\
&= \int_0^1 u'(t)\left(tu(t) - \int_0^t u(s)ds + t(u(1) - u(t))\right)dt\\
&= \int_0^1 u'(t)\left(-\int_0^t u(s)ds\right)dt = \int_0^1 u^2(t)dt.
\end{aligned}$$

More generally, if $u \in L^2(0,1)$ and (u_n) is a sequence of functions in $C_0^1(0,1)$ approximating u in the L^2 norm, by property ii) we have

$$E\left[\left(\int_0^1 u_n(t)dW_t - \int_0^1 u_m(t)dW_t\right)^2\right] = \int_0^1 (u_n(t) - u_m(t))^2 dt.$$

Therefore the sequence of integrals is a Cauchy sequence in $L^2(\Omega, P)$ and we can define

$$\int_0^1 u(t)dW_t = \lim_{n\to\infty} \int_0^1 u_n(t)dW_t.$$

We have thus constructed the stochastic integral for $u \in L^2([0,1])$ and, by passing to the limit, it is immediate to verify properties i) and ii).

Evidently this construction can be considered only an introductory step, since we are interested in defining the Brownian integral in the case u is a stochastic process. Indeed we recall that, from a financial point of view, u represents a future-investment strategy, necessarily random. On the other hand, since (4.2) seems to be a reasonable definition, in the following paragraphs we will introduce a definition of stochastic integral that agrees with the one given for the deterministic case.

4.2 Stochastic integral of simple processes

In what follows W is a real Brownian motion on the filtered probability space $(\Omega, \mathcal{F}, P, (\mathcal{F}_t))$ where the usual hypotheses hold and T is a fixed positive number.

Definition 4.1 *The stochastic process u belongs to the class $\mathbb{L}^2$ if*

i) u is progressively measurable with respect to the filtration $(\mathcal{F}_t)$;
ii) $u \in L^2([0,T] \times \Omega)$ that is

$$\int_0^T E\left[u_t^2\right] dt < \infty.$$

Condition ii) is a simple integrability condition, while i) is the property playing the crucial part in what follows. Since the definition of $\mathbb{L}^2$ depends on the given filtration $(\mathcal{F}_t)$, when it is necessary we will also write $\mathbb{L}^2(\mathcal{F}_t)$ instead of $\mathbb{L}^2$. More generally, for $p \geq 1$, we denote by $\mathbb{L}^p$ the space of progressively measurable processes in $L^p([0,T] \times \Omega)$. We note explicitly that $\mathbb{L}^p$ is a closed subspace of $L^p([0,T] \times \Omega)$.

Now we start by defining the Itô integral for a particular class of stochastic processes in $\mathbb{L}^2$.

Definition 4.2 *A process $u \in \mathbb{L}^2$ is called simple if it can be written as*

$$u_t = \sum_{k=1}^N e_k \mathbb{1}_{]t_{k-1},t_k]}(t), \qquad t \in [0,T], \tag{4.3}$$

where $0 \leq t_0 < t_1 < \cdots < t_N \leq T$ and e_k are random variables[1] *on $(\Omega, \mathcal{F}, P)$.*

[1] We assume also that

$$P(e_{k-1} = e_k) = 0, \qquad k = 2, \ldots, N,$$

so that the representation (4.3) for u is unique a.s.

Remark 4.3 It is important to observe that, since u is progressively measurable and by hypothesis (3.27) of right-continuity of the filtration, we have that e_k in (4.3) is $\mathcal{F}_{t_{k-1}}$-measurable for every $k = 1, \dots, N$. Further, $e_k \in L^2(\Omega, P)$ and we have

$$\int_0^T E\left[u_t^2\right] dt = \sum_{k=1}^N \int_0^T E\left[e_k^2\right] \mathbb{1}_{]t_{k-1},t_k]}(t) dt = \sum_{k=1}^N E\left[e_k^2\right](t_k - t_{k-1}). \quad (4.4)$$

□

If $u \in \mathbb{L}^2$ is a simple process of the form (4.3), then we define the Itô integral in the following way:

$$\int u_t dW_t = \sum_{k=1}^N e_k (W_{t_k} - W_{t_{k-1}}) \quad (4.5)$$

and also, for every $0 \le a < b \le T$,

$$\int_a^b u_t dW_t = \int u_t \mathbb{1}_{]a,b]}(t) dW_t \quad (4.6)$$

and

$$\int_a^a u_t dW_t = 0.$$

Example 4.4 Integrating the simple process $u = \mathbb{1}_{]0,t]}$, we get

$$W_t = \int_0^t dW_s.$$

Then, going back to Example 3.8, we have

$$S_t = S_0 \left(1 + \int_0^t \mu ds\right) + \int_0^t \sigma dW_s, \qquad t > 0.$$

□

The following theorem contains some important properties of the Itô integral of simple processes.

Theorem 4.5 *For all simple processes $u, v \in \mathbb{L}^2$, $\alpha, \beta \in \mathbb{R}$ and $0 \le a < b < c \le T$ the following properties hold:*

(1) linearity:

$$\int (\alpha u_t + \beta v_t) dW_t = \alpha \int u_t dW_t + \beta \int v_t \, dW_t;$$

(2) additivity:

$$\int_a^b u_t dW_t + \int_b^c u_t dW_t = \int_a^c u_t dW_t;$$

(3) null expectation:

$$E\left[\int_a^b u_t dW_t \mid \mathcal{F}_a\right] = 0, \tag{4.7}$$

and also

$$E\left[\int_a^b u_t dW_t \int_b^c v_t dW_t \mid \mathcal{F}_a\right] = 0; \tag{4.8}$$

(4) Itô isometry:

$$E\left[\int_a^b u_t dW_t \int_a^b v_t dW_t \mid \mathcal{F}_a\right] = E\left[\int_a^b u_t v_t dt \mid \mathcal{F}_a\right]; \tag{4.9}$$

(5) the stochastic process

$$X_t = \int_0^t u_s dW_s, \qquad t \in [0,T], \tag{4.10}$$

is a continuous $\mathcal{F}_t$-martingale, i.e. $X \in \mathscr{M}_c^2(\mathcal{F}_t)$, and we have

$$[\![X]\!]_T^2 = E\left[\sup_{t\in[0,T]} X_t^2\right] \le 4E\left[\int_0^T u_t^2 dt\right]. \tag{4.11}$$

Remark 4.6 Since
$$E[X] = E[E[X \mid \mathcal{F}_a]],$$
the non-conditional versions of (4.7), (4.8), (4.9) hold:

$$E\left[\int_a^b u_t dW_t\right] = 0,$$

$$E\left[\int_a^b u_t dW_t \int_b^c v_t dW_t\right] = 0,$$

$$E\left[\int_a^b u_t dW_t \int_a^b v_t dW_t\right] = E\left[\int_a^b u_t v_t dt\right].$$

The last identity for $u = v$ is equivalent to the L^2-norm equality

$$\left\| \int_a^b u_t dW_t \right\|_{L^2(\Omega)} = \|u\|_{L^2([a,b]\times\Omega)}$$

and this is why the fourth property is called "Itô isometry". □

Proof. Properties (1) and (2) are trivial. Concerning property (3), we have

$$E\left[\int_a^b u_t dW_t \mid \mathcal{F}_a\right] = \sum_{k=1}^N E\left[e_k(W_{t_k} - W_{t_{k-1}}) \mid \mathcal{F}_a\right] =$$

(since $t_0 \geq a$, e_k is $\mathcal{F}_{t_{k-1}}$-measurable by Remark 4.3 and so independent of $W_{t_k} - W_{t_{k-1}}$ and then we use Proposition A.107-(6))

$$= \sum_{k=1}^{N} E\left[e_k \mid \mathcal{F}_a\right] E\left[W_{t_k} - W_{t_{k-1}}\right] = 0.$$

To prove (4.8) we proceed analogously: if v is of the form

$$v = \sum_{h=1}^{M} d_h \mathbb{1}_{]t_{h-1}, t_h]},$$

then $E\left[\int_a^b u_t dW_t \int_b^c v_t dW_t \mid \mathcal{F}_a\right]$ is a sum of terms of the form

$$E\left[e_k d_h (W_{t_k} - W_{t_{k-1}})(W_{t_h} - W_{t_{h-1}}) \mid \mathcal{F}_a\right], \qquad \text{with } t_k \leq t_{h-1},$$

that are all equal to zero since $e_k d_h (W_{t_k} - W_{t_{k-1}})$ is $\mathcal{F}_{t_{h-1}}$-measurable and so independent of the increment $W_{t_h} - W_{t_{h-1}}$ whose expectation is null, since $a \leq t_{h-1}$.

Let us prove Itô isometry: if u and v are simple processes, we have

$$\begin{aligned} E\left[\int_a^b u_t dW_t \int_a^b v_t dW_t \mid \mathcal{F}_a\right] &= E\left[\sum_{k=1}^{N} \int_{t_{k-1}}^{t_k} e_k dW_t \sum_{h=1}^{N} \int_{t_{h-1}}^{t_h} d_h dW_t \mid \mathcal{F}_a\right] \\ &= \sum_{k=1}^{N} E\left[\int_{t_{k-1}}^{t_k} e_k dW_t \int_{t_{k-1}}^{t_k} d_k dW_t \mid \mathcal{F}_a\right] \\ &\quad + 2 \sum_{h<k} E\left[\int_{t_{k-1}}^{t_k} e_k dW_t \int_{t_{h-1}}^{t_h} d_h dW_t \mid \mathcal{F}_a\right] = \end{aligned}$$

(by (4.8) the terms in the second summation are null)

$$= \sum_{k=1}^{N} E\left[e_k d_k (W_{t_k} - W_{t_{k-1}})^2 \mid \mathcal{F}_a\right] =$$

(by Proposition A.107-(6), since $W_{t_k} - W_{t_{k-1}}$ is independent of $e_k d_k$ and of $\mathcal{F}_a$)

$$= \sum_{k=1}^{N} E\left[e_k d_k \mid \mathcal{F}_a\right] E\left[(W_{t_k} - W_{t_{k-1}})^2\right] = E\left[\sum_{k=1}^{N} e_k d_k (t_k - t_{k-1}) \mid \mathcal{F}_a\right]$$

and the claim follows by (4.4) at least for $u = v$: the general case is analogous.

Let us now prove that the stochastic process X in (4.10) is a continuous $\mathcal{F}_t$-martingale. The continuity follows directly from the definition of stochastic integral. By definition (4.5)-(4.6) and Remark 4.3, it is obvious that X is $\mathcal{F}_t$-adapted. Further, X_t is integrable since, by Hölder's inequality, we have

$$E\left[|X_t|\right]^2 \le E\left[X_t^2\right] =$$

(by Itô isometry)

$$= E\left[\int_0^t u_s^2 ds\right] < \infty$$

since $u \in \mathbb{L}^2$. Then, for $0 \le s < t$ we have

$$E\left[X_t \mid \mathcal{F}_s\right] = E\left[X_s \mid \mathcal{F}_s\right] + E\left[\int_s^t u_\tau \, dW_\tau \mid \mathcal{F}_s\right] = X_s,$$

since X_s is $\mathcal{F}_s$-measurable and (4.7) holds: therefore X is a martingale. Finally (4.11) is consequence of Doob's inequality, Theorem 3.38, and Itô isometry: indeed we have

$$[\![X]\!]_T^2 \le 4E\left[X_T^2\right] = 4E\left[\int_0^T u_t^2 dt\right].$$

Remark 4.7 The martingale property of the stochastic integral can also be written in the following meaningful way:

$$E\left[\int_0^T u_s dW_s \mid \mathcal{F}_t\right] = \int_0^t u_s dW_s, \qquad t \le T.$$

□

4.3 Integral of $\mathbb{L}^2$-processes

We extend the definition of stochastic integral to the class $\mathbb{L}^2$ of progressively measurable and square-integrable processes. Unlike the case of simple processes, the integral will be defined only modulo indistinguishability. Apart from this, all the usual properties in Theorem 4.11 carry over to this case.

To present the general idea, we consider Itô isometry

$$\left\| \int_0^T u_t dW_t \right\|_{L^2(\Omega)} = \|u\|_{L^2([0,T]\times\Omega)}. \tag{4.12}$$

This isometry plays an essential role in the construction of the stochastic integral

$$I_T(u) := \int_0^T u_t dW_t, \tag{4.13}$$

with $u \in \mathbb{L}^2$, since it guarantees that, if (u^n) is a Cauchy sequence in $L^2([0,T] \times \Omega)$, then also $(I_T(u^n))$ is a Cauchy sequence in $L^2(\Omega)$. This fact makes it possible to define the integral in $\mathbb{L}^2$ as soon as we prove that the elements in $\mathbb{L}^2$ can be approximated by simple processes.

Lemma 4.8 *For every $u \in \mathbb{L}^2$ there exists a sequence (u^n) of simple processes in $\mathbb{L}^2$ such that*

$$\lim_{n\to+\infty} E\left[\int_0^T (u_t - u_t^n)^2 dt\right] = \lim_{n\to+\infty} \left\|u - u^n\right\|^2_{L^2([0,T]\times\Omega)} = 0.$$

In particular an approximating sequence is defined by

$$u^n = \sum_{k=1}^{2^n-1} \left(\frac{1}{t_k - t_{k-1}} \int_{t_{k-1}}^{t_k} u_s ds\right) \mathbb{1}_{]t_k,t_{k+1}]}, \qquad (4.14)$$

where $t_k := \frac{kT}{2^n}$ for $0 \le k \le 2^n$: for this sequence we also have

$$\|u^n\|_{L^2([0,T]\times\Omega)} \le \|u\|_{L^2([0,T]\times\Omega)}.$$

We shall soon prove the lemma in a meaningful particular case (cf. Proposition 4.20 and Remark 4.21): for the general case we refer, for instance, to Steele [315], Theorem 6.5.

Thus we consider a sequence (u^n) of simple processes approximating $u \in \mathbb{L}^2$: since it converges, (u^n) is a Cauchy sequence in $L^2([0,T] \times \Omega)$, that is

$$\lim_{m,n\to\infty} \|u^n - u^m\|_{L^2([0,T]\times\Omega)} = 0.$$

Then, by Itô isometry, the sequence of stochastic integrals $(I_T(u^n))$ is a Cauchy sequence in $L^2(\Omega)$ and therefore it is convergent. It seems natural to define

$$\int_0^T u_t dW_t = \lim_{n\to+\infty} I_T(u^n) \quad \text{in } L^2(\Omega). \qquad (4.15)$$

Note that (4.15) defines the stochastic integral only except for a negligible event $N_T \in \mathcal{N}$. This causes problems in defining the integral as a stochastic process, i.e. as T varies. Indeed T belongs to an uncountable set and therefore the previous definition is questionable since the set $\bigcup_{T\ge 0} N_T$ might not be measurable, or if it is measurable, it might not have null probability.

On the other hand, this problem can be solved by using Doob's inequality, Theorem 3.38. Indeed, let us consider a sequence (u^n) of *simple* stochastic processes in $\mathbb{L}^2$ approximating u in $L^2([0,T] \times \Omega)$: we put

$$I_t(u^n) = \int_0^t u_s^n dW_s, \qquad t \in [0,T]. \qquad (4.16)$$

By (4.11) we obtain

$$[\![I(u^n) - I(u^m)]\!]_T \le 2\|u^n - v^n\|_{L^2([0,T]\times\Omega)},$$

and so $(I(u^n))$ is a Cauchy sequence in $\left(\mathcal{M}_c^2, [\![\cdot]\!]_T\right)$ that is a complete space by Lemma 3.43. So there exists $I(u) \in \mathcal{M}_c^2$, unique up to indistinguishability, such that

$$\lim_{n\to\infty} [\![I(u) - I(u^n)]\!]_T = 0. \qquad (4.17)$$

We observe that $I(u)$ does not depend on the approximating sequence, i.e. if v^n is another sequence of simple processes in $\mathbb{L}^2$ approximating u, we have

$$\begin{aligned}[\![I(u^n) - I(v^n)]\!]_T &\le 2\|u^n - v^n\|_{L^2([0,T]\times\Omega)} \\ &\le 2\|u^n - u\|_{L^2([0,T]\times\Omega)} + 2\|u - v^n\|_{L^2([0,T]\times\Omega)} \longrightarrow 0\end{aligned}$$

as $n \to \infty$.

Definition 4.9 *The stochastic integral of $u \in \mathbb{L}^2$ is defined (up to indistinguishability) by* (4.17)*, that is*

$$\int_0^t u_s dW_s := \lim_{n\to\infty} \int_0^t u_s^n dW_s \quad \text{in } \mathscr{M}_c^2,$$

where (u^n) is a sequence of simple processes, approximating u in $\mathbb{L}^2$.

Remark 4.10 Just as in classical functional analysis it is common practice to identify functions that are equal almost everywhere (cf., for example, Brezis [62] Chapter 4) in what follows *we will identify indistinguishable stochastic processes.* □

The following result is the natural extension of Theorem 4.5.

Theorem 4.11 *For every $u, v \in \mathbb{L}^2$, $\alpha \in \mathbb{R}$ and $0 \le a < b < c$, we have:*

(1) linearity:

$$\int_0^a (\alpha u_t + \beta v_t) dW_t = \alpha \int_0^a u_t dW_t + \beta \int_0^a v_t dW_t;$$

(2) additivity:

$$\int_a^c u_t dW_t = \int_a^b u_t dW_t + \int_b^c u_t dW_t;$$

(3) null expectation:

$$E\left[\int_a^b u_t dW_t \mid \mathcal{F}_a\right] = 0,$$

and also

$$E\left[\int_a^b u_t dW_t \int_b^c v_t dW_t \mid \mathcal{F}_a\right] = 0;$$

(4) Itô isometry:

$$E\left[\int_a^b u_t dW_t \int_a^b v_t dW_t \mid \mathcal{F}_a\right] = E\left[\int_a^b u_t v_t dt \mid \mathcal{F}_a\right];$$

(5) the process

$$X_t = \int_0^t u_s dW_s, \qquad t \in [0,T], \tag{4.18}$$

belongs to the space $\mathscr{M}_c^2$ and we have

$$[\![X]\!]_T^2 \le 4E\left[\int_0^T u_t^2 dt\right]. \tag{4.19}$$

As in Remark 4.6 the "non-conditional versions" of the identities in (3) and (4) hold.

Proof. The theorem can be proved by taking the limit in the analogous relations that hold for the integral of simple stochastic processes: the details are left as an exercise. □

Remark 4.12 An immediate but important consequence of the estimate (4.19) is that *if $u, v \in \mathbb{L}^2$ are $(m \otimes P)$-equivalent (or, in particular, if they are modifications) then their stochastic integrals coincide.* This is a fundamental consistency property of the integral (recall Example 3.27). The converse is true as well, by Corollary 4.13 below. □

Corollary 4.13 *If $u \in \mathbb{L}^2$ and for a fixed positive T we have*

$$\int_0^T u_t dW_t = 0,$$

then u is $(m \otimes P)$-equivalent to the null process on $[0,T] \times \Omega$, that is

$$\{(t,\omega) \in [0,T] \times \Omega \mid u_t(\omega) \neq 0\}$$

has null $(m \otimes P)$-measure.

Proof. The thesis follows from Itô isometry, since we have

$$0 = E\left[\left(\int_0^T u_t dW_t\right)^2\right] = E\left[\int_0^T u_t^2 dt\right]. \qquad \square$$

We wish to point out that the stochastic integral is not defined pathwise and the value of the integral in $\omega \in \Omega$ does not only depend on the paths $u(\omega)$ and $W(\omega)$ but on the entire processes u and W. For this reason the following "identity principle" for the stochastic integral will be useful later on:

Corollary 4.14 *Let $F \in \mathcal{F}$ and let $u, v \in \mathbb{L}^2$ be modifications on F, i.e. $u_t(\omega) = v_t(\omega)$ for almost all $\omega \in F$ and for every $t \in [0,T]$. If*

$$X_t = \int_0^t u_s dW_s, \qquad Y_t = \int_0^t v_s dW_s,$$

then X and Y are indistinguishable on F.

Proof. Let us consider the approximation by simple processes u^n, v^n in $\mathbb{L}^2$ defined in (4.14). By construction u^n and v^n are modifications on F for every n. Hence it follows directly that, if

$$X_t^n = \int_0^t u_s^n dW_s, \qquad Y_t^n = \int_0^t v_s^n dW_s,$$

then X^n and Y^n are modifications on F for every n.

Now, for fixed $t \in]0, T]$, we have that X_t^n, Y_t^n converge in $L^2(\Omega, P)$-norm (and pointwise a.s. after taking a subsequence) to X_t and Y_t, respectively. Therefore $X_t = Y_t$ a.s. in F and this proves that they are modifications in F. The claim follows from Proposition 3.25, since X and Y are continuous processes. □

Example 4.15 Let us consider a process of the form

$$S_t = S_0 + \int_0^t \mu(s)ds + \int_0^t \sigma(s)\, dW_s$$

where $S_0 \in \mathbb{R}$ and $\mu, \sigma \in L^2([0,T])$ are *deterministic functions*. By the previous theorem, we have

$$E[S_t] = S_0 + \int_0^t \mu(s)ds$$

and

$$\mathrm{var}(S_t) = E\left[\left(S_t - S_0 - \int_0^t \mu(s)ds\right)^2\right] =$$

(by Itô isometry)

$$= \int_0^t \sigma(s)^2 ds.$$

We will see later on that S_t *has normal distribution*: we shall prove this stronger result only after proving the Itô formula (cf. Proposition 5.13). □

Exercise 4.16 Under the hypotheses of Theorem 4.11, prove that, for every σ-algebra $\mathcal{G} \subseteq \mathcal{F}_a$, we have

$$E\left[\int_a^b u_t dW_t \mid \mathcal{G}\right] = \int_a^b E[u_t \mid \mathcal{G}]\, dW_t.$$

4.3.1 Itô and Riemann-Stieltjes integral

In this section we show that, in the case of continuous processes, the stochastic integral is the limit of Riemann sums and so it is the natural extension of the Riemann-Stieltjes integral.

Definition 4.17 *A process X is called L^2-continuous at t_0 if*

$$\lim_{t\to t_0} E\left[(X_t - X_{t_0})^2\right] = 0.$$

Example 4.18 Given $u \in \mathbb{L}^2$, the process

$$X_t = \int_0^t u_s dW_s, \qquad t \geq 0,$$

is L^2-continuous at every point. Indeed, if $t > t_0$,

$$E\left[(X_t - X_{t_0})^2\right] = E\left[\left(\int_{t_0}^t u_s dW_s\right)^2\right] =$$

(by Itô isometry)

$$= \int_{t_0}^t E\left[u_s^2\right] ds \longrightarrow 0, \qquad \text{as } t \to t_0,$$

by Lebesgue's dominated convergence theorem and the case $t < t_0$ is analogous. In particular every Brownian motion is L^2-continuous. □

Example 4.19 Let X be a continuous process such that $|X_t| \leq Y$ a.s. with $Y \in L^2(\Omega)$. Then, as an immediate consequence of the dominated convergence theorem, the process X is L^2-continuous at any point. In particular, if X is continuous and f is a bounded continuous function, then $f(X)$ is L^2-continuous. □

Proposition 4.20 *Let $u \in \mathbb{L}^2$ be an L^2-continuous process on $[0, T]$. If we put*

$$u^{(\varsigma)} = \sum_{k=1}^N u_{t_{k-1}} \mathbb{1}_{]t_{k-1},t_k]},$$

where $\varsigma = \{t_0, t_1, \dots, t_N\}$ is a partition of $[0, T]$, then $u^{(\varsigma)}$ is a simple process in $\mathbb{L}^2$ and we have

$$\lim_{|\varsigma|\to 0+} u^{(\varsigma)} = u, \qquad \text{in } L^2([0,T]\times\Omega). \tag{4.20}$$

Proof. For every $\varepsilon > 0$, there exists[2] $\delta_\varepsilon > 0$ such that, if $|\varsigma| < \delta_\varepsilon$, then we have

$$\int_0^T E\left[\left(u_t - u_t^{(\varsigma)}\right)^2\right] dt = \sum_{k=1}^N \int_{t_{k-1}}^{t_k} E\left[(u_t - u_{t_{k-1}})^2\right] dt \leq \varepsilon T. \qquad \square$$

---

[2] By the Heine-Cantor theorem, if X is L^2-continuous on the compact set $[0, T]$, then it is also uniformly L^2-continuous.

Remark 4.21 Proposition 4.20 states that $u^{(\varsigma)}$ is a simple stochastic process in $\mathbb{L}^2$ approximating u in $L^2([0,T] \times \Omega)$ for $|\varsigma| \to 0^+$. Then by definition we have

$$\lim_{|\varsigma|\to 0^+} \int_0^T u_t^{(\varsigma)} dW_t = \int_0^T u_t dW_t, \qquad \text{in } \mathscr{M}_c^2,$$

or equivalently

$$\lim_{|\varsigma|\to 0^+} \sum_{k=1}^{N} u_{t_{k-1}} (W_{t_k} - W_{t_{k-1}}) = \int_0^T u_t dW_t, \qquad \text{in } \mathscr{M}_c^2. \tag{4.21}$$

In this sense the Itô integral, being the limit of Riemann-Stieltjes sums as in (4.1), generalizes the Riemann-Stieltjes integral. □

4.3.2 Itô integral and stopping times

Some properties of the stochastic integral are similar to those of the Lebesgue integral, even though in general it is necessary to be careful: for example, let us consider the following (false) equality

$$X \int_0^T u_t dW_t = \int_0^T X u_t dW_t,$$

where $u \in \mathbb{L}^2$ and X is a $\mathcal{F}_{t_0}$-measurable random variable for some $t_0 > 0$. Although X is *constant with respect to the variable* t, the member on the right-hand side of the equality does not make sense since the integrand $Xu \notin \mathbb{L}^2$ and is not in general adapted. However, the equality

$$X \int_{t_0}^T u_t dW_t = \int_{t_0}^T X u_t dW_t \tag{4.22}$$

holds true, since (4.22) is true for every simple process u in $\mathbb{L}^2$ and can be proved in general by approximation.

The following result contains the definition of stochastic integral with *a random time as upper integration limit:* the statement might seem tautological but, in the light of the previous remark, it requires a rigorous proof.

Proposition 4.22 *Given $u \in \mathbb{L}^2(\mathcal{F}_t)$, we set*

$$X_t = \int_0^t u_s dW_s, \qquad t \in [0,T]. \tag{4.23}$$

If τ is an $(\mathcal{F}_t)$-stopping time such that $0 \le \tau \le T$ a.s. then $\left(u_t \mathbb{1}_{\{t \le \tau\}}\right) \in \mathbb{L}^2$ and

$$X_\tau = \int_0^\tau u_s dW_s = \int_0^T u_s \mathbb{1}_{\{s \le \tau\}} dW_s \quad a.s. \tag{4.24}$$

Proof. It is clear that, by definition of stopping time, the process $\left(u_t \mathbb{1}_{\{t\le\tau\}}\right)$ belongs to $\mathbb{L}^2$ and in particular is adapted. We put

$$Y = \int_0^T u_s \mathbb{1}_{\{s\le\tau\}} dW_s,$$

and we prove that

$$X_\tau = Y \quad \text{a.s.}$$

First of all, we consider the case

$$\tau = \sum_{k=1}^n t_k \mathbb{1}_{F_k} \tag{4.25}$$

with $0 < t_1 < \cdots < t_n = T$ and $F_k \in \mathcal{F}_{t_k}$ disjoint events such that

$$F := \bigcup_{k=1}^n F_k \in \mathcal{F}_0.$$

It is apparent that τ is a stopping time. Given X in (4.23), we have $X_\tau = 0$ on $\Omega \setminus F$ and

$$X_\tau = \int_0^T u_s dW_s - \int_{t_k}^T u_s dW_s, \qquad \text{on } F_k,$$

or, in other terms,

$$X_\tau = \mathbb{1}_F \int_0^T u_s dW_s - \sum_{k=1}^n \mathbb{1}_{F_k} \int_{t_k}^T u_s dW_s.$$

On the other hand, we have

$$Y = \int_0^T u_s \left(1 - \mathbb{1}_{\{s>\tau\}}\right) dW_s =$$

(by linearity)

$$= \int_0^T u_s dW_s - \int_0^T u_s \left(\mathbb{1}_{\Omega\setminus F} + \sum_{k=1}^n \mathbb{1}_{F_k} \mathbb{1}_{\{s>t_k\}} \right) dW_s$$

$$= \mathbb{1}_F \int_0^T u_s dW_s - \sum_{k=1}^n \int_{t_k}^T u_s \mathbb{1}_{F_k} dW_s,$$

and we conclude that $X_\tau = Y$ by (4.22). To use (4.22) we have written the integral from 0 to t as the difference of the integral from 0 to T and the integral from t to T.

In the case of a general stopping time τ, we adapt the approximation result of Remark 3.55 and we consider the following decreasing sequence (τ_n) of stopping times of the form (4.25):

$$\tau_n = \sum_{k=0}^{2^n} \frac{T(k+1)}{2^n} \mathbb{1}_{\left\{\frac{Tk}{2^n} < \tau \le \frac{T(k+1)}{2^n}\right\}}.$$

We have that (τ_n) converges to τ a.s. and, by continuity, X_{τ_n} converges to X_τ a.s. Further, if we put

$$Y^n = \int_0^t u_s \mathbb{1}_{\{s \le \tau_n\}} dW_s,$$

by the dominated convergence theorem, we have that Y^n converges to Y in $L^2(\Omega, P)$ and this is enough to conclude. □

The following proposition extends the usual properties of the Itô integral when the integration limit is a stopping time.

Corollary 4.23 *Let $t_0 \in [0, T[$ and $\tau \in [t_0, T]$ be a stopping time. If $u, v \in \mathbb{L}^2$ then we have*

$$E\left[\int_{t_0}^{\tau} u_t dW_t \mid \mathcal{F}_{t_0}\right] = 0,$$

$$E\left[\int_{t_0}^{\tau} u_t dW_t \int_{\tau}^{T} v_t dW_t \mid \mathcal{F}_{t_0}\right] = 0,$$

$$E\left[\int_{t_0}^{\tau} u_t dW_t \int_{t_0}^{\tau} v_t dW_t \mid \mathcal{F}_{t_0}\right] = E\left[\int_{t_0}^{\tau} u_t v_t dt \mid \mathcal{F}_{t_0}\right].$$

Proof. By (4.24) we have

$$\int_{t_0}^{\tau} u_t dW_t = \int_{t_0}^{T} u_t \mathbb{1}_{\{t \le \tau\}} dW_t$$

with $u_t \mathbb{1}_{\{t \le \tau\}} \in \mathbb{L}^2$ and so the claim follows from Theorem 4.11. □

4.3.3 Quadratic variation process

In Theorem 3.74 we computed the quadratic variation of a Brownian motion W, showing that

$$\langle W \rangle_t = t, \quad t \ge 0.$$

On the other hand, in Proposition 3.37 we proved that

$$M_t = W_t^2 - \langle W \rangle_t \tag{4.26}$$

is a martingale. Since W_t^2 is a sub-martingale (cf. Remark 3.36), this result is in line with the Doob's decomposition Theorem A.119 that states that, in discrete time, any sub-martingale can be decomposed as the sum of a martingale M and an increasing predictable process A with null initial value. Thus, in the Brownian framework, (4.26) can be interpreted as a Doob-type decomposition where the role of the process A is played by the quadratic variation $\langle W \rangle$.

In this section we aim at getting similar results for the stochastic integral process

$$X_t = \int_0^t u_s dW_s, \tag{4.27}$$

with $u \in \mathbb{L}^2$. We already proved that $X \in \mathscr{M}_c^2$. Now we introduce the quadratic variation process $\langle X \rangle$ and show that $X^2 - \langle X \rangle$ is a martingale.

Proposition 4.24 *Let X be as in (4.27) with $u \in \mathbb{L}^2$. Then for any $t > 0$, there exists the limit*

$$\lim_{\substack{|\varsigma| \to 0 \\ \varsigma \in \mathcal{P}_{[0,t]}}} \sum_{k=1}^{N} \left| X_{t_k} - X_{t_{k-1}} \right|^2 = \int_0^t u_s^2 ds \qquad \text{in } L^2(\Omega, P). \tag{4.28}$$

We set

$$\langle X \rangle_t = \int_0^t u_s^2 ds, \qquad t \in [0, T], \tag{4.29}$$

and we say that $\langle X \rangle$ is the quadratic variation process of X. We have that $X^2 - \langle X \rangle$ is a martingale.

Proof. If u is a simple $\mathbb{L}^2$-process, (4.28) can be proved by proceeding as in Theorem 3.74. In general the claim follows approximating X by integrals of simple processes.

Next we verify that $X^2 - \langle X \rangle$ is a martingale. For every $0 \le s < t$ we have

$$E\left[X_t^2 - \langle X \rangle_t \mid \mathcal{F}_s\right] = E\left[(X_t - X_s)^2 + 2X_s(X_t - X_s) + X_s^2 - \langle X \rangle_t \mid \mathcal{F}_s\right] =$$

(by (3) in Theorem 4.11)

$$= E\left[(X_t - X_s)^2 - \langle X \rangle_t \mid \mathcal{F}_s\right] + X_s^2 =$$

(by Itô isometry)

$$= E\left[\int_s^t u_\tau^2 d\tau - \langle X \rangle_t \mid \mathcal{F}_s\right] + M_s^2 = M_s^2 - \langle X \rangle_s. \qquad \square$$

Remark 4.25 Since the L^2-convergence implies convergence in probability (cf. Theorem A.136), the limit in (4.28) converges in probability as well. Moreover, by Theorem A.136, we also have that, for any sequence of partitions (ς_n) of $[0,t]$, with mesh converging to zero, there exists a subsequence (ς_{k_n}) such that

$$\lim_{n\to\infty} V_t^{(2)}(X, \varsigma_{k_n}) = \int_0^t u_s^2 ds \qquad \text{a.s.}$$

where $V_t^{(2)}$ is the quadratic variation of Definition 3.72. □

Proposition 4.24 is a particular case of the classical Doob-Meyer decomposition theorem which we state below: the interested reader can find an organic presentation of the topic, for example, in Chapter 1.4 of Karatzas-Shreve [201].

In what follows, $(\Omega, \mathcal{F}, P, (\mathcal{F}_t))$ is a filtered probability space verifying the usual hypotheses. We recall that a process A is *increasing* if almost all the paths of A are increasing functions. Moreover if $M \in \mathscr{M}^2$ then, by Jensen's inequality, $|M|^2$ is a sub-martingale.

Theorem 4.26 (Doob-Meyer decomposition theorem) *For every $M = (M_t)_{t\in[0,T]} \in \mathscr{M}_c^2(\mathcal{F}_t)$ there exists a unique (up to indistinguishability) increasing continuous process A such that $A_0 = 0$ a.s. and $|M|^2 - A$ is a $\mathcal{F}_t$-martingale. We call A the quadratic-variation process of M and we write $A_t = \langle M\rangle_t$. Moreover, for any $t \le T$ we have*

$$A_t = \lim_{\substack{|\varsigma|\to 0\\ \varsigma\in\mathcal{P}_{[0,t]}}} V_t^{(2)}(M, \varsigma) \tag{4.30}$$

in probability.

We explicitly remark that the general definition of quadratic variation agrees with that given in (4.29): indeed, for X as in (4.27), $\langle X\rangle$ in (4.29) is an increasing continuous process such that $\langle X\rangle_0 = 0$ a.s. and $|X|^2 - \langle X\rangle$ is a martingale (cf. Proposition 4.24).

It is remarkable that $\langle M\rangle$ in (4.30) does not depend on the filtration that we consider: in the case $(\mathcal{F}_t)$ is the Brownian filtration, the martingale representation Theorem 10.11 states that *any square-integrable $(\mathcal{F}_t)$-martingale can be represented as a stochastic integral of the form* (4.29)*;* as a consequence, in this particular case Theorem 4.26 follows by Proposition 4.24.

The proof of Theorem 4.26 is based on a discrete approximation procedure: we observe that, if (M_n) is a real discrete martingale, then the process (A_n) defined by $A_0 = 0$ and

$$A_n = \sum_{k=1}^{n} (M_k - M_{k-1})^2, \qquad n \ge 1,$$

is increasing and such that $M^2 - A$ is a martingale. Indeed

$$E\left[M_{n+1}^2 - A_{n+1} \mid \mathcal{F}_n\right] = M_n^2 - A_n$$

if and only if

$$E\left[M_{n+1}^2 - (M_{n+1} - M_n)^2 \mid \mathcal{F}_n\right] = M_n^2,$$

hence the claim.

The proof of (4.30) is similar to that of Theorem 3.74 and it is based on the fact that the mean of the product of increments of a martingale M over non-overlapping intervals is equal to zero[3]. More precisely, in the scalar case, for $0 \le s < t \le u < v$ we have

$$E\left[(M_v - M_u)(M_t - M_s)\right] = E\left[E\left[(M_v - M_u) \mid \mathcal{F}_u\right](M_t - M_s)\right] = 0. \quad (4.31)$$

Formula (4.31) is very simple yet useful and meaningful: for instance, (4.31) is one of the key ingredients in the construction of the stochastic integral for a general martingale.

Given $M \in \mathscr{M}_c^2$, as a consequence of Theorem 4.26, we also have that

$$E\left[|M_t|^2 - |M_s|^2 \mid \mathcal{F}_s\right] = E\left[\langle M\rangle_t - \langle M\rangle_s \mid \mathcal{F}_s\right], \qquad s \le t, \qquad (4.32)$$

that follows from the fact that $|M|^2 - \langle M\rangle$ is a martingale.

4.3.4 Martingales with bounded variation

As a consequence of the Doob-Meyer Theorem 4.26 we have that if a martingale $M \in \mathscr{M}_c^2$ has bounded variation, then it is indistinguishable from the null process: this means that almost all the paths of a non-trivial martingale M are irregular in the sense that they do not have bounded variation. More precisely, we have:

Proposition 4.27 *Let $M \in \mathscr{M}_c^2$. For almost any ω such that $\langle M\rangle_T(\omega) > 0$, the function $t \mapsto M_t(\omega)$ does not have bounded variation over $[0,T]$. Moreover, for almost any ω such that $\langle M\rangle_T(\omega) = 0$ the function $t \mapsto M_t(\omega)$ is null.*

Proof. By Theorem 4.26 there exists a sequence of partitions (ς_n) in $\mathcal{P}_{[0,T]}$, with mesh converging to zero, such that

$$\langle M\rangle_T = \lim_{n\to\infty} V_T^{(2)}(M, \varsigma_n) \qquad \text{a.s.}$$

Thus, by Proposition 3.73, the condition $\langle M\rangle_T(\omega) > 0$ is a.s. incompatible with the fact that $M(\omega)$ has bounded variation.

Concerning the second part of the claim, we set

$$\tau = \inf\{t \mid \langle M\rangle_t > 0\} \wedge T.$$

By Theorem 3.52, τ is a stopping time and since $M^2 - \langle M\rangle$ is a martingale, then, by Theorem 3.58, also[4]

$$M_{t\wedge\tau}^2 - \langle M\rangle_{t\wedge\tau} = M_{t\wedge\tau}^2$$

---

[3] For further details see, for example, Karatzas-Shreve [201], Chapter 1.5.

[4] The equality follows from the fact that $\langle M\rangle_t = 0$ for $t \le \tau$.

is a martingale. Therefore

$$E\left[M^2_{T\wedge\tau}\right] = E\left[M^2_0\right] = 0.$$

Consequently, by Doob's inequality, $(M^2_{t\wedge\tau})$ has a.s. null paths and the claim follows from the fact that $M = (M^2_{t\wedge\tau})_{t\in[0,T]}$ over $\{\langle M\rangle_T = 0\}$. □

4.3.5 Co-variation process

For the sake of simplicity, in this section we consider only real-valued processes. We remark that, by Theorem 4.26, for any $X, Y \in \mathscr{M}^2_c$ the processes

$$(X+Y)^2 - \langle X+Y\rangle, \qquad (X-Y)^2 - \langle X-Y\rangle$$

are martingales and therefore so is the following process, obtained as their difference,

$$4XY - (\langle X+Y\rangle - \langle X-Y\rangle).$$

This motivates the following:

Definition 4.28 *For any $X, Y \in \mathscr{M}^2_c$, the process*

$$\langle X, Y\rangle := \frac{1}{4}(\langle X+Y\rangle - \langle X-Y\rangle)$$

is called co-variation process of X and Y.

By Theorem 4.26, $\langle X, Y\rangle$ is the unique (up to indistinguishability) continuous adapted process with bounded variation[5] such that $\langle X, Y\rangle_0 = 0$ a.s. and $XY - \langle X, Y\rangle$ is a continuous martingale. Moreover, for any $t \le T$ we have

$$\langle X, Y\rangle_t = \lim_{\substack{|\varsigma|\to 0\\ \varsigma\in\mathcal{P}_{[0,t]}}} \sum_{k=1}^{N} (X_{t_k} - X_{t_{k-1}})(Y_{t_k} - Y_{t_{k-1}})$$

in probability. Note that $\langle X, X\rangle = \langle X\rangle$ and the following identity (that extends (4.32)) holds:

$$\begin{aligned} E\left[(X_t - X_s)(Y_t - Y_s) \mid \mathcal{F}_s\right] &= E\left[X_tY_t - X_sY_s \mid \mathcal{F}_s\right] \\ &= E\left[\langle X, Y\rangle_t - \langle X, Y\rangle_s \mid \mathcal{F}_s\right], \end{aligned}$$

for every $X, Y \in \mathscr{M}^2_c$ and $0 \le s < t$. In the following proposition we collect other straightforward properties of the co-variation process.

[5] A process has bounded variation if almost all its paths are functions with bounded variation.

Proposition 4.29 *The co-variation $\langle\cdot,\cdot\rangle$ is a bi-linear form in $\mathcal{M}_c^2$: for every $X, Y, Z \in \mathcal{M}_c^2$, $\lambda, \mu \in \mathbb{R}$ we have*

i) $\langle X, Y\rangle = \langle Y, X\rangle$;
ii) $\langle \lambda X + \mu Y, Z\rangle = \lambda\langle X, Z\rangle + \mu\langle Y, Z\rangle$;
iii) $|\langle X, Y\rangle|^2 \leq \langle X\rangle\langle Y\rangle$.

Example 4.30 A particularly important case is when

$$X_t = \int_0^t u_s dW_s, \qquad Y_t = \int_0^t v_s dW_s,$$

with $u, v \in \mathbb{L}^2$. Then, proceeding as in Proposition 4.24, we can show that

$$X_t Y_t - \int_0^t u_s v_s dW_s$$

is a martingale and therefore[6]

$$\langle X, Y\rangle_t = \int_0^t u_s v_s dW_s, \qquad t \in [0, T], \tag{4.33}$$

is the quadratic variation process of X, Y. Proceeding as in Theorem 3.74, we can also directly prove that

$$\int_0^t u_s v_s dW_s = \lim_{\substack{|\varsigma|\to 0\\ \varsigma\in\mathcal{P}_{[0,t]}}} \sum_{k=1}^{N} (X_{t_k} - X_{t_{k-1}})(Y_{t_k} - Y_{t_{k-1}})$$

where the limit is in $L^2(\Omega, P)$-norm and therefore also in probability. □

Next we recall that, by Proposition 3.79, if X is a continuous process and Y is a process with bounded variation then

$$\lim_{\substack{|\varsigma|\to 0\\ \varsigma\in\mathcal{P}_{[0,t]}}} \sum_{k=1}^{N} \left(X_{t_k}(\omega) - X_{t_{k-1}}(\omega)\right)\left(Y_{t_k}(\omega) - Y_{t_{k-1}}(\omega)\right) = 0$$

for any $t \leq T$ and $\omega \in \Omega$. Hence, if Z and V are continuous processes with bounded variation and $X, Y \in \mathcal{M}_c^2$, we formally have

$$\langle X + Z, Y + V\rangle = \langle X, Y\rangle + \underbrace{\langle Z, Y + V\rangle + \langle X + Z, V\rangle}_{=0}.$$

Therefore it seems natural to extend Definition 4.28 as follows:

[6] Note also that the process

$$I_t = \int_0^t u_s v_s dW_s, \qquad t \in [0, T],$$

has bounded variation in view of Example 3.60-iii) and $I_0 = 0$.

Definition 4.31 *Let Z and V be continuous processes with bounded variation and $X, Y \in \mathscr{M}^2_c$. We call*

$$\langle X + Z, Y + V\rangle := \langle X, Y\rangle$$

the co-variation process of $X + Z$ and $Y + V$.

Example 4.32 We go back to Example 4.15 and consider

$$S_t = S_0 + \int_0^t \mu(s)ds + \int_0^t \sigma(s)\, dW_s$$

with $\mu, \sigma \in L^2([0,T])$ *deterministic functions.* We proved that

$$\mathrm{var}(S_t) = \int_0^t \sigma(s)^2 ds.$$

Now we observe that the process $S_0 + \int_0^t \mu(s)ds$ is continuous and has bounded variation by Example 3.60-iii). Therefore, according to Definition 4.31 and formula (4.29), we have

$$\langle S\rangle_t = \mathrm{var}(S_t), \qquad t \in [0,T],$$

i.e. the quadratic variation process is deterministic and equal to the variance function. □

4.4 Integral of $\mathbb{L}^2_{\rm loc}$-processes

In this paragraph we further extend the class of processes for which the stochastic integral is defined. This generalization is necessary because simple processes like $f(W_t)$, where f is a continuous function, do not generally belong to $\mathbb{L}^2$: indeed we have

$$E\left[\int_0^T f(W_t)dt\right] = \frac{1}{\sqrt{2\pi t}}\int_0^T \int_{\mathbb{R}} \exp\left(-\frac{x^2}{2t}\right) f(x)dxdt.$$

Then, for example, $f(W_t) \notin \mathbb{L}^2$ if $f(x) = e^{x^4}$. Luckily it is not difficult to extend the construction of the Itô integral to a class of progressively measurable processes that verify an integrability condition that is weaker than in Definition 4.1-ii) and that is sufficiently general to handle most applications. However, when this generalization is made, some important properties are lost: in particular the stochastic integral is not in general a martingale.

Definition 4.33 *We denote by $\mathbb{L}^2_{\rm loc}$ the family of processes $(u_t)_{t\in[0,T]}$ that are progressively measurable with respect to the filtration $(\mathcal{F}_t)_{t\in[0,T]}$ and such that*

$$\int_0^T u_t^2\, dt < \infty \quad a.s. \tag{4.34}$$

Example 4.34 Every stochastic process that is progressively measurable and has a.s. continuous paths belongs to $\mathbb{L}^2_{\text{loc}}$. In particular $\exp(W_t^4)$, where W is a Brownian motion, belongs to $\mathbb{L}^2_{\text{loc}}$. □

It is interesting to note that the space $\mathbb{L}^2_{\text{loc}}$ is invariant with respect to changes of equivalent probability measures: if (4.34) holds and $Q \sim P$ then we have of course

$$\int_0^T u_t^2 \, dt < \infty, \qquad Q\text{-a.s.}$$

On the contrary, the space $\mathbb{L}^2$ depends on the fixed probability measure.

Now we define the stochastic integral $u \in \mathbb{L}^2_{\text{loc}}$ step by step: the rest of the paragraph can be skipped on first reading.

I) Given $u \in \mathbb{L}^2_{\text{loc}}$, the process[7]

$$A_t = \int_0^t u_s^2 ds, \qquad t \in [0, T],$$

is continuous and adapted to the filtration. Indeed it is enough to observe that u can be approximated pointwise by a sequence of simple and adapted processes.

II) For every $n \in \mathbb{N}$ we put

$$\tau_n = \inf\{t \in [0, T] \mid A_t \geq n\} \wedge T.$$

By Theorem 3.52, τ_n is a stopping time and

$$\tau_n \nearrow T \qquad \text{a.s. as } n \to \infty.$$

We have

$$F_n := \{\tau_n = T\} = \{A_T \leq n\}, \tag{4.35}$$

and so, since $u \in \mathbb{L}^2_{\text{loc}}$,

$$\bigcup_{n \in \mathbb{N}} F_n = \Omega \setminus N, \qquad N \in \mathcal{N}. \tag{4.36}$$

III) We put

$$u_t^n = u_t \mathbb{1}_{\{t \leq \tau_n\}}, \qquad t \in [0, T],$$

and note that $u^n \in \mathbb{L}^2$ since

$$E\left[\int_0^T (u_t^n)^2 \, dt\right] = E\left[\int_0^{\tau_n} u_t^2 dt\right] \leq n.$$

Therefore the process

$$X_t^n = \int_0^t u_t^n dW_t, \qquad t \in [0, T] \tag{4.37}$$

is well-defined and $X^n \in \mathcal{M}_c^2$.

[7] We put $A(\omega) = 0$ if $u(\omega) \notin L^2(0, T)$.

IV) For every $n, h \in \mathbb{N}$, we have $u^n = u^{n+h} = u$ on F_n in (4.35). So, by Corollary 4.14, the processes X^n and X^{n+h} are indistinguishable on F_n. Recalling that (F_n) is an increasing sequence for which (4.36) holds, the following definition is well-posed.

Definition 4.35 *Given $u \in \mathbb{L}^2_{\text{loc}}$, let F_n and X^n be defined as in (4.35) and (4.37), respectively. Then the stochastic integral of u is the continuous and $\mathcal{F}_t$-adapted stochastic process X that is indistinguishable from X^n on F_n, for every $n \in \mathbb{N}$. We write*

$$X_t = \int_0^t u_s dW_s, \qquad t \in [0, T].$$

Note that, by construction, we have

$$X_t = \lim_{n\to\infty} \int_0^t u_t^n dW_t, \qquad t \in [0, T], \text{ a.s.} \tag{4.38}$$

Remark 4.36 Given $p \geq 1$, we denote by $\mathbb{L}^p_{\text{loc}}$ the family of progressively measurable processes $(u_t)_{t\in[0,T]}$ such that

$$\int_0^T |u_t|^p dt < \infty \quad \text{a.s.} \tag{4.39}$$

By Hölder's inequality we have

$$\mathbb{L}^p_{\text{loc}} \subseteq \mathbb{L}^q_{\text{loc}}, \qquad p \geq q \geq 1,$$

and in particular $\mathbb{L}^2_{\text{loc}} \subseteq \mathbb{L}^1_{\text{loc}}$. Since $\mathbb{L}^p_{\text{loc}}$ depends on the filtration $(\mathcal{F}_t)$, whenever it is necessary we write more explicitly $\mathbb{L}^p_{\text{loc}}(\mathcal{F}_t)$. The space $\mathbb{L}^2_{\text{loc}}$ is the natural setting for the definition of stochastic integral: we refer to Steele [315], Paragraph 7.3, for an interesting discussion about the impossibility of defining the Itô integral of $u \in \mathbb{L}^p_{\text{loc}}$ for $1 \leq p < 2$. □

4.4.1 Local martingales

In general, the stochastic integral of a process $u \in \mathbb{L}^2_{\text{loc}}$ *is not a martingale:* however, in the sense that we are going to explain, it is not "far off to be a martingale".

Definition 4.37 *A process $M = (M_t)_{t\in[0,T]}$ is a $\mathcal{F}_t$-local martingale if there exists an increasing sequence (τ_n) of $\mathcal{F}_t$-stopping times, called localizing sequence for M, such that*

$$\lim_{n\to\infty} \tau_n = T \quad a.s. \tag{4.40}$$

and, for every $n \in \mathbb{N}$, the stochastic process $M_{t\wedge\tau_n}$ is a $\mathcal{F}_t$-martingale. We denote by $\mathscr{M}_{\text{c,loc}}$ the space of continuous local martingales such that $M_0 = 0$ a.s.

To put it simply, a local martingale is a stochastic process that can be approximated by a sequence of true martingales. Sometimes, when we want to emphasize the fact that a process M is a true martingale and not simply a local martingale, we say that M is a *strict martingale*. An interesting example of a local martingale that is not a strict martingale is given in Example 9.34.

By definition, we have that

$$M_{s\wedge\tau_n} = E\left[M_{t\wedge\tau_n} \mid \mathcal{F}_s\right], \qquad 0 \le s \le t \le T, \tag{4.41}$$

and if M is continuous, since $\tau_n \to T$ a.s., we have

$$\lim_{n\to\infty} M_{t\wedge\tau_n} = M_t \quad \text{a.s.}$$

Consequently, whenever we can take the limit inside the conditional expectation in (4.41), we have that M is a strict martingale: as particular cases, see Propositions 4.39 and 4.40 below.

Clearly every martingale is also a local martingale: it is enough to choose $\tau_n = T$ for every n. Further, we remark that every local martingale admits a right-continuous modification: indeed it is enough to note that, by Theorem 3.41, this holds true for the stopped processes $M_{t\wedge\tau_n}$. In what follows we shall always consider the right-continuous version of every local martingale.

Remark 4.38 Every continuous local martingale M admits an approximating sequence of continuous and *bounded* martingales. Indeed let (τ_n) be a localizing sequence for M and let us put

$$\sigma_n = \inf\{t \in [0,T] \mid |M_t| \ge n\} \wedge T, \qquad n \in \mathbb{N}.$$

Since M is continuous we have that σ_n satisfies (4.40) and also $(\tau_n \wedge \sigma_n)$ is a localizing sequence for M: indeed

$$M_{t\wedge(\tau_n\wedge\sigma_n)} = M_{(t\wedge\tau_n)\wedge\sigma_n}$$

and so, by Doob's Theorem 3.58, $M_t^n := M_{t\wedge(\tau_n\wedge\sigma_n)}$ is a bounded martingale such that

$$|M_t^n| \le n, \qquad t \in [0,T].$$

□

We present now some simple properties of continuous local martingales.

Proposition 4.39 *If $M \in \mathscr{M}_{c,loc}$ and*

$$\sup_{t\in[0,T]} |M_t| \in L^1(\Omega, P),$$

then M is a martingale. In particular every bounded[8] *$M \in \mathscr{M}_{c,loc}$ is a martingale.*

[8] There exists a constant c such that $|M_t| \le c$ a.s. for every $t \in [0,T]$.

Proof. The claim follows directly from (4.41), applying the dominated convergence theorem for conditional expectation. □

Proposition 4.40 *Every continuous non-negative local martingale M is also a super-martingale. Further, if*

$$E[M_T] = E[M_0] \tag{4.42}$$

then $(M_t)_{0\le t\le T}$ is a martingale.

Proof. Applying Fatou's lemma for conditional expectation to (4.41), we get

$$M_s \ge E[M_t \mid \mathcal{F}_s], \qquad 0 \le s \le t \le T, \tag{4.43}$$

and this proves the first part of the claim.

By taking the expectation in the previous relation we get

$$E[M_0] \ge E[M_t] \ge E[M_T], \qquad 0 \le t \le T.$$

By assumption (4.42), we infer that $E[M_t] = E[M_0]$ for every $t \in [0,T]$. Eventually, by (4.43), if we had $M_s > E[M_t \mid \mathcal{F}_s]$ on an event of strictly positive probability, then we would get a contradiction. □

Proposition 4.41 *If $M \in \mathscr{M}_{c,\text{loc}}$ and τ is a stopping time, then also $M_{t\wedge\tau} \in \mathscr{M}_{c,\text{loc}}$.*

Proof. If (τ_n) is a localizing sequence for M and $X_t = M_{t\wedge\tau}$, we have

$$X_{t\wedge\tau_n} = M_{(t\wedge\tau)\wedge\tau_n} = M_{(t\wedge\tau_n)\wedge\tau}.$$

Consequently, by Theorem 3.58 and since by assumption $M_{t\wedge\tau_n}$ is a continuous martingale, we have that (τ_n) is a localizing sequence for X. □

4.4.2 Localization and quadratic variation

The following theorem states that the stochastic integral of a process $u \in \mathbb{L}^2_{\text{loc}}$ is a continuous local martingale. In the whole section we use the notations

$$X_t = \int_0^t u_s dW_s, \qquad A_t = \int_0^t u_s^2 ds, \qquad t \in [0,T]. \tag{4.44}$$

Theorem 4.42 *We have:*

i) if $u \in \mathbb{L}^2$, then $X \in \mathscr{M}^2_c$;
ii) if $u \in \mathbb{L}^2_{\text{loc}}$, then $X \in \mathscr{M}_{c,\text{loc}}$ and a localizing sequence for X is given by

$$\tau_n = \inf\{t \in [0,T] \mid A_t \ge n\} \wedge T, \qquad n \in \mathbb{N}. \tag{4.45}$$

Proof. We only prove *ii*). We saw at the beginning of Paragraph 4.4 that (τ_n) in (4.45) is an increasing sequence of stopping times such that $\tau_n \to T$ a.s. for $n \to \infty$.

By Definition 4.35, on $F_k = \{A_T \le k\}$ with $k \ge n$, we have

$$X_{t\wedge\tau_n} = \int_0^{t\wedge\tau_n} u_s \mathbb{1}_{\{s\le\tau_k\}} dW_s =$$

(by Proposition 4.22, since $u_s \mathbb{1}_{\{s\le\tau_k\}} \in \mathbb{L}^2$)

$$= \int_0^t u_s \mathbb{1}_{\{s\le\tau_k\}} \mathbb{1}_{\{s\le\tau_n\}} dW_s =$$

(since $n \le k$)

$$= \int_0^t u_s \mathbb{1}_{\{s\le\tau_n\}} dW_s, \qquad \text{on } F_k.$$

By the arbitrariness of k and by (4.36), we get

$$X_{t\wedge\tau_n} = \int_0^t u_s \mathbb{1}_{\{s\le\tau_n\}} dW_s, \qquad t \in [0,T], \text{ a.s.} \tag{4.46}$$

The claim follows from the fact that $u_s \mathbb{1}_{\{s\le\tau_n\}} \in \mathbb{L}^2$ and so $X_{t\wedge\tau_n} \in \mathscr{M}_c^2$ and τ_n is a localizing sequence for X. □

Next we extend Proposition 4.24.

Proposition 4.43 *Given $u \in \mathbb{L}^2_{\text{loc}}$, let X and A be the processes in* (4.44). *Then $X^2 - A$ is a continuous local martingale: A is called quadratic variation process of X and we write $A = \langle X \rangle$.*

Proof. Let us consider the localizing sequence (τ_n) for X defined in Theorem 4.42. We proved that (cf. (4.46))

$$X_{t\wedge\tau_n} = \int_0^t u_s \mathbb{1}_{\{s\le\tau_n\}} dW_s$$

with $u_s \mathbb{1}_{\{s\le\tau_n\}} \in \mathbb{L}^2$. Therefore, by Proposition 4.24, we have that the following process is a martingale:

$$X^2_{t\wedge\tau_n} - \int_0^t u_s^2 \mathbb{1}_{\{s\le\tau_n\}} ds = X^2_{t\wedge\tau_n} - A_{t\wedge\tau_n} = \left(X^2 - A\right)_{t\wedge\tau_n}.$$

Hence $X^2 - A$ is a local martingale and τ_n is a localizing sequence for X. □

Proposition 4.43 has the following extension: for every $X, Y \in \mathscr{M}_{c,\text{loc}}$ there exists a unique (up to indistinguishability) continuous process $\langle X, Y\rangle$ with bounded variation, such that $\langle X, Y\rangle_0 = 0$ a.s. and

$$XY - \langle X, Y\rangle \in \mathscr{M}_{c,\text{loc}}.$$

We call $\langle X, Y\rangle$ the *co-variation process of* X, Y. Note that $\langle X\rangle = \langle X, X\rangle$.

Remark 4.44 If

$$X_t = \int_0^t u_s dW_s, \qquad Y_t = \int_0^t v_s dW_s,$$

with $u, v \in \mathbb{L}^2_{\text{loc}}$, then

$$\langle X, Y \rangle_t = \int_0^t u_s v_s ds. \qquad \square$$

More generally, by analogy with Definition 4.31, we give the following:

Definition 4.45 *Let Z and V be continuous processes with bounded variation and $X, Y \in \mathscr{M}_{\text{c,loc}}$. We call*

$$\langle X + Z, Y + V \rangle := \langle X, Y \rangle \tag{4.47}$$

the co-variation process of $X + Z$ and $Y + V$. In (4.47), $\langle X, Y \rangle$ is the unique (up to indistinguishability) continuous process with bounded variation, such that $\langle X, Y \rangle_0 = 0$ a.s. and $XY - \langle X, Y \rangle \in \mathscr{M}_{\text{c,loc}}$.

Proposition 4.27 can be extended as follows:

Proposition 4.46 *Let $M \in \mathscr{M}_{\text{c,loc}}$. For almost any ω such that $\langle M \rangle_T(\omega) > 0$, the function $t \mapsto M_t(\omega)$ does not have bounded variation over $[0, T]$. Moreover, for almost any ω such that $\langle M \rangle_T(\omega) = 0$ the function $t \mapsto M_t(\omega)$ is null.*

We conclude the paragraph stating[9] a classical result that claims that, for every $M \in \mathscr{M}_{\text{c,loc}}$, the expected values

$$E\left[\langle M \rangle_T^p\right] \qquad \text{and} \qquad E\left[\sup_{t \in [0,T]} |M_t|^{2p}\right]$$

are comparable, for $p > 0$. More precisely, we have

Theorem 4.47 (Burkholder-Davis-Gundy's inequalities) *For any $p > 0$ there exist two positive constants λ_p, Λ_p such that*

$$\lambda_p E\left[\langle M \rangle_\tau^p\right] \leq E\left[\sup_{t \in [0,\tau]} |M_t|^{2p}\right] \leq \Lambda_p E\left[\langle M \rangle_\tau^p\right],$$

for every $M \in \mathscr{M}_{\text{c,loc}}$ and stopping time τ.

As a consequence of Theorem 4.47 we prove a useful criterion to establish whether a stochastic integral of a process in $\mathbb{L}^2_{\text{loc}}$ is a martingale.

[9] For the proof we refer, for example, to Theorem 3.3.28 in [201].

Corollary 4.48 *If $u \in \mathbb{L}^2_{\mathrm{loc}}$ and*

$$E\left[\left(\int_0^T u_t^2 dt\right)^{\frac{1}{2}}\right] < \infty, \tag{4.48}$$

then the process

$$\int_0^t u_s dW_s, \qquad t \in [0,T],$$

is a martingale.

Proof. First of all we observe that, by Hölder's inequality, we have

$$E\left[\left(\int_0^T u_t^2 dt\right)^{\frac{1}{2}}\right] \le E\left[\int_0^T u_t^2 dt\right]^{\frac{1}{2}},$$

and so condition (4.48) is weaker than the integrability condition in the space $\mathbb{L}^2$.

By the second Burkholder-Davis-Gundy's inequality with $p = \frac{1}{2}$ and $\tau = T$, we have

$$E\left[\sup_{t\in[0,T]}\left|\int_0^t u_s dW_s\right|\right] \le \Lambda_{\frac{1}{2}} E\left[\left(\int_0^T u_t^2 dt\right)^{\frac{1}{2}}\right] < \infty$$

and so the claim follows from Proposition 4.39. □

5

Itô calculus

As for the Riemann and Lebesgue integral, the definition of stochastic integral is theoretical and it is not possible to use it directly for practical purposes, apart from some particular cases. Classical results reduce the problem of the computation of a Riemann integral to the determination of a primitive of the integrand function; in stochastic integration theory, the concept of primitive is translated into "integral terms" by the Itô-Doeblin formula[1]. This formula extends Theorem 3.70 in a probabilistic framework and lays the grounds for differential calculus for Brownian motion: as we have already seen the Brownian motion paths are generally irregular and so an integral interpretation of differential calculus for stochastic processes is natural.

In this chapter we present the fundamental connection established by the Itô formula between martingale and partial differential equation theories: this connection will be explained thoroughly in Paragraph 9.4, having at our disposal the foundations of the theory of stochastic differential equations. In Paragraph 5.2 we extend to the multi-dimensional case the main results on stochastic processes and stochastic calculus, and we dwell on the concept of correlation of processes. The last part of the chapter deals with some extensions of the Itô formula: in order to be able to apply the formula to the study of American options, we weaken the classical $C^{1,2}$ regularity assumptions of the function only requiring it belongs to a suitable Sobolev space. In the end we describe the so-called *local time of an Itô process* by which it is possible to give a direct proof of the Black-Scholes formula to price a European Call option.

[1] The formula for the "change of variable" presented in this chapter was proved by Itô [180] and is commonly known as Itô formula in the literature. Recently a posthumous 1940 paper by W. Doeblin [98] was found, where the author constructed the stochastic integral and stated the change-of-variable formula. That paper was recently published again [99] with a historical note by Bru. In what follows, for the sake of brevity we simply indicate that formula by the more common "Itô formula".

Pascucci A.: PDE and Martingale Methods in Option Pricing

5.1 Itô processes

Let W be a real Brownian motion on a filtered probability space $(\Omega, \mathcal{F}, P, (\mathcal{F}_t))$ where the usual hypotheses hold.

Definition 5.1 *An Itô process is a stochastic process X of the form*

$$X_t = X_0 + \int_0^t \mu_s ds + \int_0^t \sigma_s dW_s, \qquad t \in [0, T], \tag{5.1}$$

where X_0 is a $\mathcal{F}_0$-measurable random variable, $\mu \in \mathbb{L}^1_{\text{loc}}$ and $\sigma \in \mathbb{L}^2_{\text{loc}}$.

Formula (5.1) is usually written in the "differential form"

$$dX_t = \mu_t dt + \sigma_t dW_t. \tag{5.2}$$

The processes μ and σ are called *drift* and *diffusion* coefficients, respectively. Intuitively μ "gives the direction" to the process X, while the part of X containing σ is a (local) martingale giving only a "stochastic contribution" to the evolution of X.

On the one hand, (5.2) is shorter to write than (5.1) and so more convenient to use; on the other hand, (5.2) is more intuitive and familiar, because (just formally!) it reminds us of the usual differential calculus for functions of real variables. We remark that we have defined every single term appearing in (5.1); on the contrary (5.2) must be taken "on the whole" and it is merely a more compact notation to write (5.1). For the sake of clarity, we wish to emphasize the fact that the term dX_t, sometimes called *stochastic differential*, has not been defined and it makes sense only within formula (5.2).

The Itô process X in (5.1) is the sum of the continuous process with bounded variation

$$X_0 + \int_0^t \mu_s ds$$

with the continuous local martingale

$$\int_0^t \sigma_s dW_s.$$

Therefore, by Definition 4.45 and Remark 4.44, we have

Corollary 5.2 *If X is the Itô process in (5.1), then its quadratic variation process is given by*

$$\langle X \rangle_t = \int_0^t \sigma_s^2 ds,$$

or, in differential terms,

$$d\langle X \rangle_t = \sigma_t^2 dt.$$

The differential representation of an Itô process is unique, that is the drift and diffusion coefficients are determined uniquely. Indeed we have the following:

Proposition 5.3 *If X is the Itô process in* (5.1) *and there exist a random variable X_0', $\mu' \in \mathbb{L}^1_{\text{loc}}$ and $\sigma' \in \mathbb{L}^2_{\text{loc}}$ such that*

$$X_t = X_0' + \int_0^t \mu_s' ds + \int_0^t \sigma_s' \, dW_s, \qquad t \in [0,T],$$

then $X_0 = X_0'$ a.s. and $\mu = \mu', \sigma = \sigma'$ $(m \otimes P)$-a.e. that is

$$\{(t,\omega) \in [0,T] \times \Omega \mid \mu_t(\omega) \neq \mu_t'(\omega)\}, \quad \{(t,\omega) \in [0,T] \times \Omega \mid \sigma_t(\omega) \neq \sigma_t'(\omega)\}$$

have null $(m \otimes P)$-measure.

Proof. By assumption we have

$$M_t := \int_0^t (\mu_s - \mu_s')ds = \int_0^t (\sigma_s - \sigma_s')dW_s, \quad t \in [0,T] \text{ a.s.}$$

Therefore $M \in \mathscr{M}_{\text{c,loc}}$ and has bounded variation: hence by Proposition 4.46 M is indistinguishable from the null process. Consequently we have

$$0 = \langle M \rangle_T = \int_0^T (\sigma_t - \sigma_t')^2 dt,$$

and this proves that σ and σ' are $(m \otimes P)$-equivalent. We also have $\mu = \mu'$ $(m \otimes P)$-a.e. since it is a standard result that, if $u \in L^1([0,T])$ and

$$\int_0^t u_s ds = 0, \qquad t \in [0,T],$$

then $u = 0$ almost everywhere with respect to Lebesgue measure.

Remark 5.4 *An Itô process is a local martingale if and only if it has null drift.* More precisely, if X in (5.1) is a local martingale, then $\mu = 0$ $(m \otimes P)$-a.e. Indeed by assumption the process

$$\int_0^t \mu_s ds = X_t - X_0 - \int_0^t \sigma_s dW_s$$

would belong to $\mathscr{M}_{\text{c,loc}}$ and, at the same time, would have bounded variation, since it is a Lebesgue integral: the claim follows from Proposition 4.46. □

5.1.1 Itô formula for Brownian motion

Theorem 5.5 (Itô formula) *Let $f \in C^2(\mathbb{R})$ and let W be a real Brownian motion. Then $f(W)$ is an Itô process and we have*

$$df(W_t) = f'(W_t)dW_t + \frac{1}{2}f''(W_t)dt. \tag{5.3}$$

Proof. The great news of the Itô formula with respect to (3.53) is the presence of a "second order" term $\frac{1}{2}f''(W_t)dt$ (as we shall see, the coefficient $\frac{1}{2}$ comes from a second-order Taylor expansion) due to the fact that a Brownian motion has positive quadratic variation:

$$d\langle W\rangle_t = dt.$$

We first prove the thesis under the additional assumption that f has bounded first and second derivatives. We set

$$X_t = f(W_t) - f(W_0), \qquad Y_t = \int_0^t f'(W_s)dW_s + \frac{1}{2}\int_0^t f''(W_s)ds$$

and, for a fixed $t > 0$, we prove that $X_t = Y_t$ a.s. Given a partition $\varsigma = \{t_0, t_1, \dots, t_N\}$ of $[0,t]$, to simplify the notation we put $f_k = f(W_{t_k})$ and $\Delta_k = W_{t_k} - W_{t_{k-1}}$. We have

$$X_t = f(W_t) - f(W_0) = \sum_{k=1}^{N}(f_k - f_{k-1}) =$$

(by taking the second-order Taylor expansion with the Lagrange form of the remainder, with $t_k^* \in [t_{k-1}, t_k]$)

$$= \underbrace{\sum_{k=1}^{N} f'_{k-1}\Delta_k}_{=:I_1(\varsigma)} + \frac{1}{2}\underbrace{\sum_{k=1}^{N} f''_{k-1}\Delta_k^2}_{=:I_2(\varsigma)} + \frac{1}{2}\underbrace{\sum_{k=1}^{N}(f''(W_{t_k^*}) - f''_{k-1})\Delta_k^2}_{=:I_3(\varsigma)}.$$

Concerning $I_1(\varsigma)$, since f' is by assumption a continuous bounded function, then $f'(W)$ is a L^2-continuous process (cf. Example 4.19) and so by Remark 4.21 we have

$$\lim_{|\varsigma|\to 0+} I_1(\varsigma) = \int_0^t f'(W_s)dW_s \quad \text{in } \mathscr{M}_c^2.$$

Concerning $I_2(\varsigma)$, it is enough to proceed as in the proof of Theorem 3.74, using the fact that $\langle W\rangle_s = s$, to prove that

$$\lim_{|\varsigma|\to 0+} I_2(\varsigma) = \int_0^t f''(W_s)ds \quad \text{in } L^2(\Omega).$$

Eventually we verify that

$$\lim_{|\varsigma|\to 0+} I_3(\varsigma) = 0 \quad \text{in } L^2(\Omega). \tag{5.4}$$

Intuitively this is due to the fact that $f''(W_t)$ is a continuous process and W has finite quadratic variation: as a matter of fact the proof is based upon the same idea of Proposition 3.73, analogous result for the first variation.

Preliminarily, we observe that, for every $\varsigma = \{t_0, \dots, t_N\} \in \mathcal{P}_{[0,t]}$, $t > 0$, we have

$$t^2 = \left(\sum_{k=1}^{N} (t_k - t_{k-1}) \right)^2 = \underbrace{\sum_{k=1}^{N} (t_k - t_{k-1})^2}_{=:J_1(\varsigma)} + \underbrace{2 \sum_{h<k} (t_h - t_{h-1})(t_k - t_{k-1})}_{=:J_2(\varsigma)};$$

consequently

$$0 \le J_1(\varsigma) \le |\varsigma| \sum_{k=1}^{N} (t_k - t_{k-1}) = |\varsigma| t \xrightarrow[|\varsigma| \to 0^+]{} 0, \tag{5.5}$$

$$0 \le J_2(\varsigma) \le t, \qquad \varsigma \in \mathcal{P}_{[0,t]}. \tag{5.6}$$

Then, we have

$$E\left[(I_3(\varsigma))^2 \right] = \underbrace{\sum_{k=1}^{N} E\left[(f''(W_{t_k^*}) - f''_{k-1})^2 \Delta_k^4 \right]}_{=:L_1(\varsigma)}$$
$$+ \underbrace{2 \sum_{h<k}^{N} E\left[(f''(W_{t_h^*}) - f''_{h-1})(f''(W_{t_k^*}) - f''_{k-1}) \Delta_h^2 \Delta_k^2 \right]}_{=:L_2(\varsigma)}.$$

We have

$$L_1(\varsigma) \le 4 \sup |f''|^2 \sum_{k=1}^{N} E\left[\Delta_k^4 \right] =$$

(by (3.55))

$$= 12 \sup |f''|^2 \sum_{k=1}^{N} (t_k - t_{k-1})^2 \xrightarrow[|\varsigma| \to 0^+]{} 0$$

by (5.5). On the other hand, by Hölder's inequality, we get

$$L_2(\varsigma) \le \sum_{h<k}^{N} E\left[(f''(W_{t_h^*}) - f''_{h-1})^2 (f''(W_{t_k^*}) - f''_{k-1})^2 \right]^{\frac{1}{2}} E\left[\Delta_h^4 \Delta_k^4 \right]^{\frac{1}{2}} \le$$

(given $\varepsilon > 0$, if $|\varsigma|$ is small enough, by Lebesgue's dominated convergence theorem, since f'' is bounded and continuous)

$$\le \varepsilon \sum_{h<k}^{N} E\left[\Delta_h^4 \Delta_k^4 \right]^{\frac{1}{2}} \le$$

(by the independence of the Brownian increments)

$$\leq \varepsilon \sum_{h<k}^{N} E\left[\Delta_h^4\right]^{\frac{1}{2}} E\left[\Delta_k^4\right]^{\frac{1}{2}} = 3\varepsilon \sum_{h<k}^{N} (t_h - t_{h-1})(t_k - t_{k-1}) \leq 3\varepsilon t$$

by (5.6) and this proves (5.4).

To conclude, we remark that by Theorem A.136 the L^2-convergence implies that for any t there exists of a sequence of partitions (ς_n), with mesh converging to zero, such that

$$\lim_{n\to\infty} (I_1(\varsigma_n) + I_2(\varsigma_n) + I_3(\varsigma_n)) = \int_0^t f'(W_s)dW_s + \frac{1}{2}\int_0^t f''(W_s)ds = Y_t \quad \text{a.s.}$$

On the other hand $X_t = I_1(\varsigma_n) + I_2(\varsigma_n) + I_3(\varsigma_n)$ for any n and therefore the processes X and Y are modifications; eventually, since X and Y are continuous, we infer that they are indistinguishable as well. This concludes the proof under assumption that f has bounded first and second derivatives.

For the general case, it suffices to use a localization argument: we introduce the sequence of stopping times

$$\tau_n = \inf\{t \mid |W_t| \geq n\}, \qquad n \in \mathbb{N},$$

and consider the stopped Brownian motion $W_t^n = W_{t\wedge\tau_n}$ that takes its values in a compact set where f' and f'' are bounded. Using the same arguments as above, we prove

$$f(W_{t\wedge\tau_n}) - f(W_0) = \int_0^{t\wedge\tau_n} f'(W_s)dW_s + \frac{1}{2}\int_0^{t\wedge\tau_n} f''(W_s)ds$$

and the thesis follows since n is arbitrary (see, for instance, Steele [315] or Durrett [108] for further details). □

Example 5.6

i) Applying the Itô formula with $f(x) = x^2$ we have

$$d(W_t^2) = 2W_t dW_t + dt,$$

hence we get

$$\int_0^t W_s dW_s = \frac{W_t^2 - t}{2}.$$

ii) We compute $E\left[W_t^4\right]$: by the Itô formula we have

$$dW_t^4 = 4W_t^3 dW_t + 6W_t^2 dt,$$

i.e.

$$W_t^4 = \int_0^t 4W_s^3 dW_s + \int_0^t 6W_s^2 ds.$$

By the property of having null mean (4.7), we infer

$$E\left[W_t^4\right] = \int_0^t 6E\left[W_s^2\right] ds = \int_0^t 6s ds = 3t^2.$$

iii) Let $X_t = e^{\sigma W_t}$ with W Brownian motion and $\sigma \in \mathbb{R}$; we have

$$dX_t = \sigma X_t dW_t + \frac{1}{2}\sigma^2 X_t dt. \tag{5.7}$$

We aim at computing $E[X_t]$: since $X \in \mathbb{L}^2$, by the property of null expectation (4.7), from (5.7) we get

$$E[X_t] = \frac{\sigma^2}{2}\int_0^t E[X_s] ds.$$

In other terms, if we put $y(t) = E[X_t]$, we have that y is solution to the ordinary Cauchy problem

$$\begin{cases} y'(t) = \frac{\sigma^2}{2} y(t), \\ y(0) = 1, \end{cases}$$

and we conclude that

$$E\left[e^{\sigma W_t}\right] = e^{\frac{\sigma^2}{2}t}. \tag{5.8}$$

Formula (5.8) is also a direct consequence of the result in Exercise A.34 or can also be proved by noting that

$$E\left[e^{\sigma W_t}\right] = e^{\frac{\sigma^2}{2}t} E\left[e^{\sigma W_t - \frac{\sigma^2}{2}t}\right] = e^{\frac{\sigma^2}{2}t}$$

where the last equality follows from the the martingale property of the exponential Brownian motion, see Proposition 3.37-iii).

□

Exercise 5.7 Proceeding as in Example 5.6 and using the Itô formula, compute $E\left[W_t^6\right]$. By induction, prove that $E[W_t^n] = 0$ if n is odd and

$$E[W_t^n] = \left(\frac{t}{2}\right)^{\frac{n}{2}} \frac{n!}{(n/2)!},$$

if n is even. □

5.1.2 General formulation

Notation 5.8 *Let X be an Itô process*

$$dX_t = \mu_t dt + \sigma_t dW_t. \tag{5.9}$$

If h is a stochastic process such that $h\mu \in \mathbb{L}^1_{\text{loc}}$ and $h\sigma \in \mathbb{L}^2_{\text{loc}}$, we shorten the notation

$$dY_t = h_t \mu_t dt + h_t \sigma_t dW_t$$

by writing

$$dY_t = h_t dX_t. \tag{5.10}$$

Consistently we employ the notation

$$Y_t = Y_0 + \int_0^t h_s dX_s := Y_0 + \int_0^t h_s \mu_s ds + \int_0^t h_s \sigma_s dW_s.$$

We remark that, if $\mu \in \mathbb{L}^1_{\text{loc}}$, $\sigma \in \mathbb{L}^2_{\text{loc}}$ and h is a continuous adapted process, then $h\mu \in \mathbb{L}^1_{\text{loc}}$ and $h\sigma \in \mathbb{L}^2_{\text{loc}}$. More generally it is enough that h is progressively measurable and a.s. bounded.

We state now a more general version of the Itô formula: we will not go through the proof that is substantially analogous to that of Theorem 5.5.

Theorem 5.9 (Itô formula) *Let X be the Itô process in (5.9) and $f = f(t,x) \in C^{1,2}(\mathbb{R}^2)$. Then the stochastic process*

$$Y_t = f(t, X_t)$$

is an Itô process and we have

$$df(t, X_t) = \partial_t f(t, X_t)dt + \partial_x f(t, X_t)dX_t + \frac{1}{2}\partial_{xx} f(t, X_t)d\langle X\rangle_t. \tag{5.11}$$

Remark 5.10 Since, by Corollary 5.2, we have

$$d\langle X\rangle_t = \sigma_t^2 dt,$$

Formula (5.11) can be written more explicitly as follows

$$df = \left(\partial_t f + \mu_t \partial_x f + \frac{1}{2}\sigma_t^2 \partial_{xx} f\right) dt + \sigma_t \partial_x f dW_t, \tag{5.12}$$

where $f = f(t, X_t)$. □

Example 5.11 If $f(t,x) = tx$ and $X = W$ is a Brownian motion, we have

$$d(tW_t) = W_t dt + t dW_t.$$

We emphasize the resemblance with the classical differentiation rule for the product of two functions. In integral form we get

$$tW_t = \int_0^t W_s ds + \int_0^t s dW_s.$$

As an exercise, compute the stochastic differential of tW_t^2. □

Example 5.12 (Exponential martingale)
Given $u \in \mathbb{L}^2_{\text{loc}}$, we set

$$dY_t = u_t dW_t - \frac{1}{2} u_t^2 dt, \tag{5.13}$$

and consider the process e^Y. By the Itô formula we have

$$de^{Y_t} = e^{Y_t} dY_t + \frac{1}{2} e^{Y_t} d\langle Y \rangle_t = u_t e^{Y_t} dW_t.$$

Therefore e^Y is a local martingale, called *exponential martingale*. By Proposition 4.40, since it is a positive process, e^Y is also a super-martingale and in particular we have that

$$E\left[e^{Y_t}\right] \le E\left[e^{Y_0}\right], \qquad t \ge 0.$$

Further, if $E\left[e^{Y_T}\right] = E\left[e^{Y_0}\right]$, then $\left(e^{Y_t}\right)_{0 \le t \le T}$ is a martingale. By (5.8) this is true in particular if $u_t = \sigma$ with σ real constant (or even if $\sigma \in \mathbb{C}$): then the process

$$e^{\sigma W_t - \frac{|\sigma|^2}{2} t}$$

is a martingale. Let us remark explicitly that

$$Y_t = X_t - \frac{1}{2} \langle X \rangle_t,$$

where

$$X_t = \int_0^t u_s dW_s, \qquad \langle X \rangle_t = \int_0^t u_s^2 ds.$$

□

Proposition 5.13 *If $\mu \in L^1$ and $\sigma \in L^2$ are deterministic functions, then the process defined by*

$$dS_t = \mu(t)dt + \sigma(t)dW_t,$$

has normal distribution with

$$E[S_t] = S_0 + \int_0^t \mu(s)ds, \qquad \text{var}(S_t) = \int_0^t \sigma^2(s)ds.$$

Proof. By Theorem A.89 and recalling Example 4.15, it is enough to prove that, for every t, we have

$$E\left[e^{i\xi S_t}\right] = \exp\left(i\xi\left(S_0 + \int_0^t \mu(s)ds\right) - \frac{\xi^2}{2}\int_0^t \sigma^2(s)ds\right). \tag{5.14}$$

The proof of (5.14) is left as an exercise: it is enough to proceed as in the proof of (5.8). □

5.1.3 Martingales+and parabolic equations

We consider the Itô process X in (5.9) with constant drift and diffusion coefficients $\mu_t = \mu$ and $\sigma_t = \sigma$ and we define the parabolic differential operator with constant coefficients

$$L = \partial_t + \mu\partial_x + \frac{\sigma^2}{2}\partial_{xx}.$$

Then, for $f \in C^{1,2}(\mathbb{R}^2)$, (5.12) is equivalent to

$$df(t,X_t) = Lf(t,X_t)dt + \sigma\partial_x f(t,X_t)dW_t. \tag{5.15}$$

Corollary 5.14 *Under the previous assumptions, the process* $(f(t,X_t))_{t\in[0,T]}$ *is a local martingale if and only if* f *is a solution of* L*:*

$$Lf = 0, \qquad \text{in }]0,T[\times\mathbb{R}.$$

Proof. It is obvious that, if f is a solution, then $f(t,X_t)$ is a stochastic integral by (5.15) and so a continuous local martingale. Conversely, if $f(t,X_t)$ is a local martingale, then its drift is null by Remark 5.4, i.e. we have $Lf(t,X_t) = 0$ $(m\otimes P)$-almost everywhere. The claim follows from Proposition A.59, since X_t has strictly positive density on $\mathbb{R}$: we observe that

$$0 = \int_0^t E\left[|Lf(s,X_s)|\right] ds = \int_0^t \int_{\mathbb{R}} |Lf(s,x)|\Gamma(s,x)dxds,$$

with $\Gamma > 0$. □

We note that, if $\partial_x f(t,X_t) \in \mathbb{L}^2$, then $f(t,X_t)$ is a square integrable strict martingale, $f(t,X_t) \in \mathscr{M}_c^2$. Analogously $f(t,X_t)$ is a local super-martingale if and only if f is a supersolution[2] of L, i.e. we have that

$$Lf \le 0, \qquad \text{in }]0,T[\times\mathbb{R}. \tag{5.16}$$

5.1.4 Geometric Brownian motion

A geometric Brownian motion is a solution of the stochastic differential equation

$$dS_t = \mu S_t dt + \sigma S_t dW_t, \tag{5.17}$$

where $\mu,\sigma \in \mathbb{R}$, i.e. it is a stochastic process $S \in \mathbb{L}^2$ such that

$$S_t = S_0 + \mu\int_0^t S_s ds + \sigma\int_0^t S_s dW_s. \tag{5.18}$$

[2] See the note on p. 264.

The process S can be written explicitly in the form $S_t = f(t, W_t)$ with $f = f(t, x) \in C^{1,2}$. Indeed applying the Itô formula and imposing (5.17), we get

$$\left(\partial_t f(t, W_t) + \frac{1}{2}\partial_{xx} f(t, W_t)\right) dt + \partial_x f(t, W_t) dW_t$$
$$= \mu f(t, W_t)dt + \sigma f(t, W_t)dW_t.$$

By the uniqueness of the representation of an Itô process, (cf. Proposition 5.3) we infer[3] that, for $(t, x) \in \mathbb{R}_{>0} \times \mathbb{R}$, we have that

$$\begin{cases} \partial_x f(t, x) = \sigma f(t, x), \\ f(t, x) + \frac{1}{2}\partial_{xx} f(t, x) = \mu f(t, x). \end{cases}$$

For the first equation, there exists a function $g = g(t)$ such that

$$f(t, x) = g(t)e^{\sigma x} \tag{5.19}$$

and, plugging (5.19) into the second equation, we get

$$g' + \frac{\sigma^2}{2} g = \mu g$$

hence $g(t) = g(0)e^{\left(\mu - \frac{\sigma^2}{2}\right)t}$. In conclusion we have that

$$S_t = S_0 e^{\sigma W_t + \left(\mu - \frac{\sigma^2}{2}\right)t}, \tag{5.20}$$

and, applying the Itô formula, it is easy to verify that S in (5.20) is really a solution of equation (5.17).

Bachelier [16] was the first to use (non-geometric) Brownian motion as a model for asset prices, even though such a process can be negative with positive probability. Later on, Samuelson [295] considered geometric Brownian motion, that was then used by Black, Merton and Scholes in their classical works [49], [250] on arbitrage pricing of options. Being an exponential, if $S_0 > 0$ then the geometric Brownian motion (S_t) is a strictly positive process: more precisely, S_t has a density whose support lies in $\mathbb{R}_{\geq 0}$ and is strictly positive over $]0, +\infty[$ (see (5.22) further down).

If $\sigma = 0$, the dynamics of S is deterministic

$$S_t = S_0 e^{\mu t}$$

and this corresponds to continuous compounding with rate μ. For this reason the drift coefficient μ is usually called *expected rate of return* of S and the

[3] Here we use also the fact that, for $t > 0$, W_t has strictly positive density on $\mathbb{R}$: by Proposition A.59, if g is a continuous function such that $g(W_t) = 0$ a.s., then $g \equiv 0$.

Fig. 5.1. Graph of a path $t \mapsto S_t(\omega)$ of a geometric Brownian motion S and its mean $E[S_t]$

diffusion coefficient σ, adjusting the stochastic effect of Brownian motion, is called *volatility*. Since

$$\log(S_t) = \log(S_0) + \left(\mu - \frac{\sigma^2}{2}\right) t + \sigma W_t \sim \mathcal{N}_{\log(S_0)+\left(\mu - \frac{\sigma^2}{2}\right)t,\, \sigma^2 t}, \qquad (5.21)$$

S has log-normal distribution (cf. Example A.34). Clearly it is easy to compute

$$P(S_t \in [a, b]) = P(\log S_t \in [\log a, \log b]),$$

using, for example (A.24) to get a normal standard distribution. Alternatively we can explicitly write the density $\Psi(S_0; t, \cdot)$ of S_t: since $S_t = F(W_t)$ with $F(x) = S_0 \exp\left(\sigma x + \left(\mu - \frac{\sigma^2}{2}\right) t\right)$, by Remark A.33, we have

$$\Psi(S_0; t, x) = \frac{1}{\sigma x \sqrt{2\pi t}} \exp\left(- \frac{\left(\log\left(\frac{x}{S_0}\right) - \mu + \frac{\sigma^2 t}{2}\right)^2}{2\sigma^2 t} \right), \qquad t > 0,\ x > 0. \qquad (5.22)$$

In Figure 5.2 we depict the graph of the log-normal density.

Recalling that, by Example 5.12,

$$M_t := \exp\left(\sigma W_t - \frac{\sigma^2}{2} t\right)$$

is a martingale, we have

$$E[S_T \mid \mathcal{F}_t] = e^{\mu T} E[M_T \mid \mathcal{F}_t] = e^{\mu T} M_t = e^{\mu(T-t)} S_t,$$

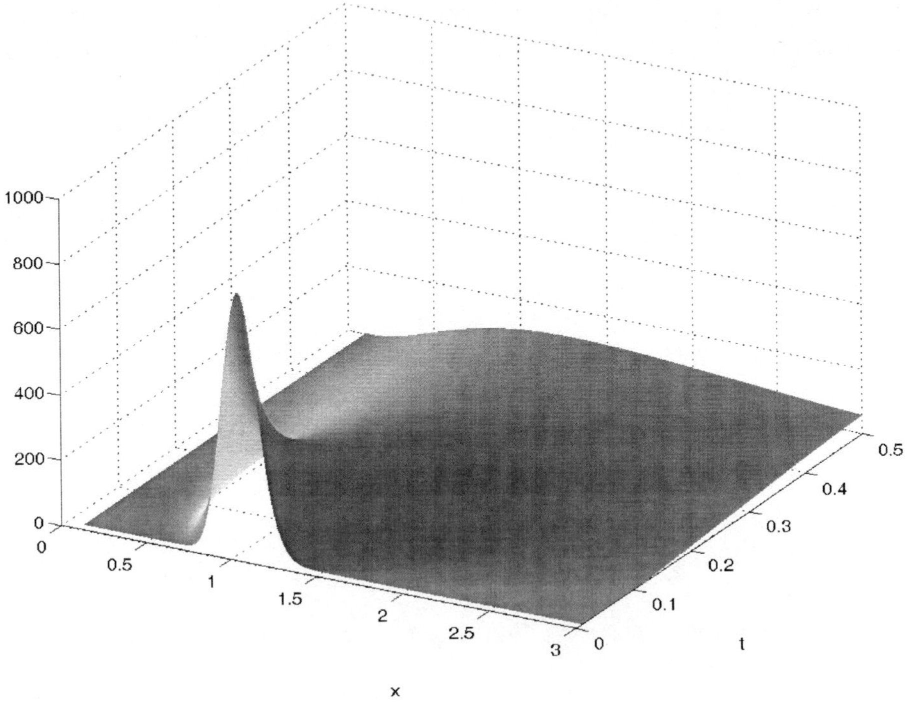

Fig. 5.2. Graph of the log-normal density $\Psi(S_0; t, x)$ with $S_0 = 1$

for every $0 \leq t \leq T$. Consequently S_t *is a sub-martingale if* $\mu \geq 0$ *and it is a martingale if and only if* $\mu = 0$. Further, if $S_0 \in \mathbb{R}$, we have

$$E[S_t] = S_0 e^{\mu t},$$

as we can verify directly by (A.27). Eventually, by (A.28), we have

$$\mathrm{var}(S_t) = S_0^2 e^{2\mu t} \left(e^{\sigma^2 t} - 1 \right).$$

5.2 Multi-dimensional Itô processes

We extend the definition of Brownian motion to the multi-dimensional case.

Definition 5.15 (d-dimensional Brownian motion) *Let* $(\Omega, \mathcal{F}, P, (\mathcal{F}_t))$ *be a filtered probability space. A d-dimensional Brownian motion is a stochastic process* $W = (W_t)_{t \in [0,+\infty[}$ *in* $\mathbb{R}^d$ *such that*

i) $W_0 = 0$ *P-a.s.;*
ii) W *is an* $\mathcal{F}_t$*-adapted and continuous stochastic process;*

iii) for $t > s \geq 0$, the random variable $W_t - W_s$ has multi-normal distribution $\mathcal{N}_{0,(t-s)I_d}$, where I_d is the $(d \times d)$-identity matrix, and it is independent of $\mathcal{F}_s$.

The following lemma contains some immediate consequences of the definition of multi-dimensional Brownian motion.

Lemma 5.16 *Let $W = (W^1, \dots, W^d)$ be a d-dimensional Brownian motion on $(\Omega, \mathcal{F}, P, (\mathcal{F}_t))$. Then, for every $i = 1, \dots, d$ we have*

1) W^i is a real Brownian motion on $(\Omega, \mathcal{F}, P, (\mathcal{F}_t))$ and so, in particular, an $\mathcal{F}_t$-martingale;
2) $W^i_{t+h} - W^i_t$ and $W^j_{t+h} - W^j_t$ are independent random variables for $i \neq j$ and $t, h \geq 0$.

Proof. The claim follows from the fact that, for $x = (x_1, \dots, x_d) \in \mathbb{R}^d$ and $h > 0$, we have

$$\Gamma(h,x) := \frac{1}{(2\pi h)^{\frac{d}{2}}} \exp\left(-\frac{|x|^2}{2h}\right) = \prod_{i=1}^{d} \frac{1}{\sqrt{2\pi h}} \exp\left(-\frac{x_i^2}{2h}\right). \tag{5.23}$$

Indeed, we prove property 1) in the case $i = 1$: it is enough to verify that

$$(W^1_{t+h} - W^1_t) \sim \mathcal{N}_{0,h}. \tag{5.24}$$

Given $H \in \mathscr{B}$ and $h > 0$, we have

$$P((W^1_{t+h} - W^1_t) \in H) = P((W_{t+h} - W_t) \in H \times \mathbb{R} \times \cdots \times \mathbb{R}) =$$

(since $(W_{t+h} - W_t) \sim \mathcal{N}_{0,tI_d}$ and (5.23) holds)

$$= \int_H \frac{1}{\sqrt{2\pi h}} \exp\left(-\frac{x_1^2}{2h}\right) dx_1 \prod_{i=2}^{d} \int_{\mathbb{R}} \frac{1}{\sqrt{2\pi h}} \exp\left(-\frac{x_i^2}{2h}\right) dx_i$$

and this proves (5.24) since all the integrals in dx_i for $i \geq 2$ evaluate to one.

Since we know the joint and marginal distributions of $W^i_{t+h} - W^i_t$ and $W^j_{t+h} - W^j_t$, property 2) is an immediate consequence of Proposition A.53 and of (5.23). □

Example 5.17 (Correlated Brownian motion) Given an $(N \times d)$-dimensional matrix α with constant real entries, we set

$$\varrho = \alpha\alpha^*. \tag{5.25}$$

Evidently $\varrho = (\varrho^{ij})$ is an $(N \times N)$-dimensional matrix, symmetric, positive semi-definite and $\varrho^{ij} = \langle \alpha^i, \alpha^j \rangle$ where α^i is the i-th row of the matrix α.

Given $\mu \in \mathbb{R}^N$ and a d-dimensional Brownian motion W, we put

$$B_t = \mu + \alpha W_t, \tag{5.26}$$

i.e.

$$dB_t = \alpha dW_t. \tag{5.27}$$

By Remark A.93, we have that

$$B_t \sim \mathcal{N}_{\mu, t\varrho}$$

and in particular

$$\mathrm{Cov}(B_t) = t\varrho \tag{5.28}$$

i.e.

$$E\left[\left(B_t^i - \mu_i\right)\left(B_t^j - \mu_j\right)\right] = t\varrho^{ij}.$$

We say that B is a *Brownian motion starting from* μ, *with (deterministic) correlation matrix* ϱ. We treat in Section 5.2.2 the case of a Brownian motion with stochastic correlation matrix.

For instance, if $N = 1$, we have $\alpha = (\alpha^{1i})_{i=1,\dots,d}$ and the random variable

$$B_t = \mu + \sum_{i=1}^{d} \alpha^{1i} W_t^i$$

has normal distribution with expectation μ and variance $|\alpha|^2 t$.

In intuitive terms, we can think of N as the number of assets traded on the market, represented by B, and of d as the number of sources of randomness. In building a stochastic model, we can suppose that the correlation matrix ϱ of the assets is observable: if ϱ is symmetric and positive definite, the Cholesky decomposition algorithm[4] allows us to determine an $(N \times N)$-dimensional lower triangular matrix α such that $\varrho = \alpha\alpha^*$, and so it is possible to obtain a representation for the risk factors in the form (5.26). □

Since the components of a d-dimensional Brownian motion W on $(\Omega, \mathcal{F}, P, (\mathcal{F}_t))$ are independent real Brownian motions, the integral

$$\int_0^t u_s dW_s^j, \qquad j = 1, \dots, d,$$

is defined in the usual way for every $u \in \mathbb{L}^2_{\mathrm{loc}}(\mathcal{F}_t)$. Moreover we have

Lemma 5.18 *For every $u, v \in \mathbb{L}^2(\mathcal{F}_t)$, $t_0 < t$ and $i \neq j$, we have*

$$E\left[\int_{t_0}^t u_s dW_s^i \int_{t_0}^t v_s dW_s^j \mid \mathcal{F}_{t_0}\right] = 0. \tag{5.29}$$

[4] See, for example, [263].

Proof. By an approximation argument it is enough to consider simple u, v: since the proof is similar to that of Theorem 4.5, we employ analogous notations. We have

$$\begin{aligned} & E\left[\int_{t_0}^t u_s dW_s^i \int_{t_0}^t v_s dW_s^j \mid \mathcal{F}_{t_0}\right] \\ =& E\left[\sum_{k=1}^N e_k(W_{t_k}^i - W_{t_{k-1}}^i)\sum_{h=1}^N \varepsilon_h(W_{t_h}^j - W_{t_{h-1}}^j) \mid \mathcal{F}_{t_0}\right] \\ =& \sum_{k=1}^N E\left[e_k\varepsilon_k(W_{t_k}^i - W_{t_{k-1}}^i)(W_{t_k}^j - W_{t_{k-1}}^j) \mid \mathcal{F}_{t_0}\right] \\ &+ 2\sum_{h<k} E\left[e_k\varepsilon_h(W_{t_k}^i - W_{t_{k-1}}^i)(W_{t_h}^j - W_{t_{h-1}}^j) \mid \mathcal{F}_{t_0}\right] \end{aligned}$$

and we conclude by using Proposition A.107-(6), by the independence of $W_{t_k}^i - W_{t_{k-1}}^i$ from $W_{t_k}^j - W_{t_{k-1}}^j$ (by Lemma 5.16), from e_k and ε_k (being $\mathcal{F}_{t_{k-1}}$-measurable random variables). □

Notation 5.19 *If u is an $(N \times d)$-matrix with components in $\mathbb{L}^2_{\text{loc}}(\mathcal{F}_t)$ (in what follows we simply write $u \in \mathbb{L}^2_{\text{loc}}$), we put*

$$\int_0^t u_s\, dW_s = \left(\sum_{j=1}^d \int_0^t u_s^{ij} dW_s^j\right)_{i=1,\dots,N}.$$

The following result extends the properties of the stochastic integral in Theorem 4.11.

Theorem 5.20 *For all $(N \times d)$-matrices $u, v \in \mathbb{L}^2$ and $0 \le a < b < c$, the following properties hold true:*

(1) null expectation:

$$E\left[\int_a^b u_s dW_s \mid \mathcal{F}_a\right] = 0, \qquad E\left[\langle \int_a^b u_t dW_t, \int_b^c v_t dW_t\rangle \mid \mathcal{F}_a\right] = 0;$$

(2) Itô isometry:

$$\begin{aligned} E\left[\langle \int_a^b u_t dW_t, \int_a^b v_t dW_t\rangle \mid \mathcal{F}_a\right] &= E\left[\int_a^b \sum_{i=1}^N \sum_{j=1}^d u_t^{ij} v_t^{ij} dt \mid \mathcal{F}_a\right] \\ &= E\left[\int_a^b \operatorname{tr}(u_t v_t^*)\, dt \mid \mathcal{F}_a\right], \end{aligned} \tag{5.30}$$

and in particular

$$E\left[\left|\int_a^b u_t dW_t\right|^2 \mid \mathcal{F}_a\right] = E\left[\int_a^b |u_t|^2\, dt \mid \mathcal{F}_a\right];$$

(3) if we put

$$X_t = \int_0^t u_s dW_s, \qquad t \in [0, T],$$

we have that $X \in \mathscr{M}_c^2$ *and*

$$[\![X]\!]_T^2 \leq 4 \int_0^T E\left[|u_t|^2\right] dt.$$

Further, the "non-conditional" versions of the equalities at points (1) and (2) hold.

Proof. We prove only (5.30):

$$E\left[\langle \int_a^b u_t dW_t, \int_a^b v_t dW_t \rangle \mid \mathcal{F}_a\right]$$
$$= \sum_{i=1}^N E\left[\left(\sum_{h=1}^d \int_a^b u_t^{ih} dW_t^h\right)\left(\sum_{k=1}^d \int_a^b v_t^{ik} dW_t^k\right) \mid \mathcal{F}_a\right]$$

(by Lemma 5.18 and Itô isometry)

$$= E\left[\int_a^b \sum_{i=1}^N \sum_{h=1}^d u_t^{ih} v_t^{ih} dt \mid \mathcal{F}_a\right].$$

□

5.2.1 Multi-dimensional Itô formula

Definition 5.21 *An N-dimensional Itô process is a stochastic process of the form*

$$X_t = X_0 + \int_0^t \mu_s ds + \int_0^t \sigma_s \, dW_s, \qquad t \in [0, T], \tag{5.31}$$

where X_0 is $\mathcal{F}_0$-measurable, W is a d-dimensional Brownian motion, $\mu \in \mathbb{L}^1_{\mathrm{loc}}$ is a $(N \times 1)$-vector and $\sigma \in \mathbb{L}^2_{\mathrm{loc}}$ is a $(N \times d)$-matrix.

Formula (5.31) can be written in the equivalent differential form

$$dX_t = \mu_t dt + \sigma_t dW_t$$

or, more explicitly

$$dX_t^i = \mu_t^i dt + \sum_{j=1}^d \sigma_t^{ij} dW_t^j, \qquad i = 1, \dots, N.$$

We recall the Definition 4.45 of co-variation process and, if

$$X = (X^1, \dots, X^N) \quad \text{and} \quad Y = (Y^1, \dots, Y^M)$$

are vector-valued processes, we put

$$\langle X, Y\rangle_t = \left(\langle X^i, Y^j\rangle_t\right)_{\substack{i=1,\dots,N \\ j=1,\dots,M}},$$

and $\langle X, X\rangle = \langle X\rangle$. The following result follows from Remark 4.44 and generalizes Corollary 5.2.

Lemma 5.22 *Consider an Itô process X of the form* (5.31) *and set*

$$\mathcal{C} = \sigma\sigma^*. \tag{5.32}$$

Then we have

$$\langle X^i, X^j\rangle_t = \int_0^t \mathcal{C}_s^{ij} ds, \qquad t \geq 0, \tag{5.33}$$

or, in differential notation,

$$d\langle X\rangle_t = \mathcal{C}_t dt.$$

In practice, given two Itô processes X, Y in $\mathbb{R}^N$, the computation of $\langle X, Y\rangle_t$ can be handled by applying the following "rule":

$$d\langle X^i, Y^j\rangle_t = dX_t^i dY_t^j,$$

where the product on the right-hand side of the previous equality can be computed using the following formal rules:

$$dtdt = dtdW^i = dW^i dt = 0, \qquad dW^i dW^j = \delta_{ij} dt,$$

and δ_{ij} denotes Kronecker's delta

$$\delta_{ij} = \begin{cases} 0, & i \neq j, \\ 1, & i = j. \end{cases}$$

Example 5.23 For $X = (X^1, X^2)$ defined by

$$\begin{aligned} dX_t^1 &= \mu_t dt + \alpha_t dW_t^1 + \beta_t dW_t^2, \\ dX_t^2 &= \nu_t dt + \gamma_t dW_t^1 + \delta_t dW_t^2, \end{aligned}$$

we have

$$\sigma = \begin{pmatrix} \alpha & \beta \\ \gamma & \delta \end{pmatrix} \quad \text{and} \quad \sigma\sigma^* = \begin{pmatrix} \alpha^2 + \beta^2 & \alpha\gamma + \beta\delta \\ \alpha\gamma + \beta\delta & \gamma^2 + \delta^2 \end{pmatrix}.$$

Then we have

$$d\langle X^1\rangle_t = d\langle X^1, X^1\rangle_t = \left(\alpha_t^2 + \beta_t^2\right) dt, \qquad d\langle X^1, X^2\rangle_t = \left(\alpha_t\gamma_t + \beta_t\delta_t\right) dt.$$

□

Example 5.24 For a correlated Brownian motion $B = \mu + \alpha W$, recalling (5.25)-(5.28), we have

$$\langle B^i, B^j \rangle_t = t(\alpha\alpha^*)^{ij} = \mathrm{cov}\left(B_t^i, B_t^j\right).$$

□

We state now the general version of Itô formula.

Theorem 5.25 *Let X be an Itô process of the form* (5.31) *and* $f = f(t,x) \in C^{1,2}(\mathbb{R} \times \mathbb{R}^N)$. *Then*

$$df = \partial_t f dt + \nabla f \cdot dX_t + \frac{1}{2} \sum_{i,j=1}^{N} \partial_{x_i x_j} f d\langle X^i, X^j \rangle_t, \tag{5.34}$$

with $f = f(t, X_t)$ *and* $\nabla f = (\partial_{x_1} f, \dots, \partial_{x_N} f)$.

In compact form, if we put $\mathcal{C} = \sigma\sigma^*$ and recall Lemma 5.22, then formula (5.34) becomes

$$\begin{aligned} df &= \left(\frac{1}{2} \sum_{i,j=1}^{N} \mathcal{C}_t^{ij} \partial_{x_i x_j} f + \mu_t \cdot \nabla f + \partial_t f \right) dt + \nabla f \cdot \sigma_t dW_t \\ &= \left(\frac{1}{2} \sum_{i,j=1}^{N} \mathcal{C}_t^{ij} \partial_{x_i x_j} f + \sum_{i=1}^{N} \mu_t^i \partial_{x_i} f + \partial_t f \right) dt + \sum_{i=1}^{N} \sum_{h=1}^{d} \partial_{x_i} f \sigma_t^{ih} dW_t^h. \end{aligned} \tag{5.35}$$

Example 5.26 (Standard Brownian motion) Let W be a d-dimensional Brownian motion and $f = f(t,x) \in C^{1,2}(\mathbb{R} \times \mathbb{R}^d)$. Then we have

$$\begin{aligned} df(t, W_t) &= \left(\partial_t f(t, W_t) + \frac{1}{2} \sum_{i=1}^{d} \partial_{x_i x_i} f(t, W_t) \right) dt + \sum_{i=1}^{d} \partial_{x_i} f(t, W_t) dW_t^i \\ &= \left(\partial_t f(t, W_t) + \frac{1}{2} \triangle f(t, W_t) \right) dt + \nabla f(t, W_t) \cdot dW_t, \end{aligned} \tag{5.36}$$

where $\triangle$ denotes the Laplace operator in $\mathbb{R}^d$. □

Example 5.27 (Correlated Brownian motion) Let $B = (B^1, \dots, B^N) = \alpha W$ be a correlated Brownian motion with correlation matrix $\varrho = \alpha\alpha^*$. We consider the Itô processes in $\mathbb{R}$

$$dX_t^i = \mu_t^i dt + \sigma_t^i dB_t^i, \qquad i = 1, \dots, N.$$

Then, for every $f = f(t,x) \in C^{1,2}(\mathbb{R} \times \mathbb{R}^N)$, we have

$$df = \left(\frac{1}{2} \sum_{i,j=1}^{N} \varrho^{ij} \sigma_t^i \sigma_t^j \partial_{x_i x_j} f + \sum_{i=1}^{N} \mu_t^i \partial_{x_i} f + \partial_t f \right) dt + \sum_{i=1}^{N} \partial_{x_i} f \sigma_t^i dB_t^i, \tag{5.37}$$

with $f = f(t, X_t)$. □

We consider further examples of application of the Itô formula that are useful to get acquainted to the multi-dimensional version.

Example 5.28 Let (W^1, W^2) be a 2-dimensional Brownian motion and

$$f(t, x_1, x_2) = x_1 x_2.$$

Then

$$d(W^1 W^2) = W^1 dW^2 + W^2 dW^1.$$

Further, for $f(t, x_1, x_2) = x_1^2 x_2$ we have

$$d\left(\left(W^1\right)^2 W^2\right) = \left(W^1\right)^2 dW^2 + 2W^1 W^2 dW^1 + W^2 dt.$$

In the case of a Brownian motion B in $\mathbb{R}^2$ with correlation matrix

$$\varrho = \begin{pmatrix} \alpha & \beta \\ \beta & \gamma \end{pmatrix}$$

and $f(t, x_1, x_2) = x_1 x_2$, we have

$$d(B_t^1 B_t^2) = B_t^1 dB_t^2 + B_t^2 dB_t^1 + \beta dt.$$

As an exercise, apply the Itô formula in the case $B = (B^1, B^2, B^3)$ and $f(B) = B^i B^j$ or $f(B) = B^i B^j B^k$. □

Example 5.29 (Integration by parts) We consider an Itô process with $N = 2$ and $d = 1$:

$$dX_t^i = \mu_t^i dt + \sigma_t^i dW_t, \qquad i = 1, 2.$$

In this case

$$\begin{aligned} d(X_t^1 X_t^2) &= X_t^1 dX_t^2 + X_t^2 dX_t^1 + \frac{1}{2}\left(d\langle X^1, X^2\rangle_t + d\langle X^2, X^1\rangle_t\right) \\ &= X_t^1 dX_t^2 + X_t^2 dX_t^1 + \sigma_t^1 \sigma_t^2 dt, \end{aligned} \tag{5.38}$$

i.e.

$$\int_0^t X_s^2 dX_s^1 = X_t^1 X_t^2 - X_0^1 X_0^2 - \int_0^t X_s^1 dX_s^2 - \int_0^t \sigma_s^1 \sigma_s^2 ds.$$

Note that it is enough that $\sigma^1 = 0$ or $\sigma^2 = 0$ in order for the usual integration-by-parts rule to hold formally true. □

Example 5.30 (Exponential martingale) Let W be a d-dimensional Brownian motion and $\sigma \in \mathbb{L}^2_{\text{loc}}$ an $(N \times d)$-matrix. We put

$$X_t = \int_0^t \sigma_s dW_s,$$

and we recall that

$$\langle X^i, X^j \rangle_t = \int_0^t \mathcal{C}_s^{ij} ds,$$

where $\mathcal{C} = \sigma\sigma^*$. Given $\xi \in \mathbb{R}^N$, we consider the process

$$\begin{aligned} Z_t^\xi &= \exp\left(\int_0^t \xi \cdot \sigma_s dW_s - \frac{1}{2}\int_0^t \langle \mathcal{C}_s \xi, \xi \rangle ds\right) \\ &= \exp\left(\xi \cdot X_t - \frac{1}{2}\sum_{i,j=1}^d \xi_i \xi_j \langle X^i, X^j \rangle_t\right). \end{aligned}$$

By the Itô formula we get

$$dZ_t^\xi = Z_t^\xi \xi \cdot dX_t = Z_t^\xi \xi \cdot \sigma_t dW_t,$$

and so Z^ξ is a positive local martingale, called exponential martingale (consistently with the 1-dimensional case, treated in Example 5.12).

In the particular case that σ is the $(d \times d)$-identity matrix, the process

$$Z_t^\xi = \exp\left(\xi \cdot W_t - \frac{|\xi|^2}{2} t\right)$$

is a martingale for every $\xi \in \mathbb{R}^d$. □

Remark 5.31 If f solves the adjoint heat equation in $\mathbb{R}^d$

$$\frac{1}{2}\triangle f + \partial_t f = 0, \tag{5.39}$$

then (5.36) becomes

$$df(t, W_t) = \nabla f(t, W_t) \cdot dW_t.$$

Hence, analogously to what we have seen in Section 5.1.3, $f(t, W_t)$ is a local martingale[5] if and only if f is solution to (5.39). In this case, if we denote the Brownian motion starting from x at time t by

$$W_T^{t,x} := x + W_T - W_t, \qquad t \le T,$$

in analogy to what we have seen in Section 3.1.2, we have:

i) $f(t,x) = E\left[f\left(T, W_T^{t,x}\right)\right]$;

---

[5] If $\nabla f(t, W_t) \in \mathbb{L}^2$ (for example, if ∇f is bounded), then

$$f(t, W_t) = f(0, W_0) + \int_0^t \nabla f(s, W_s) \cdot dW_s$$

is an $\mathcal{F}_t$-martingale.

ii) $E[f(T, W_T) \mid \mathcal{F}_t] = f(t, W_t)$

for $t \le T$ and $x \in \mathbb{R}^d$. □

We conclude the section by stating the multi-dimensional version of Proposition 5.13.

Proposition 5.32 *If $\mu \in L^1$ and $\sigma \in L^2$ are deterministic functions, then the process defined by*

$$dS_t = \mu(t)dt + \sigma(t)dW_t, \qquad S_0 = x \in \mathbb{R},$$

has multi-normal distribution with

$$E[S_t] = x + \int_0^t \mu(s)ds, \qquad \text{cov}(S_t) = \int_0^t \sigma(s)\sigma^*(s)ds.$$

The proof is analogous to that of the one-dimensional case and is therefore left as an exercise.

5.2.2 Correlated Brownian motion+and martingales

In this section we present a useful characterization of Brownian motion in terms of exponential martingales. Going back to Example 5.30, we consider the process

$$Z_t^\xi = e^{i\xi \cdot W_t + \frac{|\xi|^2}{2}t},$$

where i is the imaginary unit, W is a d-dimensional Brownian motion and $\xi \in \mathbb{R}^d$. We pointed out that Z^ξ is a local martingale and since Z^ξ is a bounded process, then it is also a strict martingale. Conversely, we have the following:

Theorem 5.33 *Let X be a continuous process in $\mathbb{R}^d$ on $(\Omega, \mathcal{F}, P, (\mathcal{F}_t))$ such that $X_0 = 0$ a.s. If for every $\xi \in \mathbb{R}^d$ the process*

$$Z_t^\xi = e^{i\xi \cdot X_t + \frac{|\xi|^2}{2}t} \tag{5.40}$$

is a martingale, then X is a Brownian motion.

Proof. We just have to verify that:

i) $X_t - X_s$ has normal distribution $\mathcal{N}_{0,(t-s)I_d}$;
ii) $X_t - X_s$ is independent of $\mathcal{F}_s$.

By (5.40) we have that

$$E\left[e^{i\xi \cdot (X_t - X_s)} \mid \mathcal{F}_s\right] = e^{-\frac{|\xi|^2}{2}(t-s)}$$

for every $\xi \in \mathbb{R}^d$ and taking the mean on both sides of the equality we get that the characteristic function of $X_t - X_s$ verifies

$$E\left[e^{i\xi \cdot (X_t - X_s)}\right] = e^{-\frac{|\xi|^2}{2}(t-s)}, \qquad \xi \in \mathbb{R}^d.$$

Then i) follows from Theorem A.89 and ii) from Proposition A.110. □

We give now a classical characterization of Brownian motion. First of all we observe that, if W is a Brownian motion in $\mathbb{R}^d$, it is immediate to verify by the Itô formula that the process

$$W^i W^j - \delta_{ij} t,$$

where δ_{ij} is Kronecker's delta, is a martingale: in terms of quadratic variation, this is tantamount to saying

$$\langle W^i, W^j \rangle_t = \delta_{ij} t.$$

It is remarkable that the quadratic variation and the martingale property characterize the Brownian motion. Indeed we have

Theorem 5.34 (Lévy's characterization of Brownian motion) *Let X be a stochastic process in $\mathbb{R}^d$ on the space $(\Omega, \mathcal{F}, P, (\mathcal{F}_t))$ such that $X_0 = 0$ a.s. Then X is a Brownian motion if and only if it is a continuous local martingale such that*

$$\langle X^i, X^j \rangle_t = \delta_{ij} t. \tag{5.41}$$

Proof. The proof is based upon Theorem 5.33 and it consists in verifying that, for every $\xi \in \mathbb{R}^N$, the exponential process

$$Z_t^\xi := \exp\left(i\xi \cdot X_t + \frac{|\xi|^2}{2} t \right)$$

is a martingale. We consider only the particular case in which X is an Itô process: for a general proof we refer, for example, to Protter [287], Theorem 39, Chapter II.

By assumption X is a local martingale, therefore its drift is null and X takes the form

$$dX_t = \sigma_t dW_t,$$

with $\sigma \in \mathbb{L}^2_{\text{loc}}$. By the Itô formula we have

$$dZ_t^\xi = Z_t^\xi \left(\frac{|\xi|^2}{2} dt + i\xi \cdot dX_t - \frac{1}{2} \sum_{i,j=1}^d \xi_i \xi_j d\langle X^i, X^j \rangle_t \right) =$$

(by (5.41))

$$= Z_t^\xi i\xi \cdot \sigma_t dW_t.$$

So Z^ξ is a local martingale but, being bounded, it is also a strict martingale. Then the claim follows from Theorem 5.33. □

Corollary 5.35 *Let $\alpha = (\alpha^1, \dots, \alpha^d)$ a progressively measurable process in $\mathbb{R}^d$ such that*

$$|\alpha_t|^2 = \sum_{i=1}^d \left(\alpha_t^i\right)^2 = 1 \qquad t \geq 0, \text{ a.s.}$$

and let W be a d-dimensional Brownian motion. Then

$$B_t = \int_0^t \alpha_s dW_s$$

is a real Brownian motion.

Proof. By assumption $\alpha \in \mathbb{L}^2$ and so B is a continuous martingale. Further, we have that

$$\langle B \rangle_t = \int_0^t |\alpha_s|^2 ds = t.$$

Therefore the hypotheses of Theorem 5.34 are verified and this concludes the proof. □

Definition 5.36 *Let us consider an $(N \times d)$-dimensional matrix α, whose components $\alpha^{ij} = \alpha_t^{ij}$ are progressively measurable processes and whose rows α^i are such that*

$$|\alpha_t^i| = 1 \qquad t \geq 0, \text{ a.s.}$$

The process

$$B_t = \int_0^t \alpha_s dW_s$$

is called correlated Brownian motion.

By Corollary 5.35, every component of B is a real Brownian motion and by Lemma 5.22

$$\langle B^i, B^j \rangle_t = \int_0^t \varrho_s^{ij} ds$$

where $\varrho_t = \alpha_t \alpha_t^*$ is called *correlation matrix of B*. Further, we have

$$\mathrm{Cov}(B_t) = \int_0^t E\left[\varrho_s\right] ds,$$

since

$$\mathrm{cov}(B_t^i, B_t^j) = E\left[B_t^i B_t^j\right] = E\left[\sum_{k=1}^d \int_0^t \alpha_s^{ik} dW_s^k \sum_{h=1}^d \int_0^t \alpha_s^{jh} dW_s^h\right] =$$

(by Lemma 5.18)

$$= E\left[\sum_{k=1}^d \int_0^t \alpha_s^{ik} dW_s^k \int_0^t \alpha_s^{jk} dW_s^k\right] =$$

(by Itô isometry)

$$= E\left[\int_0^t \sum_{k=1}^d \alpha_s^{ik} \alpha_s^{jk} ds\right] = \int_0^t E\left[\varrho_s^{ij}\right] ds.$$

If α is an orthogonal[6] matrix, then B is a standard Brownian motion, according to Definition 5.15. □

[6] I.e. such that $\alpha^* = \alpha^{-1}$. Consequently $\alpha^i \cdot \alpha^j = \delta_{ij}$ for every pair of rows.

5.3 Generalized Itô formulas

In this paragraph we are going to examine some extensions of the Itô formula (5.34): in particular we are interested in weakening the assumptions on the regularity of the function f.

The first generalization is an Itô formula for weakly differentiable functions. We will use this result in the study of American options since, as we saw in Section 2.5.5, the price of such derivatives is a function belonging to a suitable Sobolev space and does not belong, in general, to $C^{1,2}$.

Secondly, we want to extend the Itô formula to the payoff function of a Call option

$$f(x) = (x - K)^+, \qquad x \in \mathbb{R}, \tag{5.42}$$

where K is a fixed number. In this case[7] f does not possess a classical derivative in $x = K$, but it does admit first weak derivative

$$Df = \mathbb{1}_{]K,+\infty[}, \tag{5.43}$$

and has second derivative only in the distributional sense: precisely

$$D^2 f = \delta_K \tag{5.44}$$

where δ_K is Dirac's delta concentrated at K. In Section 5.3.4 we use an extension of Itô formula valid for f in (5.42) to get an interesting representation of the price of a European Call option.

5.3.1 Itô formula and+weak derivatives

The main result of the section is the following Itô formula for weakly differentiable functions. Hereafter W is an N-dimensional Brownian motion and $W^{2,p}$ (resp. $W^{2,p}_{\mathrm{loc}}$) denotes the Sobolev space of L^p (resp. L^p_{loc}) functions with first and second order weak derivatives in L^p (resp. L^p_{loc}), see Appendix A.9.2.

Theorem 5.37 *Let $f \in W^{2,p}_{\mathrm{loc}}(\mathbb{R}^N)$ with $p > 1 + \frac{N}{2}$. Then we have*

$$f(W_t) = f(0) + \int_0^t \nabla f(W_s) \cdot dW_s + \frac{1}{2}\int_0^t \triangle f(W_s) ds. \tag{5.45}$$

The proof of the theorem is based upon the following lemmas.

Lemma 5.38 *Let*

$$\Gamma(t,x) = \frac{1}{(2\pi t)^{\frac{N}{2}}} \exp\left(-\frac{|x|^2}{2t}\right), \qquad t > 0,\ x \in \mathbb{R}^N,$$

[7] We refer to the Appendix, Paragraph A.9, where we present the main results of the theory of distributions and weak derivatives.

be the density of the N-dimensional Brownian motion. Then

$$\Gamma \in L^q(]0,T[\times\mathbb{R}^N)$$

for every $q \in \left]0, 1+\frac{2}{N}\right[$ *and* $T > 0$.

Proof. We have

$$\int_0^T \int_{\mathbb{R}^N} \Gamma^q(t,x)dxdt = \int_0^T \int_{\mathbb{R}^N} \frac{1}{(2\pi t)^{\frac{Nq}{2}}} \exp\left(-\frac{q|x|^2}{2t}\right) dxdt =$$

(by the change of variables $y = \frac{x}{\sqrt{2t}}$)

$$= \frac{1}{(\pi)^{\frac{Nq}{2}}} \int_0^T \frac{1}{(2t)^{\frac{N}{2}(q-1)}} dt \int_{\mathbb{R}^N} e^{-q|y|^2} dy,$$

which is finite for $\frac{N(q-1)}{2} < 1$ and $q > 0$, i.e. $0 < q < 1 + \frac{2}{N}$. □

Lemma 5.39 *Assume that* $f \in W^{2,p}(\mathbb{R}^N)$, *with* $p > 1 + \frac{N}{2}$. *Then f is (Hölder) continuous and we have:*

i) if $p \le N$ *then* $|\nabla f|^2 \in L^q(\mathbb{R}^N)$ *for some* $q > 1 + \frac{N}{2}$;
ii) if $p > N$ *then* $\nabla f \in C(\mathbb{R}^N) \cap L^\infty(\mathbb{R}^N)$.

Proof. If $p \ge N$ the thesis follows from the Sobolev-Morrey embedding Theorem A.168. If $1 + \frac{N}{2} < p < N$ then necessarily $N > 2$ and, again by Theorem A.168, we have $\nabla f \in L^{2q}(\mathbb{R}^N)$ with

$$2q = \frac{pN}{N-p} = \frac{N}{\frac{N}{p} - 1} >$$

(since $p > 1 + \frac{N}{2}$)

$$> \frac{N}{\frac{N}{1+\frac{N}{2}} - 1} = \frac{N(N+2)}{N-2} > N + 2.$$

This proves that $|\nabla f|^2 \in L^q(\mathbb{R}^N)$ for some $q > 1 + \frac{N}{2}$ and consequently, by Theorem A.168, f is Hölder continuous. □

Proof (of Theorem 5.37). By using a localization argument as in the proof of the standard Itô formula, Theorem 5.5, it is not restrictive to assume that $f \in W^{2,p}(\mathbb{R}^N)$.

We first consider the case $N > 2$. Let (f_n) be a regularizing sequence for f, obtained by convolution with the usual mollifiers. Then, by Theorem A.176-v), $f_n \in C^\infty(\mathbb{R}^N)$ and (f_n) converges to f in $W^{2,p}$. Moreover, by the continuity of f proved in Lemma 5.39, we also have that (f_n) converges to f (uniformly on compacts) so that

$$\lim_{n\to\infty} f_n(W_t) = f(W_t)$$

for any $t \ge 0$.

By the standard Itô formula we have

$$f_n(W_t) = f_n(W_0) + \int_0^t \nabla f_n(W_s) \cdot dW_s + \frac{1}{2} \int_0^t \triangle f_n(W_s) ds.$$

Further, by Itô isometry,

$$\begin{aligned} &E\left[\left(\int_0^t (\nabla f_n(W_s) - \nabla f(W_s)) \cdot dW_s\right)^2\right] \\ &= \int_0^t E\left[|\nabla f_n(W_s) - \nabla f(W_s)|^2\right] ds \\ &= \int_0^t \int_{\mathbb{R}^N} |\nabla f_n(x) - \nabla f(x)|^2 \Gamma(s,x) dx ds =: I_n. \end{aligned}$$

If $p > N$ then, we have

$$\lim_{n\to\infty} I_n = 0$$

by the dominated convergence theorem, since by Lemma 5.39 $\nabla f \in C \cap L^\infty$, and so the integrand converges to zero pointwise and is dominated by the integrable function $\|\nabla f_n - \nabla f\|^2_{L^\infty(\mathbb{R}^N)} \Gamma$.

On the other hand, if $1 + \frac{N}{2} < p \leq N$, by Lemma 5.39 we have $|\nabla f|^2 \in L^q(\mathbb{R}^N)$ for some $q > 1 + \frac{N}{2}$. Let q' be the conjugate exponent of q: then we have

$$q' = 1 + \frac{1}{p-1} < 1 + \frac{2}{N}$$

and therefore, by Lemma 5.38, $\Gamma \in L^{q'}(]0,T[\times\mathbb{R}^N)$. By Hölder's inequality, we get

$$I_n \leq \left\| |\nabla f_n - \nabla f|^2 \right\|_{L^q(]0,T[\times\mathbb{R}^N)} \|\Gamma\|_{L^{q'}(]0,T[\times\mathbb{R}^N)} \xrightarrow[n\to\infty]{} 0.$$

Finally

$$\begin{aligned} &E\left[\left|\int_0^t (\triangle f_n(W_s) - \triangle f(W_s))\, ds\right|\right] \\ &\leq \int_0^t E\left[|\triangle f_n(W_s) - \triangle f(W_s)|\right] ds \\ &= \int_0^t \int_{\mathbb{R}^N} |\triangle f_n(x) - \triangle f(x)|\, \Gamma(s,x) dx ds \leq \end{aligned}$$

(by Hölder's inequality, with p' conjugate exponent of p)

$$\leq \|\triangle f_n - \triangle f\|_{L^p(]0,T[\times\mathbb{R}^N)} \|\Gamma\|_{L^{p'}(]0,T[\times\mathbb{R}^N)} \xrightarrow[n\to\infty]{} 0$$

since (f_n) converges to f in $W^{2,p}(\mathbb{R}^N)$ and the assumption $p > 1 + \frac{N}{2}$ implies $p' < 1 + \frac{2}{N}$: so, by Lemma 5.38, we have

$$\|\Gamma\|_{L^{p'}(]0,T[\times\mathbb{R}^N)} < \infty.$$

In conclusion, we proved that (5.45) holds a.s. for every $t > 0$, and by Proposition 3.25, this is enough to conclude.

In the case $N \leq 2$, the hypothesis $p > 1 + \frac{N}{2}$ implies $p > N$ and the thesis can be proved as before. □

Remark 5.40 The previous proof can easily be adapted to the case in which f depends also on time, i.e. the Itô formula holds for f in the parabolic Sobolev space $S^{2,p}_{\text{loc}}(\mathbb{R}^{N+1})$ with $p > 1 + \frac{N+2}{2}$. Moreover the generalized Itô formula holds can be proved for a class of processes more general than Brownian motion (see Theorem 9.47). A crucial tool in the proof of Theorem 5.37 is the integrability estimate of the transition density in Lemma 5.38: in Chapter 8 we prove an analogous estimate for a wide class of Itô processes, solutions of stochastic differential equations. In Chapter 11 we adapt the arguments used in this section to study the optimal stopping problem for American options. □

5.3.2 Tanaka formula+and local times

We consider the payoff function of a Call option

$$f(x) = (x - K)^+, \qquad x \in \mathbb{R}.$$

By applying *formally* the Itô formula to the process $f(W)$, where W is a real Brownian motion, and recalling the expression (5.43) and (5.44) of the derivatives of f, we get

$$(W_t - K)^+ = (W_0 - K)^+ + \int_0^t \mathbb{1}_{[K,+\infty[}(W_s)dW_s + \frac{1}{2}\int_0^t \delta_K(W_s)ds. \quad (5.46)$$

The relation (5.46), known as the *Tanaka formula,* besides a rigorous proof, requires also an explanation of the meaning of every term that appears in it. In particular, the last integral in (5.46), containing the distribution δ_K, must be interpreted in a formal way at this level: as we shall see, that term is indeed interesting from both a theoretical and a practical point of view, above all in financial applications. In order to give a precise meaning to that integral, it is necessary to introduce the concept of *local time of a Brownian motion,* after some preliminary considerations. In the next definition $|\cdot|$ denotes the Lebesgue measure.

Definition 5.41 (Occupation time) *Let $t \geq 0$ and $H \in \mathscr{B}$. The occupation time of H by time t of a Brownian motion W, is defined by*

$$J_t^H := |\{s \in [0,t] \mid W_s \in H\}|. \quad (5.47)$$

Intuitively, for every $\omega \in \Omega$, $J_t^H(\omega)$ measures the time W has spent in the Borel set H before t. The next properties of the occupation time follow directly from the definition:

i) we have
$$J_t^H = \int_0^t \mathbb{1}_H(W_s)ds; \tag{5.48}$$
ii) for every $H \in \mathscr{B}$, $\left(J_t^H\right)$ is an adapted and continuous stochastic process;
iii) for every $\omega \in \Omega$ and $H \in \mathscr{B}$, the function $t \mapsto J_t^H(\omega)$ is increasing and
$$0 \le J_t^H(\omega) \le t;$$
iv) for every t, ω, the map $H \mapsto J_t^H(\omega)$ is a measure on $\mathscr{B}$ and $J_t^{\mathbb{R}}(\omega) = t$;
v) by (5.48), we have
$$E(J_t^H) = \int_0^t P(W_s \in H)ds = \int_0^t \int_H \Gamma(s,x)dxds,$$
where Γ is the Gaussian density in (A.7). Consequently
$$|H| = 0 \qquad \Longleftrightarrow \qquad J_t^H = 0 \quad P\text{-a.s.} \tag{5.49}$$
In particular it follows that the occupation time of a single point in $\mathbb{R}$ by a Brownian motion is null.

Formally (5.49) suggests that $H \mapsto J_t^H$ is a measure equivalent to the Lebesgue measure, and so, by the Radon-Nikodym theorem it possesses a density:
$$J_t^H = \int_H L_t(x)dx. \tag{5.50}$$
Actually the situation is more delicate since J_t^H is a random variable: anyway, (5.50) holds true in the sense of the following:

Theorem 5.42 *There exists a two-parameter stochastic process*
$$L = \{L_t(x) = L_t(x,\omega) : \mathbb{R}_{\ge 0} \times \mathbb{R} \times \Omega \longrightarrow \mathbb{R}_{\ge 0}\}$$
with the following properties:

i) $L_t(x)$ *is* $\mathcal{F}_t$*-measurable for every* t, x*;*
ii) $(t,x) \mapsto L_t(x)$ *is an a.s. continuous function and, for every* x*,* $t \mapsto L_t(x)$ *is a.s. increasing;*
iii) (5.50) *holds for every* t *and* H *a.s.*

The process L *is called Brownian local time.*

For the proof of Theorem 5.42 we refer, for example, to Karatzas-Shreve [201], p. 207.

Remark 5.43 Combining (5.50) with (5.48) we get

$$\int_0^t \mathbb{1}_H(W_s)ds = \int_H L_t(x)dx, \qquad H \in \mathscr{B}, \quad \text{a.s.} \tag{5.51}$$

and, by Dynkin's Theorem A.9, this is equivalent to the fact that

$$\int_0^t \varphi(W_s)ds = \int_{\mathbb{R}} \varphi(x)L_t(x)dx, \qquad \text{a.s.} \tag{5.52}$$

for every bounded and measurable function φ. □

Remark 5.44 As a consequence of the a.s. continuity property of $L_t(x)$, we have that, almost surely

$$L_t(x) = \lim_{\varepsilon \to 0^+} \frac{1}{2\varepsilon} \int_{x-\varepsilon}^{x+\varepsilon} L_t(y)dy =$$

(by (5.51))

$$= \lim_{\varepsilon \to 0^+} \frac{1}{2\varepsilon} \left| \{ s \in [0,t] \mid |W_s - x| \le \varepsilon \} \right|. \tag{5.53}$$

This is the definition of local time originally introduced by P. Lévy: intuitively $L_t(x)$ measures the time (up to t) spent by W "around" the point x. □

We prove now a representation formula for Brownian local time.

Theorem 5.45 (Tanaka formula) *For every $K \in \mathbb{R}$ we have*

$$(W_t - K)^+ = (W_0 - K)^+ + \int_0^t \mathbb{1}_{[K,+\infty[}(W_s)dW_s + \frac{1}{2}L_t(K). \tag{5.54}$$

Remark 5.46 If we choose $\varphi = \varrho_n$ in (5.52), where (ϱ_n) is a regularizing sequence[8] for δ_K and if we take the limit in n, we get, by the a.s. continuity of L,

$$\lim_{n\to\infty} \int_0^t \varrho_n(W_s)ds = \lim_{n\to\infty} \int_{\mathbb{R}} \varrho_n(x)L_t(x)dx = L_t(K), \qquad \text{a.s.}$$

So it is natural to use the notation

$$\int_0^t \delta_K(W_s)ds := L_t(K). \tag{5.55}$$

Plugging (5.55) into (5.54), we get the Tanaka formula in the version given by (5.46).

We point out that the Itô formula was generalized under the only assumption of convexity of f by Meyer [254] and Wang [336]: concerning this topic, we refer the reader to Karatzas-Shreve [201], Chapter 3.6-D. □

[8] See Appendix A.9.4.

Proof (of Theorem 5.45). We construct a regularizing sequence for $f(x) = (x-K)^+$ using the mollifiers ϱ_n:

$$f_n(x) = \int_{\mathbb{R}} \varrho_n(x-y)(y-K)^+ dy.$$

We recall that, by Theorem A.176-v), we have

$$f_n'(x) = (Df)_n(x) = \int_{\mathbb{R}} \varrho_n(x-y)\mathbb{1}_{[K,+\infty[}(y)dy, \tag{5.56}$$

$$f_n''(x) = (D^2 f)_n(x) = \int_{\mathbb{R}} \varrho_n(x-y)\delta_K(dy) = \varrho_n(x-K). \tag{5.57}$$

Since $f_n \in C^\infty$, applying the Itô formula we get

$$F_n(W_t) = f_n(W_0) + \underbrace{\int_0^t f_n'(W_s)dW_s}_{=:I_t^{(1)}} + \frac{1}{2}\underbrace{\int_0^t f_n''(W_s)ds}_{=:I_t^{(2)}}.$$

By (5.57) we have

$$I_t^{(2)} = \int_0^t \varrho_n(W_s - K)ds =$$

(by (5.52))

$$= \int_{\mathbb{R}} \varrho_n(x-K)L_t(x)dx \xrightarrow[n\to\infty]{} L_t(K), \qquad \text{a.s.}$$

Further,

$$E\left[\left(I_t^{(1)} - \int_0^t \mathbb{1}_{[K,+\infty[}(W_s)dW_s\right)^2\right] =$$

(by Itô isometry)

$$= E\left[\int_0^t \left(f_n'(W_s) - \mathbb{1}_{[K,+\infty[}(W_s)\right)^2 ds\right] \xrightarrow[n\to\infty]{} 0$$

by the dominated convergence theorem, since the integrand converges to zero a.s. and it is bounded. This proves the Tanaka formula (5.54) $(m \otimes P)$-a.e.: on the other hand, by continuity, (5.54) holds true indeed for every t a.s. □

5.3.3 Tanaka+formula for Itô processes

In view of financial applications, we state the generalization of Theorem 5.42 for Itô processes in the form

$$X_t = X_0 + \int_0^t \mu_s ds + \int_0^t \sigma_s dW_s, \tag{5.58}$$

with $\mu \in \mathbb{L}^1_{\text{loc}}$ and $\sigma \in \mathbb{L}^2_{\text{loc}}$. The main difference with respect to the Brownian case lies in the fact that the local time is a continuous process in t rather than in the pair (t, x) and the term $d\langle X\rangle_t$ substitutes the term dt.

Theorem 5.47 *There exists a two-parameter stochastic process, called local time of the process* X,

$$L = \{L_t(x) = L_t(x, \omega) : \mathbb{R}_{\geq 0} \times \mathbb{R} \times \Omega \longrightarrow \mathbb{R}_{\geq 0}\}$$

with the following properties:

i) $(t, x, \omega) \mapsto L_t(x, \omega)$ *is measurable and* $L_t(x)$ *is* $\mathcal{F}_t$*-measurable for every* t, x*;*
ii) $t \mapsto L_t(x, \omega)$ *is a continuous and increasing function for every* x *a.s.;*
iii) for every $\varphi \in \mathscr{B}_b$ *we have the identity*

$$\int_0^t \varphi(X_s)d\langle X\rangle_s = \int_{\mathbb{R}} \varphi(x)L_t(x)dx, \qquad \text{a.s.}$$

Further, if we put

$$\int_0^t \delta_K(X_s)d\langle X\rangle_s := L_t(K), \qquad K \in \mathbb{R},$$

the Tanaka formula holds:

$$(X_t - K)^+ = (X_0 - K)^+ + \int_0^t \mathbb{1}_{[K,+\infty[}(X_s)dX_s + \frac{1}{2}\int_0^t \delta_K(X_s)d\langle X\rangle_s. \quad (5.59)$$

For the proof of the theorem, we refer, for example, to Karatzas-Shreve [201].

5.3.4 Local+time and Black-Scholes formula

The material in this section is partially taken from [298]. We consider a financial model in which the dynamics of the price of a risky asset is described by a geometric Brownian motion and for the sake of simplicity we assume that the expected return μ and the interest rate r are null:

$$dS_t = \sigma S_t dW_t.$$

Applying the Tanaka formula we get

$$(S_T - K)^+ = (S_0 - K)^+ + \int_0^T \mathbb{1}_{\{S_t \geq K\}}dS_t + \frac{1}{2}\int_0^T \sigma^2 S_t^2 \delta_K(S_t)dt, \quad (5.60)$$

and we have a representation for the payoff of a Call option with strike K as the sum of three terms:

- $(S_0 - K)^+$ represents the *intrinsic value* of the option;
- $\int_0^T \mathbb{1}_{\{S_t \geq K\}} dS_t$ is the final value of a self-financing strategy that consists in holding one unit of the asset when its price is greater than the strike and no units when the price is smaller than the strike. This is what we call a "stop-loss strategy";
- $\frac{1}{2}\int_0^T \sigma^2 S_t^2 \delta_K(S_t)dt$ is the local time around the strike: this non-negative term gives the error that one makes by replicating the strategy with the stop-loss strategy. Intuitively, if S does not hit the strike, the stop-loss replication is perfect. On the other hand, if S hits the strike we have to buy or sell the underlying asset. Since the paths of S are very irregular, this happens very often and in such a way that, intuitively, we are not able to buy (or sell) in the exact moment that S is worth K: in other terms we are forced to sell (buy) for a value which is a little bit smaller (greater) than the strike. This causes a replication error that is not due to transaction costs but is indeed a feature of the model based upon Brownian motion whose paths are irregular.

Taking the mean in formula (5.60) and using the property of null expectation of the stochastic integral we get

$$\begin{aligned} E\left[(S_T - K)^+\right] &= (S_0 - K)^+ + \frac{1}{2}\int_0^T E\left[\sigma^2 S_t^2 \delta_K(S_t)\right] dt \\ &= (S_0 - K)^+ + \frac{1}{2}\int_0^T \int_{\mathbb{R}} \sigma^2 S^2 \Psi(S_0; t, S)\delta_K(dS)dt \\ &= (S_0 - K)^+ + \frac{\sigma^2 K^2}{2}\int_0^T \Psi(S_0; t, K)dt, \end{aligned} \tag{5.61}$$

where $\Psi(S_0; t, \cdot)$ is the log-normal density of S_t in (5.22) with $\mu = 0$. Formula (5.61) expresses the mean of the payoff (intuitively the risk-neutral price of the Call option) as a *sum of the intrinsic value of the option with the integral with respect to the time variable of the density of the underlying asset, where the density is evaluated at K.*

Remark 5.48 Formula (5.61) can also be obtained in a easier way by using PDE arguments. We first note that $\Psi(S_0; t, x)$ is the fundamental solution of the Black-Scholes differential operator (cf. (2.113)) that, in the case $r = 0$, reads

$$L_{\mathrm{BS}} = \frac{\sigma^2 S_0^2}{2}\partial_{S_0 S_0} + \partial_t.$$

Therefore, for any $t < T$ and $S, S_0 > 0$, we have $L_{\mathrm{BS}}\Psi(S_0; T - t, S) = 0$ and

$$L_{\mathrm{BS}}^* \Psi(S_0; T - t, S) = \partial_{SS}\left(\frac{\sigma^2 S^2}{2}\Psi(S_0; T - t, S)\right) - \partial_T \Psi(S_0; T - t, S) = 0 \tag{5.62}$$

where $L_{\rm BS}^*$ is the adjoint operator of $L_{\rm BS}$: these identities follow by a direct computation. Moreover, formally we have

$$\Psi(S_0; 0, \cdot) = \delta_{S_0}$$

where δ_{S_0} denotes the Dirac's delta centered at S_0. Hence we have

$$\Psi(S_0; T, S) = \delta_{S_0}(S) + \int_0^T \partial_t \Psi(S_0; t, S) dt =$$

(by (5.62))

$$= \delta_{S_0}(S) + \int_0^T \partial_{SS} \left(\frac{\sigma^2 S^2}{2} \Psi(S_0; t, S) \right) dt.$$

Multiplying by the payoff function $(S - K)^+$ and integrating over $\mathbb{R}_{>0}$, we obtain the following representation of the Call price:

$$E\left[(S_T - K)^+ \right] =$$

$$= (S_0 - K)^+ + \int_K^{+\infty} \int_0^T (S - K)\, \partial_{SS} \left(\frac{\sigma^2 S^2}{2} \Psi(S_0; t, S) \right) dt dS =$$

(by parts)

$$= (S_0 - K)^+ - \int_K^{+\infty} \int_0^T \partial_S \left(\frac{\sigma^2 S^2}{2} \Psi(S_0; t, S) \right) dt dS =$$

$$= (S_0 - K)^+ + \frac{\sigma^2 K^2}{2} \int_0^T \Psi(S_0; t, K) dt,$$

that proves (5.61). □

Proposition 5.49 *Formula* (5.61) *is equivalent to the Black-Scholes formula* (2.108) *with interest rate* $r = 0$.

Proof. For the sake of simplicity we consider only the at-the-money case $S_0 = K$ and we leave it to the reader as an exercise to verify the general case. If C is the Black-Scholes price, by (2.108) we have

$$C = S_0 \Phi(d_1) - K e^{-rT} \Phi(d_2),$$

where d_1, d_2 are defined in (2.105) and Φ is the standard normal distribution function in (2.104). In the particular case $S_0 = K$ and $r = 0$ we have

$$C = K \left(\Phi\left(\sigma\sqrt{T}/2\right) - \Phi\left(-\sigma\sqrt{T}/2\right) \right) = 2K \int_0^{\frac{\sigma\sqrt{T}}{2}} \frac{1}{\sqrt{2\pi}} e^{-\frac{x^2}{2}} dx.$$

On the other hand, by (5.61), we have

$$E\left[(S_T - K)^+\right] = \frac{\sigma^2 K^2}{2} \int_0^T \Psi(t, K) dt =$$

(substituting the expression of Ψ given by (5.22))

$$= \frac{\sigma K}{2} \int_0^T \frac{1}{\sqrt{2\pi t}} \exp\left(-\frac{\sigma^2 t}{8}\right) dt$$

whence the claim, by the change of variables $x = \frac{\sigma\sqrt{t}}{2}$. □

We conclude by pointing out that the results in this section can be applied more generally to every model in which the underlying asset is an Itô process.

6

Parabolic PDEs with variable coefficients: uniqueness

In this chapter we consider elliptic-parabolic equations *with variable coefficients* of the form

$$L_a u := Lu - au = 0, \tag{6.1}$$

where L is the second order operator

$$L = \frac{1}{2} \sum_{j,k=1}^{N} c_{jk} \partial_{x_j x_k} + \sum_{j=1}^{N} b_j \partial_{x_j} - \partial_t, \qquad (t, x) \in \mathbb{R}^{N+1}. \tag{6.2}$$

In the whole chapter we assume:

Hypothesis 6.1 *The coefficients $c_{ij} = c_{ij}(t,x), b_i = b_i(t,x)$ and $a = a(t,x)$ are real-valued functions. The matrix $\mathcal{C}(t,x) = (c_{ij}(t,x))$ is symmetric and positive semi-definite for any (t,x). The coefficient a is bounded from below:*

$$\inf a =: a_0 \in \mathbb{R}. \tag{6.3}$$

Hereafter we use systematically the following:

Notation 6.2 *For fixed $T > 0$, we set*

$$\mathcal{S}_T =]0, T[\times \mathbb{R}^N.$$

We are interested in studying conditions that guarantee the uniqueness of the solution of the Cauchy problem

$$\begin{cases} Lu - au = f, & \text{in } \mathcal{S}_T, \\ u(0, \cdot) = \varphi, & \text{on } \mathbb{R}^N. \end{cases} \tag{6.4}$$

Such results, besides the apparent theoretical interest, are crucial in the study of the derivatives pricing problem: indeed, as we have already pointed out in the discrete-time setting, the arbitrage price of an option can be defined in terms of the solution of a problem of the form (6.4). The uniqueness for (6.4)

Pascucci A.: PDE and Martingale Methods in Option Pricing

translates into terms of absence of arbitrage opportunities and is equivalent to the uniqueness of the arbitrage price.

In general, problem (6.4) admits more than one solution: the classical Tychonov's example [327] shows that there exist classical non-null solutions of the Cauchy problem

$$\begin{cases} \frac{1}{2}\triangle u - \partial_t u = 0, & \text{in } \mathbb{R}_{>0} \times \mathbb{R}, \\ u(0,\cdot) = 0, & \text{on } \mathbb{R}. \end{cases}$$

For this reason the study of uniqueness for (6.4) consists in determining suitable families of functions in which there exists at most one classical solution: such families are usually called *uniqueness classes for L*. In what follows we specify two of them related to the main results of this chapter, Theorems 6.15 and 6.19.

In the first part, Paragraphs 6.1 and 6.2, we show a classical result, the so-called *weak maximum principle*, that allows proving the uniqueness for (6.4) within the class of functions verifying the following growth estimate at infinity:

$$|u(t,x)| \le Ce^{C|x|^2}, \qquad (t,x) \in \mathcal{S}_T, \tag{6.5}$$

for some constant C. This result, contained in Theorem 6.15, is very general and holds true under extremely weak conditions. More precisely, Hypothesis 6.1 suffices to prove uniqueness results on bounded domains (cf. Paragraph 6.1); when we study problems on unbounded domains (as the Cauchy problem), we also assume the following growth conditions on the coefficients:

Hypothesis 6.3 *There exists a constant M such that*

$$|c_{ij}(t,x)| \le M, \quad |b_i(t,x)| \le M(1+|x|), \quad |a(t,x)| \le M(1+|x|^2), \tag{6.6}$$

for every $(t,x) \in \mathcal{S}_T$ and $i,j = 1,\dots,N$.

In this chapter we only study the problem of uniqueness of the solution: we point out that Hypotheses 6.1 and 6.3 are so weak that are generally not sufficient to guarantee the existence of classical solutions. As we shall see in the following chapters, the growth condition (6.6) is usually quite natural in the financial applications.

In Paragraph 6.3 we present other more general uniqueness results, nevertheless requiring the much stronger assumption of existence of a fundamental solution of L and so, basically, the solvability of the Cauchy problem. We recall the following:

Definition 6.4 *A fundamental solution of the operator L_a, with pole in the point (s,y) in $\mathbb{R}^{N+1}$, is a function $\Gamma(\cdot,\cdot;s,y)$ defined on $]s,+\infty[\times\mathbb{R}^N$ such that, for every $\varphi \in C_b(\mathbb{R}^N)$, the function*

$$u(t,x) = \int_{\mathbb{R}^N} \Gamma(t,x;s,y)\varphi(y)dy \tag{6.7}$$

is a classical solution of the Cauchy problem

$$\begin{cases} L_a u = 0, & \text{in }]s,+\infty[\times\mathbb{R}^N, \\ u(s,\cdot) = \varphi, & \text{on } \mathbb{R}^N. \end{cases} \tag{6.8}$$

In Paragraph 6.3 we prove that the family of *non-negative* (or, more generally, bounded from below) functions is a uniqueness class for L_a: we will use this result in Chapter 7 to define the arbitrage price of a derivative; in particular we shall see that a solution of (6.8) represents the value of a self-financing strategy: the non-negativity condition translates, in economic terms, into an admissibility assumption for the strategies, necessary in order to avoid arbitrage opportunities.

We now state precisely the assumptions that will be in force in Paragraph 6.3. For a fixed positive constant λ, let

$$\Gamma_\lambda(t,x) = \frac{1}{(2\pi\lambda t)^{\frac{N}{2}}} e^{-\frac{|x|^2}{2t\lambda}}, \qquad t>0,\ x\in\mathbb{R}^N, \tag{6.9}$$

be the fundamental solution, with pole at the origin, of the heat operator in $\mathbb{R}^{N+1}$

$$\frac{\lambda}{2}\triangle - \partial_t.$$

Hypothesis 6.5 *The operator L_a has a fundamental solution Γ. Moreover, there exists $\lambda>0$ such that, for every $T>0$, $k=1,\dots,N$, $t\in]s,s+T[$ and $x,y\in\mathbb{R}^N$, the following estimates hold:*

$$\frac{1}{M}\Gamma_{\frac{1}{\lambda}}(t-s,x-y) \le \Gamma(t,x;s,y) \le M\Gamma_\lambda(t-s,x-y) \tag{6.10}$$

$$|\partial_{y_k}\Gamma(t,x;s,y)| \le \frac{M}{\sqrt{t-s}}\Gamma_\lambda(t-s,x-y), \tag{6.11}$$

with M positive constant depending on T.

Hypothesis 6.6 *The operator L_a admits the adjoint operator*

$$L_a^* u = \frac{1}{2}\sum_{j,k=1}^N c_{jk}\partial_{x_j x_k} u + \sum_{j=1}^N b_j^* \partial_{x_j} u - a^* u + \partial_t u \tag{6.12}$$

and the coefficients

$$b_i^* = -b_i + \sum_{j=1}^N \partial_{x_i} c_{ij}, \qquad a^* = a - \frac{1}{2}\sum_{i,j=1}^N \partial_{x_i x_j} c_{ij} + \sum_{j=1}^N \partial_{x_j} b_j, \tag{6.13}$$

verify growth conditions analogous to (6.6).

Remark 6.7 We note explicitly that all the previous assumptions are satisfied if L_a belongs to the class of parabolic operators with *constant coefficients* (cf. Appendix A.3) or, more generally, to the class of *uniformly parabolic operators with variable coefficients* considered in Chapter 8. □

6.1 Maximum principle and Cauchy-Dirichlet problem

In this paragraph we study the uniqueness problem on *bounded* domains. In this section we suppose that L_a in (6.1)-(6.2) verifies Hypothesis 6.1 and that Q is a bounded open set in $\mathbb{R}^N$.

Remark 6.8 Given $\alpha \in \mathbb{R}$, we set

$$v(t,x) = e^{\alpha t} u(t,x). \tag{6.14}$$

Then we have

$$L_a(e^{\alpha t} u) = e^{\alpha t}(L_a u - \alpha u)$$

that is

$$L_{(a-\alpha)} v = e^{\alpha t} L_a u. \tag{6.15}$$

In particular, if $\alpha < a_0$ then $\inf(a - \alpha) > 0$. □

For $T > 0$, we denote by

$$Q_T =]0, T[\times Q,$$

the open cylinder with basis Q and height T: moreover $\bar{Q}_T$ is the closure of Q_T and $\partial_p Q_T$ is the *parabolic boundary* defined by

$$\partial_p Q_T = \partial Q_T \setminus (\{T\} \times Q).$$

Definition 6.9 *Let $f \in C(Q_T)$ and $\varphi \in C(\partial_p Q_T)$. A classical solution of the Cauchy-Dirichlet problem for L_a in Q_T with boundary datum φ is a function $u \in C^{1,2}(Q_T) \cap C(Q_T \cup \partial_p Q_T)$ such that*

$$\begin{cases} L_a u = f, & \text{in } Q_T, \\ u = \varphi, & \text{on } \partial_p Q_T. \end{cases} \tag{6.16}$$

Theorem 6.10 (Weak maximum principle) *Let $u \in C^{1,2}(Q_T) \cap C(Q_T \cup \partial_p Q_T)$ such that $L_a u \geq 0$ on Q_T. If $u \leq 0$ on $\partial_p Q_T$, then $u \leq 0$ on Q_T.*

Proof. First of all, by Remark 2.57 it is not restrictive to assume $a_0 > 0$, since we may prove the thesis for v in (6.14) with $\alpha < a_0$ and then use the fact that u and v have the same sign.

By contradiction, we assume that $u(t_0, x_0) > 0$ at some point $(t_0, x_0) \in Q_T$: then for some $(t_1, x_1) \in \bar{Q}_{t_0} \setminus \partial_p Q_{t_0}$ we have

$$u(t_1, x_1) = \max_{\bar{Q}_{t_0}} u \geq u(t_0, x_0) > 0,$$

and

$$D^2 u(t_1, x_1) := (\partial_{x_i x_j} u(t_1, x_1)) \leq 0, \quad \partial_{x_k} u(t_1, x_1) = 0, \quad \partial_t u(t_1, x_1) \geq 0,$$

for every $k = 1, \dots, N$. Then there exists a symmetric and positive semi-definite matrix $M = (m_{ij})$ such that

$$-D^2u(t_1, x_1) = M^2 = \left(\sum_{h=1}^{N} m_{ih}m_{hj}\right)_{i,j} = \left(\sum_{h=1}^{N} m_{ih}m_{jh}\right)_{i,j}$$

and therefore we have

$$\begin{aligned} L_a u(t_1, x_1) &= -\frac{1}{2}\sum_{i,j=1}^{N} c_{ij}(t_1, x_1)\sum_{h=1}^{N} m_{ih}m_{jh} + \sum_{j=1}^{N} b_j(t_1, x_1)\partial_{x_j}u(t_1, x_1) \\ &\quad - a(t_1, x_1)u(t_1, x_1) - \partial_t u(t_1, x_1) \\ &= -\frac{1}{2}\sum_{h=1}^{N} \underbrace{\sum_{i,j=1}^{N} c_{ij}(t_1, x_1)m_{ih}m_{jh}}_{\geq 0 \text{ since } \mathcal{C}\geq 0} - a(t_1, x_1)u(t_1, x_1) - \partial_t u(t_1, x_1) \\ &\leq -a(t_1, x_1)u(t_1, x_1) < 0, \end{aligned}$$

and this contradicts the assumption $L_a u \geq 0$. □

The previous result is called *weak maximum principle* since it does not rule out the possibility that a solution may attain its maximum also inside the cylinder: the *strong maximum principle* states on the contrary that the only solution attaining its maximum internally is the constant one.

Corollary 6.11 (Comparison principle) *Let $u, v \in C^{1,2}(Q_T) \cap C(Q_T \cup \partial_p Q_T)$ such that $L_a u \leq L_a v$ in Q_T and $u \geq v$ in $\partial_p Q_T$. Then $u \geq v$ in Q_T. In particular there exists at most one classical solution of the Cauchy-Dirichlet problem* (6.16).

Proof. It suffices to apply the maximum principle to the function $v - u$. □

Now we prove an a priori[1] estimate of the maximum of a solution to (6.16).

Theorem 6.12 *Let $u \in C^{1,2}(Q_T) \cap C(Q_T \cup \partial_p Q_T)$ and let us set*

$$a_1 := \max\{0, -a_0\}.$$

Then

$$\sup_{Q_T} |u| \leq e^{a_1 T}\left(\sup_{\partial_p Q_T} |u| + T\sup_{Q_T} |L_a u|\right). \tag{6.17}$$

[1] An a priori estimate is an estimate that is valid for all the possible solutions of a family of differential equations even though the assumptions on the equations do not guarantee the existence of such solutions. In the classical theory of partial differential equations, these a priori estimates are a basic tool to prove the existence and the regularity of solutions.

Proof. We first suppose that $a_0 \geq 0$ and also that u and $L_a u$ are bounded in $\partial_p Q_T$ and Q_T respectively, otherwise there is nothing to prove. We consider the function

$$w(t,x) = \sup_{\partial_p Q_T} |u| + t \sup_{Q_T} |L_a u|;$$

we have

$$L_a w = -aw - \sup_{Q_T} |L_a u| \leq L_a u, \qquad L_a(-w) = aw + \sup_{Q_T} |L_a u| \geq L_a u,$$

and $-w \leq u \leq w$ in $\partial_p Q_T$. Then estimate (6.17), with $a_1 = 0$, follows from the comparison principle, Corollary 6.11.

On the other hand, if $a_0 < 0$ we consider v in (6.14) with $\alpha = a_0$ and by estimate (6.17) we infer

$$\sup_{Q_T} |v| \leq \sup_{\partial_p Q_T} |v| + \sup_{Q_T} |L_{(a-a_0)} v|.$$

Consequently, since $a_0 < 0$, we get

$$e^{a_0 T} \sup_{Q_T} |u| \leq \sup_{(t,x) \in Q_T} |e^{a_0 t} u(t,x)| \leq \sup_{\partial_p Q_T} |v| + \sup_{Q_T} |L_{(a-a_0)} v| \leq$$

(by (6.15))

$$\leq \sup_{(t,x) \in \partial_p Q_T} |e^{a_0 t} u(t,x)| + T \sup_{(t,x) \in Q_T} |e^{a_0 t} L_a u(t,x)| \leq$$

(since $a_0 < 0$)

$$\leq \sup_{\partial_p Q_T} |u| + T \sup_{Q_T} |L_a u|,$$

from which the thesis follows. □

Under suitable regularity assumptions, existence results for the Cauchy-Dirichlet problem can be proved by using the classical theory of Fourier series: we refer to Chapter V in DiBenedetto [97] for a clear presentation of this topic.

6.2 Maximum principle and Cauchy problem

In this paragraph we prove uniqueness results for the Cauchy problem. The standard Cauchy problem differs from the Cauchy-Dirichlet problem studied in the previous section, in that it is posed on a strip of $\mathbb{R}^{N+1}$, that is on an unbounded domain where only the initial conditions is given, but no lateral conditions. In what follows we assume that the operator L_a in (6.1)-(6.2) verifies Hypotheses 6.1 and 6.3. We recall the notation

$$\mathcal{S}_T =]0, T[\times \mathbb{R}^N.$$

Theorem 6.13 (Weak maximum principle) *Let* $u \in C^{1,2}(\mathcal{S}_T) \cap C(\overline{\mathcal{S}}_T)$ *such that*

$$\begin{cases} L_a u \le 0, & \text{in } \mathcal{S}_T, \\ u(0,\cdot) \ge 0, & \text{on } \mathbb{R}^N. \end{cases}$$

If

$$u(t,x) \ge -Ce^{C|x|^2}, \qquad (t,x) \in \mathcal{S}_T, \tag{6.18}$$

for some positive constant C*, then* $u \ge 0$ *on* $\mathcal{S}_T$*.*

Before Theorem 6.13, we prove first the following:

Lemma 6.14 *Let* $u \in C^{1,2}(\mathcal{S}_T) \cap C(\overline{\mathcal{S}}_T)$ *such that*

$$\begin{cases} L_a u \le 0, & \text{in } \mathcal{S}_T, \\ u(0,\cdot) \ge 0, & \text{on } \mathbb{R}^N, \end{cases}$$

and

$$\liminf_{|x|\to\infty} \left(\inf_{t \in]0,T[} u(t,x) \right) \ge 0. \tag{6.19}$$

Then $u \ge 0$ *on* $\mathcal{S}_T$*.*

Proof. By the same argument used in the proof of Theorem 6.10 and based on Remark 2.57, it is not restrictive to assume $a_0 \ge 0$. Then, for fixed $(t_0, x_0) \in \mathcal{S}_T$ and $\varepsilon > 0$, we have

$$\begin{cases} L_a(u+\varepsilon) \le 0, & \text{in } \mathcal{S}_T, \\ u(0,\cdot) + \varepsilon > 0, & \text{on } \mathbb{R}^N, \end{cases}$$

and, by assumption (6.19), there exists a large enough $R > |x_0|$ such that

$$u(t,x) + \varepsilon > 0, \qquad t \in]0,T[, \ |x| = R.$$

Then we can apply the maximum principle, Theorem 6.10, on the cylinder

$$Q_T =]0,T[\times \{|x| \le R\}$$

to infer that $u(t_0,x_0) + \varepsilon \ge 0$ and, by the arbitrariness of ε, $u(t_0,x_0) \ge 0$. □

Proof (of Theorem 6.13). We observe that it suffices to prove that $u \ge 0$ on a strip $\mathcal{S}_{T_0}$ with $T_0 > 0$: once we have proved this, by applying the result repeatedly we get the claim on the entire strip $\mathcal{S}_T$.

We prove first the remarkable case of the heat operator

$$L = \frac{1}{2}\triangle - \partial_t,$$

and $a = 0$, i.e. $L = L_a$. For fixed $\gamma > C$, we set $T_0 = \frac{1}{4\gamma}$ and we consider the function

$$v(t,x) = \frac{1}{(1-2\gamma t)^{\frac{N}{2}}} \exp\left(\frac{\gamma|x|^2}{1-2\gamma t}\right), \qquad (t,x) \in \mathcal{S}_{T_0}.$$

A direct computation shows that

$$Lv(t,x) = 0 \quad \text{and} \quad v(t,x) \geq e^{\gamma|x|^2} \qquad (t,x) \in \mathcal{S}_{T_0}.$$

Moreover, for every $\varepsilon > 0$, Lemma 6.14 ensures that the function

$$w = u + \varepsilon v$$

is non-negative: since ε is arbitrary, this suffices to conclude the proof.

The general case is only technically more complicated and it is based upon the crucial Hypothesis 6.3 on the growth at infinity of the coefficients of the operator. For fixed $\gamma > C$ and two parameters $\alpha, \beta \in \mathbb{R}$ to be chosen appropriately, we consider the function

$$v(t,x) = \exp\left(\frac{\gamma|x|^2}{1-\alpha t} + \beta t\right), \qquad 0 \leq t \leq \frac{1}{2\alpha}.$$

We have

$$\frac{L_a v}{v} = \frac{2\gamma^2}{(1-\alpha t)^2}\langle Cx, x\rangle + \frac{\gamma}{1-\alpha t}\operatorname{tr} C + \frac{2\gamma}{1-\alpha t}\sum_{i=1}^{N} b_i x_i - a - \frac{\alpha\gamma|x|^2}{(1-\alpha t)^2} - \beta.$$

Using Hypothesis 6.3, we see that, if α, β are large enough, then

$$\frac{L_a v}{v} \leq 0. \tag{6.20}$$

Now we consider the function $w = \frac{u}{v}$: by assumption (6.18), we have

$$\liminf_{|x|\to\infty}\left(\inf_{t\in\,]0,T[} w(t,x)\right) \geq 0;$$

furthermore w satisfies the equation

$$\frac{1}{2}\sum_{i,j=1}^{N} c_{ij}\partial_{x_i x_j} w + \sum_{i=1}^{N} \widetilde{b}_i \partial_{x_i} w - \widetilde{a} w - \partial_t w = \frac{L_a u}{v} \leq 0,$$

where

$$\widetilde{b}_i = b_i + \sum_{j=1}^{N} c_{ij}\frac{\partial_{x_j} v}{v}, \qquad \widetilde{a} = -\frac{L_a v}{v}.$$

Since $\widetilde{a} \geq 0$ by (6.20), we can apply Lemma 6.14 to infer that w (and so u) is non-negative. □

The following uniqueness result is a direct consequence of Theorem 6.13. We emphasize that L_a verifies only the very general Hypotheses 6.1 and 6.3: for example, L_a can be a first-order operator.

Theorem 6.15 *There exists at most one classical solution $u \in C^{1,2}(\mathcal{S}_T) \cap C(\overline{\mathcal{S}}_T)$ of the problem*

$$\begin{cases} L_a u = f, & \text{in } \mathcal{S}_T, \\ u(0,\cdot) = \varphi, & \text{on } \mathbb{R}^N, \end{cases}$$

such that

$$|u(t,x)| \le Ce^{C|x|^2}, \qquad (t,x) \in \mathcal{S}_T, \tag{6.21}$$

for some positive constant C.

Remark 6.16 Let us suppose that L_a also verifies Hypothesis 6.5. Then L has a fundamental solution Γ and, given $\varphi \in C(\mathbb{R}^N)$ such that

$$|\varphi(y)| \le ce^{c|y|^\gamma}, \qquad y \in \mathbb{R}^N, \tag{6.22}$$

with c, γ positive constants and $\gamma < 2$, then the function

$$u(t,x) := \int_{\mathbb{R}^N} \Gamma(t,x;0,y)\varphi(y)dy, \qquad (t,x) \in \mathcal{S}_T, \tag{6.23}$$

is a classical solution of the Cauchy problem

$$\begin{cases} L_a u = 0, & \text{in } \mathcal{S}_T, \\ u(0,\cdot) = \varphi, & \text{on } \mathbb{R}^N, \end{cases} \tag{6.24}$$

for every $T > 0$.

By using the upper estimate of Γ in (6.10), it is not hard to prove that for every $T > 0$ there exists a constant c_T such that

$$|u(t,x)| \le c_T e^{2c|x|^\gamma}, \qquad (t,x) \in \mathcal{S}_T. \tag{6.25}$$

Then, by Theorem 6.15, u *in* (6.23) *is the unique solution of the Cauchy problem* (6.24) *verifying estimate* (6.22).

Now we prove (6.25) assuming, without loss of generality, that $\gamma \ge 1$. By (6.22) we have

$$|u(t,x)| \le \frac{cM}{(2\pi\lambda t)^{\frac{N}{2}}} \int_{\mathbb{R}^N} e^{-\frac{|x-y|^2}{2t\lambda} + c|y|^\gamma} dy =$$

(by the change of variable $\eta = \frac{x-y}{\sqrt{2\lambda t}}$)

$$= \frac{cM}{\pi^{\frac{N}{2}}} \int_{\mathbb{R}^N} e^{-|\eta|^2 + c|x-\eta\sqrt{2\lambda t}|^\gamma} d\eta \le$$

(by the elementary inequality $(a+b)^\gamma \le 2^{\gamma-1}(a^\gamma + b^\gamma)$ that holds for $a, b > 0$ and $\gamma \ge 1$)

$$\le c_T e^{2c|x|^\gamma},$$

where

$$c_T = \frac{cM}{\pi^{\frac{N}{2}}} \int_{\mathbb{R}^N} e^{-\eta^2 + 2c|\eta\sqrt{2\lambda T}|^\gamma} d\eta. \qquad \square$$

Remark 6.17 Let us suppose that the operator L_a in (6.1)-(6.2) satisfies Hypotheses 6.1, 6.3 and 6.5. Then its fundamental solution Γ satisfies the *reproduction (or semigroup) property:* for every $t_0 < t < T$ and $x, y \in \mathbb{R}^N$ we have

$$\int_{\mathbb{R}^N} \Gamma(T, x; t, \eta)\Gamma(t, \eta; t_0, y) d\eta = \Gamma(T, x; t_0, y). \tag{6.26}$$

Formula (6.26) is an immediate consequence of the representation formula (6.23) and of the uniqueness of the solution of the Cauchy problem

$$\begin{cases} L_a u = 0, & \text{in }]t, T[\times\mathbb{R}^N, \\ u(t, \cdot) = \Gamma(t, \cdot; t_0, y), & \text{on } \mathbb{R}^N. \end{cases}$$

Further, if $a = 0$, then Γ is a density, i.e.

$$\int_{\mathbb{R}^N} \Gamma(T, x; t, y) dy = 1, \tag{6.27}$$

for every $t < T$ and $x, y \in \mathbb{R}^N$. Also (6.27) follows from the uniqueness of the representation (6.23) of the solution of the Cauchy problem with initial datum identically equal to one.

More generally, for a generic a, we have

$$\int_{\mathbb{R}^N} \Gamma(t, x; t_0, y) dy \le e^{-a_0(t-t_0)}. \tag{6.28}$$

Indeed (6.28) follows from the maximum principle, Theorem 6.13, applied to the function

$$u(t, x) = e^{-a_0(t-t_0)} - \int_{\mathbb{R}^N} \Gamma(t, x; t_0, y) dy, \qquad t \ge t_0, \ x \in \mathbb{R}^N.$$

Indeed we have $u(t_0, x) = 1$ and

$$L_a u(t, x) = L_a e^{-a_0(t-t_0)} = -(a(t, x) - a_0)e^{-a_0(t-t_0)} \le 0. \qquad \square$$

We conclude the paragraph proving a maximum estimate that is analogous to that of Theorem 6.12.

Theorem 6.18 *Under Hypotheses 6.1 and 6.3, let* $u \in C^{1,2}(\mathcal{S}_T) \cap C(\overline{\mathcal{S}}_T)$ *such that*

$$|u(t,x)| \leq Ce^{C|x|^2}, \qquad (t,x) \in \mathcal{S}_T,$$

for some constant C*. Then, setting*

$$a_1 := \max\{0, -a_0\},$$

we have

$$\sup_{\mathcal{S}_T} |u| \leq e^{a_1 T} \left(\sup_{\mathbb{R}^N} |u(0,\cdot)| + T \sup_{\mathcal{S}_T} |L_a u| \right). \tag{6.29}$$

Proof. If $a_0 \geq 0$, then, setting

$$w_{\pm} = \sup_{\mathbb{R}^N} |u(0,\cdot)| + t \sup_{\mathcal{S}_T} |L_a u| \pm u, \qquad \text{in } \mathcal{S}_T,$$

we have

$$\begin{cases} L_a w_{\pm} \leq -\sup\limits_{\mathcal{S}_T} |L_a u| \pm L_a u \leq 0, & \text{in } \mathcal{S}_T, \\ w_{\pm}(0,\cdot) \geq 0, & \text{on } \mathbb{R}^N, \end{cases}$$

and it is apparent that $w_{\pm}$ verify estimate (6.18) in $\mathcal{S}_T$. Therefore, by Theorem 6.13, $\omega_{\pm} \geq 0$ in $\mathcal{S}_T$ and this proves the claim.

If $a_0 < 0$, then we proceed as in the proof of Theorem 6.12. □

We will see in Chapter 12, that a priori estimates such as (6.29) play a crucial part in the proof of stability results of numerical schemes. Moreover, as a consequence of (6.29), if $u, v \in C^{1,2}(\mathcal{S}_T) \cap C(\overline{\mathcal{S}}_T)$ verify the exponential growth estimate, then

$$\sup_{\mathcal{S}_T} |u - v| \leq e^{a_1 T} \left(\sup_{\mathbb{R}^N} |u(0,\cdot) - v(0,\cdot)| + T \sup_{\mathcal{S}_T} |L_a u - L_a v| \right).$$

This formula gives an *estimate of the sensitivity of the solution of the Cauchy problem* (6.4) *with respect to variations of the initial datum and* f.

6.3 Non-negative solutions of the Cauchy problem

In this paragraph we assume that L_a has a fundamental solution and we prove that the family of non-negative functions (or, more generally, functions bounded from below) is a uniqueness class for L_a.

Theorem 6.19 *Under Hypotheses 6.1, 6.3, 6.5 and 6.6, there exists at most one function* $u \in C^{1,2}(\mathcal{S}_T) \cap C(\overline{\mathcal{S}}_T)$ *that is bounded from below and solves the problem*

$$\begin{cases} L_a u = 0, & \text{in } \mathcal{S}_T, \\ u(0,\cdot) = \varphi, & \text{on } \mathbb{R}^N. \end{cases}$$

We defer the proof to the end of the paragraph: it is based on the following result that generalizes Theorem 6.15, weakening the exponential growth condition.

Theorem 6.20 *Under Hypotheses 6.1, 6.3, 6.5 and 6.6, there exists at most one solution* $u \in C^{1,2}(\mathcal{S}_T) \cap C(\overline{\mathcal{S}}_T)$ *of the problem*

$$\begin{cases} L_a u = f, & \text{in } \mathcal{S}_T, \\ u(0,\cdot) = \varphi, & \text{on } \mathbb{R}^N, \end{cases}$$

for which there exists a constant C *such that*

$$\int_{\mathbb{R}^N} |u(t,x)| e^{-C|x|^2} dx < \infty, \tag{6.30}$$

for every $0 \le t \le T$.

Before proving the theorem, we dwell on a few remarks. First of all, it is apparent that condition (6.21) is stronger than (6.30). Moreover, Theorem 6.21 below shows that the non-negative solutions verify estimate (6.30) and consequently we have uniqueness within the class of non-negative functions. For uniformly parabolic operators this result was proven by Widder [338] for $N = 1$ and it was subsequently generalized by Kato [203] and Aronson [12], among others. The uniqueness results in Polidoro [284], Di Francesco and Pascucci [94] further examine the more general case of non-uniformly parabolic operators that arise in some financial models.

Theorem 6.21 *Under Hypotheses 6.1, 6.3 and 6.5, if* $u \in C^{1,2}(\mathcal{S}_T)$ *is a non-negative function such that* $L_a u \le 0$*, then*

$$\int_{\mathbb{R}^N} \Gamma(t,x;s,y) u(s,y) dy \le u(t,x), \tag{6.31}$$

for every $x \in \mathbb{R}^N$ *and* $0 < s < t < T$.

Proof. We consider a decreasing function $h \in C(\mathbb{R})$ such that $h(r) = 0$ for $r \ge 2$ and $h(r) = 1$ for $r \le 1$. For fixed $s \in]0,T[$, we set

$$g_n(s,y) = u(s,y) h\left(\frac{|y|}{n}\right), \qquad n \in \mathbb{N},$$

and

$$u_n(t,x) := \int_{\mathbb{R}^N} \Gamma(t,x;s,y) g_n(s,y) dy, \qquad (t,x) \in]s,T[\times \mathbb{R}^N, \quad n \in \mathbb{N}.$$

Since $y \mapsto g_n(s,y)$ is a continuous bounded function on $\mathbb{R}^N$, we have

$$\begin{cases} L_a(u - u_n) \le 0, & \text{in }]s,T[\times \mathbb{R}^N, \\ (u - u_n)(s,\cdot) = (u - g_n)(s,\cdot) \ge 0, & \text{on } \mathbb{R}^N. \end{cases}$$

Moreover, since g_n is bounded and has compact support, we have

$$\lim_{|x|\to\infty}\left(\sup_{t\in]s,T[} u_n(t,x)\right) = 0.$$

Therefore we can apply Lemma 6.14 to the function $u - u_n$ to get

$$u(t,x) \geq \int_{\mathbb{R}^N} \Gamma(t,x;s,y)g_n(s,y)dy \geq 0, \qquad (t,x) \in]s,T[\times\mathbb{R}^N,$$

for every $n \in \mathbb{N}$. Since g_n is an increasing sequence of non-negative functions tending to u, the claim follows by taking the limit in n and using Beppo Levi theorem.

We point out that in the proof we used only one part of Hypothesis 6.5, precisely the fact that L_a has a non-negative fundamental solution. □

Now we prove a corollary of Theorem 6.21 that, as we shall see in Section 7.3.2, has a very interesting financial interpretation.

Corollary 6.22 *Let Hypotheses 6.1, 6.3 and 6.5 hold and suppose that $a = 0$. If $u \in C^{1,2}(\mathcal{S}_T)$ is a function that is bounded from below such that $Lu \leq 0$, then* (6.31) *holds, i.e.*

$$\int_{\mathbb{R}^N} \Gamma(t,x;s,y)u(s,y)dy \leq u(t,x), \tag{6.32}$$

for every $x \in \mathbb{R}^N$ and $0 < s < t < T$.

Proof. Let $u_0 = \inf\limits_{\mathcal{S}_T} u$. Then, since $a = 0$, we have $L(u - u_0) = Lu \leq 0$ and, by Theorem 6.21,

$$\int_{\mathbb{R}^N} \Gamma(t,x;s,y)\left(u(s,y) - u_0\right)dy \leq u(t,x) - u_0.$$

The claim follows from (6.27). □

Proof (of Theorem 6.20). In view of the linearity of the problem, it suffices to prove that, if $L_a u = 0$ and $u(0,\cdot) = 0$, then $u = 0$. Let us consider an arbitrary point $(t_0, x_0) \in \mathcal{S}_T$ and let us show that $u(t_0,x_0) = 0$. To this end, we use the classical *Green's identity*:

$$\begin{aligned} vL_a u - uL_a^* v = &\sum_{i=1}^N \partial_{x_i}\left(\sum_{j=1}^N \left(\frac{c_{ij}}{2}\left(v\partial_{x_j}u - u\partial_{x_j}v\right) - \frac{uv}{2}\partial_{x_j}c_{ij}\right) + uvb_i\right) \\ &- \partial_t(uv), \end{aligned} \tag{6.33}$$

that follows directly from the definition of adjoint operator L_a^* in (6.12). We use identity (6.33) with

$$v(s,y) = h_R(y)\Gamma(t_0,x_0;s,y), \qquad (s,y) \in \mathcal{S}_{t_0-\varepsilon},$$

where $\varepsilon > 0$ and $h_R \in C^2(\mathbb{R}^N)$ is such that

$$0 \le h_R \le 1, \qquad h_R(y) = \begin{cases} 1, & \text{if } |y-x_0| \le R, \\ 0, & \text{if } |y-x_0| \ge 2R, \end{cases}$$

and

$$|\nabla h_R| \le \frac{1}{R}, \qquad |\partial_{y_i y_j} h_R| \le \frac{2}{R^2}, \quad i,j = 1,\dots,N. \tag{6.34}$$

Let B_R denote the ball in $\mathbb{R}^N$ with center x_0 and radius R: integrating Green's identity over the domain $]0,t_0-\varepsilon[\times B_{2R}$, by the divergence theorem we get

$$\begin{aligned} J_{R,\varepsilon} &:= \int_0^{t_0-\varepsilon}\int_{B_{2R}} u(s,y) L_a^*\left(h_R(y)\Gamma(t_0,x_0;s,y)\right) dyds \\ &= \int_{B_{2R}} h_R(y)\Gamma(t_0,x_0;t_0-\varepsilon,y)u(t_0-\varepsilon,y)dy =: I_{R,\varepsilon}. \end{aligned} \tag{6.35}$$

Here we have used the fact that $L_a u = 0$, $u(0,\cdot) = 0$ and some integral over the boundary cancels since h is zero (with its derivatives) on the boundary of B_{2R}.

Now, by condition (6.30) and by the estimate from above of Γ in (6.10), if ε is small enough we have

$$\Gamma(t_0,x_0;t_0-\varepsilon,\cdot)u(t_0-\varepsilon,\cdot) \in L^1(\mathbb{R}^N).$$

So, by the dominated convergence theorem,

$$\lim_{R\to\infty} I_{R,\varepsilon} = \int_{\mathbb{R}^N} \Gamma(t_0,x_0;t_0-\varepsilon,y)u(t_0-\varepsilon,y)dy. \tag{6.36}$$

On the other hand, since $L_a^*\Gamma(t_0,x_0;\cdot,\cdot) = 0$ in $]0,t_0-\varepsilon[\times B_{2R}$, we get

$$\begin{aligned} J_{R,\varepsilon} = \int_0^{t_0-\varepsilon}\int_{B_{2R}\setminus B_R} u(s,y)\Bigg[\sum_{i,j=1}^N \frac{c_{ij}}{2}\big(\Gamma(t_0,x_0;s,y)\partial_{y_i y_j} h_R(y) \\ +2\partial_{y_j}\Gamma(t_0,x_0;s,y)\partial_{y_i} h_R(y)\big) + \sum_{i=1}^N b_i^*\Gamma(t_0,x_0;s,y)\partial_{y_i} h_R(y)\Bigg] dyds, \end{aligned} \tag{6.37}$$

with b_i^* in (6.13). Now we use estimates (6.34) on the derivatives of h_R, the estimate from above of Γ in (6.10), estimate (6.11) of the first order derivatives of Γ and the assumption of linear growth of b^* and we obtain

$$|J_{R,\varepsilon}| \leq \text{const} \int_0^{t_0-\varepsilon} \frac{1}{t_0 - s} \int_{B_{2R}\setminus B_R} \frac{|y|}{R} \Gamma_\lambda(t_0, x_0; s, y)|u(s, y)|dyds$$
$$\leq \frac{\text{const}}{\varepsilon} \int_0^{t_0-\varepsilon} \int_{B_{2R}\setminus B_R} e^{-\frac{|y|^2}{2\lambda\varepsilon}} |u(s, y)|dyds.$$

Therefore, by condition (6.30), if $\varepsilon > 0$ is small enough, we have

$$\lim_{R\to\infty} J_{R,\varepsilon} = 0.$$

In conclusion, gathering (6.35), (6.36) and the previous result, we get

$$\int_{\mathbb{R}^N} \Gamma(t_0, x_0; t_0 - \varepsilon, y)u(t_0 - \varepsilon, y)dy = 0.$$

Taking the limit as $\varepsilon \to 0^+$ we infer that $u(t_0, x_0) = 0$. □

We conclude the paragraph with the

Proof (of Theorem 6.19). If u is non-negative, it suffices to observe that u verifies a condition analogous to (6.30) that can be obtained easily by using the estimate from below of Γ in (6.10) and by Theorem 6.21 that ensures that

$$\int_{\mathbb{R}^N} \Gamma(t, 0; s, y)u(s, y)dy < \infty.$$

If u is bounded from below, we can easily go back to the previous case by a substitution $v = u + C$: we observe that we can always perform a further substitution $v(t, x) = e^{\alpha t}u(t, x)$, so it is not restrictive to assume $a \geq 0$. □

7

Black-Scholes model

In this chapter we present some of the fundamental ideas of arbitrage pricing in continuous time, illustrating Black-Scholes theory from a point of view that is, as far as possible, elementary and close to the original ideas in the papers by Merton [250], Black and Scholes [49]. In Chapter 10 the topic will be treated in a more general fashion, fully exploiting martingale and PDEs theories.

In the Black-Scholes model the market consists of a non-risky asset, a bond B and of a risky asset, a stock S. The bond price verifies the equation

$$dB_t = rB_t dt$$

where r is the short-term (or locally risk-free) interest rate, assumed to be a constant. Therefore the bond follows a deterministic dynamics: if we set $B_0 = 1$, then

$$B_t = e^{rt}. \tag{7.1}$$

The price of the risky asset is a geometric Brownian motion, verifying the equation

$$dS_t = \mu S_t dt + \sigma S_t dW_t, \tag{7.2}$$

where $\mu \in \mathbb{R}$ is the average rate of return and $\sigma \in \mathbb{R}_{>0}$ is the volatility. In (7.2), $(W_t)_{t\in[0,T]}$ is a real Brownian motion on the probability space $(\Omega, \mathcal{F}, P, (\mathcal{F}_t))$. Recall that the explicit expression of the solution of (7.2) is

$$S_t = S_0 e^{\sigma W_t + \left(\mu - \frac{\sigma^2}{2}\right)t}. \tag{7.3}$$

In what follows we study European-style derivatives in a Markovian setting and we consider payoffs of the form $F(S_T)$, where T is the maturity and F is a function defined on $\mathbb{R}_{>0}$. The most important example is the European Call option with strike K and maturity T:

$$F(S_T) = (S_T - K)^+.$$

In Section 7.6 we study Asian-style derivatives, whose payoff depends on an average of the prices of the underlying asset.

Pascucci A.: PDE and Martingale Methods in Option Pricing

7.1 Self-financing strategies

Let us introduce some definitions that extend in a natural way the concepts analyzed in discrete time in Chapter 2.

Definition 7.1 *A strategy (or portfolio) is a stochastic process* (α_t, β_t) *where* $\alpha \in \mathbb{L}^2_{\mathrm{loc}}$ *and* $\beta \in \mathbb{L}^1_{\mathrm{loc}}$. *The value of the portfolio* (α, β) *is the stochastic process defined by*

$$V_t^{(\alpha,\beta)} = \alpha_t S_t + \beta_t B_t. \tag{7.4}$$

As usual α, β are to be interpreted as the amount of S and B held by the investor in the portfolio: let us point out that short-selling is allowed, so α, β can take negative values. Where there is no risk of ambiguity, we simply write V instead of $V^{(\alpha,\beta)}$.

Intuitively the assumption that α, β have to be progressively measurable[1] describes the fact that the investment strategy depends only on the amount of information available at that moment.

Definition 7.2 *A strategy* (α_t, β_t) *is self-financing if*

$$dV_t^{(\alpha,\beta)} = \alpha_t dS_t + \beta_t dB_t \tag{7.5}$$

holds, that is

$$V_t^{(\alpha,\beta)} = V_0^{(\alpha,\beta)} + \int_0^t \alpha_s dS_s + \int_0^t \beta_s dB_s. \tag{7.6}$$

We observe that, since S is a continuous and adapted stochastic process we have that $\alpha S \in \mathbb{L}^2_{\mathrm{loc}}$ and then the stochastic integral in (7.6) is well defined. Equation (7.5) is the continuous version[2] of the relation

$$\Delta V = \alpha \Delta S + \beta \Delta B$$

valid for discrete self-financing portfolios (cf. (2.7)): from a purely intuitive point of view, this expresses the fact that the instantaneous variation of the value of the portfolio is caused uniquely by the changes of the prices of the assets, and not by injecting or withdrawing funds from outside.

Let us now take a strategy (α, β) and define the discounted prices

$$\widetilde{S}_t = e^{-rt} S_t, \qquad \widetilde{V}_t = e^{-rt} V_t.$$

[1] In the discrete case we considered *predictable* strategies: for the sake of simplicity, in the continuous case we prefer to assume the condition (not really restrictive indeed) that α, β are progressively measurable.

[2] If α, β are Itô processes, by the two-dimensional Itô formula we have

$$dV_t^{(\alpha,\beta)} = \alpha_t dS_t + \beta_t dB_t + S_t d\alpha_t + B_t d\beta_t + d\langle \alpha, S \rangle_t,$$

and the condition that (α, β) is self-financing is equivalent to

$$S_t d\alpha_t + B_t d\beta_t + d\langle \alpha, S \rangle_t = 0.$$

The following proposition gives a remarkable characterization of the self-financing condition.

Proposition 7.3 *A strategy (α, β) is self-financing if and only if*

$$d\widetilde{V}_t^{(\alpha,\beta)} = \alpha_t d\widetilde{S}_t$$

holds, that is

$$\widetilde{V}_t^{(\alpha,\beta)} = V_0^{(\alpha,\beta)} + \int_0^t \alpha_s d\widetilde{S}_s. \tag{7.7}$$

Remark 7.4 Thanks to (7.7), *the value of a self-financing strategy (α, β) is determined uniquely by its initial value $V_0^{(\alpha,\beta)}$ and by the process α* that is the amount of risky stock held by the investor in the portfolio. The integral in (7.7) equals the difference between the final and initial discounted values and therefore represents the *gain of the strategy.*

When an initial value $V_0 \in \mathbb{R}$ and a process $\alpha \in \mathbb{L}^2_{\text{loc}}$ are given, we can construct a strategy (α, β) by putting

$$\widetilde{V}_t = V_0 + \int_0^t \alpha_s d\widetilde{S}_s, \qquad \beta_t = \frac{V_t - \alpha_t S_t}{B_t}.$$

By construction (α, β) *is a self-financing strategy with initial value* $V_0^{(\alpha,\beta)} = V_0$. In other words, a self-financing strategy can be indifferently set by specifying the processes α, β or the initial value V_0 and the process α. □

Proof (of Proposition 7.3). Given a strategy (α, β), we obviously have

$$\beta_t B_t = V_t^{(\alpha,\beta)} - \alpha_t S_t. \tag{7.8}$$

Furthermore

$$d\widetilde{S}_t = -re^{-rt} S_t dt + e^{-rt} dS_t \tag{7.9}$$

$$= (\mu - r)\widetilde{S}_t dt + \sigma \widetilde{S}_t dW_t. \tag{7.10}$$

Then (α, β) is self-financing if and only if

$$\begin{aligned} d\widetilde{V}_t^{(\alpha,\beta)} &= -r\widetilde{V}_t^{(\alpha,\beta)} dt + e^{-rt} dV_t \\ &= -r\widetilde{V}_t^{(\alpha,\beta)} dt + e^{-rt} \left(\alpha_t dS_t + \beta_t dB_t \right) = \end{aligned}$$

(since $dB_t = rB_t dt$ and by (7.8))

$$\begin{aligned} &= -r\widetilde{V}_t^{(\alpha,\beta)} dt + e^{-rt} \left(\alpha_t dS_t + rV_t^{(\alpha,\beta)} dt - r\alpha_t S_t dt \right) \\ &= e^{-rt} \alpha_t \left(dS_t - rS_t dt \right) = \end{aligned}$$

(by (7.9))

$$= \alpha_t d\widetilde{S}_t,$$

and this concludes the proof. □

Remark 7.5 Thanks to (7.10), condition (7.7) takes the more explicit form

$$\widetilde{V}_t^{(\alpha,\beta)} = \widetilde{V}_0^{(\alpha,\beta)} + (\mu - r)\int_0^t \alpha_s \widetilde{S}_s ds + \sigma \int_0^t \alpha_s \widetilde{S}_s dW_s. \tag{7.11}$$

This extends the result, proved in discrete time, according to which, if the discounted prices of the assets are martingales, then also the self-financing discounted portfolios built upon those assets are martingales.

Indeed, by (7.10), the discounted price $\widetilde{S}_t$ of the underlying asset is a martingale[3] if and only if $\mu = r$ in (7.2). Under this condition $\widetilde{S}$ is a martingale and we have

$$d\widetilde{S}_t = \sigma \widetilde{S}_t dW_t; \tag{7.12}$$

moreover (7.11) becomes

$$d\widetilde{V}_t^{(\alpha,\beta)} = \sigma \widetilde{S}_t \partial_s f(t, S_t) dW_t,$$

and therefore $\widetilde{V}^{(\alpha,\beta)}$ is a (local) martingale. □

7.2 Markovian strategies and Black-Scholes equation

Definition 7.6 *A strategy (α_t, β_t) is Markovian if*

$$\alpha_t = \alpha(t, S_t), \qquad \beta_t = \beta(t, S_t)$$

where α, β are functions in $C^{1,2}([0,T[\times \mathbb{R}_{>0})$.

The value of a Markovian strategy (α, β) is a function of time and of the price of the underlying asset:

$$f(t, S_t) := V_t^{(\alpha,\beta)} = \alpha(t, S_t) S_t + \beta(t, S_t) e^{rt}, \qquad t \in [0, T[, \tag{7.13}$$

with $f \in C^{1,2}([0,T[\times \mathbb{R}_{>0})$.

We point out that the function f in (7.13) is *uniquely determined by (α, β):* if

$$V_t^{(\alpha,\beta)} = f(t, S_t) = g(t, S_t) \quad \text{a.s.}$$

then $f = g$ in $[0,T[\times \mathbb{R}_{>0}$. This follows from Proposition A.59 and by the fact that S_t has a strictly positive (log-normal) density on $\mathbb{R}_{>0}$. As we are going to use Proposition A.59 often, for the reader's convenience we recall it here:

---

[3] In this chapter we are not going to introduce the concept of EMM: we defer the rigorous justification of the steps above to Chapter 10, where we prove the existence of a probability measure equivalent to P, under which the dynamics of S is given by (7.2) with $\mu = r$.

Proposition 7.7 *Let X be a random variable with strictly positive density on $H \in \mathscr{B}$. If $g \in m\mathscr{B}$ is such that $g(X) = 0$ a.s. ($g(X) \geq 0$ a.s.) then $g = 0$ ($g \geq 0$) almost everywhere with respect to Lebesgue measure on H. In particular if g is continuous then $g = 0$ ($g \geq 0$) on H.*

The following result characterizes the self-financing condition of a Markovian portfolio in differential terms.

Theorem 7.8 *Suppose that (α, β) is a Markovian strategy and set $f(t, S_t) = V_t^{(\alpha,\beta)}$. The following two conditions are equivalent:*

i) (α, β) is self-financing;
ii) f is solution to the partial differential equation

$$\frac{\sigma^2 s^2}{2} \partial_{ss} f(t, s) + rs\partial_s f(t, s) + \partial_t f(t, s) = rf(t, s), \tag{7.14}$$

with $(t, s) \in [0, T[\times\mathbb{R}_{>0}$, and we have that[4]

$$\alpha(t, s) = \partial_s f(t, s). \tag{7.15}$$

Equation (7.14) is called Black-Scholes partial differential equation.

We have already seen Black-Scholes partial differential equation in Section 2.3.6 as the asymptotic version of the binomial algorithm.

Theorem 7.8 relates the self-financing condition to a partial differential equation whose coefficients depend on the volatility σ of the risky asset and on the risk-free rate r, but *they do not depend on the average rate of return μ*. After examining the elementary example of Section 1.2 and the discrete case in Section 2.1, this fact should not come as a surprise: as we have already pointed out, arbitrage pricing does not depend on the subjective estimate of the future value of the risky asset.

We remark that, for a portfolio based upon formulas (7.14)-(7.15), *a inaccurate estimate of the parameters σ and r of the model might affect the self-financing property of the strategy:* for example, this means that if we change those parameters in itinere (e.g. after a re-calibration of the model), then the strategy might need more funds than the ones earmarked at the initial time. This might cause unwanted effects when we are using that strategy to hedge a derivative: if we modify the value of σ, hedging might actually cost more than expected at the beginning on the basis of the self-financing condition.

Proof (of Theorem 7.8). [$i) \Rightarrow ii)$] By the self-financing condition and expression (7.2) of S, we have that

$$dV_t^{(\alpha,\beta)} = (\alpha_t \mu S_t + \beta_t r B_t)dt + \alpha_t \sigma S_t dW_t. \tag{7.16}$$

[4] Let us recall that the expression of the process β can be obtained from α and $V_0^{(\alpha,\beta)}$, by Remark 7.4. More precisely:

$$\beta(t, s) = e^{-rt} \left(f(t, s) - s\partial_s f(t, s) \right).$$

Then, by the Itô formula and putting for brevity $f = f(t, S_t)$, we have that

$$\begin{aligned} dV_t^{(\alpha,\beta)} &= \partial_t f dt + \partial_s f dS_t + \frac{1}{2}\partial_{ss} f d\langle S\rangle_t \\ &= \left(\partial_t f + \mu S_t \partial_s f + \frac{\sigma^2 S_t^2}{2}\partial_{ss} f\right) dt + \sigma S_t \partial_s f dW_t. \end{aligned} \tag{7.17}$$

From the uniqueness of the representation of an Itô process (cf. Proposition 5.3) we infer the equality of the terms in dt and dW_t in (7.16) and (7.17). Then, concerning the terms in dW_t, since σS_t is strictly positive, we obtain

$$\alpha_t = \partial_s f(t, S_t) \quad \text{a.s.} \tag{7.18}$$

hence, by Proposition 7.7, we get relation (7.15).

Concerning now the terms in dt, by (7.18), we get

$$\partial_t f + \frac{\sigma^2 S_t^2}{2}\partial_{ss} f - r\beta_t B_t = 0 \quad \text{a.s.} \tag{7.19}$$

Substituting the expression

$$\beta_t B_t = f - S_t \partial_s f \quad \text{a.s.}$$

in (7.19), we get

$$\partial_t f(t, S_t) + r S_t \partial_s f(t, S_t) + \frac{\sigma^2 S_t^2}{2}\partial_{ss} f(t, S_t) - r f(t, S_t) = 0, \qquad \text{a.s.} \tag{7.20}$$

therefore, by Proposition 7.7, f is a solution of the deterministic differential equation (7.14).

$[ii) \Rightarrow i)]$ By the Itô formula, we have

$$dV_t^{(\alpha,\beta)} = df(t, S_t) = \partial_s f(t, S_t) dS_t + \left(\frac{\sigma^2 S_t^2}{2}\partial_{ss} f(t, S_t) + \partial_t f(t, S_t)\right) dt =$$

(since, by assumption, f is a solution of equation (7.14))

$$= \partial_s f(t, S_t) dS_t + r(f(t, S_t) - S_t \partial_s f(t, S_t)) dt = \tag{7.21}$$

(by (7.15) and since $dB_t = rB_t dt$)

$$= \alpha_t dS_t + \beta_t dB_t,$$

therefore (α, β) is self-financing. □

There is an intimate connection between the Black-Scholes equation (7.14) and the heat differential equation. To see this, let us consider the change of variables

$$t = T - \tau, \qquad s = e^{\sigma x},$$

and let us put

$$u(\tau,x)=e^{ax+b\tau}f(T-\tau,e^{\sigma x}),\qquad \tau\in[0,T],\ x\in\mathbb{R},\tag{7.22}$$

where a,b are constants to be chosen appropriately afterwards. We obtain

$$\begin{aligned}\partial_\tau u&=e^{ax+b\tau}\left(bf-\partial_t f\right),\\ \partial_x u&=e^{ax+b\tau}\left(af+\sigma e^{\sigma x}\partial_s f\right),\\ \partial_{xx} u&=e^{ax+b\tau}\left(a^2 f+2a\sigma e^{\sigma x}\partial_s f+\sigma^2 e^{\sigma x}\partial_s f+\sigma^2 e^{2\sigma x}\partial_{ss} f\right),\end{aligned}\tag{7.23}$$

hence

$$\frac{1}{2}\partial_{xx}u-\partial_\tau u=e^{ax+b\tau}\left(\frac{\sigma^2 s^2}{2}\partial_{ss}f+\left(\sigma a+\frac{\sigma^2}{2}\right)s\partial_s f+\partial_t f+\left(\frac{a^2}{2}-b\right)f\right)=$$

(if f solves (7.14))

$$=e^{ax+b\tau}\left(\left(\sigma a+\frac{\sigma^2}{2}-r\right)s\partial_s f+\left(\frac{a^2}{2}-b+r\right)f\right).$$

We have thus proved the following result.

Proposition 7.9 *Let*

$$a=\frac{r}{\sigma}-\frac{\sigma}{2},\qquad b=r+\frac{a^2}{2}.\tag{7.24}$$

Then the function f is a solution of the Black-Scholes equation (7.14) in $[0,T[\times\mathbb{R}_{>0}$ if and only if the function $u=u(\tau,x)$ defined by (7.22) satisfies the heat equation

$$\frac{1}{2}\partial_{xx}u-\partial_\tau u=0,\qquad \text{in }\ ]0,T]\times\mathbb{R}.\tag{7.25}$$

7.3 Pricing

Let us consider a European derivative with payoff $F(S_T)$. As in the discrete case, the arbitrage price equals by definition the value of a replicating strategy. In order to guarantee the well-posedness of such a definition, we ought to prove that there exists at least one replicating strategy (problem of market completeness) and that, if there exist more than one, all the replicating strategies have the same value (problem of absence of arbitrage).

In analytic terms, completeness and absence of arbitrage in the Black-Scholes model correspond to the problem of existence and uniqueness of the solution of a Cauchy problem for the heat equation. To make use of the results on differential equations, it is necessary to impose some conditions on the payoff function F (to ensure *the existence* of a solution) and narrow the family of admissible replicating strategies to a class of uniqueness for the Cauchy problem (to guarantee *the uniqueness* of the solution).

Hypothesis 7.10 *The function F is locally integrable on $\mathbb{R}_{>0}$, lower bounded and there exist two positive constants $a < 1$ and C such that*

$$F(s) \le Ce^{C|\log s|^{1+a}}, \qquad s \in \mathbb{R}_{>0}. \tag{7.26}$$

Condition (7.26) is not really restrictive: the function

$$e^{(\log s)^{1+a}} = s^{(\log s)^a}, \qquad s > 1,$$

grows, as $s \to +\infty$, less than an exponential but more rapidly than any polynomial function. This allows us to deal with the majority (if not all) of European-style derivatives actually traded on the markets.

Condition (7.26) is connected to the existence results of Appendix A.3: if we put $\varphi(x) = F(e^x)$, we obtain that φ is lower bounded and we have that

$$\varphi(x) \le Ce^{C|x|^{1+a}}, \qquad x \in \mathbb{R},$$

that is a condition analogous to (A.57).

Definition 7.11 *A strategy (α, β) is admissible if it is bounded from below, i.e. there exists a constant C such that*

$$V_t^{(\alpha,\beta)} \ge C, \qquad t \in [0, T], \ a.s. \tag{7.27}$$

We denote by $\mathcal{A}$ the family of Markovian, self-financing admissible strategies.

The financial interpretation of (7.27) is that investment strategies which request unlimited debt are not allowed. This condition is indeed realistic because banks or control institutions generally impose a limit to the investor's losses. We comment further on condition (7.27) in Section 7.3.2.

If $f(t, S_t) = V_t^{(\alpha,\beta)}$ with $(\alpha, \beta) \in \mathcal{A}$, then by Proposition 7.7, f is lower bounded so it belongs to the uniqueness class for the parabolic Cauchy problem studied in Section 6.3.

Definition 7.12 *A European derivative $F(S_T)$ is replicable if there exists an admissible portfolio $(\alpha, \beta) \in \mathcal{A}$ such that*[5]

$$V_T^{(\alpha,\beta)} = F(S_T) \ in\ \mathbb{R}_{>0}. \tag{7.28}$$

We say that (α, β) is a replicating portfolio for $F(S_T)$.

[5] Let $f(t, S_t) = V_t^{(\alpha,\beta)}$. If F is a continuous function, then (7.28) simply has to be understood in the pointwise sense: the limit

$$\lim_{(t,s)\to(T,\bar{s})} f(t, s) = F(\bar{s}),$$

exists for every $\bar{s} > 0$, which is tantamount to saying that f, defined on $[0, T[\times\mathbb{R}_{>0}$ can be prolonged by continuity on $[0, T] \times \mathbb{R}_{>0}$ and, by Proposition 7.7, $f(T, \cdot) = F$. More generally, if F is locally integrable then (7.28) is to be understood in the L^1_{loc} sense, cf. Section A.3.3.

The following theorem is the central result in Black-Scholes theory and gives the definition of arbitrage price of a derivative.

Theorem 7.13 *The Black-Scholes market model is complete and arbitrage-free, this meaning that every European derivative $F(S_T)$, with F verifying Hypothesis 7.10, is replicable in a unique way. Indeed there exists a unique strategy $h = (\alpha_t, \beta_t) \in \mathcal{A}$ replicating $F(S_T)$, that is given by*

$$\alpha_t = \partial_s f(t, S_t), \qquad \beta_t = e^{-rt} \left(f(t, S_t) - S_t \partial_s f(t, S_t) \right), \tag{7.29}$$

where f is the lower bounded solution of the Cauchy problem

$$\frac{\sigma^2 s^2}{2} \partial_{ss} f + rs\partial_s f + \partial_t f = rf, \qquad \text{in } [0, T[\times \mathbb{R}_{>0}, \tag{7.30}$$

$$f(T, s) = F(s), \qquad s \in \mathbb{R}_{>0}. \tag{7.31}$$

By definition, $f(t, S_t) = V_t^{(\alpha,\beta)}$ is the arbitrage price of $F(S_T)$.

Proof. A strategy (α, β) replicates $F(S_T)$ if and only if:

i) (α, β) *is Markovian and admissible*, so there exists $f \in C^{1,2}([0, T[\times \mathbb{R}_{>0})$ that is lower bounded and such that $V_t^{(\alpha,\beta)} = f(t, S_t)$;
ii) (α, β) *is self-financing*, so, by Theorem 7.8, f is solution of the differential equation (7.30), the first of formulas (7.29) holds and the second one follows by Remark 7.4;
iii) (α, β) *is replicating* so, by Proposition 7.7, f verifies the final condition (7.31).

To prove that (α, β) exists and is unique, let us transform problem (7.30)-(7.31) into a parabolic Cauchy problem in order to apply the results of existence and uniqueness of Appendices A.3 and 6.3. If we put

$$u(\tau, x) = e^{-r(T-\tau)} f(T - \tau, e^x), \qquad \tau \in [0, T], \ x \in \mathbb{R}, \tag{7.32}$$

we obtain that f is solution of (7.30)-(7.31) if and only if u is solution of the Cauchy problem

$$\begin{cases} \frac{\sigma^2}{2} \left(\partial_{xx} u - \partial_x u \right) + r\partial_x u - \partial_\tau u = 0, & (t, x) \in]0, T] \times \mathbb{R}, \\ u(0, x) = e^{-rT} F(e^x), & x \in \mathbb{R}. \end{cases}$$

By Hypothesis 7.10 and the lower boundedness of F, Theorem A.77 guarantees the existence of a lower bounded solution u. Furthermore, by Theorem 6.19, u is the only solution belonging to the class of lower bounded functions. Thus the existence of a replicating strategy and its uniqueness within the class of lower bounded functions follow immediately. □

Remark 7.14 The admissibility condition (7.27) can be replaced by the growth condition

$$|f(t, s)| \le C e^{C(\log s)^2}, \qquad s \in \mathbb{R}_{>0}, \ t \in]0, T[.$$

In this case, by the uniqueness of the solution guaranteed by Theorem 6.15, we obtain a result that is analogous to that of Theorem 7.13. □

Corollary 7.15 (Black-Scholes Formula) *Let us assume the Black-Scholes dynamics for the underlying asset*

$$dS_t = \mu S_t dt + \sigma S_t dW_t,$$

and let us denote by r the short rate. Then, if K is the strike price and T is the maturity, the following formulas for the price of European Call and Put options hold:

$$\begin{aligned} c_t &= S_t \Phi(d_1) - K e^{-r(T-t)} \Phi(d_2), \\ p_t &= K e^{-r(T-t)} \Phi(-d_2) - S_t \Phi(-d_1), \end{aligned} \tag{7.33}$$

where

$$\Phi(x) = \frac{1}{\sqrt{2\pi}} \int_{-\infty}^{x} e^{-\frac{y^2}{2}} dy$$

is the standard normal distribution function and

$$d_1 = \frac{\log\left(\frac{S_t}{K}\right) + \left(r + \frac{\sigma^2}{2}\right)(T-t)}{\sigma\sqrt{T-t}},$$

$$d_2 = d_1 - \sigma\sqrt{T-t} = \frac{\log\left(\frac{S_t}{K}\right) + \left(r - \frac{\sigma^2}{2}\right)(T-t)}{\sigma\sqrt{T-t}}.$$

Proof. The claim follows directly from the representation formula for the solution of the Cauchy problem (7.30)-(7.31) for the Black-Scholes equation (or for the heat equation, by transformation (7.22)). We are not going through the explicit computations, already carried out in Section 2.3.5. □

7.3.1 Dividends and time-dependent parameters

Black-Scholes pricing formulas can be adapted to treat the case of a dividend-paying underlying asset. The simplest case is when we suppose a continuous payment with constant return q, i.e. we suppose that in the amount of time dt the dividend paid equals $qS_t dt$. In this case, since dividends paid by a stock reduce its value, we assume the following dynamics

$$dS_t = (\mu - q) S_t dt + \sigma S_t dW_t. \tag{7.34}$$

Moreover we modify the self-financing condition (7.5) as follows:

$$dV_t^{(\alpha,\beta)} = \alpha_t \left(dS_t + qS_t dt\right) + \beta_t dB_t. \tag{7.35}$$

Then, proceeding as in the proof of Theorem 7.8, we obtain[6] the modified Black-Scholes equation

$$\frac{\sigma^2 s^2}{2} \partial_{ss} f(t,s) + (r-q) s \partial_s f(t,s) + \partial_t f(t,s) = r f(t,s).$$

[6] On one hand, inserting (7.34) in the self-financing condition (7.35), we get (cf. (7.16))

$$dV_t^{(\alpha,\beta)} = (\alpha_t \mu S_t + \beta_t r B_t) dt + \alpha_t \sigma S_t dW_t;$$

Therefore the Black-Scholes formula for the price of a dividend-paying Call option becomes

$$c_t = e^{-q(T-t)} S_t \Phi(\bar{d}_1) - K e^{-r(T-t)} \Phi(\bar{d}_1 - \sigma\sqrt{T-t}),$$

where

$$\bar{d}_1 = \frac{\log\left(\frac{S_t}{K}\right) + \left(r - q + \frac{\sigma^2}{2}\right)(T-t)}{\sigma\sqrt{T-t}}.$$

We can obtain explicit pricing formulas even when the parameters r, μ, σ are time-dependent deterministic functions:

$$dB_t = r(t) B_t dt,$$
$$dS_t = \mu(t) S_t dt + \sigma(t) S_t dW_t.$$

Let us suppose, for example, that r, μ, σ are continuous functions on $[0, T]$. Then we have

$$B_t = e^{\int_0^t r(s)ds},$$
$$S_t = S_0 \exp\left(\int_0^t \sigma(s)dW_s + \int_0^t \left(\mu(s) - \frac{\sigma^2(s)}{2}\right) ds\right).$$

Following the same arguments we obtain formulas that are analogous to the ones of Corollary 7.15 where the terms $r(T-t)$ and $\sigma\sqrt{T-t}$ must be replaced by

$$\int_t^T r(s)ds \quad \text{and} \quad \left(\int_t^T \sigma^2(s)ds\right)^{\frac{1}{2}},$$

respectively.

7.3.2 Admissibility and absence of arbitrage

In this section, we comment on the concept of admissibility of a strategy and on its relation with the absence of arbitrage in the Black-Scholes model.

As in the discrete case, an arbitrage is an investment strategy that requires a null initial investment, with nearly no risk, and that has the possibility of taking a future positive value. Let us formalize the concept into the following:

---

on the other hand, by the Itô formula for $V_t^{(\alpha,\beta)} = f(t, S_t)$, we have (cf. (7.17))

$$dV_t^{(\alpha,\beta)} = \left(\partial_t f + (\mu - q) S_t \partial_s f + \frac{\sigma^2 S_t^2}{2} \partial_{ss} f\right) dt + \sigma S_t \partial_s f dW_t,$$

and the modified Black-Scholes equation follows from the uniqueness of the representation of an Itô process.

Definition 7.16 *An arbitrage is a self-financing strategy* (α, β) *whose value* $V^{(\alpha,\beta)}$ *is such that*

i) $V_0^{(\alpha,\beta)} = 0$ *a.s.;*

and there exists $t_0 \in]0, T]$ *such that*

ii) $V_{t_0}^{(\alpha,\beta)} \geq 0$ *a.s.;*
iii) $P(V_{t_0}^{(\alpha,\beta)} > 0) > 0$.

In the binomial model the absence of arbitrage strategies is guaranteed under straightforward and intuitive assumptions summed up by condition (2.39) which expresses a relation between the return of the risky asset and the return of the bond. On the contrary, in the continuous-time models, the problem of existence of arbitrage opportunities is a very delicate matter. Indeed, without imposing an admissibility condition, even in the Black-Scholes market model it is possible to construct arbitrage portfolios, i.e. one can invest in the assets (7.1) and (7.3) with a self-financing strategy of null initial cost to obtain a risk-free profit.

In very loose terms[7], the idea is to use a strategy consisting in doubling the bet in case of loss: this is well known in gambling games. To fix the ideas, let us consider a coin-tossing game in which if we bet \$1 we get \$2 if the outcome is head, and nothing if the outcome is tail. In this case the doubling strategy consists in beginning by betting \$1 and keeping on gambling, doubling the bet every time one loses and then stopping the first time one wins. Thus, if one wins for the first time at the n-th game, the amount of money gained equals the difference between what one invested and lost in the game, precisely $1 + 2 + 4 + \cdots + 2^{n-1}$, and what one won at the n-th game, i.e. 2^n: so, the total wealth is positive and equals \$1. In this way one is sure to win if the following two conditions hold:

i) one can gamble an infinite number of times;
ii) one has at his/her disposal an infinite wealth.

In a discrete market with finite horizon, these strategies are automatically ruled out by i), cf. Proposition 2.12. In a continuous-time market, even in the case of finite horizon, it is necessary to impose some restrictions in order to rule out the "doubling strategies" which constitute an arbitrage opportunity: this motivates the admissibility condition of Definition 7.11.

The choice of the family of admissible strategies must be made in a suitable way: we have to be careful not to choose a family that is too wide (this might generate arbitrage opportunities), but also not too narrow (this to guarantee a certain degree of freedom in building replicating portfolios that make the market complete). In the literature different notions of admissibility can be found, not all of them being expressed in an explicit fashion: Definition 7.11 looks a simple and intuitive choice. In order to compare our notion of ad-

---

[7] For further details we refer, for example, to Steele [315], Chapter 14.

missibility to other ones, let us prove now that the class $\mathcal{A}$ does not contain arbitrage opportunities.

Proposition 7.17 (No-arbitrage principle) *The family $\mathcal{A}$ does not contain arbitrage strategies.*

Proof. The claim follows directly from Corollary 6.22. By contradiction, let $(\alpha, \beta) \in \mathcal{A}$, with $V_t^{(\alpha,\beta)} = f(t, S_t)$, be an arbitrage strategy: then f is lower bounded, it is a solution of the PDE (7.30) and we have that $f(0, S_0) = 0$. Moreover there exist $t \in]0, T]$ and $\bar{s} > 0$ such that $f(t, \bar{s}) > 0$ and $f(t, s) \geq 0$ for every $s > 0$. To use Corollary 6.22, let us transform the Black-Scholes PDE into a parabolic equation by substitution(7.32)

$$u(\tau, x) = e^{-r(T-\tau)} f(T - \tau, e^x), \qquad \tau \in [0, T], \ x \in \mathbb{R}.$$

Then u is a solution of the equation

$$\frac{\sigma^2}{2} (\partial_{xx} u - \partial_x u) + r\partial_x u - \partial_\tau u = 0, \tag{7.36}$$

and Corollary 6.22 leads to the absurd inequality:

$$0 = f(0, S_0) = u(T, \log S_0) \geq \int_{\mathbb{R}} \Gamma(T, \log S_0, T - t, y) u(T - t, y) dy > 0,$$

since $u(T - t, y) = e^{-rt} f(t, e^y) \geq 0$ for every $y \in \mathbb{R}$, $u(T - t, \log \bar{s}) = e^{-rt} f(t, \bar{s}) > 0$ and $\Gamma(T, \cdot, \tau, \cdot)$, the fundamental solution of (7.36) is strictly positive when $\tau < T$. □

7.3.3 Black-Scholes analysis: heuristic approaches

We present now some alternative ways to obtain the Black-Scholes equation (7.14). The following approaches are heuristic; their good point is that they are intuitive, while their flaw is they are not completely rigorous. Furthermore they share the fact that they assume the no-arbitrage principle as a starting point, rather than a result: we will comment briefly on this at the end of the section, in Remark 7.18. What follows is informal and not rigorous.

In the first approach, we aim at pricing a derivative H with maturity T assuming that its price at a time t in the form $H_t = f(t, S_t)$ with $f \in C^{1,2}$. To this end we consider a self-financing portfolio (α, β) and impose the replication condition

$$V_T^{(\alpha,\beta)} = H_T \quad \text{a.s.}$$

By the no-arbitrage principle, it must also hold that

$$V_t^{(\alpha,\beta)} = H_t \quad \text{a.s.}$$

for $t \leq T$. Proceeding as in the proof of Theorem 7.8, we impose that the stochastic differentials $dV_t^{(\alpha,\beta)}$ and $df(t, S_t)$ are equal to get (7.14) and the hedging strategy (7.15). The result thus obtained is formally identical: nevertheless in this way one could erroneously think that the Black-Scholes equation (7.14) is a consequence of the absence of arbitrage opportunities rather than a characterization of the self-financing condition.

Concerning the second approach, let us consider the point of view of a bank that sells an option and wants to determine a hedging strategy by investing in the underlying asset. Let us consider a portfolio consisting of a certain amount of the risky asset S_t and of a short position on a derivative with payoff $F(S_T)$ whose price, at the time t, is denoted by $f(t, S_t)$:

$$V(t, S_t) = \alpha_t S_t - f(t, S_t).$$

In order to determine α_t, we want to render V neutral with respect to the variation of S_t, or, in other terms, V immune to the variation of the price of the underlying asset by imposing the condition

$$\partial_s V(t, s) = 0.$$

By the equality $V(t, s) = \alpha_t s - f(t, s)$, we get[8]

$$\alpha_t = \partial_s f(t, s), \tag{7.37}$$

and this is commonly known as the *Delta hedging*[9] strategy. By the self-financing condition we have

$$\begin{aligned} dV(t, S_t) &= \alpha_t dS_t - df(t, S_t) \\ &= \left((\alpha_t - \partial_s f)\mu S_t - \partial_t f - \frac{\sigma^2 S_t^2}{2} \partial_{ss} f \right) dt + (\alpha_t - \partial_s f)\sigma S_t dW_t. \end{aligned}$$

Therefore the choice (7.37) wipes out the riskiness of V, represented by the term in dW_t, and cancels out also the term containing the return μ of the underlying asset. Summing up we get

$$dV(t, S_t) = -\left(\partial_t f + \frac{\sigma^2 S_t^2}{2} \partial_{ss} f \right) dt. \tag{7.38}$$

Now since the dynamics of V is deterministic, *by the no-arbitrage principle* V must have the same return of the non-risky asset:

$$dV(t, S_t) = rV(t, S_t)dt = r\left(S_t \partial_s f - f\right) dt, \tag{7.39}$$

so, equating formulas (7.38) and (7.39) we obtain again the Black-Scholes equation.

The idea that an option can be used to hedge risk is very intuitive and many arbitrage pricing techniques are based upon such arguments.

---

8 The attentive reader may wonder why, even though α_t is function of s, $\partial_s \alpha_t$ does not appear in the equation.

9 In common terminology, the derivative $\partial_s f$ is usually called *Delta*.

Remark 7.18 In the approaches we have just presented, the no-arbitrage principle, under different forms, is assumed as a hypothesis in the Black-Scholes model: this certainly helps intuition, but a rigorous justification of this might be hard to find. Indeed we have seen that in the Black-Scholes model arbitrage strategies actually exist, albeit they are pathological. In our presentation, as in other more probabilistic ones based upon the notion of EMM, all the theory is built upon the self-financing condition: in this approach, the absence of arbitrage opportunities is the natural *consequence* of the self-financing property. Intuitively this corresponds to the fact that if a strategy is adapted and self-financing, then it cannot reasonably generate a risk-free profit greater than the bond: in other words it cannot be an arbitrage opportunity. □

7.3.4 Market price of risk

Let us go back to the ideas of Section 1.2.4 and analyze the pricing and hedging of a derivative whose underlying asset is not exchanged on the market, supposing though that another derivative on the same underlying asset is traded. A noteworthy case is that of a derivative on the temperature: even though it is possible to construct a probabilistic model for the value of temperature, it is not possible to build up a replicating strategy that uses the underlying asset since this cannot be bought or sold; consequently we cannot exploit the argument of Theorem 7.13. Nevertheless, if on the market there already exists an option on the temperature, we can try to price and hedge a new derivative by means of that option.

Let us assume that the underlying asset follows the geometric Brownian motion dynamics

$$dS_t = \mu S_t dt + \sigma S_t dW_t, \tag{7.40}$$

even if the following results do not depend on the particular model considered. We suppose that a derivative on S is exchanged on the market, and that its price at time t is known. We assume also that this price can be written as $f(t, S_t)$, with $f \in C^{1,2}([0, T[\times \mathbb{R}_{>0})$. Finally we request that

$$\partial_s f \neq 0$$

and that suitable assumptions hold in order to guarantee the existence and the uniqueness of the solution of the Cauchy problem (7.49)-(7.50) below. Since we go through such conditions in Chapters 6 and 8, it seems unnecessary to recall them here.

By the Itô formula, we have

$$df(t, S_t) = Lf(t, S_t)dt + \sigma S_t \partial_s f(t, S_t)dW_t, \tag{7.41}$$

where

$$Lf(t, s) = \partial_t f(t, s) + \mu s \partial_s f(t, s) + \frac{\sigma^2 s^2}{2} \partial_{ss} f(t, s). \tag{7.42}$$

Our aim is to price a derivative with payoff $G(S_T)$. We imitate the technique of the preceding sections and build a Markovian self-financing portfolio on the bond and on the derivative f. We denote by $g(t, S_t)$ the value of such portfolio at time t,

$$g(t, S_t) = \alpha_t f(t, S_t) + \beta_t B_t, \tag{7.43}$$

and we impose the self-financing condition:

$$dg(t, S_t) = \alpha_t df(t, S_t) + \beta_t dB_t =$$

(by (7.41))

$$= (\alpha_t Lf(t, S_t) + r\beta_t B_t)\, dt + \alpha_t \sigma S_t \partial_s f(t, S_t) dW_t =$$

(since $\beta_t B_t = g(t, S_t) - \alpha_t f(t, S_t)$)

$$= (\alpha_t (Lf(t, S_t) - rf(t, S_t)) + rg(t, S_t))\, dt + \alpha_t \sigma S_t \partial_s f(t, S_t) dW_t. \tag{7.44}$$

Now we compare this expression with the stochastic differential obtained by the Itô formula

$$dg(t, S_t) = Lg(t, S_t)dt + \sigma S_t \partial_s g(t, S_t) dW_t.$$

By the uniqueness of the representation for an Itô process, we deduce the equality of the terms in dt and dW_t:

$$\alpha_t = \frac{\partial_s g(t, S_t)}{\partial_s f(t, S_t)}, \tag{7.45}$$

$$\alpha_t (Lf(t, S_t) - rf(t, S_t)) = Lg(t, S_t) - rg(t, S_t). \tag{7.46}$$

Substituting (7.45) into (7.46) and reordering the terms, we obtain

$$Lg(t, S_t) - rg(t, S_t) = \sigma S_t \lambda_f \partial_s g(t, S_t), \tag{7.47}$$

where

$$\lambda_f = \lambda_f(t, S_t) = \frac{Lf(t, S_t) - rf(t, S_t)}{\sigma S_t \partial_s f(t, S_t)}. \tag{7.48}$$

Finally, substituting expression (7.42) for L into (7.47), we have proved the following generalization of Theorems 7.8 and 7.13.

Theorem 7.19 *The portfolio given by (7.43) is self-financing if and only if g is solution of the differential equation*

$$\frac{\sigma^2 s^2}{2} \partial_{ss} g(t, s) + (\mu - \sigma \lambda_f(t, s))\, s \partial_s g(t, s) + \partial_t g(t, s) = rg(t, s), \tag{7.49}$$

with $(t, s) \in [0, T[\times \mathbb{R}_{>0}$. Under the assumptions of Theorem 7.13, there exists a unique replicating portfolio for $G(S_T)$, given by the solution of the Cauchy problem for (7.49) with terminal condition

$$g(T, s) = G(s), \qquad s \in \mathbb{R}_{>0}. \tag{7.50}$$

The value $(g(t, S_t))_{t \le T}$ is the arbitrage price of $G(S_T)$ and the replicating strategy is given by (7.45).

By Theorem 7.19, the replication of an option (and then the completeness of the market) is guaranteed even if the underlying asset is not exchanged, provided that on the market there exists another derivative on the same underlying asset.

If the underlying asset is traded, we can choose $f(t,s) = s$: in this case we simply denote $\lambda = \lambda_f$ and we observe that

$$\lambda = \frac{\mu - r}{\sigma}. \tag{7.51}$$

Substituting (7.51) into (7.49) we obtain exactly the Black-Scholes equation.

The coefficient λ represents the difference between the expected return μ and the riskless return r, that the investors request when buying S in order to take the risk represented by the volatility σ. For this very reason, λ is usually called *market price of risk* and it measures the investors' propensity to risk.

The market price of risk can be determined by the underlying asset (if exchanged) or by another derivative. Let us point out that (7.41) can be rewritten in a formally similar way to (7.40):

$$df(t,S_t) = \mu_f f(t,S_t)dt + \sigma_f f(t,S_t)dW_t,$$

where

$$\mu_f = \frac{Lf(t,S_t)}{f(t,S_t)}, \qquad \sigma_f = \frac{\sigma S_t \partial_s f(t,S_t)}{f(t,S_t)},$$

so, by definition (7.48), we have that

$$\lambda_f = \frac{\mu_f - r}{\sigma_f},$$

in analogy to (7.51).

We can now interpret the Black-Scholes differential equation (7.49) in a remarkable way: it is indeed equivalent to relation (7.47) that can be simply rewritten as

$$\lambda_f = \lambda_g. \tag{7.52}$$

To put this in another terms, the self-financing condition imposes that g and f share the same market price of risk. And since f and g are generic derivatives, (7.52) is actually a market consistency condition:

- *all the traded assets (or self-financing strategies) must have the same market price of risk.*

In the case of an incomplete market, where the only exchanged asset is the bond, the theoretical prices of the derivatives must verify a Black-Scholes equation similar to (7.49) but in this case the value of the market price of risk is not known, i.e. the coefficient λ_f that appears in the differential equation is unknown. Therefore the arbitrage price of an option is not unique, just as we have seen in the discrete case for the trinomial model.

7.4 Hedging

From a theoretical point of view the Delta-hedging strategy (7.37) guarantees a perfect replication of the payoff. So there would be no need to further study the hedging problem. However, in practice the Black-Scholes model poses some problems: first of all, the strategy (7.29) requires a continuous rebalancing of the portfolio, and this is not always possible or convenient, for example because of transition costs. Secondly, the Black-Scholes model is commonly considered too simple to describe the market realistically: the main issue lies in the hypothesis of constant volatility that appears to be definitely too strong if compared with actual data (see Paragraph 7.5).

The good point of the Black-Scholes model is that it yields explicit formulas for plain vanilla options. Furthermore, even though it has been severely criticized, it is still the reference model. At a first glance this might seem paradoxical but, as we are going to explain, it is not totally groundless.

The rest of the paragraph is structured as follows: in Section 7.4.1 we introduce the so-called sensitivities or Greeks: they are the derivatives of the Black-Scholes price with respect to the risk factors, i.e. the price of the underlying and the parameters of the model. In Section 7.4.2 we analyze the robustness of the Black-Scholes model, i.e. the effects its use might cause if it is not the "correct" model. In Section 7.4.3 we use the Greeks to get more effective hedging strategies than the mere Delta-hedging.

7.4.1 The Greeks

In the Black-Scholes model the value of a strategy is a function of several variables: the price of the underlying asset, the time to maturity and the parameters of the model, the volatility σ and the short-term rate r. From a practical point of view it is useful to be able to evaluate the sensitivity of the portfolio with respect to the variation of these factors: this means that we are able to estimate, for example, how the value of the portfolio behaves when we are getting closer to maturity or we are varying the risk-free rate or the volatility. The natural sensitivity indicators are the partial derivatives of the value of the portfolio with respect to the corresponding risk factors (price of the underlying asset, volatility, etc...). A Greek letter is commonly associated to every partial derivative, and for this reason these sensitivity measurements are usually called *the Greeks.*

Notation 7.20 *We denote by $f(t, s, \sigma, r)$ the value of a self-financing Markovian strategy in the Black-Scholes model, as a function of time t, of the price of the underlying s, of the volatility σ and of the short-term rate r. We put:*

$$\begin{aligned} \Delta &= \partial_s f && \textit{(Delta)}, \\ \Gamma &= \partial_{ss} f && \textit{(Gamma)}, \end{aligned}$$

$$\begin{aligned}
\mathcal{V} &= \partial_\sigma f && \textit{(Vega)},\\
\varrho &= \partial_r f && \textit{(Rho)},\\
\Theta &= \partial_t f && \textit{(Theta)}.
\end{aligned}$$

We say that a strategy is *neutral* with respect to one of the risk factors if the corresponding Greek is null, i.e. if the value of the portfolio is insensitive to the variation of such factor. For example, the Delta-hedging strategy is constructed in such a way that the portfolio becomes neutral to the Delta, i.e. insensitive with respect to the variation of the price of the underlying.

We can get an explicit expression for the Greeks of European Put and Call options, just by differentiating the Black-Scholes formula: some computations must be carried out, but with a little bit of shrewdness they are not particularly involved. In what follows we treat in detail only the call-option case. For the reader's convenience we recall the expression of the price at the time t of a European Call with strike K and maturity T :

$$c_t = g(d_1),$$

where g is the function defined by

$$g(d) = S_t \Phi(d) - K e^{-r(T-t)} \Phi(d - \sigma\sqrt{T-t}), \qquad d \in \mathbb{R},$$

and

$$\Phi(x) = \frac{1}{\sqrt{2\pi}} \int_{-\infty}^{x} e^{-\frac{y^2}{2}} dy, \qquad d_1 = \frac{\log\left(\frac{S_t}{K}\right) + \left(r + \frac{\sigma^2}{2}\right)(T-t)}{\sigma\sqrt{T-t}}.$$

The graph of the price of a Call option is shown in Figure 7.1. Sometimes it is convenient to use the following notation:

$$d_2 = d_1 - \sigma\sqrt{T-t} = \frac{\log\left(\frac{S_t}{K}\right) + \left(r - \frac{\sigma^2}{2}\right)(T-t)}{\sigma\sqrt{T-t}},$$

and the following lemma serves the purpose of simplifying the computations.

Lemma 7.21 *We have*

$$g'(d_1) = 0, \tag{7.53}$$

and consequently

$$S_t \Phi'(d_1) = K e^{-r(T-t)} \Phi'(d_1 - \sigma\sqrt{T-t}). \tag{7.54}$$

Proof. It is enough to observe that

$$\Phi'(x) = \frac{e^{-\frac{x^2}{2}}}{\sqrt{2\pi}}.$$

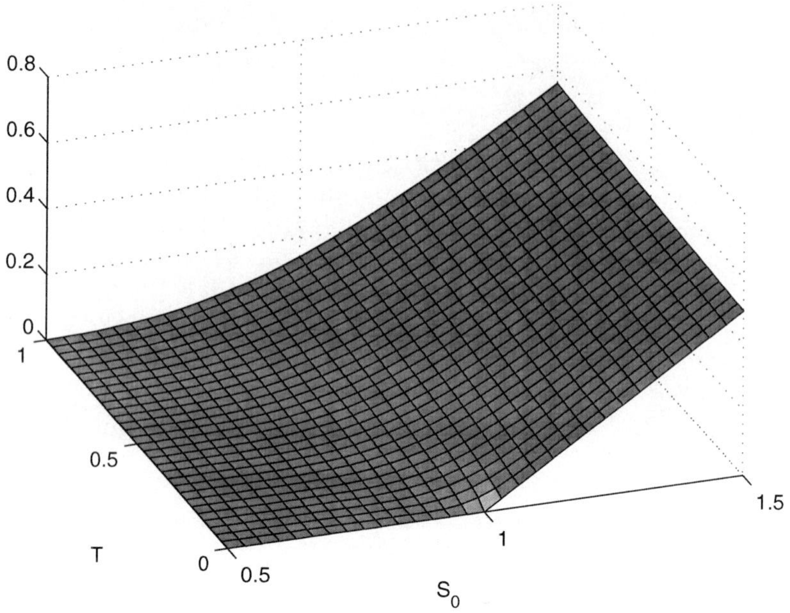

Fig. 7.1. Graph of the price of a European Call option in the Black-Scholes model, as a function of the price of the underlying asset and of time to maturity. The parameters are: strike $K = 1$, volatility $\sigma = 0.3$, risk-free rate $r = 0.05$

Then

$$
\begin{aligned}
g'(d) &= S_t \frac{e^{-\frac{d^2}{2}}}{\sqrt{2\pi}} - Ke^{-r(T-t)} \frac{e^{-\frac{(d-\sigma\sqrt{T-t})^2}{2}}}{\sqrt{2\pi}} \\
&= \frac{e^{-\frac{d^2}{2}}}{\sqrt{2\pi}} \left(S_t - Ke^{-\left(r+\frac{\sigma^2}{2}\right)(T-t)} e^{d\sigma\sqrt{T-t}} \right)
\end{aligned}
$$

and the claim follows immediately by the definition of d_1. □

Let us examine now every single Greek of a Call option.

Delta: we have

$$\Delta = \Phi(d_1). \tag{7.55}$$

Indeed

$$\Delta = \partial_s c_t = \Phi(d_1) + g'(d_1)\partial_s d_1,$$

and (7.55) follows by (7.53).
The graph of the Delta is shown in Figure 7.2. Let us point out that the Delta of the Call option is positive and less than one, because Φ is such:

$$0 < \Delta < 1.$$

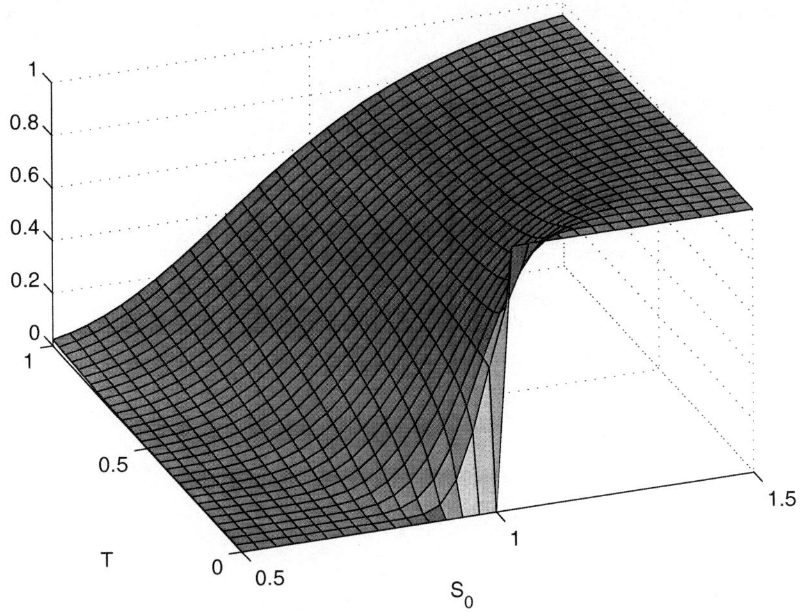

Fig. 7.2. Graph of the Delta of a European Call option in the Black-Scholes model, as a function of the price of the underlying asset and of time to maturity. The parameters are: strike $K = 1$, volatility $\sigma = 0.3$, risk-free rate $r = 0.05$

Since the Delta has to be interpreted as the amount of risky asset to be held in the Delta-hedging portfolio, this corresponds to the intuitive fact that we must buy the underlying asset in order to hedge a short position on a Call option. Let us note that

$$\lim_{s \to 0^+} d_1 = -\infty, \qquad \lim_{s \to +\infty} d_1 = +\infty,$$

so the following asymptotic expressions for price and Delta hold:

$$\lim_{s \to 0^+} c_t = 0, \qquad \lim_{s \to +\infty} c_t = +\infty,$$
$$\lim_{s \to 0^+} \Delta = 0, \qquad \lim_{s \to +\infty} \Delta = 1.$$

Gamma: we have

$$\Gamma = \frac{\Phi'(d_1)}{\sigma S_t \sqrt{T-t}}.$$

Indeed

$$\Gamma = \partial_s \Delta = \Phi'(d_1)\partial_s d_1.$$

The graph of the Gamma is shown in Figure 7.3. We note that the Gamma of a Call option is positive and therefore the price and the Delta are a

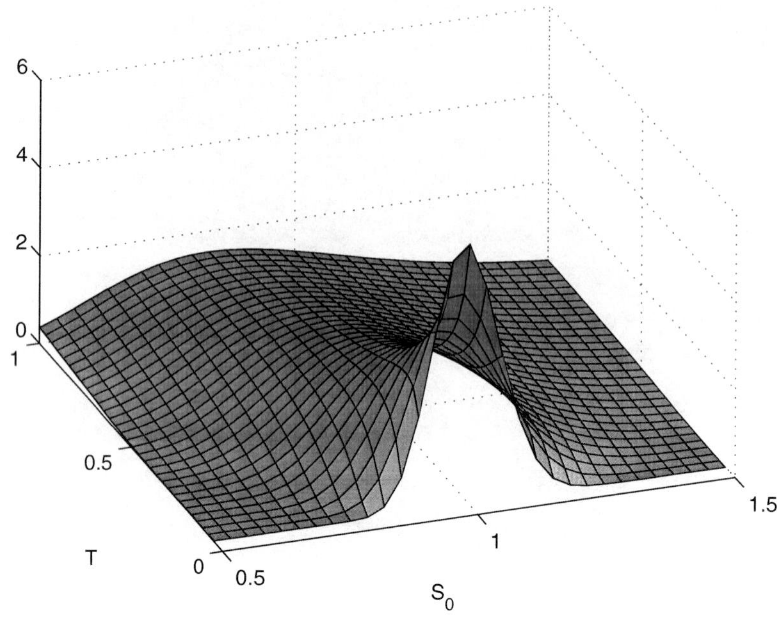

Fig. 7.3. Graph of the Gamma of a European Call option in the Black-Scholes model, as a function of the price of the underlying asset ($0.5 \le S \le 1.5$) and of time to maturity ($0.05 \le T \le 1$). The parameters are: strike $K = 1$, volatility $\sigma = 0.3$, risk-free rate $r = 0.05$

convex function and an increasing function with respect to the underlying asset, respectively. Furthermore we have that

$$\lim_{s \to 0^+} \Gamma = \lim_{s \to +\infty} \Gamma = 0.$$

Vega: we have

$$\mathcal{V} = S_t \sqrt{T - t}\, \Phi'(d_1).$$

Indeed

$$\mathcal{V} = \partial_\sigma c_t = g'(d_1)\partial_\sigma d_1 + K e^{-r(T-t)} \Phi'(d_1 - \sigma\sqrt{T-t})\sqrt{T-t} =$$

(by (7.53))

$$= K e^{-r(T-t)} \Phi'(d_1 - \sigma\sqrt{T-t})\sqrt{T-t} =$$

(by (7.54))

$$= S_t \sqrt{T-t}\, \Phi'(d_1).$$

The graph of the Vega is shown in Figure 7.4. The Vega is positive, so the price of a Call option is a strictly increasing function of the volatility (cf. Figure 7.5). Intuitively this is due to the fact that an option is a contract giving a right, not an obligation: therefore one takes advantage of

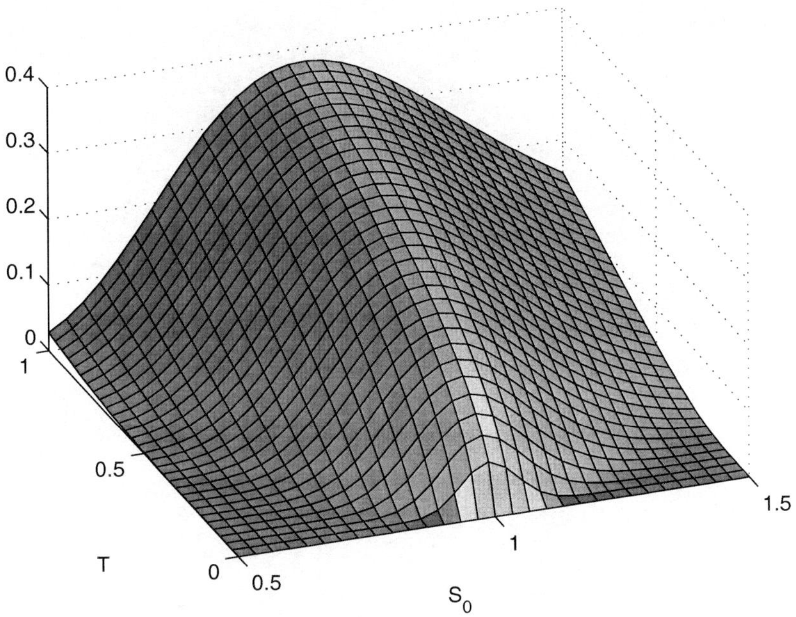

Fig. 7.4. Graph of the Vega of a European Call option in the Black-Scholes model, as a function of the price of the underlying asset and of time to maturity. The parameters are: strike $K = 1$, volatility $\sigma = 0.3$, risk-free rate $r = 0.05$

the greater riskiness of the underlying asset. It also follows that the price of the option is an *invertible* function of the volatility: in other terms, all other parameters being fixed, there is a unique value of the volatility that, plugged into the Black-Scholes formula, produces a given option price. This value is called *implied volatility*.
We show that

$$\lim_{\sigma \to 0^+} c_t = \left(S_t - Ke^{-r(T-t)}\right)^+, \qquad \lim_{\sigma \to +\infty} c_t = S_t \tag{7.56}$$

and so

$$\left(S_t - Ke^{-r(T-t)}\right)^+ < c_t < S_t,$$

in accordance with the estimates of Corollary 1.2, based upon arbitrage arguments. Indeed if we put

$$\lambda = \log\left(\frac{S_t}{K}\right) + r(T-t),$$

we have that $\lambda = 0$ if and only if

$$S_t = Ke^{-r(T-t)},$$

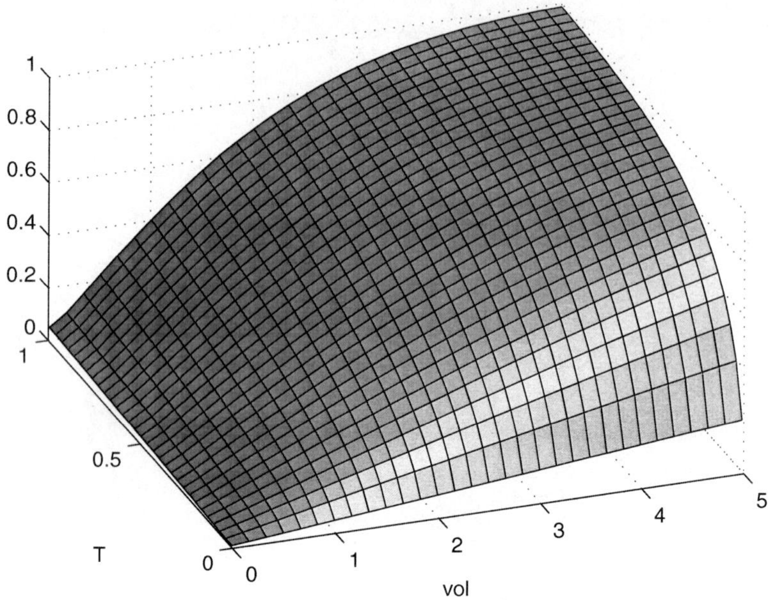

Fig. 7.5. Graph of the price of a European Call option in the Black-Scholes model, as a function of the price of the volatility ($0 \le \sigma \le 5$) and of time to maturity ($0.05 \le T \le 1$). The parameters are: $S = K = 1$, risk-free rate $r = 0.05$

and furthermore

$$\lim_{\sigma \to 0^+} d_1 = \begin{cases} +\infty, & \text{if } \lambda > 0, \\ 0, & \text{if } \lambda = 0, \\ -\infty, & \text{if } \lambda < 0. \end{cases}$$

So

$$\lim_{\sigma \to 0^+} c_t = \begin{cases} S_t - Ke^{-r(T-t)}, & \text{if } \lambda > 0, \\ 0, & \text{if } \lambda \le 0, \end{cases}$$

and this proves the first limit in (7.56). Then

$$\lim_{\sigma \to +\infty} d_1 = +\infty, \qquad \lim_{\sigma \to +\infty} d_2 = -\infty,$$

so that also the second limit in (7.56) follows easily.

Theta: we have

$$\Theta = -rKe^{-r(T-t)}\Phi(d_2) - \frac{\sigma S_t}{2\sqrt{T-t}}\Phi'(d_1). \tag{7.57}$$

Indeed

$$\Theta = \partial_t c_t = g'(d_1)\partial_t d_1 - rKe^{-r(T-t)}\Phi(d_2) - Ke^{-r(T-t)}\Phi'(d_2)\frac{\sigma}{2\sqrt{T-t}},$$

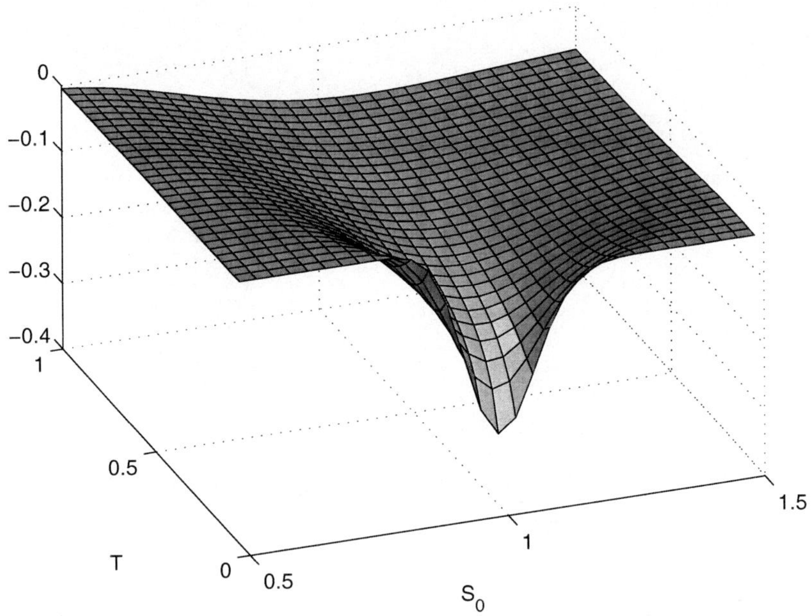

Fig. 7.6. Graph of the Theta of a European Call option in the Black-Scholes model, as a function of the price of the underlying asset ($0.5 \le S \le 1.5$) and of time to maturity ($0.05 \le T \le 1$). The parameters are: strike $K = 1$, volatility $\sigma = 0.3$, risk-free rate $r = 0.05$

and (7.57) follows from (7.54). The graph of the Theta is shown in Figure 7.6. Let us note that $\Theta < 0$ so the price of a Call option decreases when we get close to maturity: intuitively this is due to the lowering of the effect of the volatility, that is indeed multiplied in the expression for the price by a $\sqrt{T-t}$ factor.

Rho: we have

$$\varrho = K(T-t)e^{-r(T-t)}\Phi(d_2).$$

Indeed

$$\varrho = \partial_r c_t = g'(d_1)\partial_r d_1 + K(T-t)e^{-r(T-t)}\Phi(d_2),$$

and the claim follows from (7.53). The graph of the Rho is shown in Figure 7.7. Let us note that $\rho > 0$ and so the price of a Call option increases when the risk-free rate does so: this is due to the fact that if the Call is exercised, this imposes the payment of the strike K whose discounted value decreases as r increases.

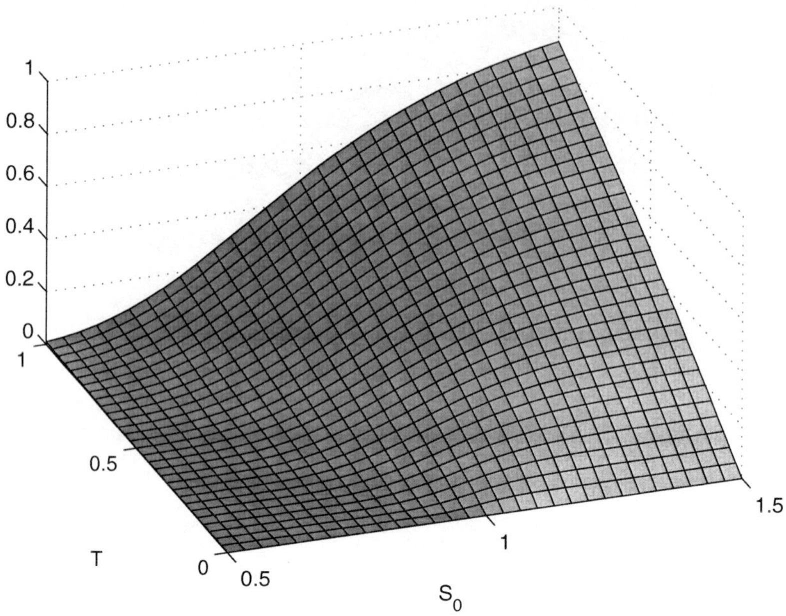

Fig. 7.7. Graph of the Rho of a European Call option in the Black-Scholes model, as a function of the price of the underlying asset and of time to maturity. The parameters are: strike $K = 1$, volatility $\sigma = 0.3$, risk-free rate $r = 0.05$

Let us mention without proof the expressions for the Greeks of a European Put option:

$$
\begin{aligned}
\Delta &= \partial_s p_t = \Phi(d_1) - 1, \\
\Gamma &= \partial_{ss} p_t = \frac{\Phi'(d_1)}{\sigma S_t \sqrt{T-t}}, \\
\mathcal{V} &= \partial_\sigma p_t = S_t \sqrt{T-t}\, \Phi'(d_1), \\
\Theta &= \partial_t p_t = rKe^{-r(T-t)} \left(1 - \Phi(d_2)\right) - \frac{\sigma S_t}{2\sqrt{T-t}} \Phi'(d_1), \\
\rho &= \partial_r p_t = K(T-t)e^{-r(T-t)} \left(\Phi(d_2) - 1\right).
\end{aligned}
$$

We point out that the Delta of a Put option is negative. Gamma and Vega have the same expression for both Put and Call options: in particular, the Vega is positive and so also the price of the Put option increases when the volatility does so. The Theta of a Put option may assume positive and negative values. The Rho of the Put is negative.

7.4.2 Robustness of the model

We assume the Black-Scholes dynamics for the underlying asset

$$dS_t = \mu S_t dt + \sigma S_t dW_t \tag{7.58}$$

where μ, σ are constant parameters and we denote by r the short-term rate. Then the price $f(t, S_t)$ of an option with payoff $F(S_T)$ is given by the solution of the Cauchy problem

$$\frac{\sigma^2 s^2}{2}\partial_{ss} f + rs\partial_s f + \partial_t f = rf, \qquad \text{in } [0, T[\times\mathbb{R}_{>0}, \tag{7.59}$$

$$f(T, s) = F(s), \qquad s \in \mathbb{R}_{>0}. \tag{7.60}$$

Moreover

$$f(t, S_t) = \alpha_t S_t + \beta_t B_t$$

is the value of the Delta-hedging strategy given by $\alpha_t = \partial_s f(t, S_t)$ and $\beta_t = f(t, S_t) - S_t \partial_s f(t, S_t)$.

Let us suppose now that the actual dynamics of the underlying asset is different from (7.58) and are described by an Itô process of the form

$$d\bar{S}_t = \mu_t \bar{S}_t dt + \sigma_t \bar{S}_t dW_t, \tag{7.61}$$

with $\mu_t \in \mathbb{L}^1_{\text{loc}}$ and $\sigma_t \in \mathbb{L}^2_{\text{loc}}$. On the basis of the final condition (7.60), the Delta-hedging strategy *replicates the payoff* $F(\bar{S}_T)$ *on any trajectory of the underlying asset.* However the fact that the actual dynamics (7.61) is different from the Black-Scholes' ones causes the *loss of the self-financing property:* in practice, this means that hedging has a different cost (possibly greater) with respect to the Black-Scholes price $f(0, \bar{S}_0)$. Indeed we have

$$df(t, \bar{S}_t) = \partial_s f d\bar{S}_t + \left(\partial_t f + \frac{\sigma_t^2 \bar{S}_t^2}{2}\partial_{ss} f\right) dt =$$

(by (7.59))

$$= \partial_s f d\bar{S}_t + \left(rf - r\bar{S}_t \partial_s f - \frac{(\sigma^2 - \sigma_t^2)\bar{S}_t^2}{2}\partial_{ss} f\right) dt$$

$$= \partial_s f d\bar{S}_t + \left(f - \bar{S}_t \partial_s f\right) dB_t - \frac{(\sigma^2 - \sigma_t^2)\bar{S}_t^2}{2}\partial_{ss} f dt. \tag{7.62}$$

More explicitly we have the following integral expression of the payoff

$$F(\bar{S}_T) = f(T, \bar{S}_T) = I_1 + I_2 + I_3$$

where

$$I_1 = f(0, \bar{S}_0)$$

is the Black-Scholes price,

$$I_2 = \int_0^T \partial_s f(t, \bar{S}_t) d\bar{S}_t + \int_0^T \left(f(t, \bar{S}_t) - \bar{S}_t \partial_s f(t, \bar{S}_t) \right) dB_t$$

is the gain of the Delta-hedging strategy,

$$I_3 = -\frac{1}{2} \int_0^T (\sigma^2 - \sigma_t^2) \bar{S}_t^2 \partial_{ss} f(t, \bar{S}_t) dt \tag{7.63}$$

is a correction term due to the erroneous specification of the model for the underlying asset. Clearly $I_3 = 0$ if $\sigma = \sigma_t$ and only in that case the strategy is self-financing.

We remark that I_3 depends only on the misspecification of the volatility term and *not on the drift*. More precisely I_3, which also represents *the replication error of the Delta-hedging strategy*, depends on the Vega which measures the convexity of the Black-Scholes price as a function of the price of the underlying asset. In particular the error is small if $\partial_{ss} f$ is small. Furthermore, if the price is convex, $\partial_{ss} f \geq 0$, as in the case of Call and Put options, then the Black-Scholes strategy (whose final value is $I_1 + I_2$) super-replicates the derivative *for any dynamics of the underlying asset* as long as we choose the volatility sufficiently large, $\sigma \geq \sigma_t$, since in this case $I_3 \leq 0$.

In this sense the Black-Scholes model is robust and, if used with all due precautions, can be effectively employed to hedge derivatives. Let us note finally that there exist options whose price is not a convex function of the underlying asset and so the Vega is not necessarily positive: this is the case of the digital option, corresponding to the Delta of a Call (see Figure 7.2), and also of some barrier options. Consequently in some cases in order to super-replicate it may be necessary to decrease the volatility.

7.4.3 Gamma and Vega-hedging

The Greeks can be used to determine more efficient hedging strategies than Delta-hedging. Here we consider the replication problem from a practical point of view: it is clear that theoretically the Delta-hedging approach offers perfect replication; nevertheless we have already mentioned some substantial problems we might have to face:

- the strategies are discrete and there are transition costs;
- the volatility is not constant.

As an example, in this section we consider the *Delta-Gamma* and *Delta-Vega-hedging* strategies whose purpose is to reduce the replication error due to the fact that rebalancing is not continuous in the first case and to the variation of the volatility in the second.

The reason why it is necessary to rebalance the Black-Scholes hedging portfolio is that the Delta changes as the underlying price varies. So, to minimize the number of times we have to rebalance (and the relative costs, of course), it seems natural to create a strategy that is neutral not only to the Delta but also to the Gamma. With all due adjustments, the procedure is similar to the Delta-hedging one in Section 7.3.3. Nevertheless in order to impose two neutrality conditions, one unknown is no longer sufficient, so it is necessary to build a portfolio with three assets. The situation is analogous to that of an incomplete market (cf. Section 2.4.1): indeed if continuous rebalancing is not allowed, not all derivatives are replicable and the Black-Scholes model loses its completeness property.

Let us suppose that we have sold a derivative $f(t, S_t)$ and we try to hedge the short position by investing on the underlying asset and on another derivative $g(t, S_t)$: the typical situation is when f is an exotic derivative and g is a plain vanilla option and we suppose it is exchanged on the market. We consider

$$V(t, S_t) = -f(t, S_t) + \alpha_t S_t + \beta_t g(t, S_t), \tag{7.64}$$

and we determine α, β by imposing the neutrality conditions

$$\partial_s V = 0, \qquad \partial_{ss} V = 0.$$

We get the system of equations

$$\begin{cases} -\partial_s f + \alpha_t + \beta_t \partial_s g = 0, \\ -\partial_{ss} f + \beta_t \partial_{ss} g = 0, \end{cases}$$

hence we deduce the Delta-Gamma-hedging strategy

$$\beta_t = \frac{\partial_{ss} f(t, S_t)}{\partial_{ss} g(t, S_t)}, \qquad \alpha_t = \partial_s f(t, S_t) - \frac{\partial_{ss} f(t, S_t)}{\partial_{ss} g(t, S_t)} \partial_s g(t, S_t).$$

We use a similar argument to reduce the uncertainty risk of the volatility parameter. The main assumption of the Black-Scholes model is that the volatility is constant, therefore the Delta-Vega-hedging strategy that we present in what follows is, in a certain sense, "beyond" the model. In this case also, the underlying asset is not sufficient and so we suppose there exists a second derivative which is exchanged on the market. Let us consider the portfolio (7.64) and let us impose the neutrality conditions

$$\partial_s V = 0, \qquad \partial_\sigma V = 0.$$

We get the system of equations

$$\begin{cases} -\partial_s f + \alpha_t + \beta_t \partial_s g = 0, \\ -\partial_\sigma f + \alpha_t \partial_\sigma S_t + \beta_t \partial_\sigma g = 0, \end{cases}$$

and then we can obtain easily the hedging strategy by observing that $\partial_\sigma S_t = S_t(W_t - \sigma t)$.

7.5 Implied volatility

In the Black-Scholes model the price of a European Call option is a function of the form

$$C_{\mathrm{BS}} = C_{\mathrm{BS}}(\sigma, S, K, T, r)$$

where σ is the volatility, S is the current price of the underlying asset, K is the strike, T is the maturity and r is the short-term rate. Actually the price can also be expressed in the form

$$C_{\mathrm{BS}} := S\varphi\left(\sigma, \frac{S}{K}, T, r\right),$$

where φ is a function whose expression can be easily deduced from the Black-Scholes formula (7.33). The number $m = \frac{S}{K}$ is usually called "moneyness" of the option: if $\frac{S}{K} > 1$, we say that that the Call option is "in the money", since we are in a situation of potential profit; if $\frac{S}{K} < 1$, the Call option is "out of the money" and has null intrinsic value; finally, if $\frac{S}{K} = 1$ i.e. $S = K$, we say that the option is "at the money".

Of all the parameters determining the Black-Scholes price, the volatility σ is the only one that is not directly observable. We recall that

$$\sigma \mapsto C_{\mathrm{BS}}(\sigma, S, K, T, r)$$

is a strictly increasing function and therefore invertible: having fixed all the other parameters, a Black-Scholes price of the option corresponds to every value of σ; conversely, a unique value of the volatility σ^* is associated to every value C^* on the interval $]0, S[$ (the interval to which the price must belong by arbitrage arguments). We set

$$\sigma^* = \mathrm{VI}(C^*, S, K, T, r),$$

where σ^* is the unique value of the volatility parameter such that

$$C^* = C_{\mathrm{BS}}(\sigma^*, S, K, T, r).$$

The function

$$C^* \mapsto \mathrm{VI}(C^*, S, K, T, r)$$

is called *implied volatility function.*

The first problem when we price an option in the Black-Scholes model is the choice of the parameter σ that, as we have already said, is not directly observable. The first idea could be to use a value of σ obtained from an estimate on the historical data on the underlying asset, i.e. the so-called *historical volatility.* Actually, the most widespread and simple approach is that of using directly, where it is available, the implied volatility of the market: we see, however, that this approach is not free from problems.

The concept of implied volatility is so important and widespread that, in financial markets, the plain vanilla options are commonly quoted in terms of implied volatility, rather than explicitly by giving their price. As a matter of fact, using the implied volatility is convenient for various reasons. First of all, since the Put and Call prices are increasing functions of the volatility, the quotation in terms of the implied volatility immediately gives the idea of the "cost" of the option. Analogously, using the implied volatility makes it easy to compare the prices of options on the same asset, but with different strikes and maturities.

For fixed S and r, and given a family of prices

$$\{C_i^* \mid i = 1, \dots M\} \tag{7.65}$$

where C_i^* denotes the price of the Call with strike K^i and maturity T^i, the *implied volatility surface* relative to (7.65) is the graph of the function

$$\left(K^i, T^i\right) \quad \mapsto \quad \mathrm{VI}\left(C_i^*, S, K^i, T^i, r\right).$$

If we assume the Black-Scholes dynamics for the underlying asset

$$dS_t = \mu S_t dt + \sigma S_t dW_t$$

and $\left(C_{\mathrm{BS}}^i\right)_{i \in I}$ is a family of Black-Scholes prices relative to the strikes K^i and maturities T^i, then the corresponding implied volatilities must obviously coincide:

$$\mathrm{VI}\left(C_{\mathrm{BS}}^i, S, K^i, T^i, r\right) = \sigma \qquad \text{for any } i \in I.$$

In other terms, the *implied volatility surface* relative to the prices obtained by the Black-Scholes model is flat and coincides with the graph of the function that is constant and equal to σ.

On the contrary, for an *empirical* implied volatility surface, inferred from quoted prices in real markets, the result is generally quite different: it is well known that the market prices of European options on the same underlying asset have implied volatilities that vary with strike and maturity. By way of example, in Figure 7.8 we depict the implied volatility surface of options on the London FTSE index on March 31st 2006.

Typically every section, with T fixed, of the implied volatility surface takes a particular form that is usually called "smile" (in the case of Figure 7.9) or "skew" (in the case of Figure 7.8). Generally we can say that market quotation tends to give more value (greater implied volatility) to the extreme cases "in" or "out of the money". This reflects that some situations in the market are perceived as more risky, in particular the case of extreme falls or rises of the quotations of the underlying asset.

Also the dependence on T, the time to maturity, is significant in the analysis of the implied volatility: this is called the *term-structure of the implied volatility*. Typically when we get close to maturity ($T \to 0^+$), we see that the smile or the skew become more marked.

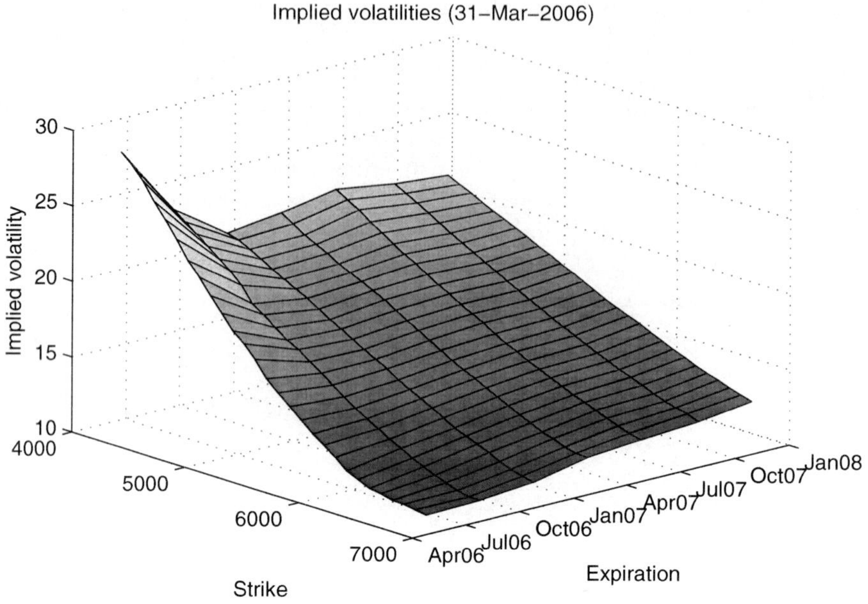

Fig. 7.8. Implied-volatility surface of options on the FTSE index on March 31st 2006

Other characteristic features make definitely different the implied volatility surface of the market from the constant Black-Scholes volatility: for example, in Figure 7.9 we show the dependence of the implied volatility of options on the S&P500 index, with respect to the so-called "deviation from trend" of the underlying asset, defined as the difference between the current price and a weighted mean of historical prices. Intuitively this parameter indicates if there have been sudden large movements of the quotation of the underlying asset.

Finally we note that the implied volatility depends also on time *in absolute terms*: indeed, it is well known that the shape of the implied volatility surface on the S&P500 index has significantly changed from the beginning of the eighties until today. The market crash of 19 October 1987 may be taken as the date marking the end of flat volatility surfaces.

This also reflects the fact that, though based on the same mathematical and probabilistic tools, the modeling of financial and, for instance, physical phenomena are essentially different: indeed, the financial dynamics strictly depends on the behaviour and beliefs of investors and therefore, differently from the general laws in physics, may vary drastically over time.

The analysis of the implied volatility surface makes it evident that the Black-Scholes model is not realistic: more precisely, we could say that nowadays Black-Scholes is the *language* of the market (since prices are quoted in

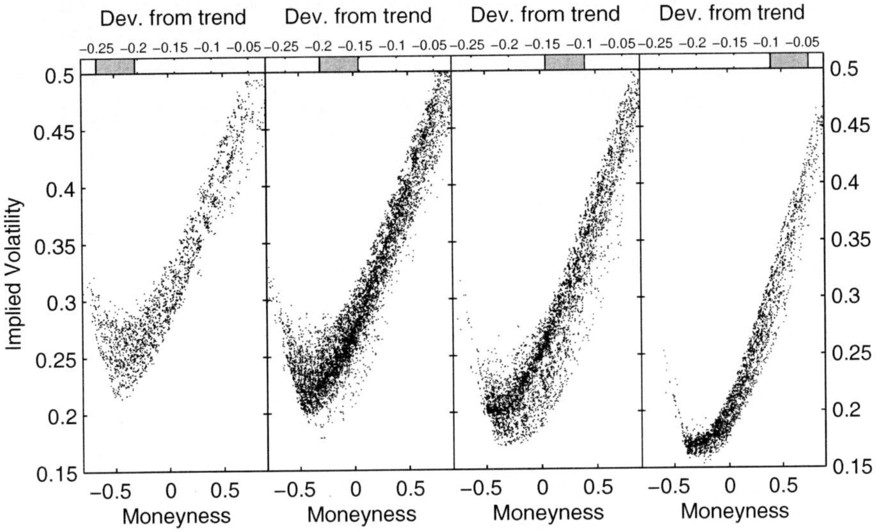

Fig. 7.9. Effect of the deviation from trend on the implied volatility. The volatility smiles for options on the S&P500 index are grouped for different values of the deviation, as indicated on top of each box

terms of implied volatility), but usually it is not the model really used by investors to price and hedge derivatives. Indeed the use of the Black-Scholes model poses some not merely theoretical problem: for instance, let us suppose that, despite all the evidence against the Black-Scholes model, we wish to use it anyway. Then we have seen that we have to face the problem of the choice of the volatility parameter for the model. If we use the historical volatility, we might get quotations that are "out of the market", especially when compared with those obtained from the market-volatility surface in the extreme "in" and "out of money" regions. On the other hand, if we want to use the implied volatility, we have to face the problem of choosing one value among all the values given by the market, since the volatility surface is not "flat". Evidently, if our goal is to price and hedge a plain vanilla option, with strike, say, K and maturity, say, T, the most natural idea is to use the implied volatility corresponding to (K,T). But the problem does not seem to be easily solvable if we are interested in the pricing and hedging of an exotic derivative: for example, if the derivative does not have a unique maturity (e.g. a Bermudan option) or if a fixed strike does not appear in the payoff (e.g., an Asian option with floating strike).

These problems make it necessary to introduce more sophisticated models than the Black-Scholes one, that can be calibrated in such a way that it is possible to price plain vanilla options in accordance with the implied volatility surface of the market. In this way such models can give prices to exotic derivatives that are consistent with the market Call and Put prices. This result is not particularly difficult and can be obtained by various models with non-constant volatility such as those in Chapter 10.5. A second goal that poses many more delicate questions and is still a research topic consists in finding a model that gives the "best" solution to the hedging problem and that is stable with respect to perturbations of the value of the parameters involved (see for instance Schoutens, Simons, and Tistaert [302] and Cont [75]).

7.6 Asian options

An Asian option is a derivative whose payoff depends on an average of the prices of the underlying asset. This kind of derivative is quite often used, for example in the currencies or commodities markets: one of the reasons to introduce this derivative is to limit speculation on plain vanilla options. Indeed it is known that the European Call and Put option prices close to maturity can be influenced by the investors through manipulations on the underlying asset.

Asian options can be classified by the payoff function and by the particular average that is used. As usual we assume that the underlying asset follows a geometric Brownian motion S verifying equation (7.2) and we denote by M_t the value of the average at time t: for an Asian option with *arithmetic average* we have

$$M_t = \frac{A_t}{t} \quad \text{with} \quad A_t = \int_0^t S_\tau d\tau; \tag{7.66}$$

for an Asian option with *geometric average* we have

$$M_t = e^{\frac{G_t}{t}} \quad \text{with} \quad G_t = \int_0^t \log(S_\tau)\, d\tau. \tag{7.67}$$

Even though arithmetic Asian options are more commonly traded in real markets, in the literature geometric Asian options have been widely studied because they are more tractable from a theoretical point of view and, under suitable conditions, they can be used to approximate the corresponding arithmetic version.

Concerning the payoff, the most common versions are the Asian Call *with fixed strike K*

$$F(S_T, M_T) = (M_T - K)^+,$$

the Asian Call *with floating strike*

$$F(S_T, M_T) = (S_T - M_T)^+,$$

and the corresponding Asian Puts.

Formally, the pricing and hedging problems for Asian options have a lot in common with their standard European counterparts: the main difference is that an Asian option depends not only on the spot price of the underlying asset but also on its entire trajectory. Nevertheless, as already mentioned in the discrete case in Section 2.3.3, it is possible to preserve the Markovian property of the model by using a technique now standard: this consists in augmenting the space by introducing an additional state variable related to the average process A_t in (7.66) or G_t in (7.67).

7.6.1 Arithmetic average

In order to make the previous ideas precise, let us examine first the arithmetic average case. We say that $(\alpha_t, \beta_t)_{t\in[0,T]}$ is a Markovian portfolio if

$$\alpha_t = \alpha(t, S_t, A_t), \qquad \beta_t = \beta(t, S_t, A_t), \qquad t \in [0, T],$$

where α, β are functions in $C^{1,2}([0,T[\times\mathbb{R}_{>0}\times\mathbb{R}_{>0})\cap C([0,T]\times\mathbb{R}_{>0}\times\mathbb{R}_{>0})$, and we denote by

$$f(t, S_t, A_t) = \alpha_t S_t + \beta_t B_t, \qquad t \in [0, T],$$

the corresponding value. The following result extends Theorems 7.8 and 7.13:

Theorem 7.22 *The following conditions are equivalent:*

i) $(\alpha_t, \beta_t)_{t\in[0,T]}$ *is self-financing, i.e. we have*

$$df(t, S_t, A_t) = \alpha_t dS_t + \beta_t dB_t;$$

ii) f *is a solution of the partial differential equation*

$$\frac{\sigma^2 s^2}{2}\partial_{ss} f(t,s,a) + rs\partial_s f(t,s,a) + s\partial_a f(t,s,a) + \partial_t f(t,s,a) = rf(t,s,a), \tag{7.68}$$

for $(t, s, a) \in [0, T[\times\mathbb{R}_{>0}\times\mathbb{R}_{>0}$, *and we have that*

$$\alpha(t, s, a) = \partial_s f(t, s, a).$$

The arbitrage price $f = f(t, S_t, A_t)$ *of an Asian arithmetic option with payoff function* F *is the solution of the Cauchy problem for equation* (7.68) *with final datum*

$$f(T, s, a) = F\left(s, \frac{a}{T}\right), \qquad s, a \in \mathbb{R}_{>0}.$$

For example, in the case of a fixed strike Asian Call, the final condition for equation (7.68) is

$$f(T, s, a) = \left(\frac{a}{T} - K\right)^+, \qquad s, a \in \mathbb{R}_{>0}. \tag{7.69}$$

For the floating strike Asian Call, the final condition becomes

$$f(T, s, a) = \left(s - \frac{a}{T}\right)^+, \qquad s, a \in \mathbb{R}_{>0}. \tag{7.70}$$

The proof of Theorem 7.22 is formally analogous to the ones of Theorems 7.8 and 7.13. Let us observe that *equation* (7.68) *cannot be transformed into a parabolic equation by a change of variables* as in the European case. In particular the results of existence and uniqueness for the Cauchy problem of Appendix A.3 and Section 6.2 are not sufficient to prove the completeness of the market and the existence and uniqueness of the arbitrage price: these results have been recently proved, for a generic payoff function, by Barucci, Polidoro and Vespri [33].

Equation (7.68) is *degenerate parabolic*, because the matrix of the second-order part of the equation is singular and only positive semi-definite: indeed, in the standard notation (A.45) of Appendix A.3, the matrix $\mathcal{C}$ corresponding to (7.68) is

$$\mathcal{C} = \begin{pmatrix} \sigma^2 s^2 & 0 \\ 0 & 0 \end{pmatrix}$$

and has rank one for every $(s,a) \in \mathbb{R}_{>0} \times \mathbb{R}_{>0}$. This does not have to come as a surprise: equation (7.68) was deduced by using the Itô formula and the second-order derivative appearing in it is "produced" by the Brownian motion of the process S. The average A brings an additional state variable in, thus augmenting the dimension of the problem, setting it in $\mathbb{R}^3$, but it does not bring a new Brownian motion in (nor second-order derivative with respect to the variable a).

In some particular cases there exists a suitable transformation to take back the problem to two dimensions. In the floating strike case, Ingersoll [178] suggests the change of variable $x = \frac{a}{s}$: if we put

$$f(t,s,a) = su\left(t, \frac{a}{s}\right) \tag{7.71}$$

we have

$$\partial_t f = s\partial_t u, \qquad \partial_s f = u - \frac{a}{s}\partial_x u, \qquad \partial_{ss} f = \frac{a^2}{s^3}\partial_{xx} u, \qquad \partial_a f = \partial_x u.$$

So f solves the Cauchy problem (7.68)-(7.70) if and only if the function $u = u(t,x)$ defined in (7.71) is a solution of the Cauchy problem in $\mathbb{R}^2$

$$\begin{cases} \frac{\sigma^2 x^2}{2}\partial_{xx} u + (1 - rx)\partial_x u + \partial_t u = 0, & t \in [0,T[,\ x > 0, \\ u(T,x) = \left(1 - \frac{x}{T}\right)^+, & x > 0. \end{cases}$$

More generally, transformation (7.71) allows to reduce the dimension of the problem in case the payoff is a homogeneous function of degree one, that is

$$F(s,a) = sF\left(1, \frac{a}{s}\right), \qquad s,a > 0.$$

For the fixed strike Asian option, Rogers and Shi [291] suggest the change of variable

$$x = \frac{\frac{a}{T} - K}{s}.$$

If we put

$$f(t,s,a) = su\left(t, \frac{\frac{a}{T} - K}{s}\right) \tag{7.72}$$

we have

$$\partial_s f = u - \frac{\frac{a}{T} - K}{s}\partial_x u, \qquad \partial_{ss} f = \frac{\left(\frac{a}{T} - K\right)^2}{s^3}\partial_{xx} u, \qquad \partial_a f = \frac{\partial_x u}{T}.$$

So f solves the Cauchy problem (7.68)-(7.69) if and only if the function $u = u(t,x)$ defined in (7.72) is a solution of the Cauchy problem in $\mathbb{R}^2$

$$\begin{cases} \frac{\sigma^2 x^2}{2}\partial_{xx} u + \left(\frac{1}{T} - rx\right)\partial_x u + \partial_t u = 0, & t \in [0,T[, \ x \in \mathbb{R}, \\ u(T,x) = x^+, & x \in \mathbb{R}. \end{cases}$$

Note that the reduction of the dimension of the problem is possible only in particular cases and assuming the Black-Scholes dynamics for the underlying asset.

7.6.2 Geometric average

We consider a geometric average Asian option: in this case the value $f = f(t,s,g)$ of the replicating portfolio is function of t, S_t and G_t in (7.67). Furthermore a result analogous to Theorem 7.22 holds, where (7.68) is replaced by the differential equation

$$\frac{\sigma^2 s^2}{2}\partial_{ss} f(t,s,g) + rs\partial_s f(t,s,g) + (\log s)\partial_g f(t,s,g) + \partial_t f(t,s,g) = rf(t,s,g), \tag{7.73}$$

with $(t,s,g) \in [0,T[\times\mathbb{R}_{>0}\times\mathbb{R}$.

Similarly to Proposition 7.9, we change the variables by putting

$$t = T - \tau, \qquad s = e^{\sigma x}, \qquad g = \sigma y,$$

and

$$u(\tau,x,y) = e^{ax+b\tau} f(T-\tau, e^{\sigma x}, \sigma y), \qquad \tau \in [0,T], \ x,y \in \mathbb{R}, \tag{7.74}$$

where a,b are constants to be determined appropriately later. Let us recall formulas (7.23) and also that

$$\partial_y u = e^{ax+b\tau}\sigma\partial_g f;$$

it follows that

$$\frac{1}{2}\partial_{xx} u + x\partial_y u - \partial_\tau u =$$

$$e^{ax+b\tau}\left(\frac{\sigma^2 s^2}{2}\partial_{ss} f + \left(\sigma a + \frac{\sigma^2}{2}\right)s\partial_s f + (\log s)\partial_g f + \partial_t f + \left(\frac{a^2}{2} - b\right)f\right) =$$

(if f solves (7.73))

$$= \left(\sigma a + \frac{\sigma^2}{2} - r\right) s\partial_s f + \left(\frac{a^2}{2} - b + r\right) f.$$

This proves the following result.

Proposition 7.23 *By choosing the constants a and b as in (7.24), the function f is a solution of the equation (7.73) in $[0,T[\times\mathbb{R}_{>0}\times\mathbb{R}$ if and only if the function $u = u(\tau, x, y)$ defined in (7.74) satisfies the equation*

$$\frac{1}{2}\partial_{xx}u + x\partial_y u - \partial_\tau u = 0, \qquad \text{in }]0,T] \times \mathbb{R}^2. \tag{7.75}$$

(7.75) is a degenerate parabolic equation, called Kolmogorov equation which will be studied in Section 9.5 and whose fundamental solution will be constructed explicitly in Example 9.53.

8

Parabolic PDEs with variable coefficients: existence

The Black-Scholes model is based upon the results of existence and uniqueness for parabolic equations with constant coefficients, in particular for the heat equation. The study of more sophisticated diffusion models requires analogous results for differential operators with variable coefficients.

In this chapter we consider a parabolic operator of the form

$$Lu = \frac{1}{2} \sum_{i,j=1}^{N} c_{ij} \partial_{x_i x_j} u + \sum_{i=1}^{N} b_i \partial_{x_i} u - au - \partial_t u, \tag{8.1}$$

where (t, x) is an element of $\mathbb{R} \times \mathbb{R}^N$ and (c_{ij}) is a symmetric matrix. We suppose that the coefficients $c_{ij} = c_{ij}(t, x)$, $b_j = b_j(t, x)$ and $a = a(t, x)$ are *bounded and Hölder continuous* functions. As already noted in Chapter 6, these assumptions are less general than those introduced in the study of the uniqueness problem.

We aim at studying on one hand the existence and the properties of the fundamental solution of L and on the other hand the free-boundary obstacle problem. The first issue is deeply connected to the solvability of the Cauchy problem and therefore to the pricing and hedging of European options. The second topic, as we had pointed out already in Section 2.5.5, has to deal with the study of American-style derivatives: in this setting, the *obstacle* function plays the part of the payoff of the option.

A thorough treatment of these topics is definitely beyond the scope of the present book: these central subjects in the theory of partial differential equations are studied in several classical monographs such as Friedman's [139], [141], Ladyzhenskaya, Ural'tseva and Solonnikov's [222], Oleĭnik and Radkevič's [274], Lieberman's [238] and Evans' [124].

Section 8.1 describes in general terms the construction of the fundamental solution by the so-called *parametrix method* introduced by E. E. Levi in [236]. In Section 8.2, using some known a priori estimates for the solutions of L in Hölder and Sobolev spaces, we give a detailed proof of the existence of strong solutions to the obstacle problem.

Pascucci A.: PDE and Martingale Methods in Option Pricing

8.1 Cauchy problem and fundamental solution

We suppose that the operator L in (8.1) is uniformly parabolic, i.e. the following condition holds:

Hypothesis 8.1 *There exists a positive constant λ such that*

$$\lambda^{-2}|\xi|^2 \leq \sum_{i,j=1}^{N} c_{ij}(t,x)\xi_i\xi_j \leq \lambda|\xi|^2, \qquad t \in \mathbb{R},\ x,\xi \in \mathbb{R}^N. \tag{8.2}$$

The prototype for the class of uniformly parabolic operators is the heat operator with constant coefficients, that has the identity matrix as (c_{ij}).

In the theory of parabolic equations, it is natural to give the time variable t "double weight" with respect to the space variables x. In order to make this concept rigorous, we define the *parabolic Hölder spaces.*

Definition 8.2 *Let $\alpha \in]0,1]$ and O be a domain in $\mathbb{R}^{N+1}$. We denote by $C_P^\alpha(O)$ the space of bounded functions u on O, such that*

$$|u(t,x) - u(s,y)| \leq C\left(|t-s|^{\frac{\alpha}{2}} + |x-y|^\alpha\right), \qquad (t,x),(s,y) \in O, \tag{8.3}$$

for some positive constant C. We define the norm

$$\|u\|_{C_P^\alpha(O)} = \sup_{(t,x)\in O} |u(t,x)| + \sup_{\substack{(t,x),(s,y)\in O \\ (t,x)\neq(s,y)}} \frac{|u(t,x)-u(s,y)|}{|t-s|^{\frac{\alpha}{2}} + |x-y|^\alpha}.$$

Moreover we denote by $C_P^{1+\alpha}(O)$ and $C_P^{2+\alpha}(O)$ the Hölder spaces defined by the following norms:

$$\|u\|_{C_P^{1+\alpha}(O)} = \|u\|_{C_P^\alpha(O)} + \sum_{i=1}^{N} \|\partial_{x_i} u\|_{C_P^\alpha(O)},$$

$$\|u\|_{C_P^{2+\alpha}(O)} = \|u\|_{C_P^{1+\alpha}(O)} + \sum_{i,j=1}^{N} \|\partial_{x_i x_j} u\|_{C_P^\alpha(O)} + \|\partial_t u\|_{C_P^\alpha(O)},$$

respectively. We write $u \in C_{P,\text{loc}}^{k+\alpha}(O)$, $k = 0,1,2$, if $u \in C_P^{k+\alpha}(M)$ for every compact domain M such that $\overline{M} \subseteq O$.

In the sequel we assume the following regularity condition on the coefficients of the operator:

Hypothesis 8.3 *The coefficients are bounded and Hölder continuous:*

$$c_{ij}, b_j, a \in C_P^\alpha(\mathbb{R}^{N+1})$$

for some $\alpha \in]0,1]$ and for every $1 \leq i,j \leq N$.

We now consider the Cauchy problem

$$\begin{cases} Lu = f, & \text{in } \mathcal{S}_T :=]0,T[\times\mathbb{R}^N, \\ u(0,\cdot) = \varphi, & \text{on } \mathbb{R}^N, \end{cases} \tag{8.4}$$

where φ and f are given functions.

Definition 8.4 *A classical solution of the Cauchy problem* (8.4) *is a function* $u \in C^{1,2}(\mathcal{S}_T) \cap C(\overline{\mathcal{S}_T})$ *that satisfies equations in* (8.4) *pointwise.*

As we have already seen in the case of the heat equation, it is natural to assume the following growth and regularity conditions:

Hypothesis 8.5 *The functions φ and f are continuous and there exist some positive constants c, γ, with $\gamma < 2$, such that*

$$|\varphi(x)| \le ce^{c|x|^\gamma}, \qquad x \in \mathbb{R}^N, \tag{8.5}$$

$$|f(t,x)| \le ce^{c|x|^\gamma}, \qquad (t,x) \in \mathcal{S}_T. \tag{8.6}$$

Moreover f is locally Hölder continuous in x, uniformly in t, i.e. for every compact set M in $\mathbb{R}^N$ we have that

$$|f(t,x) - f(t,y)| \le C|x-y|^\beta, \qquad x,y \in M,\ t \in]0,T[, \tag{8.7}$$

with $\beta \in]0,1]$ and $C > 0$.

The main result of this section is the following:

Theorem 8.6 *Under Hypotheses 8.1 and 8.3, the operator L has a fundamental solution $\Gamma = \Gamma(t,x;s,y)$ that is a positive function, defined for $x,y \in \mathbb{R}^N$ and $t > s$, such that for every φ, f verifying Hypothesis 8.5, the function u defined by*

$$u(t,x) = \int_{\mathbb{R}^N} \Gamma(t,x;0,y)\varphi(y)dy - \int_0^t \int_{\mathbb{R}^N} \Gamma(t,x;s,y)f(s,y)dyds, \tag{8.8}$$

with $(t,x) \in \mathcal{S}_T$ and by $u(0,x) = \varphi(x)$, is a classical solution of the Cauchy problem (8.4).

Remark 8.7 By Theorem 6.15, the function u in (8.8) is the unique solution of (8.4) such that

$$|u(t,x)| \le ce^{c|x|^2}, \qquad (t,x) \in \mathcal{S}_T,$$

with c positive constant.

Conditions (8.5)-(8.6) can be weakened: we could take

$$|\varphi(x)| \le c_1 \exp(c_2|x|^2), \qquad x \in \mathbb{R}^N,$$

$$|f(t,x)| \le c_1 \exp(c_2|x|^2), \qquad (t,x) \in \mathcal{S}_T,$$

with c_1, c_2 positive constants. In such a case the solution u in (8.8) is defined on $\mathcal{S}_T$ for $T < \frac{1}{2\lambda c_2}$.

Finally analogous results to those of Section A.3.3 hold if the initial datum is a locally integrable function. □

8.1.1 Levi's parametrix method

The classical proof of Theorem 8.6 is rather lengthy and involved. Here we give only the main ideas here and for further details we refer to Friedman [139]. For a more recent presentation and in a more general setting including also non-uniform parabolic operators as those that arise in the pricing problem for Asian options, we refer the reader to Di Francesco and Pascucci [94] and Polidoro [283]. For a more practical presentation, we also refer to Corielli, Foschi and Pascucci [77] where the parametrix method is used to obtain numerical approximations of the fundamental solution (and so also of the price of an option, expressed as the solution of a Cauchy problem) by an expansion of fundamental solutions of parabolic operators with constant coefficients whose explicit expression is known. In particular, analytical approximations of local volatility models are provided. Recently Gatheral, Hsu, Laurence, Ouyang and Wang [151] use a heat kernel expansion to obtain asymptotic approximations for call prices and implied volatility in local-stochastic volatility models.

In what follows we assume Hypotheses 8.1, 8.3 and for the sake of brevity we denote generic points in $\mathbb{R}^{N+1}$ by $z = (t, x)$ and $\zeta = (s, y)$. Furthermore, for fixed $w \in \mathbb{R}^{N+1}$, we denote by

$$\Gamma_w(z;\zeta)$$

the fundamental solution of the parabolic operator *with constant coefficients*

$$L_w = \frac{1}{2}\sum_{i,j=1}^{N} c_{ij}(w)\partial_{x_i x_j} - \partial_t,$$

obtained from L by freezing the second-order coefficients in w and by cancelling the lower-order terms, with the exception of the time derivative, obviously. The explicit expression of Γ_w is given in Section A.3.1.

The parametrix method is a constructive technique that allows us to prove the existence and some estimates of the fundamental solution $\Gamma(t, x; s, y)$ of L: for the sake of simplicity, in the sequel we treat only the case $s = 0$. The method is based mainly upon two ideas: the first is to approximate $\Gamma(z;\zeta)$ by the so-called parametrix defined by

$$Z(z;\zeta) = \Gamma_\zeta(z;\zeta).$$

The second idea is to suppose that the fundamental solution takes the form (let us recall that $\zeta = (0, y)$):

$$\Gamma(z;\zeta) = Z(z;\zeta) + \int_0^t \int_{\mathbb{R}^N} Z(z;w)G(w;\zeta)dw. \tag{8.9}$$

In order to find the unknown function G, we impose that Γ is the solution to the equation $L\Gamma(\cdot;\zeta) = 0$ in $\mathbb{R}_{>0} \times \mathbb{R}^N$: we wish to point out one more time, to make this totally transparent, that the operator L acts on the variable z while the point ζ is fixed. Then formally we obtain

$$\begin{aligned}0 = L\Gamma(z;\zeta) &= LZ(z;\zeta) + L\int_0^t \int_{\mathbb{R}^N} Z(z;w)G(w;\zeta)dw \\ &= LZ(z;\zeta) + \int_0^t \int_{\mathbb{R}^N} LZ(z;w)G(w;\zeta)dw - G(z;\zeta),\end{aligned}$$

hence

$$G(z;\zeta) = LZ(z;\zeta) + \int_0^t \int_{\mathbb{R}^N} LZ(z;w)G(w;\zeta)dw. \tag{8.10}$$

Therefore G is a solution of an integral equation equivalent to a fixed-point problem that can be solved by the method of successive approximations:

$$G(z;\zeta) = \sum_{k=1}^{+\infty} (LZ)_k(z;\zeta), \tag{8.11}$$

where

$$\begin{aligned}(LZ)_1(z;\zeta) &= LZ(z;\zeta), \\ (LZ)_{k+1}(z;\zeta) &= \int_0^t \int_{\mathbb{R}^N} LZ(z;w)(LZ)_k(w;\zeta)dw, \qquad k \in \mathbb{N}.\end{aligned}$$

The previous ideas are formalized by the following (cf. Proposition 4.1 in [94]):

Theorem 8.8 *There exists $k_0 \in \mathbb{N}$ such that, for all $T > 0$ and $\zeta = (0, y) \in \mathbb{R}^{N+1}$, the series*

$$\sum_{k=k_0}^{+\infty} (LZ)_k(\cdot;\zeta)$$

converges uniformly on the strip $\mathcal{S}_T$. Furthermore, the function $G(\cdot,\zeta)$ defined by (8.11) is a solution to the integral equation (8.10) in $\mathcal{S}_T$ and Γ in (8.9) is a fundamental solution to L.

Remark 8.9 The fundamental solution can be constructed in a formally analogous way also by using the backward parametrix defined by

$$Z(z;\zeta) = \Gamma_z(z;\zeta).$$

□

8.1.2 Gaussian estimates and adjoint operator

By the parametrix method it is possible to obtain also some noteworthy estimates of the fundamental solution and its derivatives in terms of the fundamental solution of the heat operator. These estimates play a basic role in several frameworks, for instance the uniqueness results of Section 6.3 and the Feynman-Kač representation Theorem 9.48. Given a positive constant λ, we denote by

$$\Gamma_\lambda(t,x) = \frac{1}{(2\pi\lambda t)^{\frac{N}{2}}} e^{-\frac{|x|^2}{2t\lambda}}, \qquad t>0,\ x\in\mathbb{R}^N,$$

the fundamental solution, with pole at the origin, of the heat operator in $\mathbb{R}^{N+1}$

$$\frac{\lambda}{2}\triangle - \partial_t.$$

Theorem 8.10 *Under Hypotheses 8.1 and 8.3, for all $T,\varepsilon>0$ there exists a positive constant C, dependent only on ε,λ,T and on the C_P^α-norm of the coefficients of the operator, such that*

$$\Gamma(t,x;s,y) \le C\,\Gamma_{\lambda+\varepsilon}(t-s,x-y), \tag{8.12}$$

$$|\partial_{x_i}\Gamma(t,x;s,y)| + |\partial_{y_i}\Gamma(t,x;s,y)| \le \frac{C}{\sqrt{t-s}}\Gamma_{\lambda+\varepsilon}(t-s,x-y), \tag{8.13}$$

$$\left|\partial_{x_ix_j}\Gamma(t,x;s,y)\right| + |\partial_t\Gamma(t,x;s,y)| + \tag{8.14}$$

$$\left|\partial_{y_iy_j}\Gamma(t,x;s,y)\right| + |\partial_s\Gamma(t,x;s,y)| \le \frac{C}{t-s}\Gamma_{\lambda+\varepsilon}(t-s,x-y), \tag{8.15}$$

for all $x,y\in\mathbb{R}^N$, $t\in]s,s+T[$ and $i,j=1,\dots,N$.

Corollary 8.11 *Under Hypotheses 8.1, 8.3 and 8.5, let u be the solution of the problem (8.4) defined in (8.8). Then there exists a positive constant C such that*

$$|u(t,x)| \le Ce^{C|x|^2}, \tag{8.16}$$

$$|\partial_{x_i}u(t,x)| \le \frac{Ce^{C|x|^2}}{\sqrt{t}}, \tag{8.17}$$

$$\left|\partial_{x_ix_j}u(t,x)\right| + |\partial_t u(t,x)| \le \frac{Ce^{C|x|^2}}{t}, \tag{8.18}$$

for all $(t,x)\in\mathcal{S}_T$ and $i,j=1,\dots,N$.

Example 8.12 Without any further regularity assumption on the initial datum φ, we have $\partial_{x_i}u(t,x) = O\left(\frac{1}{\sqrt{t}}\right)$ as $t\to0^+$, consistently with estimate (8.17). Indeed, for the 2-dimensional heat equation and initial datum $\varphi(x)=0$ when $x\ge0$ and $\varphi(x)=1$ when $x<0$, we have

$$\partial_x u(0,t) = \frac{1}{\sqrt{2\pi t}}\int_{-\infty}^0 \frac{y}{t}\exp\left(-\frac{y^2}{2t}\right)dy =$$

(by the change of variable $z=-\frac{y^2}{2t}$)

$$= -\frac{1}{\sqrt{2\pi t}}\int_{-\infty}^0 e^z dz = -\frac{1}{\sqrt{2\pi t}}.$$

□

Hereafter we assume:

Hypothesis 8.13 *The derivatives* $\partial_{x_i} c_{ij}, \partial_{x_i x_j} c_{ij}, \partial_{x_i} b_i$ *exist for all* $i, j = 1, \dots, N$ *and they belong to the space* $C_P^\alpha(\mathbb{R}^{N+1})$.

We recall the expression of the adjoint operator of L (cf. (6.12)), formally defined by the equality

$$\int_{\mathbb{R}^{N+1}} vLu = \int_{\mathbb{R}^{N+1}} uL^*v.$$

We have

$$L^*u = \frac{1}{2} \sum_{j,k=1}^{N} c_{jk} \partial_{x_j x_k} u + \sum_{j=1}^{N} b_j^* \partial_{x_j} u - a^*u + \partial_t u$$

where

$$b_i^* = -b_i + \sum_{j=1}^{N} \partial_{x_i} c_{ij}, \qquad a^* = a - \frac{1}{2} \sum_{i,j=1}^{N} \partial_{x_i x_j} c_{ij} + \sum_{j=1}^{N} \partial_{x_j} b_j.$$

The parametrix method allows to prove also the following result.

Theorem 8.14 *Under Hypotheses 8.1, 8.3, 8.5 and 8.13, there exists the fundamental solution* Γ^* *of* L^* *and we have*

$$\Gamma^*(t, x; T, y) = \Gamma(T, y; t, x),$$

when $x, y \in \mathbb{R}^N$ *and* $t < T$.

8.2 Obstacle problem

We consider problem

$$\begin{cases} \max\{Lu, \varphi - u\} = 0, & \text{in } \mathcal{S}_T =]0, T[\times \mathbb{R}^N, \\ u(0, \cdot) = \varphi, & \text{on } \mathbb{R}^N, \end{cases} \tag{8.19}$$

where L is a parabolic operator of the form (8.1) and φ is a locally Lipschitz-continuous function which is also convex in a weak sense that will be made precise later (cf. Hypothesis 8.19). In Chapter 11 we shall prove that the price of an American option with payoff φ can be expressed in terms of the solution u of (8.19).

The first equation in (8.19) asserts that $u \geq \varphi$ so the strip $\mathcal{S}_T$ is divided in two parts:

i) *the exercise region* where $u = \varphi$;
ii) *the continuation region* where $u > \varphi$ and $Lu = 0$ i.e. the price of the derivative verifies a PDE that is analogous to the Black-Scholes differential equation.

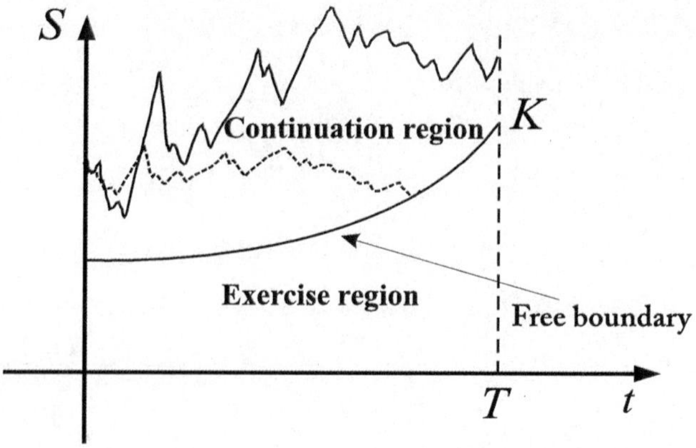

Fig. 8.1. Exercise and continuation regions of an American Put option

Problem (8.19) is equivalent[1] to:

$$\begin{cases} Lu \le 0, & \text{in } \mathcal{S}_T, \\ u \ge \varphi, & \text{in } \mathcal{S}_T, \\ (u-\varphi)\, Lu = 0, & \text{in } \mathcal{S}_T, \\ u(0,x) = \varphi(0,x), & x \in \mathbb{R}^N. \end{cases} \tag{8.20}$$

This kind of problem is usually called *obstacle problem*. The solution is a function such that:

i) it is super-solution[2] of L (i.e. $Lu \le 0$);
ii) it is greater or equal to the obstacle, represented by the function φ;
iii) it solves the equation $Lu = 0$ when $u > \varphi$;
iv) it assumes the initial condition.

[1] We use here the equivalence

$$\max\{F(x), G(x)\} = 0 \quad \Leftrightarrow \quad \begin{cases} F(x) \le 0, \\ G(x) \le 0, \\ F(x)G(x) = 0. \end{cases}$$

[2] The term "super-solution" comes from the classical theory of differential equations. More precisely, let O be a L-regular domain, that is a domain for which the Dirichlet problem for L, with boundary datum u, i.e.

$$\begin{cases} LH = 0, & \text{in } O, \\ H|_{\partial O} = u, \end{cases} \tag{8.21}$$

is solvable: in that case, we denote the solution by $H = H_u^O$. Then, by the maximum principle $Lu \le 0$ if and only if $u \ge H_u^O$ for every L-regular domain O.

Actually we can verify that u is the smallest super-solution greater than the obstacle, by analogy with the notion of Snell envelope.

One of the main features of problem (8.19) is that, in general, it does not admit a classical solution in $C^{1,2}$ even if φ is a smooth function. Therefore it is necessary to introduce a weak formulation of the problem that may be based upon different notions of generalized solution. A general theory of existence and regularity has been developed by many authors since the seventies: in the literature we can find techniques to prove the existence of solutions in the *variational* sense (cf., for example, Bensoussan and Lions [42], Kinderlehrer and Stampacchia [209]), in the *strong* sense (cf., for example, Friedman [140], [141]) and, more recently, in the *viscosity* sense (cf., for example, Barles [21], Fleming and Soner [132], Varadhan [331]). For a general presentation of the theory of optimal stopping and free-boundary problems, see also Peskir and Shiryaev [278].

The variational approach to problem (8.20) consists of looking for the solution as the minimum of a functional within a suitable space of functions admitting first order weak derivatives (we refer to [141] for a general presentation of the topic). The notions of variational solution and, above all, of viscosity solution are very weak and allow one to get existence results under very general assumptions. Strong solutions, even though requiring more restrictive assumptions (that are indeed verified in all the practical cases) seem to be preferable in the financial applications because of their better regularity properties. For this reason, we aim at studying problem (8.19) in the framework of the theory of strong solutions, following the presentation by Di Francesco, Pascucci and Polidoro [95].

8.2.1 Strong solutions

We introduce the definition of parabolic Sobolev spaces used in the study of the obstacle problem and we present some preliminary results to prove the existence of a strong solution. The proof of such results can be found, for example, in Lieberman [238]; in Appendix A.9 we briefly recall the elements of the theory of weak derivatives and Sobolev spaces.

Definition 8.15 *Let O be a domain in $\mathbb{R} \times \mathbb{R}^N$ and $1 \le p \le \infty$. We denote by $S^p(O)$ the space of the functions $u \in L^p(O)$ for which the weak derivatives*

$$\partial_{x_i} u, \partial_{x_i x_j} u, \partial_t u \in L^p(O)$$

exist for every $i, j = 1, \dots, N$. We write $u \in S^p_{\mathrm{loc}}(O)$ if $u \in S^p(O_1)$ for every bounded domain O_1 such that $\overline{O}_1 \subseteq O$.

In the definition of the parabolic Sobolev spaces, the time derivative has double weight, in the sense that Definition 8.15 involves the second order spatial derivatives of u but only the first order derivative in time. This is in line with Definition 8.2 of parabolic Hölder space.

Now we state the parabolic version of the Sobolev-Morrey imbedding Theorem A.168: in the following statements O_1, O_2 denote bounded domains in $\mathbb{R} \times \mathbb{R}^N$ with $\overline{O}_1 \subseteq O_2$.

Theorem 8.16 (Sobolev-Morrey imbedding theorem) *For every $p > N+2$ there exists a positive constant C, depending on p, N, O_1 and O_2 only, such that*

$$\|u\|_{C_P^{1+\alpha}(O_1)} \le C\|u\|_{S^p(O_2)}, \qquad \alpha = 1 - \frac{N+2}{p},$$

for all $u \in S^p(O_2)$.

Let us state now some a priori estimates

Theorem 8.17 (Interior estimates in S^p) *Let L be uniformly parabolic (Hypothesis 8.1). For every $p \in]1, \infty[$ there exists a positive constant C, depending on p, N, L, O_1 and O_2 only, such that*

$$\|u\|_{S^p(O_1)} \le C\left(\|u\|_{L^p(O_2)} + \|Lu\|_{L^p(O_2)}\right),$$

for all $u \in S^p(O_2)$.

Theorem 8.18 (Schauder interior estimates) *Under Hypotheses 8.1 and 8.3, there exists a positive constant C, depending on N, L, O_1 and O_2 only, such that*

$$\|u\|_{C_P^{2+\alpha}(O_1)} \le C\left(\sup_{O_2}|u| + \|Lu\|_{C_P^{\alpha}(O_2)}\right),$$

for all $u \in C_P^{2+\alpha}(O_2)$.

We now lay down the hypotheses on the obstacle function:

Hypothesis 8.19 *The function φ is continuous on $\overline{S}_T$, locally Lipschitz-continuous and for every bounded open set O, $\overline{O} \subseteq S_T$, there exists a constant C such that*

$$\sum_{i,j=1}^{N} \xi_i\xi_j\partial_{x_ix_j}\varphi \ge C|\xi|^2 \quad \text{in } O, \ \xi \in \mathbb{R}^N, \tag{8.22}$$

in the sense of distributions, i.e.

$$\sum_{i,j=1}^{N} \xi_i\xi_j \int_O \varphi\partial_{x_ix_j}\psi \ge C|\xi|^2 \int_O \psi,$$

for all $\xi \in \mathbb{R}^N$ and $\psi \in C_0^\infty(O)$ with $\psi \ge 0$.

Condition (8.22) gives the local lower boundedness of the matrix of the second order spatial distributional derivatives. We point out that any C^2 function verifies Hypothesis 8.19: moreover any locally Lipschitz continuous and convex function verifies Hypothesis 8.19, including the payoff functions of the Call and Put options. On the contrary the function $\varphi(x) = -x^+$ does not satisfy condition (8.22) since its second order distributional derivative is a Dirac's delta with negative sign that is "not bounded from below".

We give now the definition of strong solution.

Definition 8.20 *A strong solution of problem* (8.19) *is a function* $u \in S^1_{\text{loc}}(\mathcal{S}_T) \cap C(\overline{\mathcal{S}}_T)$ *satisfying the equation*

$$\max\{Lu, \varphi - u\} = 0$$

almost everywhere in $\mathcal{S}_T$ *and taking the initial datum pointwise. We say that* $\bar{u}$ *is a strong super-solution of* (8.19) *if* $\bar{u} \in S^1_{\text{loc}}(\mathcal{S}_T) \cap C(\overline{\mathcal{S}}_T)$ *and it verifies*

$$\begin{cases} \max\{L\bar{u}, \varphi - \bar{u}\} \leq 0, & \text{a.e. in } \mathcal{S}_T, \\ \bar{u}(0,\cdot) \geq \varphi, & \text{on } \mathbb{R}^N. \end{cases} \tag{8.23}$$

The main result of this section is the following:

Theorem 8.21 (Existence of a strong solution) *Assume Hypotheses 8.1, 8.3 and 8.19. If there exists a strong super-solution* $\bar{u}$ *to problem* (8.19), *then there exists also a strong solution* u *such that* $u \leq \bar{u}$ *in* $\mathcal{S}_T$. *Moreover* $u \in S^p_{\text{loc}}(\mathcal{S}_T)$ *for every* $p \geq 1$ *and consequently, by the imbedding Theorem 8.16,* $u \in C^{1+\alpha}_{P,\text{loc}}(\mathcal{S}_T)$ *for all* $\alpha \in]0,1[$.

Theorem 8.21 will be proved in the following section.

Remark 8.22 In typical financial applications, the obstacle corresponds to the option payoff ψ: for example, in the case of a Call option, $N = 1$ and

$$\psi(S) = (S - K)^+, \qquad S > 0.$$

In general, if ψ is a Lipschitz-continuous function, then there exists a positive constant C such that

$$|\psi(S)| \leq C(1 + S), \qquad S > 0,$$

and by the transformation

$$\varphi(t,x) = \psi(t, e^x),$$

we have that

$$|\varphi(t,x)| \leq C(1 + e^x), \qquad x \in \mathbb{R}.$$

In this case a super-solution of the obstacle problem is

$$\bar{u}(t,x) = Ce^{\gamma t}(1 + e^x), \qquad t \in [0,T], \ x \in \mathbb{R},$$

where γ is a suitable positive constant: indeed it is evident that $\bar{u} \geq \varphi$ and moreover, when $N = 1$,

$$L\bar{u} = Ce^{\gamma t}(-a - \gamma) + Ce^{x+\gamma t}\left(\frac{1}{2}c_{11} + b_1 - a - \gamma\right) \leq 0,$$

when γ is large enough. □

Remark 8.23 Theorem 8.21 gives an existence result: the uniqueness of the strong solution in the class of non-rapidly increasing functions will be proved in Section 9.4.5 as a consequence of the Feynman-Kač representation formula of Theorem 9.48. □

Remark 8.24 Concerning the regularity of the solution, we notice that on the grounds of Definition 8.2 of the space $C^{1+\alpha}_{P,\mathrm{loc}}$, the solution u is a locally Hölder continuous function, together with its first spatial derivatives $\partial_{x_1} u, \dots, \partial_{x_N}$ of exponent α for any $\alpha \in]0,1[$. Moreover, the strong solution found in Theorem 8.21 is also a solution in the weak and viscosity senses (for a proof of this claim see Di Francesco, Pascucci and Polidoro [95]). This means that the other weaker notions on generalized solution gain the stronger regularity properties of the strong solutions. □

Remark 8.25 Besides the regularity of the solution, another important theoretical issue is to determine the regularity of the free boundary, that is the boundary of the exercise region. In the classical case of a single asset following a geometric Brownian motion, that is the standard Black&Scholes [49] and Merton [250] framework, the C^∞-smoothness of the free boundary of the American put option was proved by Friedman [140] and van Moerbeke [330]. In the multi-dimensional Black&Scholes setting and for a quite general class of convex payoffs, the smoothness of the free boundary problem has been recently proved by Laurence and Salsa in [232]. In the case of variable coefficients, for the one-dimensional parabolic obstacle problem Blanchet [51], Blanchet, Dolbeault and Monneau [50] prove that the free boundary is Hölder continuous. In more general settings, only qualitative properties of the free boundary and of the exercise region are known: see for instance Jacka [182], Myneni [262], Broadie and Detemple [64], Villeneuve [334]. The asymptotic behaviour of the free boundary near maturity has been studied by Barles, Burdeau, Romano and Samsœn [23], Lamberton [224], Lamberton and Villeneuve [229], Shahgholian [306]. □

8.2.2 Penalization method

In this section we prove the existence and uniqueness of a strong solution of the obstacle problem

$$\begin{cases} \max\{Lu, \varphi - u\} = 0, & \text{in } B(T) :=]0,T[\times B, \\ u|_{\partial_P B(T)} = g, \end{cases} \tag{8.24}$$

where B is the Euclidean ball with radius R, $R > 0$ being fixed in the whole section,

$$B = \{x \in \mathbb{R}^N \mid |x| < R\},$$

and $\partial_P B(T)$ denotes the parabolic boundary of $B(T)$:

$$\partial_P B(T) := \partial B(T) \setminus (\{T\} \times B).$$

We impose a condition analogous to Hypothesis 8.19 on the obstacle:

Hypothesis 8.26 *The function φ is Lipschitz continuous on $\overline{B(T)}$ and the weak-convexity condition* (8.22) *holds with $O = B(T)$. Furthermore $g \in C(\partial_P B(T))$ and we have that $g \geq \varphi$.*

We say that $u \in S^1_{\text{loc}}(B(T)) \cap C(\overline{B(T)})$ is a strong solution of problem (8.24) if the differential equation is verified a.e. on $B(T)$ and the boundary datum is taken pointwise. The main result of this section is the following:

Theorem 8.27 *Under the Hypotheses 8.1, 8.3 and 8.26 there exists a strong solution u to the problem* (8.24). *Moreover, for every $p \geq 1$ and O such that $\overline{O} \subseteq B(T)$, there exists a positive constant c, depending only on $L, O, B(T), p$ and on the L^∞-norms of g and φ, such that*

$$\|u\|_{S^p(O)} \leq c. \tag{8.25}$$

We prove Theorem 8.27 by using a classical penalization technique. Let us consider a family $(\beta_\varepsilon)_{\varepsilon \in]0,1[}$ of functions in $C^\infty(\mathbb{R})$: for every $\varepsilon > 0$, β_ε is a bounded, increasing function with bounded first order derivative such that

$$\beta_\varepsilon(0) = 0, \quad \beta_\varepsilon(s) \leq \varepsilon, \qquad s > 0.$$

Moreover we require that

$$\lim_{\varepsilon \to 0} \beta_\varepsilon(s) = -\infty, \qquad s < 0.$$

When $\delta \in]0,1[$, we denote by φ^δ the regularization of φ obtained with the usual mollifiers (cf. Appendix A.9.4). Since $g \geq \varphi$ on $\partial_P B(T)$, we have that

$$g^\delta := g + \lambda\delta \geq \varphi^\delta, \qquad \text{in } \partial_P B(T),$$

where λ is the Lipschitz constant of φ.

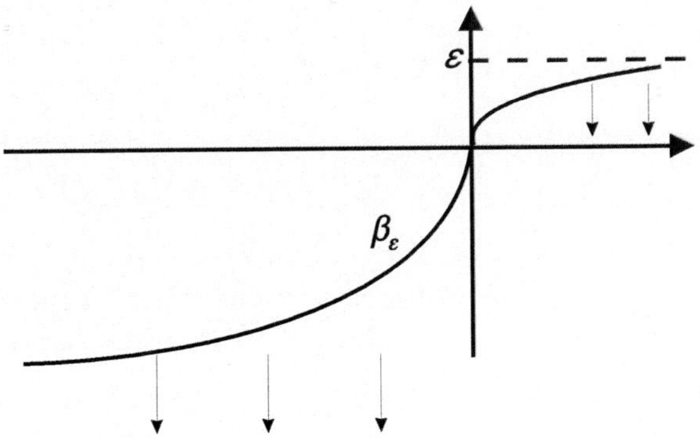

Fig. 8.2. Penalization function β_ε

Let us consider the *penalized problem*

$$\begin{cases} Lu = \beta_\varepsilon(u - \varphi^\delta), & \text{in } B(T), \\ u|_{\partial_P B(T)} = g^\delta, \end{cases} \tag{8.26}$$

and at a first stage we prove that it admits a classical solution. The proof consists in determining the solution of the non-linear differential equation internally and then in verifying that it is a continuous function up to the boundary. To study the behaviour of the solution close to the boundary, we use a standard tool in PDE theory, the *barrier functions.*

Definition 8.28 *Given a point $(t, x) \in \partial_P B(T)$, a barrier function for L at (t, x) is a function $w \in C^2(V \cap \overline{B(T)}; \mathbb{R})$, where V is a neighborhood of (t, x), such that*

i) $Lw \leq -1$ in $V \cap B(T)$;
ii) $w > 0$ in $V \cap \overline{B(T)} \setminus \{(t, x)\}$ and $w(t, x) = 0$.

Lemma 8.29 *There exists a barrier function for L at any point $(t, x) \in \partial_P B(T)$.*

Proof. If the point belongs to the basis of the cylinder $B(T)$, i.e. it is of the form $(0, \bar{x})$, then a barrier function is given by

$$w(t, x) = e^{t\|a\|_\infty} \left(|x - \bar{x}|^2 + Ct\right),$$

with C a sufficiently large constant.

If the point belongs to the lateral boundary of the cylinder, $(\bar{t}, \bar{x}) \in \partial_P B(T)$ with $\bar{t} \in]0, T[$, then we put

$$w(t, x) = Ce^{t\|a\|_\infty} \left(\frac{1}{|\bar{x} - \widetilde{x}|^p} - \frac{1}{R^p}\right),$$

where $(\bar{t}, \widetilde{x})$ is the centre of a sphere which is externally tangent to the cylinder in $(\bar{t}, \bar{x})$ and

$$R = \left(|x - \widetilde{x}|^2 + (t - \bar{t})^2\right)^{\frac{1}{2}}.$$

Then we have

$$\begin{aligned} Lw = \frac{Cp}{R^{p+4}} e^{\|a\|_\infty t} \Bigg(&-\frac{p+2}{2} \sum_{i,j=1}^N c_{ij}(x_i - \widetilde{x}_i)(x_j - \widetilde{x}_j) \\ &+ \frac{R^2}{2} \sum_{i=1}^N c_{ii} + R^2 \sum_{i=1}^N b_i(x_i - \widetilde{x}_i) - (t - \bar{t})R^2 \Bigg) + (a - \|a\|_\infty)w. \end{aligned}$$

Since L is uniformly parabolic, the expression within the parentheses is negative when p is large enough and then $Lw < 0$: by taking a suitably large C, we prove property i) and we conclude that w is a barrier function. □

Theorem 8.30 *Assume Hypotheses 8.1 and 8.3. If* $g \in C(\partial_P B(T))$ *and* $h = h(z,u) \in \mathrm{Lip}\left(\overline{B(T)} \times \mathbb{R}\right)$, *then there exists a classical solution* $u \in C_P^{2+\alpha}(B(T)) \cap C(\overline{B(T)})$ *of the problem*

$$\begin{cases} Lu = h(\cdot, u), & \text{in } B(T), \\ u|_{\partial_P B(T)} = g. \end{cases}$$

Moreover there exists a positive constant c, *depending on* h *and* $B(T)$ *only, such that*

$$\sup_{B(T)} |u| \le e^{cT}(1 + \|g\|_{L^\infty}). \tag{8.27}$$

Proof. It is not restrictive to take $a = 0$ since, after regularizing it, we can always include this term in the function h. We use a monotone iteration technique based upon the maximum principle. Let us put

$$u_0(x,t) = e^{ct}(1 + \|g\|_{L^\infty}) - 1,$$

where c is a positive constant such that

$$|h(t,x,u)| \le c(1 + |u|), \qquad (t,x,u) \in \overline{B(T)} \times \mathbb{R}.$$

Then we define recursively the sequence $(u_j)_{j \in \mathbb{N}}$ by means of

$$\begin{cases} Lu_j - \lambda u_j = h(\cdot, u_{j-1}) - \lambda u_{j-1}, & \text{in } B(T), \\ u_j|_{\partial_P B(T)} = g, \end{cases} \tag{8.28}$$

where λ is the Lipschitz constant of the function h. Here we use the classical theory (cf. for example, Chapter 3 in Friedman [139]) which assures that the *linear* problem (8.28) possesses a unique solution $C_P^{2,\alpha}(B(T)) \cap C(\overline{B(T)})$ for every $\alpha \in]0,1]$.

Now we prove by induction that (u_j) is a decreasing sequence. By the maximum principle, Theorem 6.10, we have $u_1 \le u_0$: indeed (recalling that $a = 0$)

$$L(u_1 - u_0) - \lambda(u_1 - u_0) = h(\cdot, u_0) - Lu_0 = h(\cdot, u_0) + c(1 + u_0) \ge 0,$$

and $u_1 \le u_0$ on $\partial_P B(T)$. For a fixed $j \in \mathbb{N}$, let us assume the inductive hypothesis $u_j \le u_{j-1}$; then, recalling that λ is the Lipschitz constant of h, we have that

$$L(u_{j+1} - u_j) - \lambda(u_{j+1} - u_j) = h(\cdot, u_j) - h(\cdot, u_{j-1}) - \lambda(u_j - u_{j-1}) \ge 0.$$

Furthermore $u_{j+1} = u_j$ on $\partial_P B(T)$ and so the maximum principle implies that $u_{j+1} \le u_j$. With an analogous argument we show that u_j is bounded from below by $-u_0$. Summing up, for $j \in \mathbb{N}$, we have

$$-u_0 \le u_{j+1} \le u_j \le u_0. \tag{8.29}$$

We denote by u the pointwise limit of the sequence (u_j) on $\overline{B(T)}$. Since u_j is solution to (8.28) and by the uniform estimate (8.29), we can apply the a priori estimates in S^p and the imbedding theorems, Theorems 8.17 and 8.16, in order to prove that, on every open set O included with its closure in $B(T)$ and for every $\alpha \in]0,1[$, the norm $\|u_j\|_{C_P^{1+\alpha}(O)}$ is bounded by a constant depending on L, $B(T)$, O, α and λ only. Then by the Schauder estimates, Theorem 8.18, we infer that the norm $\|u_j\|_{C_P^{2+\alpha}(O)}$ is uniformly bounded with respect to $j \in \mathbb{N}$. It follows that, by the Ascoli-Arzelà theorem, there exists a subsequence of $(u_j)_{j\in\mathbb{N}}$ (that, for the sake of simplicity, we still denote by $(u_j)_{j\in\mathbb{N}}$) converging locally in $C_P^{2+\alpha}$. Taking the limit in (8.28) as $j \to \infty$, we get

$$Lu = h(\cdot, u), \qquad \text{in } B(T),$$

and $u|_{\partial_p B(T)} = g$.

Finally, to prove that $u \in C(\overline{B(T)})$, we use the barrier functions. Given $\bar{z} = (\bar{t}, \bar{x}) \in \partial_P B(T)$ and $\varepsilon > 0$, we consider an open neighborhood V of $\bar{z}$ such that

$$|g(z) - g(\bar{z})| \le \varepsilon, \qquad z = (t, x) \in V \cap \partial_P B(T),$$

and suppose there exists a barrier function w for L in $V \cap B(T)$. We put

$$v^{\pm}(z) = g(\bar{z}) \pm (\varepsilon + k_\varepsilon w(z))$$

where k_ε is a sufficiently large constant, not depending on j, such that

$$L(u_j - v^+) \ge h(\cdot, u_{j-1}) - \lambda\,(u_{j-1} - u_j) + k_\varepsilon \ge 0,$$

and $u_j \le v^+$ on $\partial(V \cap B(T))$. By the maximum principle we have that $u_j \le v^+$ on $V \cap B(T)$.

Analogously we have $u_j \ge v^-$ on $V \cap B(T)$ and, when $j \to \infty$, we get

$$g(\bar{z}) - \varepsilon - k_\varepsilon w(z) \le u(z) \le g(\bar{z}) + \varepsilon + k_\varepsilon w(z), \qquad z \in V \cap B(T).$$

Then

$$g(\bar{z}) - \varepsilon \le \liminf_{z \to \bar{z}} u(z) \le \limsup_{z \to \bar{z}} u(z) \le g(\bar{z}) + \varepsilon, \qquad z \in V \cap B(T),$$

and this proves the claim by the arbitrariness of ε. Finally the estimate (8.27) can be justified by the maximum principle and by (8.29). □

Proof (of Theorem 8.27). We apply Theorem 8.30 with

$$h(\cdot, u) = \beta_\varepsilon(u - \varphi^\delta),$$

in order to infer the existence of a classical solution $u_{\varepsilon,\delta} \in C_P^{2+\alpha}(B(T)) \cap C(\overline{B(T)})$ of the penalized problem (8.26). After the simple change of variable $v(t, x) = e^{t\|a\|_\infty} u(t, x)$, we can always assume that $a \ge 0$.

First of all we prove that we have

$$|\beta_\varepsilon(u_{\varepsilon,\delta} - \varphi^\delta)| \leq \widetilde{c} \tag{8.30}$$

for some constant $\widetilde{c}$ not depending on ε and δ. Since $\beta_\varepsilon \leq \varepsilon$ we have to prove only the estimate from below. We denote by ζ a minimum point of the function $\beta_\varepsilon(u_{\varepsilon,\delta} - \varphi^\delta) \in C(\overline{B(T)})$ and we suppose that $\beta_\varepsilon(u_{\varepsilon,\delta}(\zeta) - \varphi^\delta(\zeta)) \leq 0$, otherwise there is nothing to prove. If $\zeta \in \partial_P B(T)$ then

$$\beta_\varepsilon(g^\delta(\zeta) - \varphi^\delta(\zeta)) \geq \beta_\varepsilon(0) = 0.$$

Conversely, if $\zeta \in B(T)$, then, since β_ε is an increasing function, also $u_{\varepsilon,\delta} - \varphi^\delta$ assumes the (negative) minimum at ζ and therefore

$$(L + a)u_{\varepsilon,\delta}(\zeta) - (L + a)\varphi^\delta(\zeta) \geq 0 \geq a(\zeta)\left(u_{\varepsilon,\delta}(\zeta) - \varphi^\delta(\zeta)\right),$$

that is

$$Lu_{\varepsilon,\delta}(\zeta) \geq L\varphi^\delta(\zeta). \tag{8.31}$$

Now, by Hypothesis 8.26, $L\varphi^\delta(\zeta)$ is bounded from below by a constant not depending on δ. So by (8.31) we get

$$\beta_\varepsilon(u_{\varepsilon,\delta}(\zeta) - \varphi^\delta(\zeta)) = Lu_{\varepsilon,\delta}(\zeta) \geq L\varphi^\delta(\zeta) \geq \widetilde{c},$$

with $\widetilde{c}$ independent on ε, δ thus proving the estimate (8.30).

By the maximum principle, Theorem 6.12, we have

$$\|u_{\varepsilon,\delta}\|_\infty \leq \|g\|_{L^\infty} + T\widetilde{c}. \tag{8.32}$$

Then by the a priori estimates in S^p, Theorems 8.17, and the estimates (8.30), (8.32) we infer that the norm $\|u_{\varepsilon,\delta}\|_{S^p(O)}$ is uniformly bounded with respect to ε and δ, for every open set O included with its closure in $B(T)$ and for every $p \geq 1$. It follows that there exists a subsequence of $(u_{\varepsilon,\delta})$ weakly convergent as $\varepsilon, \delta \to 0$ in S^p (and in $C_P^{1+\alpha}$) on compact subsets of $B(T)$ to a function u. Furthermore,

$$\limsup_{\varepsilon,\delta \to 0} \beta_\varepsilon(u_{\varepsilon,\delta} - \varphi^\delta) \leq 0,$$

so that $Lu \leq 0$ a.e. in $B(T)$. Finally, $Lu = 0$ a.e. on the set $\{u > \varphi\}$.

We can eventually conclude that $u \in C(\overline{B(T)})$ and $u = g$ on $\partial_P B(T)$ by using the argument of the barrier functions, just as in the proof of Theorem 8.30. □

We now prove a comparison principle for the obstacle problem.

Proposition 8.31 *Let u be a strong solution to the problem* (8.24) *and v a super-solution, i.e. $v \in S^1_{\text{loc}}(B(T)) \cap C(\overline{B(T)})$. If*

$$\begin{cases} \max\{Lv, \varphi - v\} \leq 0, & \text{a.e. in } B(T), \\ v|_{\partial_P B(T)} \geq g, \end{cases}$$

then $u \leq v$ in $B(T)$. In particular the solution to (8.24) *is unique.*

Proof. By contradiction, we suppose that the open set defined by

$$D := \{z \in B(T) \mid u(z) > v(z)\}$$

is not empty. Then, since $u > v \geq \varphi$ in D, we have that

$$Lu = 0, \quad Lv \leq 0 \qquad \text{in } D,$$

and $u = v$ on ∂D. The maximum principle implies $u \leq v$ in D and we get a contradiction. □

Proof (of Theorem 8.21). We prove the thesis by solving a sequence of obstacle problems on a family of cylinders that cover the strip $\mathcal{S}_T$, namely

$$B_n(T) =]0, T[\times\{|x| < n\}, \qquad n \in \mathbb{N}.$$

For every $n \in \mathbb{N}$, we consider a function $\chi_n \in C(\mathbb{R}^N; [0,1])$ such that $\chi_n(x) = 1$ if $|x| \leq n - \frac{1}{2}$ and $\chi_n(x) = 0$ if $|x| \geq n$, and we set

$$g_n(t,x) = \chi_n(x)\varphi(t,x) + (1 - \chi_n(x))\bar{u}(t,x), \qquad (t,x) \in \mathcal{S}_T.$$

By Theorem 8.27, for every $n \in \mathbb{N}$, there exists a strong solution u_n of problem

$$\begin{cases} \max\{Lu, \varphi - u\} = 0, & \text{in } B_n(T), \\ u|_{\partial_P B_n(T)} = g_n. \end{cases}$$

By Proposition 8.31 we have

$$\varphi \leq u_{n+1} \leq u_n \leq \bar{u}, \qquad \text{in } B_n(T),$$

and we can conclude the proof by using again the arguments of Theorems 8.27 and 8.30, based on the a priori estimates in S^p_{loc} and the barrier functions. □

9

Stochastic differential equations

In this chapter we present some basic results on stochastic differential equations, hereafter shortened to SDEs, and we examine the connection to the theory of parabolic partial differential equations.

We consider $Z \in \mathbb{R}^N$ and two measurable functions

$$b = b(t, x) : [0, T] \times \mathbb{R}^N \longrightarrow \mathbb{R}^N, \qquad \sigma = \sigma(t, x) : [0, T] \times \mathbb{R}^N \longrightarrow \mathbb{R}^{N \times d}.$$

In the sequel, we refer to b and σ as the *drift and diffusion coefficient*, respectively.

Definition 9.1 *Let W a d-dimensional Brownian motion on the filtered probability space $(\Omega, \mathcal{F}, P, (\mathcal{F}_t))$ on which the usual hypotheses hold. A solution relative to W of the SDE with coefficients Z, b, σ is a $\mathcal{F}_t$-adapted continuous process $(X_t)_{t \in [0,T]}$ such that*

i) $b(t, X_t) \in \mathbb{L}^1_{\text{loc}}$ and $\sigma(t, X_t) \in \mathbb{L}^2_{\text{loc}}$;
ii) we have that

$$X_t = Z + \int_0^t b(s, X_s)ds + \int_0^t \sigma(s, X_s)dW_s, \qquad t \in [0, T], \tag{9.1}$$

that is

$$dX_t = b(t, X_t)dt + \sigma(t, X_t)dW_t, \qquad X_0 = Z.$$

Next we introduce the notions of solution to (9.1).

Definition 9.2 *The SDE with coefficients Z, b, σ is solvable in the weak sense if a standard Brownian motion exists with respect to which the SDE admits a solution.*

The SDE with coefficients Z, b, σ is solvable in the strong sense if, for every fixed standard Brownian motion W, there exists a solution relative to W.

On the basis of the previous definition, a strong solution is constructed on a given probability space and with respect to a given Brownian motion W.

On the contrary, for weak solutions, the Brownian motion and the probability space are not assigned a priori and actually they are part of the solution rather than the statement of the problem.

Also for the concept of uniqueness, it is natural to introduce two different notions depending on whether we consider a strong or a weak solution.

Definition 9.3 *For the SDE with coefficients Z, b, σ we have uniqueness*

- *in the weak sense (or in law) if two solutions are equivalent processes, i.e. they have the same law;*
- *in the strong sense (or pathwise) if two solutions defined on the same probability space are indistinguishable.*

In general it is possible to assign a stochastic initial datum. When we consider strong solutions and we suppose that we have fixed a priori the probability space with filtration $(\mathcal{F}_t)$, we assume that the initial datum Z is a $\mathcal{F}_0$-measurable random variable: by (9.1), we have $X_0 = Z$. When we study the solvability in the weak sense, we merely assign the initial distribution μ of the solution: $X_0 \sim \mu$, i.e. if the solution is defined on the space $(\Omega, \mathcal{F}, P)$ we have that

$$P(X_0 \in H) = \mu(H), \qquad H \in \mathscr{B}(\mathbb{R}^N).$$

9.1 Strong solutions

In the case $\sigma = 0$ and $Z \in \mathbb{R}^N$, equation (9.1) reduces to the deterministic Volterra equation

$$X_t = Z + \int_0^t b(s, X_s) ds, \tag{9.2}$$

and assuming that b is a continuous function, (9.2) is equivalent to the ordinary Cauchy problem

$$\frac{d}{dt} X_t = b(t, X_t), \qquad X_0 = Z.$$

In the theory of existence and uniqueness for strong solutions of SDEs, many results are analogous to those for ordinary differential equations. In particular, it is known that, in order to obtain results of existence and uniqueness for the solution of (9.2) it is necessary to assume some regularity assumption on the coefficient b: typically it is assumed that $b = b(t, x)$ is locally Lipschitz continuous with respect to the variable x. For example, the equation

$$X_t = \int_0^t |X_s|^\alpha ds \tag{9.3}$$

has as unique solution the null function if $\alpha \geq 1$, while if $\alpha \in]0,1[$ there exist infinitely many solutions[1] of the form

$$X_t = \begin{cases} 0, & 0 \leq t \leq s, \\ \left(\frac{t-s}{\beta}\right)^{\beta}, & s \leq t \leq T, \end{cases}$$

where $\beta = \frac{1}{1-\alpha}$ and $s \in [0,T]$.

Furthermore it is known that, in order to guarantee the *global* existence of a solution it is necessary to impose conditions on the growth of the coefficient $b(t,x)$ as $|x| \to \infty$: typically it is assumed a linear growth. For example, for fixed $x > 0$, the equation

$$X_t = x + \int_0^t X_s^2 ds$$

has a (unique) solution $X_t = \frac{x}{1-xt}$ which diverges for $t \to \frac{1}{x}$.

On the grounds of these examples, we introduce the so-called "standard hypotheses" for a SDE. Since we are interested in studying strong solutions, in this section we assume that a d-dimensional Brownian motion W is fixed on the filtered probability space $(\Omega, \mathcal{F}, P, (\mathcal{F}_t))$.

Definition 9.4 *The SDE*

$$dX_t = b(t, X_t)dt + \sigma(t, X_t)dW_t, \qquad X_0 = Z,$$

verifies the standard hypotheses if

i) $Z \in L^2(\Omega, P)$ *and it is* $\mathcal{F}_0$*-measurable;*
ii) b, σ *are locally Lipschitz continuous in* x *uniformly with respect to* t*, i.e. for every* $n \in \mathbb{N}$ *there exists a constant* K_n *such that*

$$|b(t,x) - b(t,y)|^2 + |\sigma(t,x) - \sigma(t,y)|^2 \leq K_n |x-y|^2, \tag{9.5}$$

for $|x|, |y| \leq n$, $t \in [0,T]$*;*
iii) b, σ *have at most linear growth in* x*, i.e.*

$$|b(t,x)|^2 + |\sigma(t,x)|^2 \leq K(1 + |x|^2) \qquad x \in \mathbb{R}^N, \ t \in [0,T], \tag{9.6}$$

for a positive constant K.

[1] On the other hand, for the problem with stochastic perturbation

$$X_t = Z + \int_0^t b(X_s)ds + W_t, \tag{9.4}$$

where W is a Brownian motion, a remarkable result was proved by Zvonkin [350] and Veretennikov [333]: they proved pathwise uniqueness for (9.4) when b is only bounded and measurable.

9.1.1 Uniqueness

A classical tool for the study of the properties of differential equations is the following:

Lemma 9.5 (Gronwall's lemma) *Let $\varphi \in C([0,T])$ be such that*

$$\varphi(t) \le a + \int_0^t f(s)\varphi(s)ds, \qquad t \in [0,T],$$

where $a \in \mathbb{R}$ and f is a continuous, non-negative function. Then we have

$$\varphi(t) \le a e^{\int_0^t f(s)ds}, \qquad t \in [0,T].$$

Proof. We put

$$F(t) = a + \int_0^t f(s)\varphi(s)ds.$$

By assumption, $\varphi \le F$ and since f is non-negative we have

$$\frac{d}{dt}\left(e^{-\int_0^t f(s)ds} F(t)\right) = e^{-\int_0^t f(s)ds}\left(-f(t)F(t) + f(t)\varphi(t)\right) \le 0.$$

Integrating we get

$$e^{-\int_0^t f(s)ds} F(t) \le a,$$

hence the claim:

$$\varphi(t) \le F(t) \le a e^{\int_0^t f(s)ds}.$$

□

As in the case of deterministic equations, the uniqueness of the solution is consequence of the Lipschitz continuity on the coefficients. More precisely we have

Theorem 9.6 *If the standard conditions i) and ii) hold, then the solution of the SDE*

$$dX_t = b(t, X_t)dt + \sigma(t, X_t)dW_t, \qquad X_0 = Z,$$

is pathwise unique, i.e. two strong solutions are indistinguishable.

Proof. Let $X, \widetilde{X}$ be strong solutions with initial datum Z and $\widetilde{Z}$, respectively. For $n \in \mathbb{N}$ and $\omega \in \Omega$, we put

$$s_n(\omega) = T \wedge \inf\{t \in [0,T] \mid |X_t(\omega)| \ge n\}$$

and we define $\widetilde{s}_n$ analogously. By Theorem 3.52, $s_n, \widetilde{s}_n$ are stopping times. Therefore also

$$\tau_n := s_n \wedge \widetilde{s}_n$$

is a stopping time and

$$\lim_{n\to\infty} \tau_n(\omega) = T, \qquad \text{a.s.}$$

Recalling Proposition 4.22, which defines the Itô integral with a stopping time as integration limit, we have

$$\begin{aligned} X_{t\wedge\tau_n} - \widetilde{X}_{t\wedge\tau_n} = Z - \widetilde{Z} + &\int_0^{t\wedge\tau_n} (b(s, X_s) - b(s, \widetilde{X}_s))ds \\ + &\int_0^{t\wedge\tau_n} (\sigma(s, X_s) - \sigma(s, \widetilde{X}_s))dW_s. \end{aligned}$$

By the elementary inequality $(a+b+c)^2 \le 3(a^2+b^2+c^2)$ (cf. (9.9)), we have

$$\begin{aligned} E\left[\left|X_{t\wedge\tau_n} - \widetilde{X}_{t\wedge\tau_n}\right|^2\right] \le\ & 3E\left[|Z-\widetilde{Z}|^2\right] \\ &+ 3E\left[\left|\int_0^{t\wedge\tau_n} (b(s, X_s) - b(s, \widetilde{X}_s)ds\right|^2\right] \\ &+ 3E\left[\left|\int_0^{t\wedge\tau_n} (\sigma(s, X_s) - \sigma(s, \widetilde{X}_s))dW_s\right|^2\right] \le \end{aligned}$$

(by Hölder's inequality and Itô isometry, Corollary 4.23, since $(\sigma(s, X_s) - \sigma(s, \widetilde{X}_s))\mathbb{1}_{\{s\le t\wedge\tau_n\}} \in \mathbb{L}^2$)

$$\begin{aligned} \le 3E\left[|Z-\widetilde{Z}|^2\right] + 3tE&\left[\int_0^{t\wedge\tau_n} |b(s, X_s) - b(s, \widetilde{X}_s)|^2 ds\right] \\ +3E&\left[\int_0^{t\wedge\tau_n} |\sigma(s, X_s) - \sigma(s, \widetilde{X}_s)|^2 ds\right] \le \end{aligned}$$

(by the assumption of Lipschitz continuity of the coefficients)

$$\le 3\left(E\left[|Z-\widetilde{Z}|^2\right] + K_n(T+1)\int_0^t E\left[\left|X_{s\wedge\tau_n} - \widetilde{X}_{s\wedge\tau_n}\right|^2\right]ds\right).$$

By applying Gronwall's inequality, we infer that

$$E\left[\left|X_{t\wedge\tau_n} - \widetilde{X}_{t\wedge\tau_n}\right|^2\right] \le 3E\left[|Z-\widetilde{Z}|^2\right] e^{3K_n(T+1)t}.$$

In particular, if $Z = \widetilde{Z}$ a.s., then by Fatou's lemma we have

$$\begin{aligned} E\left[\left|X_t - \widetilde{X}_t\right|^2\right] &= E\left[\lim_{n\to\infty}\left|X_{t\wedge\tau_n} - \widetilde{X}_{t\wedge\tau_n}\right|^2\right] \\ &\le \liminf_{n\to\infty} E\left[\left|X_{t\wedge\tau_n} - \widetilde{X}_{t\wedge\tau_n}\right|^2\right] = 0, \end{aligned}$$

and therefore $X, \widetilde{X}$ are modifications. Finally, since $X, \widetilde{X}$ are continuous processes, it follows that they are indistinguishable, by Proposition 3.25. □

9.1.2 Existence

As in the deterministic case, existence of a solution of an SDE can be reduced to a fixed-point problem: formally the process X is solution of the SDE

$$dX_t = b(t, X_t)dt + \sigma(t, X_t)dW_t, \qquad X_0 = Z, \tag{9.7}$$

if and only if it is a fixed point for the functional Ψ defined by

$$\Psi(X)_t = Z + \int_0^t b(s, X_s)ds + \int_0^t \sigma(s, X_s)dW_s, \qquad t \in [0, T]. \tag{9.8}$$

To present the proper setting in which we look for the strong solution of the SDE (9.7), we introduce the following:

Notation 9.7 *$\mathcal{A}_c$ is the space of continuous $\mathcal{F}_t$-adapted processes $(X_t)_{t\in[0,T]}$ such that*

$$[\![X]\!]_T^2 := E\left[\sup_{0\le t\le T} |X_t|^2\right] < \infty.$$

The following result can be proved just as Lemma 3.43.

Lemma 9.8 *$(\mathcal{A}_c, [\![\cdot]\!]_T)$ is a semi-normed complete space.*

In what follows we repeatedly use the following inequalities:

Lemma 9.9 *For all $n \in \mathbb{N}$ and $a_1, \dots, a_n \in \mathbb{R}$ we have*

$$(a_1 + \dots + a_n)^2 \le n(a_1^2 + \dots + a_n^2). \tag{9.9}$$

For all $X \in \mathcal{A}_c$ we have that

$$E\left[\sup_{0\le s\le t}\left|\int_0^s X_u du\right|^2\right] \le t\int_0^t [\![X]\!]_s^2 ds, \tag{9.10}$$

$$E\left[\sup_{0\le s\le t}\left|\int_0^s X_u dW_u\right|^2\right] \le 4\int_0^t [\![X]\!]_s^2 ds. \tag{9.11}$$

Proof. We have

$$\begin{aligned}(a_1 + \dots + a_n)^2 &= a_1^2 + \dots + a_n^2 + 2\sum_{i<j} a_i a_j \\ &\le a_1^2 + \dots + a_n^2 + \sum_{i<j}(a_i^2 + a_j^2) \\ &= n(a_1^2 + \dots + a_n^2).\end{aligned}$$

By Hölder's inequality we get

$$\begin{aligned}E\left[\sup_{0\le s\le t}\left|\int_0^s X_u du\right|^2\right] &\le E\left[\sup_{0\le s\le t} s\int_0^s |X_u|^2 du\right] \\ &= tE\left[\int_0^t |X_u|^2 du\right] \le t\int_0^t [\![X]\!]_u^2 du.\end{aligned}$$

Finally (9.11) is a consequence of Doob's inequality and Itô isometry. □

Lemma 9.10 *Under the standard hypotheses i) and iii) of Definition 9.4, the functional Ψ in (9.8) is well defined from $\mathcal{A}_c$ to $\mathcal{A}_c$. Further, there exists a constant C_1 depending on T and K only, such that*

$$[\![\Psi(X)]\!]_t^2 \leq C_1 \left(1 + E\left[|Z|^2\right] + \int_0^t [\![X]\!]_s^2 ds\right), \qquad t \in [0,T]. \tag{9.12}$$

Proof. By the assumption of linear growth on the coefficients, we have

$$E\left[\sup_{0\leq s\leq t} |b(s,X_s)|^2\right] + E\left[\sup_{0\leq s\leq t} |\sigma(s,X_s)|^2\right] \leq K(1 + [\![X]\!]_t^2) \qquad t \in [0,T], \tag{9.13}$$

and so $b(t,X_t), \sigma(t,X_t) \in \mathcal{A}_c$ when $X \in \mathcal{A}_c$. Then we get

$$[\![\Psi(X)]\!]_t^2 = E\left[\sup_{0\leq s\leq t}\left|Z + \int_0^s b(u,X_u)du + \int_0^s \sigma(u,X_u)du\right|^2\right] \leq$$

(by (9.9), (9.10) and (9.11))

$$\leq 3\Bigg(E\left[|Z|^2\right] + t\int_0^t E\left[\sup_{0\leq u\leq s}|b(u,X_u)|^2\right]ds$$
$$+ 4\int_0^t E\left[\sup_{0\leq u\leq s}|\sigma(u,X_u)|^2\right]ds\Bigg) \leq$$

(by (9.13))

$$\leq 3\left(E\left[|Z|^2\right] + K(4+t)\left(t + \int_0^t [\![X]\!]_s^2 ds\right)\right).$$

□

The following classical theorem gives sufficient conditions for the existence of a unique strong solution of the SDE (9.1): although it is not the most general result, it is satisfactory for many applications.

Theorem 9.11 *Under the standard hypotheses of Definition 9.4 the SDE*

$$X_t = Z + \int_0^t b(s,X_s)ds + \int_0^t \sigma(s,X_s)dW_s, \qquad t \in [0,T], \tag{9.14}$$

has a strong solution in the space $\mathcal{A}_c$. Such a solution is unique modulo indistinguishability and it verifies the estimate

$$[\![X]\!]_t^2 \leq C(1 + E\left[|Z|^2\right])e^{Ct}, \qquad t \in [0,T], \tag{9.15}$$

where C is a constant depending on K in (9.6) and T only.

Proof. The uniqueness of the solution has already been proved in Theorem 9.6. Concerning its existence, for the sake of simplicity we consider the standard hypotheses with $K_n \equiv K$, not depending on n: the general case can be proved by a localization argument, just as in the proof of Theorem 9.6.

As in the deterministic case, the proof is based upon the Banach-Caccioppoli fixed-point theorem: we have already proved that Ψ is well-defined from $\mathcal{A}_c$ to $\mathcal{A}_c$ (cf. Lemma 9.10) and that $(\mathcal{A}_c, [\![\cdot]\!]_T)$ is a semi-normed complete space (cf. Lemma 3.43). What is left to prove is then the existence of an integer $n \in \mathbb{N}$ such that

$$\Psi^n = \underbrace{\Psi \circ \cdots \circ \Psi}_{n \text{ times}}$$

is a contraction, that is

$$[\![\Psi^n(X) - \Psi^n(Y)]\!]_T \le C_0 [\![X - Y]\!]_T, \qquad X, Y \in \mathcal{A}_c,$$

for a constant $C_0 \in]0,1[$. More precisely, we prove by induction that for every $n \in \mathbb{N}$ we have that

$$[\![\Psi^n(X) - \Psi^n(Y)]\!]_t^2 \le \frac{(C_2 t)^n}{n!} [\![X - Y]\!]_t^2, \qquad X, Y \in \mathcal{A}_c,\ t \in [0,T], \quad (9.16)$$

where $C_2 = 2K(T+4)$.

We have

$$[\![\Psi^{n+1}(X) - \Psi^{n+1}(Y)]\!]_t^2 = E\left[\sup_{0 \le s \le t} \left| \int_0^s (b(u, \Psi^n(X)_u) - b(u, \Psi^n(Y)_u))du \right.\right.$$
$$\left.\left. + \int_0^s (\sigma(u, \Psi^n(X)_u) - \sigma(u, \Psi^n(Y)_u))dW_u \right|^2 \right] \le$$

(by (9.9), (9.10) and (9.11))

$$\le 2t \int_0^t E\left[\sup_{0 \le u \le s} |b(u, \Psi^n(X)_u) - b(u, \Psi^n(Y)_u)|^2 \right] ds$$
$$+ 8 \int_0^t E\left[\sup_{0 \le u \le s} |\sigma(u, \Psi^n(X)_u) - \sigma(u, \Psi^n(Y)_u)|^2 \right] ds \le$$

(by the hypothesis of Lipschitz continuity)

$$\le C_2 \int_0^t [\![\Psi^n(X) - \Psi^n(Y)]\!]_s^2 ds \le$$

(by the inductive hypothesis)

$$\le C_2^{n+1} \int_0^t \frac{s^n}{n!} ds [\![X - Y]\!]_t^2,$$

hence (9.16) follows. We infer that Ψ admits a unique fixed point X in $\mathcal{A}_c$. Since $X = \Psi(X)$ the estimate (9.12) becomes

$$[\![X]\!]_t^2 \le C_1 \left(1 + E\left[|Z|^2\right] + \int_0^t [\![X]\!]_s^2 ds \right), \qquad t \in [0,T],$$

and, by applying Gronwall's inequality, we directly obtain (9.15). □

Remark 9.12 The previous proof implicitly contains a uniqueness result independent of Theorem 9.6: indeed, since Ψ^n is a contraction, it admits a unique fixed point in the space $\mathcal{A}_c$. On the other hand the uniqueness result in Theorem 9.6 is stronger because it gives the uniqueness not only within the class $\mathcal{A}_c$: in particular under the standard hypotheses, every solution to (9.14) belongs to the class $\mathcal{A}_c$. □

Remark 9.13 As in the deterministic case, the solution of a SDE can be determined by successive approximations. More precisely, let (X_n) be the sequence in $\mathcal{A}_c$ defined by

$$\begin{cases} X_0 = Z, \\ X_n = \Psi(X_{n-1}), & n \in \mathbb{N}, \end{cases}$$

where Ψ is the functional in (9.8). Then, under the standard hypotheses and denoting by X the solution, we have that

$$\lim_{n\to\infty} [\![X - X_n]\!]_T = 0.$$ □

9.1.3 Properties of solutions

In this section we prove some remarkable growth estimates and results on regularity, comparison and dependence on the data for the solution of a SDE. This kind of estimates plays a crucial role, for instance, in the study of the numerical solution of stochastic equations.

Theorem 9.14 *Let X be solution of the SDE*

$$X_t = X_0 + \int_0^t b(s, X_s)ds + \int_0^t \sigma(s, X_s)dW_s, \qquad t \in [0, T]. \tag{9.17}$$

If the standard hypotheses of Definition 9.4 hold and $E\left[|X_0|^{2p}\right]$ is finite for some $p \geq 1$, then there exists a constant C depending only on T, p and K in (9.6), such that

$$E\left[\sup_{t_0 \leq s \leq t} |X_s|^{2p}\right] \leq C\left(1 + E\left[|X_{t_0}|^{2p}\right]\right) e^{C(t-t_0)}, \tag{9.18}$$

$$E\left[\sup_{t_0 \leq s \leq t} |X_s - X_{t_0}|^{2p}\right] \leq C\left(1 + E\left[|X_{t_0}|^{2p}\right]\right) (t-t_0)^p, \tag{9.19}$$

with $0 \leq t_0 < t \leq T$.

Proof. We prove the claim in the case $p = 1$, $N = 1$ and $t_0 = 0$. The case $p > 1$ is analogous and can be proved by using the fact that X^{2p} is a solution of the SDE

$$X_t^{2p} = X_0^{2p} + \int_0^t \left(2pX_s^{2p-1}b(s,X_s) + p(2p-1)X_s^{2p-2}\sigma^2(s,X_s)\right) ds$$
$$+ \int_0^t 2pX_s^{2p-1}\sigma(s,X_s)dW_s.$$

For further details we refer, for example, to Kloeden and Platen [210], Theorem 4.5.4.

The inequality (9.18) for $p = 1$ is equivalent to (9.15) of Theorem 9.11. Concerning (9.19), by Lemma 9.9 and the condition of linear growth of the coefficients, we have

$$[\![X - X_0]\!]_t^2 \leq 2K(t+4)\int_0^t (1 + [\![X]\!]_s^2)ds \leq$$

(by (9.18))

$$\leq Ct\left(1 + E\left[|X_0|^2\right]\right).$$

□

We now prove a result on the continuous dependence on the parameters of a SDE. First of all we introduce the following:

Notation 9.15 *We put*

$$\mathcal{L}_{t_0,t}X := X_t - X_{t_0} - \int_{t_0}^t b(s,X_s)ds - \int_{t_0}^t \sigma(s,X_s)dW_s, \qquad t \in [t_0,T], \quad (9.20)$$

and, for the sake of simplicity, $\mathcal{L}_{0,t}X = \mathcal{L}_tX$. Further, when we write $\mathcal{L}_{t_0,t}X$ we implicitly assume that $(X_t)_{t\in[t_0,T]}$ is an adapted process such that

$$b(t,X_t) \in \mathbb{L}^1_{\mathrm{loc}} \quad \textit{and} \quad \sigma(t,X_t) \in \mathbb{L}^2_{\mathrm{loc}}.$$

Clearly X is a solution to the SDE (9.17) if $\mathcal{L}_tX = 0$.

Theorem 9.16 *Consider $\mathcal{L}_{t_0,t}$ in (9.20) for $0 \leq t_0 < t \leq T$ and assume that the coefficients of the SDE are Lipschitz continuous in x uniformly with respect to t, that is*

$$|b(t,x) - b(t,y)|^2 + |\sigma(t,x) - \sigma(t,y)|^2 \leq K|x-y|^2, \quad t \in [t_0,T],\ x,y \in \mathbb{R}^N,$$

for some positive constant K. Then there exists a constant C depending on K, T and $p \geq 1$ only, such that for every pair of processes X,Y, we have that

$$E\left[\sup_{t_0\leq s\leq t}|X_s - Y_s|^{2p}\right] \leq Ce^{C(t-t_0)}\Bigg(E\left[|X_{t_0} - Y_{t_0}|^{2p}\right]$$
$$+E\left[\sup_{t_0\leq s\leq t}|\mathcal{L}_{t_0,s}X - \mathcal{L}_{t_0,s}Y|^{2p}\right]\Bigg). \quad (9.21)$$

Proof. We only consider the case $p = 1$ and $t_0 = 0$. Using Lemma 9.9 we get

$$[\![X - Y]\!]_t^2 \leq 4\Big(E\left[(X_0 - Y_0)^2\right] + t\int_0^t [\![b(\cdot, X) - b(\cdot, Y)]\!]_s^2 ds \\ +4\int_0^t [\![\sigma(\cdot, X) - \sigma(\cdot, Y)]\!]_s^2 ds + [\![\mathcal{L}X - \mathcal{L}Y]\!]_t^2\Big) \leq$$

(by the assumption of Lipschitz continuity on the coefficients)

$$\leq 4\left(E\left[(X_0 - Y_0)^2\right] + K(t+4)\int_0^t [\![X - Y]\!]_s^2 ds + [\![\mathcal{L}X - \mathcal{L}Y]\!]_t^2\right).$$

The claim follows by Gronwall's Lemma. □

Remark 9.17 If X, Y are solutions of the SDE (9.17), then by (9.21) we have that

$$[\![X - Y]\!]_t^2 \leq 4E\left[|X_0 - Y_0|^2\right] e^{Ct}.$$

By examining the proof, we can improve the previous estimate by using an elementary inequality such as

$$(a + b)^2 \leq (1 + \varepsilon)a^2 + (1 + \varepsilon^{-1})b^2, \qquad \varepsilon > 0,$$

and so we get

$$[\![X - Y]\!]_t^2 \leq (1 + \varepsilon)E\left[|X_0 - Y_0|^2\right] e^{\bar{C}t}, \tag{9.22}$$

with $\bar{C}$ depending on ε, K and T. Inequality (9.22) provides us with a sensitivity (or stability) estimate of the solution in terms of dependence on the initial datum. Formula (9.22) can be useful if one wants to estimate the error made by an erroneous specification of the initial datum due, for instance, to incomplete information. □

We conclude this section mentioning a comparison result for solutions of SDE: for the proof we refer, for example, to [201], Theorem 5.2.18.

Theorem 9.18 *Let X^1, X^2 be solutions of the SDEs*

$$X_t^j = Z^j + \int_0^t b_j(s, X_s^j)ds + \int_0^t \sigma(s, X_s^j)dW_s, \qquad t \in [0, T],\ j = 1, 2,$$

with the coefficients verifying the standard hypotheses. If

i) $Z^1 \leq Z^2$ *a.s.;*
ii) $b_1(t, x) \leq b_2(t, x)$ *for every* $x \in \mathbb{R}$ *and* $t \in [0, T]$;

then

$$P(X_t^1 \leq X_t^2,\ t \in [0, T]) = 1.$$

9.2 Weak solutions

In this section we present some classical theorems on existence and uniqueness of weak solutions of SDEs with *continuous and bounded coefficients.* The material in this section summarizes very classical results: a more detailed exposition can be found, for instance, in the monographs by Stroock and Varadhan [321], Karatzas and Shreve [201].

We begin by presenting a SDE solvable in the weak sense but not in the strong sense. The following example shows also that a SDE can have solutions that are equivalent in law but not indistinguishable: in this sense uniqueness in law does not imply pathwise uniqueness.

9.2.1 Tanaka's example

The following example is due to Tanaka [324] (see also Zvonkin [350]). Let us consider the scalar SDE ($N = d = 1$) with coefficients $Z = 0 = b$ and

$$\sigma(x) = \operatorname{sgn}(x) = \begin{cases} 1 & x \geq 0, \\ -1 & x < 0. \end{cases}$$

First of all we prove that, for such a SDE, we have uniqueness in the weak sense. Indeed, if X is a solution relative to a Brownian motion W, then

$$X_t = \int_0^t \operatorname{sgn}(X_s) dW_s,$$

and by Corollary 5.35, X is a Brownian motion. Therefore we have uniqueness in law. On the other hand, $-X$ is a solution relative to W as well, we do not have pathwise uniqueness.

Let us now prove the existence of a weak solution. We consider a standard Brownian motion W on the probability space $(\Omega, \mathcal{F}, P, (\mathcal{F}_t))$ and we put

$$B_t = \int_0^t \operatorname{sgn}(W_s) dW_s.$$

Again by Corollary 5.35, B is a Brownian motion on $(\Omega, \mathcal{F}, P, (\mathcal{F}_t))$. Further, we have that

$$dW_t = (\operatorname{sgn}(W_t))^2 \, dW_t = \operatorname{sgn}(W_t) dB_t,$$

i.e. W is solution relative to the Brownian motion B.

Eventually we prove that the SDE does not admit a strong solution. By contradiction, let X be a solution relative to a Brownian motion W defined on $(\Omega, \mathcal{F}, P, \mathcal{F}_t^W)$ where $(\mathcal{F}_t^W)$ denotes the standard filtration[2] of W. Then we have

$$dW_t = (\operatorname{sgn}(X_t))^2 \, dW_t = \operatorname{sgn}(X_t) dX_t. \tag{9.23}$$

[2] Theorem 3.47, p. 118.

Since X is a Brownian motion on $(\Omega, \mathcal{F}, P, \mathcal{F}_t^W)$, applying Tanaka's formula[3], we obtain

$$|X_t| = \int_0^t \operatorname{sgn}(X_s)dX_s + 2L_t^X(0) \tag{9.24}$$

where, by (5.53),

$$L_t^X(0) = \lim_{\varepsilon\to 0^+} \frac{1}{2\varepsilon} |\{s \in [0,t] \mid |X_s| \le \varepsilon\}|$$

is the local time of X at zero. Putting (9.23) and (9.24) together we get

$$W_t = |X_t| - 2L_t^X(0)$$

and this implies that W is adapted to the standard filtration $\mathcal{F}_t^{|X|}$ of $|X|$. On the other hand, by its very definition, X is $\mathcal{F}_t^W$-adapted: so the following inclusion holds

$$\mathcal{F}_t^X \subseteq \mathcal{F}_t^{|X|},$$

where $\mathcal{F}_t^X$ is the standard filtration of X, and this is absurd.

We would like to mention also Barlow's work [24] where the author gives an example of a SDE *with continuous coefficients* which does not admit a strong solution.

9.2.2 Existence: the martingale problem

In this section we give an overview of the classical results of Stroock and Varadhan [319; 320] on the existence and uniqueness of weak solutions to SDEs with *bounded and continuous coefficients.* Instead of confronting the question of solvability directly, Stroock and Varadhan formulate and solve an equivalent problem, called *the martingale problem.*

To introduce the martingale problem, let us consider a SDE with bounded and continuous coefficients

$$b \in C_b(\mathbb{R}_{\ge 0} \times \mathbb{R}^N; \mathbb{R}^N), \qquad \sigma \in C_b(\mathbb{R}_{\ge 0} \times \mathbb{R}^N; \mathbb{R}^{N\times d}).$$

We suppose there exists a solution X to the SDE

$$dX_t = b(t, X_t)dt + \sigma(t, X_t)dW_t, \tag{9.25}$$

relative to a d-dimensional Brownian motion W defined on the probability space $(\Omega, \mathcal{F}, P, (\mathcal{F}_t))$.

Applying the Itô formula (5.34), for every $f \in C_0^2(\mathbb{R}^N)$ we have

$$df(X_t) = \mathcal{A}_t f(X_t)dt + \nabla f(X_t) \cdot \sigma(t, X_t)dW_t,$$

[3] Formula (5.54), p. 196 con $K = 0$, recalling that $|X| = X^+ + (-X)^+$.

where

$$\mathcal{A}_t f(x) := \frac{1}{2} \sum_{i,j=1}^{N} c_{ij}(t,x)\partial_{x_i x_j} f(x) + \sum_{j=1}^{N} b_j(t,x)\partial_{x_j} f(x), \tag{9.26}$$

and $(c_{ij}) = \sigma\sigma^*$.

Definition 9.19 *The operator $\mathcal{A}_t$ is called characteristic operator of the SDE (9.25).*

Since by assumption ∇f and σ are bounded, we have that $\nabla f(X_t)\sigma(t, X_t) \in \mathbb{L}^2$ and consequently the process

$$M_t^f := f(X_t) - f(X_0) - \int_0^t \mathcal{A}_s f(X_s) ds \tag{9.27}$$

is a $\mathcal{F}_t$-continuous martingale.

Now, in order to state the martingale problem, instead of considering the stochastic equation we start directly from a differential operator of the form (9.26), we assume that the coefficients $c_{ij}, b_j \in C_b(\mathbb{R}_{\geq 0} \times \mathbb{R}^N)$ and that the matrix (c_{ij}) is symmetric and positive semi-definite.

We recall briefly the results of Section 3.2.1: on the space

$$C(\mathbb{R}_{\geq 0}) = C(\mathbb{R}_{\geq 0}; \mathbb{R}^N)$$

endowed with the Borel σ-algebra $\mathscr{B}(C(\mathbb{R}_{\geq 0}))$, we define the "canonical" process

$$\mathbb{X}_t(w) = w(t), \qquad w \in C(\mathbb{R}_{\geq 0}),$$

and the related standard filtration[4] $\mathscr{B}_t(C(\mathbb{R}_{\geq 0}))$. In the following we prefer to use the more intuitive notation $w(t)$ instead of $\mathbb{X}_t(w)$.

Definition 9.20 *A solution of the martingale problem associated to operator $\mathcal{A}_t$ is a probability measure P on the space*

$$(C(\mathbb{R}_{\geq 0}), \mathscr{B}(C(\mathbb{R}_{\geq 0})))$$

such that, for every $f \in C_0^2(\mathbb{R}^N)$, the process

$$M_t^f(w) = f(w(t)) - f(w(0)) - \int_0^t \mathcal{A}_s f(w(s)) ds,$$

is a P-martingale, with respect to the filtration $\mathscr{B}_t(C(\mathbb{R}_{\geq 0}))$.

[4] Obtained by completing the natural filtration $\widetilde{\mathscr{B}}_t(C(\mathbb{R}_{\geq 0}))$ in (3.17), in accordance with Definition 3.45.

If the SDE (9.25) has a solution X, then the martingale problem for $\mathcal{A}_t$ in (9.26) is solvable: a solution is the law of X. Actually, it turns out that the problems are equivalent since, according to Theorem 9.22, the existence of a solution of the martingale problem implies the solvability in the weak sense of the associated SDE.

Let us point out that the SDE appears only indirectly in the formulation of the martingale problem, i.e. only through the coefficients of the equation defining the operator $\mathcal{A}_t$. The martingale-problem approach turns out to give a great deal of advantage in the study of SDEs for many reasons: for instance, one can use the results of convergence for Markov chains to diffusion processes which play a crucial part in the proof of the existence of the solution. With these techniques it is possible to prove weak existence results under mild assumptions. The question of uniqueness in law is in general more delicate: in Section 9.2.3 we present a theorem based upon the results of existence for the parabolic Cauchy problem of Chapter 8.

In order to state the equivalence between the martingale problem and SDEs we have to introduce the notion of extension of a probability space.

Remark 9.21 (Extension of a probability space) Let X be an adapted process on the space $(\Omega, \mathcal{F}, P, (\mathcal{F}_t))$. In general it is not possible to construct a Brownian motion on Ω, since the space could not be sufficiently "rich" to support it. On the other hand, if W is a Brownian motion on the space $(\widetilde{\Omega}, \widetilde{\mathcal{F}}, \widetilde{P}, \widetilde{\mathcal{F}}_t)$, we can consider the product space

$$\left(\Omega \times \widetilde{\Omega}, \mathcal{F} \otimes \widetilde{\mathcal{F}}, P \otimes \widetilde{P}\right)$$

endowed with the standard filtration $\bar{\mathcal{F}}_t$ obtained from $\mathcal{F}_t \otimes \widetilde{\mathcal{F}}_t$, and extend in a natural fashion the processes X and W by putting

$$\bar{X}(\omega, \widetilde{\omega}) = X(\omega), \qquad \bar{W}(\omega, \widetilde{\omega}) = W(\widetilde{\omega}).$$

Then we have that, on the product space, $\bar{W}$ is a $\bar{\mathcal{F}}_t$-Brownian motion independent of $\bar{X}$. □

The following result, that we merely mention, establishes the equivalence between the martingale problem and the weak formulation of the associated SDE. The proof is based upon the representation of continuous martingales in terms of Brownian integrals: we refer to, for example, Karatzas and Shreve [201], Proposition 5.4.11 and Corollary 5.4.9.

Theorem 9.22 *Let ζ be a distribution on $\mathbb{R}^N$. There exists a solution P of the martingale problem associated to $\mathcal{A}_t$ with initial datum ζ (i.e. such that $P(w(0) \in H) = \zeta(H)$ for every $H \in \mathscr{B}(\mathbb{R}^N)$) if and only if there exists a d-dimensional Brownian motion W, defined on an extension of*

$$(C(\mathbb{R}_{\geq 0}), \mathscr{B}(C(\mathbb{R}_{\geq 0})), P, \mathscr{B}_t(C(\mathbb{R}_{\geq 0}))),$$

such that the extension of the process $\mathbb{X}_t(w) = w(t)$ *is a solution of the SDE* (9.25) *relative to* W *with initial datum* ζ.

Further, the uniqueness of the solution of the martingale problem with initial datum ζ *is equivalent to the uniqueness in law for the SDE with initial datum* ζ.

We conclude the section stating the main existence result. The proof is based on the discretization of the SDE and on a limiting procedure for the sequence (P_n) of solutions of the martingale problem associated to the discrete SDEs (we refer, for instance, to Stroock and Varadhan [321], Theorem 6.1.7, or Karatzas and Shreve [201], Theorem 5.4.22).

Theorem 9.23 *Let us consider the SDE*

$$dX_t = b(t, X_t)dt + \sigma(t, X_t)dW_t, \tag{9.28}$$

with continuous coefficients, satisfying the linear growth condition

$$|b(t,x)|^2 + |\sigma(t,x)|^2 \le K(1 + |x|^2) \qquad x \in \mathbb{R}^N, \ t \in [0,T],$$

for a positive constant K. *Then, for any* $x \in \mathbb{R}^N$, (9.28) *admits at least one weak solution with initial datum* x.

9.2.3 Uniqueness

As already mentioned, the weak uniqueness is generally more involved to deal with, compared to the mere existence. It is enough to consider the deterministic equation (9.3) to notice that the only assumption of continuity and boundedness of the coefficients is not sufficient to guarantee this property. In this section we show that the formulation in terms of the martingale problem allows one to obtain a very natural condition for uniqueness: the existence of a solution of the Cauchy problem relative to the elliptic-parabolic operator $\mathcal{A}_t + \partial_t$. As we have seen in Chapter 8, under suitable assumptions, for such an operator a well-established theory is available.

Let us recall that[5] two measures P, Q on $(C(\mathbb{R}_{\ge 0}), \mathscr{B}(C(\mathbb{R}_{\ge 0})))$ are equal if and only they have the same finite-dimensional distributions, i.e. if

$$P\left(w(t_1) \in H_1, \dots, w(t_n) \in H_n\right) = Q\left(w(t_1) \in H_1, \dots, w(t_n) \in H_n\right)$$

for every $n \in \mathbb{N}$, $0 \le t_1 < \dots < t_n$ and $H_1, \dots, H_n \in \mathscr{B}(\mathbb{R}^N)$.

The following result gives a sufficient condition for which two solutions P and Q of the martingale problem with the same initial datum have the same *one-dimensional* distributions, i.e.

$$P\left(w(t) \in H\right) = Q\left(w(t) \in H\right)$$

for every $t \ge 0$ and $H \in \mathscr{B}(\mathbb{R}^N)$.

[5] Proposition 3.22, p. 105.

Proposition 9.24 *Let P, Q be solutions of the martingale problem associated to $\mathcal{A}_t$ with initial datum $x_0 \in \mathbb{R}^N$, i.e. such that*

$$P(w(0) = x_0) = Q(w(0) = x_0) = 1.$$

Suppose that for every $T > 0$ and for every $\varphi \in C_b(\mathbb{R}^N)$ there exists a bounded classical solution

$$u \in C^{1,2}(]0, T[\times\mathbb{R}^N) \cap C_b([0, T] \times \mathbb{R}^N),$$

to the Cauchy problem with final datum

$$\begin{cases} \mathcal{A}_t u(t, x) + \partial_t u(t, x) = 0, & \text{in }]0, T[\times\mathbb{R}^N, \\ u(T, \cdot) = \varphi, & \text{on } \mathbb{R}^N. \end{cases} \tag{9.29}$$

Then P and Q have the same one-dimensional distributions.

Proof. By Theorem 9.22, the process $\mathbb{X}_t(w) = w(t)$ is solution to the SDE (9.25) on some extension of the space of continuous functions endowed with the probability measure P and the same result holds for Q. It follows that, if u is a solution of the problem (9.29), then the process $u(t, w(t))$ is a local martingale, by the Itô formula. On the other hand, $u(t, w(t))$ is a strict martingale because u is bounded and therefore we have

$$\begin{aligned} E^P[\varphi(w(T))] = E^P[u(T, w(T))] &= u(0, x_0) \\ &= E^Q[u(T, w(T))] = E^Q[\varphi(w(T))]. \end{aligned} \tag{9.30}$$

Now it is fairly easy to conclude by using Dynkin's Theorem: indeed if H is a bounded open set in $\mathbb{R}^N$, we construct the increasing sequence of non-negative, continuous and bounded functions

$$\varphi_n(x) = n \min\left\{\frac{1}{n}, \inf_{y \notin H} |x - y|\right\},$$

approximating the characteristic function of H as n tends to infinity. By the theorem of monotone convergence and (9.30), we get

$$P(w(T) \in H) = Q(w(T) \in H),$$

and the claim follows easily by Proposition A.6. □

Now we are interested to go from the uniqueness of the one-dimensional distributions to the uniqueness of all finite-dimensional distributions. We mention the following result, due to Stroock and Varadhan [321], Theorem 6.2.3.

Proposition 9.25 *Suppose that the solutions of the martingale problem associated to $\mathcal{A}_t$ with initial condition $x_0 \in \mathbb{R}^N$ have the same one-dimensional distributions. Then the solution of the martingale problem with initial datum x_0 is unique.*

Remark 9.26 A similar result is proved in Karatzas and Shreve [201], Proposition 5.4.27, by using the Markov property in a way not dissimilar, for example, to the ideas in the proof of Proposition 3.23, which characterized the finite-dimensional distributions of Brownian motion. Nevertheless, this approach requires the hypothesis of *autonomous coefficients*, $b = b(x)$ and $\sigma = \sigma(x)$, since it has to be proven first that P, Q have the Markov property. □

We can eventually state a remarkable result of weak uniqueness for SDEs.

Theorem 9.27 *Consider a SDE with measurable and bounded coefficients b and σ. As usual, we denote by $\mathcal{A}_t$ the related differential operator defined in* (9.26)*. If, for all $T > 0$ and for all $\varphi \in C_b(\mathbb{R}^N)$, there exists a bounded classical solution of the Cauchy problem* (9.29)*, then for the SDE we have uniqueness in law.*

Sufficient conditions for the solvability of problem (9.29), as requested in Theorem 9.27, were given in Chapter 8. If the coefficients c_{ij}, b_j are Hölder continuous bounded functions and the matrix (c_{ij}) is uniformly positive definite, then the operator $\mathcal{A}_t + \partial_t$ has a fundamental solution Γ such that

$$u(t, x) = \int_{\mathbb{R}^N} \Gamma(t, x; T, y)\varphi(y)dy$$

is a classical solution of the Cauchy problem (9.29). Further, u is the only bounded solution:

$$|u(t, x)| \le \|\varphi\|_\infty \int_{\mathbb{R}^N} \Gamma(t, x; T, y)dy = \|\varphi\|_\infty.$$

In Section 9.5.2 we treat also the case of *non-uniformly parabolic PDEs* that appear in some financial models: in the case of constant coefficients, the prototype of such a class is the Kolmogorov equation (7.75)

$$\partial_{xx} + x\partial_y + \partial_t, \qquad (t, x, y) \in \mathbb{R}^3,$$

introduced in the study of Asian options.

9.3 Maximal estimates

We consider the solution X of a SDE. In order to prove some fundamental results, as the Feynman-Kač formula of Section 9.4.2 on unbounded domains, it is necessary to estimate "how far from the starting point the process X has gone in a given time interval". We loosely use the adjective "maximal" to denote an estimate of the upper bound

$$\sup_{0 \le t \le T} X_t.$$

9.3.1 Maximal estimates for martingales

We have already seen in the proof of Doob's inequality, Theorem 3.38, that for martingales it is possible to obtain uniform estimates with respect to time. The following result is the natural "uniform in t" version of the Markov inequality.

Theorem 9.28 (Maximal martingale inequalities) *Let X be a right-continuous super-martingale. For every $\lambda > 0$ we have*

$$P\left(\sup_{0\le t\le T} X_t \ge \lambda\right) \le \frac{E[X_0] + E\left[X_T^-\right]}{\lambda}, \tag{9.31}$$

$$P\left(\inf_{0\le t\le T} X_t \le -\lambda\right) \le \frac{E[|X_T|]}{\lambda}, \tag{9.32}$$

where $X_T^- = \max\{-X_T, 0\}$. In particular

$$P\left(\sup_{0\le t\le T} |X_t| \ge \lambda\right) \le \frac{E[X_0] + 2E[|X_T|]}{\lambda}. \tag{9.33}$$

Proof. We use the notation

$$\hat{X}_t = \sup_{0\le s\le t} X_s,$$

and, for fixed $\lambda > 0$, we put

$$\tau(\omega) = \inf\{t \ge 0 \mid X_t(\omega) \ge \lambda\} \wedge T, \qquad \omega \in \Omega.$$

Then τ is a bounded stopping time and, by Theorem 3.56, we have that

$$\begin{aligned} E[X_0] \ge E[X_\tau] &= \int_{\{\hat{X}_T \ge \lambda\}} X_\tau dP + \int_{\{\hat{X}_T < \lambda\}} X_T dP \\ &\ge \lambda P\left(\hat{X}_T \ge \lambda\right) - E\left[X_T^-\right], \end{aligned}$$

and this proves (9.31).

Now we put

$$\check{X}_t = \inf_{0\le s\le t} X_s,$$

and

$$\tau = \inf\{t \ge 0 \mid X_t \le -\lambda\} \wedge T.$$

By Theorem 3.56 we have

$$\begin{aligned} E[X_T] \le E[X_\tau] &= \int_{\{\check{X}_T \le -\lambda\}} X_\tau dP + \int_{\{\check{X}_T > -\lambda\}} X_\tau dP \\ &= \lambda P\left(\check{X}_T \le -\lambda\right) + \int_{\{\check{X}_T > -\lambda\}} X_T dP, \end{aligned}$$

hence (9.32) follows. Finally, (9.33) follows because we have

$$P\left(\sup_{0\le s\le t} |X_s| \ge \lambda\right) \le P\left(\sup_{0\le s\le t} X_s \ge \lambda\right) + P\left(\inf_{0\le s\le t} X_s \le -\lambda\right). \qquad \square$$

Now we use Theorem 9.28 to get a maximal estimate for integral processes.

Corollary 9.29 (Exponential inequality) *Let W be a real Brownian motion and $\sigma \in \mathbb{L}^2$ such that*

$$\int_0^T \sigma_s^2 ds \le k \qquad a.s.$$

for a constant k. Then, if we put

$$X_t = \int_0^t \sigma_s dW_s,$$

for every $\lambda > 0$ we have that

$$P\left(\sup_{0\le t\le T} |X_t| \ge \lambda\right) \le 2e^{-\frac{\lambda^2}{2k}}. \tag{9.34}$$

Proof. We consider the quadratic variation process

$$\langle X\rangle_t = \int_0^t \sigma_s^2 ds,$$

and we recall that[6]

$$Z_t^{(\alpha)} = \exp\left(\alpha X_t - \frac{\alpha^2}{2}\langle X\rangle_t\right)$$

is a continuous super-martingale for every $\alpha \in \mathbb{R}$. Further, we point out that, for every $\lambda, \alpha > 0$, we have

$$\begin{aligned}\{X_t \ge \lambda\} &= \{\exp(\alpha X_t) \ge \exp(\alpha\lambda)\}\\ &\subseteq \left\{Z_t^{(\alpha)} \ge \exp\left(\alpha\lambda - \frac{\alpha^2 k}{2}\right)\right\}.\end{aligned}$$

Then, by applying the maximal inequality (9.31), we get

$$P\left(\sup_{0\le t\le T} X_t \ge \lambda\right) \le P\left(\sup_{0\le t\le T} Z_t^{(\alpha)} \ge e^{\alpha\lambda - \frac{\alpha^2 k}{2}}\right) \le e^{-\alpha\lambda + \frac{\alpha^2 k}{2}}.$$

By choosing $\alpha = \frac{\lambda}{k}$ we maximize the last term of the previous inequality and we get

$$P\left(\sup_{0\le t\le T} X_t \ge \lambda\right) \le e^{-\frac{\lambda^2}{2k}}.$$

An analogous argument applied to the process $-X$ gives the estimate

$$P\left(\inf_{0\le t\le T} X_t \le -\lambda\right) \le e^{-\frac{\lambda^2}{2k}};$$

hence the claim follows. □

[6] Cf. Example 5.12.

Remark 9.30 With the technique of Corollary 9.29 we can also prove the following inequality: let W be a d-dimensional Brownian motion and $\sigma \in \mathbb{L}^2$ an $(N \times d)$-matrix such that

$$\int_0^T \langle \sigma_s \sigma_s^* \theta, \theta \rangle ds \leq k \tag{9.35}$$

for some $\theta \in \mathbb{R}^N$, $|\theta| = 1$, and a constant k. Then, if we put

$$X_t = \int_0^t \sigma_s dW_s,$$

for every $\lambda > 0$ we have that

$$P\left(\sup_{0 \leq t \leq T} |\langle \theta, X_t \rangle| \geq \lambda \right) \leq 2e^{-\frac{\lambda^2}{2k}}. \tag{9.36}$$

□

Now we prove the multi-dimensional version of Corollary 9.29.

Corollary 9.31 *Let W be a d-dimensional Brownian motion and $\sigma \in \mathbb{L}^2$ an $(N \times d)$-matrix such that*[7]

$$\int_0^T |\sigma_s \sigma_s^*| ds \leq k$$

for a constant k. Then, if we put

$$X_t = \int_0^t \sigma_s dW_s,$$

for every $\lambda > 0$ we have that

$$P\left(\sup_{0 \leq t \leq T} |X_t| \geq \lambda \right) \leq 2Ne^{-\frac{\lambda^2}{2kN}}.$$

Proof. Let us notice that, if

$$\sup_{0 \leq t \leq T} |X_t(\omega)| \geq \lambda,$$

then

$$\sup_{0 \leq t \leq T} \left| X_t^i(\omega) \right| \geq \frac{\lambda}{\sqrt{N}}$$

[7] We recall that, if $A = (a_{ij})$ is a matrix, we have that

$$|A| := \sqrt{\sum_{i,j} a_{ij}^2} \geq \max_{|\theta|=1} |A\theta| =: \|A\|.$$

for some $i = 1, \dots, N$, where X^i denotes the i-th component of the vector X. Consequently

$$P\left(\sup_{0\le t\le T} |X_t| \ge \lambda\right) \le \sum_{i=1}^{N} P\left(\sup_{0\le t\le T} |X_t^i| \ge \frac{\lambda}{\sqrt{N}}\right) \le 2Ne^{-\frac{\lambda^2}{2kN}},$$

where the last inequality follows from (9.36), by choosing θ among the vectors of the canonical basis. □

9.3.2 Maximal estimates for diffusions

The following maximal estimates play a crucial part in the proof of the representation formulas for the Cauchy problem of Section 9.4.4 that extend, by a localization technique, the results of Section 9.4.2. In this section we prove maximal estimates for solutions of SDE with bounded diffusion coefficient (cf. Theorem 9.32) or with diffusion coefficient growing at most linearly (cf. Theorem 9.33).

Theorem 9.32 *Let us consider the SDE in* $\mathbb{R}^N$

$$X_t = x_0 + \int_0^t b(s, X_s)ds + \int_0^t \sigma(s, X_s)dW_s. \tag{9.37}$$

We suppose that σ *is a bounded and measurable* $(N \times d)$*-matrix: in particular we have that*

$$|\sigma\sigma^*(t,x)| \le k, \qquad t \in [0,T],\ x \in \mathbb{R}^N; \tag{9.38}$$

further, we suppose that b *is measurable with at most linear growth,*

$$|b(t,x)| \le K(1 + |x|), \qquad t \in [0,T],\ x \in \mathbb{R}^N. \tag{9.39}$$

Then there exists a positive constant α *depending only on* k, K, T *and* N *such that, if* X *is a solution of* (9.37)*, then we have*

$$E\left[e^{\alpha \bar{X}_T^2}\right] < \infty, \tag{9.40}$$

where

$$\bar{X}_T = \sup_{0\le t\le T} |X_t|.$$

Proof. By Proposition A.56 we have

$$E\left[e^{\alpha \bar{X}_T^2}\right] = 1 + \int_0^{+\infty} 2\alpha\lambda e^{\alpha\lambda^2} P\left(\bar{X}_T \ge \lambda\right) d\lambda,$$

so it is enough to have a suitable estimate of $P\left(\bar{X}_T \ge \lambda\right)$ when $\lambda \gg 1$. If we put

$$M_t = \int_0^t \sigma(s, X_s)dW_s,$$

by Corollary 9.31 we have that

$$P\left(\sup_{0\le t\le T}|M_t|\ge R\right)\le 2Ne^{-\frac{R^2}{2kNT}},\qquad R>0.$$

On the other hand on the event

$$\left\{\sup_{0\le t\le T}|M_t|<R\right\},$$

we have

$$|X_t|\le|x_0|+\int_0^t K(1+|X_s|)ds+R$$

hence, by Gronwall's lemma, we get

$$|X_t|\le(|x_0|+KT+R)\,e^{KT},\qquad t\in[0,T].$$

Summing up

$$P\left(\bar X_T\ge(|x_0|+KT+R)\,e^{KT}\right)\le 2Ne^{-\frac{R^2}{2kNT}},$$

that is, for λ large enough,

$$P\left(\bar X_T\ge\lambda\right)\le 2N\exp\left(-\frac{\left(e^{-KT}\lambda-|x_0|-KT\right)^2}{2kNT}\right),\tag{9.41}$$

hence the claim by choosing

$$\alpha<\frac{e^{-2KT}}{2kNT}.$$

□

If the diffusion coefficients have linear growth, we can get a result of maximal integrability of polynomial type: the following result generalizes the estimate (9.18).

Theorem 9.33 *Suppose that the coefficients of the SDE* (9.37) *are measurable and satisfy the estimate* (9.6) *of linear growth. Then if X is a solution of* (9.37), *for every $p\ge 1$ we have*

$$E\left[\sup_{0\le t\le T}|X_t|^p\right]<\infty.\tag{9.42}$$

Proof. We resort to a trick to go back to the case of a SDE with bounded coefficients. We consider the function $f(x)=\log(1+|x|^2)$ and compute the derivatives of first and second order:

$$\partial_{x_i}f(x)=\frac{2x_i}{1+|x|^2},\qquad \partial_{x_ix_j}f(x)=\frac{2\delta_{ij}}{1+|x|^2}-\frac{4x_ix_j}{(1+|x|^2)^2}.$$

Since

$$\partial_{x_i} f(x) = \mathrm{O}\left(|x|^{-1}\right), \quad \partial_{x_i x_j} f(x) = \mathrm{O}\left(|x|^{-2}\right), \qquad \text{as } |x| \to +\infty,$$

by the assumption of linear growth on the coefficients, it is immediate to verify, by applying the Itô formula (5.35), that the coefficients of the stochastic differential of the process

$$Y_t = \log\left(1 + |X_t|^2\right)$$

are bounded. Therefore by proceeding as in the proof of Theorem 9.32 we get

$$P\left(\sup_{0\le t\le T} Y_t \ge \lambda\right) \le c e^{-c\lambda^2}, \qquad \lambda > 0,$$

for some positive constant depending on x_0, T, N and on the growth constant K in (9.6): this is tantamount to writing that

$$\begin{aligned} P\left(\sup_{0\le t\le T} |X_t| \ge \lambda\right) &= P\left(\sup_{0\le t\le T} Y_t \ge \log(1+\lambda^2)\right) \\ &\le c e^{-c\log^2(1+\lambda^2)} \le \frac{c}{\lambda^{c\log\lambda}}, \qquad \lambda > 0. \end{aligned} \tag{9.43}$$

The claim follows from Proposition A.56, since

$$E\left[\sup_{0\le t\le T} |X_t|^p\right] = \int_0^\infty p\lambda^{p-1} P\left(\sup_{0\le t\le T} |X_t| \ge \lambda\right) d\lambda,$$

and the last integral converges by the estimate (9.43). □

9.4 Feynman-Kač representation formulas

In this section we examine the deep connection between SDEs and PDEs, where the trait d'union is the Itô formula. To face the problem in a systematic way, we treat first the stationary[8] (or ellipic) case, which does not have direct financial applications but is nevertheless introductory to the study of evolution (or parabolic) problems that typically arise in the study of American and European derivatives.

Let us fix some notations and assumptions that will hold in the entire section. We consider the SDE in $\mathbb{R}^N$

$$dX_t = b(t, X_t)dt + \sigma(t, X_t)dW_t, \tag{9.44}$$

we denote by D a bounded domain[9] in $\mathbb{R}^N$ and we assume that:

[8] The coefficients do not depend on time.
[9] Open and connected set.

i) the coefficients are locally bounded: $b, \sigma \in L^{\infty}_{\mathrm{loc}}(\mathbb{R}_{\geq 0} \times \mathbb{R}^N)$;
ii) for every $t \geq 0$ and $x \in D$ there exists a solution $X^{t,x}$ of (9.44) such that $X_t^{t,x} = x$, relative to a d-dimensional Brownian motion W on the space $(\Omega, \mathcal{F}, P, (\mathcal{F}_t))$.

In the following, $\tau_{(t,x)}$ denotes the first exit time of $X^{t,x}$ from D: for the sake of simplicity, we write $X^{0,x} = X^x$ and $\tau_{(0,x)} = \tau_x$. Further, putting $(c_{ij}) = \sigma\sigma^*$,

$$\mathcal{A}_t f(x) := \frac{1}{2} \sum_{i,j=1}^{N} c_{ij}(t,x) \partial_{x_i x_j} f(x) + \sum_{j=1}^{N} b_j(t,x) \partial_{x_j} f(x) \tag{9.45}$$

denotes the characteristic operator of the SDE (9.44).

The main results of this section, commonly known as Feynman-Kač theorems, give a representation of the solution u of the Cauchy-Dirichlet, Cauchy and obstacle problems relative to (9.45) in terms of expectation of $u(t, X_t)$. For example, let us consider $u \in C^2(\mathbb{R}^{N+1})$, solution of the equation

$$\mathcal{A}_t u + \partial_t u = 0. \tag{9.46}$$

By the Itô formula we have

$$u(T, X_T^{t,x}) = u(t,x) + \int_t^T \nabla u(s, X_s^{t,x}) \cdot \sigma(s, X_s^{t,x}) dW_s, \tag{9.47}$$

and *if the stochastic integral on the right hand side is a martingale*, by taking expectations we get

$$u(t,x) = E\left[u(T, X_T^{t,x})\right]. \tag{9.48}$$

This formula has a remarkable financial meaning, since it shows the connection between the notions of risk-neutral price and arbitrage price of a derivative. As a matter of fact, on one hand (9.48) is the usual risk-neutral pricing formula for a financial instrument, for example a European option with payoff $u(T, X_T^{t,x})$. On the other hand, if u represents the value of an investment strategy, the PDE (9.46) expresses the self-financing condition (cf. Section 7.1) that, put together with the final replicating condition, determines the arbitrage price of a European derivative as the solution of the corresponding Cauchy problem.

Note however that the stochastic integral on the right hand side of (9.47) is in general only a *local martingale*, as the following remarkable example shows (cf., for instance, Example 7.19 in [259]).

Example 9.34 Consider the radially symmetric function defined on $\mathbb{R}^d \setminus \{0\}$

$$u(x) = \begin{cases} \log|x| & \text{for } d = 2, \\ |x|^{-d+2} & \text{for } d \geq 3, \end{cases}$$

which is a *harmonic* function, that is a solution of the Laplace equation

$$\triangle u(x) = \sum_{i=1}^{d} \partial_{x_i x_i} u(x) = 0, \qquad x \in \mathbb{R}^d \setminus \{0\}.$$

The function u is usually called *fundamental solution* of the Laplace equation since it plays a role analogous to the Gaussian function for the heat equation. For a given a d-dimensional Brownian motion W and $x_0 \in \mathbb{R}^d \setminus \{0\}$, we set $B = W + x_0$ and denote by τ the first time when B hits the origin: it is known (cf. Corollary 2.26 in [259]) that, for $d \geq 2$, we have

$$P(\{\omega \mid B_s = 0 \text{ for some } s \leq t\}) = 0, \qquad t > 0, \tag{9.49}$$

that is, a Brownian motion in $\mathbb{R}^d$, with $d \geq 2$, does not hit almost surely singletons and therefore $\tau = \infty$ a.s. Consequently the process $X_t = u(B_t)$ is well defined and $X_t = u(B_{t\wedge\tau})$ a.s. Now we show that X is a local, but not strict, martingale.

First of all, we consider an increasing sequence (K_n) of compacts whose union is $\mathbb{R}^d \setminus \{0\}$ and a sequence of functions $u_n \in C_0^\infty(\mathbb{R}^d)$. We also denote by τ_n the exit time of B from K_n. Then by Itô formula and assuming that $u_n = u$ on K_n, we have

$$X_{t\wedge\tau_n} = u_n(B_{t\wedge\tau_n}) = u_n(x_0) + \int_0^{t\wedge\tau_n} \nabla u_n(B_s) \cdot dW_s.$$

Since $\nabla u_n(B) \in \mathbb{L}^2$, by the Optional sampling theorem $(X_{t\wedge\tau_n})$ is a martingale for any n and this proves that X is a local martingale.

Next we show that X is not a strict martingale by proving that its expectation is not constant in time. We only consider the case $d \geq 3$: we have

$$E[X_t] = \int_{\mathbb{R}^d} \frac{1}{|x|^{d-2}(2\pi t)^{\frac{d}{2}}} \exp\left(-\frac{|x-x_0|^2}{2t}\right) dx =$$

(by the change of variable $y = \frac{x-x_0}{2t}$)

$$= \frac{1}{\pi^{\frac{d}{2}}} \int_{\mathbb{R}^d} \frac{e^{-|y|^2}}{|2ty + x_0|^{d-2}} dy$$

and therefore, by the dominated convergence theorem, $E[X_t]$ tends to zero as $t \to \infty$. □

The rest of the paragraph is structured as follows: in the first three sections we study the representation of the solution of the Cauchy-Dirichlet problem on a bounded domain. Section 9.3.2 is devoted to the proof of some preliminary estimates, necessary to the study of the Cauchy problem in Section 9.4.4. In Section 9.4.5 we represent the solution of the obstacle problem in terms of the solution of an optimal stopping problem.

9.4.1 Exit time from a bounded domain

In this section we study some simple conditions that guarantee that the first exit time from a bounded domain D

$$\tau_x = \inf\{t \mid X_t^x \notin D\}$$

of the solution of the SDE (9.44) is integrable and therefore, in particular, finite a.s.

Proposition 9.35 *If there exists a function $f \in C^2(\mathbb{R}^N)$, non-negative over D and such that*

$$\mathcal{A}_t f \leq -1, \qquad \text{in } D,\ t \geq 0, \tag{9.50}$$

then $E[\tau_x]$ is finite for every $x \in D$.

Before proving the proposition, we examine a remarkable example. Let us suppose there exists $\lambda > 0$ such that

$$c_{11}(t, \cdot) \geq \lambda, \qquad \text{over } D,\ t \geq 0. \tag{9.51}$$

Then there exists $f \in C^2(\mathbb{R}^N)$, non-negative over D, such that (9.50) holds: indeed it is enough to put

$$f(x) = \alpha(e^{\beta R} - e^{\beta x_1})$$

where α, β are suitable positive constants and R is large enough to make the Euclidean ball of radius R, centered in the origin include D. Indeed f is non-negative over D and we have

$$\begin{aligned}\mathcal{A}_t f(x) &= -\alpha e^{\beta x_1}\left(\frac{1}{2}c_{11}(t,x)\beta^2 + b_1(t,x)\beta\right)\\ &\leq -\alpha\beta e^{-\beta R}\left(\frac{\lambda\beta}{2} - \|b\|_{L^\infty(D)}\right)\end{aligned}$$

hence the claim, by choosing α, β large enough.

Condition (9.51) ensures that $\mathcal{A}_t$ is not totally degenerate and is obviously fulfilled when (c_{ij}) is uniformly positive definite.

Proof (of Proposition 9.35). For fixed t, by the Itô formula we have that

$$f(X^x_{t\wedge\tau_x}) = f(x) + \int_0^{t\wedge\tau_x} \mathcal{A}_s f(X^x_s)ds + \int_0^{t\wedge\tau_x} \nabla f(X^x_s)\cdot\sigma(s, X^x_s)dW_s.$$

Since ∇f and $\sigma(s,\cdot)$ are bounded in D when $s \leq t$, the stochastic integral has null expectation and by (9.50) we have

$$E\left[f(X^x_{t\wedge\tau_x})\right] \leq f(x) - E\left[t\wedge\tau_x\right];$$

hence, since $f \geq 0$, we have

$$E\left[t\wedge\tau_x\right] \leq f(x).$$

Finally, taking the limit as $t \to \infty$, by Beppo Levi's theorem we get

$$E\left[\tau_x\right] \leq f(x). \qquad \square$$

Remark 9.36 Similar arguments yield a condition on the first-order (drift) term: if $|b_1(t,\cdot)| \geq \lambda$ in D for $t \geq 0$, with λ positive constant, then $E[\tau_x]$ is finite. Indeed, let us suppose for example that $b_1(t,x) \geq \lambda$ (the case $b_1(t,x) \leq -\lambda$ is analogous): then by applying the Itô formula to the function $f(x) = x_1$ we have

$$\left(X^x_{t\wedge\tau_x}\right)_1 = x_1 + \int_0^{t\wedge\tau_x} b_1(s, X^x_s)ds + \sum_{i=1}^d \int_0^{t\wedge\tau_x} \sigma_{1i}(s, X^x_s)dW^i_s,$$

and in mean

$$E\left[\left(X^x_{t\wedge\tau_x}\right)_1\right] \geq x_1 + \lambda E\left[t \wedge \tau_x\right],$$

hence the claim, taking the limit as $t \to \infty$. □

9.4.2 Elliptic-parabolic equations and Dirichlet problem

In this section we assume that the coefficients of the SDE (9.44) are autonomous, i.e. $b = b(x)$ and $\sigma = \sigma(x)$. In many cases this assumption is not restrictive since time-dependent problems can be treated analogously by including time in the state variables (cf. Example 9.42). In addition to the assumptions that we stated at the beginning of the paragraph, we suppose that $E[\tau_x]$ is finite for every $x \in D$ and we denote the characteristic operator of (9.44) by

$$\mathcal{A} := \frac{1}{2}\sum_{i,j=1}^N c_{ij}\partial_{x_i x_j} + \sum_{j=1}^N b_j \partial_{x_j}. \tag{9.52}$$

The following result gives a representation formula (and so, in particular, *a uniqueness result*) for the classical solutions of the Dirichlet problem relative to the elliptic-parabolic operator $\mathcal{A}$:

$$\begin{cases} \mathcal{A}u - au = f, & \text{in } D, \\ u|_{\partial D} = \varphi, \end{cases} \tag{9.53}$$

where f, a, φ are given functions.

Theorem 9.37 *Let $f \in L^\infty(D)$, $\varphi \in C(\partial D)$ and $a \in C(D)$ such that $a \geq 0$. If $u \in C^2(D) \cap C(\bar{D})$ is solution to the Dirichlet problem (9.53) then, for fixed $x \in D$ and writing for the sake of simplicity $\tau = \tau_x$, we have*

$$u(x) = E\left[e^{-\int_0^\tau a(X^x_t)dt}\varphi(X^x_\tau) - \int_0^\tau e^{-\int_0^t a(X^x_s)ds} f(X^x_t)dt\right]. \tag{9.54}$$

Proof. For $\varepsilon > 0$ small enough, let D_ε be a domain such that

$$x \in D_\varepsilon, \qquad \bar{D}_\varepsilon \subseteq D, \qquad \operatorname{dist}(\partial D_\varepsilon, \partial D) \leq \varepsilon.$$

We denote the exit time of X^x from D_ε by τ_ε and we observe that, since X^x is continuous,

$$\lim_{\varepsilon \to 0} \tau_\varepsilon = \tau.$$

We put

$$Z_t = e^{-\int_0^t a(X_s^x)ds},$$

and we notice that, by assumption, $Z_t \in]0,1]$. Further, if $u_\varepsilon \in C_0^2(\mathbb{R}^N)$ is such that $u_\varepsilon = u$ in D_ε, by the Itô formula we get

$$d(Z_t u_\varepsilon(X_t^x)) = Z_t \left((\mathcal{A}u_\varepsilon - au_\varepsilon)(X_t^x)dt + \nabla u_\varepsilon(X_t^x) \cdot \sigma(X_t^x)dW_t \right)$$

hence

$$Z_{\tau_\varepsilon} u(X_{\tau_\varepsilon}^x) = u(x) + \int_0^{\tau_\varepsilon} Z_t f(X_t^x)dt + \int_0^{\tau_\varepsilon} Z_t \nabla u(X_t^x) \cdot \sigma(X_t^x)dW_t.$$

Since ∇u and σ are bounded in D, by taking expectations we get

$$u(x) = E\left[Z_{\tau_\varepsilon} u(X_{\tau_\varepsilon}^x) - \int_0^{\tau_\varepsilon} Z_t f(X_t^x)dt \right].$$

The claim follows by passing to the limit as $\varepsilon \to 0$, by the dominated convergence theorem: indeed, recalling that $Z_t \in]0,1]$, we get

$$\left| Z_{\tau_\varepsilon} u(X_{\tau_\varepsilon}^x) \right| \le \|u\|_{L^\infty(D)}, \qquad \left| \int_0^{\tau_\varepsilon} Z_t f(X_t^x)dt \right| \le \tau \|f\|_{L^\infty(D)},$$

where, by assumption, τ is integrable. □

By the techniques of Section 9.4.5 it is possible to get a similar result for the *strong solutions* of the Dirichlet problem, i.e. for solutions $u \in W_{\mathrm{loc}}^{2,p}(D) \cap C(\bar{D})$ that satisfy the equation $\mathcal{A}u - au = f$ almost everywhere.

From the numerical point of view, formula (9.54) is relevant since it allows the use of Monte Carlo-type methods for approximating the solution of the Dirichlet problem (9.53).

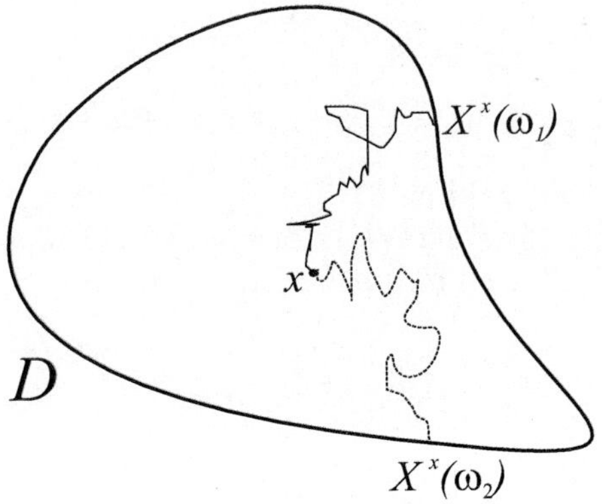

Fig. 9.1. Dirichlet problem and paths of the corresponding SDE

Remark 9.38 The hypothesis $a \geq 0$ is essential: the function

$$u(x, y) = \sin x \sin y$$

is a solution of the problem

$$\begin{cases} \Delta u + 2u = 0, & \text{in } D =]0, 2\pi[\times]0, 2\pi[, \\ u|_{\partial D} = 0, \end{cases}$$

but it does not satisfy (9.54). □

Existence results for problem (9.53) are well known in the *uniformly elliptic* case: we mention the following classical theorem (we refer, e.g., to Gilbarg and Trudinger [157], Theorem 6.13).

Theorem 9.39 *Under the following assumptions*

i) $\mathcal{A}$ is a uniformly elliptic operator, i.e. there exists a constant $\lambda > 0$ such that

$$\sum_{i,j=1}^{N} c_{ij}(x)\xi_i\xi_j \geq \lambda|\xi|^2, \qquad x \in D, \ \xi \in \mathbb{R}^N;$$

ii) the coefficients are Hölder-continuous functions, $c_{ij}, b_j, a, f \in C^\alpha(D)$. Further, the functions c_{ij}, b_j, f are bounded and $a \geq 0$;
iii) for every $y \in \partial D$ there exists[10] *an Euclidean ball B contained in the complement of D and such that $y \in \bar{B}$;*
iv) $\varphi \in C(\partial D)$;

there exists a classical solution $u \in C^{2+\alpha}(D) \cap C(\bar{D})$ of problem (9.53).

Let us now consider some remarkable examples.

Example 9.40 (Expectation of an exit time) If the problem

$$\begin{cases} \mathcal{A}u = -1, & \text{in } D, \\ u|_{\partial D} = 0, \end{cases}$$

has a solution, then by (9.54) we have that $u(x) = E[\tau_x]$. □

Example 9.41 (Poisson kernel) If $a = f = 0$, (9.54) can be rewritten in terms of a mean value formula. More precisely, we denote the distribution of the random variable $X^x_{\tau_x}$ by μ^x: then μ^x is a probability measure on ∂D and by (9.54) we have

$$u(x) = E\left[u(X^x_{\tau_x})\right] = \int_{\partial D} u(y)\mu^x(dy).$$

[10] This is a regularity condition of the boundary of D, verified if, for example, ∂D is a C^2-manifold.

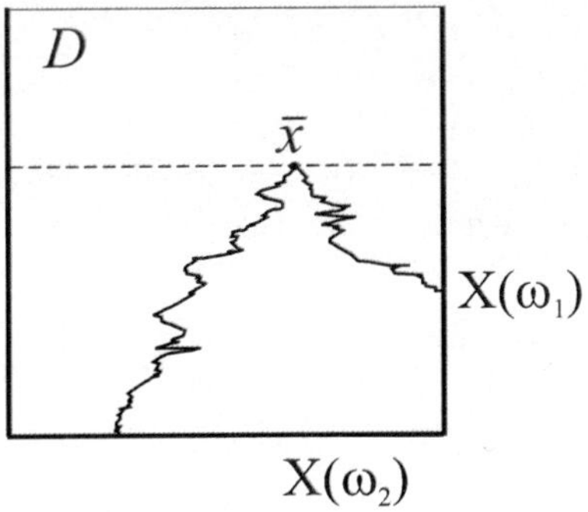

Fig. 9.2. Cauchy-Dirichlet problem and paths of the corresponding SDE

The law μ^x is usually called *harmonic measure* of $\mathcal{A}$ over ∂D. In particular, let us consider the case of a Brownian motion X^x with initial point $x \in \mathbb{R}^N$: then $\mathcal{A} = \frac{1}{2}\Delta$ and if $D = B(0, R)$ is the Euclidean ball with radius R, μ^x has a density (with respect to the surface measure) whose explicit expression is known and is given by the so-called *Poisson kernel*

$$\frac{1}{R\omega_N} \frac{R - |x|^2}{|x - y|^N},$$

where ω_N is the measure of the unit sphere in $\mathbb{R}^N$. □

Example 9.42 (Heat equation) The process $X_t = (W_t, -t)$, where W is a real Brownian motion, is solution of the SDE

$$\begin{cases} dX_t^1 = dW_t, \\ dX_t^2 = -dt, \end{cases}$$

and the corresponding characteristic operator

$$\mathcal{A} = \frac{1}{2}\partial_{x_1 x_1} - \partial_{x_2}$$

is the heat operator in $\mathbb{R}^2$. Let us consider formula (9.54) in a rectangular domain

$$D =]a_1, b_1[\times]a_2, b_2[.$$

By examining the explicit expression of the paths of X (see also Figure 9.2), it is clear that the value $u(\bar{x}_1, \bar{x}_2)$ of a solution of the heat equation depends only on the values of u on the portion of the boundary of D contained in $\{x_2 < \bar{x}_2\}$. In general the value of u in D depends only on the values of u over the *parabolic boundary* of D, defined as

$$\partial_p D = \partial D \setminus (]a_1, b_1[\times\{b_2\}).$$

This fact is consistent with the results on the Cauchy-Dirichlet problem of Section 6.1. $\partial_p D$ is also called *regular boundary* of D since it is the part of ∂D where it is allowed to impose the boundary condition of the Dirichlet problem for the heat equation: indeed, if $\mathcal{A}u = 0$ in D then by (9.54) the values of u on $]a_1, b_1[\times\{b_2\}$ are determined by the values of u on $\partial_p D$ and cannot be imposed arbitrarily. □

Example 9.43 (Method of characteristics) If $\sigma = 0$, the characteristic operator is a first-order differential operator

$$\mathcal{A} = \sum_{i=1}^{N} b_i \partial_{x_i}.$$

The corresponding SDE is actually deterministic and becomes

$$X_t^x = x + \int_0^t b(X_s^x)ds,$$

i.e. X is an integral curve of the vector field b:

$$\frac{d}{dt}X_t = b(X_t).$$

Note that

$$\frac{d}{dt}u(X_t) = \langle b(X_t), \nabla u(X_t)\rangle = \mathcal{A}u(X_t)$$

and therefore a solution of $\mathcal{A}u = 0$ is constant along the integral curves of b.

By Theorem 9.37, if the exit time of X from D is finite (cf. Remark 9.36), we have the representation

$$u(x) = e^{-\int_0^{\tau_x} a(X_t^x)dt}\varphi(X_{\tau_x}^x) - \int_0^{\tau_x} e^{-\int_0^t a(X_s^x)ds} f(X_t^x)dt, \tag{9.55}$$

for the solution of the problem

$$\begin{cases} \langle b, \nabla u\rangle - au = f & \text{in } D, \\ u|_{\partial D} = \varphi. \end{cases}$$

Formula (9.55) is related to the classical it method of characteristics which can be used to solve the initial value problem for general first order (only contain first order partial derivatives) PDEs: for a description of the method we refer, for instance, to Evans [124], Chapter 3.2.

For example, let us consider $b(x, y) = (1, -x)$ on the (unbounded) domain $D = \mathbb{R} \times \mathbb{R}_{>0}$. For a fixed point $(x, y) \in D$, the integral curve of b starting from (x, y) is given by

$$\begin{cases} \frac{d}{dt}X_t = 1, & X_0 = x, \\ \frac{d}{dt}Y_t = -X_t, & Y_0 = y, \end{cases}$$

that is $(X_t, Y_t) = \left(x + t, y - xt - \frac{t^2}{2}\right)$. Putting $Y_t = 0$ we find the time t at which the curve reaches the boundary of D: we have $t = -x \pm \sqrt{x^2 + 2y}$ and imposing $t > 0$, we obtain $\bar{t} = -x + \sqrt{x^2 + 2y}$. Thus

$$(X_{\bar{t}}, Y_{\bar{t}}) = (\sqrt{x^2 + 2y}, 0)$$

is the point at which the integral curve starting from (x, y) reaches the boundary. Note that $(X_{\bar{t}}, Y_{\bar{t}}) \in \mathbb{R}_{>0} \times \{0\}$ for any $(x, y) \in D$ and therefore $\mathbb{R}_{>0} \times \{0\}$ is the *regular boundary* of D for the operator $\mathcal{A} = \partial_x - x\partial_y$. Moreover, by (9.55), the solution of the problem

$$\begin{cases} \partial_x u(x, y) - x\partial_y u(x, y) = 0, & \text{in } \mathbb{R} \times \mathbb{R}_{>0}, \\ u(x, 0) = \varphi(x), & x > 0, \end{cases}$$

is $u(x, y) = \varphi(\sqrt{x^2 + 2y})$ at least if φ is a sufficiently regular function: note indeed that, in the case of a first order equation, the solution in (9.55) inherits the regularity of the boundary datum and there is no smoothing effect due to the expectation (i.e. convolution with the density) as in the diffusive case (9.54). □

9.4.3 Evolution equations and Cauchy-Dirichlet problem

In this section we state the parabolic version of Theorem 9.37. Let the assumptions we imposed at the beginning of the paragraph hold and let us denote the characteristic operator of the SDE (9.44) by

$$\mathcal{A}u(t, x) = \frac{1}{2} \sum_{i,j=1}^{N} c_{ij}(t, x)\partial_{x_i x_j} u(t, x) + \sum_{j=1}^{N} b_j(t, x)\partial_{x_j} u(t, x). \tag{9.56}$$

Further, we consider the cylinder

$$Q =]0, T[\times D,$$

whose backward[11] parabolic boundary is defined by

$$\partial_p Q = \partial Q \setminus (\{0\} \times D).$$

The following theorem gives a representation formula for the classical solutions of the Cauchy-Dirichlet problem:

$$\begin{cases} \mathcal{A}u - au + \partial_t u = f, & \text{in } Q, \\ u|_{\partial_p Q} = \varphi, \end{cases} \tag{9.57}$$

where f, a, φ are given functions.

[11] In this section we consider backward operators such as $\triangle + \partial_t$.

Theorem 9.44 *Let $f \in L^\infty(Q)$, $\varphi \in C(\partial_p Q)$ and $a \in C(Q)$ such that*

$$a_0 := \inf a$$

is finite. If $u \in C^2(Q) \cap C(\bar{Q})$ is a solution of the problem (9.57) then, for any $(t,x) \in Q$, we have

$$\begin{aligned} u(t,x) =& E\left[e^{-\int_t^{\tau\wedge T} a(s,X_s)ds}\varphi(\tau\wedge T, X_{\tau\wedge T})\right] \\ &-E\left[\int_t^{\tau\wedge T} e^{-\int_t^s a(r,X_r)dr} f(s,X_s)ds\right], \end{aligned}$$

where, for the sake of simplicity, we put $X = X^{t,x}$ and $\tau = \tau_{(t,x)}$.

Proof. The proof is analogous to that of Theorem 9.37. □

9.4.4 Fundamental solution and transition density

In this section we prove a representation formula for the classical solution of the Cauchy problem

$$\begin{cases} \mathcal{A}u - au + \partial_t u = f, & \text{in } \mathcal{S}_T :=]0,T[\times\mathbb{R}^N, \\ u(T,\cdot) = \varphi, \end{cases} \tag{9.58}$$

where f, a, φ are given functions, $(c_{ij}) = \sigma\sigma^*$ and

$$\mathcal{A} = \frac{1}{2}\sum_{i,j=1}^N c_{ij}\partial_{x_i x_j} + \sum_{j=1}^N b_j \partial_{x_j} \tag{9.59}$$

is the characteristic operator of the SDE

$$dX_t = b(t,X_t)dt + \sigma(t,X_t)dW_t. \tag{9.60}$$

We assume that

i) the coefficients b, σ are measurable and have at most linear growth in x;
ii) for every $(t,x) \in \mathcal{S}_T$, there exists a solution $X^{t,x}$ of the SDE (9.60) relative to a d-dimensional Brownian motion W on the space $(\Omega, \mathcal{F}, P, (\mathcal{F}_t))$.

Theorem 9.45 (Feynman-Kač formula) *Let $u \in C^2(\mathcal{S}_T) \cap C(\bar{\mathcal{S}}_T)$ be a solution of the Cauchy problem (9.58) where $a \in C(\mathcal{S}_T)$ is such that $a_0 = \inf a > -\infty$. Assume that i), ii) and at least one of the following conditions are in force:*

1) there exist two positive constants M, p such that

$$|u(t,x)| + |f(t,x)| \le M(1+|x|^p), \qquad (t,x) \in \mathcal{S}_T;$$

2) the matrix σ is bounded and there exist two positive constants M and α, with α small enough[12], *such that*

$$|u(t,x)| + |f(t,x)| \le M e^{\alpha |x|^2}, \qquad (t,x) \in \mathcal{S}_T.$$

Then for every $(t,x) \in \mathcal{S}_T$, we have the representation formula

$$u(t,x) = E\left[e^{-\int_t^T a(s,X_s)ds} \varphi(X_T) - \int_t^T e^{-\int_t^s a(r,X_r)dr} f(s,X_s)ds \right],$$

where, for the sake of simplicity, $X = X^{t,x}$.

Proof. If τ_R denotes the exit time of X from the Euclidean ball with radius R, by Theorem 9.44 we have

$$\begin{aligned} u(t,x) =& E\left[e^{-\int_t^{T\wedge\tau_R} a(s,X_s)ds} u(T\wedge\tau_R, X_{T\wedge\tau_R}) \right] \\ &- E\left[\int_t^{T\wedge\tau_R} e^{-\int_t^s a(r,X_r)dr} f(s,X_s)ds \right]. \end{aligned} \tag{9.61}$$

Since

$$\lim_{R\to\infty} T\wedge\tau_R(\omega) = T,$$

for every $\omega \in \Omega$, the claim follows by taking the limit in R in (9.61) by the dominated convergence theorem. Indeed we have pointwise convergence of the integrands and under condition 1), we have that

$$e^{-\int_t^{T\wedge\tau_R} a(s,X_s)ds} |u(T\wedge\tau_R, X_{T\wedge\tau_R})| \le M e^{|a_0|T}\left(1+\bar{X}_T^p\right),$$

$$\left| \int_t^{T\wedge\tau_R} e^{-\int_t^s a(r,X_r)dr} f(s,X_s)ds \right| \le T e^{|a_0|T} M \left(1+\bar{X}_T^p\right),$$

where

$$\bar{X}_T = \sup_{0\le t\le T} |X_t|$$

is integrable by the maximal estimate of Theorem 9.33.

Under condition 2) we can proceed in an analogous way, by using the integrability estimate (9.40) in Theorem 9.32. □

The Feynman-Kač representation formula allows us to generalize the results of Paragraph 3.1.1 on the transition density of Brownian motion. More

[12] It is sufficient to take

$$\alpha < \frac{e^{-2KT}}{2kNT},$$

where $|\sigma\sigma^*| \le k$ and K is the growth constant in (9.39), so that we can apply Theorem 9.32.

precisely, if[13] the operator $\mathcal{A}+\partial_t$ has a fundamental solution $\Gamma(t,x;T,y)$ then, for every $\varphi \in C_b(\mathbb{R}^N)$, the function

$$u(t,x) = \int_{\mathbb{R}^N} \varphi(y)\Gamma(t,x;T,y)dy$$

is the classical bounded solution of the Cauchy problem (9.58) with $a = f = 0$ and so, by the Feynman-Kač formula, we have that

$$E\left[\varphi(X_T^{t,x})\right] = \int_{\mathbb{R}^N} \varphi(y)\Gamma(t,x;T,y)dy.$$

By the arbitrariness of φ, this means that, for fixed $x \in \mathbb{R}^N$ and $t < T$, the function

$$y \mapsto \Gamma(t,x;T,y)$$

is the density of the random variable $X_T^{t,x}$: we express this fact by saying that Γ is the *transition density* of the SDE (9.60). This fundamental result unveils the deep connection between PDEs and SDEs:

Theorem 9.46 *If there exists the fundamental solution of the differential operator $\mathcal{A}+\partial_t$ with $\mathcal{A}$ in* (9.59), *then it is equal to the transition density of the SDE* (9.60).

9.4.5 Obstacle problem and optimal stopping

In this section we prove a representation formula for the strong solution of the obstacle problem

$$\begin{cases} \max\{\mathcal{A}u - au + \partial_t u, \varphi - u\}, & \text{in } \mathcal{S}_T :=]0,T[\times\mathbb{R}^N, \\ u(T,\cdot) = \varphi, \end{cases} \tag{9.62}$$

where a and φ are given functions and, if we put $(c_{ij}) = \sigma\sigma^*$,

$$\mathcal{A} = \frac{1}{2}\sum_{i,j=1}^{N} c_{ij}\partial_{x_i x_j} + \sum_{j=1}^{N} b_j \partial_{x_j} \tag{9.63}$$

is the characteristic operator of the SDE

$$dX_t = b(t,X_t)dt + \sigma(t,X_t)dW_t. \tag{9.64}$$

We assume that the operator

$$Lu := \mathcal{A}u - au + \partial_t u$$

is uniformly parabolic (Hypothesis 8.1) and has bounded and Hölder continuous coefficients (Hypothesis 8.3). We sum up here some of the main consequences of those hypotheses:

---

[13] As we have seen in Chapter 8, typical conditions that guarantee the existence of the fundamental solution are the uniform parabolicity, the boundedness and the Hölder continuity of the coefficients.

- by Theorem 8.6, L has a fundamental solution Γ;
- by Theorem 8.10, the Gaussian estimate

$$\Gamma(t,x;s,y) \le C\,\Gamma_0(t,x;s,y), \qquad s \in]t,T[,$$

holds for every $T > t$ and $x, y \in \mathbb{R}^N$, where Γ_0 is a Gaussian function, fundamental solution of a suitable parabolic operator with constant coefficients. In particular, as a consequence of Lemma 5.38, we have

$$\Gamma(t,x;\cdot,\cdot) \in L^{\bar{q}}(]t,T[\times\mathbb{R}^N), \qquad \bar{q} \in [1, 1+2/N[, \tag{9.65}$$

for every $(t,x) \in \mathbb{R}^{N+1}$ and $T > t$;
- by Theorem 9.27, for every $(t,x) \in \mathcal{S}_T$, there exists a unique solution $X^{t,x}$ of the SDE (9.64), with initial datum $X_t = x \in \mathbb{R}^N$, relative to a d-dimensional Brownian motion W on the space $(\Omega, \mathcal{F}, P, (\mathcal{F}_t))$. By Theorem 9.46, $\Gamma(t,x;\cdot,\cdot)$ is the transition density of $X^{t,x}$;
- under Hypothesis 8.19 on the regularity of the function φ, Theorem 8.21 guarantees that the obstacle problem (9.62) has a strong solution $u \in S^p_{\text{loc}}(\mathcal{S}_T) \cap C(\overline{\mathcal{S}}_T)$ for every $p \ge 1$; in particular $u \in C^{1+\alpha}_{P,\text{loc}}(\mathcal{S}_T)$ for every $\alpha \in]0,1[$. We recall that the Hölder spaces C^α_P and the Sobolev spaces S^p were introduced in Definitions 8.2 and 8.15, respectively.

Moreover, the following weak Itô formula can be proved by using the same arguments as in the proof of Theorem 5.37.

Theorem 9.47 (Itô formula) *If $f = f(t,x) \in S^p([0,T]\times\mathbb{R}^N)$ and $(\nabla f)^2 \in L^q([0,T]\times\mathbb{R}^N)$ with $p, q > 1 + \frac{N}{2}$, then we have*

$$f(t,X_t) = f(0,X_0) + \int_0^t Lf(s,X_s)ds + \int_0^t \nabla f(s,X_s)\cdot\sigma(s,X_s)dW_s.$$

The main result of this section is a representation theorem for the obstacle problem in terms of a solution of the optimal stopping problem for the diffusion X. Recalling the results in discrete time in Paragraph 2.5, the connection with the problem of pricing American options is evident: in the continuous time case, this connection will be made precise in Chapter 11.

Theorem 9.48 (Feynman-Kač formula) *Under Hypotheses 8.1 and 8.3, let u be a strong solution of the obstacle problem (9.62) and let us assume that there exist two positive constants C and λ, with λ small enough, such that*

$$|u(t,x)| \le Ce^{\lambda|x|^2}, \qquad (t,x) \in \mathcal{S}_T. \tag{9.66}$$

Then, for every $(t,x) \in \mathcal{S}_T$, we have the representation formula

$$u(t,x) = \sup_{\tau \in \mathcal{T}_{t,T}} E\left[e^{-\int_t^\tau a(s,X_s^{t,x})ds}\varphi(\tau, X_\tau^{t,x})\right],$$

where $\mathcal{T}_{t,T}$ denotes the family of stopping times with values in $[t,T]$.

Proof. As for the standard Feynman-Kač formula, the proof is based on the Itô formula: since a strong solution is generally not in C^2, then we have to apply the generalized Itô formula of Theorem 9.47 by means of a localization argument. For more clarity, we only treat the case $a = 0$.

We set $B_R = \{x \in \mathbb{R}^N \mid |x| < R\}$, $R > 0$, and for a fixed $x \in B_R$ we denote by τ_R the first exit time of $X^{t,x}$ from B_R. Under our assumptions, it is well-known that $E[\tau_R]$ is finite.

We show that for any $(t, x) \in]0, T[\times B_R$ and $\tau \in \mathcal{T}_{t,T}$ such that $\tau \leq \tau_R$ a.s., it holds

$$u(t, x) = E\left[u(\tau, X_\tau^{t,x}) - \int_t^\tau Lu(s, X_s^{t,x})ds\right]. \tag{9.67}$$

Since $u \in S^p_{\text{loc}}(\mathcal{S}_T)$ for any $p \geq 1$ then, for any positive and suitably small ε, there exists a function $u^{\varepsilon,R}$ such that $u^{\varepsilon,R} \in S^p(\mathbb{R}^{N+1})$ for any $p \geq 1$ and $u^{\varepsilon,R} = u$ in $]t, T - \varepsilon[\times B_R$.

We next apply Itô formula to $u^{\varepsilon,R}$ and using the fact that $u^{\varepsilon,R} = u$ in $]t, T - \varepsilon[\times B_R$, we get

$$u(\tau, X_\tau^{t,x}) = u(t, x) + \int_t^\tau Lu(s, X_s^{t,x})ds + \int_t^\tau \nabla u(s, X_s^{t,x}) \cdot \sigma(s, X_s^{t,x})dW_s, \tag{9.68}$$

for any $\tau \in \mathcal{T}_{t,T}$ such that $\tau \leq \tau_R \wedge (T - \varepsilon)$. Since $u \in C^{1+\alpha}_{P,\text{loc}}$ then $(\nabla u)\sigma$ is a bounded function on $]t, T - \varepsilon[\times B_R$ so that

$$E\left[\int_t^\tau \nabla u(s, X_s^{t,x}) \cdot \sigma(s, X_s^{t,x})dW_s\right] = 0.$$

Thus, taking expectations in (9.68), we conclude the proof of formula (9.67), since $\varepsilon > 0$ is arbitrary.

Next we recall that $Lu \leq 0$ a.e.: since the law of $X^{t,x}$ is absolute continuous with respect to the Lebesgue measure, we have

$$E\left[\int_t^\tau Lu(s, X_s^{t,x})ds\right] \leq 0, \qquad \tau \in \mathcal{T}_{t,T},$$

so that from (9.67) we deduce

$$u(t, x) \geq E\left[u(\tau \wedge \tau_R, X_{\tau\wedge\tau_R}^{t,x})\right], \qquad \tau \in \mathcal{T}_{t,T}. \tag{9.69}$$

Now we pass to the limit as $R \to +\infty$: it holds

$$\lim_{R\to+\infty} \tau \wedge \tau_R = \tau$$

and, by the growth condition (9.66), we have

$$\left|u(\tau \wedge \tau_R, X_{\tau\wedge\tau_R}^{t,x})\right| \leq C \exp\left(\lambda \sup_{t\leq s\leq T} |X_s^{t,x}|^2\right).$$

By the maximal estimates of Theorem 9.32, the random variable on the right hand side is integrable, thus by the dominated convergence theorem, passing to the limit in (9.69) as $R \to +\infty$, we infer

$$u(t,x) \geq E\left[u(\tau, X_\tau^{t,x})\right] \geq E\left[\varphi(\tau, X_\tau^{t,x})\right].$$

This proves that

$$u(t,x) \geq \sup_{\tau \in \mathscr{T}_{t,T}} E\left[\varphi(\tau, X_\tau^{t,x})\right].$$

We conclude the proof by setting

$$\tau_0 = \inf\{s \in [t,T] \mid u(s, X_s^{t,x}) = \varphi(s, X_s^{t,x})\}.$$

Since $Lu = 0$ a.e. on $\{u > \varphi\}$, it holds

$$E\left[\int_t^{\tau_0 \wedge \tau_R} Lu(s, X_s^{t,x}) ds\right] = 0,$$

so that by (9.67) we have

$$u(t,x) = E\left[u(\tau_0 \wedge \tau_R, X_{\tau_0 \wedge \tau_R}^{t,x})\right].$$

Using the previous argument to pass to the limit as $R \to +\infty$, we finally deduce

$$u(t,x) = E\left[u(\tau_0, X_{\tau_0}^{t,x})\right] = E\left[\varphi(\tau_0, X_{\tau_0}^{t,x})\right].$$

□

From the Feynman-Kač representation it is possible to obtain useful information about the solution of the obstacle problem under more specific assumptions. For example, if we suppose that the function φ is Lipschitz continuous in x uniformly in t, i.e. there exists a constant C such that

$$|\varphi(t,x) - \varphi(t,y)| \leq C|x-y|, \qquad (t,x),(t,y) \in \mathcal{S}_T,$$

then we can prove that the spatial gradient ∇u is bounded over $\mathcal{S}_T$. More precisely we have the following:

Proposition 9.49 *Under the assumptions of Theorem 9.48, suppose that the function φ and the coefficients of the SDE* (9.64) *are Lipschitz continuous in x uniformly with respect to t over $\mathcal{S}_T$. Further, let the coefficient a be a constant or φ be bounded. Then the strong solution u of the obstacle problem* (9.62) *satisfies*

$$\nabla u \in L^\infty(\mathcal{S}_T).$$

Proof. Let us first consider the case a is a constant. The claim follows from the general inequality

$$\left|\sup_\tau F(\tau) - \sup_\tau G(\tau)\right| \leq \sup_\tau |F(\tau) - G(\tau)|,$$

that holds true for every function F, G. By the Feynman-Kač representation formula we have

$$|u(t,x) - u(t,y)| \le \sup_{\tau \in \mathscr{T}_{t,T}} E\left[e^{-a(\tau - t)} \left|\varphi(\tau, X_\tau^{t,x}) - \varphi(\tau, X_\tau^{t,y})\right|\right] \le$$

(by the assumption of Lipschitz continuity, for a suitable positive constant c)

$$\le c \sup_{\tau \in \mathscr{T}_{t,T}} E\left[\left|X_\tau^{t,x} - X_\tau^{t,y}\right|\right] \le$$

(by the result on the dependence on the initial datum, Theorem 9.16)

$$\le c_1 |x - y|,$$

where the constant c_1 depends only on T and on the Lipschitz constants of φ and of the coefficients.

If φ is bounded, the claim follows in an analogous way by using the fact that the product of two bounded Lipschitz-continuous functions

$$(t,x) \mapsto e^{-\int_t^\tau a(s, X_s^{t,x}) ds} \varphi(\tau, X_\tau^{t,x})$$

is itself a Lipschitz-continuous function. □

9.5 Linear equations

In this paragraph we study the simplest and most important class of stochastic equations, namely those whose coefficients are linear functions of the solution, and we introduce the corresponding class of second-order differential operators, the Kolmogorov operators. Such operators arise in some classical physical and financial models, and mostly possess all the good properties of the heat operator, even though they are not, in general, uniformly parabolic.

Let us consider the following linear SDE in $\mathbb{R}^N$

$$dX_t = (B(t)X_t + b(t))dt + \sigma(t)dW_t \tag{9.70}$$

where b, B and σ are L^∞_{loc}-functions with values in the space of $(N \times 1)$, $(N \times N)$ and $(N \times d)$-dimensional matrices respectively, and W is a d-dimensional Brownian motion with $d \le N$. Since the standard hypotheses of Definition 9.4 hold, a strong solution of (9.70) exists and is unique. Further, just as in the case of deterministic linear equations, it is also possible to obtain the explicit expression for the solution.

Let us denote by $\Phi = \Phi(t)$ the solution of the ordinary Cauchy problem

$$\begin{cases} \Phi'(t) = B(t)\Phi(t), \\ \Phi(t_0) = I_N, \end{cases}$$

where I_N is the $(N \times N)$ identity matrix.

Proposition 9.50 *The solution of the SDE (9.70) with initial condition $X_0^x = x$ is given by*

$$X_t^x = \Phi(t)\left(x + \int_0^t \Phi^{-1}(s)b(s)ds + \int_0^t \Phi^{-1}(s)\sigma(s)dW_s\right). \tag{9.71}$$

Further, X_t^x has multi-normal distribution with mean

$$E\left[X_t^x\right] = \Phi(t)\left(x + \int_0^t \Phi^{-1}(s)b(s)ds\right) \tag{9.72}$$

and covariance matrix

$$\operatorname{cov}(X_t^x) = \Phi(t)\left(\int_0^t \Phi^{-1}(s)\sigma(s)\left(\Phi^{-1}(s)\sigma(s)\right)^* ds\right)\Phi^*(t). \tag{9.73}$$

Proof. Hereafter we use the notation

$$m_x(t) = E\left[X_t^x\right], \qquad \mathcal{C}(t) = \operatorname{cov}(X_t^x). \tag{9.74}$$

To prove that X^x in (9.71) is the solution, we merely have to use the Itô formula: we put

$$Y_t = x + \int_0^t \Phi^{-1}(s)b(s)ds + \int_0^t \Phi^{-1}(s)\sigma(s)dW_s$$

and we have

$$dX_t^x = d(\Phi(t)Y_t) = \Phi'(t)Y_t dt + \Phi(t)dY_t = (B(t)X_t^x + b(t))dt + \sigma(t)dW_t.$$

Since X_t^x is the sum of integrals of deterministic functions, by Proposition 5.32 we have that X_t has multi-normal distribution with mean and covariance given by (9.72) and (9.73) respectively. For the sake of clarity, we repeat the computation of the covariance matrix: we have

$$\begin{aligned}\operatorname{cov}(X_t^x) &= E\left[(X_t - m_x(t))(X_t - m_x(t))^*\right] \\ &= \Phi(t)E\left[\int_0^t \Phi^{-1}(s)\sigma(s)dW_s\left(\int_0^t \Phi^{-1}(s)\sigma(s)dW_s\right)^*\right]\Phi^*(t)\end{aligned}$$

and the thesis follows by Itô isometry. □

We explicitly note that, since $d \le N$, in general *the matrix $\mathcal{C}(t)$ is only positive semi-definite.* The case $\mathcal{C}(t) > 0$ is particularly important: indeed in this case X_t^x has density $y \mapsto \Gamma(0, x; t, y)$, where

$$\Gamma(0, x; t, y) = \frac{(2\pi)^{-\frac{N}{2}}}{\sqrt{\det \mathcal{C}(t)}} \exp\left(-\frac{1}{2}\langle \mathcal{C}^{-1}(t)(y - m_x(t)), (y - m_x(t))\rangle\right)$$

for $x, y \in \mathbb{R}^N$ and $t > 0$. Moreover, by the results in Section 9.4.4, Γ *is the fundamental solution of the differential operator in* $\mathbb{R}^{N+1}$ *associated to the linear SDE:*

$$\begin{aligned} L &= \frac{1}{2}\sum_{i,j=1}^{N} c_{ij}(t)\partial_{x_i x_j} + \sum_{i=1}^{N} b_i(t)\partial_{x_i} + \sum_{i=1}^{N} B_{ij}(t)x_i\partial_{x_j} + \partial_t \\ &= \frac{1}{2}\sum_{i,j=1}^{N} c_{ij}(t)\partial_{x_i x_j} + \langle b(t) + B(t)x, \nabla\rangle + \partial_t \end{aligned} \tag{9.75}$$

where $(c_{ij}) = \sigma\sigma^*$ and $\nabla = (\partial_{x_1}, \dots, \partial_{x_N})$.

Remark 9.51 The case constant coefficients, $b(t) \equiv b$, $B(t) \equiv B$ and $\sigma(t) \equiv \sigma$, is utterly important. First of all, let us recall that in this case we have $\Phi(t) = e^{tB}$ where

$$e^{tB} = \sum_{n=0}^{\infty} \frac{(tB)^n}{n!}. \tag{9.76}$$

Note that the series in (9.76) is absolutely convergent, since

$$\sum_{n=0}^{\infty} \frac{\|t^n B^n\|}{n!} \le \sum_{n=0}^{\infty} \frac{|t|^n}{n!}\|B\|^n = e^{|t|\|B\|}.$$

Moreover we have

$$\left(e^{tB}\right)^* = e^{tB^*}, \qquad e^{tB}e^{sB} = e^{(t+s)B}, \qquad t, s \in \mathbb{R}.$$

In particular, e^{tB} is not degenerate and we have that

$$\left(e^{tB}\right)^{-1} = e^{-tB}.$$

Then, by Proposition 9.50, the solution of the linear SDE

$$dX_t = (b + BX_t)dt + \sigma dW_t \tag{9.77}$$

with initial datum x, is given by

$$X_t^x = e^{tB}\left(x + \int_0^t e^{-sB}bds + \int_0^t e^{-sB}\sigma dW_s\right)$$

and we have

$$m_x(t) = E\left[X_t^x\right] = e^{tB}\left(x + \int_0^t e^{-sB}bds\right) = e^{tB}x + \int_0^t e^{sB}bds \tag{9.78}$$

and

$$\mathcal{C}(t) = \operatorname{cov}(X_t^x) = e^{tB}\int_0^t e^{-sB}\sigma\left(e^{-sB}\sigma\right)^* ds\, e^{tB^*} = \int_0^t \left(e^{sB}\sigma\right)\left(e^{sB}\sigma\right)^* ds. \tag{9.79}$$

In this case we also have that

$$X_t^{t_0,x} = X_{t-t_0}^x, \qquad t \geq t_0, \tag{9.80}$$

is the solution of (9.77) with initial condition $X_{t_0}^{t_0,x} = x$. □

Example 9.52 If $N = d$, $B = 0$ and b, σ are constant with σ non degenerate, L in (9.75) is the parabolic operator with constant coefficients

$$L = \frac{1}{2} \sum_{i,j=1}^{N} (\sigma\sigma^*)_{ij} \partial_{x_i x_j} + \sum_{i=1}^{N} b_i \partial_{x_i} + \partial_t.$$

Then, since e^{tB} is the identity matrix, by (9.78)-(9.79) we have

$$m_x(t) = x + tb, \qquad \mathcal{C}(t) = t\sigma\sigma^*,$$

and by (9.80) the fundamental solution of L is given by (see also Appendix A.3.2)

$$\Gamma(t, x; T, y) = \frac{(2\pi(T-t))^{-\frac{N}{2}}}{|\det \sigma|} \exp\left(-\frac{\left|\sigma^{-1}(y - x - (T-t)b)\right|^2}{2(T-t)} \right) \tag{9.81}$$

for $x, y \in \mathbb{R}^N$ and $t < T$. □

Example 9.53 The SDE in $\mathbb{R}^2$

$$\begin{cases} dX_t^1 = dW_t, \\ dX_t^2 = X_t^1 dt, \end{cases}$$

is the simplified version of the Langevin equation [231] that describes the motion of a particle in the phase space: X_t^1 and X_t^2 represent the velocity and the position of the particle, respectively. In this case $d = 1 < N = 2$ and we have

$$B = \begin{pmatrix} 0 & 0 \\ 1 & 0 \end{pmatrix}, \qquad \sigma = \begin{pmatrix} 1 \\ 0 \end{pmatrix}.$$

Since $B^2 = 0$, the matrix B is nilpotent and

$$e^{tB} = \begin{pmatrix} 1 & 0 \\ t & 1 \end{pmatrix}.$$

Moreover, if we put $x = (x_1, x_2)$, using notation (9.78)-(9.79) we have

$$m_x(t) = e^{tB}x = (x_1, x_2 + tx_1),$$

and

$$\mathcal{C}(t) = \int_0^t e^{sB}\sigma\sigma^* e^{sB^*} ds = \int_0^t \begin{pmatrix} 1 & 0 \\ s & 1 \end{pmatrix} \begin{pmatrix} 1 & 0 \\ 0 & 0 \end{pmatrix} \begin{pmatrix} 1 & s \\ 0 & 1 \end{pmatrix} ds = \begin{pmatrix} t & \frac{t^2}{2} \\ \frac{t^2}{2} & \frac{t^3}{3} \end{pmatrix}.$$

We note that $\mathcal{C}(t)$ is positive definite for every $t > 0$ and so the associated differential operator

$$L = \frac{1}{2}\partial_{x_1 x_1} + x_1 \partial_{x_2} + \partial_t \tag{9.82}$$

has the following fundamental solution

$$\begin{aligned}\Gamma(t,x;T,y) =& \frac{\sqrt{3}}{\pi(T-t)^2}\cdot \\ &\cdot \exp\left(-\frac{1}{2}\langle \mathcal{C}^{-1}(T-t)(y - e^{(T-t)B}x), (y - e^{(T-t)B}x)\rangle\right)\end{aligned}$$

for $x, y \in \mathbb{R}^2$ and $t < T$, where

$$\mathcal{C}^{-1}(t) = \begin{pmatrix} \frac{4}{t} & -\frac{6}{t^2} \\ -\frac{6}{t^2} & \frac{12}{t^3} \end{pmatrix}.$$

More explicitly we have

$$\begin{aligned}\Gamma(t,x;T,y) =& \frac{\sqrt{3}}{\pi(T-t)^2}\cdot \\ &\cdot \exp\left(-\frac{(y_1 - x_1)^2}{2(T-t)} - \frac{3(2y_2 - 2x_2 - (T-t)(y_1 + x_1))^2}{2(T-t)^3}\right).\end{aligned} \tag{9.83}$$

We emphasize that L *in* (9.82) *it is not a uniformly parabolic operator* since the matrix of the second-order part of L

$$\sigma\sigma^* = \begin{pmatrix} 1 & 0 \\ 0 & 0 \end{pmatrix}$$

is degenerate. Nevertheless L has a Gaussian fundamental solution as the heat equation. Kolmogorov [214] was the first to determine the fundamental solution of L in (9.82): for further details we refer to the introduction in Hörmander's paper [170].

From a financial point of view, the operator L arises in the problem of pricing Asian options with geometric average and Black-Scholes dynamics (cf. Section 7.6.2). □

9.5.1 Kalman condition

The distribution of X_t, solution of a linear SDE, is multi-normal and in general it is degenerate. In this section we give some necessary and sufficient conditions that make the covariance matrix of X_t positive definite so that X_t has a density.

For the sake of simplicity and clarity of exposition, in the sequel we take B and σ constant. To avoid trivial situations, we also suppose that the matrix

σ has maximum rank d: then, after a suitable linear transformation, we can assume that the columns of σ are the first d elements of the canonical basis, i.e. σ assumes the block form

$$\sigma = \begin{pmatrix} I_d \\ 0 \end{pmatrix},$$

where I_d is the $(d \times d)$ identity matrix. As usual B is a $(N \times N)$-dimensional generic matrix. Note that by (9.73) the co-variance matrix of X_t does not depend on b.

The first result that we present gives a condition in terms of controllability in the setting of linear-system theory; for further details we refer, for instance, to Lee and Markus [233] or Zabczyk [341]. We first recall the following classical:

Definition 9.54 *The pair (B, σ) is controllable over $[0, T]$ if for every $x, y \in \mathbb{R}^N$ there exists a function $v \in C([0, T]; \mathbb{R}^d)$ such that the problem*

$$\begin{cases} \gamma'(t) = B\gamma(t) + \sigma v(t), & t \in]0, T[, \\ \gamma(0) = x, \qquad \gamma(T) = y, \end{cases}$$

has a solution. The function v is called a control for (B, σ).

Theorem 9.55 *Given $T > 0$, the matrix*

$$\mathcal{C}(T) = \int_0^T \left(e^{tB}\sigma\right)\left(e^{tB}\sigma\right)^* dt \tag{9.84}$$

is positive definite if and only if the pair (B, σ) is controllable over $[0, T]$. In that case a control is given by

$$v(t) = G^*(t) M^{-1}(T)(e^{-TB} y - x), \qquad t \in [0, T], \tag{9.85}$$

where

$$M(T) = \int_0^T G(t) G^*(t) dt. \tag{9.86}$$

Before proving the theorem, we dwell on some remarks. First of all we introduce the notation

$$G(t) = e^{-tB}\sigma$$

that will be used systematically in what follows. Then, for fixed $x \in \mathbb{R}^N$, as a particular case of formula (9.71), we have that

$$\gamma(t) = e^{tB}\left(x + \int_0^t G(s) v(s) ds\right) \tag{9.87}$$

is the solution of the linear Cauchy problem

$$\begin{cases} \gamma'(t) = B\gamma(t) + \sigma v(t), & t \in]0, T[, \\ \gamma(0) = x. \end{cases} \tag{9.88}$$

If (B, σ) is controllable over $[0, T]$, then for every $y \in \mathbb{R}^N$ there exists a control v such that the trajectory γ in (9.87) hits the target y at time T. The existence of a control is not guaranteed in general since v in (9.88) multiplies the matrix σ that "reduces" the influence of the control: this is obvious in the case $\sigma = 0$. In general, the differential equation in (9.88) can be rewritten in the following way:

$$\gamma' = B\gamma + \sum_{i=1}^{d} v_i \sigma^i,$$

where the vectors σ^i, $i = 1, \dots, d$, denote the columns of σ, i.e. the first d vectors of the canonical basis of $\mathbb{R}^N$. The physical interpretation is that the "speed" γ' equals $B\gamma$ plus a linear combination of the vectors σ^i, $i = 1, \dots, d$: the coefficients of this linear combination are the components of the control v. Therefore v *allows us to control the speed of the trajectory* γ *in* $\mathbb{R}^N$ *only in the first* d *directions.* Evidently if σ is the identity matrix, then the columns of σ constitute the canonical basis of $\mathbb{R}^N$ and (B, σ) is controllable for any matrix B. Nevertheless there are cases in which the contribution of B is crucial, as in the following:

Example 9.56 Let B and σ be as in Example 9.53: then v has real values and problem (9.88) becomes

$$\begin{cases} \gamma_1'(t) = v(t), \\ \gamma_2'(t) = \gamma_1(t), \\ \gamma(0) = x. \end{cases} \tag{9.89}$$

The control v acts directly only on the first component of γ, but influences also γ_2 through the second equation: in this case we can directly verify that (B, σ) is controllable over $[0, T]$ for every positive T by using the control in (9.85) (see Figure 9.3). □

Proof (of Theorem 9.55). We recall that by (9.79) we have

$$\mathcal{C}(T) = e^{TB} M(T) e^{TB^*},$$

with M as in (9.86). Since the exponential matrices are non-degenerate, $\mathcal{C}(T)$ is positive definite if and only if $M(T)$ is as such.

We suppose $M(T) > 0$ and prove that (B, σ) is controllable over $[0, T]$. For fixed $x \in \mathbb{R}^N$, we consider the curve γ in (9.87), solution of problem (9.88): given $y \in \mathbb{R}^N$, we have that $\gamma(T) = y$ if and only if

$$\int_0^T G(t) v(t) dt = e^{-TB} y - x =: z, \tag{9.90}$$

and therefore, using the non-degeneracy assumption on $M(T)$, a control is simply given as in (9.85) by

$$v(t) = G^*(t) M^{-1}(T) z, \qquad t \in [0, T].$$

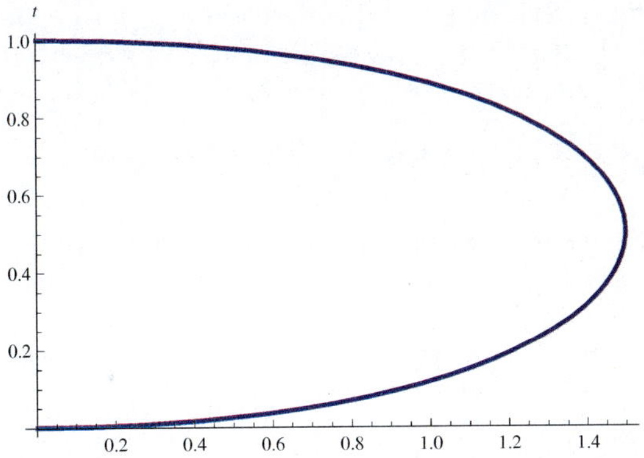

Fig. 9.3. Graph of the optimal trajectory $\gamma(t) = (6(t-t^2), 3t^3 - 2t^3)$ solution of the problem (9.89) with $\gamma(0) = x = (0,0)$ and such that it satisfies the final condition $\gamma(1) = y = (0,1)$

Conversely, let (B, σ) be controllable over $[0,T]$ and, by contradiction, let us suppose that $M(T)$ is degenerate. Then there exists $w \in \mathbb{R}^N \setminus \{0\}$ such that

$$0 = \langle M(T)w, w\rangle = \int_0^T |w^* G(t)|^2 dt,$$

and consequently we have

$$w^* G(t) = 0, \qquad t \in [0,T].$$

By assumption, (B, σ) is controllable over $[0,T]$ and so for every $x, y \in \mathbb{R}^N$ there exists a suitable control v such that (9.90) holds. Multiplying by w^*, we have

$$w^* z = \int_0^T w^* G(s) v(s) ds = 0,$$

and this leads to a contradiction. □

Remark 9.57 The control v in (9.85) is optimal in the sense that it minimizes the "cost functional"

$$U(v) := \|v\|^2_{L^2([0,T])} = \int_0^T v(t)^* v(t) dt.$$

This can be proved by the Lagrange-Ljusternik theorem (cf. for instance [322]) that extends to the functional setting the standard Lagrange multipliers theorem. Indeed, in order to minimize U subject to the constraint (9.90), we consider the Lagrange functional

$$\mathcal{L}(v, \lambda) = \|v\|^2_{L^2([0,T])} - \lambda^* \left(\int_0^T G(t) v(t) dt - z \right),$$

where $\lambda \in \mathbb{R}^N$ is the Lagrange multiplier. Taking the Fréchet differential of $\mathcal{L}$, we impose that v is a stationary point of $\mathcal{L}$ and we get

$$\partial_v \mathcal{L}(u) = 2\int_0^T v(t)^* u(t)dt - \lambda^* \int_0^T G(t)u(t)dt = 0, \qquad u \in L^2([0,T]).$$

Then we find $v = \frac{1}{2}\lambda G^*$ where λ is determined by the constraint (9.90), $\lambda = 2M^{-1}(T)z$, according with (9.85). □

The following result gives a practical criterion to check whether the covariance matrix is non-degenerate.

Theorem 9.58 (Kalman rank condition) *The matrix $\mathcal{C}(T)$ in (9.84) is positive definite for $T > 0$ if and only if the pair (B, σ) verifies the Kalman condition, i.e. the $(N \times (Nd))$-dimensional block matrix, defined by*

$$\begin{pmatrix} \sigma & B\sigma & B^2\sigma & \cdots & B^{N-1}\sigma \end{pmatrix}, \tag{9.91}$$

has maximum rank, equal to N.

We point out explicitly that the Kalman condition does not depend on T and consequently $\mathcal{C}(T)$ is positive definite for some positive T if and only if it is positive definite *for every positive* T.

Example 9.59 In Example 9.53, we have

$$\sigma = \begin{pmatrix} 1 \\ 0 \end{pmatrix}, \qquad B\sigma = \begin{pmatrix} 0 & 0 \\ 1 & 0 \end{pmatrix}\begin{pmatrix} 1 \\ 0 \end{pmatrix} = \begin{pmatrix} 0 \\ 1 \end{pmatrix},$$

therefore $(\sigma \;\, B\sigma)$ is the identity matrix and evidently the Kalman condition is satisfied. □

Proof (of Theorem 9.58). We recall the Cayley-Hamilton theorem: let

$$p(\lambda) = \det(A - \lambda I_N) = \lambda^N + a_1\lambda^{N-1} + \cdots + a_{N-1}\lambda + a_N$$

be the characteristic polynomial of an $(N \times N)$-dimensional matrix A. Then we have $p(A) = 0$ and so every power A^k with $k \geq N$ can be expressed as a linear combination of $I_N, A, \dots, A^{N-1}$.

Now we observe that the matrix (9.91) does not have maximum rank if and only if there exists $w \in \mathbb{R}^N \setminus \{0\}$ such that

$$w^*\sigma = w^*B\sigma = \cdots = w^*B^{N-1}\sigma = 0. \tag{9.92}$$

Then, assuming that the matrix (9.91) does not have maximum rank, by (9.92) and the Cayley-Hamilton theorem, we have

$$w^*B^k\sigma = 0, \qquad k \in \mathbb{N}_0,$$

hence we infer

$$w^* e^{tB} \sigma = 0, \qquad t \geq 0.$$

Consequently

$$\langle \mathcal{C}(T)w, w \rangle = \int_0^T \left| w^* e^{tB} \sigma \right|^2 dt = 0, \tag{9.93}$$

and $\mathcal{C}(T)$ is degenerate for every $T > 0$.

Conversely, if $\mathcal{C}(T)$ is degenerate for some $T > 0$ then there exists $w \in \mathbb{R}^N \setminus \{0\}$ such that (9.93) holds, hence

$$f(t) := w^* e^{tB} \sigma = 0, \qquad t \in [0, T].$$

Differentiating we get

$$0 = \frac{d^k}{dt^k} f(t) \mid_{t=0} = w^* B^k \sigma, \qquad k \in \mathbb{N}_0,$$

so we infer that the matrix (9.91) does not have maximum rank, by (9.92).□

9.5.2 Kolmogorov equations and Hörmander condition

Let us consider the linear SDE

$$dX_t = (BX_t + b)dt + \sigma dW_t, \tag{9.94}$$

with B, b, σ constant, σ given by

$$\sigma = \begin{pmatrix} I_d \\ 0 \end{pmatrix},$$

and let us assume that the Kalman condition holds true:

$$\operatorname{rank} \begin{pmatrix} \sigma & B\sigma & B^2\sigma & \cdots & B^{N-1}\sigma \end{pmatrix} = N.$$

Definition 9.60 *We say that the differential operator in* $\mathbb{R}^{N+1}$

$$L = \frac{1}{2} \triangle_{\mathbb{R}^d} + \langle b + Bx, \nabla \rangle + \partial_t, \tag{9.95}$$

associated to the SDE (9.94), is a Kolmogorov-type operator with constant coefficients. Here we use the notation

$$\triangle_{\mathbb{R}^d} = \sum_{i=1}^d \partial_{x_i x_i}.$$

We recall the definition of $\mathcal{C}$ and m_x in (9.78)-(9.79). By the Kalman condition $\mathcal{C}(t)$ is positive definite for $t > 0$, hence it follows that L has a fundamental solution whose explicit expression is

$$\Gamma(t,x;T,y) = \frac{(2\pi)^{-\frac{N}{2}}}{\sqrt{\det \mathcal{C}(T-t)}} e^{-\frac{1}{2}\langle \mathcal{C}^{-1}(T-t)(y-m_x(T-t)),(y-m_x(T-t))\rangle},$$

for $x, y \in \mathbb{R}^N$ and $t < T$.

Now we prove that the Kalman condition is equivalent to the Hörmander condition which is a non-degeneracy criterion, well known in the theory of partial differential equations. By convention, we identify every first-order differential operator Z in $\mathbb{R}^N$ given by

$$Zf(x) = \sum_{k=1}^{N} \alpha_k(x)\partial_{x_k} f(x),$$

with the vector field of its coefficients and so we also write

$$Z = (\alpha_1, \dots, \alpha_N).$$

The commutator of Z with

$$U = \sum_{k=1}^{N} \beta_k \partial_{x_k}$$

is defined by

$$[Z,U] = ZU - UZ = \sum_{k=1}^{N} (Z\beta_k - U\alpha_k)\,\partial_{x_k}.$$

Hörmander's theorem [170] (see also Stroock [318] for a more recent exposition) is a very general result that, in the particular case of the Kolmogorov operator with constant coefficients in (9.95), states that L has a fundamental solution if and only if, at any point $x \in \mathbb{R}^N$, the vector space spanned by the differential operators (vector fields)

$$\partial_{x_1}, \dots, \partial_{x_d} \qquad \text{and} \qquad Y := \langle Bx, \nabla \rangle,$$

and by their commutators of every order, computed at x, coincides with $\mathbb{R}^N$. This is the so-called *Hörmander condition.*

Example 9.61 □

i) If the operator is parabolic we have $d = N$ therefore the Hörmander condition is obviously satisfied, without resorting to commutators, since $\partial_{x_1}, \dots, \partial_{x_N}$ form the canonical basis of $\mathbb{R}^N$.

ii) In Example 9.53 we simply have $Y = x_1\partial_{x_2}$. So

$$\partial_{x_1} \sim (1,0) \qquad \text{and} \qquad [\partial_{x_1}, Y] = \partial_{x_2} \sim (0,1)$$

span $\mathbb{R}^2$.

iii) Let us consider the differential operator

$$\partial_{x_1x_1} + x_1\partial_{x_2} + x_2\partial_{x_3} + \partial_t.$$

Here $N = 3$, $d = 1$ and $Y = x_1\partial_{x_2} + x_2\partial_{x_3}$: also in this case the Hörmander condition is verified since

$$\partial_{x_1}, \qquad [\partial_{x_1}, Y] = \partial_{x_2}, \qquad [[\partial_{x_1}, Y], Y] = \partial_{x_3},$$

span $\mathbb{R}^3$.

□

Proposition 9.62 *Kalman and Hörmander conditions are equivalent.*

Proof. It is enough to notice that, for $i = 1, \dots, d$,

$$[\partial_{x_i}, Y] = \sum_{k=1}^{N} b_{ki}\partial_{x_k}$$

is the i-th column of the matrix B. Further, $[[\partial_{x_i}, Y], Y]$ is the i-th column of the matrix B^2 and an analogous representation holds for higher-order commutators.

On the other hand, for $k = 1, \dots, N$, $B^k\sigma$ in (9.91) is the $(N \times d)$-dimensional matrix whose columns are the first d columns of B^k. □

Let us now introduce the definition of Kolmogorov operator with variable coefficients. We consider the SDE in $\mathbb{R}^N$

$$dX_t = (BX_t + b(t, X_t))dt + \sigma(t, X_t)dW_t, \tag{9.96}$$

where as usual W is a d-dimensional Brownian motion and we assume:

i) the matrix σ takes the form

$$\sigma = \begin{pmatrix} \sigma_0 \\ 0 \end{pmatrix},$$

where $\sigma_0 = \sigma_0(t, x)$ is a $d \times d$-dimensional matrix such that $(c_{ij}) = \sigma_0\sigma_0^*$ is uniformly positive definite, i.e. there exists a positive constant Λ such that

$$\sum_{i,j=1}^{d} c_{ij}(t, x)\eta_i\eta_j \geq \Lambda^2|\eta|^2, \qquad \eta \in \mathbb{R}^d, \ (t, x) \in \mathbb{R}^{N+1};$$

ii) B and $\begin{pmatrix} I_d \\ 0 \end{pmatrix}$ verify the Kalman condition or, in other terms,

$$\frac{1}{2}\triangle_{\mathbb{R}^d} + \langle Bx, \nabla \rangle + \partial_t$$

is a Kolmogorov operator with constant coefficients;

iii) $b_{d+1}, \dots, b_N$ are functions of the variable t only.

The first condition weakens the uniform parabolicity in (8.2). The second assumption is a non-degeneracy condition that makes up for the possible absence of parabolicity: if the coefficients are constant, this guarantees the existence of a fundamental solution. The third condition aims at preserving the second one: if b could be a generic function, then the linear term BX_t in the stochastic equation would be superfluous. In particular the Kalman condition, which is based upon the particular structure of the matrix B, would be lost.

Definition 9.63 *We say that the differential operator in* $\mathbb{R}^{N+1}$

$$L = \frac{1}{2}\sum_{i,j=1}^{d} c_{ij}(t,x)\partial_{x_i}\partial_{x_j} + \sum_{i=1}^{d} b_i(t,x)\partial_{x_i} + \langle Bx, \nabla \rangle + \partial_t,$$

associated to the SDE (9.96)*, is a Kolmogorov-type operator with variable coefficients.*

A theory, analogous to the classical one for uniformly parabolic operators, presented in Chapter 8, has been developed by various authors for the general class of Kolmogorov operators with variable coefficients: we mention the results in Lanconelli and Polidoro [230], Polidoro [283], [284], [285], Di Francesco and Pascucci [94]. Recently in [95] and [276] the obstacle problem for Kolmogorov operators was studied, and with this the corresponding optimal stopping problem that arises in the pricing problem for American Asian options.

9.5.3 Examples

We examine a couple of interesting examples of linear SDEs.

Example 9.64 (Brownian bridge) Let $b \in \mathbb{R}$. We consider the 1-dimensional SDE

$$dB = \frac{b - B_t}{1 - t} dt + dW_t,$$

whose solution, at least for $t < 1$, is given by

$$B_t = B_0(1-t) + bt + (1-t)\int_0^t \frac{dW_s}{1-s}.$$

Then we have

$$E[B_t] = B_0(1-t) + bt,$$

and, by Itô isometry,

$$\operatorname{var}(B_t) = (1-t)^2 \int_0^t \frac{ds}{(1-s)^2} = t(1-t).$$

We note that

$$\lim_{t\to 1^-} E\left[B_t\right] = b, \qquad \text{and} \qquad \lim_{t\to 1^-} \operatorname{var}(B_t) = 0.$$

As a matter of fact, we can prove that

$$\lim_{t\to 1^-} B_t = b, \qquad \text{a.s.}$$

since, for $t < 1$, we have

$$\begin{aligned}
&E\left[(B_t - b)^2\right] \\
&= (1-t)^2 \left((b - B_0)^2 - 2(b - B_0) \underbrace{E\left[\int_0^t \frac{dW_s}{1-s}\right]}_{=0} + E\left[\left(\int_0^t \frac{dW_s}{1-s}\right)^2\right] \right) = \\
&= (1-t)^2 \left((b - B_0)^2 + \int_0^t \frac{ds}{(1-s)^2} \right) = \\
&= (1-t)^2 \left((b - B_0)^2 + \frac{1}{1-t} - 1 \right) \xrightarrow[t\to 1^-]{} 0.
\end{aligned}$$

□

Example 9.65 (Ornstein and Uhlenbeck [328], Langevin [231]) We consider the following model for the motion of a particle with friction: speed and position are described by the pair $X_t = (V_t, P_t)$, solution of the linear SDE

$$\begin{cases} dV_t = -\mu V_t dt + \sigma dW_t \\ dP_t = V_t dt, \end{cases}$$

where W is a real Brownian motion, μ and σ are the positive friction and diffusion coefficients. Equivalently we have

$$dX_t = BX_t dt + \bar{\sigma} dW_t$$

where

$$B = \begin{pmatrix} -\mu & 0 \\ 1 & 0 \end{pmatrix}, \qquad \bar{\sigma} = \begin{pmatrix} \sigma \\ 0 \end{pmatrix}.$$

We can easily check that the Kalman condition is verified. Further, it is immediate to prove by induction that, for every $n \in \mathbb{N}$, we have that

$$B^n = \begin{pmatrix} (-\mu)^n & 0 \\ (-\mu)^{n-1} & 0 \end{pmatrix},$$

and so

$$e^{tB} = I_2 + \sum_{n=1}^{N} \frac{(tB)^n}{n!} = \begin{pmatrix} e^{-\mu t} & 0 \\ \frac{1-e^{-\mu t}}{\mu} & 1 \end{pmatrix}.$$

We know that X_t has normal distribution. To conclude, we compute the expectation and covariance matrix:

$$E[X_t] = \begin{pmatrix} E[V_t] \\ E[P_t] \end{pmatrix} = e^{tB} \begin{pmatrix} V_0 \\ P_0 \end{pmatrix} = \begin{pmatrix} V_0 e^{-\mu t} \\ P_0 + \frac{V_0}{\mu}(1 - e^{-\mu t}) \end{pmatrix};$$

further,

$$\begin{aligned} \mathcal{C}(t) &= \begin{pmatrix} \text{var}(V_t) & \text{cov}(V_t, P_t) \\ \text{cov}(V_t, P_t) & \text{var}(P_t) \end{pmatrix} = \int_0^t \left(e^{sB}\bar{\sigma}\bar{\sigma}^*\right) e^{sB^*} ds \\ &= \sigma^2 \int_0^t \begin{pmatrix} e^{-\mu s} & 0 \\ \frac{1-e^{-\mu s}}{\mu} & 0 \end{pmatrix} \begin{pmatrix} e^{-\mu s} & \frac{1-e^{-\mu s}}{\mu} \\ 0 & 1 \end{pmatrix} ds \\ &= \sigma^2 \int_0^t \begin{pmatrix} e^{-2\mu s} & \frac{e^{-\mu s} - e^{-2\mu s}}{\mu} \\ \frac{e^{-\mu s} - e^{-2\mu s}}{\mu} & \left(\frac{1-e^{-\mu s}}{\mu}\right)^2 \end{pmatrix} ds \\ &= \sigma^2 \begin{pmatrix} \frac{1}{2\mu}\left(1 - e^{-2\mu t}\right) & \frac{1}{2\mu^2}\left(1 - 2e^{-\mu t} + e^{-2\mu t}\right) \\ \frac{1}{2\mu^2}\left(1 - 2e^{-\mu t} + e^{-2\mu t}\right) & \frac{1}{\mu^3}\left(\mu t + 2e^{-\mu t} - \frac{e^{-2\mu t} - 3}{2}\right) \end{pmatrix}. \qquad \square \end{aligned}$$

10

Continuous market models

In this chapter we present the theory of derivative pricing and hedging for continuous-time diffusion models. As in the discrete-time case, the concept of martingale measure plays a central role: we prove that any equivalent martingale measure (EMM) is associated to a market price of risk and determines a risk-neutral price for derivatives, that avoids the introduction of arbitrage opportunities. In this setting we generalize the theory in discrete time of Chapter 2 and extend the Markovian formulation of Chapter 7, based upon parabolic equations.

Our presentation follows essentially the probabilistic approach introduced in the papers by Harrison and Kreps [163], Harrison and Pliska [164]. In the first two paragraphs we give the theoretical results on the change of probability measure and on the representation of Brownian martingales. Then, we introduce the market models in continuous time and we study the existence of an EMM and its relation with the absence of arbitrage opportunities. At first we discuss pricing and hedging of options in a general framework; afterwards we treat the Markovian case that is based upon the parabolic PDE theory developed in the previous chapters: this case is particularly significant, since it allows the use of efficient numerical methods to determine the price and the hedging strategy of a derivative. We next give a coincise description of the well-known technique of the change of numeraire: in particular, we examine some remarkable applications to the fixed-income markets and prove a quite general pricing formula.

10.1 Change of measure

10.1.1 Exponential martingales

We consider a d-dimensional Brownian motion $(W_t)_{t\in[0,T]}$ on the space $(\Omega, \mathcal{F}, P, (\mathcal{F}_t))$. Let $\lambda \in \mathbb{L}^2_{\text{loc}}$ be a d-dimensional process: we define the *ex-*

Pascucci A.: PDE and Martingale Methods in Option Pricing

ponential martingale associated to λ (cf. Example 5.12) as

$$Z_t^\lambda = \exp\left(-\int_0^t \lambda_s \cdot dW_s - \frac{1}{2}\int_0^t |\lambda_s|^2 ds\right), \qquad t \in [0,T]. \tag{10.1}$$

We recall that the symbol "·" denotes the scalar product in $\mathbb{R}^d$. By the Itô formula we have

$$dZ_t^\lambda = -Z_t^\lambda \lambda_t \cdot dW_t, \tag{10.2}$$

so that Z^λ is a local martingale. Since Z^λ is positive, by Proposition 4.40 it is also a super-martingale:

$$E\left[Z_t^\lambda\right] \le E\left[Z_0^\lambda\right] = 1, \qquad t \in [0,T],$$

and $(Z_t^\lambda)_{t\in[0,T]}$ is a strict martingale if and only if $E\left[Z_T^\lambda\right] = 1$.

Lemma 10.1 *If there exists a constant C such that*

$$\int_0^T |\lambda_t|^2 dt \le C \qquad a.s. \tag{10.3}$$

then Z^λ in (10.1) *is a martingale such that*

$$E\left[\sup_{0\le t\le T} \left(Z_t^\lambda\right)^p\right] < \infty, \qquad p \ge 1. \tag{10.4}$$

In particular $Z^\lambda \in \mathbb{L}^p(\Omega, P)$ for every $p \ge 1$.

Proof. We put

$$\hat{Z}_T = \sup_{0\le t\le T} Z_t^\lambda.$$

For every $\zeta > 0$, we have

$$P\left(\hat{Z}_T \ge \zeta\right) \le P\left(\sup_{0\le t\le T} \exp\left(-\int_0^t \lambda_s \cdot dW_s\right) \ge \zeta\right)$$
$$= P\left(\sup_{0\le t\le T}\left(-\int_0^t \lambda_s \cdot dW_s\right) \ge \log\zeta\right) \le$$

(by Corollary 9.31, using condition (10.3) with c_1, c_2 positive constants)

$$\le c_1 e^{-c_2(\log\zeta)^2}.$$

Then, by Proposition A.56 we have

$$E\left[\hat{Z}_T^p\right] = p\int_0^\infty \zeta^{p-1} P\left(\hat{Z}_T \ge \zeta\right) d\zeta < \infty.$$

In particular for $p = 2$ we have that $\lambda Z^\lambda \in \mathbb{L}^2$ and so, by (10.2), that Z^λ is a martingale. □

Remark 10.2 Given $z = (z_1, \dots, z_d) \in \mathbb{C}^d$, we set

$$z^2 = \sum_{k=1}^{d} z_k^2 \quad \text{and} \quad |z|^2 = \sum_{k=1}^{d} |z_k|^2.$$

If $\lambda = (\lambda^1, \dots, \lambda^d)$ takes complex values, $\lambda_t^k \in \mathbb{C}$, $k = 1, \dots, d$, then by proceeding as in the proof of Lemma 10.1, we can prove that, if

$$\int_0^T |\lambda_t|^2 \, dt \le C \qquad P\text{-a.s.},$$

then

$$Z_t^\lambda := \exp\left(- \int_0^t \lambda_s \cdot dW_s - \frac{1}{2} \int_0^t \lambda_s^2 ds \right), \qquad t \in [0, T],$$

is a (complex) martingale in $\mathbb{L}^p(\Omega, P)$ for every $p \ge 1$. □

We now suppose that Z^λ in (10.1) is a martingale and define the measure Q on $(\Omega, \mathcal{F})$ by

$$\frac{dQ}{dP} = Z_T^\lambda, \tag{10.5}$$

i.e.

$$Q(F) = \int_F Z_T^\lambda dP, \qquad F \in \mathcal{F}.$$

We recall Bayes' formula, Theorem A.113: for every $X \in L^1(\Omega, Q)$ we have

$$E^Q [X \mid \mathcal{F}_t] = \frac{E^P \left[X Z_T^\lambda \mid \mathcal{F}_t \right]}{E^P \left[Z_T^\lambda \mid \mathcal{F}_t \right]} \qquad t \in [0, T]. \tag{10.6}$$

Consequently we get the following:

Lemma 10.3 *Assume that Z^λ in (10.1) is a P-martingale and Q is the probability measure defined in (10.5). Then a process $(M_t)_{t \in [0,T]}$ is a Q-martingale if and only if $(M_t Z_t^\lambda)_{t \in [0,T]}$ is a P-martingale.*

Proof. Since Z^λ is strictly positive and adapted, it is clear that M is adapted if and only if $M Z^\lambda$ is adapted. Moreover, since Z^λ is a P-martingale, M is Q-integrable if and only if $M Z^\lambda$ is P-integrable: indeed

$$E^Q [|M_t|] = E^P \left[|M_t| Z_T^\lambda \right] = E^P \left[E^P \left[|M_t| Z_T^\lambda \mid \mathcal{F}_t \right] \right] =$$

(since M is adapted)

$$= E^P \left[|M_t| E^P \left[Z_T^\lambda \mid \mathcal{F}_t \right] \right] = E^P \left[|M_t| Z_t^\lambda \right].$$

Analogously, for $s \le t$ we have

$$E^P \left[M_t Z_T^\lambda \mid \mathcal{F}_s \right] = E^P \left[E^P \left[M_t Z_T^\lambda \mid \mathcal{F}_t \right] \mid \mathcal{F}_s \right] = E^P \left[M_t Z_t^\lambda \mid \mathcal{F}_s \right].$$

Then by (10.6) with $X = M_t$ we have

$$E^Q[M_t \mid \mathcal{F}_s] = \frac{E^P\left[M_t Z_T^\lambda \mid \mathcal{F}_s\right]}{E^P\left[Z_T^\lambda \mid \mathcal{F}_s\right]} = \frac{E^P\left[M_t Z_t^\lambda \mid \mathcal{F}_s\right]}{Z_s^\lambda},$$

hence the claim. □

Remark 10.4 Under the assumptions of Lemma 10.3, the process

$$\left(Z_t^\lambda\right)^{-1} = \exp\left(\int_0^t \lambda_s \cdot dW_s + \frac{1}{2}\int_0^t |\lambda_s|^2 ds\right)$$

is a Q-martingale since $Z^\lambda \left(Z^\lambda\right)^{-1}$ is clearly a P-martingale. Further, for every integrable random variable X, we have

$$E^P[X] = E^P\left[X\left(Z_T^\lambda\right)^{-1} Z_T^\lambda\right] = E^Q\left[X\left(Z_T^\lambda\right)^{-1}\right]$$

and so

$$\frac{dP}{dQ} = \left(Z_T^\lambda\right)^{-1}.$$

In particular P, Q *are equivalent measures* since they reciprocally have strictly positive densities.

Finally, by proceeding as in Lemma 10.1, we can prove that, if condition (10.3) holds, then $\left(Z^\lambda\right)^{-1} \in \mathbb{L}^p(\Omega, P)$ for every $p \geq 1$. □

10.1.2 Girsanov's theorem

Girsanov's theorem shows that it is possible to substitute "arbitrarily" the drift of an Itô process by modifying appropriately the considered probability measure and Brownian motion, while keeping unchanged the diffusion coefficient. In this section $(W_t)_{t\in[0,T]}$ denotes a d-dimensional Brownian motion on the space $(\Omega, \mathcal{F}, P, (\mathcal{F}_t))$. The main result is the following:

Theorem 10.5 (Girsanov's theorem) *Let Z^λ in (10.1) be the exponential martingale associated to the process $\lambda \in \mathbb{L}^2_{\text{loc}}$. We assume that Z^λ is a P-martingale and we consider the measure Q defined by*

$$\frac{dQ}{dP} = Z_T^\lambda. \tag{10.7}$$

Then the process

$$W_t^\lambda = W_t + \int_0^t \lambda_s ds, \qquad t \in [0,T], \tag{10.8}$$

is a Brownian motion on $(\Omega, \mathcal{F}, Q, (\mathcal{F}_t))$.

Proof. We use the Brownian motion characterization Theorem 5.33. We have to prove that, for every $\xi \in \mathbb{R}^d$, the process

$$Y_t^\xi = e^{i\xi \cdot W_t^\lambda + \frac{|\xi|^2}{2} t}, \qquad t \in [0, T],$$

is a Q-martingale or, equivalently, by Lemma 10.3, that the process

$$\begin{aligned} Y_t^\xi Z_t &= \exp\left(i\xi \cdot W_t + i \int_0^t \xi \cdot \lambda_s ds + \frac{|\xi|^2 t}{2} - \int_0^t \lambda_s \cdot dW_s - \frac{1}{2} \int_0^t |\lambda_s|^2 ds \right) \\ &= \exp\left(- \int_0^t (\lambda_s - i\xi) \cdot dW_s - \frac{1}{2} \sum_{k=1}^d \int_0^t \left(\lambda_s^k - i\xi^k\right)^2 ds \right) \end{aligned}$$

is a P-martingale. If

$$\int_0^T |\lambda_t|^2 dt \le C \qquad P\text{-a.s.},$$

then the claim follows from Lemma 10.1 that holds true also for complex-valued processes and in particular for $\lambda - i\xi$ (cf. Remark 10.2).

In general we have to use a localization argument: we consider the sequence of stopping times

$$\tau_n = T \wedge \inf \left\{ t \mid \int_0^t |\lambda_s|^2 ds \ge n \right\}, \qquad n \in \mathbb{N}.$$

By Lemma 10.1, the process $(Y_{t\wedge\tau_n}^\xi Z_{t\wedge\tau_n})$ is a P-martingale and we have

$$E^P \left[Y_{t\wedge\tau_n}^\xi Z_{t\wedge\tau_n} \mid \mathcal{F}_s \right] = Y_{s\wedge\tau_n}^\xi Z_{s\wedge\tau_n}, \qquad s \le t, \ n \in \mathbb{N}.$$

Therefore, in order to prove that $Y^\xi Z$ is a martingale, it is enough to show that $(Y_{t\wedge\tau_n}^\xi Z_{t\wedge\tau_n})$ converges to $(Y_t^\xi Z_t)$ in L^1-norm as n tends to infinity. Since

$$\lim_{n\to\infty} Y_{t\wedge\tau_n}^\xi = Y_t^\xi \qquad \text{a.s.}$$

and $0 \le Y_{t\wedge\tau_n}^\xi \le e^{\frac{|\xi|^2 T}{2}}$, it is enough to prove that

$$\lim_{n\to\infty} Z_{t\wedge\tau_n} = Z_t \qquad \text{in } L^1(\Omega, P).$$

Putting

$$M_n = \min\{Z_{t\wedge\tau_n}, Z_t\},$$

we have $0 \le M_n \le Z_t$ and, by the dominated convergence theorem

$$\lim_{n\to\infty} E[M_n] = E[Z_t].$$

On the other hand,

$$E[|Z_t - Z_{t\wedge\tau_n}|] = E[Z_t - M_n] + E[Z_{t\wedge\tau_n} - M_n] =$$

(since $E[Z_t] = E[Z_{t\wedge\tau_n}] = 1$)

$$= 2E[Z_t - M_n]$$

hence the claim. □

The main assumption of the Girsanov's theorem is the martingale property of the process Z^λ. In financial applications we frequently assume that λ is a bounded process: in that case the martingale property of Z^λ follows from Lemma 10.1. Nevertheless in general the fact that λ is bounded may not be verified directly, so the following Novikov condition [266] can be very useful: we merely state it here.

Theorem 10.6 (Novikov condition) *If $\lambda \in \mathbb{L}^2_{\text{loc}}$ is such that*

$$E\left[\exp\left(\frac{1}{2}\int_0^T |\lambda_s|^2 ds\right)\right] < \infty$$

then the exponential martingale Z^λ in (10.1) is a strict martingale.

10.1.3 Representation of Brownian martingales

Let $(W_t)_{t\in[0,T]}$ be a d-dimensional Brownian motion on the space $(\Omega, \mathcal{F}, P)$ endowed with the Brownian filtration $\mathcal{F}^W = (\mathcal{F}_t^W)_{t\in[0,T]}$. We know (cf. Theorem 5.20) that, for every d-dimensional process $u \in \mathbb{L}^2(\mathcal{F}^W)$ and $M_0 \in \mathbb{R}$, the real-valued integral process

$$M_t = M_0 + \int_0^t u_s \cdot dW_s, \qquad t \in [0,T], \tag{10.9}$$

is a $\mathcal{F}^W$-martingale. In this paragraph we prove that, conversely, every real $\mathcal{F}^W$-martingale can be represented in the form (10.9).

Theorem 10.7 *For every real random variable $X \in L^2(\Omega, \mathcal{F}_T^W)$ there exists a unique*[1] *$u \in \mathbb{L}^2(\mathcal{F}^W)$ such that*

$$X = E[X] + \int_0^T u_t \cdot dW_t. \tag{10.10}$$

For the sake of simplicity, we consider only the 1-dimensional case $d = 1$ even though the arguments that follow can be adapted to the general case. The proof of Theorem 10.7 is based upon the following preliminary results.

Lemma 10.8 *The collection of random variables of the form*

$$\varphi(W_{t_1}, \dots, W_{t_n})$$

with $\varphi \in C_0^\infty(\mathbb{R}^n)$, $t_k \in [0,T]$ for $k = 1, \dots, n$ and $n \in \mathbb{N}$, is dense in $L^2(\Omega, \mathcal{F}_T^W)$.

[1] In the sense of the $(m \otimes P)$-equivalence, cf. Definition 3.28.

Proof. We consider a countable dense subset $\{t_n\}_{n\in\mathbb{N}}$ of $[0,T]$ and we define the discrete filtration

$$\mathcal{F}_n := \sigma(W_{t_1},\dots,W_{t_n}), \qquad n\in\mathbb{N};$$

we observe that $\mathcal{F}_T^W = \sigma(\mathcal{F}_n, n\in\mathbb{N})$. Given $X\in L^2(\Omega,\mathcal{F}_T^W)$, we consider the discrete martingale defined by

$$X_n = E\left[X \mid \mathcal{F}_n\right], \qquad n\in\mathbb{N}.$$

By Corollary A.134 we have

$$\lim_{n\to\infty} X_n = X, \qquad \text{in } L^2;$$

further, by Corollary A.10, for every $n\in\mathbb{N}$ there exists a measurable function $\varphi^{(n)}$ such that

$$X_n = \varphi^{(n)}(W_{t_1},\dots,W_{t_n}).$$

By density, $\varphi^{(n)}$ can be approximated in $L^2(\mathbb{R}^n)$ by a sequence $(\varphi_k^{(n)})_{k\in\mathbb{N}}$ in $C_0^\infty(\mathbb{R}^n)$: it follows that

$$\lim_{k\to\infty} \varphi_k^{(n)}(W_{t_1},\dots,W_{t_n}) = X_n, \qquad \text{in } L^2(\Omega,P),$$

and this concludes the proof. □

Lemma 10.9 *The space of the linear combinations of random variables of the form*

$$Z^\lambda = \exp\left(-\int_0^T \lambda(t)\cdot dW_t - \frac{1}{2}\int_0^T |\lambda(t)|^2 dt\right),$$

where λ is a function in $L^\infty([0,T];\mathbb{R}^d)$, is dense in $L^2(\Omega,\mathcal{F}_T^W,P)$.

Proof. We prove the claim by verifying that, if

$$\langle X, Z^\lambda\rangle_{L^2(\Omega)} = \int_\Omega X Z^\lambda dP = 0, \tag{10.11}$$

for every $\lambda\in L^\infty([0,T])$, then $X=0$ a.s. As before, we consider only the case $d=1$.

By choosing a suitable piecewise constant function λ, from (10.11) we infer

$$F(\xi) := \int_\Omega e^{\xi_1 W_{t_1}+\dots+\xi_n W_{t_n}} X dP = 0, \tag{10.12}$$

for every $\xi\in\mathbb{R}^n$, $t_1,\dots,t_n\in[0,T]$ and $n\in\mathbb{N}$. Now we consider the extension of F on $\mathbb{C}^n$:

$$F(z) = \int_\Omega e^{z_1 W_{t_1}+\dots+z_n W_{t_n}} X dP, \qquad z\in\mathbb{C}^n,$$

and we observe that, by the analytic continuation principle and (10.12), $F \equiv 0$. Then, on the grounds of the inverse Fourier transform Theorem A.65, for every $\varphi \in C_0^\infty(\mathbb{R}^n)$ we have

$$\begin{aligned}\int_\Omega \varphi(W_{t_1},\dots,W_{t_n}) X dP &= \int_\Omega \left(\frac{1}{(2\pi)^n} \int_{\mathbb{R}^n} e^{\xi_1 W_{t_1}+\dots+\xi_n W_{t_n}} \hat{\varphi}(-\xi) d\xi \right) X dP \\ &= \frac{1}{(2\pi)^n} \int_{\mathbb{R}^n} \hat{\varphi}(-\xi) \int_\Omega e^{\xi_1 W_{t_1}+\dots+\xi_n W_{t_n}} X dP d\xi = 0,\end{aligned}$$

and, by Lemma 10.8, this proves the claim. □

Proof (of Theorem 10.7). Concerning the uniqueness, if $u, v \in \mathbb{L}^2$ satisfy (10.10), then

$$\int_0^T (u_t - v_t) \cdot dW_t = 0$$

and therefore, by Corollary 4.13, u and v are $(m \otimes P)$-equivalent.

Concerning the existence, first of all we consider the case in which X is of the form

$$X = Z_T^\lambda = \exp\left(- \int_0^T \lambda(t) \cdot dW_t - \frac{1}{2} \int_0^T |\lambda(t)|^2 dt \right) \tag{10.13}$$

with $\lambda \in L^\infty([0,T])$ deterministic function. By the Itô formula we have

$$dZ_t^\lambda = -Z_t^\lambda \lambda(t) \cdot dW_t$$

hence

$$X = 1 - \int_0^T Z_t^\lambda \lambda(t) \cdot dW_t.$$

Further, by Lemma 10.1, since λ is a bounded function we have $\lambda Z^\lambda \in \mathbb{L}^2$; this proves (10.10) for X in (10.13).

Now, by Lemma 10.9 every $X \in L^2(\Omega, \mathcal{F}_T^W, P)$ can be approximated in L^2 by a sequence (X_n) of linear combinations of random variables of the type (10.13): therefore we have the representation

$$X_n = E[X_n] + \int_0^T u_t^n \cdot dW_t \tag{10.14}$$

with $u^n \in \mathbb{L}^2$. By Itô isometry we have

$$E\left[(X_n - X_m)^2\right] = (E[X_n - X_m])^2 + E\left[\int_0^T |u_t^n - u_t^m|^2 \, dt\right],$$

hence (u^n) is a Cauchy sequence in $\mathbb{L}^2(\mathcal{F}^W)$ and therefore it is convergent. Taking the limit in (10.14) as $n \to \infty$ we have the claim. □

Remark 10.10 By using Malliavin calculus, in Section 16.2.1 we obtain the expression of the process u in (10.10) in terms of conditional expectation of the stochastic derivative of X. □

Theorem 10.11 *Let $(M_t)_{t\in[0,T]}$ be a $\mathcal{F}^W$-martingale such that*

$$M_T \in L^2(\Omega, \mathcal{F}_T^W).$$

Then there exists a unique (up to $(m \otimes P)$-equivalence) process $u \in \mathbb{L}^2(\mathcal{F}^W)$ such that

$$M_t = M_0 + \int_0^t u_s \cdot dW_s \qquad a.s. \tag{10.15}$$

for any $t \in [0, T]$. In particular, every square integrable $\mathcal{F}^W$-martingale admits a continuous modification.

Proof. Since $M_T \in L^2(\Omega, \mathcal{F}_T^W)$, by Theorem 10.7 there exists $u \in \mathbb{L}^2(\mathcal{F}^W)$ such that

$$M_T = M_0 + \int_0^T u_s \cdot dW_s.$$

For a fixed $t \le T$, taking the conditional expectation, we have

$$M_t = E\left[M_T \mid \mathcal{F}_t^W\right] = M_0 + \int_0^t u_s \cdot dW_s, \qquad t \in [0, T].$$

□

Theorem 10.12 *Let $(M_t)_{t\in[0,T]}$ be a $\mathcal{F}^W$-local martingale. Then there exists a unique (up to $(m \otimes P)$-equivalence) process $u \in \mathbb{L}^2_{\text{loc}}(\mathcal{F}^W)$ such that*

$$M_t = M_0 + \int_0^t u_s \cdot dW_s, \qquad t \in [0, T]. \tag{10.16}$$

Proof. Uniqueness follows from Proposition 5.3, that is from the uniqueness of the representation of an Itô process. Regarding the existence, we assume at first that M is continuous: by Remark 4.38, there exists a localizing sequence (τ_n) such that (M^{τ_n}) is a sequence of continuous and bounded martingales. Then, by Theorem 10.11, there exists a sequence (u^n) in $\mathbb{L}^2(\mathcal{F}^W)$ such that

$$M_t^{\tau_n} = M_{t\wedge\tau_n} = M_0 + \int_0^t u_s^n \cdot dW_s, \qquad t \in [0, T]. \tag{10.17}$$

Now

$$M_t^{\tau_n} = M_t^{\tau_{n+1}} \qquad \text{on } \{t \le \tau_n\},$$

and so, by the uniqueness result of the Theorem 10.11 and by using an argument analogous to that in Paragraph 4.4, the definition

$$u_t \mathbb{1}_{\{t\le\tau_n\}} = u_t^n, \qquad t \in [0, T]$$

is well-posed; furthermore $u \in \mathbb{L}^2_{\text{loc}}$ and by (10.17) we get (10.16).

Now we prove that every local martingale M admits a continuous modification. Initially we consider the case of a martingale M: since $M_T \in L^1(\Omega, P)$ and $L^2(\Omega, P)$ is dense in $L^1(\Omega, P)$, there exists a sequence (X_n) of $\mathcal{F}_T^W$-measurable and square-integrable random variables such that

$$\|X_n - M_T\|_{L^1} \leq \frac{1}{2^n}, \qquad n \in \mathbb{N}.$$

By Theorem 10.11 the sequence of martingales

$$M_t^n := E\left[X_n \mid \mathcal{F}_t^W\right], \qquad t \in [0, T],$$

admits a continuous modification. By the maximal inequality (cf. Theorem 9.28), applied to the super-martingale $-|M_t - M_t^n|$, we have

$$P\left(\sup_{t\in[0,T]} |M_t - M_t^n| \geq \frac{1}{k}\right) \leq kE\left[|M_T - X_n|\right] \leq \frac{k}{2^n},$$

and so by the Borel-Cantelli lemma[2] we infer that (M_n) converges uniformly a.s. to M which therefore is continuous a.s.

Finally, if M is a local martingale, we consider a localizing sequence (τ_n): as we have just seen, M^{τ_n} admits a continuous modification, hence

$$M = M^{\tau_n} \qquad \text{on } \{t \leq \tau_n\},$$

is continuous and, since $n \in \mathbb{N}$ is arbitrary, we have proved the claim. □

We conclude the paragraph by proving that the representation result for Brownian martingales holds true also after a Girsanov type change of measure.

Theorem 10.13 *Under the assumptions of Girsanov's Theorem 10.5, if M is a local martingale in $(\Omega, \mathcal{F}, Q, (\mathcal{F}_t^W))$, then there exists a unique (up to $(m \otimes P)$-equivalence) $u \in \mathbb{L}^2_{\text{loc}}(\mathcal{F}^W)$ such that*

$$M_t = M_0 + \int_0^t u_s \cdot dW_s^\lambda, \qquad t \in [0, T],$$

where W^λ is the Q-Brownian motion defined in (10.8).

---

[2] Given a sequence (A_n) of events and putting

$$A = \bigcap_{n\geq 1} \bigcup_{k \geq n} A_k,$$

if

$$\sum_{n\geq 1} P(A_n) < \infty,$$

then $P(A) = 0$.

Proof. As usual we can always use the localization argument as in the proof of Theorem 10.12, so it is enough to consider the case M is a martingale. We note that, since M is a Q-martingale with respect to $\mathcal{F}^W$ which is the natural filtration for W and not for W^λ, we cannot apply Theorem 10.12 directly.

By Lemma 10.3, the process $Y := MZ$, where $Z = Z^\lambda$ is the exponential martingale defining Q, is a P-martingale and so

$$Y_t = M_0 + \int_0^t v_s \cdot dW_s, \qquad t \in [0,T],$$

where $v \in \mathbb{L}^2_{\text{loc}}$. We observe that[3]

$$\begin{aligned} dZ_t^{-1} &= d\exp\left(\int_0^t \lambda_s \cdot dW_s + \frac{1}{2}\int_0^t |\lambda_s|^2 ds\right) \\ &= Z_t^{-1}\left(\lambda_t \cdot dW_t + |\lambda_t|^2 dt\right) = Z_t^{-1}\lambda_t \cdot dW_t^\lambda, \end{aligned} \tag{10.18}$$

and so by the Itô formula we have

$$\begin{aligned} dM_t &= d\left(Y_t Z_t^{-1}\right) = Y_t dZ_t^{-1} + Z_t^{-1} dY_t + d\langle Y, Z^{-1}\rangle_t \\ &= Z_t^{-1}\left(Y_t\lambda_t \cdot dW_t^\lambda + v_t \cdot dW_t + v_t \cdot \lambda_t dt\right) \\ &= Z_t^{-1}\left(Y_t\lambda_t + v_t\right) \cdot dW_t^\lambda. \end{aligned}$$

Therefore we have proved the claim with

$$u = Z^{-1}\left(Y\lambda + v\right). \qquad \square$$

10.1.4 Change of drift

Let W be a d-dimensional standard Brownian motion on a probability space $(\Omega, \mathcal{F}, P)$ endowed with the Brownian filtration $\left(\mathcal{F}_t^W\right)$. We combine the results of the previous sections and examine how a change of measure from P to an equivalent measure Q, modifies the coefficients of an Itô process.

Theorem 10.14 (Change of drift) *Let Q be a probability measure equivalent to P. The Radon-Nikodym derivative of Q with respect to P is an exponential martingale*

$$\frac{dQ}{dP}|_{\mathcal{F}_t^W} = Z_t^\lambda, \qquad dZ_t^\lambda = -Z_t^\lambda \lambda_t \cdot dW_t,$$

with $\lambda \in \mathbb{L}^2_{\text{loc}}$ and the process W^λ, defined by

$$dW_t = dW_t^\lambda - \lambda_t dt, \tag{10.19}$$

is a Brownian motion on $(\Omega, \mathcal{F}, Q, (\mathcal{F}_t^W))$.

---

[3] Note the analogy between formulas (10.18) and (10.2).

Proof. We denote by Z the density process of Q with respect to P (cf. Example 3.34):

$$Z_t = E^P\left[\frac{dQ}{dP} \mid \mathcal{F}_t^W\right] = \frac{dQ}{dP}|_{\mathcal{F}_t^W}, \qquad t \in [0,T].$$

Since $Q \sim P$, the process Z is a positive P-martingale. Then, by the martingale representation Theorem 10.12, there exists a unique d-dimensional process $u \in \mathbb{L}^2_{\mathrm{loc}}(\mathcal{F}^W)$ such that

$$dZ_t = u_t \cdot dW_t,$$

or equivalently

$$dZ_t = -Z_t\lambda_t \cdot dW_t,$$

where λ is the process defined by

$$\lambda_t = -\frac{u_t}{Z_t}, \qquad t \in [0,T]. \tag{10.20}$$

Note that λ belongs to $\mathbb{L}^2_{\mathrm{loc}}$ because $u \in \mathbb{L}^2_{\mathrm{loc}}$ and Z is positive and continuous. Hence Z is the exponential martingale associated to λ. Moreover, since by construction Z is a strict martingale, by Girsanov's theorem we infer that W^λ in (10.19) is a Brownian motion on $(\Omega, \mathcal{F}, Q, (\mathcal{F}_t^W))$. □

Remark 10.15 Let X be an N-dimensional Itô process of the form

$$dX_t = b_t dt + \sigma_t dW_t.$$

Under the assumptions of Theorem 10.14, the Q-dynamics of X is given by

$$dX_t = (b_t - \sigma_t\lambda_t)\, dt + \sigma_t dW_t^\lambda. \tag{10.21}$$

We emphasize the fundamental feature of the changes of measure: *a change of measure only affects the drift coefficient of the process X; the diffusion coefficient (or volatility) does not vary.* □

10.2 Arbitrage theory

In this paragraph we study the problem of pricing European derivatives in a continuous-time market model. First of all, we fix the assumptions that are going to hold in the rest of the chapter: we consider a market with N risky assets and d sources of risk that are represented by a d-dimensional *correlated* Brownian motion $W = (W^1, \dots, W^d)$ on the probability space $(\Omega, \mathcal{F}, P)$, endowed with the Brownian filtration $(\mathcal{F}_t^W)$. For simplicity, we only consider the case of constant correlation matrix even if all the following results can be extended to the more general case of stochastic correlation (see Remark 10.23 below). Thus, we assume that

$$W_t = A\bar{W}_t \tag{10.22}$$

where $\bar{W}$ is a standard d-dimensional Brownian motion and $A = \left(A^{ij}\right)_{i,j=1,\dots,d}$ is a *non-singular* $d \times d$ constant matrix. We denote by $\varrho = AA^*$ the correlation matrix and, for any $i = 1, \dots, d$, we assume that

$$\varrho_t^{ii} = \sum_{j=1}^{d} \left(A_t^{ij}\right)^2 = 1, \qquad t \in [0, T] \quad \text{a.s.}$$

Then, by Corollary 5.35,

$$W_t^i = \sum_{j=1}^{d} A_t^{ij} \bar{W}_t^j, \qquad i = 1, \dots, d,$$

is a standard 1-dimensional Brownian motion and the covariance processes are given by

$$d\langle W^i, W^j \rangle_t = \varrho^{ij} dt, \qquad i, j = 1, \dots, d.$$

Example 10.16 In the case $d = 2$, we typically assume

$$A = \begin{pmatrix} 1 & 0 \\ \bar{\varrho} & \sqrt{1 - \bar{\varrho}^2} \end{pmatrix}$$

where $\bar{\varrho} \in]-1, 1[$. Then $W_t = A\bar{W}_t$ is a correlated Brownian motion with non-singular correlation matrix

$$\varrho = \begin{pmatrix} 1 & \bar{\varrho} \\ \bar{\varrho} & 1 \end{pmatrix}.$$

In general, for $d \geq 3$, A can be obtained from ϱ by the Cholesky decomposition algorithm (see, for example, [263]). □

We assume that the number N of risky assets is less or equal to the number d of risk factors, that is

$$N \leq d.$$

We give grounds for this last choice in Example 10.36 and the argument preceding it. Intuitively the idea is that, if $N > d$ then we have two possibilities: the market admits arbitrage opportunities or some assets are "redundant", i.e. they can be replicated by using only d "primitive" assets among the N traded assets.

We denote by $S = \left(S^1, \dots, S^N\right)$ the price process where S_t^i is the price at time $t \in [0, T]$ of the i-th risky asset. We suppose that

$$S_t^i = e^{X_t^i}$$

where X^i is an Itô process of the form

$$dX_t^i = b_t^i dt + \sigma_t^i dW_t^i, \qquad i = 1, \dots, N, \tag{10.23}$$

with $b \in \mathbb{L}^1_{\text{loc}}$ and σ^i is a positive process in $\mathbb{L}^2_{\text{loc}}$.

Remark 10.17 By assumption, the processes b and σ are progressively measurable with respect to the Brownian filtration $\mathcal{F}^W$: in particular, the dynamics of the N assets depends on the d-dimensional Brownian motion W and, apart from trivial cases, the filtration $\mathcal{F}^S$ of the assets coincides with $\mathcal{F}^W$. □

Equation (10.23) can be rewritten in compact form as

$$dX_t = b_t dt + \sigma_t dW_t$$

where $b = (b^1, \dots, b^N)$ and σ is the $(N \times d)$-matrix valued process

$$\sigma_t = \begin{pmatrix} \sigma_t^1 & 0 & \cdots & 0 & 0 & 0 & \cdots & 0 \\ 0 & \sigma_t^2 & \cdots & 0 & 0 & 0 & \cdots & 0 \\ \vdots & \vdots & \ddots & \vdots & \vdots & \vdots & \ddots & \vdots \\ 0 & 0 & \cdots & \sigma_t^{N-1} & 0 & 0 & \cdots & 0 \\ 0 & 0 & \cdots & 0 & \sigma_t^N & 0 & \cdots & 0 \end{pmatrix}. \tag{10.24}$$

By the Itô formula we have

$$dS_t^i = \mu_t^i S_t^i dt + \sigma_t^i S_t^i dW_t^i, \qquad i = 1, \dots, N, \tag{10.25}$$

where $\mu^i = b^i + \frac{(\sigma^i)^2}{2}$, or equivalently

$$S_t^i = S_0^i \exp\left(\int_0^t \sigma_s^i dW_s^i + \int_0^t \left(\mu_s^i - \frac{\left(\sigma_s^i\right)^2}{2} \right) ds \right). \tag{10.26}$$

Concerning the locally non-risky asset B, we suppose it satisfies the equation

$$dB_t = r_t B_t dt, \qquad B_0 = 1,$$

with $r \in \mathbb{L}^1_{\text{loc}}$, that is we have

$$B_t = e^{\int_0^t r_s ds}, \qquad t \in [0, T]. \tag{10.27}$$

We remark that, although B represents the "non-risky" asset, it is a stochastic process because r is $\mathcal{F}^W$-progressively measurable: however B has bounded variation (cf. Example 3.60-iii)) and null quadratic variation so, intuitively, it possesses a smaller degree of randomness with respect to the other risky assets.

To simplify the exposition, we also assume some further integrability condition on the coefficients: we assume that r and σ^i, $i = 1, \dots, N$, verify the estimate (10.3); more explicitly, we suppose that

$$\int_0^T r_t^2 dt + \sum_{i=1}^N \int_0^T \left(\sigma_t^i\right)^2 dt \le C \qquad \text{a.s.} \tag{10.28}$$

for some positive constant C.

Remark 10.18 Under condition (10.28), the process B is bounded and strictly positive since, by Hölder's inequality, we have

$$\left| \int_0^t r_s ds \right| \le \sqrt{t} \left(\int_0^t r_s^2 ds \right)^{\frac{1}{2}} \le \sqrt{CT}, \qquad t \in [0,T], \quad \text{a.s.} \qquad \square$$

Summing up, we consider a market model where:

Hypothesis 10.19

- *There are N risky assets $S^1, \dots, S^N$ and one no-risky asset B whose dynamics is given by (10.25) and (10.27) respectively;*
- *W is a d-dimensional Brownian motion with constant correlation matrix ϱ and $d \ge N$;*
- *the processes σ, r satisfy the integrability condition (10.28) and the processes σ^i are positive.*

10.2.1 Change of drift with correlation

We provide some results on the change of measure that are preliminary to the discussion of the concept of equivalent martingale measure. Specifically, we extend the results of Section 10.1.4 on the change of drift, to the case of the *correlated* Brownian motion in (10.22).

Theorem 10.20 (Change of drift with correlation) *For any probability measure Q equivalent to P there exists a process $\lambda \in \mathbb{L}^2_{\text{loc}}$ such that*

$$\frac{dQ}{dP}|_{\mathcal{F}_t^W} = Z_t \qquad \text{and} \qquad dZ_t = -Z_t \lambda_t \cdot dW_t.$$

Moreover the process W^λ defined by

$$dW_t = dW_t^\lambda - \varrho \lambda_t dt \tag{10.29}$$

is a Brownian motion on $(\Omega, \mathcal{F}, Q, (\mathcal{F}_t^W))$ with correlation matrix ϱ.

Proof. By the martingale representation Theorem 10.12 for the standard Brownian motion $\bar{W}$, there exists a d-dimensional process $\bar{\lambda} \in \mathbb{L}^2_{\text{loc}}(\mathcal{F}^W)$ such that $\frac{dQ}{dP}|_{\mathcal{F}_t^W} = Z_t$ and

$$dZ_t = -Z_t \bar{\lambda}_t \cdot d\bar{W}_t = -Z_t \bar{\lambda}_t \cdot \left(A^{-1} dW_t\right) = -Z_t \lambda_t \cdot dW_t,$$

where[4] $\lambda_t = \left(A^{-1}\right)^* \bar{\lambda}_t$. We remark that, by the Itô formula we have

$$Z_t = \exp\left(-\int_0^t \lambda_s \cdot dW_s - \frac{1}{2} \int_0^t \langle \varrho \lambda_s, \lambda_s \rangle ds \right). \tag{10.30}$$

[4] Note that

$$\langle \bar{\lambda}_t, A^{-1} dW_t \rangle = \langle \left(A^{-1}\right)^* \bar{\lambda}_t, dW_t \rangle.$$

By Girsanov's theorem, the process $\bar{W}^{\bar{\lambda}}$ defined by

$$d\bar{W}_t^{\bar{\lambda}} = d\bar{W}_t + \bar{\lambda}_t dt, \qquad t \in [0,T],$$

is a standard Q-Brownian motion: since $\bar{\lambda} = A^*\lambda$, multiplying by A in the previous equation we infer that

$$dW_t^{\lambda} := A d\bar{W}_t^{\bar{\lambda}} = dW_t + \varrho\lambda_t dt$$

is a correlated Q-Brownian motion with correlation matrix ϱ. □

Remark 10.21 Under the assumptions of Theorem 10.20, let X be an N-dimensional Itô process of the form

$$dX_t = b_t dt + \sigma_t dW_t. \tag{10.31}$$

Then the Q-dynamics of X is given by

$$dX_t = (b_t - \sigma_t \varrho\lambda_t)\, dt + \sigma_t dW_t^{\lambda}. \tag{10.32}$$

□

Remark 10.22 It is sometimes useful to rephrase Theorem 10.20 as follows: *if Q is a probability measure equivalent to P then there exists a process $\lambda \in \mathbb{L}^2_{\text{loc}}$ such that*

$$\frac{dQ}{dP}|_{\mathcal{F}_t^W} = Z_t \qquad \textit{and} \qquad dZ_t = -Z_t \left(\varrho^{-1}\lambda_t\right) \cdot dW_t.$$

Moreover the process W^{λ} defined by

$$dW_t = dW_t^{\lambda} - \lambda_t dt \tag{10.33}$$

is a Brownian motion on $(\Omega, \mathcal{F}, Q, (\mathcal{F}_t^W))$ with correlation matrix ϱ. □

Remark 10.23 It is possible to extend the previous results to the case of a stochastic correlation matrix: specifically, if we assume that ϱ has bounded variation, that is

$$d\varrho_t = \theta_t dt \tag{10.34}$$

with $\theta \in \mathbb{L}^1_{\text{loc}}$, then we have the following formula for the change of drift

$$dW_t = dW_t^{\lambda} - \left(\varrho_t\lambda_t + \theta_t \int_0^t \lambda_s ds\right) dt \tag{10.35}$$

that generalizes formula (10.29). In particular, for the Itô process X in (10.31), we have

$$dX_t = \left(b_t - \sigma_t\left(\varrho_t\lambda_t + \theta_t \int_0^t \lambda_s ds\right)\right) dt + \sigma_t dW_t^{\lambda}.$$ □

10.2.2 Martingale measures and market prices of risk

The concept of EMM plays a central role in the theory of financial derivatives. As already seen in the discrete case, it gives a characterization of arbitrage-free markets and allows us to introduce the risk-neutral or arbitrage price of replicable derivatives (see Section 10.2.5).

Definition 10.24 *An equivalent martingale measure Q with numeraire B is a probability measure on $(\Omega, \mathcal{F})$ such that*

i) Q is equivalent to P;
ii) the process of discounted prices

$$\widetilde{S}_t = e^{-\int_0^t r_s ds} S_t, \qquad t \in [0,T],$$

is a strict[5] Q-martingale. In particular, the risk-neutral pricing formula

$$S_t = E^Q\left[e^{-\int_t^T r_s ds} S_T \mid \mathcal{F}_t^W\right], \qquad t \in [0,T], \tag{10.36}$$

holds.

Now we consider an EMM Q and we use Theorem 10.20, in the form of Remark 10.22, to find the Q-dynamics of the price process: we recall that there exists a process $\lambda = \left(\lambda^1, \dots, \lambda^d\right) \in \mathbb{L}^2_{\text{loc}}$ such that

$$\frac{dQ}{dP}|_{\mathcal{F}_t^W} = Z_t \tag{10.37}$$

where Z solves

$$dZ_t = -Z_t \left(\varrho^{-1}\lambda_t\right) \cdot dW_t, \qquad Z_0 = 1. \tag{10.38}$$

Moreover the process $W^\lambda = \left(W^{\lambda,1}, \dots, W^{\lambda,d}\right)$ defined by

$$dW_t = dW_t^\lambda - \lambda_t dt \tag{10.39}$$

is a Q-Brownian motion with correlation matrix ϱ. Therefore, for $i = 1, \dots, N$ we have

$$\begin{aligned} d\widetilde{S}_t^i &= \left(\mu_t^i - r_t\right) \widetilde{S}_t^i dt + \sigma_t^i \widetilde{S}_t^i dW_t^i \\ &= \left(\mu_t^i - r_t\right) \widetilde{S}_t^i dt + \sigma_t^i \widetilde{S}_t^i \left(dW_t^{\lambda,i} - \lambda_t^i dt\right) \\ &= \left(\mu_t^i - r_t - \sigma_t^i \lambda_t^i\right) \widetilde{S}_t^i dt + \sigma_t^i \widetilde{S}_t^i dW_t^{\lambda,i}. \end{aligned} \tag{10.40}$$

---

[5] We assume that $\widetilde{S}$ is a *strict*, not simply a *local*, Q-martingale. This subtle distinction is necessary because pathologies may arise when the discounted prices process is a local martingale: we refer to Sin [312] where interesting examples are analyzed.

Now we recall that an Itô process is a local martingale if and only if it has null drift (cf. Remark 5.4); since Q is an EMM, the following *drift condition* necessarily holds[6]

$$\lambda_t^i = \frac{\mu_t^i - r_t}{\sigma_t^i}, \qquad i = 1, \dots, N. \tag{10.41}$$

Due to the strong analogy with the concept of market price of risk introduced in Section 7.3.4 (see in particular formula (7.51)), we give the following:

Definition 10.25 *A market price of risk is a d-dimensional process $\lambda \in \mathbb{L}^2_{\mathrm{loc}}$ such that:*

i) the first N components of λ are given by (10.41)*;*
ii) the solution Z to the SDE (10.38) *is a strict P-martingale.*

The drift condition (10.41) fixes the first N components of the d-dimensional process λ: when $N < d$, it is clear that the market price of risk is not uniquely determined. The following result shows the strict relationship between EMMs and market prices of risk.

Theorem 10.26 *Formulas* (10.37)-(10.38) *establish a one-to-one correspondence between EMMs and market prices of risk. The dynamics, under an EMM Q, of the asset prices is given by*

$$dS_t^i = r_t S_t^i dt + \sigma_t^i S_t^i dW_t^{\lambda,i}, \tag{10.42}$$

where $W^\lambda = \left(W^{\lambda,1}, \dots, W^{\lambda,d}\right)$ is the Q-Brownian motion in (10.39)*. Moreover*

$$E^Q\left[\sup_{0 \le t \le T} |S_t|^p\right] < \infty, \tag{10.43}$$

for every $p \ge 1$.

Proof. We have already proved, by using Theorem 10.20, that any EMM Q defines a market price of risk λ such that (10.42) holds.

Conversely, if λ is a market price of risk, we consider the process Z in (10.38) and using that Z is a P-martingale, we define the measure Q by putting $\frac{dQ}{dP} = Z_T$. Then Q is an EMM: indeed, by Girsanov's theorem for correlated Brownian motions (cf. Remark 10.22), W^λ in (10.39) is a correlated Brownian motion on $(\Omega, \mathcal{F}, Q, (\mathcal{F}_t^W))$ and, by Remark 10.4, P and Q are equivalent. Further, by (10.41) we directly get (10.42) and therefore $\widetilde{S}^i$ is the exponential martingale

$$\widetilde{S}_t^i = \exp\left(\int_0^t \sigma_s^i dW_s^{\lambda,i} - \int_0^t \frac{\left(\sigma_s^i\right)^2}{2} ds\right). \tag{10.44}$$

[6] More precisely, we have

$$\lambda_t^i(\omega) = \frac{\mu_t^i(\omega) - r_t(\omega)}{\sigma_t^i(\omega)}$$

for almost all $(t, \omega) \in [0, T] \times \Omega$.

Since σ^i and r verify the integrability condition (10.28), by Lemma 10.1 and Remark 10.18, we infer that $\widetilde{S}^i$ is a strict martingale and verifies (10.43). $\square$

Theorem 10.26 shows that the notions of EMM and market price of risk are equivalent. However, while the EMM is mainly a theoretical concept, on the contrary the market price of risk is of essential importance from the practical point of view: indeed, by the change of drift formula (10.39), it determines explicitly the change of measure and the dynamics of the assets as Itô processes. Moreover, the market price of risk is also an essential tool in the proof of the existence of an EMM. In the following theorem we give a simple condition on the coefficients μ, σ, r which guarantees the existence of a market price of risk and of the related EMM.

Theorem 10.27 (Existence of an EMM) *Assume that the processes*

$$\lambda_t^i = \frac{\mu_t^i - r_t}{\sigma_t^i}, \qquad i = 1, \dots, N,$$

verify the integrability condition (10.3), *that is*

$$\int_0^T \left|\lambda_t^i\right|^2 dt \le C \qquad a.s., \tag{10.45}$$

for some positive constant C. Then an EMM Q exists.

Proof. By Theorem 10.26, in order to show that an EMM exists, it suffices to construct a market price of risk. Let $\lambda \in \mathbb{L}^2$ be any d-dimensional process with the first N components defined by (10.41) and such that estimate (10.45) holds for any $i = 1, \dots, d$. By Lemma 10.1, Z in (10.38) is a strict P-martingale and therefore λ is a market price of risk. $\square$

Notation 10.28 *We denote by $\mathcal{Q}$ the family of EMMs whose corresponding process of market prices of risk λ verify estimate* (10.45) *for any $i = 1, \dots, d$.*

Remark 10.29 The class $\mathcal{Q}$ will play a central role in the sequel (see, for instance, Proposition 10.41). We remark that if $Q \in \mathcal{Q}$, then *the Radon-Nikodym derivative of Q with respect to P belongs to $\mathbb{L}^p(\Omega, P)$ for every $p \ge 1$*: more precisely, let us recall that

$$\frac{dQ}{dP} = Z_T$$

where

$$dZ_t = -Z_t \left(\varrho^{-1}\lambda_t\right) \cdot dW_t, \qquad Z_0 = 1,$$

and λ is the market price of risk associated to Q. If λ is is such that

$$\int_0^T |\lambda_t|^2 dt \le C \qquad \text{a.s.}$$

then, by Lemma 10.1, Z is a P-martingale and

$$E\left[\sup_{0\le t\le T} Z_t^p\right] < \infty, \qquad p \ge 1.$$

Clearly, under the assumptions of Theorem 10.27, the class $\mathcal{Q}$ is not empty.□

Remark 10.30 The following condition is stronger than (10.45), but simpler to check: the process r is bounded and the processes σ^i, $i = 1, \dots, N$, are bounded and uniformly positive, i.e. there exists $C \in \mathbb{R}_{>0}$ such that

$$|r_t| \le C, \qquad \frac{1}{C} \le |\sigma_t^i| \le C, \qquad i = 1, \dots, N, \quad t \in [0, T],$$

almost surely. □

10.2.3 Examples

Example 10.31 In the Black-Scholes market model $N = d = 1$ and the coefficients r, μ, σ are constant. In this case the market price of risk is uniquely determined by equation (10.41) and we have

$$\lambda = \frac{\mu - r}{\sigma}$$

which corresponds to the value found in Section 7.3.4. By Theorem 10.26, the process

$$W_t^\lambda = W_t + \lambda t, \qquad t \in [0, T],$$

is a Brownian motion under the measure Q defined by

$$\frac{dQ}{dP} = \exp\left(-\lambda W_T - \frac{\lambda^2}{2} T\right), \tag{10.46}$$

and the dynamics of the risky asset is

$$dS_t = rS_t dt + \sigma S_t dW_t^\lambda.$$

Moreover the discounted price process $\widetilde{S}_t = e^{-rt} S_t$ is a Q-martingale and we have

$$S_t = e^{-r(T-t)} E^Q\left[S_T \mid \mathcal{F}_t^W\right], \qquad t \in [0, T].$$

□

Example 10.32 In a market model where the number of risky assets is equal to the dimension of the Brownian motion, i.e. $N = d$, the drift condition (10.41) determines the process λ univocally. Therefore, under the assumptions of Theorem 10.27 we have that *the EMM Q exists and is unique*. As usual the Q-dynamics of the discounted prices is

$$d\widetilde{S}_t^i = \sigma_t^i \widetilde{S}_t^i dW_t^{\lambda,i}, \qquad i = 1, \dots, N,$$

where W^λ is the Q-Brownian motion defined by $dW_t = dW_t^\lambda - \lambda_t dt$. □

Example 10.33 In the Heston stochastic volatility model [165], there is an underlying asset ($N = 1$) whose volatility is a stochastic process that is driven by a second real Brownian motion ($d = 2$). More precisely, we assume that

$$dS_t = \mu S_t dt + \sqrt{\nu_t} S_t dW_t^1, \tag{10.47}$$

$$d\nu_t = k(\bar{\nu} - \nu_t)dt + \eta\sqrt{\nu_t} dW_t^2, \tag{10.48}$$

where $\mu, k, \bar{\nu}, \eta$ are constant parameters and W is a two-dimensional correlated Brownian motion. As in Example 10.16, we set

$$A = \begin{pmatrix} 1 & 0 \\ \bar{\varrho} & \sqrt{1-\bar{\varrho}^2} \end{pmatrix}$$

and

$$W_t^1 = \bar{W}_t^1, \qquad W_t^2 = \bar{\varrho}\bar{W}_t^1 + \sqrt{1-\bar{\varrho}^2}\bar{W}_t^2,$$

where $(\bar{W}^1, \bar{W}^2)$ is a two-dimensional standard Brownian motion and $\bar{\varrho} \in]-1, 1[$ is the (constant) correlation parameter. The interest rate r is supposed to be constant.

Equation (10.48) was previously suggested by Cox, Ingersoll and Ross [80] as a model for the short rate dynamics in a fixed-income market (cf. Section 10.3.1): here the process ν represents the variance of S. By the Itô formula, the solution of (10.47) is

$$S_t = S_0 \exp\left(\int_0^t \sqrt{\nu_s} dW_s^1 + \int_0^t \left(\mu - \frac{\nu_s}{2}\right) ds \right).$$

On the other hand, we remark that the existence and uniqueness results for SDEs in Chapter 9 do not apply to (10.48) because the diffusion coefficient is only Hölder continuous. However we have the following result (see, for instance, [177] p.168).

Theorem 10.34 *For any $\nu_0 \geq 0$, there exists a unique non-negative strong solution to* (10.48) *starting from ν_0.*

A solution to (10.48) is called a *mean reverting square root process*. For $k > 0$, the drift is positive if $\nu_t < \bar{\nu}$ and it is negative if $\nu_t > \bar{\nu}$ and so the process ν_t is "pushed" towards the value $\bar{\nu}$ that can be interpreted as a long-term mean. The other parameters represent respectively: μ the drift of S_t, k the speed of mean reversion and η the volatility of the variance. We remark that in general the solution ν can reach the origin: more precisely, let us denote by τ the stopping time defined by

$$\tau = \inf\{t \geq 0 \mid \nu_t = 0\}$$

and $\tau(\omega) = \infty$ if $\nu_t(\omega) > 0$ for any t. Then we have (see Feller [129] or Proposition 6.2.3 in [226]):

Proposition 10.35 *Assume that $\nu_0 > 0$. Then we have:*

- *if $k\bar{\nu} \geq \frac{\eta^2}{2}$ then $\tau = \infty$ a.s.;*
- *if $0 \leq k\bar{\nu} < \frac{\eta^2}{2}$ and $k < 0$ then $0 < P(\tau < \infty) < 1$;*
- *if $0 \leq k\bar{\nu} < \frac{\eta^2}{2}$ and $k \geq 0$ then $\tau < \infty$ a.s.*

In the notations introduced at the beginning of the paragraph, σ_t in (10.24) is the (1×2)-dimensional matrix

$$\sigma_t = \begin{pmatrix} \sqrt{\nu_t} & 0 \end{pmatrix}.$$

A market price of risk is a two-dimensional process $\lambda = (\lambda^1, \lambda^2) \in \mathbb{L}^2_{\text{loc}}$ such that, by (10.41),

$$\lambda^1_t = \frac{\mu - r}{\sqrt{\nu_t}}, \tag{10.49}$$

while there is no restriction on the second component λ^2 except for the fact that Z in (10.38) must be a martingale. If this is the case, we consider the corresponding EMM Q with respect to which the process W^λ, defined by (cf. (10.39))

$$dW_t = dW^\lambda_t - \lambda_t dt = dW^\lambda_t - \begin{pmatrix} \frac{\mu - r}{\sqrt{\nu_t}} \\ \lambda^2_t \end{pmatrix} dt,$$

is a 2-dimensional Brownian motion: then the Q-dynamics of the risky asset is given by

$$dS_t = rS_t dt + \sqrt{\nu_t} S_t dW^{\lambda,1}_t, \tag{10.50}$$

$$d\nu_t = \left(k(\bar{\nu} - \nu_t) - \eta\sqrt{\nu_t}\lambda^2_t\right) dt + \eta\sqrt{\nu_t} dW^{\lambda,2}_t. \tag{10.51}$$

This is the dynamics to be used in order to compute a risk neutral price in the form $E^Q[f(S_T)]$, for instance by Monte Carlo simulation (cf. Section 12.4): clearly, the result depends on the choice of the process λ^2, that is on the market price of risk. We also remark that, by taking the process λ^2 of the form $\lambda^2_t = \frac{a\nu_t + b}{\sqrt{\nu_t}}$ with $a, b \in \mathbb{R}$, then (10.51) reduces to

$$d\nu_t = \widetilde{k}(\widetilde{\nu} - \nu_t)dt + \eta\sqrt{\nu_t} dW^{\lambda,2}_t, \tag{10.52}$$

where

$$\widetilde{k} = k + \eta a, \qquad \widetilde{\nu} = \frac{k\bar{\nu} - \eta b}{k + \eta a}, \tag{10.53}$$

and therefore ν is a square root process under Q as well.

Since the log-characteristic function of S

$$E\left[e^{i\xi \log S_T}\right]$$

can be computed explicitly, analytical approximations of the price of European Calls and Puts are available by Fourier inversion techniques: this will be discussed in Examples 15.15 and 15.20.

From the economical point of view, *the price of risk* λ *is determined by the market:* in other terms, λ must be chosen on the basis of observations, by calibrating the parameters of the model to the available data. Once λ and the corresponding EMM Q have been selected, the risk neutral price of a derivative on S is defined as in Section 10.2.5 by a risk neutral formula under Q. Note that it is not possible in general to construct a hedging strategy based only on the underlying asset and the bond and therefore the Heston model is incomplete. □

By Theorem 10.26, the existence of an EMM implies that a market price of risk λ exists and verifies the drift condition (10.41): since λ is a d-dimensional process and (10.41) gives N constraints on λ, it is natural to assume $N \leq d$. If this is not the case, the market might admit arbitrage opportunities: the proof of this claim in a general setting goes beyond the scope of this exposition. This result, which is part of the first fundamental theorem of asset pricing, has been proved by many authors and under different hypotheses: we mention, among others, Stricker [317], Ansel and Stricker [10], Delbaen [85], Schweizer [304], Lakner [223], Delbaen and Schachermayer [86; 87; 88; 89; 90; 91], Frittelli and Lakner [143]. Here we confine ourselves to examine a simple example in which the arbitrage strategy can be constructed explicitly.

Example 10.36 We consider a market which consists of two geometric Brownian motions

$$dS_t^i = \mu^i S_t^i dt + \sigma^i S_t^i dW_t, \qquad i = 1, 2,$$

where W is a real Brownian motion: in this case $N = 2 > d = 1$. The drift condition (10.41) takes the form:

$$\begin{cases} \sigma^1 \lambda = \mu^1 - r, \\ \sigma^2 \lambda = \mu^2 - r, \end{cases}$$

and the system is solvable if and only if

$$\frac{\mu^1 - r}{\sigma^1} = \frac{\mu^2 - r}{\sigma^2}. \tag{10.54}$$

This is in line with what we had observed in Section 7.3.4, in particular with formula (7.52) which states that, in an arbitrage-free market, *all the assets must have the same market price of risk.* If (10.54) is not satisfied, the market admits arbitrage opportunities: indeed, let us suppose that

$$k := \frac{\mu^1 - r}{\sigma^1} - \frac{\mu^2 - r}{\sigma^2} > 0,$$

and let us consider the self-financing (cf. Definition 10.37 and Corollary 10.40) portfolio $(\alpha^1, \alpha^2, \beta)$ with null initial value, defined by

$$\alpha^i = \frac{1}{S_t^i \sigma^i}, \qquad i = 1, 2.$$

Then the value V of the portfolio verifies

$$\begin{aligned} dV_t &= \alpha_t^1 dS_t^1 + \alpha_t^2 dS_t^2 + r\left(V_t - \alpha_t^1 S_t^1 - \alpha_t^2 S_t^2\right) dt \\ &= \frac{\mu^1 - r}{\sigma^1} dt + dW_t - \frac{\mu^2 - r}{\sigma^2} dt - dW_t + rV_t dt \\ &= (rV_t + k)\, dt, \end{aligned}$$

and therefore it gives rise to an arbitrage opportunity since it produces a certain profit that is strictly greater than the bond. □

10.2.4 Admissible strategies and arbitrage opportunities

We consider a standard market (S, B) under Hypothesis 10.19, where

$$dS_t^i = \mu_t^i S_t^i dt + S_t^i \sigma_t^i dW_t^i, \qquad i = 1, \dots, N,$$

with W d-dimensional correlated Brownian motion. Moreover the coefficients μ, σ satisfy condition (10.45) of Theorem 10.27 that guarantees the existence of an EMM.

We introduce the family of self-financing strategies.

Definition 10.37 *A strategy (or portfolio) is a process (α, β) where $\alpha, \beta \in \mathbb{L}^1_{\text{loc}}$ have values in $\mathbb{R}^N$ and in $\mathbb{R}$, respectively. The value of the strategy (α, β) is the real-valued process*

$$V_t^{(\alpha,\beta)} = \alpha_t \cdot S_t + \beta_t B_t = \sum_{i=1}^{N} \alpha_t^i S_t^i + \beta_t B_t, \qquad t \in [0, T].$$

A strategy (α, β) is self-financing if

$$dV_t = \alpha_t \cdot dS_t + \beta_t dB_t. \tag{10.55}$$

Remark 10.38 In the definition of self-financing strategy (α, β), we implicitly assume that

$$\alpha^i \sigma^i \in \mathbb{L}^2_{\text{loc}}, \qquad i = 1, \dots, N. \tag{10.56}$$

This condition ensures that the stochastic integrals

$$\alpha_t \cdot dS_t = \sum_{i=1}^{N} \alpha_t^i \sigma_t^i S_t^i dW_t^i$$

are well-defined. □

The following result extends a useful characterization of self-financing strategies already proved in the Black-Scholes framework. As usual, we denote the discounted values by $\widetilde{S}$ and $\widetilde{V}$.

Proposition 10.39 *A strategy (α, β) is self-financing if and only if*

$$d\widetilde{V}_t^{(\alpha,\beta)} = \alpha_t \cdot d\widetilde{S}_t. \tag{10.57}$$

Moreover, a self-financing strategy (α, β) is determined by its initial value V_0 and by the process α of the amount of risky assets held in portfolio: indeed, for any $V_0 \in \mathbb{R}$ and $\alpha \in \mathbb{L}^1_{\text{loc}}$ verifying (10.56), there exists a unique self-financing strategy (α, β) such that $V_0^{(\alpha,\beta)} = V_0$.

Proof. We have

$$d\widetilde{V}_t^{(\alpha,\beta)} = e^{-\int_0^t r_s ds} \left(-r_t V_t^{(\alpha,\beta)} dt + dV_t^{(\alpha,\beta)} \right) =$$

(by the self-financing property (10.55))

$$= e^{-\int_0^t r_s ds} \left(-r_t V_t^{(\alpha,\beta)} dt + \alpha_t \cdot dS_t + r_t \beta_t B_t dt \right) =$$

(since $V_t^{(\alpha,\beta)} - \beta_t B_t = \alpha_t \cdot S_t$)

$$= e^{-\int_0^t r_s ds} \left(-r_t \alpha_t \cdot S_t dt + \alpha_t \cdot dS_t \right) = \alpha_t \cdot d\widetilde{S}_t,$$

and this proves (10.57).

To prove the second part of the thesis we proceed as in the Black-Scholes case and, for any $V_0 \in \mathbb{R}$ and $\alpha \in \mathbb{L}^1_{\text{loc}}$ such that (10.56) holds, we define the processes V and β by putting

$$e^{-\int_0^t r_s ds} V_t = V_0 + \int_0^t \alpha_s \cdot d\widetilde{S}_s, \qquad \beta_t = B_t^{-1} \left(V_t - \alpha_t \cdot S_t \right), \qquad t \in [0, T].$$

Then, by (10.57), (α, β) is a self-financing strategy such that $V_t^{(\alpha,\beta)} = V_t$ for $t \in [0, T]$. □

A simple consequence of Proposition 10.39 is the following:

Corollary 10.40 *Let Q be an EMM with associated Q-Brownian motion $W^\lambda = \left(W^{\lambda,1}, \dots, W^{\lambda,d} \right)$ defined by (10.37)-(10.39). For any self-financing strategy (α, β), we have*

$$\widetilde{V}_t^{(\alpha,\beta)} = V_0^{(\alpha,\beta)} + \sum_{i=1}^N \int_0^t \alpha_s^i \widetilde{S}_s^i \sigma_s^i dW_s^{\lambda,i}, \tag{10.58}$$

and in particular $\widetilde{V}^{(\alpha,\beta)}$ is a local Q-martingale.

Proof. The thesis follows from (10.57) and Theorem 4.42. □

The following proposition gives very natural conditions under which $\widetilde{V}^{(\alpha,\beta)}$ is a *strict* Q-martingale. The result is non trivial since the integrability condition (10.59), which replaces (10.56), is given *under the objective probability P* and therefore is independent on the selected EMM. Let us first recall Notation 10.28 of the family $\mathcal{Q}$ of EMMs.

Proposition 10.41 *If Q is an EMM in $\mathcal{Q}$ and (α,β) be a self-financing strategy such that*

$$\alpha^i\sigma^i \in \mathbb{L}^2(\Omega,P), \qquad i=1,\dots,N, \tag{10.59}$$

then $\widetilde{V}^{(\alpha,\beta)}$ is a strict Q-martingale. In particular the following risk-neutral pricing formula holds:

$$V_t^{(\alpha,\beta)} = E^Q\left[e^{-\int_t^T r_s ds} V_T^{(\alpha,\beta)} \mid \mathcal{F}_t^W\right], \qquad t\in[0,T]. \tag{10.60}$$

Proof. By (10.58) and Corollary 4.48, if

$$E^Q\left[\left(\int_0^T \left(\alpha_t^i\sigma_t^i\widetilde{S}_t^i\right)^2 dt\right)^{\frac{1}{2}}\right] < \infty,$$

for every $i=1,\dots,N$, then $\widetilde{V}^{(\alpha,\beta)}$ is a Q-martingale. We recall that under our main Hypothesis 10.19, B and B^{-1} are bounded processes (cf. Remark 10.18); therefore in order to prove the thesis it is enough to verify that

$$E^Q\left[Y_i^{\frac{1}{2}}\right] < \infty, \qquad i=1,\dots,N,$$

where

$$Y_i = \int_0^T \left(\alpha_t^i\sigma_t^iS_t^i\right)^2 dt.$$

Now we use the fact that $Q\in\mathcal{Q}$ and therefore, by Remark 10.29, the Radon-Nikodym derivative Z of Q with respect to P belongs to $\mathbb{L}^p(\Omega,P)$ for every $p\geq 1$. Given two conjugate exponents q,q' with $1<q<2$, by Hölder's inequality we get

$$E^Q\left[Y_i^{\frac{1}{2}}\right] = E^P\left[Y_i^{\frac{1}{2}}Z_T\right] \leq E^P\left[Y_i^{\frac{q}{2}}\right]^{\frac{1}{q}} E^P\left[Z_T^{q'}\right]^{\frac{1}{q'}},$$

and we conclude by verifying that

$$E^P\left[Y_i^{\frac{q}{2}}\right] < \infty.$$

We have

$$E^P\left[Y_i^{\frac{q}{2}}\right] \leq E^P\left[\left(\int_0^T \left(\alpha_t^i\sigma_t^i\right)^2 dt\right)^{\frac{q}{2}} \sup_{t\in[0,T]}|S_t|^q\right] \leq$$

(by Hölder's inequality)

$$\leq E^P\left[\int_0^T \left(\alpha_t^i\sigma_t^i\right)^2 dt\right]^{\frac{q}{2}} E^P\left[\sup_{t\in[0,T]}|S_t|^{\frac{2q}{2-q}}\right]^{\frac{2-q}{2}} < \infty$$

by the assumption on α and estimate (10.43). □

Definition 10.42 *A self-financing strategy* (α, β) *such that* $\widetilde{V}^{(\alpha,\beta)}$ *is a* Q*-martingale for every* $Q \in \mathcal{Q}$*, is called an admissible strategy. We denote by* $\mathcal{A}$ *the collection of all admissible strategies.*

Proposition 10.41 guarantees that the family $\mathcal{A}$ is not empty: indeed, any self-financing strategy (α, β) verifying condition (10.59) is admissible.

As in the discrete case, it is immediate to verify that an admissible strategy cannot be an arbitrage: in particular, the collection of self-financing strategies (α, β) with $\alpha \in \mathbb{L}^2(\Omega, P)$ does not contain arbitrage portfolios. Indeed we have the following version of the no-arbitrage principle.

Corollary 10.43 (No-arbitrage principle) *If an EMM in* $\mathcal{Q}$ *exists and* $(\alpha, \beta), (\alpha', \beta')$ *are admissible self-financing strategies such that*

$$V_T^{(\alpha,\beta)} = V_T^{(\alpha',\beta')} \qquad P\text{-a.s.}$$

then $V^{(\alpha,\beta)}$ *and* $V^{(\alpha',\beta')}$ *are indistinguishable.*

Proof. If $Q \in \mathcal{Q}$ exists and $(\alpha, \beta), (\alpha', \beta')$ are admissible, then $\widetilde{V}^{(\alpha,\beta)}, \widetilde{V}^{(\alpha',\beta')}$ are Q-martingales with the same final value Q-a.s. (because $Q \sim P$) and the thesis follows. □

Remark 10.44 Consider the following alternative definition of admissibility: a self-financing strategy (α, β) is admissible if there exists *at least one* $Q \in \mathcal{Q}$ such that $\widetilde{V}^{(\alpha,\beta)}$ is a Q-martingale. With this definition, we are not able to prove the thesis of Corollary 10.43: indeed, in general $\widetilde{V}^{(\alpha,\beta)}, \widetilde{V}^{(\alpha',\beta')}$ are only local martingales with respect to a generic EMM but it could be the case that they are not strict martingales with respect to the *same* EMM. This explains why we adopted the stronger Definition 10.42 and spent some effort in the proof of Proposition 10.41. □

10.2.5 Arbitrage pricing

We consider a standard market (S, B) under Hypothesis 10.19 and we assume condition (10.45) of Theorem 10.27 for the existence of an EMM in $\mathcal{Q}$. By arguments that are substantially analogous to those used in discrete time (cf. Section 2.1), we analyze the problem of pricing of a European derivative.

Definition 10.45 *A European derivative* X *with maturity* T *is a* $\mathcal{F}_T^W$*-measurable random variable such that* $X \in L^p(\Omega, P)$ *for some* $p > 1$*. A derivative* X *is called replicable if there is an admissible strategy* $(\alpha, \beta) \in \mathcal{A}$ *such that*

$$X = V_T^{(\alpha,\beta)} \qquad P\text{-a.s.} \tag{10.61}$$

The random variable X represents the payoff of the derivative. The $\mathcal{F}_T^W$-measurability condition describes the fact that X depends on the risk factors given by $(W_t)_{t \le T}$: note that replicable payoffs are necessarily $\mathcal{F}_T^W$-measurable since so is $V_T^{(\alpha,\beta)}$. An admissible strategy (α, β) such that (10.61) holds, is called a *replicating strategy* for X.

Definition 10.46 *The risk-neutral price of a European derivative X with respect to the EMM $Q \in \mathcal{Q}$, is defined as*

$$H_t^Q = E^Q \left[e^{-\int_t^T r_s ds} X \mid \mathcal{F}_t^W \right], \qquad t \in [0,T]. \tag{10.62}$$

Remark 10.47 The assumption $X \in L^p(\Omega, P)$ for some $p > 1$, guarantees that X is Q-integrable for any $Q \in \mathcal{Q}$, so that definition (10.62) is well-posed. Indeed, for a fixed $Q \in \mathcal{Q}$, let us denote by Z the Radon-Nikodym derivative of Q with respect to P and recall that (cf. Remark 10.29) $Z \in \mathbb{L}^q(\Omega, P)$ for every $q \geq 1$. Then, by Hölder's inequality, we have

$$E^Q [|X|] = E^P [|X| Z_T] \leq \|X\|_{L^p(\Omega,P)} \|Z_T\|_{L^q(\Omega,P)} < \infty$$

where p, q are conjugate exponents and this shows that $X \in L^1(\Omega, Q)$. On the other hand, by Remark 10.18 the discount factor appearing in (10.62) is a bounded process. □

Next we introduce the collections of super and sub-replicating strategies:

$$\mathcal{A}_X^+ = \{(\alpha,\beta) \in \mathcal{A} \mid V_T^{(\alpha,\beta)} \geq X, \ P\text{-a.s.}\},$$
$$\mathcal{A}_X^- = \{(\alpha,\beta) \in \mathcal{A} \mid V_T^{(\alpha,\beta)} \leq X, \ P\text{-a.s.}\}.$$

For a given $(\alpha,\beta) \in \mathcal{A}_X^+$ (resp. $(\alpha,\beta) \in \mathcal{A}_X^-$), the value $V_0^{(\alpha,\beta)}$ represents the initial wealth sufficient to build a strategy that super-replicates (resp. sub-replicates) the payoff X at maturity. The following result confirms the natural consistency relation among the initial values of the sub and super-replicating strategies and the risk-neutral price: this relation must necessarily hold true in any arbitrage-free market.

Lemma 10.48 *Let X be a European derivative. For every EMM $Q \in \mathcal{Q}$ and $t \in [0,T]$ we have*

$$\sup_{(\alpha,\beta)\in\mathcal{A}_X^-} V_t^{(\alpha,\beta)} \leq E^Q \left[e^{-\int_t^T r_s ds} X \mid \mathcal{F}_t^W \right] \leq \inf_{(\alpha,\beta)\in\mathcal{A}_X^+} V_t^{(\alpha,\beta)}.$$

Proof. If $(\alpha,\beta) \in \mathcal{A}_X^-$, then $\widetilde{V}^{(\alpha,\beta)}$ is a Q-martingale for any $Q \in \mathcal{Q}$: thus we have

$$V_t^{(\alpha,\beta)} = E^Q \left[e^{-\int_t^T r_s ds} V_T^{(\alpha,\beta)} \mid \mathcal{F}_t^W \right] \leq$$

(since $V_T^{(\alpha,\beta)} \leq X, \ P$-a.s.)

$$\leq E^Q \left[e^{-\int_t^T r_s ds} X \mid \mathcal{F}_t^W \right]$$

and an analogous estimate holds for $(\alpha,\beta) \in \mathcal{A}_X^+$. □

Lemma 10.48 ensures that any risk-neutral price does not give rise to arbitrage opportunities since it is greater than the price of every sub-replicating strategy and smaller than the price of every super-replicating strategy. By definition, H^Q depends on the selected EMM Q; however, this is not the case if X is replicable. Indeed the following result shows that the risk-neutral price of a replicable derivative is uniquely defined and independent of $Q \in \mathcal{Q}$.

Theorem 10.49 *Let X be a replicable European derivative. For every replicating strategy $(\alpha, \beta) \in \mathcal{A}$ and for every EMM $Q \in \mathcal{Q}$, we have*

$$H_t := V_t^{(\alpha,\beta)} = E^Q \left[e^{-\int_t^T r_s ds} X \mid \mathcal{F}_t^W \right]. \tag{10.63}$$

The process H is called risk-neutral (or arbitrage) price of X.

Proof. If $(\alpha, \beta) \in \mathcal{A}$ replicates X, then $(\alpha, \beta) \in \mathcal{A}_X^- \cap \mathcal{A}_X^+$ and by Lemma 10.48 we have

$$E^Q \left[e^{-\int_t^T r_s ds} X \mid \mathcal{F}_t^W \right] = V_t^{(\alpha,\beta)}, \qquad t \in [0, T],$$

for every EMM $Q \in \mathcal{Q}$. □

10.2.6 Complete markets

We consider a standard market (S, B) and, as usual, we assume Hypothesis 10.19 and condition (10.45) of Theorem 10.27 for the existence of an EMM.

In this section we show that, if the number of risky assets is equal to the dimension of the underlying Brownian motion, i.e. $N = d$, then the market is complete and the martingale measure is unique. Roughly speaking, in a complete market every European derivative X is replicable and by Theorem 10.49 it can be priced in a unique way by arbitrage arguments: the price of X coincides with the value of any replicating strategy and with the risk-neutral price under the unique EMM.

Theorem 10.50 *When $N = d$, the market model (S, B) in* (10.25)-(10.27) *is complete, that is every European derivative is replicable. Moreover there exists only one EMM.*

Proof. The uniqueness of the EMM has been already pointed out in Example 10.32: it follows from the fact that, when $N = d$, the drift condition (10.41) determines uniquely the market price of risk.

Next we denote by Q the EMM and by W^λ the associated Q-Brownian motion. We define the Q-martingale[7]

$$M_t = E^Q \left[e^{-\int_0^T r_t dt} X \mid \mathcal{F}_t^W \right], \qquad t \in [0, T].$$

[7] Let us recall that $X \in L^1(\Omega, Q)$ by Remark 10.47.

By Theorem 10.13 we have the representation

$$M_t = E^Q \left[e^{-\int_0^T r_t dt} X \right] + \int_0^t u_s \cdot dW_s^\lambda$$

with $u \in \mathbb{L}^2_{\mathrm{loc}}(\mathcal{F}^W)$: in other terms we have

$$\begin{aligned} M_t &= E^Q \left[e^{-\int_0^T r_t dt} X \right] + \sum_{i=1}^N \int_0^t u_s^i dW_s^{\lambda,i} \\ &= E^Q \left[e^{-\int_0^T r_t dt} X \right] + \sum_{i=1}^N \int_0^t \alpha_s^i \sigma_s^i \widetilde{S}_s^i dW_s^{\lambda,i}, \end{aligned}$$

where

$$\alpha^i = \frac{u^i}{\sigma^i \widetilde{S}^i}, \qquad i = 1, \dots, N.$$

Note that $\alpha^i \sigma^i = \frac{u^i}{\widetilde{S}^i} \in \mathbb{L}^2_{\mathrm{loc}}$ because S^i is positive and continuous, so that condition (10.56) is fulfilled. By Proposition 10.39, α and M_0 define a self-financing strategy (α, β) such that

$$\widetilde{V}_t^{(\alpha,\beta)} = M_t, \qquad t \in [0, T].$$

The strategy (α, β) is admissible, $(\alpha, \beta) \in \mathcal{A}$, because M is a Q-martingale. Moreover we have

$$\widetilde{V}_T^{(\alpha,\beta)} = M_T = e^{-\int_0^T r_t dt} X,$$

and therefore (α, β) is a replicating strategy for X. □

10.2.7 Parity formulas

By the risk-neutral pricing formula (10.62), the price of a derivative is defined as the expectation of the discounted payoff and therefore it depends linearly on the payoff. Let us denote by H^X the risk-neutral price of a derivative X, under a fixed EMM Q: then we have

$$H^{c_1 X^1 + c_2 X^2} = c_1 H^{X^1} + c_2 H^{X^2}, \tag{10.64}$$

for every $c_1, c_2 \in \mathbb{R}$. This fact may be useful to decompose complex payoffs in simpler ones: for instance, the payoff of a *straddle* on the underlying asset S, with strike K and maturity T, is given by

$$X = \begin{cases} (S_T - K), & \text{if } S_T \geq K, \\ (K - S_T), & \text{if } 0 < S_T < K. \end{cases}$$

By (10.64), we simply have $H^X = c + p$ where c and p denote the prices of a European Call and a Put option, respectively, with the same strike and maturity.

Using (10.64), we can also obtain a generalization of the Put-Call parity formula of Corollary 1.1. Indeed, let us consider the following payoffs:

$$\begin{aligned} X^1 &= (S_T - K)^+, && \text{(Call option)},\\ X^2 &= 1, && \text{(bond)},\\ X^3 &= S_T, && \text{(underlying asset)}. \end{aligned}$$

Then we have

$$H_t^{X^2} = E^Q\left[e^{-\int_t^T r_s ds} \mid \mathcal{F}_t^W\right], \qquad H_t^{X^3} = S_t.$$

Now we observe that the payoff of a Put option is a linear combination of X^1, X^2 and X^3:

$$(K - S_T)^+ = KX^2 - X^3 + X^1.$$

Then, by (10.64), we get the Put-Call parity formula

$$p_t = KE^Q\left[e^{-\int_t^T r_s ds} \mid \mathcal{F}_t^W\right] - S_t + c_t, \qquad t \in [0,T], \tag{10.65}$$

that is obviously equivalent to (1.4) if the short rate is deterministic.

10.3 Markovian models: the PDE approach

In this section we examine a typical Markovian realization of the general market model analyzed in Paragraph 10.2. Specifically, we consider a model of the form

$$dS_t^i = \mu_t^i S_t^i dt + S_t^i \sigma_t^i dW_t^i, \qquad i = 1, \dots, N, \tag{10.66}$$

$$d\nu_t^j = m_t^j dt + \eta_t^j dW_t^{N+j}, \qquad j = 1, \dots, d-N, \tag{10.67}$$

where $W = (W^1, \dots, W^d)$ is a d-dimensional correlated Brownian motion, $S = (S^1, \dots, S^N)$ is the stochastic process of the risky assets that are supposed to be traded on the market and $\nu = (\nu^1, \dots, \nu^{d-N})$ is the vector of additional stochastic factors (e.g. the stochastic volatility in the Heston model of Example 10.33). We assume that

$$\mu_t = \mu(t, S_t, \nu_t), \quad \sigma_t = \sigma(t, S_t, \nu_t), \quad m_t = m(t, S_t, \nu_t), \quad \eta_t = \eta(t, S_t, \nu_t)$$

where μ, σ, m, η are deterministic functions:

- $\mu, \sigma : [0,T] \times \mathbb{R}^d \longrightarrow \mathbb{R}^N$ are the drift and volatility functions, respectively, of the assets;
- $m, \eta : [0,T] \times \mathbb{R}^d \longrightarrow \mathbb{R}^{d-N}$ are the drift and volatility functions, respectively, of the stochastic factors.

The locally non-risky asset is given by

$$B_t = e^{\int_0^t r_s ds}, \qquad t \in [0, T],$$

where $r_t = r(t, S_t, \nu_t)$ for some (deterministic) bounded function r.

Under suitable conditions on μ, σ, m, η, the results in Chapter 9 guarantee that a unique solution to the system of SDEs (10.66)-(10.67) exists: for every $(t, \bar{s}, \bar{\nu}) \in [0, T[\times\mathbb{R}^d$ we denote by $(S^{t,\bar{s},\bar{\nu}}, \nu^{t,\bar{s},\bar{\nu}})$ the solution of (10.66)-(10.67) such that $S_t^{t,\bar{s},\bar{\nu}} = \bar{s}$ and $\nu_t^{t,\bar{s},\bar{\nu}} = \bar{\nu}$. Moreover there exists a d-dimensional process of the market price of risk (cf. Definition 10.25) of the form

$$\lambda_t = \lambda(t, S_t, \nu_t).$$

Let Q be the EMM associated to λ and W^λ denote the correlated Q-Brownian motion defined by

$$dW_t = dW_t^\lambda - \lambda_t dt.$$

Then we have the following risk-neutral dynamics under Q:

$$dS_t^i = r_t S_t^i dt + S_t^i \sigma_t^i dW_t^{\lambda,i}, \qquad i = 1, \dots, N, \tag{10.68}$$

$$d\nu_t^j = \left(m_t^j - \eta_t^j \lambda_t^j\right) dt + \eta_t^j dW_t^{\lambda,N+j}, \qquad j = 1, \dots, d - N. \tag{10.69}$$

Next we consider a derivative of the form $X = F(S_T, \nu_T)$, where F is the deterministic payoff function. In this Markovian setting, the risk-neutral price of X (cf. (10.62)), under the selected EMM Q, is equal to $H_t^Q = f(t, S_t, \nu_t)$ where

$$f(t, s, \nu) = E^Q\left[e^{-\int_0^t r\left(u, S_u^{t,s,\nu}, \nu_u^{t,s,\nu}\right)du} F\left(S_T^{t,s,\nu}, \nu_T^{t,s,\nu}\right)\right].$$

Under the hypotheses of the Feynman-Kač representation Theorem 9.45, the price function f is the solution of the Cauchy problem for the differential operator associated to the system of SDEs (10.68)-(10.69), with final condition $f(T, s, \nu) = F(s, \nu)$: this is in perfect analogy with what we proved in the Black-Scholes framework. On the other hand, we emphasize that the non-uniqueness of the risk-neutral price reflects on the fact that the pricing differential operator depends on the fixed EMM: indeed, the market price of risk λ enters as a coefficient of the differential operator and therefore we have *a different pricing PDE for each EMM.*

To facilitate a deeper comprehension of these facts, we now examine three remarkable examples:

- the Heston model where $N = 1$ and $d = 2$;
- a model for interest rate derivatives where $N = 0$ and $d \geq 1$;
- a general complete model where $N = d \geq 1$: in this case the PDE approach has also the great advantage of providing the hedging strategy.

Example 10.51 We consider the Heston stochastic volatility model of Example 10.33. The price of the risky asset S is given by the system of SDEs

(10.47)-(10.48). In this case the market price of risk is a 2-dimensional process $\lambda = (\lambda^1, \lambda^2)$ with λ_t^1 determined by the drift condition (10.41):

$$\lambda_t^1 = \frac{\mu - r}{\sqrt{\nu_t}}. \tag{10.70}$$

As already mentioned, λ is generally not unique and a natural choice for the second component of the market price of risk is

$$\lambda_t^2 = \frac{a\nu_t + b}{\sqrt{\nu_t}} \tag{10.71}$$

with $a, b \in \mathbb{R}$: then the risk-neutral dynamics, under the EMM Q related to λ, is given by equations (10.50)-(10.52), that is

$$\begin{aligned} dS_t &= rS_t dt + \sqrt{\nu_t} S_t dW_t^{\lambda,1}, \\ d\nu_t &= m_t dt + \eta_t dW_t^{\lambda,2}, \end{aligned}$$

where W^λ is a 2-dimensional Brownian motion with correlation

$$d\langle W^{\lambda,1}, W^{\lambda,2}\rangle = \varrho dt, \qquad \varrho \in]-1,1[,$$

and

$$m_t = m(t, S_t, \nu_t) = \widetilde{k}\left(\widetilde{\nu} - \nu_t\right), \qquad \eta_t = \eta(t, S_t, \nu_t) = \eta\sqrt{\nu_t},$$

where $\widetilde{k}, \widetilde{\nu}, \eta$ are real constants (cf. (10.53)).

The Q-risk-neutral price of the derivative $F(S_T, \nu_T)$ is equal to $f(t, S_t, \nu_t)$ where $f = f(t, s, \nu)$ is solution to the Cauchy problem

$$\begin{cases} L^\lambda f - rf = 0, & \text{in }]0,T[\times \mathbb{R}^2_{>0}, \\ f(T, s, \nu) = F(s, \nu), & (s, \nu) \in \mathbb{R}^2_{>0}, \end{cases} \tag{10.72}$$

and L^λ is the pricing operator related to λ in (10.70)-(10.71):

$$L^\lambda f = \frac{\nu s^2}{2}\partial_{ss} f + \eta\varrho s\nu \partial_{s\nu} f + \frac{\eta^2 \nu}{2}\partial_{\nu\nu} f + rs\partial_s f + \widetilde{k}\left(\widetilde{\nu} - \nu\right)\partial_\nu f + \partial_t f.$$

As already mentioned, in the Heston model semi-analytical approximations for the price of European Calls and Puts are available (cf. Section 15): these formulas are generally preferable due to precision and computational efficiency with respect to the solution of problem (10.72) by standard numerical techniques. □

10.3.1 Martingale models for the short rate

We consider a Markovian market model in the very particular case when $N = 0$ (no risky asset) and $d \geq 1$ (d risk factors). In this case the money market account

$$B_t = e^{\int_0^t r_s ds}, \qquad t \in [0, T],$$

is the only traded asset. If the function $r = r(t,\nu)$ is smooth, then the short rate $r_t = r(t,\nu_t)$ is an Itô process whose stochastic differential can be easily computed by the Itô formula starting from the dynamics (10.67) of the risk factor ν. For simplicity, we consider the case $d = 1$ and directly assume that r solves the SDE

$$dr_t = \mu(t,r_t)dt + \sigma(t,r_t)dW_t, \tag{10.73}$$

where W is a standard 1-dimensional Brownian motion. This kind of model can be used to describe the primary objects of a so-called *fixed-income market.*

Definition 10.52 *A zero coupon bound with maturity T (or, simply, a T-bond) is a contract paying to its holder one unit of cash at time T. We denote by $p(t,T)$ the price at time t of the T-bond.*

Since the final value (payoff) of a T-bond is known, $p(T,T) = 1$, it seems natural to view bonds as *interest rate derivatives*, that is derivatives with "underlying" the short rate r. However, since r *is not an asset traded in the market,* the corresponding market model is incomplete. Indeed, even if we can invest in the money market account B, in general we cannot hope to replicate a T-bond with certainty. On one hand, this is due to the fact that a self-financing strategy involving only the asset B is necessarily constant; on the other hand, we remark that r is only "locally" riskless, but in the long run it is stochastic. Indeed, the wealth needed at time t to replicate a T-bond at maturity is equal to the *discount factor*

$$D(t,T) = e^{-\int_t^T r_s ds},$$

that is unknown at time t since r is a progressively measurable stochastic process: mathematically, $D(t,T)$ is a $\mathcal{F}_T^W$-random variable. Note the conceptual difference between $p(t,T)$ and $D(t,T)$: at maturity they have the same value $p(T,T) = D(T,T) = 1$; but while $p(t,T)$ is a price and, as such, it is observable at time t (i.e. $\mathcal{F}_t^W$-measurable), on the contrary the discount factor is $\mathcal{F}_T^W$-measurable and unobservable at time $t < T$.

By the definition of risk-neutral price of a T-bond, under the selected EMM Q, we have

$$p(t,T) = E^Q\left[e^{-\int_t^T r_s ds} p(T,T) \mid \mathcal{F}_t^W\right] = E^Q\left[D(t,T) \mid \mathcal{F}_t^W\right]. \tag{10.74}$$

Remark 10.53 By definition (10.74), the Put-Call parity formula (10.65) becomes

$$p_t = Kp(t,T) - S_t + c_t, \qquad t \in [0,T].$$

□

In this framework, the existence of an EMM Q is a trivial fact: indeed, the only traded asset that, when discounted, has to be a Q-martingale is the money market account B. But obviously $\widetilde{B} \equiv 1$ and therefore *any probability measure Q, that is equivalent to P, is an EMM.* Analogously, in order for a process λ to be a market price of risk (cf. Definition 10.25) it is sufficient that the exponential martingale defined by $dZ_t = -Z_t\lambda_t dW_t$ is a strict P-martingale: indeed, the martingale (or drift adjustment) condition (10.41) is automatically satisfied because there are no traded risky assets. This is the reason why, instead of specifying μ and λ under the real-world probability measure P, it is common practice to specify the dynamics of the short rate *directly under the martingale measure* Q: in other terms, we suppose that (10.73) describes the dynamics of r under the EMM Q. This procedure is known as *martingale modeling.* In the literature there is a large number of proposed models for the Q-dynamics of r: Vasiček [332], Cox, Ingersoll and Ross [80], Black, Derman and Toy [48], Ho and Lee [168], Hull and White [176] were among the first proposed models. We refer to Bjork [47] for an effective introduction to the subject and to Brigo and Mercurio [63], Part II, for a deep and exhaustive account of short rate modeling.

Assuming the Q-dynamics (10.73) for the short rate r, by the pricing formula (10.74) and Feynman-Kač representation, we obtain that $p(t,T) = F(t, r_t; T)$ where $(t,r) \mapsto F(t,r;T)$ solves the so-called *term structure equation*

$$\begin{cases} \frac{\sigma^2(t,r)}{2}\partial_{rr}F + \mu(t,r)\partial_r F + \partial_t F - rF = 0, \\ F(T,r;T) = 1. \end{cases} \tag{10.75}$$

Contrary to the Black-Scholes equation, the PDE in (10.75) depends on the drift coefficient μ of the underlying process, because it takes into account the fixed EMM. Therefore, in the framework of martingale modeling, the selection of the EMM among all probability measures equivalent to P is an important task: essentially, it can be considered a problem equivalent to the *calibration* of the model, that is the problem of determining the coefficients μ and σ of the SDE (10.73). More precisely, since we know that the diffusion coefficient remains unchanged through a change of measure à la Girsanov, selecting Q is equivalent to estimating μ. Note however that μ represents the drift coefficient in the risk-neutral dynamics (i.e. under an EMM) and therefore we cannot adopt standard statistical procedures to find μ starting from the historical dynamics of the process r: indeed, the historical data describe the real-world dynamics and not the risk-neutral ones that we are trying to estimate.

Alternatively, this problem is typically approached by calibrating the model to the set of today's prices of the T-bonds and other liquid interest rate derivatives: more precisely, we assume that the so called *empirical term structure,* that is the set $\{p^*(0,T) \mid T > 0\}$ of the initial prices of T-bonds, is observable. Then we can estimate μ and σ by imposing that the theoretical prices $p(0,T)$, $T > 0$, given by the model via the term structure equation (10.75), agree with the empirical data, that is

$$p(0,T) = p^*(0,T), \qquad T > 0.$$

This procedure is particularly efficient for the so-called *affine models*, where $\mu(t,r) = \alpha(t)r + \beta(t)$ and $\sigma(t,r) = \sqrt{\gamma(t)r + \delta(t)}$ with $\alpha, \beta, \gamma, \delta$ deterministic functions. These models are of particular interest since the term structure equation can be solved semi-analytically in terms of first order differential equations of Riccati type: for more details we refer to Bjork [47], Filipović [131] and the more comprehensive studies by Duffie, Filipović and Schachermayer [103] and Keller-Ressel [204].

The short rate models examined so far, are also called one-factor models since the risk factor ν is one-dimensional ($d = 1$). As a matter of fact, it is straightforward to extend the previous analysis to models with two or more risk factors: these models are more appealing for the practical applications and still widely used. We refer, for instance, to [63], Chapter 4, for a practice-oriented account of two-factors short rate models.

10.3.2 Pricing and hedging in a complete model

We have already commented on the fact that Theorem 10.50 is indeed an interesting result from a theoretical point of view but it is not constructive and it does not provide the expression of the hedging strategy for the derivative. By using Malliavin calculus, in Section 16.2.1 we will prove the Clark-Ocone formula that, under suitable assumptions, expresses the replicating strategy in terms of the so-called stochastic derivative of the payoff. Without using the advanced tools of Malliavin calculus, the most interesting and general results can be obtained in the Markovian setting, employing the theory of parabolic PDEs.

In this section we consider a Markovian model of the form (10.66)-(10.67) which, under the assumption $N = d$, reduces to

$$dS_t^i = \mu_t^i S_t^i dt + S_t^i \sigma_t^i dW_t^i, \qquad i = 1, \dots, N, \tag{10.76}$$

where $\mu_t = \mu(t, S_t)$, $\sigma_t = \sigma(t, S_t)$ and $W = \left(W^1, \dots, W^N\right)$ is Brownian motion with constant correlation matrix ϱ:

$$d\langle W^i, W^j \rangle_t = \varrho^{ij} dt.$$

We also assume that $r_t = r(t, S_t)$ and the following:

Hypothesis 10.54 *The coefficients μ, σ, r are Hölder continuous and bounded functions. The matrix $(c_{ij}) = \left(\varrho^{ij}\sigma^i\sigma^j\right)$ is uniformly positive-definite: there exists a constant $C > 0$ such that*

$$\sum_{i,j=1}^{N} c_{ij}(t, s)\xi_i\xi_j \geq C|\xi|^2, \qquad t \in [0, T], \ s \in \mathbb{R}^N_{>0}, \ \xi \in \mathbb{R}^N.$$

The arguments presented in Chapter 7 for the study of the Black-Scholes model can be easily adapted to the general case of a market with N risky assets. In particular it is possible to characterize the self-financing property in terms of a parabolic PDE of Black-Scholes type: moreover, the price and the hedging strategy are given in terms of the solution of a suitable Cauchy problem. Without going through the details that were already seen in Chapter 7, we can directly establish the connection between PDE and arbitrage pricing theories, by using the Feynman-Kač representation Theorem 9.45.

We recall that the results in Paragraphs 9.2 and 8.1 guarantee the existence of a weak solution S of (10.76). Further, by Theorem 10.27 (see also Remark 10.30), there exists a unique EMM Q: we denote by $W^Q = (W^{Q,1}, \dots, W^{Q,N})$ the Brownian motion associated to Q, under which the risk-neutral dynamics of the prices is

$$dS_t^i = r_t S_t^i dt + S_t^i \sigma_t^i dW_t^{Q,i}, \qquad i = 1, \dots, N. \tag{10.77}$$

As usual, for every $(t, s) \in [0, T[\times \mathbb{R}^N_{>0}$, we denote by $S^{t,s}$ the solution of (10.77) such that

$$S_t^{t,s} = s.$$

We consider a European derivative with payoff $F(S_T)$, where F is a locally integrable function on $\mathbb{R}^N_{>0}$ such that

$$|F(s)| \le Ce^{C|\log s|^\gamma}, \qquad s \in \mathbb{R}^N_{>0},$$

with C, γ positive constants and $\gamma < 2$.

Theorem 10.55 *Let f be the solution of the Cauchy problem*

$$\begin{cases} Lf = 0, & \text{in }]0, T[\times \mathbb{R}^N_{>0}, \\ f(T, \cdot) = F, & \text{on } \mathbb{R}^N_{>0}, \end{cases}$$

where

$$\begin{aligned} Lf(t,s) =& \frac{1}{2} \sum_{i,j=1}^N c_{ij}(t,s) s_i s_j \partial_{s_i s_j} f(t,s) \\ &+ r(t,s) \sum_{j=1}^N s_j \partial_{s_j} f(t,s) + \partial_t f(t,s) - r(t,s) f(t,s). \end{aligned}$$

Then

$$f(t,s) = E^Q \left[e^{-\int_t^T r(a, S_a^{t,s}) da} F(S_T^{t,s}) \right], \qquad (t,s) \in [0,T] \times \mathbb{R}^N_{>0}, \tag{10.78}$$

and $H_t = f(t, S_t)$ is the arbitrage price, at time t, of the derivative $F(S_T)$. Further, a replicating strategy (α, β) is given by[8]

$$V_t^{(\alpha,\beta)} = f(t, S_t), \qquad \alpha_t = \nabla f(t, S_t), \qquad t \in [0, T]. \tag{10.79}$$

[8] Here
$$\nabla f = (\partial_{s_1} f, \dots, \partial_{s_N} f).$$

Proof. The claim is a consequence of the existence results for the Cauchy problem in Paragraph 8.1 and of the Feynman-Kač formula: they can be applied directly after the transformation $s = e^x$. More precisely, for $i = 1, \dots, N$, we set

$$X_t^i = \log S_t^i, \quad \tilde{\sigma}(t,x) = \sigma\left(t, e^x\right), \quad \tilde{r}(t,x) = r\left(t, e^x\right), \quad \tilde{c}(t,x) = c\left(t, e^x\right).$$

Then, by the Itô formula, we have

$$dX_t^i = \left(\tilde{r}(t, X_t) - \frac{\left(\tilde{\sigma}^i(t, X_t)\right)^2}{2}\right) dt + \tilde{\sigma}^i(t, X_t) dW_t^{Q,i}, \qquad i = 1, \dots, N. \tag{10.80}$$

The characteristic operator associated to the system of SDEs (10.80) is

$$Au(t,x) = \frac{1}{2} \sum_{i,j=1}^{N} \tilde{c}_{ij}(t,x) \partial_{x_i x_j} u(t,x) + \sum_{j=1}^{N} \left(\tilde{r}(t,x) - \frac{\left(\tilde{\sigma}^i(t,x)\right)^2}{2}\right) \partial_{x_j} u(t,x).$$

Since, by assumption, $A + \partial_t$ is a uniformly parabolic operator, Theorem 8.6 ensures the existence of a classical smooth solution of the Cauchy problem

$$\begin{cases} Au - \tilde{r}u + \partial_t u = 0, & \text{in }]0,T[\times \mathbb{R}^N, \\ u(T,x) = F(e^x), & \text{on } \mathbb{R}^N, \end{cases} \tag{10.81}$$

and (10.78) follows by the Feynman-Kač formula. By definition f is the arbitrage price of the derivative; moreover, by Proposition 10.39, formula (10.79) defines an admissible strategy (α, β) that, by construction, replicates the derivative. □

10.4 Change of numeraire

We consider a market model (S, B) of the form (10.25)-(10.27), introduced in Paragraph 10.2. Throughout the paragraph we assume that the class $\mathcal{Q}$ (cf. Notation 10.28) of EMMs is not empty.

Definition 10.56 *Let $Q \in \mathcal{Q}$ be an EMM with numeraire B. A process U is called a "Q-price process" if:*

i) U is strictly positive;
ii) the discounted process defined by $\widetilde{U}_t = \frac{U_t}{B_t}$, $t \le T$, is a strict Q-martingale.

In economic terms, a Q-price process is a stochastic process with the main features of a "true price": it is positive and it verifies the risk-neutral pricing formula under the EMM Q. Indeed, the martingale property is equivalent to

$$U_t = E^Q\left[D(t,T) U_T \mid \mathcal{F}_t^W\right], \qquad t \in [0,T],$$

where

$$D(t,T) = e^{-\int_t^T r_s ds}$$

is the standard discount factor. By definition of martingale measure, any risky asset S^i, $i = 1, \dots, N$, is a Q-price process; further, the value of any admissible (cf. Definition 10.42) and positive self-financing strategy (α, β) is a Q-price process.

We already noted in the framework of discrete market models (see in particular Section 2.1.5) that it is sometimes convenient to use a numeraire different from the standard money market account. The Q-price processes of Definition 10.56 are exactly the kind of processes which we might choose as numeraire, i.e. as a basic standard by which prices of all other assets are measured.

Definition 10.57 *Let U be a Q-price process. A probability measure Q^U on $(\Omega, \mathcal{F})$ is called an EMM with numeraire U if:*

i) Q^U is equivalent to P;
ii) the processes of the U-discounted prices $\frac{S_t}{U_t}$ and $\frac{B_t}{U_t}$ are strict Q^U-martingales. In particular, the risk-neutral pricing formulas

$$S_t = E^{Q^U}\left[D^U(t,T)S_T \mid \mathcal{F}_t^W\right],$$
$$B_t = E^{Q^U}\left[D^U(t,T)B_T \mid \mathcal{F}_t^W\right], \qquad t \in [0,T],$$

hold, where $D^U(t,T) = \frac{U_t}{U_T}$ is the U-discount factor.

Next we prove the first basic result on the change of numeraire: any Q-price process U can be chosen as numeraire, that is for any Q-price process U there exists an EMM Q^U with numeraire U. Moreover we have an explicit formula for the change of measure between EMMs in terms of their Radon-Nikodym derivative. The following result is essentially equivalent to Lemma 10.3.

Theorem 10.58 *Let Q be an EMM with numeraire B and let U be a Q-price process. Consider the probability measure Q^U on $(\Omega, \mathcal{F})$ defined by*[9]

$$\frac{dQ^U}{dQ} = \frac{U_T B_0}{B_T U_0}. \tag{10.82}$$

Then for any $X \in L^1(\Omega, Q)$ we have

$$E^Q\left[D(t,T)X \mid \mathcal{F}_t^W\right] = E^{Q^U}\left[D^U(t,T)X \mid \mathcal{F}_t^W\right], \qquad t \in [0,T]. \tag{10.83}$$

Consequently Q^U is an EMM with numeraire U and the Q-risk-neutral price of a European derivative X is also equal to

$$E^{Q^U}\left[D^U(t,T)X \mid \mathcal{F}_t^W\right], \qquad t \in [0,T]. \tag{10.84}$$

---

[9] Or equivalently, by

$$D(0,T)dQ = D^U(0,T)dQ^U.$$

Proof. We first prove that for any $X \in L^1\left(\Omega, Q^U\right)$ we have

$$E^{Q^U}\left[X \mid \mathcal{F}_t^W\right] = E^Q\left[\frac{D(t,T)}{D^U(t,T)}X \mid \mathcal{F}_t^W\right], \qquad t \in [0,T]. \tag{10.85}$$

We set

$$Z_t = \frac{U_t B_0}{B_t U_0}, \qquad t \in [0,T].$$

Since U is a Q-price process, Z is a strictly positive Q-martingale. By the Bayes' formula in Theorem A.113, we have

$$E^{Q^U}\left[X \mid \mathcal{F}_t^W\right] = \frac{E^Q\left[XZ_T \mid \mathcal{F}_t^W\right]}{E^Q\left[Z_T \mid \mathcal{F}_t^W\right]} =$$

(since Z is a Q-martingale)

$$= E^Q\left[X\frac{Z_T}{Z_t} \mid \mathcal{F}_t^W\right] = E^Q\left[X\frac{D(t,T)}{D^U(t,T)} \mid \mathcal{F}_t^W\right],$$

where in the last equality we have used the following identity:

$$\frac{Z_T}{Z_t} = \frac{U_T B_0}{U_0 B_T}\frac{U_0 B_t}{U_t B_0} = \frac{U_T B_t}{U_t B_T} = \frac{D(t,T)}{D^U(t,T)}.$$

This proves (10.85) and now (10.83) simply follows from

$$E^Q\left[D(t,T)X \mid \mathcal{F}_t^W\right] = E^Q\left[\frac{D(t,T)}{D^U(t,T)}\left(D^U(t,T)X\right) \mid \mathcal{F}_t^W\right]$$

(by (10.85))

$$= E^{Q^U}\left[D^U(t,T)X \mid \mathcal{F}_t^W\right].$$

Moreover, $Q^U \sim Q$ because $\frac{dQ^U}{dQ} > 0$ and therefore Q^U is equivalent to P. Finally

$$S_t = E^Q\left[D(t,T)S_T \mid \mathcal{F}_t^W\right] = \tag{10.86}$$

(by (10.83))

$$= E^{Q^U}\left[D^U(t,T)S_T \mid \mathcal{F}_t^W\right], \tag{10.87}$$

and an analogous result holds for B. Thus Q^U is an EMM with numeraire U and this concludes the proof. □

As a consequence of the previous result, we also have the following useful:

Corollary 10.59 *Let U, V be Q-price processes with corresponding EMMs Q^U and Q^V. Then we have*

$$\frac{dQ^V}{dQ^U}\Big|_{\mathcal{F}_t^W} = \frac{V_t U_0}{U_t V_0}. \tag{10.88}$$

As the following example shows, the change of numeraire is a very powerful tool when dealing with stochastic interest rates: Geman [152] and Jamshidian [186] first put it to a systematic use for facilitating the computation of option prices.

Example 10.60 (Forward measure) As in Section 10.3.1, we denote by $p(t, T)$ the price at time t, under a fixed EMM Q with numeraire B, of the zero coupon bond with maturity T (cf. (10.74)):

$$p(t, T) = E^Q\left[e^{-\int_t^T r_s ds} \mid \mathcal{F}_t^W\right], \qquad t \le T.$$

Clearly $p(\cdot, T)$ is a Q-price process and we may consider the associated EMM Q^T, usually called *T-forward measure*. By (10.84) the risk neutral price H of a European derivative X is equal to

$$H_t = E^{Q^T}\left[\frac{p(t,T)}{p(T,T)} X \mid \mathcal{F}_t^W\right] = p(t,T) E^{Q^T}\left[X \mid \mathcal{F}_t^W\right]. \tag{10.89}$$

In this formula the price is given in terms of the Q^T-expectation of X and therefore, at least formally, it appears much simpler than the standard expression

$$E^Q\left[e^{-\int_t^T r_s ds} X \mid \mathcal{F}_t^W\right]$$

where the Q-expectation also involves the *stochastic discount factor*. On the other hand, in order to use formula (10.89) (for instance, for a Monte Carlo simulation) we need to determine the distribution of X under Q^T. Let us mention how this can be done: to fix ideas, we assume that $X = X_T$ is the final value of an Itô process and recall that a change of measure simply affects the drift of X_t (not the diffusion coefficient). Now Theorem 10.20 gives us the expression of the change of drift in terms of the Radon-Nikodym derivative $\frac{dQ^T}{dQ}$; in turn, this Radon-Nikodym derivative is known explicitly by Theorem 10.58:

$$\frac{dQ^T}{dQ} = \frac{B_0}{p(0,T) B_T}.$$

We will come back to this matter in a more comprehensive way in Section 10.4.2 where we will give more direct formulas for the change of measure/drift for Itô processes. □

10.4.1 LIBOR market model

As a second meaningful example of application of Theorem 10.58, we analyze the construction of the so-called LIBOR market model that is a widespread model in fixed-income markets introduced by Miltersen [257], Brace [58], Jamshidian [187]. At first our discussion will be informal, trying to determine the main feature of the model we aim at setting up.

We denote as usual by $p(t,T)$ the price at time t of the T-bond. The *simple forward rate* (or, more commonly, LIBOR rate) $L = L(t;T,S)$ is defined by the formula

$$p(t,T) = p(t,S)\left(1 + (S-T)L\right), \qquad t \le T \le S.$$

Then $L(t;T,S)$ is the simple interest rate, contracted at t, for the period $[T,S]$, which agrees with the quoted prices of zero-coupon bonds.

We consider a fixed set of increasing maturities $T_0, T_1, \dots, T_N$ and we put $\delta_i = T_i - T_{i-1}$ for $i = 1, \dots, N$. Moreover, to shorten notations we set $p_t^i = p(t,T_i)$ and we denote by $L_t^i = L(t;T_{i-1},T_i)$ the LIBOR rates corresponding to the given maturities: then we have

$$L_t^i = \frac{1}{\delta_i}\left(\frac{p_t^{i-1}}{p_t^i} - 1\right), \qquad i = 1, \dots, N. \tag{10.90}$$

Now we aim at constructing a mathematical model for the fixed-income market where an EMM Q exists (i.e. the model is arbitrage-free) and the prices p^i of T-bonds are Q-price processes (i.e. they are positive processes that can be used as numeraire). In this model, we denote by Q^i the EMM with numeraire p^i and we remark that the processes $\frac{p^j}{p^i}$ are Q^i-martingales for any $j \le N$: consequently, by formula (10.90), L^i is a Q^i-martingale as well, i.e. it is a driftless process.

Keeping in mind the previous considerations, we look for the processes L^i in the form

$$dL_t^i = \mu_t^i L_t^i dt + \sigma_t^i L_t^i dW_t^{N,i}, \qquad i = 1, \dots, N, \tag{10.91}$$

where $W^N = \left(W^{N,1}, \dots, W^{N,d}\right)$, with $N \le d$, is a *correlated* d-dimensional Brownian motion with correlation matrix ϱ and the (scalar) volatilities σ_t^i may be positive constants such as in the standard LIBOR market model or positive processes depending on some additional stochastic factors driven by the last $d - N$ components of W^N, such as in a typical stochastic volatility model. We suppose that (10.91) gives the dynamics of the LIBOR rates under the T_N-forward measure Q^N (cf. Example 10.60) and we try to determine, if they exist, the processes μ^i such that L^i is a martingale under the forward measure Q^i with numeraire p^i, for any $i = 1, \dots, N$: more precisely,

$$dL_t^i = \sigma_t^i L_t^i dW_t^{i,i}, \qquad i = 1, \dots, N, \tag{10.92}$$

where $W^{i,j}$ denotes the j-th component of the Q^i-Brownian motion W^i, for $i = 1, \dots, N$ and $j = 1, \dots, d$, in agreement with the fact that L^i must be a Q^i-martingale.

As in [47], in order to find μ^i, *we perform the change of measure from Q^N to Q^i and impose that the Q^i-drift is null*. More precisely, by Corollary 10.59, we have

$$\frac{dQ^{i-1}}{dQ^i}|_{\mathcal{F}_t} = \frac{p_t^{i-1}}{p_t^i}\frac{p_0^i}{p_0^{i-1}} := \gamma_t^i, \tag{10.93}$$

and we note that, by (10.90),

$$\gamma_t^i = \frac{p_0^i}{p_0^{i-1}}\left(1 + \delta_i L_t^i\right). \tag{10.94}$$

Therefore, assuming the dynamics (10.91), we get

$$d\gamma_t^i = \frac{p_0^i}{p_0^{i-1}}\delta_i L_t^i \sigma_t^i dW_t^{N,i} = \gamma_t^i \sigma_t^i \frac{\delta_i L_t^i}{1 + \delta_i L_t^i} dW_t^{N,i}.$$

Now that we have obtained explicitly the Radon-Nikodym derivative $\frac{dQ^{i-1}}{dQ^i}$ as an exponential martingale, we can apply directly the result about the change of drift with correlation, Theorem 10.20 (see in particular formula (10.29)), to get

$$dW_t^{i,j} = dW_t^{i-1,j} + \varrho^{ji}\sigma_t^i \frac{\delta_i L_t^i}{1 + \delta_i L_t^i} dt.$$

Applying this inductively, we also obtain

$$dW_t^{N,j} = dW_t^{i,j} + \sum_{k=i+1}^{N} \varrho^{jk}\sigma_t^k \frac{\delta_k L_t^k}{1 + \delta_k L_t^k} dt, \qquad i < N. \tag{10.95}$$

Plugging (10.95) into (10.91), we conclude that, in order to get (10.92), we necessarily have to put

$$\mu_t^i = -\sigma_t^i \sum_{k=i+1}^{N} \varrho^{jk}\sigma_t^k \frac{\delta_k L_t^k}{1 + \delta_k L_t^k}, \qquad i < N, \tag{10.96}$$

and obviously $\mu^N = 0$ because L^N is a Q^N-martingale.

We conclude that, if a LIBOR market model with all the desired features exists, it must be of the form (10.91)-(10.96). Actually, the argument can be now reverted and we have the following existence result.

Theorem 10.61 *Let W^N be a correlated d-dimensional Brownian motion on a space $(\Omega, \mathcal{F}, Q^N)$ endowed with the Brownian filtration $(\mathcal{F}_t^W)$. We define the processes $L^1, \dots, L^N$ by*

$$dL_t^i = -L_t^i \sum_{k=i+1}^{N} \varrho^{jk}\sigma_t^i \sigma_t^k \frac{\delta_k L_t^k}{1 + \delta_k L_t^k} dt + \sigma_t^i L_t^i dW_t^{N,i}, \qquad i = 1, \dots, N-1,$$

$$dL_t^N = \sigma_t^N L_t^N dW_t^{N,N},$$

where $\sigma^i \in \mathbb{L}^2_{\text{loc}}$ *are positive processes and* ϱ *is the correlation matrix of* W^N. *Then, for any* $i = 1, \dots, N$, *the process* L^i *is a positive martingale under the forward measure* Q^i *defined by* (10.93)-(10.94) *and it satisfies equation* (10.92).

10.4.2 Change of numeraire for Itô processes

In this section we examine the change of measure induced by numeraires that are Itô processes. As usual Q is a fixed EMM with numeraire B. As a preliminary result, we compute the diffusion coefficient of the ratio of two Itô processes: since we are only interested in the diffusion coefficient, we use the symbol $(\cdots)$ to denote a generic drift term.

Lemma 10.62 *Let* U, V *be two positive Itô processes of the form*

$$dU_t = (\cdots)dt + \sigma_t^U \cdot dW_t,$$
$$dV_t = (\cdots)dt + \sigma_t^V \cdot dW_t,$$

where W *is a correlated* d*-dimensional Brownian motion and* $\sigma^U, \sigma^V \in \mathbb{L}^2_{\text{loc}}$ *are the* $\mathbb{R}^d$*-valued diffusion coefficients. Then* $\frac{V}{U}$ *is an Itô process of the form*

$$d\frac{V_t}{U_t} = (\cdots)dt + \frac{V_t}{U_t}\left(\frac{\sigma_t^V}{V_t} - \frac{\sigma_t^U}{U_t}\right) \cdot dW_t. \tag{10.97}$$

Proof. The thesis follows directly by the Itô formula, since we have

$$d\frac{V_t}{U_t} = \frac{dV_t}{U_t} - \frac{V_t dU_t}{U_t^2} + \frac{V_t}{U_t^3} d\langle U, U\rangle_t - \frac{1}{U_t^2} d\langle U, V\rangle_t. \qquad \square$$

Remark 10.63 Since W is a correlated Brownian motion we may assume (cf. (10.22)) that it takes the form $W = A\bar{W}$ where $\bar{W}$ is a standard d-dimensional Brownian motion. Then $\varrho = AA^*$ is the correlation matrix of W. Under the assumptions of Lemma 10.62, we set

$$\hat{\sigma}_t = A^* \left(\frac{\sigma_t^V}{V_t} - \frac{\sigma_t^U}{U_t}\right),$$

and from (10.97) we get

$$\begin{aligned} d\frac{V_t}{U_t} &= (\cdots)dt + \frac{V_t}{U_t}\hat{\sigma}_t \cdot d\bar{W}_t \\ &= (\cdots)dt + \frac{V_t}{U_t}|\hat{\sigma}_t|\, d\hat{W}_t, \end{aligned} \tag{10.98}$$

where

$$d\hat{W}_t = \frac{\hat{\sigma}_t}{|\hat{\sigma}_t|} \cdot d\bar{W}_t.$$

We remark explicitly that, by Corollary 5.35, $\hat{W}$ is a *one-dimensional standard Brownian motion.* Note also that

$$|\hat{\sigma}_t| = \sqrt{\langle \varrho \left(\frac{\sigma_t^V}{V_t} - \frac{\sigma_t^U}{U_t} \right), \left(\frac{\sigma_t^V}{V_t} - \frac{\sigma_t^U}{U_t} \right) \rangle}. \tag{10.99}$$

□

The following example is due to Margrabe [247] who first used explicitly a change of numeraire in order to value an exchange option.

Example 10.64 (Exchange option) We consider an exchange option whose payoff is

$$\left(S_T^1 - S_T^2\right)^+$$

where the two stocks S^1, S^2 are modeled as geometric Brownian motions:

$$dS_t^i = \mu^i S_t^i dt + \sigma^i S_t^i dW_t^i, \qquad i = 1, 2.$$

Here $W = \left(W^1, W^2\right)$ is a 2-dimensional Brownian motion with correlation $\bar{\varrho}$:

$$d\langle W^1, W^2 \rangle_t = \bar{\varrho} dt, \qquad \bar{\varrho} \in]-1, 1[.$$

By the results in Section 10.2.6 the market is complete, the martingale measure is unique and by the pricing formula (10.84) of Theorem 10.58, the arbitrage price H of the exchange option under the EMM Q^2 with numeraire S^2, is given by

$$H_t = E^{Q^2} \left[\frac{S_t^2}{S_T^2} \left(S_T^1 - S_T^2\right)^+ \mid \mathcal{F}_t^W \right] = S_t^2 E^{Q^2} \left[\left(\frac{S_T^1}{S_T^2} - 1 \right)^+ \mid \mathcal{F}_t^W \right].$$

Hence the price of the exchange option is equal to the price of a Call option with underlying asset $Y = \frac{S^1}{S^2}$ and strike 1.

Next we find the Q^2-law of the process Y: we first remark that Y is driftless since it is a Q^2-martingale and therefore *its dynamics is only determined by the diffusion coefficient that is independent on the change of measure.* Therefore, by (10.97) we have[10]

$$dY_t = Y_t \left(\sigma^1 dW_t^1 - \sigma^2 dW_t^2\right) =$$

(by (10.98))

$$= Y_t \sigma d\hat{W}_t,$$

[10] In the notations of (10.97) we have

$$V = S^1, \quad U = S^2, \quad \sigma_t^V = \left(\sigma^1 S_t^1, 0\right), \quad \sigma_t^U = \left(0, \sigma^2 S_t^2\right).$$

where $\hat{W}$ is a one-dimensional standard Brownian motion and, as in (10.99),

$$\sigma = \sqrt{(\sigma^1)^2 + (\sigma^2)^2 - 2\bar{\varrho}\sigma^1\sigma^2}.$$

We have thus proved that Y is a geometric Brownian motion with volatility σ and therefore an explicit Black-Scholes type formula for the exchange option holds. □

Combining Corollary 10.59 with Lemma 10.62 and Theorem 10.20, we get the following:

Theorem 10.65 (Change of numeraire) *Let U, V be Q-price processes of the form*

$$\begin{aligned} dU_t &= (\cdots)dt + \sigma_t^U \cdot dW_t, \\ dV_t &= (\cdots)dt + \sigma_t^V \cdot dW_t, \end{aligned} \tag{10.100}$$

where W is a d-dimensional Brownian motion with correlation matrix ϱ. Let Q^U, Q^V be the EMMs related to U, V respectively and W^U, W^V be the related Brownian motions. Then the following formula for the change of drift holds:

$$dW_t^U = dW_t^V + \varrho\left(\frac{\sigma_t^V}{V_t} - \frac{\sigma_t^U}{U_t}\right) dt. \tag{10.101}$$

Example 10.66 In the Black-Scholes model, let us denote by W^B and W^S the Brownian motions with numeraires B and S respectively. Then by (10.101) we have

$$dW_t^B = dW_t^S + \sigma dt.$$

In particular the dynamics of S under W^S is given by

$$\begin{aligned} dS_t &= rS_t dt + \sigma S_t dW_t^B \\ &= rS_t dt + \sigma S_t \left(dW_t^S + \sigma dt\right) \\ &= \left(r + \sigma^2\right) S_t dt + \sigma S_t dW_t^S. \end{aligned}$$

□

10.4.3 Pricing with stochastic interest rate

The aim of this section is to give a fairly general formula for the pricing of a European call option in a model with *stochastic interest rate*. This formula is particularly suitable for the use of Fourier inversion techniques in the case r is not deterministic. The following result is a special case of a systematic study of general changes of numeraire that has been carried out by Geman, El Karoui and Rochet [154]. As already mentioned, in the context of interest rate theory, changes of numeraires were previously used by Geman [152] and Jamshidian [186]; this technique was also used by Harrison and Kreps [163], Harrison and Pliska [164] and even earlier by Merton [250].

We consider a general market model of the type introduced in Paragraph 10.2 with $N = 1$ (only one asset S) and $d \geq 1$. We assume the existence of an EMM Q with numeraire B. The main idea is to write the Q-neutral price C as follows:

$$C_0 = E^Q \left[e^{-\int_0^T r_s ds} (S_T - K)^+ \right]$$
$$= E^Q \left[e^{-\int_0^T r_s ds} (S_T - K) \mathbb{1}_{\{S_T \geq K\}} \right] = I^1 + I^2$$

where, by (10.83) of Theorem 10.58, we have

$$I^1 = E^Q \left[e^{-\int_0^T r_s ds} S_T \mathbb{1}_{\{S_T \geq K\}} \right] = S_0 E^{Q^S} \left[\mathbb{1}_{\{S_T \geq K\}} \right],$$
$$I^2 = K E^Q \left[e^{-\int_0^T r_s ds} \mathbb{1}_{\{S_T \geq K\}} \right] = K p(0,T) E^{Q^T} \left[\mathbb{1}_{\{S_T \geq K\}} \right],$$

by the change the numeraire where in the first term above we use the measure Q^S with numeraire S, and for the second term we use the T-forward measure Q^T with numeraire the T-bond $p(\cdot, T)$. Thus we get the following general pricing formula.

Theorem 10.67 *The Q-risk neutral price of a Call option with underlying S, strike K and maturity T is given by*

$$C_0 = S_0 Q^S (S_T \geq K) - K p(0,T) Q^T (S_T \geq K), \tag{10.102}$$

where Q^S and Q^T denote the EMMs obtained from Q by the change of numeraire S and $p(\cdot, T)$ respectively.

For the practical use of this formula we have to determine the distribution of S under the new martingale measures. We first recall the dynamics of S under the EMM Q with numeraire B and related Brownian motion $W^Q = (W^{Q,1}, \dots, W^{Q,d})$ with correlation matrix ϱ:

$$dS_t = r_t S_t dt + \sigma_t S_t dW_t^{Q,1}.$$

Moreover we introduce the d-dimensional process $\bar{\sigma}_t = (\sigma_t, 0, \dots, 0)$, and we assume that

$$dp(t,T) = r_t p(t,T) dt + p(t,T) \sigma_t^T \cdot dW_t^Q,$$

where σ^T denotes the d-dimensional volatility process of $p(\cdot, T)$. Then by Theorem 10.65 we have that the processes W^S and W^T, defined by

$$dW_t^S = dW_t^Q - \varrho \bar{\sigma}_t dt$$

and

$$dW_t^T = dW_t^Q - \varrho \sigma_t^T dt,$$

are Brownian motions with correlation matrix ϱ, under the measures Q^S and Q^T respectively. Hence we have

$$dS_t = \left(r_t + \sigma_t^2\right) S_t dt + \sigma_t S_t dW_t^{S,1}$$

and

$$dS_t = \left(r_t + \sigma_t \left(\varrho\sigma_t^T\right)^1\right) S_t dt + \sigma_t S_t dW_t^{T,1},$$

where $\left(\varrho\sigma_t^T\right)^1$ denotes the first component of the vector $\varrho\sigma_t^T$.

Remark 10.68 When the short rate $r = r(t)$ is deterministic, as in most models for equity derivatives, we have $p(t,T) = e^{-\int_t^T r(s)ds}$ and $\sigma^T = 0$: thus the dynamics of S under Q and Q^T coincide. □

Finally, we also give the expression of the Radon-Nikodym derivatives for the changes of measure: by Corollary 10.59, for $t \in [0,T]$ we have

$$\frac{dQ^S}{dQ}|_{\mathcal{F}_t^W} = \frac{S_t B_0}{B_t S_0} = \frac{e^{-\int_0^t r_s ds} S_t}{S_0},$$

and

$$\frac{dQ^T}{dQ}|_{\mathcal{F}_t^W} = \frac{p(t,T) B_0}{B_t p(0,T)} = \frac{e^{-\int_0^t r_s ds}}{p(0,t)}.$$

10.5 Diffusion-based volatility models

This section is devoted to the analysis of volatility risk with special focus on the most popular extensions of the Black-Scholes models in the diffusion-based framework, from local to stochastic volatility models. In order to explain the systematic differences between the market prices and the theoretical Black-Scholes prices (cf. Paragraph 7.5), various approaches to model volatility have been introduced. The general idea is to modify the dynamics of the underlying asset, thus obtaining a stochastic process that is more flexible than the standard geometric Brownian motion. Broadly speaking, the models with non-constant volatility can be divided in two groups:

- in the first one, the volatility is endogenous, i.e. it is described by a process that depends on the same risk factors of the underlying asset. In this case, the completeness of the market is generally preserved;
- in the second one, the volatility is exogenous, i.e. it is described by a process that is driven by some additional risk factors: for example other Brownian motions and/or jump processes. In this case the corresponding market model is generally incomplete.

10.5.1 Local and path-dependent volatility

Among the models with endogenous volatility, the most popular ones are the so-called *local-volatility* models for which σ is assumed to be function of time and of the price of the underlying asset: the dynamics of the underlying asset is simply that of a diffusion process

$$dS_t = \mu(t, S_t)S_t dt + \sigma(t, S_t)S_t dW_t. \tag{10.103}$$

Under the assumptions of Section 10.3.2 such a model is complete and it is possible to determine the price and the hedging strategy by solving numerically a Cauchy problem as in Theorem 10.55.

Actually the dependence of σ on S_t does not seem to be easily justified from an intuitive point of view. Nevertheless, local-volatility models have enough flexibility to give the theoretical price of an option in accordance (at least approximately) with the implied volatility surface of the market. In order to replicate an implied volatility surface, the model must be calibrated; in other terms, a so-called *inverse problem* must be solved: this consists in determining the function $\sigma = \sigma(t, S)$ such that the theoretical prices match the quoted market prices. The calibration of the local volatility is an extremely delicate matter that made several authors in the literature question the effectiveness and the validity of the model: we refer, for example, to Dumas, Fleming and Whaley [106] and Cont [75].

Starting from Breeden and Litzenberger's work [59], Dupire [107] has shown how, at least theoretically, it is possible to solve the inverse problem for a local volatility model. In what follows, for the sake of simplicity we consider the one-dimensional case, with $r = 0$ and we denote by $\Gamma(0, S; T, \cdot)$ the transition density of the process of the underlying asset, with initial value S at time 0. In view of the risk-neutral pricing formula, we have that the price $C = C(0, S, T, K)$ of a European Call option with strike K and maturity T is equal to

$$C(0, S, T, K) = E^Q\left[(S_T - K)^+\right] = \int_{\mathbb{R}_{>0}} (s - K)^+ \Gamma(0, S; T, s)ds.$$

Now, the second-order *distributional* derivative, with respect to K, of the payoff is

$$\partial_{KK}(s - K)^+ = \delta_K(s),$$

where δ_K is the Dirac's delta and so, at least formally, we get

$$\partial_{KK}C(0, S, T, K) = \Gamma(0, S; T, K). \tag{10.104}$$

By (10.104), as soon as we know all the market prices of the Call options for all the strikes and maturities, it is theoretically possible to obtain the density of S_T: in other terms, knowing exactly the implied volatility surface means knowing the transition density of the underlying asset.

Now let us recall (cf. Theorem 9.46) that the transition density, as a function of T, K, satisfies the adjoint PDE associated to the SDE (10.103) and so

$$\partial_T \Gamma(0, S; T, K) = \frac{1}{2}\partial_{KK}\left(\sigma^2(T, K)K^2\Gamma(0, S; T, K)\right). \tag{10.105}$$

Substituting (10.104) into (10.105) we have

$$\partial_{TKK} C(0, S; T, K) = \frac{1}{2}\partial_{KK}\left(\sigma^2(T, K)K^2\partial_{KK}C(0, S; T, K)\right)$$

and integrating in K we get

$$\partial_T C(0, S; T, K) - \frac{1}{2}\sigma^2(T, K)K^2\partial_{KK}C(0, S; T, K) = A(T)K + B(T), \tag{10.106}$$

where A, B are arbitrary functions of T. Since, at least formally, the right-hand side of (10.106) tends to zero as $K \to +\infty$, we must have $A = B = 0$ and so

$$\partial_T C(0, S; T, K) = \frac{1}{2}\sigma^2(T, K)K^2\partial_{KK}C(0, S; T, K), \qquad K, T > 0. \tag{10.107}$$

In principle, $\partial_T C(0, S; T, K)$ and $\partial_{KK}C(0, S; T, K)$ can be computed once the implied volatility surface is known: therefore from (10.107) we get

$$\sigma^2(T, K) = \frac{2\partial_T C(0, S; T, K)}{K^2\partial_{KK}C(0, S; T, K)}, \tag{10.108}$$

which is the expression of the volatility function to plug as a coefficient into the SDE (10.103) in order for the local-volatility model to replicate the observed volatility surface.

Unfortunately formula (10.108) cannot be used in practice since the implied volatility surface is known only at a finite number of strikes and maturities: more precisely, the computation of the derivatives $\partial_T C, \partial_{KK}C$ strongly depends on the interpolation scheme used to build a continuous surface starting from discrete data, this scheme being necessary to compute the derivatives of the price. This makes formula (10.108) and the corresponding volatility surface highly unstable.

The true interest in equation (10.107) lies in the fact that, by solving the Cauchy problem for (10.107) with initial datum $C(0, S; 0, K) = (S - K)^+$, it is possible to obtain the prices of the Call options for all the strikes and maturities in one go.

A variant of the local volatility is the so-called path-dependent volatility, introduced by Hobson and Rogers [169] and generalized by Foschi and Pascucci [135]. Path-dependent volatility describes the dependence of the volatility on the movements of the asset in terms of deviation from trend (cf.

Figure 7.9). The model is very simple: we consider a function ψ that is non-negative, piecewise continuous and integrable over $]-\infty, T]$. We assume that ψ is strictly positive on $[0, T]$ and we set

$$\Psi(t) = \int_{-\infty}^{t} \psi(s)ds.$$

We define the weighted average process of the underlying asset as

$$M_t = \frac{1}{\Psi(t)} \int_{-\infty}^{t} \psi(s) Z_s ds, \qquad t \in]0, T],$$

where $Z_t = \log(e^{-rt} S_t)$ is the logarithm of the discounted price. The Hobson-Rogers model corresponds to the choice $\psi(t) = e^{\alpha t}$ with α a positive parameter. By the Itô formula we have

$$dM_t = \frac{\varphi(t)}{\Phi(t)} (Z_t - M_t)\, dt.$$

Assuming the following dynamics for the logarithm of the price

$$dZ_t = \mu(Z_t - M_t)dt + \sigma(Z_t - M_t)dW_t,$$

with suitable functions μ, σ, we get the pricing PDE

$$\frac{\sigma^2(z-m)}{2} (\partial_{zz} f - \partial_z f) + \frac{\varphi(t)}{\Phi(t)}(z-m)\partial_m f + \partial_t f = 0, \quad (t, z, m) \in]0, T[\times \mathbb{R}^2. \tag{10.109}$$

Equation (10.109) is a Kolmogorov equation, similar to the ones that arise in the pricing problem for Asian options: as we have seen in Section 9.5.2, for such equations an existence and uniqueness theory for the Cauchy problem, similar to that for uniformly parabolic PDEs, is available. Further, since no additional risk factor is introduced, the path-dependent volatility model is complete. As shown by Foschi and Pascucci [136], a suitable choice of the function σ allows to replicate the volatility surface of the market and reproduce some typical feature like the rapid increase of implied volatility corresponding to large decreases of the underlying asset.

10.5.2 CEV model

We consider a local volatility model where the volatility depends on the underlying asset according to a power law

$$\sigma(t, S_t) = \sigma(t) S_t^{\beta - 1} \tag{10.110}$$

with $\sigma(t)$ deterministic function of time and $0 < \beta < 1$. This model was first proposed in Cox [79] where it was called the constant elasticity of variance (CEV) model.

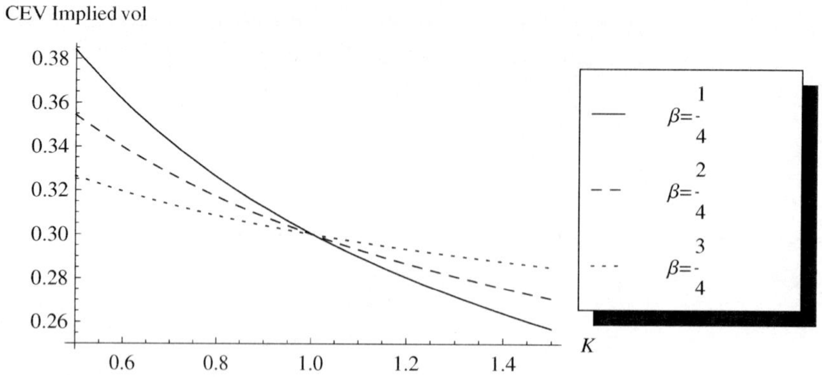

Fig. 10.1. CEV-implied volatilities for $T-t=\frac{1}{2}$, $r=5\%$, $\sigma(t)\equiv 30\%$ and different values of β

The function in (10.110) does not satisfy the standard assumption for the coefficient of a SDE which guarantee the strong uniqueness of the solution: more precisely, by the classical results on one-dimensional SDEs (see, for instance, Section 5.5.5 in Karatzas and Shreve [201]), it is known that there exists a pathwise unique solution of the SDE

$$dS_t = \sigma(t) S_t^{\beta} dW_t \tag{10.111}$$

for any $\beta \geq \frac{1}{2}$, while uniqueness fails to hold for any $\beta \in]0, 1/2[$; further, the point 0 is an attainable state for the process S. On the other hand, Delbaen and Shirakawa [92] studied the arbitrage free option pricing problem and proved that, for any $0 < \beta < 1$, there exists a unique EMM measure under which the risk-neutral dynamics of the asset is given by

$$dS_t = rS_t dt + \sigma(t) S_t^{\beta} dW_t.$$

Thus the model is complete and the arbitrage price of a European Call option, with strike K and maturity T, is equal to $C(t, S_t) = e^{-r(T-t)} u(t, S_t)$ where

$$u(t,s) = E\left[\left(S_T^{t,s} - K\right)^+\right]$$

and, by the Feynman-Kac formula, u solves the Cauchy problem

$$\begin{cases} \partial_t u(t,s) + \frac{\sigma^2(t)s^{2\beta}}{2} \partial_{ss} u(t,s) + rs\partial_s u(t,s) = 0, & t \in]0,T[, \ s > 0, \\ u(T,s) = (s-K)^+, & s > 0. \end{cases} \tag{10.112}$$

Notice that for $\beta = 1$, (10.112) yields the standard Black-Scholes model.

A main feature of the CEV model is that volatility changes inversely with the price, and this reproduces a well documented characteristic of actual price movements, recognized by several econometric studies. From the point of view of volatility modeling, the CEV model introduces the skew pattern that is commonly observed in some markets (see Figure 10.1) but it seems incapable of reproducing realistic smiles. Moreover, a time dependent parameter $\sigma(t)$ is required to accommodate the observed term structure of implied volatility.

The transition density of the price in the CEV model can be explicitly represented in terms of special functions: in particular, Cox [79] expressed the price of a Call option as the sum of a series of Gamma cumulative distribution functions. It is known that these formulae give a good *local* (at-the-money) approximation of the option price. For instance, Figure 10.2 shows the Cox option prices in the case $\beta = \frac{3}{4}$ and $T = \frac{1}{3}$ with a number n of terms in the series expansion equal to $n = 400, 420, 440, 460$: it is evident that for far from the money options this approximation gives wrong prices unless we consider a high number of terms in the series expansion. This is particularly sensible for short times to maturity.

On the other hand, the approximation by Shaw [307] (see also Schroder [303] and Lipton [239]) expresses the payoff random variable in terms of Bessel functions and then uses numerical integration to provide the option price. Since it is an adaptive method, the representation of prices is valid globally even if the method may become computationally expensive when we have to compute deep out-of or in-the money option prices.

Here we present another approach due to Hagan and Woodward [160] who employed singular perturbation techniques to obtain an analytical approximation formula for the implied volatility in the CEV model. Singular perturbation methods were largely originated by workers in fluid dynamics but their use has spread throughout other fields of applied mathematics: Kevorkian and Cole [205], [206] are reference texts on the subject. We present this approach in some detail since it is quite general and can be applied in different settings (an example is given by the SABR stochastic volatility model).

We recall that the implied volatility is defined as the value of the volatility to be inserted in the classical Black-Scholes formula to obtain a given Call or Put option price (cf. Section 7.5). The main result in [160] is the following:

Theorem 10.69 *The implied volatility generated by the CEV model* (10.111), *with* $\beta \in]0,1[$, *is approximated by the following formula:*

$$\sigma_{\text{CEV}}(S_t, T, K) = \frac{\sqrt{\alpha_{t,T}}}{F_t^{1-\beta}} \left(1 + \frac{(1-\beta)(2+\beta)}{6} \left(\frac{F_t - K}{F_t} \right)^2 + \frac{(1-\beta)^2 (T-t) \alpha_{t,T}}{24 F_t^{2(1-\beta)}} \right) \tag{10.113}$$

where

$$\alpha_{t,T} = \frac{1}{T-t} \int_t^T e^{2r(T-\tau)(1-\beta)} \sigma^2(\tau) d\tau$$

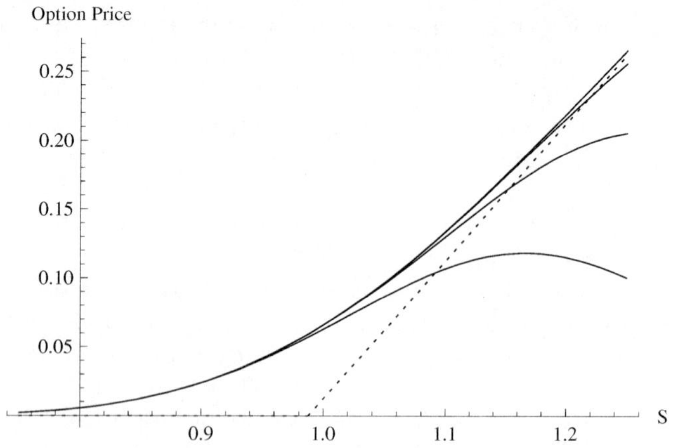

Fig. 10.2. CEV-expansion option price by Cox [79], in the case $\beta = \frac{3}{4}$ and $T = \frac{1}{3}$ with a number n of terms in the series expansion equal to $n = 400, 420, 440, 460$

and

$$F_t = \frac{e^{r(T-s)}S_t + K}{2}.$$

Proof. The proof proceeds in some steps.

First step. We consider the pricing problem

$$\begin{cases} \partial_t u(t,s) + \frac{\sigma^2(t)A^2(s)}{2}\partial_{ss}u(t,s) + rs\partial_s u(t,s) = 0, & t \in]0,T[, \ s > 0, \\ u(T,s) = (s-K)^+, & s > 0, \end{cases} \tag{10.114}$$

where, for greater convenience, we put $A(s) = s^\beta$. We also set $\varepsilon = A(K)$ and

$$\tau(t) = \int_t^T \left(e^{r(T-\varrho)}\sigma(\varrho)A\left(e^{-r(T-\varrho)}\right)\right)^2 d\varrho, \quad x(t,s) = \frac{e^{r(T-t)}s - K}{\varepsilon}. \tag{10.115}$$

By the change of variable

$$u(t,s) = \varepsilon Q(\tau(t), x(t,s)), \tag{10.116}$$

problem (10.114) is equivalent to

$$\begin{cases} \partial_\tau Q(\tau,x) - \frac{A^2(K+\varepsilon x)}{2\varepsilon^2}\partial_{xx}Q(\tau,x) = 0, & \tau > 0, \ x > -\frac{K}{\varepsilon}, \\ Q(0,x) = x^+, & x > -\frac{K}{\varepsilon}. \end{cases} \tag{10.117}$$

Next we consider the Taylor expansion

$$A(K + \varepsilon x) = A(K)\left(1 + \varepsilon x \nu_1 + \frac{1}{2}\varepsilon^2 x^2 \nu_2 + \mathrm{O}(\varepsilon^3)\right) \tag{10.118}$$

as $\varepsilon \to 0$, where

$$\nu_1 = \frac{A'(K)}{A(K)}, \qquad \nu_2 = \frac{A''(K)}{A(K)}. \tag{10.119}$$

Plugging (10.118) into (10.117), we get

$$\begin{cases} \partial_\tau Q - \frac{1}{2}\partial_{xx} Q = \varepsilon x \nu_1 \partial_{xx} Q + \frac{\varepsilon^2(\nu_1^2+\nu_2)}{2}\partial_{xx} Q + \mathrm{O}(\varepsilon^3), & \tau > 0,\ x > -\frac{K}{\varepsilon}, \\ Q(0,x) = x^+, \qquad x > -\frac{K}{\varepsilon}. \end{cases} \tag{10.120}$$

Second step. Let

$$\Gamma(t,x) = \frac{1}{\sqrt{2\pi t}} e^{-\frac{x^2}{2t}}$$

be the fundamental solution of the heat equation. We set

$$\widetilde{Q} = G + \varepsilon G_1 + \varepsilon^2 G_2$$

where

$$G(t,x) = x \int_{-\infty}^{x} \Gamma(t,y)dy + t\Gamma(t,x) \tag{10.121}$$

is the positive solution to the Cauchy problem

$$\begin{cases} \partial_t G(t,x) - \frac{1}{2}\partial_{xx} G(t,x) = 0, & t > 0,\ x \in \mathbb{R}, \\ G(0,x) = x^+, & x \in \mathbb{R}, \end{cases} \tag{10.122}$$

while the functions G_1, G_2 are defined recursively by

$$\begin{cases} \partial_t G_1 - \frac{1}{2}\partial_{xx} G_1 = \nu_1 x \partial_{xx} G, & t > 0,\ x \in \mathbb{R}, \\ G_1(0,x) = 0, & x \in \mathbb{R}, \end{cases} \tag{10.123}$$

$$\begin{cases} \partial_t G_2 - \frac{1}{2}\partial_{xx} G_2 = \nu_1 x \partial_{xx} G_1 + \frac{1}{2}\left(\nu_1^2 + \nu_2\right) x^2 \partial_{xx} G, & t > 0,\ x \in \mathbb{R}, \\ G_2(0,x) = 0, & x \in \mathbb{R}. \end{cases} \tag{10.124}$$

A direct computation shows that

$$\begin{cases} \partial_t \widetilde{Q} - \frac{1}{2}\partial_{xx} \widetilde{Q} = \varepsilon x \nu_1 \partial_{xx} \widetilde{Q} + \frac{\varepsilon^2(\nu_1^2+\nu_2)}{2}\partial_{xx} \widetilde{Q} + \mathrm{O}(\varepsilon^3), & \tau > 0,\ x > -\frac{K}{\varepsilon}, \\ \widetilde{Q}(0,x) = x^+, \qquad x > -\frac{K}{\varepsilon}, \end{cases}$$

and therefore, by comparison with problem (10.120), we take $\widetilde{Q}$ as an approximation of the solution Q.

Third step. Now we derive the explicit expression of $\widetilde{Q}$. We first remark that, as a simple consequence of the following identities

$$\partial_x \Gamma(t,x) = -\frac{x}{t}\Gamma(t,x), \tag{10.125}$$

$$\partial_t \Gamma(t,x) = \frac{1}{2}\partial_{xx}\Gamma(t,x) = \frac{x^2-t}{2t^2}\Gamma(t,x), \tag{10.126}$$

we have

$$\partial_t G(t,x) = \frac{1}{2}\Gamma(t,s), \quad \partial_x G(t,x) = \int_{-\infty}^{x} \Gamma(t,y)dy, \tag{10.127}$$

$$\partial_{tt} G(t,x) = \frac{x^2-t}{4t^2}\Gamma(t,x), \quad \partial_{xt} G(t,x) = -\frac{x}{2t}\Gamma(t,x), \quad \partial_{xx} G(t,x) = \Gamma(t,x), \tag{10.128}$$

$$\partial_{ttt} G(t,x) = \frac{x^4-6x^2t+3t^2}{8t^4}\Gamma(t,x). \tag{10.129}$$

Now we prove that

$$G_1(t,x) = \nu_1 tx\partial_t G(t,x). \tag{10.130}$$

Indeed by the classical representation formula (A.67) for solutions of the Cauchy problem, we have

$$G_1(t,x) = \nu_1 \int_0^t \int_{\mathbb{R}} \Gamma(t-s, x-y) y \partial_{yy} G(s,y) dyds =$$

(by (10.128))

$$= -\nu_1 \int_0^t \int_{\mathbb{R}} \Gamma(t-s, x-y) s \partial_y \Gamma(s,y) dyds =$$

(integrating by parts and since $\partial_y \Gamma(\cdot, x-y) = -\partial_x \Gamma(\cdot, x-y)$)

$$= -\nu_1 \int_0^t s\partial_x \int_{\mathbb{R}} \Gamma(t-s, x-y)\Gamma(s,y)dyds =$$

(by the reproduction property (6.26) and then again by (10.128))

$$= -\frac{\nu_1 t^2}{2}\partial_x \Gamma(t,x) = \nu_1 tx\partial_t G(t,x)$$

and this proves (10.130). A similar argument yields

$$G_2 = \nu_1^2\left(t^4\partial_{ttt}G + \frac{8t^3}{3}\partial_{tt}G + \frac{t^2}{2}\partial_t G\right) + \nu_2\left(\frac{2t^3}{2}\partial_{tt}G + \frac{t^2}{2}\partial_t G\right) =$$

(by (10.128) and (10.129))

$$= \frac{t}{12}\left(x^2(4\nu_2 + \nu_1^2) + t(2\nu_2 - \nu_1^2)\right)\partial_t G + \frac{\nu_1^2 t^2 x^2}{2}\partial_{tt} G. \tag{10.131}$$

Summing up, we have

$$\begin{aligned}\widetilde{Q}(\tau, x) =& G(\tau, x) + \varepsilon G_1(\tau, x) + \varepsilon^2 G_2(\tau, x)\\ =& G(\tau, x) + \left(\varepsilon\nu_1\tau x + \frac{\varepsilon^2\tau}{12}\left(x^2(4\nu_2 + \nu_1^2) + \tau(2\nu_2 - \nu_1^2)\right)\right)\partial_\tau G(\tau, x)\\ &+ \frac{\varepsilon^2\nu_1^2\tau^2 x^2}{2}\partial_{\tau\tau} G(\tau, x),\end{aligned}$$

and therefore

$$\widetilde{Q}(\tau, x) = G(\widetilde{\tau}, x) + \mathrm{O}(\varepsilon^3), \qquad \text{as } \varepsilon \to 0,$$

where

$$\widetilde{\tau} = \tau\left(1 + \varepsilon\nu_1 x + \frac{\varepsilon^2}{12}\left(x^2(4\nu_2 + \nu_1^2) + \tau(2\nu_2 - \nu_1^2)\right)\right).$$

Thus, ignoring the errors of order greater than or equal to three, by (10.116) we get the following approximation formula for the price of the Call option:

$$C(t, S_t) = e^{-r(T-t)}u(t, S_t) = e^{-r(T-t)}A(K)G\left(\widetilde{\tau}, \frac{e^{r(T-t)}S_t - K}{A(K)}\right) =$$

(since $\varepsilon G(t, x) = G(\varepsilon^2 t, \varepsilon x)$ for $\varepsilon > 0$)

$$= e^{-r(T-t)}G\left(A^2(K)\widetilde{\tau}, e^{r(T-t)}S_t - K\right) \tag{10.132}$$

with G as in (10.121) and

$$\begin{aligned}\widetilde{\tau} =& \tau\Bigg(1 + \nu_1\left(e^{r(T-t)}S_t - K\right) + \frac{1}{12}\left(e^{r(T-t)}S_t - K\right)^2(4\nu_2 + \nu_1^2)\\ &+ \tau\frac{A^2(K)}{12}(2\nu_2 - \nu_1^2)\Bigg),\end{aligned} \tag{10.133}$$

where

$$\tau = \int_t^T \left(e^{r(T-\varrho)}\sigma(\varrho)A\left(e^{-r(T-\varrho)}\right)\right)^2 d\varrho.$$

Fourth step. We finally prove the approximation formula (10.113) for the implied volatility. For the special case of the Black-Scholes model, that is with $A(s) = s$ and $\sigma(t) = \sigma_{\text{impl}}$, the approximation formula (10.132)-(10.133) reads

$$C_{\text{BS}}(t, S_t) = e^{-r(T-t)}G\left(K^2\tau_{\text{BS}}, e^{r(T-t)}S_t - K\right)$$

with G as in (10.121) and

$$\begin{aligned}\tau_{\rm BS} = &\sigma_{\rm impl}^2(T-t)\Bigg(1+\frac{1}{K}\left(e^{r(T-t)}S_t - K\right)\\ &+\frac{1}{12K^2}\left(e^{r(T-t)}S_t - K\right)^2 - \frac{1}{12}\sigma_{\rm impl}^2(T-t)\Bigg).\end{aligned} \tag{10.134}$$

Since $G(\tau, x)$ is an increasing function of τ, the Black-Scholes price matches the CEV price if and only if $\widetilde{\tau}$ in (10.133) equals τ in (10.134): solving this equation with respect to $\sigma_{\rm impl}$, we find formula (10.113) for the implied volatility in the CEV model. □

10.5.3 Stochastic volatility and the SABR model

Local volatility, and more generally complete market models, are commonly considered unrealistic and unable to hedge the volatility risk. Hagan, Kumar, Lesniewski and Woodward [159] argued that local volatility models have an inherent flaw of predicting the wrong dynamics of the implied volatility. Moreover several studies support the fact that the stock price volatility should be modeled by an autonomous stochastic process; the presence of jumps in the dynamics of risky assets and/or their volatilities is another well documented characteristic (see, for instance, the accounts given by Musiela and Rutkowski [261], Cont and Tankov [76] and the bibliographies therein).

Typically, a stochastic volatility model is a Markovian model of the form examined in Section 10.3: in the one-dimensional case, the stock price S is given by

$$dS_t = \mu(t, S_t)S_t dt + \sigma_t S_t dW^1$$

where σ_t is a stochastic process, solution to the SDE

$$d\sigma_t = a(t, \sigma_t)dt + b(t, \sigma_t)dW_t^2,$$

and (W^1, W^2) is a two-dimensional correlated Brownian motion. Many different specifications of stochastic volatility have been proposed in the literature: among others we mention Hull and White [175], Johnson and Shanno [189], Stein and Stein [316], Heston [165], Ball and Roma [18], Renault and Touzi [290]. The Heston model, that is one of the classical and most widely used models, was already presented in Example 10.33: the related numerical issues will be discussed in Chapter 15.

Another popular model, also used in the modelling of fixed income markets, is the so called SABR model proposed and analyzed by Hagan, Kumar, Lesniewski and Woodward [159]. The SABR model is the natural extension of the classical CEV model to stochastic volatility: the risk-neutral dynamics of the forward price $F_t = e^{r(T-t)}S_t$ is given by

$$\begin{aligned} dF_t &= \alpha_t F_t^{\beta} dW_t^1, \\ d\alpha_t &= \nu\alpha_t dW_t^2, \end{aligned}$$

where (W^1, W^2) is a Brownian motion with constant correlation ϱ. The singular perturbation techniques presented in Section 10.5.2 can be employed to prove the following approximation formula of the implied volatility in the SABR model:

$$\sigma(K,T,F_0,r) = \frac{\alpha_0}{(F_0K)^{\frac{1-\beta}{2}}\left(1+\frac{(1-\beta)^2}{24}\log^2\left(\frac{F_0}{K}\right)+\frac{(1-\beta)^4}{1920}\log^4\left(\frac{F_0}{K}\right)\right)}\frac{z}{x(z)}\cdot$$
$$\cdot\left(1+\left(\frac{(1-\beta)^2\alpha_0^2}{24(F_0K)^{1-\beta}}+\frac{\rho\beta\nu\alpha_0}{4(F_0K)^{(1-\beta)/2}}+\frac{(2-3\rho^2)\nu^2}{24}\right)T\right),$$

where

$$z = \frac{\nu}{\alpha_0}(F_0K)^{(1-\beta)/2}\log\frac{F_0}{K}$$

and

$$x(z) = \log\frac{\sqrt{1-2\varrho z+z^2}+z-\varrho}{1-\varrho}.$$

We mention that Kahl and Jackel [193], Rogers and Veraart [292] have recently proposed stochastic-local volatility models alternative to SABR, in which option prices for European plain vanilla options have accurate closed-form expressions. Moreover, Benhamou, Gobet and Miri in [39] and [40] proposed a recent methodology based on stochastic analysis and Malliavin calculus to derive tractable approximations of option prices in various stochastic volatility models.

A common feature of stochastic volatility models is the market incompleteness: typically, in a market model with stochastic volatility and/or jumps, it is not possible to replicate all the payoffs and the arbitrage price is not unique because it depends on the market price of risk. On the other hand, in practice these models can be effectively used by employing a procedure of market completion, analogous to that we presented in the Gamma and Vega hedging (cf. Section 7.4.3). The parameters of the model are usually calibrated to market data in order to determine the market price of risk: then, in some cases a hedging strategy for an exotic derivative can be constructed by using a plain vanilla option, besides the bond and the underlying assets.

It is well known that stochastic volatility models account for long term smiles and skews but they cannot give rise to realistic short-term implied volatility patterns. To cope with these and other problems, more recently models with jumps have become increasingly popular. Financial modeling with jump processes will be discussed in Chapters 13, 14 and 15.

11

American options

We present the main results on the pricing and hedging of American derivatives by extending to continuous time the ideas introduced in the discrete-market setting in Section 2.5. Even in the simplest case of the Black-Scholes market model, the hedging and pricing problems for American options need very refined mathematical tools. In the complete-market setting, Bensoussan [41] and Karatzas [198], [199] developed a probabilistic approach based upon the notion of Snell envelope in continuous time and upon the Doob-Meyer decomposition. The problem was also studied by Jaillet, Lamberton and Lapeyre [185] who employed variational techniques, and by Oksendal and Reikvam [273], Gatarek and Świech [149] in the framework of the theory of viscosity solutions. American options for models with jumps were studied among others by Zhang [345], Mulinacci [260], Pham [279], Levendorskii [235], Ekström [119], Ivanov [181], Lamberton and Mikou [227], Bayraktar and Xing [36].

In this chapter we present an analytical Markovian approach, based on the existence results for the obstacle problem proved in Section 8.2 and on the Feynman-Kač representation Theorem 9.48. In order to avoid technicalities and to show the main ideas in a clear fashion, we consider first the Black-Scholes model and then in Section 11.3 we treat the case of a complete market model with d risky assets.

11.1 Pricing and hedging in the Black-Scholes model

We consider the Black-Scholes model with risk-free rate r on a bounded time interval $[0, T]$. Since in the theory of American options dividends play an essential role, we assume the following risk-neutral dynamics for the underlying asset under the EMM Q:

$$dS_t = (r - q)S_t dt + \sigma S_t dW_t, \tag{11.1}$$

where as usual, σ is the volatility parameter, $q \geq 0$ is the dividend yield and W is a real Brownian motion on the filtered space $(\Omega, \mathcal{F}, Q, \mathcal{F}_t)$. Let us point out that the discounted price $\widetilde{S}_t = e^{-rt} S_t$ follows the dynamics

$$d\widetilde{S}_t = -q\widetilde{S}_t dt + \sigma \widetilde{S}_t dW_t. \tag{11.2}$$

Definition 11.1 *An American option is a process of the form*

$$(\psi(t, S_t))_{t \in [0,T]}$$

where ψ is a convex Lipschitz continuous function on $[0,T] \times \mathbb{R}_{>0}$: $\psi(t, S_t)$ represents the premium obtained by exercising the option at time t.

An early-exercise strategy is a stopping time on $(\Omega, \mathcal{F}, Q, \mathcal{F}_t)$ taking values in $[0,T]$: we denote by $\mathscr{T}_T$ the family of all exercise strategies. We say that $\tau_0 \in \mathscr{T}_T$ is an optimal strategy if we have that

$$E^Q \left[e^{-r\tau_0} \psi(\tau_0, S_{\tau_0}) \right] = \sup_{\tau \in \mathscr{T}_T} E^Q \left[e^{-r\tau} \psi(\tau, S_\tau) \right].$$

The following result relates the parabolic obstacle problem to the corresponding problem for the Black-Scholes differential operator

$$L_{\mathrm{BS}} f(t, S) = \frac{\sigma^2 S^2}{2} \partial_{SS} f(t, S) + (r - q) S \partial_S f(t, S) + \partial_t f(t, S) - r f(t, S).$$

We recall Definition 8.20 of *strong solution* of an obstacle problem and that any strong solution belongs to the parabolic Hölder space $C^{1+\alpha}_{P,\mathrm{loc}}$, for any $\alpha \in]0, 1[$.

Theorem 11.2 *There exists a unique strong solution f of the obstacle problem*

$$\begin{cases} \max\{L_{\mathrm{BS}} f, \psi - f\} = 0, & \text{in }]0,T[\times \mathbb{R}_{>0}, \\ f(T, \cdot) = \psi(T, \cdot), & \text{on } \mathbb{R}_{>0}, \end{cases} \tag{11.3}$$

satisfying the following properties:

i) for every $(t, y) \in [0, T[\times \mathbb{R}_{>0}$, we have

$$f(t, y) = \sup_{\substack{\tau \in \mathscr{T}_T \\ \tau \in [t,T]}} E^Q \left[e^{-r(\tau - t)} \psi(\tau, S_\tau^{t,y}) \right], \tag{11.4}$$

where $S^{t,y}$ is a solution of the SDE (11.1) with initial condition $S_t = y$;

ii) f admits first-order partial derivative with respect to S in the classical sense and we have

$$\partial_S f \in C \cap L^\infty(]0, T[\times \mathbb{R}_{>0}). \tag{11.5}$$

Proof. With the change of variables

$$u(t, x) = f(t, e^x), \qquad \varphi(t, x) = \psi(t, e^x)$$

problem (11.3) is equivalent to the obstacle problem

$$\begin{cases} \max\{Lu, \varphi - u\} = 0, & \text{in }]0, T[\times\mathbb{R}, \\ u(T, \cdot) = \varphi(T, \cdot), & \text{on } \mathbb{R}, \end{cases}$$

for the parabolic operator with constant coefficients

$$Lu = \frac{\sigma^2}{2}\partial_{xx}u + \left(r - q - \frac{\sigma^2}{2}\right)\partial_x u + \partial_t u - ru.$$

The existence of a strong solution is guaranteed by Theorem 8.21 and Remark 8.22. Furthermore, again by Remark 8.22, u is bounded from above by a super-solution and from below by φ so that an exponential-growth estimate similar to (9.66) holds: then we can apply the Feynman-Kač representation theorem, Theorem 9.48, which justifies formula (11.4). Finally, the uniqueness of the solution follows from (11.4). Moreover, by proceeding as in the proof of Proposition 9.49, we get the global boundedness of the gradient. □

We now consider a strategy $h = (\alpha_t, \beta_t)$, with $\alpha \in \mathbb{L}^2_{\text{loc}}$ and $\beta \in \mathbb{L}^1_{\text{loc}}$, with value

$$V_t(h) = \alpha_t S_t + \beta_t B_t.$$

We recall that h is self-financing if and only if

$$dV_t(h) = \alpha_t (dS_t + qS_t dt) + \beta_t dB_t.$$

If we set

$$\widetilde{V}_t(h) = e^{-rt}V_t(h),$$

we have:

Proposition 11.3 *A strategy $h = (\alpha, \beta)$ is self-financing if and only if*

$$d\widetilde{V}_t(h) = \alpha_t \left(d\widetilde{S}_t + q\widetilde{S}_t dt\right),$$

i.e.

$$\begin{aligned} \widetilde{V}_t(h) &= V_0(h) + \int_0^t \alpha_s d\widetilde{S}_s + \int_0^t \alpha_s q\widetilde{S}_s ds \\ &= V_0(h) + \int_0^t \alpha_s \sigma \widetilde{S}_s dW_s. \end{aligned} \tag{11.6}$$

In particular every self-financing strategy is uniquely determined by its initial value and its α-component. Furthermore, $\widetilde{V}(h)$ is a Q-local martingale.

Proof. The proof is analogous to that of Proposition 7.3, the only difference being the term due to the dividend. Formula (11.6) follows from (11.2). □

On the grounds of the previous proposition, the discounted value of any self-financing strategy is a Q-*local* martingale. In what follows we are interested in the strategies whose value is a strict martingale. Then we denote by $\mathcal{A}$ the family of self-financing strategies $h = (\alpha, \beta)$ such that $\alpha \in \mathbb{L}^2(P)$: a noteworthy example is represented by the strategies where α is a bounded process. We recall that, by Proposition 10.41, the discounted value of every $h \in \mathcal{A}$ is a Q-martingale. Let us now prove a version of the no-arbitrage principle.

Lemma 11.4 *Let $h^1, h^2 \in \mathcal{A}$ be two self-financing strategies such that*

$$V_\tau(h^1) \le V_\tau(h^2) \tag{11.7}$$

for some $\tau \in \mathscr{T}_T$. Then

$$V_0(h^1) \le V_0(h^2).$$

Proof. The claim is an immediate consequence of (11.7), of the martingale property of both $\widetilde{V}(h^1)$ and $\widetilde{V}(h^2)$ and of Doob's optional sampling theorem, Theorem 3.56. □

Just as in the discrete case, we define the rational price of an American option by comparing it from above and below with the value of suitable self-financing strategies. This kind of argument is necessary because, differently from the European case, the payoff $\psi(t, S_t)$ of an American option is not replicable in general, i.e. no self-financing strategy assumes the same value of the payoff at every single time. Indeed by Proposition 11.3 the discounted value of a self-financing strategy is a local martingale (or, in analytical terms, a solution of a parabolic PDE) while $\psi(t, S_t)$ is a generic process.

Let us denote by

$$\mathcal{A}_\psi^+ = \{h \in \mathcal{A} \mid V_t(h) \ge \psi(t, S_t),\ t \in [0, T] \text{ a.s.}\},$$

the family of self-financing strategies that super-replicate the payoff $\psi(t, S_t)$. Intuitively, in order to avoid arbitrage opportunities, *the initial price of the American option must be less than or equal to the initial value $V_0(h)$ for every $h \in \mathcal{A}_\psi^+$.*

Furthermore, we set

$$\mathcal{A}_\psi^- = \{h \in \mathcal{A} \mid \text{there exists } \tau \in \mathscr{T}_T \text{ s.t. } \psi(\tau, S_\tau) \ge V_\tau(h) \text{ a.s.}\}.$$

We can think of $h \in \mathcal{A}_\psi^-$ as a strategy on which we assume a short position to obtain funds to invest in the American option. In other words, $V_0(h)$ represents the amount that we can initially borrow to buy the option that has to be exercised, exploiting the early-exercise feature, at time τ to obtain the payoff $\psi(\tau, S_\tau)$ which is greater or equal to $V_\tau(h)$, amount necessary to close the short position on the strategy h. To avoid arbitrage opportunities, intuitively *the initial price of the American option must be greater than or equal to $V_0(h)$ for all $h \in \mathcal{A}_\psi^-$.*

These remarks are formalized by the following results. In particular, as an immediate consequence of Lemma 11.4, we have the following:

Proposition 11.5 *If $h_1 \in \mathcal{A}_\psi^-$ and $h_2 \in \mathcal{A}_\psi^+$ then we have*

$$V_0(h^1) \le V_0(h^2).$$

Further, for every $h_1, h_2 \in \mathcal{A}_\psi^- \cap \mathcal{A}_\psi^+$ we have

$$V_0(h^1) = V_0(h^2).$$

By Theorem 11.7, there exists a strategy $\bar{h} \in \mathcal{A}_\psi^+ \cap \mathcal{A}_\psi^-$: then, by Proposition 11.5 the following definition is well-posed.

Definition 11.6 *The arbitrage price of the American option $\psi(t, S_t)$ is the initial value of any strategy $\bar{h} \in \mathcal{A}_\psi^+ \cap \mathcal{A}_\psi^-$:*

$$V_0(\bar{h}) = \inf_{h \in \mathcal{A}_\psi^+} V_0(h) = \sup_{h \in \mathcal{A}_\psi^-} V_0(h).$$

Theorem 11.7 *Let f be the strong solution to the obstacle problem (11.3). The self-financing strategy $h = (\alpha, \beta)$ defined by*

$$V_0(h) = f(0, S_0), \qquad \alpha_t = \partial_S f(t, S_t)$$

belongs to $\mathcal{A}_\psi^+ \cap \mathcal{A}_\psi^-$. Consequently $f(0, S_0)$ is the arbitrage price of $\psi(t, S_t)$. Furthermore an optimal exercise strategy is defined by

$$\tau_0 = \inf\{t \in [0, T] \mid f(t, S_t) = \psi(t, S_t)\}, \tag{11.8}$$

and we have that

$$V_0(h) = E^Q\left[e^{-r\tau_0}\psi(\tau_0, S_{\tau_0})\right] = \sup_{\tau \in \mathscr{T}_T} E^Q\left[e^{-r\tau}\psi(\tau, S_\tau)\right],$$

where

$$S_t = S_0 e^{\sigma W_t + \left(r - q - \frac{\sigma^2}{2}\right)t}$$

is the solution of the SDE (11.1) with initial condition S_0.

Proof. The idea is to use the Itô formula to compute the stochastic differential of $f(t, S_t)$ and to separate the martingale part from the drift part of the process. We recall that, by definition of strong solution, (cf. Definition 8.20), $f \in S^p_{\text{loc}}([0,T] \times \mathbb{R}_{>0})$ and so it does not belong in general to C^2. Consequently we have to use a weak version of the Itô formula: however, since we do not have a global estimate of the second derivatives[1] of f (and therefore of $L_{\text{BS}} f$),

[1] It is possible to prove (cf. for example [225]) the following global estimate

$$\|\partial_t f(t,\cdot)\|_{L^\infty(\mathbb{R}_{>0})} + \|\partial_{SS} f(t,\cdot)\|_{L^\infty(\mathbb{R}_{>0})} \le \frac{C}{\sqrt{T-t}}$$

and then we could use it as in Remark 5.40 in order to prove the validity of the Itô formula for f.

but only a local one, we cannot use Theorem 5.37 directly, and we must use a localization argument. After fixing $R > 0$, let us consider the stopping time

$$\tau_R = T \wedge \inf\{t \mid S_t \in]0, 1/R[\cup]R, +\infty[\}.$$

By the standard regularization argument used in the proof of Theorems 5.37 and 9.48, we can prove that, for all $\tau \in \mathcal{T}_T$, we have

$$\begin{aligned} e^{-r(\tau\wedge\tau_R)} f(\tau \wedge \tau_R, S_{\tau\wedge\tau_R}) = f(0, S_0) + & \int_0^{\tau\wedge\tau_R} \sigma \widetilde{S}_s \partial_S f(s, S_s) dW_s \\ + & \int_0^{\tau\wedge\tau_R} e^{-rs} L_{\text{BS}} f(s, S_s) ds \end{aligned} \tag{11.9}$$

or equivalently, by (11.6),

$$e^{-r(\tau\wedge\tau_R)} f(\tau \wedge \tau_R, S_{\tau\wedge\tau_R}) = \widetilde{V}_{\tau\wedge\tau_R} + \int_0^{\tau\wedge\tau_R} e^{-rs} L_{\text{BS}} f(s, S_s) ds, \tag{11.10}$$

where $\widetilde{V}$ is the discounted value of the self-financing strategy $\bar{h} = (\alpha, \beta)$ defined by the initial value $f(0, S_0)$ and $\alpha_t = \partial_S f(t, S_t)$. Let us point out the analogy with the hedging strategy and the Delta of a European option (cf. Theorem 7.13).

We observe that $\widetilde{V}$ is a martingale since $\partial_S f$ is a continuous, bounded function by (11.5), and therefore $\bar{h} \in \mathcal{A}$. Let us prove that, for all $\tau \in \mathcal{T}_T$, we have that

$$\lim_{R\to\infty} \widetilde{V}_{\tau\wedge\tau_R} = \widetilde{V}_\tau. \tag{11.11}$$

Indeed

$$\begin{aligned} & E\left[\left(\int_{\tau\wedge\tau_R}^{\tau} \sigma \widetilde{S}_s \partial_S f(s, S_s) dW_s\right)^2\right] \\ =& E\left[\left(\int_0^T \sigma \widetilde{S}_t \partial_S f(t, S_t) \mathbb{1}_{\{\tau\wedge\tau_R \le t \le \tau\}} dW_t\right)^2\right] = \end{aligned}$$

(by Itô isometry, since the integrand belongs to $\mathbb{L}^2$)

$$= E\left[\int_0^T \left(\sigma \widetilde{S}_t \partial_S f(t, S_t) \mathbb{1}_{\{\tau\wedge\tau_R \le t \le \tau\}}\right)^2 dt\right] \xrightarrow[R\to\infty]{} 0$$

by the dominated convergence theorem, being $\partial_S f \in L^\infty$.

Now we can prove that $\bar{h} \in \mathcal{A}_\psi^+ \cap \mathcal{A}_\psi^-$. First of all, since $L_{\text{BS}} f \le 0$ a.e. and S_t has positive density, by (11.10), we have

$$V_{t\wedge\tau_R} \ge f(t \wedge \tau_R, S_{t\wedge\tau_R})$$

for all $t \in [0, T]$ and $R > 0$. Taking the limit in R, by (11.11) and the continuity of f, we get

$$V_t \geq f(t, S_t) \geq \psi(t, S_t), \qquad t \in [0, T],$$

and this proves that $\bar{h} \in \mathcal{A}_\psi^+$.

Secondly, since $L_{\mathrm{BS}} f(t, S_t) = 0$ a.s. on $\{\tau_0 \geq t\}$ with τ_0 defined by (11.8), again by (11.10) we have

$$V_{\tau_0 \wedge \tau_R} = f(\tau_0 \wedge \tau_R, S_{\tau_0 \wedge \tau_R})$$

for all $R > 0$. Taking the limit in R as above, we get

$$V_{\tau_0} = f(\tau_0, S_{\tau_0}) = \psi(\tau_0, S_{\tau_0}).$$

This proves that $\bar{h} \in \mathcal{A}_\psi^-$ and concludes the proof. □

11.2 American Call and Put options in the Black-Scholes model

By Theorem 11.7 we have the following expressions for the prices of Call and Put American options in the Black-Scholes model, with risk-neutral dynamics (11.1) for the underlying asset:

$$C(T, S_0, K, r, q) = \sup_{\tau \in \mathcal{T}_T} E\left[e^{-r\tau} \left(S_0 e^{\sigma W_\tau + \left(r - q - \frac{\sigma^2}{2}\right)\tau} - K \right)^+ \right],$$

$$P(T, S_0, K, r, q) = \sup_{\tau \in \mathcal{T}_T} E\left[e^{-r\tau} \left(K - S_0 e^{\sigma W_\tau + \left(r - q - \frac{\sigma^2}{2}\right)\tau} \right)^+ \right].$$

In the preceding expressions, $C(T, S_0, K, r, q)$ and $P(T, S_0, K, r, q)$ represent the prices at time 0 of Call and Put American options respectively, with maturity T, initial price of the underlying asset S_0, strike K, interest rate r and dividend yield q. For American options explicit formulas as in the European case are not known, and in order to compute the prices and the hedging strategies we have to resort to numerical methods.

The following result establishes a symmetry relation between the prices of American Call and Put options.

Proposition 11.8 *We have*

$$C(T, S_0, K, r, q) = P(T, K, S_0, q, r). \tag{11.12}$$

Proof. We set

$$Z_t = e^{\sigma W_t - \frac{\sigma^2}{2} t},$$

and recall that Z is a Q-martingale with unitary mean. Moreover, the process

$$\widetilde{W}_t = W_t - \sigma t$$

is a Brownian motion with respect to the measure $\widetilde{Q}$ defined by

$$\frac{d\widetilde{Q}}{dQ} = Z_T.$$

Then we have

$$\begin{aligned}
C(T, S_0, K, r, q) &= \sup_{\tau \in \mathscr{T}_T} E^Q \left[Z_\tau e^{-q\tau} \left(S_0 - Ke^{-\sigma W_\tau + \left(q - r + \frac{\sigma^2}{2}\right)\tau} \right)^+ \right] \\
&= \sup_{\tau \in \mathscr{T}_T} E^Q \left[Z_T e^{-q\tau} \left(S_0 - Ke^{-\sigma W_\tau + \left(q - r + \frac{\sigma^2}{2}\right)\tau} \right)^+ \right] \\
&= \sup_{\tau \in \mathscr{T}_T} E^{\widetilde{Q}} \left[e^{-q\tau} \left(S_0 - Ke^{-\sigma \widetilde{W}_\tau + \left(q - r - \frac{\sigma^2}{2}\right)\tau} \right)^+ \right].
\end{aligned}$$

The claim follows because, by symmetry, $-\widetilde{W}$ is a $\widetilde{Q}$-Brownian motion. □

Now we state a Put-Call parity formula for American options, analogous to Corollaries 1.1 and 1.2. The proof is a general consequence of the absence of arbitrage opportunities and is left as an exercise.

Proposition 11.9 (Put-Call parity for American options) *Let C, P be the arbitrage prices of American Call and Put options, respectively, with strike K and maturity T. The following relations hold:*

$$S_t - K \le C_t - P_t \le S_t - Ke^{-r(T-t)}, \tag{11.13}$$

and

$$(K - S_t)^+ \le P_t \le K. \tag{11.14}$$

Now we study some qualitative properties of the prices: in view of Proposition 11.8 it is enough to consider the case of the American Put option. In the following result we denote by

$$P(T, S) = \sup_{\tau \in \mathscr{T}_T} E \left[e^{-r\tau} \left(K - Se^{\sigma W_\tau + \left(r - q - \frac{\sigma^2}{2}\right)\tau} \right)^+ \right], \tag{11.15}$$

the price of the American Put option.

Proposition 11.10 *The following properties hold:*

i) for all $S \in \mathbb{R}_{>0}$, the function $T \mapsto P(T, S)$ is increasing. In other words, if we fix the parameters of the option, the price of the Put option decreases as we get closer to maturity;
ii) for all $T \in [0, T]$, the function $S \mapsto P(T, S)$ is decreasing and convex;
iii) for all $(T, S) \in [0, T[\times \mathbb{R}_{>0}$ we have that

$$-1 \le \partial_S P(T, S) \le 0.$$

Proof. i) is trivial. ii) is an immediate consequence of (11.15), of the monotonicity and convexity properties of the payoff function and of the fact that those properties are preserved by taking the supremum over all stopping times: indeed, if (g_τ) is a family of increasing and convex functions then also

$$g := \sup_\tau g_\tau$$

is increasing and convex.

Then $\partial_S P(T, S) \leq 0$ since $S \mapsto P(T, S)$ is decreasing. Furthermore, if $\psi(S) = (K - S)^+$ then we have

$$|\psi(S) - \psi(S')| \leq |S - S'|,$$

and so the third property follows from the argument used in the proof of Proposition 9.49, by observing that

$$\left| E\left[e^{-r\tau}\psi\left(S_0 e^{\sigma W_\tau + \left(r - q - \frac{\sigma^2}{2}\right)\tau} \right) - e^{-r\tau}\psi\left(S_0' e^{\sigma W_\tau + \left(r - q - \frac{\sigma^2}{2}\right)\tau} \right) \right] \right|$$
$$\leq |S_0 - S_0'| E\left[e^{\sigma W_\tau - \left(q + \frac{\sigma^2}{2}\right)\tau} \right] \leq$$

(since $q \geq 0$)

$$\leq |S_0 - S_0'| E\left[e^{\sigma W_\tau - \frac{\sigma^2}{2}\tau} \right] =$$

(since the exponential martingale has unitary mean)

$$= |S_0 - S_0'|.$$

□

In the last part of this section, we study the relation between the prices of the European Put option and American Put option by introducing the concept of *early exercise premium*. In the sequel we denote by $f = f(t, S)$ the solution of the obstacle problem (11.3) relative to the payoff function of the Put option

$$\psi(t, S) = (K - S)^+.$$

For $t \in [0, T]$, we define

$$S^*(t) = \inf\{S > 0 \mid f(t, S) > \psi(t, S)\}.$$

$S^*(t)$ is called *critical price at time* t and it corresponds to the point where f "touches" the payoff ψ.

Lemma 11.11 *For all $(t, S) \in [0, T[\times\mathbb{R}_{>0}$, we have that*

$$L_{\mathrm{BS}} f(t, S) = (qS - rK)\mathbb{1}_{\{S \leq S^*(t)\}}. \tag{11.16}$$

In particular $L_{\mathrm{BS}} f$ is a bounded function.

Proof. First of all let us observe that $S^*(t) < K$: indeed if we had $S^*(t) \geq K$ then it should hold that

$$f(t,K) = \psi(t,K) = 0$$

and this is absurd (since $f > 0$ by (11.4)). Then by the convexity of $S \mapsto f(t,S)$ (that follows from Proposition 11.10-ii)) we infer that

i) $f(t,S) = K - S$ for $S \leq S^*(t)$;
ii) $f(t,S) > \psi(t,S)$ for $S > S^*(t)$.

So we have that

$$L_{\mathrm{BS}}f(t,S) = \begin{cases} (qS - rK), & \text{for } S \leq S^*(t), \\ 0, & \text{a.e. for } S > S^*(t). \end{cases}$$ □

Now we go back to formula (11.9) with $\tau = T$: since $L_{\mathrm{BS}}f$ is bounded, we can take the limit as $R \to +\infty$ and then get

$$e^{-rT}f(T,S_T) = f(0,S_0) + \int_0^T e^{-rt}L_{\mathrm{BS}}f(t,S_t)dt + \int_0^T \sigma\widetilde{S}_t\partial_S f(t,S_t)dW_t,$$

and taking expectation, by (11.16),

$$p(T,S_0) = P(T,S_0) + \int_0^T e^{-rt}E^Q\left[(qS_t - rK)\mathbb{1}_{\{S_t \leq S^*(t)\}}\right]dt, \qquad (11.17)$$

where $p(T,S_0)$ and $P(T,S_0)$ denote the price at time 0 of the European and American options respectively, with maturity T. The expression (11.17) gives the difference $P(T,S_0) - p(T,S_0)$, usually called *early exercise premium:* it quantifies the value of the possibility of exercising before maturity. Formula (11.17) has been proved originally by Kim [208].

11.3 Pricing and hedging in a complete market

Let us consider a market model consisting of d risky assets S_t^i, $i = 1,\dots,d$, and one non-risky asset B_t, $t \in [0,T]$. We suppose that

$$S_t^i = e^{X_t^i}, \qquad i = 1,\dots,d,$$

where $X = (X^1,\dots,X^d)$ is solution of the system of SDEs

$$dX_t^i = b^i(t,X_t)dt + \sigma^i(t,X_t)dW_t^i, \qquad i = 1,\dots,d, \qquad (11.18)$$

and $W = (W^1,\dots,W^d)$ is Brownian motion on the space $(\Omega,\mathcal{F},P,(\mathcal{F}_t))$, with constant correlation matrix ϱ:

$$d\langle W^i,W^j\rangle_t = \varrho^{ij}dt.$$

We will specify in Hypothesis 11.12 the regularity conditions on the coefficients that guarantee the existence of a strong solution to (11.18). By the Itô formula we have

$$dS_t^i = \mu_t^i S_t^i dt + \sigma_t^i S_t^i dW_t^i, \qquad i = 1, \dots, d, \tag{11.19}$$

where $\mu_t^i = b^i(t, X_t) + \frac{(\sigma^i(t,X_t))^2}{2}$. We also assume that the price of the non-risky asset is

$$B_t = e^{\int_0^t r_s ds}, \qquad t \in [0, T],$$

where $r_t = r(t, X_t)$, with r a suitable function, and that the i-th asset pays continuous dividends at the rate $q_t^i = q^i(t, X_t)$.

Hypothesis 11.12 *The functions b, σ, r and q are bounded and locally Hölder continuous on $]0, T[\times \mathbb{R}^d$. The matrix $(c_{ij}) = \left(\varrho^{ij}\sigma^i\sigma^j\right)$ is uniformly positive definite: there exists a positive constant Λ such that*

$$\Lambda^{-1}|\xi|^2 \le \sum_{i,j=1}^d c_{ij}(t, x)\xi_i\xi_j \le \Lambda|\xi|^2, \qquad t \in]0, T[, \ x, \xi \in \mathbb{R}^d.$$

Under these conditions, by Theorem 10.27 (see also Remark 10.30), there exists a unique EMM Q: we denote by $W^Q = \left(W^{Q,1}, \dots, W^{Q,d}\right)$ the Brownian motion associated to Q, under which the risk-neutral dynamics of the discounted prices

$$\widetilde{S}_t^i = e^{-\int_0^t r_s ds} S_t^i$$

is given by

$$d\widetilde{S}_t^i = -q_t^i \widetilde{S}_t^i dt + \widetilde{S}_t^i \sigma_t^i dW_t^{Q,i}, \qquad i = 1, \dots, d. \tag{11.20}$$

The definitions of American option and exercise strategy are analogous to the ones in the Black-Scholes case. An American option is a process of the form

$$(\psi(t, S_t))_{t \in [0,T]}$$

where ψ *is a convex Lipschitz continuous function* on $[0, T] \times \mathbb{R}_{>0}^d$.

We denote by $\mathscr{T}_T$ the family of the $\mathcal{F}_t$-stopping times with values in $[0, T]$ and we say that $\tau \in \mathscr{T}_T$ is an *early-exercise strategy.* Furthermore $\tau_0 \in \mathscr{T}_T$ is an *optimal strategy* if we have

$$E^Q\left[e^{-\int_0^{\tau_0} r_s ds}\psi(\tau_0, S_{\tau_0})\right] = \sup_{\tau \in \mathscr{T}_T} E^Q\left[e^{-\int_0^{\tau} r_s ds}\psi(\tau, S_\tau)\right].$$

The following result generalizes Theorem 11.2. In the following statement (t, S) is an element of $[0, T] \times \mathbb{R}_{>0}^d$ and L is the parabolic operator associated to the process (S_t):

$$Lf = \frac{1}{2}\sum_{i,j=1}^d \widetilde{c}_{ij} S^i S^j \partial_{S^i S^j} f + \sum_{i=1}^d (\widetilde{r} - \widetilde{q}^i) S^i \partial_{S^i} f + \partial_t f - \widetilde{r} f,$$

where, after setting $\log S = (\log S^1, \dots, \log S^d)$,

$$\widetilde{c}_{ij} = c_{ij}(t, \log S), \quad \widetilde{r} = r(t, \log S) \quad \text{and} \quad \widetilde{q}^i = q^i(t, \log S).$$

Theorem 11.13 *There exists a unique strong solution f of the obstacle problem*

$$\begin{cases} \max\{Lf, \psi - f\} = 0, & \text{in }]0, T[\times \mathbb{R}^d_{>0}, \\ f(T, \cdot) = \psi(T, \cdot), & \text{on } \mathbb{R}^d_{>0}, \end{cases} \tag{11.21}$$

satisfying the following properties:

i) for all $(t, y) \in [0, T[\times \mathbb{R}^d_{>0}$, we have

$$f(t, y) = \sup_{\substack{\tau \in \mathcal{T}_T \\ \tau \in [t,T]}} E^Q \left[e^{-\int_t^\tau r_s^{t,y} ds} \psi(\tau, S_\tau^{t,y}) \right],$$

where $S^{t,y}$ is the price process with initial value $S_t = y$, and $r_s^{t,y} = r(s, \log S_s^{t,y})$;

ii) f admits spatial gradient $\nabla f = (\partial_{S^1} f, \dots, \partial_{S^d} f)$ in the classical sense and

$$\nabla f \in C(]0, T[\times \mathbb{R}^d_{>0}).$$

Furthermore, assume that the functions b, σ and q are globally Lipschitz continuous in x, uniformly with respect to t, on $]0, T[\times \mathbb{R}^d$: if ψ is bounded or the coefficient r is constant[2] *then*

$$\nabla f \in L^\infty(]0, T[\times \mathbb{R}^d_{>0}).$$

Proof. By the change of variable $S = e^x$, the claim is direct consequence of Theorems 8.21, 9.48 and of Proposition 9.49. □

Let us now consider a strategy $h = (\alpha_t, \beta_t)$, $\alpha \in \mathbb{L}^2_{\text{loc}}$ and $\beta \in \mathbb{L}^1_{\text{loc}}$, with value

$$V_t(h) = \alpha_t \cdot S_t + \beta_t B_t,$$

and let us recall the self-financing condition:

$$dV_t(h) = \sum_{i=1}^d \alpha_t^i \left(dS_t^i + q_t^i S_t^i dt \right) + \beta_t dB_t.$$

If we set

$$\widetilde{V}_t(h) = e^{-\int_0^t r_s ds} V_t(h),$$

the following holds:

[2] In general, without the assumptions on the boundedness of ψ or the fact that r is constant, proceeding as in Proposition 9.49 we can prove that ∇f has at most linear growth in S.

Proposition 11.14 *The strategy $h = (\alpha, \beta)$ is self-financing if and only if*

$$\begin{aligned}\widetilde{V}_t(h) &= V_0(h) + \sum_{i=1}^{d} \int_0^t \alpha_s^i d\widetilde{S}_s^i + \sum_{i=1}^{d} \int_0^t \alpha_s^i q_s^i \widetilde{S}_s^i ds \\ &= V_0(h) + \sum_{i=1}^{d} \int_0^t \alpha_s^i \widetilde{S}_s^i \sigma_s^i dW_t^{Q,i}.\end{aligned} \tag{11.22}$$

Proof. The proof is analogous to that of Proposition 7.3. The second equality in (11.22) follows from (11.20). □

The definition of arbitrage price of the American Option is based upon the same arguments already used in the Black-Scholes market setting.

Notation 11.15 *We recall Definition 10.42 and we set*

$$\mathcal{A}_\psi^+ = \{h \in \mathcal{A} \mid V_t(h) \geq \psi(t, S_t),\ t \in [0, T]\ a.s.\},$$
$$\mathcal{A}_\psi^- = \{h \in \mathcal{A} \mid \text{there exists } \tau_0 \in \mathscr{T}_T \text{ s.t. } \psi(\tau_0, S_{\tau_0}) \geq V_{\tau_0}(h)\ a.s.\}.$$

$\mathcal{A}_\psi^+$ and $\mathcal{A}_\psi^-$ denote the families of self-financing strategies super- and sub-replicating respectively. By the martingale property, it follows that

$$V_0(h^-) \leq V_0(h^+)$$

for any $h^- \in \mathcal{A}_\psi^-$ and $h^+ \in \mathcal{A}_\psi^+$. Furthermore, in order not to introduce arbitrage opportunities, the price of the American option $\psi(t, S_t)$ must be less or equal to the initial value $V_0(h)$ for all $h \in \mathcal{A}_\psi^+$ and greater or equal to the initial value $V_0(h)$ for all $h \in \mathcal{A}_\psi^-$.

The following result, analogous to Theorem 11.7, gives the definition of the arbitrage price of the American option by showing that

$$\inf_{h \in \mathcal{A}_\psi^+} V_0(h) = \sup_{h \in \mathcal{A}_\psi^-} V_0(h).$$

Theorem 11.16 *Let f be the solution to the obstacle problem (11.21). The self-financing strategy $\bar{h} = (\alpha, \beta)$ defined by*

$$V_0(\bar{h}) = f(0, S_0), \qquad \alpha_t = \nabla f(t, S_t),$$

belongs to $\mathcal{A}_\psi^+ \cap \mathcal{A}_\psi^-$. By definition

$$f(0, S_0) = V_0(\bar{h}) = \inf_{h \in \mathcal{A}_\psi^+} V_0(h) = \sup_{h \in \mathcal{A}_\psi^-} V_0(h)$$

is the arbitrage price of $\psi(t, S_t)$. Furthermore, an optimal exercise strategy is defined by

$$\tau_0 = \inf\{t \in [0, T] \mid f(t, S_t) = \psi(t, S_t)\},$$

and we have that

$$V_0(\bar{h}) = E^Q\left[e^{-\int_0^{\tau_0} r_s ds}\psi(\tau_0, S_{\tau_0})\right] = \sup_{\tau \in \mathscr{T}_T} E^Q\left[e^{-\int_0^{\tau} r_s ds}\psi(\tau, S_\tau)\right].$$

12

Numerical methods

In this chapter we present some methods for the numerical solution of deterministic and stochastic differential equations. The numerical approximation is necessary when it is not possible to determine explicitly the solution of an equation (i.e. nearly always).

What underlies many numerical methods for differential equations is simply the approximation of the derivatives (the integrals) with incremental ratios (with sums). We will follow this approach in the whole chapter, trying to present the methods for different kinds of equations (ordinary, with partial derivatives, deterministic or stochastic) in the same common setting. Broadly speaking, the main ingredients we will use in order for a solution X of a differential equation $LX = 0$ to be approximated by a solution X^δ of the "discretized" equation are three:

- *the regularity* of the solution X, that is derived from the properties of the differential equation and is in general a consequence of the regularity hypotheses on the coefficients;
- *the consistency* of the discretization (or numerical scheme), i.e. the fact that $L - L^\delta \xrightarrow[\delta \to 0+]{} 0$ in an appropriate sense: this is in general a consequence of the approximation by a Taylor series expansion and of the regularity of the solution in the previous point;
- *the stability* of the numerical scheme, in general a consequence of a maximum principle for L^δ that gives an estimate of a function (or a process) Y in terms of the initial datum Y_0 and of $L^\delta Y$.

12.1 Euler method for ordinary equations

Let us consider the ordinary differential equation

$$\frac{dX_t}{dt} = \mu(t, X_t), \qquad t \in [0, T], \tag{12.1}$$

where

$$\mu : [0, T] \times \mathbb{R} \longrightarrow \mathbb{R}$$

is a continuous function. For the sake of simplicity and clarity, we confine ourselves to the 1-dimensional case, but the following results can be extended without difficulties. We assume the linear-growth condition

$$|\mu(t, x)| \leq K(1 + |x|), \qquad x \in \mathbb{R},\ t \in [0, T], \tag{12.2}$$

and, further, we assume Lipschitz continuity with respect to both variables (so a slightly stronger assumption than the standard Lipschitz continuity in x):

$$|\mu(t, x) - \mu(s, y)| \leq K(|t - s| + |x - y|), \qquad x, y \in \mathbb{R},\ t, s \in [0, T]. \tag{12.3}$$

For fixed $N \in \mathbb{N}$, we divide the interval $[0, T]$ into N intervals $[t_{n-1}, t_n]$ whose length is $\delta := \frac{T}{N}$, so that $t_n = n\delta$ for $n = 0, \dots, N$. We denote by δ the discretization step. By approximating the derivative in (12.1) by the incremental ratio relative to it (or, equivalently, by truncating the Taylor expansion of the function X with initial point t_n to the first order), we get the following discretization of (12.1):

$$X^{\delta}_{t_{n+1}} = X^{\delta}_{t_n} + \mu(t_n, X^{\delta}_{t_n})\delta, \qquad n = 1, \dots, N. \tag{12.4}$$

By imposing $X^{\delta}_0 = X_0$, (12.4) defines recursively the values of $X^{\delta}_{t_n}$ for $n = 1, \dots, N$, giving an algorithm for the determination of an approximation of the solution X.

It can be useful to consider the equivalent integral version of (12.1):

$$L_t X = 0, \qquad t \in [0, T], \tag{12.5}$$

where L_t is the operator defined by

$$L_t X := X_t - X_0 - \int_0^t \mu(s, X_s)ds, \qquad t \in [0, T]. \tag{12.6}$$

For fixed t_n as before, equation (12.5) can be discretized by making the integrand $\mu(s, X_s) \simeq \mu(t_{n-1}, X_{t_{n-1}})$ constant over the interval $[t_{n-1}, t_n]$. More precisely, we define the discretized operator L^{δ} by putting

$$L^{\delta}_t X := X_t - X_0 - \int_0^t \sum_{n=1}^{N} \mu(t_{n-1}, X_{t_{n-1}}) \mathbb{1}_{]t_{n-1}, t_n]}(s)ds, \qquad t \in [0, T]. \tag{12.7}$$

The equation

$$L^{\delta}_t X^{\delta} = 0, \qquad t \in [0, T],$$

is equivalent to

$$X^{\delta}_t = X_0 - \int_0^t \sum_{n=1}^{N} \mu(t_{n-1}, X^{\delta}_{t_{n-1}}) \mathbb{1}_{]t_{n-1}, t_n]}(s)ds, \qquad t \in [0, T], \tag{12.8}$$

and it defines recursively the same (at the points $t = t_n$) approximation X^δ of the solution X introduced earlier with formula (12.4): more precisely the function X^δ is defined by linear interpolation of the values $X^\delta_{t_n}$, $n = 0, \dots, N$.

In order to study the convergence of the Euler numerical scheme, first we prove an a-priori estimate of the regularity of the solutions to the differential equation.

Proposition 12.1 (Regularity) *The solution X in (12.6) is such that*

$$|X_t - X_s| \le K_1|t - s|, \qquad t, s \in [0, T], \tag{12.9}$$

with K_1 depending on K in (12.2), T and X_0 only.

Proof. By definition, if $s < t$, we have

$$|X_t - X_s| = \left| \int_s^t \mu(u, X_u)du \right| \le (t - s) \max_{u \in [0,T]} |\mu(u, X_u)|.$$

The claim follows by the assumption of linear growth on μ and from the following estimate

$$|X_t| \le e^{Kt} \left(|X_0| + KT \right), \qquad t \in [0, T], \tag{12.10}$$

that can be proved using Gronwall's Lemma and the inequality

$$|X_t| \le |X_0| + \int_0^t |\mu(s, X_s)| \, ds \le$$

(by the linear-growth assumption on μ)

$$\le |X_0| + KT + K \int_0^t |X_s| \, ds.$$

□

Now we verify the consistency of the discretized operator L^δ with L.

Proposition 12.2 (Consistency) *Let Y be a Lipschitz continuous function on $[0, T]$ with Lipschitz constant K_1. For every $t \in [0, T]$*

$$\left| L_t Y - L^\delta_t Y \right| \le C\delta, \tag{12.11}$$

where the constant C depends only on K, K_1 and T.

Proof. It suffices to consider the case $t = t_n$. We have

$$\left| L_{t_n} Y - L^\delta_{t_n} Y \right| = \left| \sum_{k=1}^n \int_{t_{k-1}}^{t_k} \left(\mu(s, Y_s) - \mu\left(t_{k-1}, Y_{t_{k-1}}\right) \right) ds \right| \le$$

(by the Lipschitz-continuity assumption on μ)

$$\le K\sum_{k=1}^{n}\int_{t_{k-1}}^{t_k}\left(s-t_{k-1}+|Y_s-Y_{t_{k-1}}|\right)ds$$

(by (12.9))

$$\begin{aligned}&\le K(1+K_1)\sum_{k=1}^{n}\int_{t_{k-1}}^{t_k}(s-t_{k-1})\,ds\\&\le K(1+K_1)T\delta.\end{aligned}$$

□

The third step is the proof of a maximum principle for the discrete operator L^δ.

Proposition 12.3 (Stability - Maximum principle) *Let X, Y be continuous functions on $[0,T]$. Then*

$$\max_{t\in[0,T]}|X_t-Y_t|\le e^{KT}\left(|X_0-Y_0|+\max_{t\in[0,T]}|L_t^\delta X-L_t^\delta Y|\right). \qquad (12.12)$$

Proof. Since

$$\begin{aligned}X_t-Y_t=&X_0-Y_0+L_t^\delta X-L_t^\delta Y\\&+\int_0^t\sum_{n=1}^{N}\left(\mu(t_{n-1},X_{t_{n-1}})-\mu(t_{n-1},Y_{t_{n-1}})\right)\mathbb{1}_{]t_{n-1},t_n]}(s)ds\end{aligned}$$

by the Lipschitz-continuity assumption on μ, we have

$$\max_{s\in[0,t]}|X_s-Y_s|\le|X_0-Y_0|+\max_{s\in[0,T]}\left|L_s^\delta X-L_s^\delta Y\right|+K\int_0^t\max_{u\in[0,s]}|X_u-Y_u|\,ds.$$

The claim follows from Gronwall's Lemma. □

Remark 12.4 The previous result is sometimes called "maximum principle" because, if $Y_t\equiv 0$ and the differential equation is linear and homogeneous, i.e. of the form $\mu(t,x)=a(t)x$, then (12.12) becomes

$$\max_{t\in[0,T]}|X_t|\le e^{KT}\left(|X_0|+\max_{t\in[0,T]}|L_tX|\right),$$

and this expresses the fact that the maximum of the solution to the equation $L_tX=f$ can be estimated in terms of the initial value X_0 and the known function f. This kind of result guarantees *the stability* of a numerical scheme: this means that, for two solutions X^δ, Y^δ of $L^\delta X=0$, (12.12) becomes

$$\max_{t\in[0,T]}|X_t^\delta-Y_t^\delta|\le e^{Kt}|X_0^\delta-Y_0^\delta|$$

and this gives an estimate of the sensitivity of the solution with respect to some perturbation of the initial datum. □

Now we prove that the order of convergence of the Euler discretization scheme is one.

Theorem 12.5 *Let X and X^δ be the solutions of $L_t X = 0$ and $L_t^\delta X^\delta = 0$, respectively, with the same initial datum $X_0 = X_0^\delta$. There exists a constant C depending only on T, K in (12.2) and X_0 such that*

$$\max_{t\in[0,T]} \left| X_t - X_t^\delta \right| \le C\delta. \tag{12.13}$$

Proof. By the maximum principle we have

$$\max_{t\in[0,T]} \left| X_t - X_t^\delta \right| \le e^{KT} \max_{t\in[0,T]} \left| L_t^\delta X - L_t^\delta X^\delta \right| = e^{KT} \max_{t\in[0,T]} \left| L_t^\delta X - L_t X \right| \le$$

(by the consistency results, Proposition 12.2, and by the regularity results, Proposition 12.1)

$$\le C\delta$$

where C depends only on T, K and X_0. □

12.1.1 Higher order schemes

The Euler discretization is extremely simple and intuitive, nevertheless it gives satisfying results only if the coefficient μ can be well-approximated by linear functions. In general it is preferable to use higher order numerical schemes. We briefly touch upon the main ideas. In what follows we assume that the coefficient μ is sufficiently regular and we consider the equation

$$X_t' = \mu(t, X_t), \qquad t \in [0, T].$$

Differentiating the previous equation and omitting the arguments of the function μ and of its derivatives, we get

$$\begin{aligned} X'' &= \mu_t + \mu_x X' = \mu_t + \mu_x \mu, \\ X''' &= \mu_{tt} + 2\mu_{tx} X' + \mu_{xx} (X')^2 + \mu_x X'', \end{aligned}$$

where μ_t, μ_x denote the partial derivatives of the function $\mu = \mu(t, x)$. Substituting these expressions in the Taylor expansion of the p-th order, we have

$$X_{t_{n+1}} = X_{t_n} + X_{t_n}' \delta + \dots + \frac{1}{p!} X_{t_n}^{(p)} \delta^p$$

and we obtain the *p-th order Euler scheme*. For example, the second order scheme is

$$X_{t_{n+1}}^\delta = X_{t_n}^\delta + \mu(t_n, X_{t_n}^\delta)\delta + \frac{\delta^2}{2} \left(\mu_t(t_n, X_{t_n}^\delta) + \mu_x(t_n, X_{t_n}^\delta) \mu(t_n, X_{t_n}^\delta) \right).$$

Under suitable regularity assumptions on the coefficient μ, it is possible to prove that the order of convergence of the *p-th order Euler scheme is p*, i.e.

$$\max_{t\in[0,T]} \left| X_t - X_t^\delta \right| \le C\delta^p.$$

12.2 Euler method for stochastic differential equations

We examine the problem of numerical approximation of a stochastic differential equation. We refer to the monographs by Kloeden and Platen [210], Bouleau and Lépingle [54] for the presentation of the general theory.

We use the notations of Paragraph 9.1 and we define the operator

$$L_t X := X_t - X_0 - \int_0^t \mu(s, X_s) ds - \int_0^t \sigma(s, X_s) dW_s, \qquad t \in [0, T], \quad (12.14)$$

where X_0 is a given $\mathcal{F}_0$-measurable random variable in $L^2(\Omega, P)$ and the coefficients

$$\mu = \mu(t,x) : [0,T] \times \mathbb{R} \longrightarrow \mathbb{R}, \qquad \sigma = \sigma(t,x) : [0,T] \times \mathbb{R} \longrightarrow \mathbb{R},$$

verify the following assumption

$$|\mu(t,x) - \mu(s,y)|^2 + |\sigma(t,x) - \sigma(t,y)|^2 \le K\left(|t-s| + |x-y|^2\right), \qquad (12.15)$$

for $x, y \in \mathbb{R}$ and $t, s \in [0, T]$. Notice that (12.15) is slightly stronger than the standard assumptions (cf. Definition 9.4), since it is equivalent to global Lipschitz continuity in the x variable and to Hölder continuity with exponent $1/2$ in t: in particular (12.15) contains the usual linear-growth condition. Under these assumptions, Theorem 9.11 ensures the existence of a strong solution $X \in \mathcal{A}_c$ of the equation

$$L_t X = 0, \qquad t \in [0, T].$$

We recall (cf. Notation 9.7) that $\mathcal{A}_c$ denotes the space of continuous $\mathcal{F}_t$-adapted processes such that

$$[\![X]\!]_T = \sqrt{E\left[\sup_{0 \le t \le T} X_t^2\right]}$$

is finite.

We divide the interval $[0, T]$ into N intervals $[t_{n-1}, t_n]$ whose length is $\delta := \frac{T}{N}$ and we define the discretized operator L^δ obtained by making the integrands in (12.14) piecewise constant:

$$\begin{aligned} L_t^\delta X := X_t - X_0 - & \int_0^t \sum_{n=1}^N \mu(t_{n-1}, X_{t_{n-1}}) \mathbb{1}_{]t_{n-1}, t_n]}(s) ds \\ & - \int_0^t \sum_{n=1}^N \sigma(t_{n-1}, X_{t_{n-1}}) \mathbb{1}_{]t_{n-1}, t_n]}(s) dW_s, \end{aligned} \qquad (12.16)$$

for $t \in [0, T]$. The equation

$$L_t^\delta X^\delta = 0, \qquad t \in [0, T], \qquad (12.17)$$

defines the discretized process X^δ: for $t = t_n$ (12.17) is equivalent to the formula

$$X^\delta_{t_{n+1}} = X^\delta_{t_n} + \mu(t_n, X^\delta_{t_n})\delta + \sigma(t_n, X^\delta_{t_n}) \left(W_{t_{n+1}} - W_{t_n}\right), \tag{12.18}$$

that determines the discretized process X^δ recursively, starting from the initial datum $X^\delta_0 = X_0$.

The first tool we need in order to prove the convergence of the Euler scheme is the following result on the regularity of the solutions of the stochastic equation, contained in Theorem 9.14:

Proposition 12.6 (Regularity) *The solution X of $L_t X = 0$ is such that*

$$E\left[\sup_{s\in[t,t']} |X_s - X_t|^2\right] \le K_1(t'-t), \qquad 0 \le t < t' \le T, \tag{12.19}$$

where K_1 is a constant that depends only on T, $E\left[X_0^2\right]$ and K in (12.15).

The second step consists in verifying the consistency of the discretized operator L^δ with L: the next result is analogous to Proposition 12.2.

Proposition 12.7 (Consistency) *Let $Y \in \mathcal{A}_c$ such that*

$$E\left[\sup_{s\in[t,t']} |Y_s - Y_t|^2\right] \le K_1(t'-t), \qquad 0 \le t < t' \le T. \tag{12.20}$$

Then

$$[\![LY - L^\delta Y]\!]_T \le C\sqrt{\delta}, \tag{12.21}$$

where the constant C depends only on K, K_1 and T.

Proof. We have

$$L_t Y - L^\delta_t Y = \int_0^t \underbrace{\sum_{n=1}^N \left(\mu(t_{n-1}, Y_{t_{n-1}}) - \mu(s, Y_s)\right) \mathbb{1}_{]t_{n-1},t_n]}(s)}_{:=Z^\mu_s} ds$$

$$+ \int_0^t \underbrace{\sum_{n=1}^N \left(\sigma(t_{n-1}, Y_{t_{n-1}}) - \sigma(s, Y_s)\right) \mathbb{1}_{]t_{n-1},t_n]}(s)}_{:=Z^\sigma_s} dW_s,$$

and so, by Lemma 9.9,

$$[\![LY - L^\delta Y]\!]_T^2 \le 2\int_0^T \left(T[\![Z^\mu]\!]_t^2 + 4[\![Z^\sigma]\!]_t^2\right) dt.$$

To conclude the proof, we observe that

$$[\![Z^\mu]\!]_t^2 = E\left[\sup_{s\le t}\sum_{n=1}^{N}\left(\mu(s,Y_s)-\mu(t_{n-1},Y_{t_{n-1}})\right)^2 \mathbb{1}_{]t_{n-1},t_n]}(s)\right] \le$$

(by the Lipschitz-continuity assumption (12.15))

$$\le KE\left[\sup_{s\le t}\sum_{n=1}^{N}\left(|s-t_{n-1}|+|Y_s-Y_{t_{n-1}}|^2\right)\mathbb{1}_{]t_{n-1},t_n]}(s)\right]\le C\delta$$

in view of the regularity assumption (12.20), and an analogous estimate holds for $[\![Z^\mu]\!]_t^2$. □

The third tool is a maximum principle for the discrete operator L^δ.

Proposition 12.8 (Stability - Maximum principle) *There exists a constant C_0, depending only on K and T such that, for every pair of processes $X, Y \in \mathcal{A}_c$, we have*

$$[\![X-Y]\!]_T^2 \le C_0\left(E\left[|X_0-Y_0|^2\right]+[\![L^\delta X - L^\delta Y]\!]_T^2\right). \tag{12.22}$$

Proof. Since

$$\begin{aligned}X_t - Y_t =& L_t^\delta X - L_t^\delta Y + X_0 - Y_0\\ &+\int_0^t \underbrace{\sum_{n=1}^{N}\left(\mu(t_{n-1},X_{t_{n-1}})-\mu(t_{n-1},Y_{t_{n-1}})\right)\mathbb{1}_{]t_{n-1},t_n]}(s)}_{:=Z_s^\mu}\,ds\\ &+\int_0^t \underbrace{\sum_{n=1}^{N}\left(\sigma(t_{n-1},X_{t_{n-1}})-\sigma(t_{n-1},Y_{t_{n-1}})\right)\mathbb{1}_{]t_{n-1},t_n]}(s)}_{:=Z_s^\sigma}\,dW_s,\end{aligned}$$

using Lemma 9.9 we get

$$\begin{aligned}[\![X-Y]\!]_t^2 \le& 4\Big(E\left[(X_0-Y_0)^2\right]+[\![L^\delta X - L^\delta Y]\!]_T^2\\ &+t\int_0^t [\![Z^\mu]\!]_s^2 ds + 4\int_0^t [\![Z^\sigma]\!]_s^2 ds\Big).\end{aligned}$$

On the other hand we have

$$[\![Z^\mu]\!]_t^2 = E\left[\sup_{s\le t}\sum_{n=1}^{N}\left(\mu(t_{n-1},X_{t_{n-1}})-\mu(t_{n-1},Y_{t_{n-1}})\right)^2\mathbb{1}_{]t_{n-1},t_n]}(s)\right]\le$$

(by the Lipschitz-continuity assumption on μ)

$$\leq KE\left[\sup_{s\leq t}\sum_{n=1}^{N}|X_{t_{n-1}}-Y_{t_{n-1}}|^2\mathbb{1}_{]t_{n-1},t_n]}(s)\right]\leq K[\![X-Y]\!]_t^2,$$

and an analogous estimate holds for $[\![Z^\sigma]\!]_t^2$. So putting the previous estimates together, we get

$$[\![X-Y]\!]_t^2\leq 4E\left[(X_0-Y_0)^2\right]+4[\![L^\delta X-L^\delta Y]\!]_T^2+4K(T+4)\int_0^t[\![X-Y]\!]_s^2ds$$

and the claim follows from Gronwall's Lemma. □

We can now prove the following result that states that *the order of strong convergence of the Euler scheme is* $\frac{1}{2}$.

Theorem 12.9 *There exists a constant C depending only on K,T and $E\left[X_0^2\right]$, such that*

$$[\![X-X^\delta]\!]_T\leq C\sqrt{\delta}.$$

Proof. By the maximum principle, Proposition 12.8, we have

$$[\![X-X^\delta]\!]_T^2\leq C_0[\![L^\delta X-L^\delta X^\delta]\!]_T^2=C_0[\![L^\delta X-LX]\!]_T^2\leq C\delta$$

where C depends only on T,K and $E\left[X_0^2\right]$, and the last inequality follows from the consistency and regularity results, Propositions 12.7 and 12.6. □

12.2.1 Milstein scheme

Analogously to the deterministic case, it is possible to introduce higher-order schemes for the discretization of stochastic equations. One of the simplest is the Milstein scheme, which uses a first-order approximation of the diffusion term with respect to the variable x:

$$\int_{t_n}^{t_{n+1}}\sigma(t,X_t)dW_t\sim\int_{t_n}^{t_{n+1}}\left(\sigma(t_n,X_{t_n})+\partial_x\sigma(t_n,X_{t_n})(W_t-W_{t_n})\right)dW_t.$$

By simple computation we get

$$\int_{t_n}^{t_{n+1}}(W_t-W_{t_n})dW_t=\frac{\left(W_{t_{n+1}}-W_{t_n}\right)^2-(t_{n+1}-t_n)}{2}.$$

Then, putting $\delta=t_{n+1}-t_n$ and denoting a standard Normal random variable by Z, we get the natural extension of the iterative scheme (12.18):

$$X_{t_{n+1}}=X_{t_n}+\mu(t_n,X_{t_n})\delta+\sigma(t_n,X_{t_n})\sqrt{\delta}Z+\partial_x\sigma(t_n,X_{t_n})\frac{\delta(Z^2-1)}{2}.$$

It is known that the order of strong convergence of the Milstein scheme is one.

By way of example, for the discretization of a geometric Brownian motion

$$dS_t = \mu S_t dt + \sigma S_t dW_t,$$

we have

$$S_{t_{n+1}} = S_{t_n} \left(1 + \delta \left(\mu + \frac{\sigma^2}{2}(Z^2 - 1)\right) + \sigma\sqrt{\delta}Z\right).$$

12.3 Finite-difference methods for parabolic equations

In this section we present some simple finite-difference schemes for parabolic differential operators in $\mathbb{R}^2$. Finite-difference methods give superior results, in terms of accuracy and speed of computation of the price and the Greeks of an option, with respect to other numerical schemes (binomial and Monte Carlo), even though their application is limited to low dimensional problems.

Among the monographs that study in depth finite-difference schemes applied to financial problems, we mention Zhu, Wu and Chern [349], Tavella and Randall [325]. The monographs by Mitchell and Griffiths [258], Raviart and Thomas [289], Smith [314], Hall and Porsching [161] investigate finite-difference methods for partial differential equations on a more advanced and general level.

Let us consider an operator of the form $A + \partial_t$, where

$$Au(t, x) := a(t, x)\partial_{xx}u(t, x) + b(t, x)\partial_x u(t, x) - r(t, x)u(t, x), \qquad (12.23)$$

and $(t, x) \in \mathbb{R}^2$. We suppose that A verifies the standard hypotheses in Paragraph 8.1: the coefficients a, b and r are bounded Hölder continuous functions and there exists a positive constant μ such that

$$\mu^{-1} \le a(t, x) \le \mu, \qquad (t, x) \in \mathbb{R}^2.$$

If we assume the dynamics

$$dX_t = \mu(t, X_t)dt + \sigma(t, X_t)dW_t, \qquad (12.24)$$

for the logarithm of the price of a risky asset S and if r is the short-term rate, in Section 10.3.2 we have expressed the arbitrage price of a derivative with payoff $F(S_T)$ in terms of a solution of the Cauchy problem

$$\begin{cases} \partial_t u(t, x) + Au(t, x) = 0, & (t, x) \in]0, T[\times\mathbb{R}, \\ u(T, x) = \varphi(x), & x \in \mathbb{R}, \end{cases} \qquad (12.25)$$

where A is the differential operator in (12.23) with

$$a = \frac{\sigma^2}{2}, \qquad b = r - \frac{\sigma^2}{2},$$

and $\varphi(x) = F(e^x)$.

12.3.1 Localization

In order to construct a discretization scheme and to implement it, the first step is to localize the problem (12.25) on a bounded domain. More precisely, for fixed $R > 0$, we introduce the Cauchy-Dirichlet problem

$$\begin{cases} \partial_t u + Au = 0, & \text{in }]0,T[\times]-R,R[, \\ u(t,-R) = \varphi_{-R}(t), \ u(t,R) = \varphi_R(t), & t \in [0,T], \\ u(T,x) = \varphi(x), & |x| < R, \end{cases} \tag{12.26}$$

where $\varphi_{\pm R}$ are functions that express the data on the lateral boundary and have to be chosen in a suitable way: the simplest choice is $\varphi_{\pm R} = 0$ and other typical choices are

$$\varphi_{\pm R}(t) = \varphi(\pm R), \qquad \text{or} \qquad \varphi_{\pm R}(t) = e^{-\int_t^T r(s,\pm R)ds}\varphi(\pm R), \qquad t \in [0,T].$$

Instead of lateral Cauchy-Dirichlet-type conditions, it is possible to assign Neumann-type ones: for example, in the case of a Put option,

$$\partial_x u(t,-R) = \partial_x u(t,R) = 0, \qquad t \in [0,T].$$

By using the Feynman-Kač probabilistic representation of Theorems 9.44 and 9.45, it is possible to easily obtain an estimate of the difference between the solution u_R of (12.26) and u. For the sake of simplicity we consider only the case $\varphi_{\pm R} = 0$: an analogous result can be proved without major difficulties when $\varphi_{\pm R}$ are bounded functions. We have

$$u(t,x) = E\left[e^{-\int_t^T r(s,X_s^{t,x})ds}\varphi(X_T^{t,x})\right],$$
$$u_R(t,x) = E\left[e^{-\int_t^T r(s,X_s^{t,x})ds}\varphi\left(X_T^{t,x}\right)\mathbb{1}_{\{\tau_x \geq T\}}\right],$$

where τ_x is the exit time of the process $X^{t,x}$, solution of the SDE (12.24) with $\mu = b$, from the interval $]-R,R[$. Then

$$|u(t,x) - u_R(t,x)| \leq E\left[e^{-\int_t^T r(s,X_s^{t,x})ds}\left|\varphi\left(X_T^{t,x}\right)\right|\mathbb{1}_{\{\tau_x < T\}}\right] \leq$$

(since r is bounded)

$$\leq e^{\|r\|_{L^\infty}(T-t)}\|\varphi\|_{L^\infty} P\left(\sup_{t\leq s\leq T}\left|X_s^{t,x}\right| \geq R\right) \leq$$

(by the maximal estimate (9.41))

$$\leq 2\|\varphi\|_{L^\infty}\exp\left(-\frac{\left(e^{-K(T-t)}R - |x| - K(T-t)\right)^2}{2k(T-t)} + \|r\|_{L^\infty}(T-t)\right), \tag{12.27}$$

where k, K are positive constants depending explicitly on the coefficients of the stochastic equations (cf. (9.38)-(9.39)). Formula (12.27) proves that u_R converges uniformly on compact sets to u for $R \to +\infty$ and gives an explicit estimate, quite coarse indeed, of the approximation error. More precise sup-norm estimates for the truncation error in terms of the maximal error at the boundary are given by Kangro and Nicolaides [197].

12.3.2 θ-schemes for the Cauchy-Dirichlet problem

For a fixed discretization step $\delta > 0$, we introduce the following first-order finite differences:

$$D_\delta^+ v(y) = \frac{v(y+\delta) - v(y)}{\delta}, \qquad \text{"forward"},$$

$$D_\delta^- v(y) = \frac{v(y) - v(y-\delta)}{\delta}, \qquad \text{"backward"},$$

$$D_\delta v(y) = \frac{1}{2}\left(D_\delta^+ v(y) + D_\delta^- v(y)\right) = \frac{v(y+\delta) - v(y-\delta)}{2\delta}, \qquad \text{"central"}.$$

Further, we define the second-order central ratio

$$D_\delta^2 v(y) = \frac{D_\delta^+ v(y) - D_\delta^- v(y)}{\delta} = \frac{v(y+\delta) - 2v(y) + v(y-\delta)}{\delta^2}.$$

We prove the consistency of the previous finite differences with the corresponding derivatives: the order of approximation of the backward and forward differences is one, whilst for the central differences the order is two.

Lemma 12.10 *If v is four times differentiable in a convex neighborhood of the point y, then the following estimates hold:*

$$\left|D_\delta^+ v(y) - v'(y)\right| \le \delta \frac{\sup |v''|}{2}, \tag{12.28}$$

$$\left|D_\delta^- v(y) - v'(y)\right| \le \delta \frac{\sup |v''|}{2}, \tag{12.29}$$

$$\left|D_\delta v(y) - v'(y)\right| \le \delta^2 \frac{\sup |v'''|}{3} + \delta^3 \frac{\sup |v''''|}{12}, \tag{12.30}$$

$$\left|D_\delta^2 v(y) - v''(y)\right| \le \delta^2 \frac{\sup |v''''|}{12}. \tag{12.31}$$

Proof. Taking the Taylor series expansion of v with initial point y, we get

$$v(y+\delta) = v(y) + v'(y)\delta + \frac{1}{2} v''(\hat{y})\delta^2, \tag{12.32}$$

$$v(y-\delta) = v(y) - v'(y)\delta + \frac{1}{2} v''(\check{y})\delta^2, \tag{12.33}$$

with $\hat{y}, \check{y} \in]y-\delta, y+\delta[$. Then (12.28) follows from (12.32) and (12.29) follows from (12.33).

Now let us consider the fourth-order expansion:

$$v(y+\delta) = v(y) + v'(y)\delta + \frac{1}{2}v''(y)\delta^2 + \frac{1}{3!}v'''(y)\delta^3 + \frac{1}{4!}v''''(\hat{y})\delta^4, \qquad (12.34)$$

$$v(y-\delta) = v(y) - v'(y)\delta + \frac{1}{2}v''(y)\delta^2 - \frac{1}{3!}v'''(y)\delta^3 + \frac{1}{4!}v''''(\check{y})\delta^4 \qquad (12.35)$$

with $\hat{y}, \check{y} \in]y-\delta, y+\delta[$. Summing (subtracting) (12.34) and (12.35), we obtain immediately (12.30) ((12.31)). □

For fixed $M, N \in \mathbb{N}$, we define the space-discretization and the time-discretization steps

$$\delta = \frac{2R}{M+1}, \qquad \tau = \frac{T}{N},$$

and on the domain $[0,T] \times [-R,R]$ we construct the grid of points

$$G^{(\tau,\delta)} = \{(t_n, x_i) = (n\tau, -R+i\delta) \mid n = 0,\dots,N,\ i = 0,\dots,M+1\}. \qquad (12.36)$$

For every function $g = g(t,x)$ and for every $t \in [0,T]$, we denote by $g(t)$ the $\mathbb{R}^M$-vector with components

$$g_i(t) = g(t, x_i), \qquad i = 1,\dots,M; \qquad (12.37)$$

further, on the grid $G^{(\tau,\delta)}$ we define the function

$$g_{n,i} = g_i(t_n) = g(t_n, x_i),$$

for $n = 0,\dots,N$ and $i = 0,\dots,M+1$.

Using the finite differences described earlier, we introduce the discretization in the space variable of the Cauchy-Dirichlet problem: we define the linear operator $A_\delta = A_\delta(t)$ approximating A in (12.23) and acting on $u(t)$, vector in $\mathbb{R}^M$ defined as in (12.37), in the following way

$$\begin{aligned}(A_\delta u)_i(t) :=& a_i(t)\frac{u_{i+1}(t) - 2u_i(t) + u_{i-1}(t)}{\delta^2} \\ &+ b_i(t)\frac{u_{i+1}(t) - u_{i-1}(t)}{2\delta} - r_i(t)u_i(t) \\ =& \alpha_i(t)u_{i-1}(t) - \beta_i(t)u_i(t) + \gamma_i(t)u_{i+1}(t), \qquad i = 1,\dots,M,\end{aligned}$$

where

$$\alpha_i(t) = \frac{a_i(t)}{\delta^2} - \frac{b_i(t)}{2\delta}, \qquad \beta_i(t) = \frac{2a_i(t)}{\delta^2} + r_i(t), \qquad \gamma_i(t) = \frac{a_i(t)}{\delta^2} + \frac{b_i(t)}{2\delta}.$$

In other terms, incorporating the null Dirichlet condition at the boundary, the operator $A_\delta(t)$ is represented by the tridiagonal matrix

$$A_\delta(t) = \begin{pmatrix} \beta_1(t) & \gamma_1(t) & 0 & 0 & \cdots & 0 \\ \alpha_2(t) & \beta_2(t) & \gamma_2(t) & 0 & \cdots & 0 \\ 0 & \alpha_3(t) & \beta_3(t) & \gamma_3(t) & \cdots & 0 \\ \vdots & \vdots & \ddots & \ddots & \ddots & \vdots \\ 0 & 0 & \cdots & \alpha_{M-1}(t) & \beta_{M-1}(t) & \gamma_{M-1}(t) \\ 0 & 0 & \cdots & 0 & \alpha_M(t) & \beta_M(t) \end{pmatrix}$$

and the discretized version of problem (12.26) with null lateral-boundary data is

$$\begin{cases} \frac{d}{dt}u(t) + A_\delta u(t) = 0, & t \in]0,T[, \\ u_i(T) = \varphi(x_i), & i = 1, \dots, M. \end{cases} \tag{12.38}$$

Finally we approximate the time derivative by a forward finite difference:

$$\frac{d}{dt}u_i(t_n) \sim \frac{u_{n+1,i} - u_{n,i}}{\tau}.$$

Definition 12.11 *For fixed $\theta \in [0,1]$, the finite-difference θ-scheme for the problem (12.26) with null lateral-boundary data consists of the final condition*

$$u_{N,i}(t) = \varphi(x_i), \qquad i = 1, \dots, M, \tag{12.39}$$

associated to the following equation to be solved iteratively for n decreasing from $n = N-1$ to $n = 0$:

$$\frac{u_{n+1,i} - u_{n,i}}{\tau} + \theta(A_\delta u)_{n,i} + (1-\theta)(A_\delta u)_{n+1,i} = 0, \qquad i = 1, \dots, M. \tag{12.40}$$

The finite-difference θ-scheme is called *explicit* if $\theta = 0$: in this case the computation of $(u_{n,i})$ starting from $(u_{n+1,i})$ in (12.40) is immediate, since

$$u_{n,i} = u_{n+1,i} + \tau(A_\delta u)_{n+1,i}, \qquad i = 1, \dots, M.$$

In general, we note that (12.40) is equivalent to

$$((I - \tau\theta A_\delta)u)_{n,i} = ((I + \tau(1-\theta)A_\delta)u)_{n+1,i}, \qquad i = 1, \dots, M.$$

So, if $\theta > 0$, in order to solve this equation it is necessary to invert the matrix $I - \tau\theta A_\delta$: algorithms to solve tri-diagonal linear systems can be found, for example, in Press, Teukolsky, Vetterling and Flannery [286]. For $\theta = 1$ we say that the scheme is *totally implicit,* whilst for $\theta = \frac{1}{2}$ it is called *Crank-Nicolson scheme* [81]. It is evident that the simplest choice seems $\theta = 0$, nevertheless the greater complexity of implicit schemes gives better convergence results (cf. Remark 12.14).

We also give the expression of the operator $A_\delta(t)$ if null Neumann-type conditions are assumed at the boundary:

$$A_\delta(t) = \begin{pmatrix} \alpha_1(t)+\beta_1(t) & \gamma_1(t) & 0 & 0 & \cdots & 0 \\ \alpha_2(t) & \beta_2(t) & \gamma_2(t) & 0 & \cdots & 0 \\ 0 & \alpha_3(t) & \beta_3(t) & \gamma_3(t) & \cdots & 0 \\ \vdots & \vdots & \ddots & \ddots & \ddots & \vdots \\ 0 & 0 & \cdots & \alpha_{M-1}(t) & \beta_{M-1}(t) & \gamma_{M-1}(t) \\ 0 & 0 & \cdots & 0 & \alpha_M(t) & \beta_M(t)+\gamma_M(t) \end{pmatrix}.$$

In this case (12.38) is the discretized version of the problem

$$\begin{cases} \partial_t u + Au = 0, & \text{in }]0,T[\times]-R,R[, \\ \partial_x u(t,-R) = \partial_x u(t,R) = 0, & t \in [0,T], \\ u(T,x) = \varphi(x), & |x| < R. \end{cases}$$

By way of example, now we study the explicit scheme ($\theta = 0$) for an equation with constant coefficients: more simply, with a change of variables as in (7.22), we can consider the heat equation directly, i.e. $a = 1$ and $b = r = 0$ in (12.23). Setting

$$Lu = \partial_t u + \partial_{xx} u,$$
$$(L^{(\tau,\delta)}u)_{n,i} = \frac{u_{n+1,i} - u_{n,i}}{\tau} + \frac{u_{n+1,i+1} - u_{n+1,i} + u_{n+1,i-1}}{\delta^2},$$

the Cauchy-Dirichlet problem

$$\begin{cases} Lu = 0, & \text{in }]0,T[\times]-R,R[, \\ u(t,-R) = \varphi_{-R}(t), \ u(t,R) = \varphi_R(t), & t \in [0,T], \\ u(T,x) = \varphi(x), & |x| < R, \end{cases} \tag{12.41}$$

is discretized by the system of equations

$$\begin{cases} L^{(\tau,\delta)}u = 0, & \text{in } G^{(\tau,\delta)}, \\ u_{n,0} = \varphi_{-R}(t_n), \ u_{n,M+1} = \varphi_R(t_n), & n = 0,\dots,N, \\ u_{N,i} = \varphi(x_i), & i = 1,\dots,M. \end{cases} \tag{12.42}$$

The next result extends the weak maximum principle proved in Paragraph 6.1.

Proposition 12.12 (Discrete maximum principle) *Let g be a function defined on the grid $G^{(\tau,\delta)}$ such that*

$$\begin{cases} L^{(\tau,\delta)} g \geq 0, & \text{on } G^{(\tau,\delta)}, \\ g_{n,0} \leq 0, \; g_{n,M+1} \leq 0, & n = 0, \dots, N, \\ g_{N,i} \leq 0, & i = 1, \dots, M. \end{cases}$$

If the condition

$$\tau \leq \frac{\delta^2}{2} \tag{12.43}$$

is satisfied, then $g \leq 0$ on $G^{(\tau,\delta)}$.

Proof. We observe that $L^{(\tau,\delta)} g \geq 0$ on $G^{(\tau,\delta)}$ if and only if

$$g_{n,i} \leq g_{n+1,i} \left(1 - \frac{2\tau}{\delta^2}\right) + (g_{n+1,i+1} + g_{n+1,i-1}) \frac{\tau}{\delta^2}$$

for $n = 0, \dots, N-1$ and $i = 1, \dots, M$. Then the claim follows from the fact that, in view of condition (12.43), the coefficients on the right-hand side of the previous inequality are non-negative: consequently, since the boundary data are less or equal to zero, we have that $g_{n+1,i} \leq 0$ implies $g_{n,i} \leq 0$. □

The following theorem proves that the explicit finite-difference scheme converges with speed proportional to δ^2.

Theorem 12.13 *Let u be a solution of problem* (12.41) *and let us suppose that $\partial_{xxxx} u$ and $\partial_{tt} u$ are bounded. If condition* (12.43) *holds, then there exists a positive constant C such that, for every $\delta > 0$*

$$\max_{G^{(\tau,\delta)}} |u - u^{(\tau,\delta)}| \leq C\delta^2,$$

where $u^{(\tau,\delta)}$ is the solution of the discretized problem (12.42).

Proof. Firstly we observe that, by Lemma 12.10 combined with condition (12.43), we have

$$|(L^{(\tau,\delta)} u)_{n,i}| = |(L^{(\tau,\delta)} u)_{n,i} - (Lu)_{n+1,i}| \leq C\delta^2, \tag{12.44}$$

with $C = \frac{\|\partial_{tt} u\|_\infty}{4} + \frac{\|\partial_{xxxx} u\|_\infty}{12}$.

Then, on the grid $G^{(\tau,\delta)}$ we define the functions

$$w^+ = u - u^{(\tau,\delta)} - C(T-t)\delta^2, \qquad w^- = u^{(\tau,\delta)} - u - C(T-t)\delta^2,$$

and we observe that $w^\pm \leq 0$ on the parabolic boundary of $G^{(\tau,\delta)}$. Further, since $L^{(\tau,\delta)} t = 1$,

$$L^{(\tau,\delta)} w^+ = L^{(\tau,\delta)} (u - u^{(\tau,\delta)}) + C\delta^2 = L^{(\tau,\delta)} u + C\delta^2 \geq 0,$$

by the estimate (12.44). Analogously we have $L^{(\tau,\delta)} w^- \geq 0$ and so, by Proposition 12.12, we get $w^\pm \leq 0$ on $G^{(\tau,\delta)}$, hence the claim. □

Remark 12.14 Inequality (12.43) is usually called *stability condition* and it is in general necessary for the convergence of a θ-scheme if $0 \le \theta < \frac{1}{2}$: if (12.43) is not satisfied, the claim of the previous proposition does not hold true. For this reason the θ-schemes are called *conditionally convergent* for $\theta < \frac{1}{2}$. On the contrary, if $\frac{1}{2} \le \theta \le 1$, the θ-schemes are called *unconditionally convergent* because they converge when τ, δ tend to zero. □

We conclude the section by stating a general convergence result for finite-difference θ-schemes; for the proof we refer to Raviart and Thomas [289].

Theorem 12.15 *Let u and $u^{(\tau,\delta)}$ be the solutions of problem* (12.26) *and of the corresponding problem discretized by a θ-scheme, respectively. Then*

- *if $0 \le \theta < \frac{1}{2}$ and the stability condition*

$$\lim_{\tau,\delta\to 0^+} \frac{\tau}{\delta^2} = 0$$

holds, we have

$$\lim_{\tau,\delta\to 0^+} u^{(\tau,\delta)} = u, \qquad \text{in } L^2(]0,T[\times]-R,R[);$$

- *if $\frac{1}{2} \le \theta \le 1$, we have*

$$\lim_{\tau,\delta\to 0^+} u^{(\tau,\delta)} = u, \qquad \text{in } L^2(]0,T[\times]-R,R[).$$

12.3.3 Free-boundary problem

The finite-difference schemes can be easily adapted to the pricing problem of options with early exercise. Basically the idea consists in approximating an American option with the corresponding Bermudan option that admits a finite number of possible exercise dates. Using the notations in the preceding section and in particular having fixed the time discretization

$$t_n = n\tau, \qquad \tau = \frac{T}{N},$$

we employ the usual θ-scheme in (12.40) to compute the "European" price $(\widetilde{u}_{n,i})_{i=1,\dots,M}$ at time t_n starting from $(u_{n+1,i})_{i=1,\dots,M}$ and then we apply the early-exercise condition

$$u_{n,i} = \max\{\widetilde{u}_{n,i}, \varphi(t_n, x_i)\}, \qquad i = 1, \dots, M,$$

to determine the approximation of the price of the American option. In this way also an approximation of the free boundary is obtained.

In the last years many numerical methods for American options have been proposed in the literature. Brennan and Schwartz [61] were the first to use analytic methods (i.e. based on the solution of the corresponding obstacle problem) for options with early exercise: Jaillet, Lamberton and Lapeyre [185] and Han-Wu [162] gave a rigorous justification of the method and Zhang [346] studied its convergence and its extension to jump models. In Barraquand and Martineau [31], Barraquand and Pudet [32], Dempster and Hutton [93] the previous techniques have been improved to price exotic options. Among the other methods that have been proposed, we mention the finite elements in Achdou and Pironneau [1], the methods ADI in Villeneuve and Zanette [335] and the methods based on wavelets in Matache, Nitsche and Schwab [248]. MacMillan [242], Barone-Adesi and Whaley [30], Carr and Faguet [66], Jourdain and Martini [190; 191] give semi-explicit approximation formulas for the price of American derivatives.

12.4 Monte Carlo methods

The Monte Carlo method is a simple technique of numerical approximation of the mean of a random variable X. It is used in many circumstances in mathematical finance and in particular in the pricing problem and in the computation of the Greeks of derivatives. More generally, the Monte Carlo method allows approximating the value of an integral numerically: indeed we recall that, if $Y \sim \text{Unif}_{[0,1]}$ is uniformly distributed on $[0,1]$ and $X = f(Y)$, then we have

$$E[X] = \int_0^1 f(x)dx.$$

The Monte Carlo method is based on the strong law of large numbers (cf. Section A.7.1): if (X_n) is a sequence of integrable i.i.d. random variables and such that $E[X_1] = E[X]$, then

$$\lim_{n\to\infty} \frac{1}{n}\sum_{k=1}^{n} X_k = E[X] \qquad \text{a.s.}$$

Consequently, if we are able to draw samples $\bar{X}_1, \dots, \bar{X}_n$ from X in an independent way, then the mean

$$\frac{1}{n}\sum_{k=1}^{n} \bar{X}_k$$

gives an a.s. approximation of $E[X]$.

In order to analyze some of the main features of this technique, we consider the problem of numerical approximation of the following integral over the unitary cube in $\mathbb{R}^d$:

$$\int_{[0,1]^d} f(x)dx. \tag{12.45}$$

The most natural way to approximate the value of the integral consists in considering a discretization by Riemann sums: for fixed $n \in \mathbb{N}$, on $[0,1]^d$ we build a grid of points with coordinates of the form $\frac{k}{n}$, $k = 0, \dots, n$. Then we rewrite the integral in the form

$$\int_{[0,1]^d} f(x)dx = \sum_{k_1=0}^{n-1} \cdots \sum_{k_d=0}^{n-1} \int_{\frac{k_1}{n}}^{\frac{k_1+1}{n}} \cdots \int_{\frac{k_d}{n}}^{\frac{k_d+1}{n}} f(x_1, \dots, x_d)dx_1 \cdots dx_d$$

and we approximate the right-hand side by

$$\sum_{k_1=0}^{n-1} \cdots \sum_{k_d=0}^{n-1} \int_{\frac{k_1}{n}}^{\frac{k_1+1}{n}} \cdots \int_{\frac{k_d}{n}}^{\frac{k_d+1}{n}} f\left(\frac{k_1}{n}, \dots, \frac{k_d}{n}\right) dx_1 \cdots dx_d \tag{12.46}$$

$$= \frac{1}{n^d} \sum_{k_1=0}^{n-1} \cdots \sum_{k_d=0}^{n-1} f\left(\frac{k_1}{n}, \dots, \frac{k_d}{n}\right) =: S_n(f).$$

If f is Lipschitz continuous, with Lipschitz constant L, then

$$\left| \int_{[0,1]^d} f(x)dx - S_n(f) \right| \le \frac{L}{n}.$$

Further, if $f \in C^q([0,1]^d)$, we can easily obtain an n^{-q}-order scheme, by substituting $f\left(\frac{k_1}{n}, \dots, \frac{k_d}{n}\right)$ in (12.46) with the q-th order Taylor expansion of f with initial point $\left(\frac{k_1}{n}, \dots, \frac{k_d}{n}\right)$.

In principle, this kind of approximation gives better results than the Monte Carlo method. However, we wish to highlight the following aspects concerning the regularity assumption and the computational complexity:

1) the convergence of the scheme depends heavily on the regularity of f. For example, the measurable function

$$f(x) = \mathbb{1}_{[0,1]^d \setminus \mathbb{Q}^d} \tag{12.47}$$

has integral equal to 1, but $S_n(f) = 0$ for every $n \in \mathbb{N}$;
2) the computation of the approximation term $S_n(f)$ necessary to get an error of the order of $\frac{1}{n}$ involves the valuation of f in n^d points; so the number of points increases exponentially with the dimension of the problem. It follows that, in practice, only if d is small enough it is possible to implement the method in an effective way. In other terms, if the number of points taken in the discretization is fixed, say, equal to n, the quality of the approximation gets worse when the dimension d of the problem increases: the order of the error is $n^{-\frac{1}{d}}$.

Now we consider the approximation with the Monte Carlo method. If (Y_n) is a sequence of i.i.d random variables with uniform distribution on $[0,1]^d$, we have

$$\int_{[0,1]^d} f(x)dx = E\left[f(Y_1)\right] = \lim_{n\to\infty} \frac{1}{n} \sum_{k=1}^{n} f(Y_k) \qquad \text{a.s.} \tag{12.48}$$

We observe that, in order for the integral to converge, it suffices that f is integrable on $[0,1]^d$ and no further regularity assumption is required: for example, for the function f in (12.47) we have $f(Y_k) = 1$ a.s. and so the approximation is correct a.s.

Concerning the computational complexity, we can give a first estimate of the error of the Monte Carlo method directly by the Markov inequality proceeding as in Remark A.144. We consider a sequence of real i.i.d. random variables (X_n) with $\mu = E[X_1]$ and $\sigma^2 = \text{var}(X_1)$ finite. Furthermore, we set

$$M_n = \frac{1}{n}\sum_{k=1}^{n} X_k.$$

By Markov's inequality, for every $\varepsilon > 0$, we have

$$P(|M_n - \mu| \geq \varepsilon) \leq \frac{\text{var}(M_n)}{\varepsilon^2} =$$

(by the independence)

$$= \frac{n\text{var}\left(\frac{X_1}{n}\right)}{\varepsilon^2} = \frac{\sigma^2}{n\varepsilon^2},$$

that can be rewritten in a more appealing way as follows:

$$P(|M_n - \mu| \leq \varepsilon) \geq p, \qquad \text{where } p := 1 - \frac{\sigma^2}{n\varepsilon^2}. \tag{12.49}$$

First of all we note that, being the technique based on the generation of random numbers, *the result and the error of the Monte Carlo method are random variables.* Formula (12.49) gives an estimate of the error in terms of three parameters:

i) n, the number of samples, i.e. how many random numbers we have generated;
ii) ε, the maximum approximation error;
iii) p, the minimum probability that the approximated value M_n belongs to the confidence interval $[\mu - \varepsilon, \mu + \varepsilon]$.

According to (12.49), for fixed $n \in \mathbb{N}$ and $p \in]0,1[$, the maximum approximation error of the Monte Carlo method is

$$\varepsilon = \frac{\sigma}{\sqrt{n(1-p)}}. \tag{12.50}$$

In the example of the computation of the integral (12.45), we have $X = f(Y)$ with Y uniformly distributed on $[0,1]^d$ and in this case the maximum error of the method can be estimated by

$$\sqrt{\frac{\text{var}(f(Y))}{n(1-p)}}.$$

In other terms *the error is of the order of* $\frac{1}{\sqrt{n}}$ *regardless of the dimension of the problem:* for comparison, we recall that the order of the error of the deterministic scheme that we examined earlier was $n^{-\frac{1}{d}}$.

Summing up, if the dimension is low and some suitable regularity assumptions are verified, then it is not difficult to implement deterministic algorithms performing better than Monte Carlo. However, when the dimension of the problem increases, these deterministic algorithms become burdensome and the Monte Carlo method is, for now, the only viable alternative.

We also observe that, by (12.50), the standard deviation σ is directly proportional to the approximation error: as a matter of fact, from a computational point of view σ is a crucial parameter which influences significantly the efficiency of the approximation. Typically σ is not known; nevertheless it is possible to use the random numbers that we have generated to construct an estimator of σ:

$$\sigma_n^2 := \frac{1}{n-1}\sum_{k=1}^{n}(X_k - \mu_n)^2, \qquad \mu_n := \frac{1}{n}\sum_{k=1}^{n} X_k.$$

In other words, we can use the realizations of X to have simultaneously the approximation of $E[X]$ and of the error that we commit, in terms of confidence intervals. Evidently σ_n is just an approximation of σ, even though in general it is sufficiently accurate to estimate the error satisfactorily.

Usually, in order to improve the effectiveness of the Monte Carlo method, *variance-reduction methods* are used. These techniques, elementary in some cases, employ the specific features of the problem to reduce the value of σ_n and consequently increase the speed of convergence: for the description of such techniques, we refer to Chapter 4 in [158].

In the next sections we will briefly deal with some questions on the implementation of the Monte Carlo method. In Section 12.4.1 we analyze the problem of the simulation of X, i.e. how to generate independent realizations of X; further, we discuss the application of the Monte Carlo method to the problem of derivative pricing. In Section 12.4.2 we present some techniques to compute the Greeks. Finally in Section 12.4.3 we go back to the problem of the error analysis and of the determination of confidence intervals by making use of the central limit theorem. A complete presentation of Monte Carlo methods and their applications to mathematical finance, can be found in the reference text by Glasserman [158].

12.4.1 Simulation

The first step to approximate $E[X]$ by the Monte Carlo method consists in generating n independent realizations of the random variable X: this poses some practical problems.

First of all n must be large enough and so the generation of the simulations cannot be made by hand (for example, by tossing a coin): therefore we must use the power of a computer to perform the computation. This rather obvious remark introduces the first serious problem: a computer can generate "random" values only by using deterministic algorithms. So, in order to implement the Monte Carlo method, actually we have at our disposal only "pseudo-random" numbers, i.e. numbers that have the same statistical properties as the actual random values but, when the number of times we simulate increases, are not generated in a really independent way. This translates into an additional error that cannot be easily estimated in the approximated result. Therefore it should always be borne in mind the fact that the quality of the random-number generator influences the numerical result significantly.

After shedding some light on this first matter, for the vast majority of the well-known distributions, and in particular for the Normal standard distribution, it is not difficult to find a pseudo-random number generator. Having this at our disposal, pricing of a European option with payoff F is indeed an easy task. For example, in the Black-Scholes model, where the final price of the underlying asset is[1]

$$S_T = S_0 \exp\left(\sigma W_T + \left(r - \frac{\sigma^2}{2}\right)T\right),$$

the procedure is as follows:

(A.1) we draw n independent samples $\bar{Z}_1, \dots, \bar{Z}_n$, from the standard Normal distribution;

(A.2) we consider the corresponding realizations of the final value of the underlying asset

$$\bar{S}_T^{(k)} = S_0 \exp\left(\sigma\sqrt{T}\bar{Z}_k + \left(r - \frac{\sigma^2}{2}\right)T\right);$$

(A.3) we compute the approximation of the price of the derivative

$$\frac{e^{-rT}}{n}\sum_{k=1}^{n} F\left(\bar{S}_T^{(k)}\right) \approx e^{-rT}E\left[F(S_T)\right].$$

Because of its easy applications to a wide range of problems, the Monte Carlo is one of the most popular numerical methods. Now we see how it can be used in conjunction with the Euler scheme. We consider a local-volatility model in which the dynamics of the underlying asset under the EMM is given by

$$dS_t = rS_t dt + \sigma(t, S_t)dW_t.$$

In this case the distribution of the final price S_T is not known explicitly. In order to obtain some realizations of S_T we use a Euler-type scheme: it is clear that, in this way, the discretization error of the SDE must be added to the error of the Monte Carlo method. The procedure is as follows:

[1] Here σ is, as usual, the volatility coefficient.

(B.1) we produce nm independent realizations $\bar{Z}_{k,i}$, for $k = 1, \dots, n$ and $i = 1, \dots, m$, of the Normal standard distribution $\mathcal{N}_{0,1}$;

(B.2) using the iterative formula

$$\bar{S}_{t_i}^{(k)} = \bar{S}_{t_{i-1}}^{(k)}(1 + r(t_i - t_{i-1})) + \sigma(t_{i-1}, \bar{S}_{t_{i-1}}^{(k)})\sqrt{t_i - t_{i-1}}\bar{Z}_{k,i}$$

we determine the corresponding realizations of the final value of the underlying asset $\bar{S}_T^{(1)}, \dots, \bar{S}_T^{(n)}$;

(B.3) we compute the approximation of the price of the derivative as in (A.3).

Finally we consider an Up&Out contract with barrier B and payoff

$$H_T = F(S_T)\mathbb{1}_{\left\{\max\limits_{0 \le t \le T} S_t \le B\right\}}.$$

Since it is a path-dependent option, also in this case the Euler-Monte Carlo method is suitable in order to simulate the full path of the underlying asset and not only the final price. For the sake of simplicity we set $r = 0$. In this case the steps are as follows:

(C.1) as in (B.1);

(C.2) using (B.2) we determine the realizations of the final value of the underlying asset $\bar{S}_T^{(k)}$ and of the maximum $\bar{M}^{(k)} := \max\limits_{i=1,\dots,m} \bar{S}_{t_i}^{(k)}$;

(C.3) we compute the approximation of the price of the derivative

$$\frac{1}{n}\sum_{k=1}^{n} F\left(\bar{S}_T^{(k)}\right)\mathbb{1}_{[0,B]}\left(\bar{M}^{(k)}\right) \approx E\left[H_T\right].$$

12.4.2 Computation of the Greeks

With some precautions, the Monte Carlo method can be used also to compute the sensitivities. We consider in particular the problem of computing the Delta of an option: we denote by $S_t(x)$ the price of the underlying asset with initial value x and F the payoff function of a European derivative. In the sequel the mean is computed with respect to an EMM.

The simplest approach to compute the Delta

$$\Delta = e^{-rT}\partial_x E\left[F(S_T(x))\right]$$

consists in approximating the derivative by an incremental ratio:

$$\Delta \approx e^{-rT} E\left[\frac{F(S_T(x+h)) - F(S_T(x))}{h}\right]. \tag{12.51}$$

The mean in (12.51) can be approximated by the Monte Carlo method, by choosing a suitable h and *taking care to use the same realizations of the Normal standard variables to simulate $S_T(x+h)$ and $S_T(x)$*. It is often preferable to use a central incremental ratio or also a higher-order one to get a more accurate result. Nevertheless it is important to note that this approach is efficient only if F is sufficiently regular: in general extra care must be taken.

We also touch upon an alternative method that will be presented in greater detail in Chapter 16. The following technique uses the fact that, in the Black-Scholes model, we have an explicit expression of the density of the underlying asset as a function of the initial price x:

$$S_T(x) = e^Y, \qquad Y \sim \mathcal{N}_{\log x + \left(r - \frac{\sigma^2}{2}\right)T, \sigma^2 T}.$$

The price of the option is

$$H(x) = e^{-rT} E\left[F\left(S_T(x)\right)\right] = e^{-rT} \int_{\mathbb{R}} F(e^y)\Gamma(x,y)dy,$$

where

$$\Gamma(x,y) = \frac{1}{\sqrt{2\pi\sigma^2 T}} \exp\left(-\frac{\left(y - \log x - \left(r - \frac{\sigma^2}{2}\right)T\right)^2}{2\sigma^2 T}\right).$$

Then, under suitable assumptions justifying the exchange of the derivative-integral sign, we have

$$\begin{aligned}
\Delta = \partial_x H(x) &= e^{-rT} \int_{\mathbb{R}} F(e^y)\partial_x \Gamma(x,y)dy \\
&= e^{-rT} \int_{\mathbb{R}} F(e^y)\Gamma(x,y) \frac{y - \log x - \left(r - \frac{\sigma^2}{2}\right)T}{\sigma^2 T x} dy \\
&= \frac{e^{-rT}}{\sigma^2 T x} E\left[F\left(S_T(x)\right)\left(\log S_T(x) - \log x - \left(r - \frac{\sigma^2}{2}\right)T\right)\right] && (12.52) \\
&= \frac{e^{-rT}}{\sigma T x} E\left[F(S_T)W_T\right]. && (12.53)
\end{aligned}$$

We observe that (12.52) expresses the Delta in terms of the price of a new option: the same result can be obtained by means of Malliavin calculus techniques (cf. (16.19)). What is peculiar about formula (12.52) is that *the derivative of F does not appear anymore:* indeed the partial derivative ∂_x was applied directly to the density of the underlying asset. The advantage from a numerical point of view is remarkable above all if F is not regular: the typical case is that of the digital option, in which the derivative of the payoff function $F = \mathbb{1}_{[K,+\infty[}$ is (in the distributional sense) a Dirac delta.

With the same technique it is also possible to get similar expressions of the other Greeks in the Black-Scholes model. For more general models, or when the explicit expression of the density is not known, analogous results can be proved by using the more sophisticated tools of Malliavin calculus, which will be presented in Chapter 16.

12.4.3 Error analysis

We consider a sequence (X_n) of i.i.d. random variables with finite expectation and variance:

$$\mu = E[X_1], \qquad \sigma^2 = \text{var}(X_1).$$

By the strong Law of large numbers, the sequence

$$M_n := \frac{X_1 + \cdots + X_n}{n}$$

converges a.s. to μ. Now we show that the central limit theorem provides an estimate of the speed of convergence and the error distribution. Indeed, by Theorem A.146 we have

$$\sqrt{n}\left(\frac{M_n - \mu}{\sigma}\right) \xrightarrow[n\to\infty]{} Z \sim \mathcal{N}_{0,1},$$

and so, *asymptotically* for $n \to \infty$, for every $x \in \mathbb{R}$ we have

$$P\left(\sqrt{n}\left(\frac{M_n - \mu}{\sigma}\right) \le x\right) \approx \Phi(x),$$

where Φ is the standard Normal distribution function as in (A.25). Consequently, for every $x > 0$,

$$P\left(M_n \in \left[\mu - \frac{\sigma x}{\sqrt{n}}, \mu + \frac{\sigma x}{\sqrt{n}}\right]\right) \approx p, \qquad \text{where } p = 2\Phi(x) - 1. \tag{12.54}$$

Therefore, for a fixed $p \in]0,1[$, the distance between the exact value and the approximated one is with probability p (asymptotically) less than

$$\frac{\sigma}{\sqrt{n}} \Phi^{-1}\left(\frac{p+1}{2}\right).$$

For example, $\Phi^{-1}(\frac{p+1}{2}) \approx 1,96$ for $p = 95\%$.

From a theoretical point of view, it is apparent that the previous estimates are inconsistent, since they hold asymptotically, for $n \to \infty$ and we cannot control the speed of convergence. However, in practice they give a more accurate estimate than (12.49). This fact can be justified rigorously by the Berry-Esseen Theorem. This result gives the speed of convergence in the central limit theorem, thus allowing us to obtain rigorous estimates for the confidence intervals. In the next statement we assume, for the sake of simplicity, that $E[X] = 0$: we can always make this assumption satisfied by substituting X with $X - E[X]$.

Theorem 12.16 (Berry-Esseen) *Let (X_n) be a sequence of i.i.d. random variables such that $E[X_1] = 0$ and $\sigma^2 = \mathrm{var}(X_1)$, $\varrho = E\left[|X_1|^3\right]$ are finite. If Φ_n is the distribution function of $\frac{\sqrt{n}M_n}{\sigma}$, then*

$$|\Phi_n(x) - \Phi(x)| \leq \frac{\varrho}{\sigma^3\sqrt{n}}$$

for every $x \in \mathbb{R}$.

For the proof we refer to, for example, Durrett [109].

13

Introduction to Lévy processes

(with Rossella Agliardi)

The classical Black-Scholes model employs the Brownian motion as the driving stochastic process of asset prices. Empirical evidence has pointed out that such an assumption does not provide an accurate description of financial data and has promoted the development of more flexible models. This chapter presents the fundamentals of Lévy processes and option pricing under such stochastic processes. Since this chapter is intended as an informal introduction to Lévy processes, many of the proofs are omitted: for a complete treatment of the theory we refer to the classical monographs by Bertoin [44], Sato [297], Jacod and Shiryaev [184].

13.1 Beyond Brownian motion

The classical Black-Scholes model assumes that the price S of the risky underlying asset follows a geometric Brownian motion. In other words, the asset price returns behave according to a normal distribution and the paths of asset prices are continuous functions of time: more precisely, by (7.3), we have

$$S_t = S_0 e^{X_t} \quad \text{where} \quad X_t \sim \mathcal{N}_{\left(\mu - \frac{\sigma^2}{2}\right)t, \sigma^2 t}.$$

However, empirical evidence has brought to the light several stylized facts which are in contrast with this simple assumption and are now recognized as essential ingredients of any mathematical model of financial assets. A list of such empirical facts and their implications on the modeling of market moves is included in the monograph by Cont and Tankov [76] (see also the recent account by Tankov and Voltchkova [115]) and here we give a short outline.

The presence of jumps in the stock price trajectories, which appear as discontinuities in the price path, is a well-documented evidence. One remedy is proposed by Merton [251] who adopted a jump-diffusion model, that is a mixture of independent Brownian motion and Poisson processes. A closed

Pascucci A.: PDE and Martingale Methods in Option Pricing

form solution to the option pricing problem is available under specific assumptions on the distribution of jumps. However some criticisms have been raised towards the use of jump-diffusions: tractable jump-diffusion models require special assumptions, they do not allow for too many jumps, empirical analysis of high-frequency data supports purely discontinuous processes, etc.

Another important feature that has led to adopting more flexible distributions than the normal one is the widely recognized non-Gaussian behaviour of the log-returns. As we know, the normal distribution has zero skewness (cf. (A.79)). Moreover the kurtosis is 3 for the normal distribution, while it is greater (less) than 3 for the distributions with a higher (flatter) peak. Excess kurtosis is related to fat tails, that is, large movements are more frequent than a normal distribution may predict. The empirical analysis of stock returns typically exhibits some significant (usually negative) skewness and an excess kurtosis. For example, the kurtosis of S&P 500 index in the period 1970-2001 is 7.17 if the pathological data of the crash of October 19, 1987 are expelled, and is 43.36 if the complete dataset is considered (see Schoutens [301]).

While extreme stock market movements are relatively rare (for example, the −20% stock market crash of October 19, 1987 was the only daily movement in the post World War II era to exceed 10% in magnitude), persistent deviations from the normal distribution have been noticed, especially after the October 1987 crash. Indeed the classical models tend to under-estimate the probability of large drops of stock prices, thus leading to under-pricing of financial risk. The use of alternative distributions to capture outliers dates back to Mandelbrot [245] who pointed out that "the empirical distributions of price changes are usually too peaked to be relative to samples from Gaussian populations" and the adoption of the stable Paretian distribution was proposed as an alternative. The Gaussian assumption was rejected also in Fama [125]. Stable distributions (cf. Section 13.4.2) are defined in terms of an index $\alpha \in]0,2]$: the case $\alpha = 2$ corresponds to the normal distribution, which is the only stable distribution with a finite variance. Since stable distributions with $\alpha < 2$ are more peaked and have fatter tails than the normal one, they have been proposed as a more realistic description of price returns. A stable random variable X satisfies:

$$P(|X| > x) = \mathrm{O}\left(x^{-\alpha}\right), \qquad \text{as } x \to \infty,$$

while

$$P(|X| > x) = \mathrm{O}\left(x^{-1} e^{-\frac{x^2}{2}}\right), \qquad \text{as } x \to \infty,$$

in the standard Gaussian case. Therefore the tails are heavier when a stable non-normal distribution is employed. However further empirical studies suggested a faster decrease at infinity than predicted in this framework. For example, Officer [271] found evidence of thinner tails for large sums of daily returns; moreover, he found an increase in α for daily returns and concluded that it would be more appropriate to consider a modified model with a finite second moment distribution and *semi-heavy tails*, that is such that

$$P(|X| > x) = \mathrm{O}\left(|x|^{\alpha_\pm} e^{-\lambda_\pm |x|}\right), \qquad \text{as } x \longrightarrow \pm\infty,$$

for some positive constants $\lambda_\pm$.

Madan and Seneta [243] introduced an alternative distribution, the Variance-Gamma (VG), and tested the goodness-of-fit on several Australian stocks: the best performance was attained by the VG model, while the stable model out-performed in two cases out of 19 and the normal distribution in none. Several other models have been proposed to overcome the inconsistency of the Gaussian assumption with empirical data. Examples of such models are the truncated Lévy flights proposed by Mantegna and Stanley [246], the Normal Inverse Gaussian (NIG) employed by Barndorff-Nielsen [25], the Hyperbolic distribution adopted by Eberlein and Keller [113] and the more general Generalized Hyperbolic (GH) model (see Eberlein [110], Eberlein and Prause [117], the Meixner process (see Schoutens [300]), the CGMY model introduced by Carr, Geman, Madan and Yor [67] and generalized to a six-parameter model in [68].

The above listed processes belong to the wide family of the Lévy processes that will be the object of the following sections. The Lévy class has gained increasing favor in the financial literature thanks to its flexibility which allows to capture some features of the empirical distributions as sharp peaks and semi-heavy tails. Another qualitative features of empirical price trajectories which is not captured in a purely Brownian framework is the non-self-similarity, that is, dramatic changes occur in the distributions of returns if one looks at them on different time scales, as already pointed out in Fama [126]. On the contrary, a Wiener process has the self-similarity property: $W_{\lambda^2 t} = \lambda W_t$ for any scaling factor $\lambda > 0$. The only self-similar Lévy processes are the Brownian motion (without drift) and the symmetric α-stable Lévy process. The deviation from the normal distribution is investigated also in Eberlein and Keller [113] in details. Note that it is significant especially if prices are considered on a daily or an intraday time grid, which has become of major interest for the nowadays trading on electronic platforms and availability of high-frequency data.

Another interesting property of some Lévy processes proposed in the financial literature is that they can be obtained as time-changed Brownian motions: that is, $X_t = W_{S_t}$ where S_t is a stochastic time change or a "stochastic clock". As Geman [153] points out, such a "stochastic time change is in fact a measure of the economic activity" and accounts for the asset price reaction to the arrival of information. "Some days, very little news, good or bad, is released; trading is typically slow and prices barely fluctuate. In contrast, when new information arrives and traders adjust their expectation accordingly, trading becomes brisk and the price evolution accelerates". This point of view is related to the problem of handling stochastic volatility. In [68] a stochastic volatility effect is obtained by letting the price process be subordinated by a second stochastic process, a "stochastic clock": periods with high volatility are obtained letting time run faster than in periods with low volatility. As Cont and Tankov [76] emphasize, "unlike the Brownian model where realized volatility has a deterministic continuous-time limit, model based on Lévy processes lead to a realized volatility which remains stochastic when computed on fine time grids" and a stochastic volatility effect may be achieved without employing any additional random factors.

Another major issue that has driven beyond the classical Black-Scholes theory is the calibration to the market option prices. The discrepancy between the Black-Scholes prices and the empirical prices results in the shape of the implied volatility surface. The problem of fitting of the implied volatility and the study of the stability of the solution to this inverse problem has generated several diffusion-based models, either of the level dependent volatility type or in the stochastic volatility approach. The main drawback of stochastic volatility models is their inability to reproduce the variation in moneyness of the volatility surfaces at shorter term. As we shall see in Chapter 14, the combination of stochastic volatility models with jump processes seems to offer more powerful tools to face the problem, both in terms of greater flexibility in generating smile/skew patterns and in giving a convincing explanation.

13.2 Poisson process

The Poisson process is a fundamental example of stochastic process with discontinuous paths and serves as the basic building block for jump processes. To construct a Poisson process, we consider a sequence $(\tau_n)_{n\geq 1}$ of independent random variables with exponential distribution with parameter $\lambda > 0$:

$$\tau_n \sim \text{Exp}_\lambda, \qquad n \geq 1.$$

We refer to Example A.29 for a review of the main properties of the exponential distribution. We consider a model where jumps occur randomly and τ_n denotes the time distance of the n-th jump from the preceding one: thus, the first jump occurs at time τ_1, the second jump occurs τ_2 time units after τ_1 and so forth: then, for any $n \in \mathbb{N}$,

$$T_n := \sum_{k=1}^{n} \tau_k \tag{13.1}$$

is the time of the n-th jump. We remark that

$$E\left[T_n - T_{n-1}\right] = E\left[\tau_n\right] = \frac{1}{\lambda}, \qquad n \in \mathbb{N},$$

that is $\frac{1}{\lambda}$ is the average distance among subsequent jumps: this can also be expressed by saying that λ jumps are expected in a unit time interval; for this reason, λ is also called the *intensity* parameter.

Lemma 13.1 *For any $n \in \mathbb{N}$, the random variable T_n has probability density*

$$f_{T_n}(t) = \lambda e^{-\lambda t} \frac{(\lambda t)^{n-1}}{(n-1)!} \mathbb{1}_{\mathbb{R}_{\geq 0}}(t), \qquad t \in \mathbb{R}. \tag{13.2}$$

Proof. We prove the thesis by induction. For $n = 1$ it is obvious. Next we assume that T_n has probability density given by (13.2): by the independence of the variables $\{\tau_n\}$ and Corollary A.54, we have

$$\begin{aligned} f_{T_{n+1}}(t) = f_{T_n+\tau_{n+1}}(t) &= \int_{\mathbb{R}} f_{T_n}(s) f_{\tau_{n+1}}(t-s) ds \\ &= \int_0^{\infty} \lambda e^{-\lambda s} \frac{(\lambda s)^{n-1}}{(n-1)!} \lambda e^{-\lambda(t-s)} \mathbb{1}_{\{t-s>0\}} ds \\ &= \frac{\lambda^{n+1} e^{-\lambda t}}{(n-1)!} \int_0^t s^{n-1} ds \end{aligned}$$

and the thesis easily follows. □

Definition 13.2 (Poisson process) *The Poisson process with intensity λ is the process*

$$N_t = \sum_{n \geq 1} n \mathbb{1}_{[T_n, T_{n+1}[}(t), \qquad t \in \mathbb{R}_{\geq 0},$$

with T_n as in (13.1).

The Poisson process N_t *counts the number of jumps that occur at or before time t*. In particular N_t takes only non-negative integer values. Figure 13.1 shows a path of the Poisson process. Notice that by definition the trajectories of N are *right-continuous functions:*

$$N_t = N_{t+} := \lim_{s \downarrow t} N_s, \qquad t \geq 0.$$

The following proposition shows some other important properties of the Poisson process and also helps in understanding the difference between the notions of *pathwise* continuous and *stochastically* continuous process.

Proposition 13.3 *Let $(N_t)_{t \geq 0}$ be a Poisson process. Then:*

i) the trajectories $t \mapsto N_t(\omega)$ are right continuous with finite left limits, that is N is a càdlàg[1] *process;*

ii) for any positive t, almost all trajectories are continuous at t, that is

$$N_t = N_{t-} = \lim_{s \uparrow t} N_s \qquad a.s.;$$

iii) N is stochastically continuous that is, for all $\varepsilon > 0$ and for all $t \geq 0$, we have

$$\lim_{h \to 0} P(|N_{t+h} - N_t| \geq \varepsilon) = 0.$$

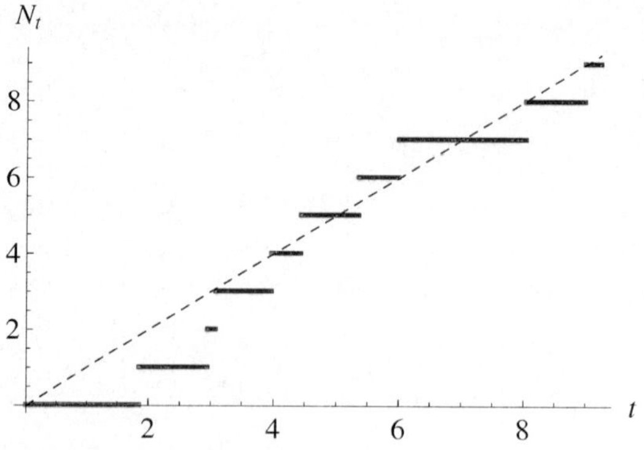

Fig. 13.1. One path of a Poisson process with $\lambda = 1$. The dashed line is the expected value $E[N_t] = \lambda t$

Proof. Property *i)* follows from the definition of Poisson process. Secondly, the discontinuities of N are at the jump times T_n, $n \in \mathbb{N}$: however, by Lemma 13.1 for any $t > 0$ we have

$$P(T_n = t) = 0,$$

and therefore, with probability one, t is not a discontinuity point. This proves *ii)*. Finally *iii)* follows directly from *ii)* because the almost sure convergence implies the convergence in probability (cf. Theorem A.136). □

Next we study the distribution of the Poisson process.

Proposition 13.4 *Let* $(N_t)_{t \geq 0}$ *be a Poisson process with intensity* λ. *Then:*

i) for any $t \geq 0$, N_t *has the distribution*

$$P(N_t = n) = e^{-\lambda t} \frac{(\lambda t)^n}{n!}, \qquad n \in \mathbb{N}, \tag{13.3}$$

and in particular

$$E[N_t] = \lambda t, \qquad \mathrm{var}(N_t) = \lambda t;$$

ii) N *has independent increments, that is for any* $0 \leq t_1 < \cdots < t_n$ *the random variables* N_{t_1}, $N_{t_2} - N_{t_1}, \ldots, N_{t_n} - N_{t_{n-1}}$ *are independent;*

iii) N *has stationary increments, that is*

$$N_t - N_s \overset{d}{=} N_{t-s}, \qquad t \geq s \geq 0.$$

Proof. We only prove *i)*: for the other properties, that are consequence of the absence of memory of the exponential distribution (cf. Example A.29), we

[1] Càdlàg is the French shortening for "right-continuous with finite left limits at all t" (continue à droite et limité à gauche).

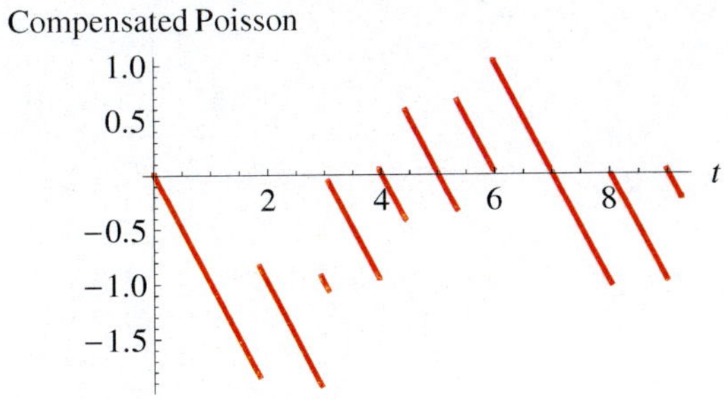

Fig. 13.2. One path of a compensated Poisson process with $\lambda = 1$

refer for instante to [76]. We first observe that by (13.2) we have

$$P(t \geq T_{n+1}) = \int_0^t \lambda e^{-\lambda s} \frac{(\lambda s)^n}{n!} ds =$$

(by parts)

$$= -\left[e^{-\lambda s} \frac{(\lambda s)^n}{n!} \right]_{s=0}^{s=t} + \int_0^t \lambda e^{-\lambda s} \frac{(\lambda s)^{n-1}}{(n-1)!} ds$$
$$= -e^{-\lambda t} \frac{(\lambda t)^n}{n!} + P(t \geq T_n).$$

Hence we have

$$P(N_t = n) = P(t \geq T_n) - P(t \geq T_{n+1}) = e^{-\lambda t} \frac{(\lambda t)^n}{n!}.$$

□

Remark 13.5 Notice that, by Proposition 13.4-*i*) and *iii*), we have

$$E[N_{t+1} - N_t] = E[N_1] = \sum_{n \geq 1} n P(N_1 = n) = e^{-\lambda} \sum_{n \geq 1} \frac{\lambda^n}{(n-1)!} = \lambda,$$

which confirms that the intensity λ represents the number of jumps expected in a unit time interval. □

Remark 13.6 It is known that any counting process with stationary independent increments is a Poisson process (see, for instance, Protter [287]). □

Remark 13.7 Let $(\mathcal{F}_t)$ be the filtration generated by a Poisson process N. By the independence of increments, for any $t > s \geq 0$ we have

$$E[N_t \mid \mathcal{F}_s] = E[N_t - N_s] + N_s = \lambda(t - s) + N_s.$$

As a consequence the process $N_t - \lambda t$ is a martingale and is usually called *compensated Poisson process*. Figure 13.2 shows a path of a compensated Poisson process: note that $N_t - \lambda t$ is not an integer valued process. □

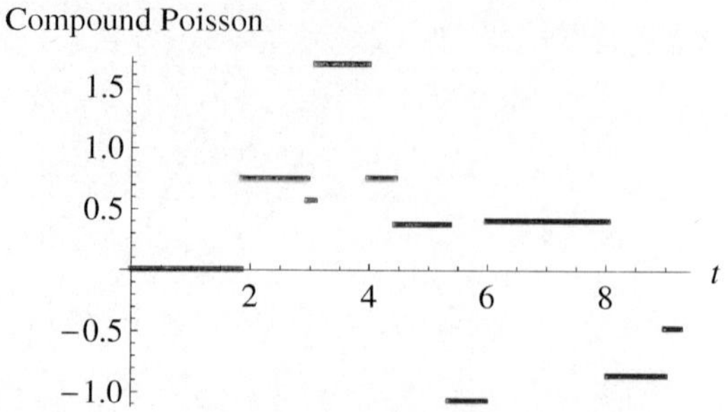

Fig. 13.3. One path of a compound Poisson process with $\lambda = 1$ and $\eta = \mathcal{N}_{0,1}$

Definition 13.8 (Compound Poisson process) *Let N be a Poisson process with intensity λ and assume that $Z = (Z_n)$ is a sequence of i.i.d. random variables in $\mathbb{R}^d$ with distribution η, i.e. $Z_n \sim \eta$ for $n \geq 1$, and which are independent of N. The compound Poisson process is defined as*[2]

$$X_t = \sum_{n=1}^{N_t} Z_n, \qquad t \geq 0. \tag{13.4}$$

The jumps of the compound Poisson process X in (13.4) occur at the same times as the jumps of N and X is a càdlàg process: however, while the jumps of N are of fixed size equal to one, the jumps of X are of random size with distribution η. Figure 13.3 shows a path of a 1-dimensional compound Poisson process with $\lambda = 1$ and $\eta = \mathcal{N}_{0,1}$.

A compound Poisson process has independent and stationary increments. Moreover, if we set[3]

$$m = E[Z_1] \in \mathbb{R}^d, \tag{13.5}$$

then we have

$$E[X_t] = \sum_{n \geq 1} E\left[\mathbb{1}_{\{N_t = n\}} \sum_{k=1}^{n} Z_k\right] =$$

(by the independence of N and Z)

$$= \sum_{n \geq 1} n E[Z_1] P(N_t = n) =$$

[2] By convention, $X_t = 0$ when $N_t = 0$.

[3] Hereafter, we shall always implicitly assume that

$$E[Z_1] = \int_{\mathbb{R}^d} x \eta(dx) < \infty.$$

(by (13.3))

$$= me^{-\lambda t} \sum_{n \geq 1} \frac{(\lambda t)^n}{(n-1)!} = m\lambda t.$$

Definition 13.9 (Compensated compound Poisson process) *Let X be a compound Poisson process with intensity λ and distribution of jumps η. The process*

$$\widetilde{X}_t = X_t - E[X_t] = X_t - m\lambda t,$$

where

$$m = \int_{\mathbb{R}^d} x\, \eta(dx) = E[Z_1],$$

is called compensated compound Poisson process.

A compensated compound Poisson process is a martingale with respect to the filtration generated by N and Z.

13.3 Lévy processes

In this section Lévy processes are introduced and their main properties are presented. They are named after the French mathematician Paul Lévy (1886-1971) who is one of the founding fathers of the theory of stochastic processes. The class of Lévy processes includes Brownian motion and Poisson process and retains the property of the independence and stationarity of the increments. An important consequence is the infinite divisibility of distributions, which implies that X_t at a fixed time, say $t = 1$, can be expressed as the sum of a finite number of i.i.d. random variables: this provides a motivation for modeling price changes as resulting from a large number of shocks in the economy. The Brownian motion is a very special example, since it is the only Lévy process with continuous trajectories; on the other hand, the presence of jumps is one main motivation that has led to consider Lévy processes in finance. A part of this section is devoted to introducing the characteristic exponent which is a convenient concept for handling Lévy processes. Moreover we revise the most popular Lévy processes that have been used in finance. Hereafter, we assume given a filtered probability space $(\Omega, \mathcal{F}, P, (\mathcal{F}_t))$ satisfying the usual hypotheses (cf. Section 3.3.3).

Definition 13.10 *A Lévy process is an adapted stochastic process $X = (X_t)_{t\geq 0}$ defined on the space $(\Omega, \mathcal{F}, P, (\mathcal{F}_t))$ with values in $\mathbb{R}^d$, such that $X_0 = 0$ a.s. and*

L-i) X has increments independent of the past, that is $X_t - X_s$ is independent of $\mathcal{F}_s$ for $0 \leq s < t$;

L-ii) X has stationary increments, that is $X_t - X_s$ has the same distribution of X_{t-s}:

$$X_t - X_s \stackrel{d}{=} X_{t-s}, \qquad 0 \le s < t;$$

L-iii) X is stochastically continuous that is, for all $\varepsilon > 0$ and for all $t \ge 0$, we have

$$\lim_{h \to 0} P(|X_{t+h} - X_t| \ge \varepsilon) = 0.$$

Remark 13.11 In Protter [287], Chapter 1, it is shown that every Lévy process has a unique càdlàg modification that is itself a Lévy process. Therefore we will incorporate such property in the definition of Lévy process: we shall assume that sample paths of a Lévy process are almost surely continuous from the right and have finite limits from the left. Note that a càdlàg function

$$f : \mathbb{R}_{\ge 0} \longrightarrow \mathbb{R}^d$$

can only have jump discontinuities: if t is a discontinuity point then we denote by

$$\Delta f(t) := f(t) - f(t-) \in \mathbb{R}^d \setminus \{0\} \tag{13.6}$$

the jump of f at t; we also set $\Delta f(0) = 0$. Intuitively the value $f(t)$ is not foreseeable by an observer approaching t from the past, which amounts to saying that càdlàg paths have unpredictable jumps. □

Lemma 13.12 *Let f be a càdlàg function defined on a compact interval $[0, T]$. Then, for any $n \in \mathbb{N}$, the number of jumps of f of size greater than $\frac{1}{n}$ is finite:*

$$\#\{t \in]0, T] \mid |\Delta f(t)| \ge \frac{1}{n}\} < \infty.$$

In particular, f has at most a countable number of jumps.

Proof. By contradiction, assume that for some $n \in \mathbb{N}$ the number of jumps of size greater than $\frac{1}{n}$ is infinite: then, since the domain is compact, there exists a sequence (t_k) in $[0, T]$, strictly increasing or decreasing, which converges to $\bar{t} \in [0, T]$ as $k \to \infty$ and is such that

$$|\Delta f(t_k)| \ge \frac{1}{n}, \qquad k \in \mathbb{N}. \tag{13.7}$$

We only consider the case when (t_k) is strictly increasing, the other case being analogous. Since f has *finite* left limits, we have that $f(t_k)$ converges to $f(\bar{t}-)$ as $k \to \infty$ and this contradicts (13.7). □

Remark 13.13 Let X be a Lévy process and consider $T > 0$ and $H \in \mathscr{B}(\mathbb{R}^d)$ such that $0 \notin \overline{H}$ so that

$$\operatorname{dist}(H, 0) = \inf\{|x| \mid x \in H\} > 0.$$

As a consequence of Lemma 13.12 we have that, with probability one, $(X_t)_{t\in[0,T]}$ has only a finite number of jumps of size belonging to H: we express this by saying that *a Lévy process can only have a finite number of large jumps.* On the other hand, X can have infinitely (countably) many *small jumps.*

If a Lévy process has only a finite number of jumps in any bounded time interval (like the compound Poisson process) we say that it is a *finite activity* Lévy process, otherwise we say that it has *infinite activity.* □

Remark 13.14 Theorem 3.47 on the completion of the Brownian filtration extends to the case of Lévy processes. More precisely, let X be a Lévy process and consider the natural filtration of X completed by the negligible events:

$$\mathcal{F}_t^X = \sigma\left(\widetilde{\mathcal{F}}_t^X \cup \mathcal{N}\right).$$

Then $\left(\mathcal{F}_t^X\right)$ is right-continuous and therefore it satisfies the usual hypotheses (see, for instance, Theorem I-31 in [287]); moreover X is a Lévy process with respect to $\mathcal{F}^X$. □

13.3.1 Infinite divisibility and characteristic function

Since most option pricing theory under Lévy processes builds on Fourier transform methods, now we examine the characteristic function (cf. Appendix A.7.1)

$$\varphi_{X_t}(\xi) = E\left[e^{i\xi\cdot X_t}\right], \qquad \xi \in \mathbb{R}^d,\ t \geq 0,$$

of a Lévy process X. A remarkable property of the characteristic function of a Lévy process is that it can be expressed in the form $e^{t\psi(\xi)}$ for some continuous function ψ. More precisely, we have

Theorem 13.15 *If X is a Lévy process, then there exists and is unique a function $\psi \in C\left(\mathbb{R}^d, \mathbb{C}\right)$ such that $\psi(0) = 0$ and*

$$\varphi_{X_t}(\xi) = e^{t\psi(\xi)}, \qquad t \geq 0,\ \xi \in \mathbb{R}^d. \tag{13.8}$$

The function ψ is called the characteristic (or Lévy) exponent of X.

Let us recall that the distribution of a random variable is determined by its characteristic function: a consequence of Theorem 13.15 is that the law of X_t is only determined by the characteristic exponent ψ or equivalently by the law of X_1. Therefore *in order to specify the distribution of a Lévy process X, it is sufficient to specify the distribution of X_t for a single time.*

To give a sketch of the proof of Theorem 15.16, we show some preliminary result and introduce the notion of infinite divisibility.

Definition 13.16 *A random value Y is said to be infinitely divisible if, for any $n \geq 2$, there exist i.i.d. random variables $Y_1^{(n)}, \dots, Y_n^{(n)}$ such that*

$$Y \stackrel{d}{=} Y_1^{(n)} + \cdots + Y_n^{(n)}.$$

In other words, an infinitely divisible random variable can be decomposed into the sum of an arbitrary number of i.i.d. variables: for example, if $Y \sim \mathcal{N}_{\mu,\sigma^2}$ then we can take independent variables $Y_i^{(n)} \sim \mathcal{N}_{\frac{\mu}{n},\frac{\sigma^2}{n}}$, $i = 1, \dots, n$.

Lemma 13.17 *If X is Lévy process, then X_t is infinitely divisible for each $t \geq 0$ and we have*

$$\varphi_{X_t}(\xi) = \left(\varphi_{X_{\frac{t}{n}}}(\xi)\right)^n, \qquad t \geq 0, \ n \in \mathbb{N}. \tag{13.9}$$

Proof. The thesis follows by the properties *L-i)* and *L-ii)* of Definition 13.10: indeed, for any $n \geq 2$ we set

$$Y_i^{(n)} := X_{\frac{it}{n}} - X_{\frac{(i-1)t}{n}} \overset{d}{=} X_{\frac{t}{n}}, \qquad i = 1, \dots, n, \tag{13.10}$$

and we remark that $Y_i^{(n)}$ are i.i.d. by *L-ii)*. Then it suffices to observe that

$$X_t = Y_1^{(n)} + \dots + Y_n^{(n)}. \tag{13.11}$$

Formula (13.9) follows from (13.10)-(13.11) and the independence of the variables $Y_i^{(n)}$, $i = 1, \dots, n$. □

Lemma 13.18 *If $(X_t)_{t \geq 0}$ is stochastically continuous, then the map $t \mapsto \varphi_{X_t}(\xi)$ is continuous for each $\xi \in \mathbb{R}^d$.*

Proof. Let $\xi \in \mathbb{R}^d$ be fixed: for any $\varepsilon > 0$ we consider $\delta_\varepsilon > 0$ such that

$$\sup_{|y| \leq \delta_\varepsilon} \left| e^{i\xi \cdot y} - 1 \right| < \frac{\varepsilon}{2}.$$

If X is stochastically continuous, there exists $\delta'_\varepsilon > 0$ such that

$$P(|X_t - X_s| > \delta_\varepsilon) < \frac{\varepsilon}{4}$$

whenever $|t - s| \leq \delta'_\varepsilon$. Hence we have

$$\begin{aligned}
|\varphi_{X_t}(\xi) - \varphi_{X_s}(\xi)| =& E\left[\left|e^{i\xi\cdot(X_t - X_s)} - 1\right|\right] \\
=& E\left[\left|e^{i\xi\cdot(X_t - X_s)} - 1\right| \mathbb{1}_{\{|X_t - X_s| \leq \delta_\varepsilon\}}\right] \\
&+ E\left[\left|e^{i\xi\cdot(X_t - X_s)} - 1\right| \mathbb{1}_{\{|X_t - X_s| > \delta_\varepsilon\}}\right] \\
\leq& \frac{\varepsilon}{2} + 2P\left(|X_t - X_s| > \delta_\varepsilon\right) \leq \varepsilon
\end{aligned}$$

for $|t - s| \leq \delta'_\varepsilon$. □

The following non trivial lemma is another consequence of the infinite divisibility property of Lévy processes.

Lemma 13.19 *If $(X_t)_{t\geq 0}$ is a Lévy process, then $\varphi_{X_t}(\xi) \neq 0$ for any $\xi \in \mathbb{R}^d$ and $t \geq 0$.*

Proof. See Sato [297], Lemma 7.5. □

Lemma 13.20 *Let $\varphi \in C(\mathbb{R}^d, \mathbb{C})$ such that $\varphi(0) = 1$ and $\varphi(\xi) \neq 0$ for any $\xi \in \mathbb{R}^d$. Then there exists a unique continuous function $g \in C(\mathbb{R}^d, \mathbb{C})$ such that $g(0) = 0$ and $\varphi(\xi) = e^{g(\xi)}$ for any $\xi \in \mathbb{R}^d$.*

Proof. See Sato [297], Lemma 7.6. □

We improperly write
$$g(\xi) = \log \varphi(\xi)$$
and we call g the *distinguished complex logarithm* of φ. We recall that the complex logarithm is a multi-valued function (see also Section 15.1); therefore it is important to remark that g *is not the composition of φ with a fixed branch of the complex logarithm function:* in particular, $\varphi(\xi) = \varphi(\xi')$ does not imply $g(\xi) = g(\xi')$.

We are now in position to prove Theorem 15.16.

Proof (of Theorem 13.15). Let us denote by g_t the distinguished complex logarithm of φ_{X_t} that is well-defined for any Lévy process by Lemmas 13.19 and 13.20. By (13.9) with $t = m \in \mathbb{N}$, we have

$$e^{g_m(\xi)} = \varphi_{X_m}(\xi) = \left(\varphi_{X_{\frac{m}{n}}}(\xi)\right)^n = e^{n g_{\frac{m}{n}}(\xi)}$$

and this implies

$$g_m(\xi) = n g_{\frac{m}{n}}(\xi) + 2\pi i k(\xi), \qquad \xi \in \mathbb{R}^d,$$

where $\xi \mapsto k(\xi)$ is, by definition, a continuous function with values in $\mathbb{Z}$ and therefore is a constant function: more precisely, $k(\xi) = k(0) = 0$ because $g_t(0) = 0$ for any t. Thus we have

$$g_m = n g_{\frac{m}{n}}, \qquad m, n \in \mathbb{N}. \tag{13.12}$$

By (13.12), we also have

$$g_{\frac{m}{n}} = \frac{1}{n} g_m =$$

(by (13.12) with $n = m$)

$$= \frac{m}{n} g_1, \qquad m, n \in \mathbb{N}. \tag{13.13}$$

By Lemma 13.18, $t \mapsto g_t(\xi)$ is a continuous function for any $\xi \in \mathbb{R}^d$: therefore, if $t \geq 0$ and (q_n) is a sequence of positive rational numbers approximating t, we have

$$g_t(\xi) = \lim_{n\to\infty} g_{q_n}(\xi) =$$

(by (13.13))

$$= \lim_{n\to\infty} q_n g_1(\xi) = t g_1(\xi), \qquad \xi \in \mathbb{R}^d,\ t \geq 0.$$

Therefore the thesis is proved with $\psi = g_1$. □

Example 13.21 (Brownian motion with drift) Let $X_t = \mu t + \sigma W_t$ where W is a standard real Brownian motion: as a consequence of Example A.60 and Lemma A.70, we have

$$E\left[e^{i\xi X_t}\right] = e^{i\mu t\xi} E\left[e^{i\xi\sigma W_t}\right] = e^{i\mu t\xi + \frac{1}{2}(i\xi\sigma)^2 t},$$

that is

$$\psi(\xi) = i\mu\xi - \frac{\sigma^2\xi^2}{2}.$$

In the d-dimensional case when $\mu \in \mathbb{R}^d$ and σ is an $d \times N$ constant matrix, we have

$$\psi(\xi) = i\mu\cdot\xi - \frac{1}{2}\langle \mathcal{C}\xi, \xi\rangle,$$

where $\mathcal{C} = \sigma\sigma^*$. □

Example 13.22 (Poisson process) Let N_t denote a Poisson process with intensity λ (cf. Definition 13.2). We have

$$\varphi_{N_t}(\xi) = E\left[e^{i\xi N_t}\right] = \sum_{n\geq 0} E\left[e^{i\xi n}\mathbb{1}_{\{N_t=n\}}\right] =$$

(by (13.3))

$$= e^{-\lambda t}\sum_{n\geq 0}\frac{\left(e^{i\xi}\lambda t\right)^n}{n!} = e^{-\lambda t}e^{\lambda t e^{i\xi}}.$$

Hence in this case the characteristic exponent is given by

$$\psi(\xi) = \lambda\left(e^{i\xi} - 1\right). \tag{13.14}$$

□

Example 13.23 (Compound Poisson process) Let

$$X_t = \sum_{n=1}^{N_t} Z_n, \qquad t \geq 0,$$

be a d-dimensional compound Poisson process (cf. Definition 13.8) with intensity λ and distribution of jumps η. We denote by

$$\hat{\eta}(\xi) = E\left[e^{i\xi \cdot Z_1}\right] = \int_{\mathbb{R}^d} e^{i\xi \cdot x} \eta(dx), \qquad \xi \in \mathbb{R}^d,$$

the characteristic function of Z_1. Then we have

$$\varphi_{X_t}(\xi) = \sum_{n \geq 0} E\left[e^{i\xi \cdot \sum_{k=1}^{n} Z_k} \mathbb{1}_{\{N_t = n\}}\right] =$$

(by the independence of $N_t, Z_1, \dots, Z_n$, $n \in \mathbb{N}$)

$$= \sum_{n \geq 0} \left(E\left[e^{i\xi \cdot Z_1}\right]\right)^n P(N_t = n) =$$

(by (13.3))

$$= e^{-\lambda t} \sum_{n \geq 0} \frac{(\lambda t \hat{\eta}(\xi))^n}{n!} = e^{-\lambda t} e^{\lambda t \hat{\eta}(\xi)}.$$

Hence the characteristic exponent is given by

$$\psi(\xi) = \lambda \left(\hat{\eta}(\xi) - 1\right) = \int_{\mathbb{R}^d} \left(e^{i\xi \cdot x} - 1\right) \lambda \eta(dx). \tag{13.15}$$

Note that the measure $\lambda\eta$ is equal to the product of the intensity λ, that is the expected number of jumps in a unit time interval, with the distribution of the size of the jumps η: hence, for any Borel set H, $\lambda\eta(H)$ can be interpreted as the *intensity of jumps with size in* H, that is $\lambda\eta(H)$ is the number of jumps, with size in H, that are expected in a unit time interval. We also remark that (13.14) is a special case of (13.15) with $\eta = \delta_1$ (the Dirac distribution centered at 1). □

Example 13.24 (Compensated compound Poisson process) Let

$$\widetilde{X}_t = X_t - m\lambda t, \qquad m = \int_{\mathbb{R}^d} x \eta(dx),$$

be a compensated compound Poisson process with intensity λ and distribution of jumps η (cf. Definition 13.9). Then we have

$$\varphi_{\widetilde{X}_t}(\xi) = \varphi_{X_t}(\xi) e^{-i\lambda t \xi \cdot m};$$

therefore the characteristic exponent has the form

$$\psi(\xi) = \lambda\left(\hat{\eta}(\xi) - 1 - i\xi \cdot m\right) = \int_{\mathbb{R}^d} \left(e^{i\xi\cdot x} - 1 - i\xi \cdot x\right) \lambda\eta(dx). \tag{13.16}$$

□

Remark 13.25 Let X and Y be independent Lévy processes with characteristic exponents ψ_X and ψ_Y respectively. Then the sum $X + Y$ is a Lévy process with characteristic exponent $\psi_X + \psi_Y$. Indeed, by independence we have

$$e^{t\psi_{X+Y}(\xi)} = E\left[e^{i\xi\cdot(X_t+Y_t)}\right] = \varphi_X(\xi)\varphi_Y(\xi) = e^{t(\psi_X(\xi)+\psi_Y(\xi))}.$$ □

Example 13.26 (Jump-diffusion process) Consider the jump-diffusion process

$$X_t = \mu t + B_t + \sum_{n=1}^{N_t} Z_n$$

which is the independent sum of a correlated Brownian motion with drift and covariance matrix $\mathcal{C}$, and a compound Poisson process. Then

$$\psi(\xi) = i\mu \cdot \xi - \frac{1}{2}\langle \mathcal{C}\xi, \xi\rangle + \int_{\mathbb{R}^d} (e^{ix\cdot\xi} - 1)\lambda\eta(dx),$$

where λ is the intensity and η is the distribution of jumps. In particular, if Z_1 is one-dimensional and normal, $Z_1 \sim \mathcal{N}_{\alpha,\delta^2}$ then

$$\psi(\xi) = i\mu\xi - \frac{1}{2}\sigma^2\xi^2 + \lambda\left(e^{i\alpha\xi - \frac{1}{2}\delta^2\xi^2} - 1\right).$$

The first model for risky assets in finance which employed a jump-diffusion is the Merton model [251]. □

13.3.2 Jump measures of compound Poisson processes

In this section we consider a jump-diffusion process X in $\mathbb{R}^d$ of the form

$$X_t = \mu t + B_t + \sum_{n=1}^{N_t} Z_n \tag{13.17}$$

where $\mu \in \mathbb{R}^d$, B is a d-dimensional correlated Brownian motion with correlation matrix[4] $\mathcal{C}$, N is a Poisson process with intensity λ and $(Z_n)_{n\geq 1}$ are i.i.d. random variables in $\mathbb{R}^d$ with distribution η. As usual, the Brownian and Poisson components are independent.

[4] In terminology used here $B_t \sim \mathcal{N}_{0,t\mathcal{C}}$.

We denote by (T_n) the increasing sequence of jump times. For any $I \times H \in \mathscr{B}([0,+\infty[\times\mathbb{R}^d)$, we put

$$J(I \times H) := \sum_{n\geq 1} \delta_{T_n}(I)\delta_{Z_n}(H) \tag{13.18}$$

where δ denotes the Dirac delta. In particular, note that for $I = [0,t]$ we have

$$J([0,t] \times H) = \sum_{n=1}^{N_t} \delta_{Z_n}(H).$$

In other terms, $J([0,t] \times H)$ counts the number of jumps occurring in the time interval $[0,t]$ and such that their size is in H. Since, with probability one, only a finite number of jumps occurs on any bounded time interval, the sum in (13.18) has a finite number of terms and J is well-defined. Moreover, J is a finite sum of Dirac deltas and therefore it is a σ-finite measure on $\mathscr{B}([0,+\infty[\times\mathbb{R}^d)$ taking values in the set of non-negative integers $\mathbb{N}_0$: notice that J also depends on $\omega \in \Omega$ and thus it is a measure taking random values, i.e. it is a *random measure*. We refer to Kallenberg [194] or Jacod and Shiryaev [184], Chapter II, for a presentation of the general theory of random measures.

Definition 13.27 *The random measure J in* (13.18) *is called jump measure of X.*

The expectation of J is given by

$$\begin{aligned}
E\left[J\left([0,t] \times H\right)\right] &= E\left[\sum_{k=1}^{N_t} \delta_{Z_k}(H)\right] \\
&= \sum_{n\geq 1} E\left[\sum_{k=1}^{n} \delta_{Z_k}(H)\mathbb{1}_{\{N_t=n\}}\right] \\
&= \sum_{n\geq 1} P(N_t = n) \sum_{k=1}^{n} P\left(\delta_{Z_k}(H)\right) \\
&= e^{-\lambda t} \sum_{n\geq 1} \frac{(\lambda t)^n}{n!} n\eta(H) = t\lambda\eta(H).
\end{aligned}$$

In particular we have

$$E\left[J\left([0,t] \times H\right)\right] = tE\left[J\left([0,1] \times H\right)\right],$$

and

$$\nu(H) := E\left[J\left([0,1] \times H\right)\right] = \lambda\eta(H), \qquad H \in \mathscr{B}(\mathbb{R}^d), \tag{13.19}$$

defines a finite measure on $\mathscr{B}(\mathbb{R}^d)$ such that

$$\nu(\mathbb{R}^d) = \lambda.$$

Definition 13.28 *The measure ν in* (13.19) *is called the intensity (or Lévy) measure of X.*

The intensity measure $\nu(H)$ determines the expected number, per unit time, of jumps of X whose amplitude belongs to H. Note that, differently from J, ν is not integer-valued.

Remark 13.29 We recall (13.15) of Example 13.23 and note that the characteristic exponent of the jump part of X in (13.17) (i.e. the compound Poisson process $\sum_{n=1}^{N_t} Z_n$), can be expressed in terms of the Lévy measure as follows:

$$\psi(\xi) = \int_{\mathbb{R}^d} \left(e^{i\xi\cdot x} - 1\right) \nu(dx), \qquad \xi \in \mathbb{R}^d.$$

Since the characteristic exponent determines the distribution of the process, we have that *the Lévy measure characterizes the jump part of X*. More generally, the Lévy process X in (13.17) is completely identified by the triplet $(\mu, \mathcal{C}, \nu)$ where:

- μ is the coefficient of the drift part;
- $\mathcal{C}$ is the covariance matrix of the diffusion part;
- ν is the intensity measure of the jump part.

As we shall see later, a similar characterization holds for the general class of Lévy processes. □

Next we prove a crucial result which allows to obtain a representation of X in terms of its jump measure J.

Theorem 13.30 *Let X be the jump-diffusion process in* (13.17) *with jump measure J and Lévy measure ν. For any function $f = f(t,x)$ we have*

$$\sum_{\substack{0<s\le t \\ \Delta X_s \neq 0}} f(s, \Delta X_s) = \int_0^t \int_{\mathbb{R}^d} f(s,x) J(ds, dx). \tag{13.20}$$

Further, assume that $f \in L^1\left([0,+\infty[\times\mathbb{R}^d, ds \otimes \nu\right)$ and let

$$\begin{aligned} M_t &= \int_0^t \int_{\mathbb{R}^d} f(s,x) \widetilde{J}(ds, dx) \\ &= \int_0^t \int_{\mathbb{R}^d} f(s,x) J(ds, dx) - \int_0^t \int_{\mathbb{R}^d} f(s,x) \nu(dx) ds \end{aligned} \tag{13.21}$$

where

$$\widetilde{J}(dt, dx) := J(dt, dx) - dt\nu(dx) \tag{13.22}$$

is called the compensated jump measure of X. Then M is a martingale and $E[M_t] = 0$, that is

$$E\left[\int_0^t \int_{\mathbb{R}^d} f(s,x) J(ds, dx)\right] = \int_0^t \int_{\mathbb{R}^d} f(s,x) \nu(dx) ds. \tag{13.23}$$

Moreover, if $f \in L^2\left([0,+\infty[\times\mathbb{R}^d, ds\otimes\nu\right)$ *then we have*

$$\operatorname{var}(M_t) = E\left[\left(\int_0^t\int_{\mathbb{R}^d} f(s,x)\widetilde{J}(ds,dx)\right)^2\right] = \int_0^t\int_{\mathbb{R}^d} f^2(s,x)\nu(dx)ds. \tag{13.24}$$

Proof. For simplicity, we only consider the case $f = f(x)$. We first remark that, for a jump-diffusion process X, we have

$$\sum_{\substack{0<s\le t\\ \Delta X_s\neq 0}} f(\Delta X_s) = \sum_{n=1}^{N_t} f(Z_n), \tag{13.25}$$

where, as usual, $\Delta X_s = X_s - X_{s-}$. Clearly both sums in (13.25) have the same *finite* number of terms, that is N_t, and therefore they are finite. Moreover we have

$$\sum_{n=1}^{N_t} f(Z_n) = \sum_{n\ge 1}\int_0^t\int_{\mathbb{R}^d} f(x)\delta_{Z_n}(dx)\delta_{T_n}(ds) = \int_0^t\int_{\mathbb{R}^d} f(x)J(ds,dx),$$

which proves (13.20).

Next we remark that

$$\begin{aligned}
E\left[\sum_{n=1}^{N_t} f(Z_n)\right] &= \sum_{n\ge 1} E\left[\sum_{k=1}^{n} f(Z_k)\mathbb{1}_{\{N_t=n\}}\right]\\
&= \sum_{n\ge 1} P(N_t=n)\sum_{k=1}^{n} E\left[f(Z_k)\right]\\
&= e^{-\lambda t}\sum_{n\ge 1}\frac{(\lambda t)^n}{n!} nE\left[f(Z_1)\right]\\
&= \lambda t E\left[f(Z_1)\right] = \left[N_t f(Z_1)\right].
\end{aligned} \tag{13.26}$$

Now assume that $f \in L^1\left(\mathbb{R}^d,\nu\right)$: then M in (13.21) is integrable because

$$E\left[\left|\int_0^t\int_{\mathbb{R}^d} f(x)J(ds,dx)\right|\right] \le E\left[\sum_{n=1}^{N_t}\left|f(Z_n)\right|\right]$$

(by (13.26))

$$\begin{aligned}
&= E\left[N_t\left|f(Z_1)\right|\right]\\
&= \lambda t E\left[\left|f(Z_1)\right|\right] = t\int_{\mathbb{R}^d}\left|f(x)\right|\nu(dx).
\end{aligned}$$

Moreover, M is adapted by definition and we have

$$E\left[\int_s^t \int_{\mathbb{R}^d} f(x) J(ds, dx) \mid \mathcal{F}_s\right] = E\left[\sum_{N_s < n \le N_t} f(Z_n)\right]$$

(by (13.26))

$$\begin{aligned}
&= E\left[(N_t - N_s) f(Z_1)\right] \\
&= \lambda(t-s) E\left[f(Z_1)\right] = (t-s)\int_{\mathbb{R}^d} f(x)\nu(dx),
\end{aligned}$$

and this proves that M is a martingale.

Finally, assume that $f \in L^2\left(\mathbb{R}^d, \nu\right)$: we have

$$\begin{aligned}
&E\left[\left(\sum_{k=1}^{N_t} f(Z_k) - t\int_{\mathbb{R}^d} f(x)\nu(dx)\right)^2\right] \\
&= \sum_{n\ge 1} E\left[\left(\sum_{k=1}^{n} f(Z_k) - t\int_{\mathbb{R}^d} f(x)\nu(dx)\right)^2 \mathbb{1}_{\{N_t = n\}}\right] \\
&= e^{-\lambda t}\sum_{n\ge 1}\frac{(\lambda t)^n}{n!}\left(E\left[\left(\sum_{k=1}^{n} f(Z_k)\right)^2\right] - 2t\int_{\mathbb{R}^d} f(x)\nu(dx) E\left[\sum_{k=1}^{n} f(Z_k)\right]\right. \\
&\quad \left. + \left(t\int_{\mathbb{R}^d} f(x)\nu(dx)\right)^2\right) \\
&= e^{-\lambda t}\sum_{n\ge 1}\frac{(\lambda t)^n}{n!} E\left[\left(\sum_{k=1}^{n} f(Z_k)\right)^2\right] - \left(t\int_{\mathbb{R}^d} f(x)\nu(dx)\right)^2 \\
&= e^{-\lambda t}\sum_{n\ge 1}\frac{(\lambda t)^n}{n!}\left(E\left[\sum_{k=1}^{n} f^2(Z_k)\right] + E\left[\sum_{h\neq k} f(Z_k) f(Z_h)\right]\right) \\
&\quad - \left(t\int_{\mathbb{R}^d} f(x)\nu(dx)\right)^2 \\
&= e^{-\lambda t}\sum_{n\ge 1}\frac{(\lambda t)^n}{n!}\left(nE\left[f^2(Z_1)\right] + n(n-1)E\left[f(Z_1)\right]^2\right) - \left(t\int_{\mathbb{R}^d} f(x)\nu(dx)\right)^2 \\
&= \lambda t E\left[f^2(Z_1)\right] = t\int_{\mathbb{R}^d} f^2(x)\nu(dx),
\end{aligned}$$

and this concludes the proof of (13.24). □

Remark 13.31 Combining (13.20) with (13.23), we have

$$E\left[\sum_{\substack{0<s\leq t\\ \Delta X_s\neq 0}} f(s,\Delta X_s)\right]=\int_0^t\int_{\mathbb{R}^d} f(s,x)\nu(dx)ds. \tag{13.27}$$

Combining (13.23) with (13.24), we obtain the interesting isometry

$$E\left[\left(\int_0^t\int_{\mathbb{R}^d} f(s,x)\widetilde{J}(ds,dx)\right)^2\right]=E\left[\int_0^t\int_{\mathbb{R}^d} f^2(s,x)J(ds,dx)\right]. \tag{13.28}$$

□

By (13.20) with $f(x)=x$, we get the following remarkable representation of the process X in (13.17):

$$X_t=\mu t+B_t+\int_0^t\int_{\mathbb{R}^d} xJ(ds,dx); \tag{13.29}$$

this is a particular case of the fundamental Lévy-Itô decomposition of Theorem 13.35. Moreover, if $f(x)=x$ is η-integrable (and therefore ν-integrable) then we have

$$E\left[X_t\right]=t\left(\mu+\int_{\mathbb{R}^d} x\nu(dx)\right). \tag{13.30}$$

Remark 13.32 We consider a compound Poisson process

$$X_t=\sum_{n=1}^{N_t} Z_n. \tag{13.31}$$

By Theorem 13.30, we have the following representations of X and of its compensated version (cf. Definition 13.9):

$$X_t=\int_0^t\int_{\mathbb{R}^d} xJ(ds,dx), \tag{13.32}$$

$$\widetilde{X}_t=X_t-E\left[X_t\right]=\int_0^t\int_{\mathbb{R}^d} x\widetilde{J}(ds,dx). \tag{13.33}$$

Moreover by taking $f(x)=|x|$ in Theorem 13.30, we deduce that the first variation[5] of X on $[0,t]$ is given by

$$V_{[0,t]}(X)=\sum_{n=1}^{N_t}|Z_n|=\int_0^t\int_{\mathbb{R}^d}|x|J(ds,dx);$$

---

[5] See Definition 3.59.

taking the expectation, we also have

$$E\left[V_{[0,t]}(X)\right] = t\int_{\mathbb{R}^d} |x|\nu(dx)$$

which may be infinite. Concerning the quadratic variation (cf. Definition 3.72), taking $f(x) = |x|^2$ in Theorem 13.30, we get

$$V_t^2(X) = \sum_{n=1}^{N_t} |Z_n|^2 = \int_0^t \int_{\mathbb{R}^d} |x|^2 J(ds, dx), \tag{13.34}$$

$$E\left[V_t^2(X)\right] = t\int_{\mathbb{R}^d} |x|^2\nu(dx) \le \infty. \tag{13.35}$$

We emphasize that the quadratic variation process $V_t^2(X)$ in (13.34) is well-defined and finite for any compound Poisson process X: indeed identity (13.20) holds *for any function* f. On the contrary, the expected quadratic variation $E\left[V_t^2(X)\right]$ in (13.35) is finite if and only if $f(x) = |x|^2$ is ν-integrable. □

13.3.3 Lévy-Itô decomposition

In Section 13.3.2 we saw that every jump-diffusion process can be represented as

$$X_t = \mu t + B_t + \int_0^t \int_{\mathbb{R}^d} xJ(ds, dx), \tag{13.36}$$

where J is the jump measure of X. We recall that $J([0,t] \times H)$ takes only non-negative integer values and counts the number of jumps of X occurring in the time interval $[0,t]$ and whose amplitude belongs to $H \in \mathscr{B}(\mathbb{R}^d)$: for any jump-diffusion process X, the measure $J([0,t] \times \mathbb{R}^d)$ is finite a.s. because X has a finite number of jumps in any bounded time interval.

We may ask if every Lévy process X admits a representation of the form (13.36). Actually, the first question concerns the definition of the jump measure J: for a compound Poisson process the definition is well-posed because the sum in (13.18) has only a finite number of terms. However, if X has infinite activity (cf. Remark 9.85), i.e. X has infinitely many jumps in finite time, then J may become infinite. On the other hand, as already noted in Remark 13.13, a Lévy process X can only have a finite number of "large" jumps: more precisely, if $H \in \mathscr{B}(\mathbb{R}^d)$ and $0 \notin \overline{H}$, where $\overline{H}$ denotes the closure of H, then X has only a *finite* number of jumps with size in H. This allows to define $J(I \times H)$ for any $I \times H \in \mathscr{B}([0,+\infty[\times\mathbb{R}^d)$ with I bounded and H such that $0 \notin \overline{H}$:

$$J(I \times H) := \#\{t \in I \mid \Delta X_t \in H\}.$$

Then, by the general results of measure theory, J can be extended to a σ-finite (in general, *not finite*) random measure on[6] $\mathscr{B}\left([0,+\infty[\times\mathbb{R}^d \setminus \{0\}\right)$. Once we

[6] J can be extended to $\mathscr{B}\left([0,+\infty[\times\mathbb{R}^d\right)$ by setting $J(\{0\} \times \mathbb{R}^d) = 0$.

have defined the jump measure of a generic Lévy process X, we may also consider its *Lévy measure*:

$$\nu(H) := E\left[J\left([0,1] \times H\right)\right], \qquad H \in \mathscr{B}(\mathbb{R}^d).$$

The Lévy measure $\nu(H)$ gives the expected number, per unit time, of jumps of X whose amplitude belongs to H. We remark that ν is a measure on $\mathscr{B}(\mathbb{R}^d)$ but it is not a probability measure nor it is necessarily finite[7].

Lemma 13.33 *Let X be a Lévy process with jump measure J and Lévy measure ν. Then*

i) if $H \in \mathscr{B}(\mathbb{R}^d)$ is such that $0 \notin \overline{H}$ then the process

$$t \mapsto J_t(H) := J([0,t] \times H) = \#\{s \in]0,t] \mid \Delta X_s \in H\} \tag{13.37}$$

is a Poisson process with intensity $\nu(H)$ and the compensated process

$$t \mapsto \widetilde{J}_t(H) = J_t(H) - t\nu(H)$$

is a martingale;

ii) if $H \in \mathscr{B}(\mathbb{R}^d)$ is such that $0 \notin \overline{H}$ and f is a measurable function, then the process

$$t \mapsto J_t(H,f) := \int_0^t \int_H f(s,x) J(ds,dx) = \sum_{0<s\le t} f(s,\Delta X_s)\mathbb{1}_H(\Delta X_s) \tag{13.38}$$

is a compound Poisson process;

iii) if f, g are measurable functions and H, K are disjoint Borel sets such that $0 \notin \overline{H} \cup \overline{K}$, then the processes $J_t(H,f), J_t(K,g)$ are independent.

Proof. Part i) can be verified directly using the definition of Poisson process: see, for instance, Protter [287], p. 26. Part ii) can be proved via the characteristic function: see Corollary 13.42 below or Theorem 2.3.9 in Applebaum [11]. Concerning part iii), we refer for instance to Kallenberg [195], Lemma 13.6.□

Let X be a Lévy process with jump measure J and Lévy measure ν. For any Borel function $f = f(t,x)$ on $\mathbb{R}_{\ge 0} \times \mathbb{R}^d$ one can construct, ω by ω, the integral

$$\int_0^t \int_{\mathbb{R}^d} f(s,x) J(ds,dx) \tag{13.39}$$

with respect to the random measure J proceeding as in the deterministic case (see Appendix A.1.4 and also Cont and Tankov [76], Section 2.6.4): this definition coincides with that given by (13.20) or (13.38) in case the number of jumps is finite.

The following result generalizes Theorem 13.30 and is one of the fundamental results about Lévy processes.

[7] However it is σ-finite.

Theorem 13.34 *Let $(X_t)_{t\geq 0}$ be a d-dimensional Levy process with Levy measure ν and jump measure J. For any measurable function f such that*

$$\int_0^t \int_{|x|\leq\varepsilon} |f(s,x)|\nu(dx)ds < \infty \tag{13.40}$$

for some $\varepsilon > 0$, we have

$$\int_0^t \int_{\mathbb{R}^d} f(s,x)J(ds,dx) = \sum_{\substack{0<s\leq t \\ \Delta X_s \neq 0}} f(s, \Delta X_s) < \infty \qquad \text{a.s.} \tag{13.41}$$

Moreover, if $f \in L^1\left([0,+\infty[\times\mathbb{R}^d, ds \otimes \nu\right)$, then the process

$$\begin{aligned} M_t &= \int_0^t \int_{\mathbb{R}^d} f(s,x)\widetilde{J}(ds,dx) \\ &= \int_0^t \int_{\mathbb{R}^d} f(s,x)\left(J(ds,dx) - \nu(dx)ds\right) \end{aligned} \tag{13.42}$$

is a martingale and $E[M_t] = 0$ or equivalently

$$E\left[\int_0^t \int_{\mathbb{R}^d} f(s,x)J(ds,dx)\right] = \int_0^t \int_{\mathbb{R}^d} f(s,x)\nu(dx)ds. \tag{13.43}$$

If $f \in L^2\left([0,+\infty[\times\mathbb{R}^d, ds \otimes \nu\right)$ then $M_t \in L^2$ and we have

$$\operatorname{var}(M_t) = E\left[\left(\int_0^t \int_{\mathbb{R}^d} f(s,x)\widetilde{J}(ds,dx)\right)^2\right] = \int_0^t \int_{\mathbb{R}^d} f^2(s,x)\nu(dx)ds. \tag{13.44}$$

Proof. Formulas (13.43) and (13.44) follows from (13.23)-(13.24) by limit arguments (for further details, see Section 2.4 in Applebaum [11] and Section I-4 in Protter [287]).

Now assume that f satisfies condition (13.40). The idea is that "large jumps" contribute with only a finite number of terms in the series in (13.41), while "small jumps" can be infinitely many but the local integrability condition (13.40) on f and formula (13.43) guarantee that the series is absolutely convergent. Indeed, by (13.43) we have

$$E\left[\int_0^t \int_{|x|\leq\varepsilon} |f(s,x)|J(ds,dx)\right] = \int_0^t \int_{|x|\leq\varepsilon} f(s,x)\nu(dx)ds$$

which is finite by assumption. □

Next we state the first fundamental result on the structure of the paths of a Lévy process (for the proof, consult Applebaum [11] Section 2.4, Bertoin [44] or Jacod and Shiryaev [184]).

Theorem 13.35 (Lévy-Itô decomposition) *Let $(X_t)_{t\geq 0}$ be a d-dimensional Lévy process with jump measure J_t and Lévy measure ν. Then the Lévy measure ν satisfies*

$$\int_{|x|\geq 1} \nu(dx) < \infty, \tag{13.45}$$

$$\int_{|x|<1} |x|^2 \nu(dx) < \infty. \tag{13.46}$$

Moreover, there exists a d-dimensional correlated Brownian motion B and, for any $R > 0$, there exists $\mu_R \in \mathbb{R}^d$ such that

$$X_t = \mu_R t + B_t + X_t^R + M_t^R \tag{13.47}$$

where

$$X_t^R = \int_0^t \int_{|x|\geq R} x J(ds, dx), \tag{13.48}$$

$$M_t^R = \int_0^t \int_{|x|<R} x \widetilde{J}(ds, dx), \tag{13.49}$$

and $\widetilde{J}$ denotes the compensated jump measure (cf. (13.22)). The terms in (13.47) are independent.

The first two terms in (13.47) correspond to a Brownian motion with drift and form the continuous part of X. The other two terms are discontinuous processes incorporating the jumps of X and only depend on the jump measure. In particular, by (13.41)

$$X_t^R = \sum_{0<s\leq t} \Delta X_s \mathbb{1}_{\{|\Delta X_s|\geq R\}}, \tag{13.50}$$

and by Lemma 13.33-ii), X^R *is a compound Poisson process* that is responsible for the large jumps of X: indeed, X^R has a finite number of jumps in $[0, t]$, which correspond to the jumps of X with absolute value larger than R.

Analogously, by (13.46) and Theorem 13.34, M^R is a L^2*-martingale* which is responsible for the small jumps: indeed,

$$M_t^R = \lim_{\varepsilon\to 0+} \widetilde{X}_t^{R,\varepsilon}$$

where

$$\begin{aligned} \widetilde{X}_t^{\varepsilon,R} &= \int_0^t \int_{\varepsilon\leq|x|<R} x\,(J(ds, dx) - \nu(dx)ds) \\ &= \sum_{\substack{0<s\leq t \\ \varepsilon\leq|\Delta X_s|<R}} \Delta X_s - tE\left[\Delta X_1 \mathbb{1}_{\{\varepsilon\leq|\Delta X_1|<R\}}\right] \end{aligned} \tag{13.51}$$

is the compensated compound Poisson process of the jumps of X with size between ε and R. As ε goes to zero, the number of jumps may become infinite and this prevents us from considering directly the limit of

$$X_t^{\varepsilon,R} = \int_0^t \int_{\varepsilon \le |x| < R} xJ(ds, dx)$$

in the Lévy-Itô decomposition: indeed, in order to guarantee the convergence of the last term, we have to adjust the drift of $X^{\varepsilon,R}$ by considering its compensated version that is a L^2-martingale by (13.46) and Theorem 13.34; then the isometry (13.44) can be applied to show convergence. Intuitively, this adjustment does not affect the convergence to the original process X because it is balanced by the change in the drift coefficient μ_R.

Let us remark explicitly that, for $0 < S \le R$, we have

$$\mu_S = \mu_R - \int_{S<|x|\le R} x\nu(dx). \tag{13.52}$$

Indeed, by (13.47) we have

$$\mu_R t + X_t^R + \widetilde{X}_t^{S,R} = \mu_S t + X_t^S$$

that is

$$\mu_R t + \int_0^t \int_{S\le|x|<R} x\widetilde{J}(ds, dx) = \mu_S t + \int_0^t \int_{S<|x|\le R} xJ(ds, dx),$$

and (13.52) follows by taking expectations, using (13.27) and the fact the last term in the left-hand side is a martingale with null expectation.

By the Lévy-Itô decomposition, every Lévy process is determined by the triplet $(\mu_R, \mathcal{C}, \nu)$ where μ_R is the drift coefficient in (13.47), $\mathcal{C}$ is the covariance matrix of the Brownian motion and ν is the Lévy measure.

Definition 13.36 $(\mu_R, \mathcal{C}, \nu)$ *is called the characteristic R-triplet of X.*

The specification of R is necessary to avoid ambiguities caused by the fact that the drift coefficient μ_R depends on the truncation range R which is an arbitrary positive number. Indeed at least the following three different choices of R will be used in the sequel:

- the choice $R = 1$ is common in the literature: in this case $(\mu_1, \mathcal{C}, \nu)$ is simply called the characteristic triplet of X;
- if[8]

$$\int_{|x|\le 1} |x|\nu(dx) < \infty, \tag{13.53}$$

[8] We shall show in Proposition 13.43 that condition (13.53) amounts to assume that the jump part of X has bounded variation.

such as for the compound Poisson process, we can pass to the limit and take $S = 0$ in (13.52). Then X has 0-triplet $(\mu_0, \mathcal{C}, \nu)$ where

$$\mu_0 = \mu_1 - \int_{|x|\leq 1} x\nu(dx), \tag{13.54}$$

and we get the following Lévy-Itô decomposition

$$\begin{aligned} X_t &= \mu_0 t + B_t + \int_0^t \int_{\mathbb{R}^d} xJ(ds, dx) \\ &= \mu_0 t + B_t + \sum_{0<s\leq t} \Delta X_s. \end{aligned} \tag{13.55}$$

The second equality follows by letting R go to zero in (13.48)-(13.50). Note that the last term in (13.55) is a pure jump process containing all the jumps of X and has its own drift and martingale parts;

- if[9]

$$\int_{|x|\geq 1} |x|\nu(dx) < \infty, \tag{13.56}$$

we can let R go to infinity in (13.52) and we get

$$\mu_\infty := \lim_{R\to\infty} \mu_R = \mu_S + \int_{|x|>S} x\nu(dx). \tag{13.57}$$

Then X has ∞-triplet $(\mu_\infty, \mathcal{C}, \nu)$ and the alternative Lévy-Itô decomposition

$$\begin{aligned} X_t &= \mu_\infty t + B_t + \int_0^t \int_{\mathbb{R}^d} x\widetilde{J}(ds, dx) \\ &= \mu_\infty t + B_t + \left(\sum_{0<s\leq t} \Delta X_s - tE\left[\Delta X_1\right] \right). \end{aligned} \tag{13.58}$$

Note that in this case the last term in (13.58) is a martingale (it is a process with a jump part that is compensated by a continuous part) and therefore the drift of X is entirely contained in the term $\mu_\infty t$: in Remark 13.46 we shall see that $E[X_t] = t\mu_\infty$ and in particular $\mu_\infty = E[X_1]$.

To sum up:

- by using the 0-triplet, *we separate the jump part from the continuous part* of the process;
- by the ∞-triplet, *we separate the martingale from the drift part.*

---

[9] We shall show in Proposition 13.45 that condition (13.56) amounts to assume that X has finite expectation or, more precisely, that $E[|X_t|] < \infty$ for any t. This condition is satisfied in almost all the financial applications.

Example 13.37 Let

$$X_t = \mu t + B_t + \sum_{n=1}^{N_t} Z_n$$

be a jump-diffusion process: $\mu \in \mathbb{R}^d$, B is d-dimensional correlated Brownian motion with correlation matrix $\mathcal{C}$, N is a Poisson process with intensity λ and $(Z_n)_{n\geq 1}$ are i.i.d. random variables in $\mathbb{R}^d$ with distribution η. Then conditions (13.53) and (13.56) are satisfied and we have:

- the 0-triplet of X is $(\mu, \mathcal{C}, \lambda\eta)$;
- 1-triplet of X is $(\mu_1, \mathcal{C}, \lambda\eta)$ where

$$\mu_1 = \mu + \lambda \int_{|x|\leq 1} x\eta(dx);$$

- the ∞-triplet of X is $(\mu + \lambda, \mathcal{C}, \lambda\eta)$ (see also (13.73)).

□

We remark that the second and fourth terms in (13.47) (i.e. Brownian motion and compensated small jumps) form the martingale part of X, while the first and third terms (i.e. drift term and large jumps) govern the drift of the process. More precisely, it is always possible to split a Lévy process into the sum of a martingale with bounded jumps and a process with bounded variation.

Corollary 13.38 *Let X be a Lévy process. Then $X = M + Z$ where M and Z are Lévy processes, M is a martingale such that $M_t \in L^p(\Omega)$ for any $p \geq 1$ and Z has (locally in time) bounded variation.*

Proof. By the Lévy-Ito decomposition (13.47), it is suffices to set

$$Z_t = \mu_R t + X_t^R = \mu_R t + \int_0^t \int_{|x|\geq R} xJ(ds, dx),$$

$$M_t = B_t + M_t^R = B_t + \int_0^t \int_{|x|<R} x\widetilde{J}(ds, dx).$$

Clearly the process Z has bounded variation because X^R is a compound Poisson process. On the other hand, M is a martingale by Theorem 13.35: moreover $B_t \in L^p(\Omega)$ for any $p \geq 1$ and, since M^R has jumps bounded by one, by Proposition 13.45, we also have $M_t^R \in L^p(\Omega)$ for any $p \geq 1$. □

Remark 13.39 It is also possible to show, but this is a very deep result (cf. Theorem III-29 in Protter [287]), that any local martingale X can be written $X = M + Z$ where M is a local martingale with bounded jumps and Z has (locally in time) bounded variation. □

Now we consider the function $f(x) = |x|^2$: by the general property (13.46) of Lévy measures, f is ν-integrable in a neighborhood of the origin and therefore by (13.41) in Theorem 13.34 we have

$$\sum_{0<s\le t} |\Delta X_s|^2 = \int_0^t \int_{\mathbb{R}^d} |x|^2 J(ds, dx) < \infty \qquad \text{a.s.} \tag{13.59}$$

Note that, for a pure jump process, the sum in (13.59) represents the quadratic variation (cf. Section 14.2.2). Indeed, having a closer look to the properties of ν stated in Theorem 13.35, we see that (13.45) is a consequence of the fact that X can only have a *finite* number of large jumps because it is càdlàg (cf. Lemma 13.12). Property (13.46) is more subtle and is related to the fact that the quadratic variation process of X is well-defined: we recall that, for a Brownian motion W, the quadratic variation process coincides with the variance, $\langle W\rangle_t = \text{var}(W_t) = t$. On the contrary, for a jump process, the variance is not necessarily finite[10] but the quadratic variation process is always well-defined by (13.59). Basically, the main difficulties about Lévy processes come from the fact that it is possible to have

$$\sum_{0<s\le t} |\Delta X_s| = \infty \qquad \text{a.s.}$$

however, many of these difficulties are overcome by using the fact that

$$\sum_{0<s\le t} |\Delta X_s|^2 < \infty \qquad \text{a.s.}$$

Finally, we also remark that by the Lévy-Itô decomposition, every Lévy process can be approximated with arbitrary precision by a jump diffusion process which is an independent sum of a Brownian motion with drift and a compound Poisson process. This fact has important practical implications, for instance in the simulation of Lévy processes.

13.3.4 Lévy-Khintchine representation

From Theorem 13.35 we obtain the most general form for the characteristic exponent of a Lévy process: this is provided by the celebrated Lévy-Khintchine formula.

Theorem 13.40 (Lévy-Khintchine representation) *Let X be a Lévy process in $\mathbb{R}^d$ with characteristic triplet $(\mu_1, \mathcal{C}, \nu)$. Then we have*

$$\varphi_{X_t}(\xi) = E\left[e^{i\xi\cdot X_t}\right] = e^{t\psi_X(\xi)}$$

[10] Because of the contribution of large jumps which are almost surely a finite number but whose expectation may be infinite such as in the case of a stable process in Example 13.47.

where the characteristic exponent ψ_X *is equal to*

$$\psi_X(\xi) = i\mu_1 \cdot \xi - \frac{1}{2}\langle \mathcal{C}\xi, \xi \rangle + \int_{\mathbb{R}^d} \left(e^{i\xi \cdot x} - 1 - i\xi \cdot x \mathbb{1}_{\{|x|<1\}} \right) \nu(dx). \quad (13.60)$$

Proof. By the Lévy-Itô decomposition X_t can be represented as the independent sum of is the a.s. limit of the sum of $\mu_1 t + B_t$, X_t^1 and $\widetilde{X}_t^{\varepsilon,1}$ in (13.51), as $\varepsilon \to 0^+$. Since these terms are independent, by Remark 13.25, Examples 13.23 and 13.24, we have

$$\begin{aligned} \psi_{\mu_1 t + B_t + X_t^1 + \widetilde{X}_t^{\varepsilon,1}}(\xi) =& \psi_{\mu_1 t + B_t}(\xi) + \psi_{X_t^1}(\xi) + \psi_{\widetilde{X}_t^{\varepsilon,1}}(\xi) \\ =& i\mu_1 \cdot \xi - \frac{1}{2}\langle \mathcal{C}\xi, \xi \rangle + \int_{|x| \geq 1} \left(e^{i\xi \cdot x} - 1 \right) \nu(dx) \\ &+ \int_{\varepsilon \leq |x| < 1} \left(e^{i\xi \cdot x} - 1 - i\xi \cdot x \right) \nu(dx). \end{aligned} \quad (13.61)$$

By the integrability condition (13.46) and the fact that, for fixed ξ,

$$e^{i\xi \cdot x} - 1 - i\xi \cdot x = \mathrm{O}\left(|x|^2\right), \qquad \text{as } |x| \to 0,$$

we infer that the expression in (13.61) converges to the characteristic exponent in (13.60) as $\varepsilon \to 0^+$. We conclude recalling that the a.s. convergence implies the convergence of the corresponding characteristic functions (cf. Theorems A.136 and A.141). □

An equivalent Lévy-Khintchine representation may be obtained by using the Lévy-Itô decomposition with a generic $R > 0$:

$$\begin{aligned} \psi_X(\xi) = i\mu_R \cdot \xi - \frac{1}{2}\langle \mathcal{C}\xi, \xi \rangle &+ \int_{|x| \geq R} \left(e^{i\xi \cdot x} - 1 \right) \nu(dx) \\ &+ \int_{|x| < R} \left(e^{i\xi \cdot x} - 1 - i\xi \cdot x \right) \nu(dx), \end{aligned} \quad (13.62)$$

where

$$\mu_R = \mu_1 + \int_{\mathbb{R}^d} x \left(\mathbb{1}_{|x| \leq R} - \mathbb{1}_{|x| \leq 1} \right) \nu(dx). \quad (13.63)$$

Remark 13.41 Under the integrability condition

$$\int_{|x| \leq 1} |x| \nu(dx) < \infty, \quad (13.64)$$

we can avoid to truncate the small jumps and we may represent X in terms of its 0-triplet. Indeed, by (13.53) and (13.45), we can let R go to 0 in (13.62) and we get the following simplified Lévy-Khintchine representation

$$\psi_X(\xi) = i\mu_0 \cdot \xi - \frac{1}{2}\langle \mathcal{C}\xi, \xi \rangle + \int_{\mathbb{R}^d} \left(e^{i\xi \cdot x} - 1 \right) \nu(dx). \quad (13.65)$$

□

Formula (13.65) was obtained in Example 13.26 in the particular case of jump-diffusion processes: in that case, $\mu_0 = \mu$ and $\nu = \lambda\eta$ where λ is the intensity of the Poisson process and η is the distribution of jumps. Actually, we are now able to show that any finite activity process is a jump-diffusion.

Corollary 13.42 *Let X be a Lévy process with characteristic triplet $(\mu_1, \mathcal{C}, \nu)$ and Lévy measure ν such that*

$$\nu(\mathbb{R}^d) < \infty. \tag{13.66}$$

Then X is a jump-diffusion process with intensity $\lambda = \nu(\mathbb{R}^d)$ and distribution of jumps $\eta = \lambda^{-1}\nu$.

Proof. Under condition (13.66), we can let R go to zero in the Lévy-Khintchine representation (13.62): we get

$$\psi_X(\xi) = i\mu_0 \cdot \xi - \frac{1}{2}\langle \mathcal{C}\xi, \xi\rangle + \int_{\mathbb{R}^d} \left(e^{i\xi\cdot x} - 1\right) \nu(dx)$$

which is the characteristic exponent of a jump-diffusion process with intensity $\lambda = \nu(\mathbb{R}^d)$ and distribution of jumps $\eta = \lambda^{-1}\nu$ (cf. Example 13.26). □

Condition (13.64) is of particular importance because it characterizes the Lévy process that (up the Brownian term) have the trajectories with bounded variation. Indeed, we have:

Proposition 13.43 *Let X be a Lévy process with triplet $(\mu_1, \mathcal{C}, \nu)$. Then X has (locally in time) bounded variation if and only if*

$$\mathcal{C} = 0 \qquad \textit{and} \qquad \int_{|x|\le 1} |x|\nu(dx) < \infty.$$

Proof. We only prove the "if" part. By the Lévy-Itô decomposition we have

$$X_t = \mu t + X_t^1 + \lim_{\varepsilon \to 0^+} \widetilde{X}_t^\varepsilon$$

where

$$X_t^1 = \int_{|x|\ge 1} xJ(ds, dx), \qquad \widetilde{X}_t^\varepsilon = \int_{\varepsilon \le |x| < 1} x\widetilde{J}_t(dx).$$

We first note that the drift term μt and the processes X^1 and $\widetilde{X}^\varepsilon$ have bounded variation[11]. Thus it is sufficient to prove that the variation of $\widetilde{X}^\varepsilon$ is bounded uniformly in $\varepsilon > 0$: by Remark 13.32, we have

$$E\left[V_{[0,t]}\left(\widetilde{X}^\varepsilon\right)\right] \le 2t \int_{\varepsilon \le |x| < 1} |x|\nu(dx) \le 2t \int_{|x|<1} |x|\nu(dx) < \infty, \qquad \varepsilon > 0,$$

by the integrability hypothesis on ν and this concludes the proof. □

[11] Any compound Poisson process has bounded variation since it has piecewise constant trajectories.

By the previous results, we have the following classification of Lévy processes:

- Lévy processes with finite activity and bounded variation: by Corollary 13.42 and Proposition 13.43, these are necessarily compound Poisson processes with drift;
- Lévy processes with finite activity and unbounded variation: by Corollary 13.42, these are necessarily jump-diffusions;
- Lévy processes with infinite activity and bounded variation: for instance, a α-stable process (cf. Example 13.47) with $\alpha \in]0,1[$;
- Lévy processes with infinite activity and unbounded variation for instance, a α-stable process with $\alpha \in [1,2[$;

We close this section by collecting the main results on Lévy processes with bounded variation. Incidentally we observe that, for a Lévy process X with bounded variation, often it is more convenient to use the 0-triplet $(\mu_0, \mathcal{C}, \nu)$ rather than standard triplet $(\mu_1, \mathcal{C}, \nu)$.

Corollary 13.44 *Let X be a Lévy process with (locally in time) bounded variation and characteristic triplet $(\mu_1, 0, \nu)$. Then we have the Lévy-Itô decomposition*

$$X_t = \mu_0 t + \int_{\mathbb{R}^d} xJ(ds, dx),$$

where

$$\mu_0 = \mu_1 - \int_{|x|\le 1} x\nu(dx). \tag{13.67}$$

Moreover the characteristic exponent takes the form

$$\psi_X(\xi) = i\mu_0 \cdot \xi + \int_{\mathbb{R}^d} \left(e^{i\xi\cdot x} - 1\right) \nu(dx).$$

13.3.5 Cumulants and Lévy martingales

The Lévy-Khintchine formula allows to compute easily the cumulants of a one dimensional Lévy process: let us recall (cf. (A.78)) that the n-cumulant is defined as

$$c_n(X_t) = t\frac{d^n}{d\xi^n}\psi(-i\xi)|_{\xi=0} = \frac{t}{i^n}\frac{d^n}{d\xi^n}\psi(\xi)|_{\xi=0} \tag{13.68}$$

where ψ is the characteristic exponent of X. Thus if X is a real Lévy process with characteristic triplet (μ_1, σ^2, ν), differentiating (13.60), formally we get

$$c_1(X_t) = E[X_t] = t\left(\mu_1 + \int_{|x|\ge 1} x\nu(dx)\right), \tag{13.69}$$

$$c_2(X_t) = \text{var}(X_t) = t\left(\sigma^2 + \int_{|x|\ge 1} x^2\nu(dx)\right), \tag{13.70}$$

and more generally

$$c_n(X_t) = t \int_{|x|\geq 1} x^n \nu(dx), \qquad n \geq 3. \tag{13.71}$$

More precisely, it turns out that the finiteness of the moments depends only on the large jumps of the process, i.e. on the tail behaviour of the Lévy measure. Indeed, we have (for the proof see, for instance, Sato [297] Section 25):

Proposition 13.45 *Let X be a Lévy process on $\mathbb{R}$ with characteristic triplet (μ_1, σ^2, ν). The n-absolute moment $E[|X_t|^n]$ is finite if and only if*

$$\int_{|x|\geq 1} |x|^n \nu(dx) < \infty.$$

In this case, formulas (13.69), (13.70) *and* (13.71) *hold true and in particular we have*

$$c_n(X_t) = t\, c_n(X_1), \qquad n \geq 1. \tag{13.72}$$

Remark 13.46 Let $(X_t)_{t\geq 0}$ be a real Lévy process with triplet (μ_1, σ^2, ν) and finite expectation, that is by Proposition 13.45

$$\int_{|x|\geq 1} |x| \nu(dx) < \infty.$$

Then, letting R go to infinity in (13.52), we get (cf. the definition of μ_∞ in (13.57))

$$\mu_\infty := \lim_{R\to\infty} \mu_R = \mu_1 + \int_{|x|\geq 1} x\nu(dx) = E[X_1], \tag{13.73}$$

where the last equality follows from (13.69). Thus we get the alternative Lévy-Khintchine representation in terms of the ∞-triplet $(\mu_\infty, \sigma^2, \nu)$:

$$\psi_X(\xi) = i\mu_\infty \xi - \frac{\sigma^2 \xi^2}{2} + \int_{\mathbb{R}} \left(e^{i\xi x} - 1 - i\xi x\right) \nu(dx), \tag{13.74}$$

with $\mu_\infty = E[X_1]$ and more generally $t\mu_\infty = E[X_t]$.

In particular, if X has also bounded variation then by Corollary 13.44 we also have

$$X_t = \mu_0 t + \int_{\mathbb{R}^d} xJ(ds, dx),$$

$$\psi_X(\xi) = i\mu_0 \xi + \int_{\mathbb{R}} \left(e^{i\xi\cdot x} - 1\right) \nu(dx),$$

where

$$\mu_0 = \mu_\infty - \int_{\mathbb{R}} x\nu(dx) = E[X_1] - \int_{\mathbb{R}} x\nu(dx),$$

which follows combining (13.73) with (13.67). In other words, μ_0 is the drift coefficient which only takes into account the deterministic drift part of the process; while μ_∞ is a drift coefficient which contains the contributions of the deterministic and jump parts of the process. □

Example 13.47 The Lévy measure of a stable distribution is of the form:

$$\nu(dx) = \left(\frac{C_1}{x^{1+\alpha}} \mathbb{1}_{\{x>0\}} + \frac{C_2}{(-x)^{1+\alpha}} \mathbb{1}_{\{x<0\}} \right) dx,$$

where $C_1, C_2 > 0$. By conditions (13.45) and (13.46), we necessarily have $\alpha \in]0, 2[$. Moreover

$$\int_{|x|\geq 1} |x|^n \nu(dx) < \infty$$

if and only if $n < \alpha$: this entails that the corresponding Lévy process has no finite variance and the expectation is finite for $1 < \alpha < 2$. □

Remark 13.48 Proposition 13.45 shows that *the contribution to infinite moments may come only from large jumps.* We also remark that the skewness and kurtosis (cf. (A.79)) are given by

$$s(X_t) = \frac{c_3(X_t)}{c_2(X_t)^{\frac{3}{2}}} = \frac{s(X_1)}{\sqrt{t}}, \qquad k(X_t) = \frac{c_4(X_t)}{c_2(X_t)^2} = \frac{k(X_1)}{t}.$$

Therefore Lévy processes typically exhibit skewness and kurtosis (i.e. fat tails): however these parameters decay as time increases at the rates of $t^{-\frac{1}{2}}$ and t^{-1} respectively. □

We also state an important result on the exponential moments (for the proof see, for instance, Sato [297] Theorem 25.17).

Proposition 13.49 *Let X be a Lévy process on $\mathbb{R}$ with characteristic triplet (μ_1, σ^2, ν). The exponential moment $E\left[e^{\xi X_t}\right]$, $\xi \in \mathbb{R}$, is finite if and only if*

$$\int_{|x|\geq 1} e^{\xi x} \nu(dx) < \infty.$$

In this case

$$E\left[e^{\xi X_t}\right] = e^{t\psi(-i\xi)}, \tag{13.75}$$

where ψ is the characteristic exponent of X.

We close this section by showing some martingales that can be constructed from Lévy processes.

Theorem 13.50 *Let X be a real valued Lévy process with characteristic triplet (μ_1, σ^2, ν). We have:*

i) if $E[|X_1|] < \infty$ then $(X_t - E[X_t])_{t\geq 0}$ is a martingale;
ii) X is a martingale if and only if

$$\int_{|x|\geq 1} |x| \nu(dx) < \infty$$

and

$$\mu_1 + \int_{|x|\geq 1} x \nu(dx) = 0;$$

iii) if $E\left[e^{\xi X_1}\right] < \infty$ *for some* $\xi \in \mathbb{R}$ *then* $\left(\frac{e^{\xi X_t}}{E[e^{\xi X_t}]}\right)_{t\geq 0}$ *is a martingale;*

iv) $\left(e^{X_t}\right)_{t\geq 0}$ *is a martingale if and only if*

$$\int_{|x|\geq 1} e^x \nu(dx) < \infty$$

and

$$\mu_1 + \frac{\sigma^2}{2} + \int_{\mathbb{R}} \left(e^x - 1 - x\mathbb{1}_{\{|x|\leq 1\}}\right) \nu(dx) = 0.$$

Proof. i) Since $E[X_t] = tE[X_1]$ (cf. (13.72)), if $E[|X_1|] < \infty$ then $X_t - E[X_t]$ is integrable. Moreover, by the independence of increments we have

$$E[X_t - X_s \mid \mathcal{F}_s] = E[X_t] - E[X_s],$$

and therefore $X_t - E[X_t]$ is a martingale.

ii) By i), X is a martingale if and only if $E[X_1] = 0$: then the thesis follows from (13.69).

iii) By (13.75), we have that the process

$$M_t := \frac{e^{\xi X_t}}{E\left[e^{\xi X_t}\right]} = e^{\xi X_t - t\psi(-i\xi)}$$

is integrable: here ψ denotes the characteristic exponent of X. Moreover we have

$$E[M_t - M_s \mid \mathcal{F}_s] = M_s E\left[e^{\xi(X_t - X_s) - (t-s)\psi(-i\xi)} - 1 \mid \mathcal{F}_s\right] =$$

(by the independence and stationarity of the increments of X)

$$= M_s\left(e^{-(t-s)\psi(-i\xi)} E\left[e^{\xi X_{t-s}}\right] - 1\right) = 0.$$

iv) By iii), e^{X_t} is a martingale if and only if $1 = E\left[e^{X_t}\right] = e^{t\psi(-i)}$, i.e. if $\psi(-i) = 0$. The thesis follows from formula (13.60) of the Lévy-Khintchine representation. □

13.4 Examples of Lévy processes

In this section we examine several examples of Lévy processes used in the financial modeling both with finite or infinite activity (i.e. with a finite or infinite number of jumps in every time interval). By Corollary 13.42, finite activity models are based on jump-diffusion processes that are independent sums of a Brownian motion with drift and a compound Poisson process: in this case, the jumps are "rare" events and the evolution of the process is similar to that of a diffusion. On the contrary, for an infinite activity process, we have $\nu(\mathbb{R}^d) = \infty$, where ν is the Lévy measure of the process, and it is known that the *set of jumps times of every trajectory is countable and dense in* $\mathbb{R}_{\geq 0}$ (cf., for instance, Sato [297]): in this case, jumps arrive infinitely often and the dynamics of jumps can be considered rich enough to avoid the introduction of the Brownian component.

Concerning the construction of Lévy processes, the simplest way to define a Lévy process is via the Lévy-Khintchine representation, that is by giving the characteristic triplet of the process. Alternatively, one can directly specify the distribution of the process (see, for instance, Section 13.4.5 on hyperbolic Lévy processes) even if in this way the structure of jumps is less transparent because the Lévy measure is not given explicitly. We shall also examine another important technique for constructing new Lévy processes from known ones, called the subordination method: a new process is built out of two independent stochastic processes where the first one is a Lévy process $(Y_t)_{t\geq 0}$ and the time t is made stochastic by employing another stochastic process $(S_t)_{t\geq 0}$ (a "stochastic clock"). The new process is defined as $X_t = Y_{S_t}$. In other words, the first process is subordinated (i.e. time-changed) by a stochastic clock, which is called the subordinator.

13.4.1 Jump-diffusion processes

A Lévy jump-diffusion process X in $\mathbb{R}$ has the form

$$X_t = \mu t + \sigma W_t + \sum_{n=1}^{N_t} Z_n \tag{13.76}$$

where $\mu \in \mathbb{R}$, W is a standard real Brownian motion, N is a Poisson process with intensity λ and (Z_n) are i.i.d. real random variables with distribution η. Recalling formula (13.71), we see that the cumulants, and therefore the tail behaviour of the process, depend on the Lévy measure $\nu = \lambda\eta$.

Example 13.51 (Merton model) In the Merton model [251], the log-price is modeled by a process of the form (13.76) where $\eta = \mathcal{N}_{m,\delta^2}$. Thus the 0-triplet is (μ, σ^2, ν) with Lévy measure

$$\nu(dx) = \frac{\lambda}{\sqrt{2\pi\delta^2}} \exp\left(-\frac{(x-m)^2}{2\delta^2}\right) dx.$$

As already proved in Example 13.26, the characteristic exponent is

$$\psi(\xi) = i\mu\xi - \frac{1}{2}\sigma^2\xi^2 + \lambda\left(e^{im\xi - \frac{1}{2}\delta^2\xi^2} - 1\right).$$

In this case, the density of the process admits a series expansion: indeed, we have

$$P(X_t \in H) = \sum_{n\geq 0} P(X_t \in H \mid N_t = n) P(N_t = n), \qquad H \in \mathscr{B},$$

and since, by independence, $\mu t + \sigma W_t + \sum_{k=1}^{n} Z_k \sim \mathcal{N}_{\mu t+nm,\sigma^2 t+n\delta^2}$, the density Φ_{X_t} is given by

$$\Phi_{X_t}(x) = e^{\lambda t} \sum_{n=0}^{\infty} \frac{(\lambda t)^n}{n!\sqrt{2\pi(\sigma^2 t + n\delta^2)}} e^{-\frac{(x-\mu t-nm)^2}{2(\sigma^2 t+n\delta^2)}}.$$

Analogously, a series expansion in terms of Black&Scholes prices is available for plain vanilla options. By (13.68), the first four cumulants are equal to

$$\begin{aligned}
c_1(X_t) &= E[X_t] = t(\mu + \lambda m), \\
c_2(X_t) &= \mathrm{var}(X_t) = t\left(\left(m^2 + \delta^2\right)\lambda + \sigma^2\right), \\
c_3(X_t) &= m\lambda t\left(m^2 + 3\delta^2\right), \\
c_4(X_t) &= \lambda t\left(m^4 + 6m^2\delta^2 + 3\delta^4\right).
\end{aligned}$$

□

Example 13.52 (Kou model) In the Kou model [217], the distribution of jumps is defined in terms of an asymmetric double exponential density: more precisely, we have

$$\eta(dx) = \left(p\lambda_1 e^{-\lambda_1 x}\mathbb{1}_{\{x>0\}} + (1-p)\lambda_2 e^{\lambda_2 x}\mathbb{1}_{\{x<0\}}\right) dx,$$

where $\lambda_1, \lambda_2 > 0$ and $p \in [0,1]$ represents the probability of upward jumps. Since, for $\nu = \lambda\eta$, we have

$$\int_{|x|\le 1} |x|\nu(dx) < \infty,$$

by Corollary 13.44, we have the simplified Lévy-Khintchine representation:

$$\begin{aligned}
\psi_X(\xi) &= i\mu\xi - \frac{\sigma^2\xi^2}{2} + \int_{\mathbb{R}} \left(e^{i\xi x} - 1\right)\nu(dx) \\
&= i\mu\xi - \frac{\sigma^2\xi^2}{2} + i\lambda\xi\left(\frac{p}{\lambda_1 - i\xi} - \frac{1-p}{\lambda_2 + i\xi}\right).
\end{aligned}$$

Moreover the first cumulants are given by

$$\begin{aligned}
c_1(X_t) &= E[X_t] = t\left(\mu + \frac{p\lambda}{\lambda_1} - \frac{(1-p)\lambda}{\lambda_2}\right), \\
c_2(X_t) &= \mathrm{var}(X_t) = t\left(\frac{2p\lambda}{\lambda_1^2} + \frac{2(1-p)\lambda}{\lambda_2^2} + \sigma^2\right), \\
c_3(X_t) &= 6t\lambda\left(\frac{p}{\lambda_1^3} - \frac{1-p}{\lambda_2^3}\right), \\
c_4(X_t) &= 24t\lambda\left(\frac{p}{\lambda_1^4} + \frac{1-p}{\lambda_2^4}\right).
\end{aligned}$$

The density f_{X_t} of the Kou process is not known in closed form; however, it is known that it has semi-heavy tails and more precisely

$$f_{X_t}(x) = \mathrm{O}\left(e^{-\lambda_1 x}\right), \qquad \text{as } x \to +\infty,$$
$$f_{X_t}(x) = \mathrm{O}\left(e^{-\lambda_2 |x|}\right), \qquad \text{as } x \to -\infty. \qquad \square$$

13.4.2 Stable processes

The stable distribution was suggested as a potential model for stock returns in Mandelbrot [245]. A random variable X in $\mathbb{R}^d$ is *stable* if for each $n \geq 2$ there exist $c_n > 0$, $d_n \in \mathbb{R}^d$ such that

$$X_1 + \cdots + X_n \overset{d}{=} c_n X + d_n \tag{13.77}$$

where $X_1, \dots, X_n$ are i.i.d. copies of X. For example, if $X \sim \mathcal{N}_{0,\sigma^2}$ then

$$X_1 + \cdots + X_n \sim \mathcal{N}_{0,n\sigma^2} \sim \sqrt{n}X.$$

In terms of characteristic functions, the stability property is equivalent to

$$(\varphi_X(\xi))^n = e^{i\xi \cdot d_n} \varphi_X(c_n \xi).$$

If (13.77) holds with $d_n = 0$ the term *strictly stable* is used. Due to the particular form (13.8) of the characteristic function, a Lévy process $(X_t)_{t \geq 0}$ is (strictly) stable if and only if X_1 is (strictly) stable.

It can be shown (see Sato [297], Theorems 13.11 and 13.15) that for any stable random variable there exists a constant $\alpha \in]0, 2]$ such that $c_n = n^{\frac{1}{\alpha}}$; thus the class of Lévy stable processes is defined in terms of the exponent α and the expression α-stable process is used to denote a stable process with index α. The Lévy triplet of a stable process can be obtained from the following result (see Sato [297], Theorem 14.3).

Proposition 13.53 *Let X be a non-trivial Lévy process with characteristic triplet $(\mu_1, \mathcal{C}, \nu)$ and let $\alpha \in]0, 2[$. Then X is α-stable if and only if $\mathcal{C} = 0$ and*

$$\nu(H) = r^{-\alpha} \nu(r^{-1} H)$$

for every $H \in \mathscr{B}$ and $r > 0$.

In the special case $d = 1$, we have the following explicit representation of the Lévy measure of an α-stable Lévy process:

$$\nu(dx) = \frac{C_1}{x^{1+\alpha}} \mathbb{1}_{\{x>0\}} dx + \frac{C_2}{|x|^{1+\alpha}} \mathbb{1}_{\{x<0\}} dx \tag{13.78}$$

for some positive constants C_1, C_2.

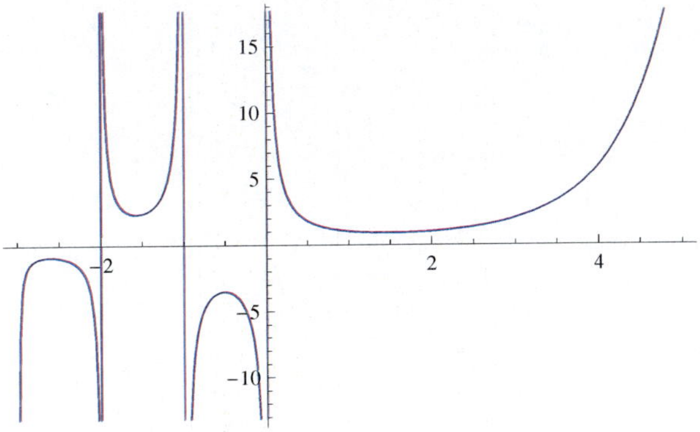

Fig. 13.4. Graph of the Euler Gamma function in (13.79) for $\alpha \in [-3, 5]$

Let us compute the characteristic exponent of a α-stable process with $C_2 = 0$ in (13.78). If $\alpha \in]0,1[$ the process has bounded variation and by Corollary 13.44 we only need to compute:

$$\int_0^\infty \left(e^{i\xi x} - 1\right) \frac{1}{x^{1+\alpha}} dx = |\xi|^\alpha \Gamma(-\alpha) \left(\cos\frac{\pi\alpha}{2} - i\text{sgn}(\xi)\sin\frac{\pi\alpha}{2}\right),$$

where

$$\Gamma(\alpha) = \int_0^\infty t^{\alpha-1} e^{-t} dt \tag{13.79}$$

is the Euler Gamma function (see Figure 13.4) and $\text{sgn}(x) = \frac{|x|}{x}$, $x \neq 0$. Thus the characteristic exponent is given by

$$\psi(\xi) = i\mu\xi - C_\alpha |\xi|^\alpha \left(1 - i\text{sgn}(\xi)\tan\frac{\pi\alpha}{2}\right)$$

where

$$C_\alpha = -C_1 \Gamma(-\alpha)\cos\frac{\pi\alpha}{2} > 0.$$

If $\alpha \in]1,2[$ one has to calculate

$$\int_0^1 \left(e^{ix\xi} - 1 - ix\xi\right)\nu(dx) = C_1 \int_0^1 \left(e^{ix\xi} - 1 - ix\xi\right)\frac{1}{x^{1+\alpha}} dx,$$

and this can be done explicitly integrating by parts. We omit further details and give the general expression of the characteristic exponent of a real-valued α-stable Lévy process:

$$\begin{aligned} \psi(\xi) &= i\mu\xi - \sigma^\alpha |\xi|^\alpha \left(1 - i\theta\text{sgn}(\xi)\tan\left(\frac{\pi\alpha}{2}\right)\right), && \text{if } \alpha \neq 1, \\ \psi(\xi) &= i\mu\xi - \sigma|\xi|\left(1 + i\theta\text{sgn}(\xi)\log|\xi|\right), && \text{if } \alpha = 1, \end{aligned} \tag{13.80}$$

where σ is positive and θ is real and such that $|\theta| \leq 1$. A stable distribution in this parameterization is usually denoted by $\mathcal{S}_\alpha(\sigma, \theta, \mu)$.

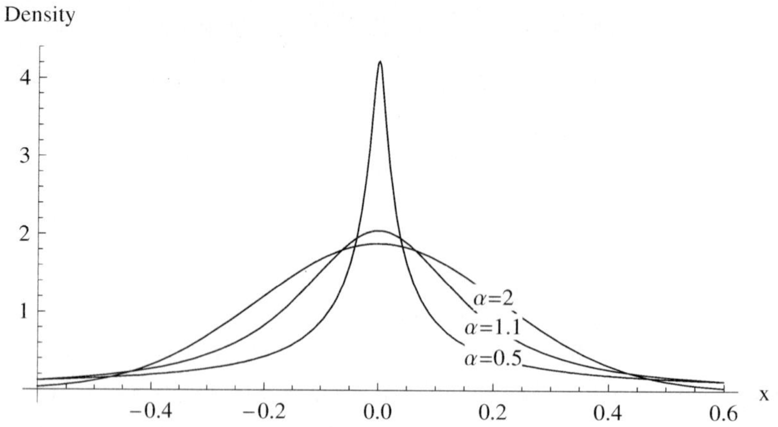

Fig. 13.5. Density of the α-stable distribution for $\alpha = 0.5, 1.1, 2$ and $\sigma = 0.15, \mu = 0, \theta = 0$

If $X \sim \mathcal{S}_\alpha(\sigma, \theta, \mu)$ then $X + b \sim \mathcal{S}_\alpha(\sigma, \theta, \mu + b)$ and we have

$$\begin{aligned} &\lambda X \sim \mathcal{S}_\alpha\left(|\lambda|\sigma, \operatorname{sgn}(\lambda)\theta, \lambda\mu\right) && \text{if } \alpha \neq 1, \\ &\lambda X \sim \mathcal{S}_1\left(|\lambda|\sigma, \operatorname{sgn}(\lambda)\theta, \lambda\mu - \frac{2}{\pi}\lambda\sigma\theta\log|\lambda|\right) && \text{if } \alpha = 1. \end{aligned} \tag{13.81}$$

Therefore μ is a shift parameter, while σ is a scale parameter. Moreover, θ is a skewness parameter, because the density is symmetric if and only if $\theta = 0$ (see also Figure 13.6). The distribution $\mathcal{S}_0(\sigma, 0, 0)$, that is with $\mu = \theta = 0$, is called *symmetric stable distribution*: in this case, the characteristic exponent is given by

$$\psi(\xi) = -\sigma^\alpha |\xi|^\alpha. \tag{13.82}$$

We point out that in the case $\alpha = 2$ and $\theta = 0$ the expression for $\psi(\xi)$ collapses into $i\mu\xi - \sigma^2\xi^2$ which is the characteristic exponent of the Gaussian distribution, that is

$$\mathcal{S}_2(\sigma, 0, \mu) = \mathcal{N}_{\mu, 2\sigma^2}.$$

The explicit expression of the Lévy measure and Proposition 13.45 show that α-stable distributions on $\mathbb{R}$ never admit a second moment, and they only admit a first moment if $\alpha > 1$. Moreover a real α-stable Lévy process has bounded variation only if $\alpha < 1$.

Finally we point out that explicit formulas for the densities of stable distributions are known only in few cases:

- the normal distribution $\mathcal{S}_2(\sigma, 0, \mu)$ with density

$$\frac{1}{2\sigma\sqrt{\pi}} \exp\left(-\frac{(x-\mu)^2}{4\sigma^2}\right), \qquad x \in \mathbb{R};$$

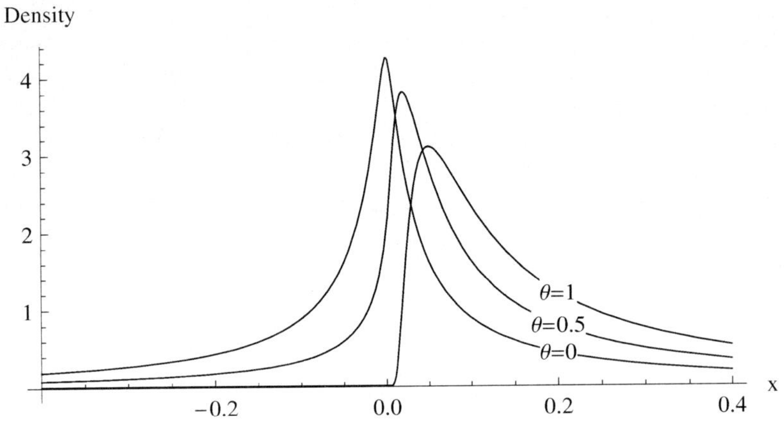

Fig. 13.6. Density of the α-stable distribution for $\theta = 0, \frac{1}{2}, 1$ and $\alpha = 0.5, \mu = 0, \sigma = 0.15$

- the Cauchy distribution $\mathcal{S}_1(\sigma, 0, \mu)$ (see Example 14.51) with density

$$\frac{\sigma}{\pi((x-\mu)^2+\sigma^2)}, \qquad x \in \mathbb{R};$$

- the Lévy-Smirnov distribution $\mathcal{S}_{\frac{1}{2}}(\sigma, 1, \mu)$ with density

$$\frac{\sqrt{\sigma}}{\sqrt{2\pi}(x-\mu)^{\frac{3}{2}}} \exp\left(-\frac{\sigma}{2(x-\mu)}\right), \qquad x > \mu.$$

Figures 13.5 and 13.6 show the α-stable densities for different choices of the parameters α and θ: these graphs have been plotted by using the Fourier-cosine approximation method illustrated in Section 15.3.

13.4.3 Tempered stable processes

A tempered[12] stable process X is obtained from a one-dimensional stable process by "tempering" the large jumps, i.e. by damping exponentially the tails of the Lévy measure. This class of Lévy processes was introduced by Koponen [216]. The characteristic triplet of X is of the form $(\mu_1, 0, \nu)$ where

$$\nu(dx) = C_1 \frac{e^{-\lambda_1 x}}{x^{1+\alpha_1}} \mathbb{1}_{\{x>0\}} dx + C_2 \frac{e^{-\lambda_2 |x|}}{|x|^{1+\alpha_2}} \mathbb{1}_{\{x<0\}} dx, \tag{13.83}$$

where $C_1, C_2 \geq 0$, $\lambda_1, \lambda_2 > 0$ and $\alpha_1, \alpha_2 < 2$. The parameters λ_1 and λ_2 govern the exponential decay of the tails for the distribution of positive and negative jump sizes. The parameters C_1 and C_2 determine the overall frequency of positive and negative jumps respectively. Due to the presence of

[12] The term "*generalized* tempered stable processes" is also used in the literature.

the exponential damping and contrary to the case of stable processes, ν in (13.83) is a well-defined Lévy measure for all $\alpha_1, \alpha_2 < 2$, and by Proposition 13.45 the cumulants of any order exist. However, negative values of α_1, α_2 lead to a compound Poisson process by Corollary 13.42 and therefore the interesting case is when $\alpha_1, \alpha_2 \in [0, 2[$, that is when the process has infinite activity. Notice that small jumps have a stable-like behaviour; however in (13.83) we allow a greater flexibility by a possibly asymmetric distribution of positive and negative jumps.

Now we give the characteristic exponent of a tempered stable process X. We remark that

$$\mu_\infty = E[X_1] = \mu_1 + \int_{|x|\geq 1} x\nu(dx)$$

is finite because of the exponential decay of the tails of the Lévy measure. Therefore we can avoid considering separately the bounded ($0 \leq \alpha < 1$) and unbounded ($1 \leq \alpha < 2$) variation cases, by using the Lévy-Khintchine representation in the form (13.74), i.e. in terms of the ∞-triplet $(\mu_\infty, 0, \nu)$:

$$\psi_X(\xi) = i\mu_\infty\xi + \int_{\mathbb{R}} \left(e^{i\xi x} - 1 - i\xi x\right)\nu(dx).$$

We have

- if $\alpha_1, \alpha_2 \notin \{0, 1\}$ then

$$\begin{aligned}\psi_X(\xi) = i\mu_\infty\xi &+ C_1\left((\lambda_1 - i\xi)^{\alpha_1} - \lambda_1^{\alpha_1 - 1}(\lambda_1 - i\alpha_1\xi)\right)\Gamma(-\alpha_1)\\ &+ C_2\left((\lambda_2 + i\xi)^{\alpha_2} - \lambda_2^{\alpha_2 - 1}(\lambda_2 + i\alpha_2\xi)\right)\Gamma(-\alpha_2);\end{aligned}$$

- if $\alpha_1 = \alpha_2 = 1$ then

$$\begin{aligned}\psi_X(\xi) = i\mu_\infty\xi &+ C_1\left(i\xi + (\lambda_1 - i\xi)\log\left(1 - \frac{i\xi}{\lambda_1}\right)\right)\\ &+ C_2\left(-i\xi + (\lambda_2 + i\xi)\log\left(1 + \frac{i\xi}{\lambda_2}\right)\right);\end{aligned}$$

- if $\alpha_1 = \alpha_2 = 0$ then

$$\begin{aligned}\psi_X(\xi) = i\mu_\infty\xi &- C_1\left(\frac{i\xi}{\lambda_1} + \log\left(1 - \frac{i\xi}{\lambda_1}\right)\right)\\ &+ C_2\left(\frac{i\xi}{\lambda_2} - \log\left(1 + \frac{i\xi}{\lambda_2}\right)\right).\end{aligned}$$

Concerning the cumulants, in general we have $c_1(X_t) = E(X_t) = t\mu_\infty$. Moreover, for $\alpha_1 \neq 0$ and $\alpha_2 \neq 0$, we have

$$\begin{aligned}c_2(X_t) &= \mathrm{var}(X_t) = tC_1\lambda_1^{-2+\alpha_1}\Gamma(2-\alpha_1) + C_2\lambda_2^{-2+\alpha_2}\Gamma(2-\alpha_2),\\ c_3(X_t) &= C_1\lambda_1^{-3+\alpha_1}\Gamma(3-\alpha_1) - C_2\lambda_2^{-3+\alpha_2}\Gamma(3-\alpha_2)),\\ c_4(X_t) &= C_1\lambda_1^{-4+\alpha_1}\Gamma(4-\alpha_1) + C_2\lambda_2^{-4+\alpha_2}\Gamma(4-\alpha_2).\end{aligned}$$

For $\alpha_1 = \alpha_2 = 1$, the previous expressions simplify to

$$c_2(X_t) = \frac{C_1}{\lambda_1} + \frac{C_2}{\lambda_2}, \qquad c_3(X_t) = \frac{C_1}{\lambda_1^2} - \frac{C_2}{\lambda_2^2}, \qquad c_4(X_t) = \frac{2C_1}{\lambda_1^3} + \frac{2C_2}{\lambda_2^3},$$

and, for $\alpha_1 = \alpha_2 = 0$, to

$$c_2(X_t) = \frac{C_1}{\lambda_1^2} + \frac{C_2}{\lambda_2^2}, \qquad c_3(X_t) = \frac{2C_1}{\lambda_1^3} - \frac{2C_2}{\lambda_2^3}, \qquad c_4(X_t) = \frac{6C_1}{\lambda_1^4} + \frac{6C_2}{\lambda_2^4}.$$

Example 13.54 (CGMY model) The CGMY model is based on a particular tempered stable process with Lévy measure as in (13.83) and parameters

$$C := C_1 = C_2, \quad G := \lambda_2, \quad M := \lambda_1, \quad Y := \alpha_1 = \alpha_2.$$

This model was proposed by Carr, Geman, Madan and Yor [67] who denoted the parameters according the four authors' names. We remark that, contrary to the general class of tempered stable processes, the CGMY processes have symmetric small jumps. For a further analysis of the model we refer to Example 15.22. □

13.4.4 Subordination

As already mentioned, a general method for constructing new Lévy processes from known ones is the subordination method. A new process is built out of two independent stochastic processes: the first is a Lévy process $(Y_t)_{t\geq 0}$ and the time t is made stochastic by employing another stochastic process $(S_t)_{t\geq 0}$, a "stochastic clock" also called subordinator. Since S provides a random model of time evolution, it needs to be non-negative and increasing a.s.

Definition 13.55 *A 1-dimensional Lèvy process $(S_t)_{t\geq 0}$ is called a subordinator if it is a.s increasing, that is*

$$t_1 \leq t_2 \quad \Longrightarrow \quad S_{t_1} \leq S_{t_2} \ a.s.$$

Since increasing processes have bounded variation and cannot have negative jumps, by using Proposition 13.43, it is not difficult prove the following characterization of subordinators.

Proposition 13.56 *A Lévy process S is a subordinator if and only if its* 0*-triplet $(b, \mathcal{C}, \varrho)$ satisfies*

$$b \geq 0, \qquad \mathcal{C} = 0, \qquad \varrho(\mathbb{R}_{\leq 0}) = 0, \qquad \int_0^1 x\varrho(dx) < \infty.$$

In this case the characteristic exponent of S takes the form

$$\psi_S(\xi) = ib\xi + \int_0^\infty \left(e^{i\xi x} - 1\right) \varrho(dx).$$

Example 13.57 A Poisson processes is a subordinator. A compound Poisson process is a subordinator if and only if all the Z_n (cf. Definition 13.8) take only non-negative values. □

Since a subordinator S takes only non-negative values, it is convenient to characterize it by the Laplace transform instead of the Fourier transform: if ψ_S denotes as usual the characteristic exponent of S, by Proposition 13.49 the Laplace exponent

$$\ell_S(\xi) := \psi_S(-i\xi) \tag{13.84}$$

is well-defined for $\xi \le 0$ and we have

$$E\left[e^{\xi S_t}\right] = e^{t\ell_S(\xi)}, \qquad \xi \le 0.$$

More explicitly, we have

$$\ell_S(\xi) = b\xi + \int_0^\infty \left(e^{\xi x} - 1\right) \varrho(dx).$$

We also note that the cumulants of S (see (13.68)), when they exist, are defined in term of the Laplace exponent by

$$c_n(S_1) = \frac{d^n}{d\xi^n} \ell_S(\xi)|_{\xi=0}.$$

Next we state the fundamental result on subordination of Lévy processes (see Sato [297], Theorem 30.1).

Theorem 13.58 (Lévy subordination) *Let Y be a Lévy process with triplet $(\mu_1, \mathcal{C}, \nu)$, characteristic exponent ψ_Y and density function f_{Y_t}, $t \ge 0$. Let S be a subordinator with 0-triplet $(b, 0, \varrho)$ and Laplace exponent ℓ_S. Then the process $X_t := Y_{S_t}$ is a Lévy process with characteristic exponent*

$$\psi_X(\xi) = \ell_S\left(\psi_X(\xi)\right), \qquad \xi \in \mathbb{R}^d,$$

and triplet $(\mu_1^X, \mathcal{C}^X, \nu^X)$ where

$$\begin{aligned}
\mu_1^X &= b\mu_1 + \int_0^\infty \int_{|x|\le 1} x f_{Y_t}(dx)dt, \\
\mathcal{C}^X &= b\mathcal{C}, \\
\nu^X(H) &= b\nu(H) + \int_0^\infty f_{Y_t}(H)\varrho(dt), \qquad H \in \mathscr{B}(\mathbb{R}^d).
\end{aligned}$$

Example 13.59 (Stable subordinators) A stable subordinator S is an α-stable process with $\alpha \in]0,1[$, and 0-triplet $(b, 0, \varrho)$ with $b \ge 0$ and Lévy measure ϱ such that $\varrho(\mathbb{R}_{\le 0}) = 0$: by Proposition 13.56, these conditions on the parameters guarantee that S is a subordinator. The Lévy measure is of the form

$$\varrho(dx) = \frac{C_1}{x^{1+\alpha}} \mathbb{1}_{\{x>0\}} dx$$

and the Laplace exponent is given by

$$\ell(\xi) = b\xi + C_1 \int_0^{\infty} \left(e^{x\xi} - 1\right) \frac{1}{x^{1+\alpha}} dx = b\xi + C_1 \Gamma(-\alpha)(-\xi)^{\alpha}, \qquad \xi < 0.$$

We remark explicitly that if Y is symmetric β-stable (cf. (13.82)) with characteristic exponent $\psi_Y(\xi) = -\sigma^{\beta}|\xi|^{\beta}$ and S is an α-stable subordinator with null drift coefficient, $b = 0$, then the subordinate process $X_t = Y_{S_t}$ is $\alpha\beta$-stable: indeed, by Theorem 13.58, we have

$$\psi_X(\xi) = C_1 \Gamma(-\alpha) \left(\sigma^{\beta}|\xi|^{\beta}\right)^{\alpha} = -C|\xi|^{\alpha\beta}$$

where $C = -C_1 \sigma^{\alpha\beta} \Gamma(-\alpha)$. A remarkable example is given by the Brownian motion ($\beta = 2$): in this case the subordinate process is 2α-stable. □

The main disadvantage of stable subordinators is that they don't have finite cumulants: for this reason their tempered version is much more commonly used in the literature. Let S be a general tempered stable process with triplet $(\beta, 0, \varrho)$ and Lévy measure of the form (13.83): by Proposition 13.56, for S to be a subordinator it is necessary that $\beta \geq 0$, $C_2 = 0$ and $\alpha := \alpha_1 < 1$; hereafter we only consider the infinite activity case $\alpha \in [0, 1[$. Thus

$$\varrho(dx) = C \frac{e^{-\lambda x}}{x^{1+\alpha}} \mathbb{1}_{\{x>0\}} dx,$$

with C and λ positive parameters, and the Laplace exponent is

$$\psi_S(\xi) = \beta\xi + C \int_0^{\infty} \left(e^{\xi x} - 1\right) \frac{e^{-\lambda x}}{x^{1+\alpha}} dx = \left(-\lambda^{\alpha} + (\lambda - \xi)^{\alpha}\right) \Gamma(-\alpha),$$

for $\xi < 0$ and $\alpha \in]0, 1[$. Since the probability density of tempered stable subordinators is known for $\alpha = 0$ (Gamma subordinator) and $\alpha = \frac{1}{2}$ (inverse Gaussian subordinator), we examine in details these two cases.

Example 13.60 (Gamma subordinator) We use slightly different notations and consider a Lévy process S with 0-triplet $(0, 0, \varrho)$ and Lévy measure

$$\varrho(dx) = \frac{ae^{-bx}}{x} \mathbb{1}_{\{x>0\}} dx,$$

where a, b are positive parameters: we call S a *Gamma subordinator*. In this case the characteristic function is given by

$$\varphi_{S_t}(\xi) = \left(1 - \frac{i\xi}{b}\right)^{-at},$$

and the Laplace exponent is

$$\ell(\xi) = -a \log\left(1 - \frac{\xi}{b}\right), \qquad \xi < 0.$$

The density of S_t can be recovered from the characteristic function by Fourier inversion:

$$f_{S_t}(x) = \frac{1}{2\pi} \int_{\mathbb{R}} e^{-ix\xi} \varphi_{S_t}(\xi) d\xi = \frac{e^{-bx}(bx)^{at}}{x\Gamma(at)}, \qquad x > 0,$$

which shows that S_t has a Gamma distribution. By (13.68), the first four cumulants are equal to[13]

$$c_1(S_1) = E[S_1] = \frac{a}{b}, \qquad c_2(S_1) = \mathrm{var}(S_1) = \frac{a}{b^2},$$
$$c_3(S_1) = \frac{2a}{b^3}, \qquad c_4(S_1) = \frac{6a}{b^4}.$$

Due to the scaling property of stable processes (cf. (13.81)), it is not restrictive to consider stable subordinators such that $E[S_t] = t$: for the Gamma subordinator, imposing $c_1(S_1) = 1$ and $c_2(S_1) = v$, we find $a = b = \frac{1}{v}$. Thus a Gamma subordinator can be parameterized in terms of its variance v. □

Example 13.61 (Variance-Gamma process) By subordinating a Brownian motion with drift μ and volatility σ by a Gamma process S with variance v (and unitary mean), we obtain the so-called Variance-Gamma (VG) process

$$X_t = \mu S_t + \sigma W_{S_t}.$$

This is a three-parameter process: the variance v of the subordinator, the drift μ and the volatility σ of the Brownian motion. By Theorem 13.58, the characteristic exponent of X is

$$\psi_X(\xi) = -\frac{1}{v} \log\left(1 - iv\mu\xi + \frac{v\xi^2\sigma^2}{2}\right).$$

By (13.68), the first four cumulants are equal to

$$c_1(X_1) = E[X_1] = \mu, \qquad c_2(X_1) = \mathrm{var}(X_1) = v\mu^2 + \sigma^2,$$
$$c_3(X_1) = v\mu\left(2v\mu^2 + 3\sigma^2\right), \qquad c_4(X_1) = 3v\left(2v^2\mu^4 + 4v\mu^2\sigma^2 + \sigma^4\right). \tag{13.85}$$

Examining the characteristic exponent, one recognizes that *a VG process is a particular tempered stable process* (cf. Section 13.4.3): its 0-triplet is $(0, 0, \nu)$ with Lévy measure

$$\nu(dx) = \frac{1}{v}\frac{e^{-\lambda_1 x}}{x}\mathbb{1}_{\{x>0\}}(x)dx + \frac{1}{v}\frac{e^{\lambda_2 x}}{|x|}\mathbb{1}_{\{x<0\}}(x)dx,$$

where

$$\lambda_1 = \left(\sqrt{\frac{\mu^2 v^2}{4} + \frac{\sigma^2 v}{2}} + \frac{\mu v}{2}\right)^{-1}, \qquad \lambda_2 = \left(\sqrt{\frac{\mu^2 v^2}{4} + \frac{\sigma^2 v}{2}} - \frac{\mu v}{2}\right)^{-1}. \tag{13.86}$$

[13] We recall that by (13.72) it is sufficient to give the expression of $c_n(S_t)$ for $t = 1$.

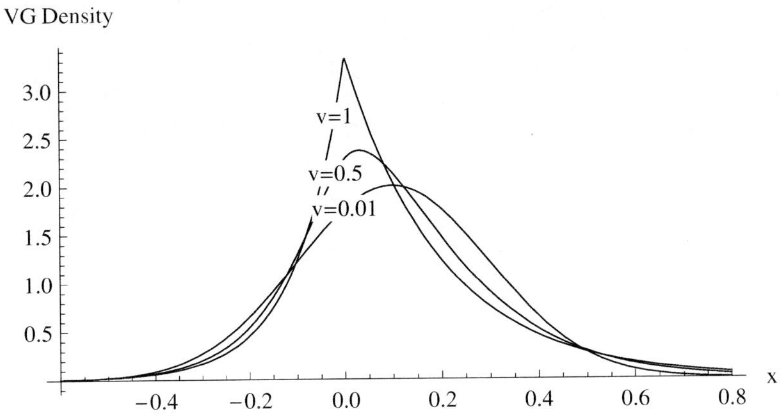

Fig. 13.7. VG density with $\sigma = 0.2, \mu = 0.1$ and different values of the variance of the subordinator: $v = 0.01$, $v = 0.5$ and $v = 1$

Notice that the VG process has bounded variation and infinite activity. The characteristic exponent can be expressed in terms of λ_1 and λ_2, as follows

$$\psi_X(\xi) = -\frac{1}{v}\left(\log\left(1 - \frac{i\xi}{\lambda_1}\right) + \log\left(1 + \frac{i\xi}{\lambda_2}\right)\right).$$

The density function of X_t is

$$f_{X_t}(x) = \frac{C_0}{\Gamma\left(\frac{t}{v}\right)}|x|^{\frac{t}{v}-\frac{1}{2}} e^{\frac{(\lambda_1+\lambda_2)x}{2}} \mathbf{K}_{\frac{t}{v}-\frac{1}{2}}\left(\frac{\lambda_2 - \lambda_1}{2}|x|\right)$$

where

$$C_0 = \left(\left(\mu^2 v + 2\sigma^2\right)^{\frac{1}{4}-\frac{\mu}{2v}}\right)\sqrt{\frac{\sigma^2 v}{2\pi}},$$

and $\mathbf{K}$ is the modified Bessel function of the second kind[14]. In Figure 13.7 we plot the probability density of a VG distribution for different choices of the variance v of the subordinator. □

Example 13.62 (Inverse Gaussian subordinator) The Inverse Gaussian (IG) subordinator S is a tempered stable subordinator with $\alpha = \frac{1}{2}$: thus S has 0-triplet $(0, 0, \varrho)$ and Lévy measure

$$\varrho(dx) = \frac{ae^{-bx}}{x^{\frac{3}{2}}}\mathbb{1}_{\{x>0\}}dx,$$

where a, b are positive parameters. The characteristic function is given by

$$\varphi_{S_t}(\xi) = e^{2at\sqrt{\pi}\left(\sqrt{b}-\sqrt{b-i\xi}\right)},$$

[14] See, for instance,
`mathworld.wolfram.com/ModifiedBesselFunctionoftheSecondKind.html`

and the Laplace exponent is

$$\ell(\xi) = 2a\sqrt{\pi}\left(\sqrt{b} - \sqrt{b-\xi}\right), \qquad \xi < 0.$$

The density of S_t can be recovered from the characteristic function by Fourier inversion:

$$f_{S_t}(x) = \frac{1}{2\pi}\int_{\mathbb{R}} e^{-ix\xi}\varphi_{S_t}(\xi)d\xi = \frac{at}{x^{3/2}}\exp\left(-\frac{\left(at\sqrt{\pi} - x\sqrt{b}\right)^2}{x}\right), \qquad x > 0.$$

The first four cumulants are equal to

$$c_1(S_1) = E[S_1] = \frac{a\sqrt{\pi}}{\sqrt{b}}, \qquad c_2(S_1) = \mathrm{var}(S_1) = \frac{a\sqrt{\pi}}{2b^{3/2}},$$
$$c_3(S_1) = \frac{3a\sqrt{\pi}}{4b^{5/2}}, \qquad c_4(S_1) = \frac{15a\sqrt{\pi}}{8b^{7/2}}.$$

As in the case of the Gamma subordinator, also the IG subordinator with unitary mean can be parameterized in terms of its variance: imposing $c_1(S_1) = 1$ and $c_2(S_1) = v$, we find

$$a = \frac{1}{\sqrt{2\pi v}}, \qquad b = \frac{1}{2v}. \tag{13.87}$$

□

Example 13.63 (Normal inverse Gaussian process) The Normal Inverse Gaussian (NIG) distribution has been introduced by Barndorff-Nielsen [25]. It is obtained by subordinating a Brownian motion with drift through an IG process S with variance v and unitary mean (i.e. with a, b as in (13.87)):

$$X_t = \mu S_t + \sigma W_{S_t}.$$

The three parameters of the model are the variance v of the subordinator, the drift μ and the volatility σ of the Brownian motion. By Theorem 13.58, the characteristic exponent of X is[15]

$$\psi_X(\xi) = \frac{1 - \sqrt{1 + v\xi\left(-2i\mu + \xi\sigma^2\right)}}{v}, \tag{13.88}$$

and by (13.68), the first four cumulants are equal to

$$\begin{aligned} c_1(X_1) &= E[X_1] = \mu, & c_2(X_1) &= \mathrm{var}(X_1) = v\mu^2 + \sigma^2, \\ c_3(X_1) &= 3v\mu\left(v\mu^2 + \sigma^2\right), & c_4(X_1) &= 3v\left(v\mu^2 + \sigma^2\right)\left(5v\mu^2 + \sigma^2\right). \end{aligned} \tag{13.89}$$

[15] Often the NIG characteristic exponent is written in the form (equivalent to (13.88))

$$\psi_X(\xi) = \delta\left(\sqrt{\alpha^2 - \beta^2} - \sqrt{\alpha^2 - (\beta + i\xi)^2}\right)$$

where $\beta = \frac{\mu}{\sigma^2}$, $\delta = \frac{\sigma}{\sqrt{v}}$ and $\alpha^2 = \frac{1}{v\sigma^2} + \beta^2$.

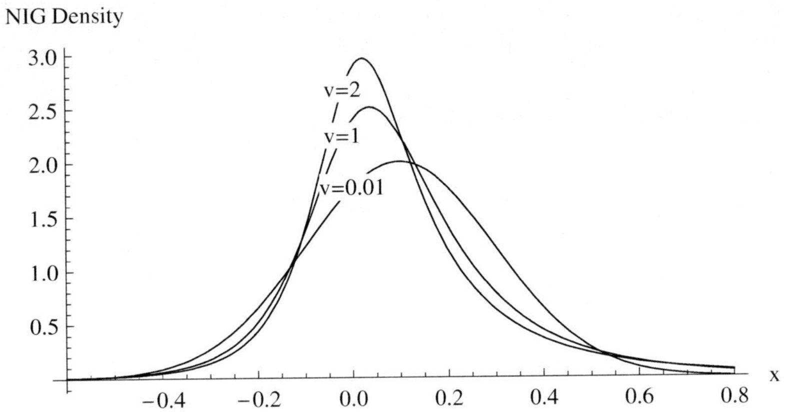

Fig. 13.8. NIG density with $\sigma = 0.2$, $\mu = 0.1$ and different values of the variance of the subordinator: $v = 0.01$, $v = 1$ and $v = 2$

The Lévy measure of a NIG process is

$$\nu(dx) = \frac{C}{|x|} e^{\lambda x} \mathbf{K}_1 \left(\gamma |x|\right) dx,$$

where $\mathbf{K}$ is the modified Bessel function of the second kind and

$$C = \frac{\sqrt{\mu^2 + \frac{\sigma^2}{v}}}{2\pi\sigma\sqrt{v}}, \qquad \lambda = \frac{\mu}{\sigma^2}, \qquad \gamma = \frac{\sqrt{\mu^2 + \frac{\sigma^2}{v}}}{\sigma^2}.$$

From the known properties of $\mathbf{K}$, we have the following asymptotic behaviour of the Lévy measure:

$$\nu(x) = \mathrm{O}\left(\frac{1}{x^2}\right) \qquad \text{as } x \to 0,$$

that is, the NIG process has unbounded variation and infinite activity (1-stable like behaviour of small jumps); moreover

$$\nu(x) = \mathrm{O}\left(\frac{e^{-\lambda_\pm |x|}}{x^{\frac{3}{2}}}\right) \qquad \text{as } x \to \pm\infty,$$

where

$$\lambda_\pm = \frac{1}{\sigma^2}\left(\sqrt{\mu^2 + \frac{\sigma^2}{v}} \mp \mu\right),$$

that is, the NIG process has asymmetric semi-heavy tails. Finally, the density function of X_t is

$$f_{X_t}(x) = \frac{t}{\pi}\sqrt{\frac{\mu^2}{v\sigma^2} + \frac{1}{v^2}} \frac{e^{\lambda x + \frac{t}{v}}}{\sqrt{x^2 + \frac{t^2\sigma^2}{v}}} \mathbf{K}_1 \left(\gamma\sqrt{x^2 + \frac{t^2\sigma^2}{v}}\right).$$

In Figure 13.8 we plot the probability density of a NIG distribution for different choices of the variance v of the subordinator. □

13.4.5 Hyperbolic processes

The hyperbolic distribution was first introduced by Barndorff-Nielsen [26] to model the distribution of grain size in wind-blown sand deposits. It was employed in finance by Eberlein and Keller [113]. It takes the attribute "hyperbolic" from the shape of the logarithm of the density function which is an hyperbola, in contrast to the Gaussian case, which follows a parabola. The density function of the hyperbolic distribution is of the form:

$$\frac{\sqrt{\alpha^2-\beta^2}}{2\alpha\delta\mathbf{K}_1(\delta\sqrt{\alpha^2-\beta^2})}\exp\left(-\alpha\sqrt{\delta^2+(x-\mu)^2}+\beta(x-\mu)\right)$$

where $\mathbf{K}_\nu$ denotes the modified Bessel function of the second kind and the parameters satisfy: $\mu \in \mathbb{R}$, $\delta \geq 0$, $|\beta| < \alpha$. Here μ is a shift parameter, δ a scale parameter and β is related to the skewness since the distribution is symmetric about μ for $\beta = 0$. Barndorff-Nielsen and Halgren [26] proved that the hyperbolic distribution is infinitely divisible. Therefore one can define Lèvy processes whose increments follow the hyperbolic distribution. The hyperbolic distribution can be viewed as a special case of the *generalized hyperbolic distribution* (GH) that has been studied by Eberlein, Keller and Prause [114], Eberlein and Prause [117] (see also Eberlein [110]). Its density function is:

$$\frac{(\alpha^2-\beta^2)^{\frac{\nu}{2}}(\delta^2+(x-\mu)^2)^{\frac{\nu}{2}-\frac{1}{4}}}{\sqrt{2\pi}\alpha^{\nu-\frac{1}{2}}\delta^\nu\mathbf{K}_\nu(\delta\sqrt{\alpha^2-\beta^2})}e^{\beta(x-\mu)}\mathbf{K}_{\nu-\frac{1}{2}}\left(\alpha\sqrt{\delta^2+(x-\mu)^2}\right)$$

where $\mu \in \mathbb{R}$ and

$$\begin{aligned} &\delta \geq 0, \quad |\beta| < \alpha \quad \text{if } \nu > 0, \\ &\delta > 0, \quad |\beta| < \alpha \quad \text{if } \nu = 0, \\ &\delta > 0, \quad |\beta| \leq \alpha \quad \text{if } \nu < 0. \end{aligned} \tag{13.90}$$

Note that the hyperbolic distribution is a particular case of the GH distribution with $\nu = 1$, in view of the property: $K_{\frac{1}{2}}(x) = \sqrt{\frac{\pi}{2x}}e^{-x}$. The tails of the density of GH distributions behave like

$$|x|^{\nu-1}e^{(\beta\pm\alpha)x} \qquad \text{as } x \longrightarrow \mp\infty$$

that is, they possess the semi-heaviness property which is in keeping with a stylized feature observed in financial data. Also the GH distribution is infinitely divisible and its characteristic exponent is:

$$\psi(x) = i\mu\xi + \frac{\nu}{2}\log\frac{\alpha^2-\beta^2}{\alpha^2-(\beta+i\xi)^2} + \log\frac{\mathbf{K}_\nu\left(\delta\sqrt{\alpha^2-(\beta+i\xi)^2}\right)}{\mathbf{K}_\nu\left(\delta\sqrt{\alpha^2-\beta^2}\right)}.$$

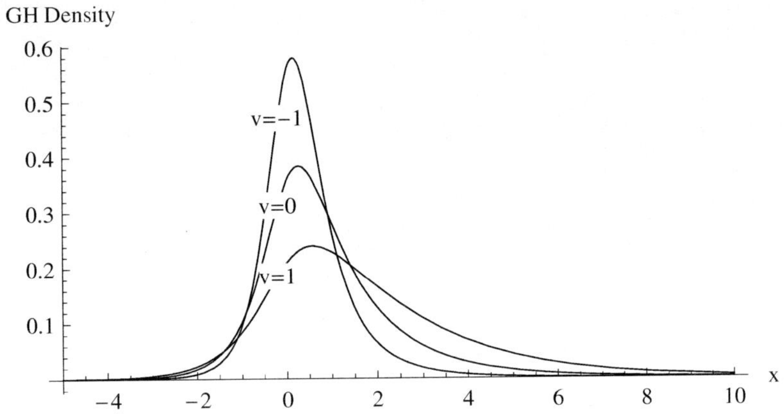

Fig. 13.9. GH density with $\mu = 0, \delta = 1, \alpha = 1, \beta = 0.5$ and different values of the parameter ν: $\nu = -1$, $\nu = 0$ and $\nu = 1$

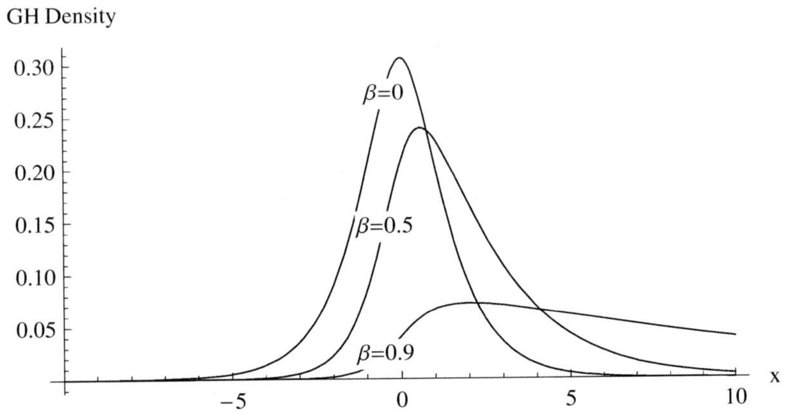

Fig. 13.10. GH density with $\mu = 0, \delta = 1, \alpha = 1, \nu = 1$ and different values of the parameter β: $\beta = 0$, $\beta = 0.5$ and $\beta = 0.9$

Put $\zeta = \delta\sqrt{\alpha^2 - \beta^2}$: then the mean of the GH distribution is

$$\mu + \frac{\beta\delta}{\sqrt{\alpha^2 - \beta^2}} \frac{\mathbf{K}_{\nu+1}(\zeta)}{\mathbf{K}_{\nu}(\zeta)}$$

and the variance is

$$\delta^2 \left(\frac{\mathbf{K}_{\nu+1}(\zeta)}{\zeta \mathbf{K}_{\nu}(\zeta)} + \frac{\beta^2}{\alpha^2 - \beta^2} \left(\frac{\mathbf{K}_{\nu+2}(\zeta)}{\mathbf{K}_{\nu}(\zeta)} - \frac{\mathbf{K}^2_{\nu+1}(\zeta)}{\mathbf{K}^2_{\nu}(\zeta)} \right) \right).$$

The other cumulants can be computed explicitly as well, but they have a quite complicated expression. Some known processes are special cases or limiting cases of the GH process. The NIG process is obtained with $\nu = -\frac{1}{2}$; the

Student-t process is obtained for $\nu < 0$, $\alpha = \beta = \mu = 0$; the VG process is a limiting case for $\delta \to 0$ and $\alpha = \sqrt{\beta^2 + \frac{2}{\nu}}$; the normal distribution is a limiting case $\delta \to \infty$ and $\frac{\delta}{\alpha} \to \sigma^2$.

A disadvantage of GH Lévy processes is that the Lévy measure is only known in implicit form (see, for instance, Schoutens [301] Section 5.3.11) and therefore the structure of jumps is unclear and it may not be easy to see, for instance, whether the process has finite/infinite activity. Moreover the sum of independent GH variables is not a GH variables (that is, the class of GH distributions is not closed under convolution).

In Figures 13.9 and 13.10 we plot the probability density of a GH distribution for different choices of the parameters ν and β.

13.5 Option pricing under exponential Lévy processes

The aim of this section is to introduce the basic notions of option pricing theory for contingent claims of European type when the underlying asset is assumed to follow a Lévy process. The classical modeling of stock prices as a geometric Brownian motion is replaced by a geometric Lévy motion, that is, the stock price S_t is e^{X_t}, where X_t is a Lévy process.

13.5.1 Martingale modeling in Lévy markets

A general Lévy market is a model of financial market with a deterministic saving account $B_t = e^{rt}$, $r \geq 0$, and $d \geq 1$ risky assets (say stocks) with stochastic price process $S_t^{(j)} = S_0^{(j)} e^{X_t^{(j)}}$ where $X = (X^{(1)}, \dots, X^{(d)})$ is a Lévy process. The riskless account is used for discounting. As usual, the information flow is given by a filtration $(\mathcal{F}_t)_{t\in[0,T]}$. In what follows we will restrict ourselves to the case $d = 1$, because our main concern will be the pricing of options on a single asset.

Since our aim is to price contingent claims on a stock, it is convenient to adopt the martingale modeling approach of Section 10.3.1. Thus, we assume that the dynamics of the risky assets is given directly under an equivalent martingale measure (EMM) Q: this means that, by definition, the discounted price process is a martingale under Q. Moreover, if $F(X_T)$ denotes the terminal payoff (at time T) of a contingent claim on the stock, following Section 10.2.5, we define the risk-neutral price of the contingent claim as follows:

$$H_t^Q = e^{-r(T-t)} E^Q \left[F(X_T) \mid \mathcal{F}_t \right] \qquad t \in [0, T].$$

In general, if the risky asset is driven by a Lévy process then Q is not unique and the market is incomplete: from the practical point of view, this corresponds to the fact that there are several possible choices for the values of the parameters of the model under consideration. It is important to note that the martingale condition imposes some restrictions on the possible specifications of the parameters: indeed, our first result is a characterization of the EMMs in terms of a condition on the characteristic exponent of the process (and therefore on the parameters of the model).

Proposition 13.64 *Assume that X is a Lévy process with characteristic exponent ψ_Q under Q. The discounted price process $\widetilde{S}_t = S_0 e^{X_t - rt}$ is a Q-martingale if and only if*

$$E^Q[S_t] = E^Q\left[S_0 e^{X_T}\right] < \infty \tag{13.91}$$

and the following drift condition holds:

$$\psi_Q(-i) = r. \tag{13.92}$$

Proof. It suffices to recall that, by Theorem 13.50-iii), $\widetilde{S}_t = S_0 e^{X_t - rt}$ is a martingale if and only if $1 = e^{-rt} E^Q\left[e^{X_t}\right] = e^{t(\psi_Q(-i) - r)}$. □

If X is a Lévy process with triplet (μ_1^Q, σ^2, ν) under Q, the drift condition (13.92) can be rewritten more explicitly as follows

$$\mu_1^Q = r - \frac{\sigma^2}{2} - \int_{\mathbb{R}} \left(e^x - 1 - x\mathbb{1}_{\{|x|<1\}}\right) \nu(dx). \tag{13.93}$$

Moreover, under assumption

$$E^Q[|X_t|] < \infty,$$

condition (13.93) is also equivalent to

$$\mu_\infty^Q = r - \frac{\sigma^2}{2} - \int_{\mathbb{R}} (e^x - 1 - x)\, \nu(dx). \tag{13.94}$$

For greater convenience, now we give the risk neutral dynamics and a coincise summary of some tractable example of exponential Lévy models. In Chapter 15 we discuss the related numerical issues and examine the implied volatility surfaces generated by several Lévy models. As usual, we assume that the asset process is of the form

$$S_t = S_0 e^{X_t}$$

where X is a Lévy process verifying the integrability condition (13.91). Since in all the models considered, X has also finite expectation, that is condition (13.56) is satisfied, we will directly give the ∞-triplet of the process: we recall the expression of μ_∞^Q in (13.94) and that the characteristic exponent of X is equal to

$$\psi(\xi) = i\mu_\infty^Q \xi - \frac{\sigma^2 \xi^2}{2} + \int_{\mathbb{R}} \left(e^{i\xi x} - 1 - i\xi x\right) \nu(dx).$$

If, in addition, the jump part of X has bounded variation, i.e. if (13.53) holds, then we also give the drift coefficient μ_0^Q of the 0-triplet. The generic R-triplet can be derived from the relation

$$\mu_S^Q = \mu_R^Q - \int_{S<|x|\le R} x\nu(dx), \qquad 0 \le S \le R.$$

- **Black&Scholes:** the characteristic triplet is $(\mu^Q, \sigma^2, 0)$ with

$$\mu^Q = r - \frac{\sigma^2}{2},$$

  and the characteristic exponent is

$$\psi_Q(\xi) = i\mu^Q\xi - \frac{\sigma^2\xi^2}{2}. \tag{13.95}$$

  There is only one parameter, the volatility σ, that is positive.
- **Merton:** the ∞-triplet is $(\mu_\infty^Q, \sigma^2, \nu)$ where

$$\begin{aligned}\nu(dx) &= \frac{\lambda}{\sqrt{2\pi\delta^2}} \exp\left(-\frac{(x-m)^2}{2\delta^2}\right) dx,\\ \mu_0^Q &= r - \frac{\sigma^2}{2} - \lambda\left(e^{m+\frac{\delta^2}{2}} - 1\right),\\ \mu_\infty^Q &= \mu_0^Q + \lambda m.\end{aligned}$$

  The characteristic exponent is

$$\psi_Q(\xi) = i\mu_\infty^Q\xi - \frac{\sigma^2\xi^2}{2} + \lambda\left(e^{im\xi - \frac{\delta^2\xi^2}{2}} - 1 - im\xi\right). \tag{13.96}$$

  The parameters are: σ (diffusion coefficient, non-negative), λ (jump intensity, positive), m (mean jump size, real), δ (standard deviation of jump size, positive).
- **Kou:** the ∞-triplet is $(\mu_\infty^Q, \sigma^2, \nu)$ where

$$\nu(dx) = \lambda\left(p\lambda_1 e^{-\lambda_1 x}\mathbb{1}_{\{x>0\}} + (1-p)\lambda_2 e^{\lambda_2 x}\mathbb{1}_{\{x<0\}}\right) dx.$$

  The expectation $E\left[e^{X_t}\right]$ is finite if $\lambda_1 > 1$ and in that case

$$\begin{aligned}\mu_0^Q &= r - \frac{\sigma^2}{2} + \lambda\left(\frac{p}{1-\lambda_1} + \frac{1-p}{1+\lambda_2}\right),\\ \mu_\infty^Q &= \mu_0^Q + \lambda\left(\frac{p}{\lambda_1} + \frac{p-1}{\lambda_2}\right).\end{aligned}$$

  The characteristic exponent is

$$\psi_Q(\xi) = i\mu_\infty^Q\xi - \frac{\sigma^2\xi^2}{2} + \lambda\xi^2\left(\frac{p-1}{\lambda_2(\lambda_2 + i\xi)} - \frac{p}{\lambda_1^2 - i\lambda_1\xi}\right).$$

The parameters are: σ (diffusion coefficient, non-negative), λ (jump intensity, positive), p (probability of upward jump, $p \in]0,1[$), λ_1, λ_2 (decay parameters of the jump distribution, $\lambda_1 > 1$, $\lambda_2 > 0$).

- **Generalized tempered stable:** the ∞-triplet is $(\mu_\infty^Q, \sigma^2, \nu)$ where

$$\nu(dx) = C_1 \frac{e^{-\lambda_1 x}}{x^{1+\alpha_1}} \mathbb{1}_{\{x>0\}} dx + C_2 \frac{e^{-\lambda_2 |x|}}{|x|^{1+\alpha_2}} \mathbb{1}_{\{x<0\}} dx, \tag{13.97}$$

and

◇ if $\alpha_1, \alpha_2 \notin \{0,1\}$ then

$$\begin{aligned}\mu_\infty^Q = r - \frac{\sigma^2}{2} &+ C_1 \left(\lambda_1^{\alpha_1} - (\lambda_1 - 1)^{\alpha_1} - \alpha_1 \lambda_1^{\alpha_1 - 1}\right) \Gamma(-\alpha_1) \\ &+ C_2 \left(\lambda_2^{\alpha_2 - 1}(\alpha_2 + \lambda_2) - (\lambda_2 + 1)^{\alpha_2}\right) \Gamma(-\alpha_2),\end{aligned}$$

where Γ denotes the Euler Gamma function;

◇ if $\alpha_1 = \alpha_2 = 0$ then

$$\begin{aligned}\mu_\infty^Q = r - \frac{\sigma^2}{2} &+ C_1 \left(\frac{1}{\lambda_1} + \log\left(1 - \frac{1}{\lambda_1}\right)\right) \\ &+ C_2 \left(-\frac{1}{\lambda_2} + \log\left(1 + \frac{1}{\lambda_2}\right)\right);\end{aligned}$$

◇ if $\alpha_1 = \alpha_2 = 1$ then

$$\begin{aligned}\mu_\infty^Q = r - \frac{\sigma^2}{2} &- C_1 \left(1 + (\lambda_1 - 1) \log\left(1 - \frac{1}{\lambda_1}\right)\right) \\ &+ C_2 \left(1 - (\lambda_2 + 1) \log\left(1 + \frac{1}{\lambda_2}\right)\right).\end{aligned}$$

The characteristic exponent is given in Section 13.4.3. The parameters are such that $C_1, C_2 \geq 0$, $\lambda_1 > 1$, $\lambda_2 > 0$ and $\alpha_1, \alpha_2 \in [0, 2[$.

- **Variance Gamma (VG):** it is a particular tempered stable process with Lévy measure (cf. Example 13.61)

$$\nu(dx) = \frac{1}{v} \left(\frac{e^{-\lambda_1 x}}{x} \mathbb{1}_{\{x>0\}} dx + \frac{e^{-\lambda_2 |x|}}{|x|} \mathbb{1}_{\{x<0\}} dx \right),$$

and risk neutral drift

$$\mu_0^Q = r - \frac{\sigma^2}{2} + \frac{1}{v} \log\left(1 - \frac{1}{\lambda_1}\right)\left(1 + \frac{1}{\lambda_2}\right),$$

$$\mu_\infty^Q = \mu_0^Q + \frac{1}{v}\left(\frac{1}{\lambda_1} - \frac{1}{\lambda_2}\right).$$

The characteristic exponent is

$$\psi_Q(\xi) = i\xi\mu_\infty^Q + \frac{i\xi}{v}\left(\frac{1}{\lambda_2} - \frac{1}{\lambda_1}\right) - \frac{\sigma^2\xi^2}{2} - \frac{1}{v}\log\left(1 - \frac{i\xi}{\lambda_1}\right)\left(1 + \frac{i\xi}{\lambda_2}\right).$$

The parameters are: v (the variance of the subordinator, positive), λ (jump intensity, positive) and λ_1, λ_2 (decay parameters of the jump distribution, $\lambda_1 > 1$, $\lambda_2 > 0$): λ_1, λ_2 can be expressed in terms of the drift μ and volatility σ of the subordinated Brownian motion by formula (13.86).

- **CGMY:** it is a particular tempered stable process with Lévy measure

$$\nu(dx) = C\left(\frac{e^{-Mx}}{x^{1+Y}}\mathbb{1}_{\{x>0\}}dx + \frac{e^{-G|x|}}{|x|^{1+Y}}\mathbb{1}_{\{x<0\}}dx\right),$$

and risk neutral drift

$$\begin{aligned}\mu_\infty^Q = r + C\Gamma(-Y)\Big(&G^Y - (1+G)^Y + M^Y - (M-1)^Y \\ &+ Y\left(G^{-1+Y} - M^{-1+Y}\right)\Big),\end{aligned} \tag{13.98}$$

where Γ denotes the Euler Gamma function. The characteristic exponent is

$$\begin{aligned}\psi_Q(\xi) = i\mu_\infty^Q\xi + C\Big(&(M - i\xi)^Y - M^Y + (G + i\xi)^Y - G^Y \\ &+ i\xi Y\left(M^{Y-1} - G^{Y-1}\right)\Big)\Gamma(-Y).\end{aligned} \tag{13.99}$$

The parameters verify the conditions: $C > 0$, $M > 1$, $G > 0$, $Y < 2$.

- **Normal Inverse Gaussian (NIG):** a NIG process X is a Brownian motion with drift and volatility $\sigma > 0$, subordinated by an Inverse Gaussian process with variance $v > 0$ and unitary mean. The characteristic exponent is

$$\psi_Q(\xi) = \frac{1 - \sqrt{1 + v\xi\left(-2i\mu + \xi\sigma^2\right)}}{v},$$

where the drift coefficient $\mu = \mu_\infty^Q = E^Q[X_1]$ is determined by the martingale condition $\psi_Q(-i) = r$: under the assumption $vr \leq 1$, we get

$$\mu_\infty^Q = r - \frac{r^2 v}{2} - \frac{\sigma^2}{2}.$$

The Lévy measure is equal to

$$\nu(dx) = \frac{C}{|x|}e^{\lambda x}\mathbf{K}_1(\gamma|x|)\,dx,$$

where $\mathbf{K}$ is the modified Bessel function of the second kind and

$$C = \frac{\sqrt{\left(\mu_\infty^Q\right)^2 + \frac{\sigma^2}{v}}}{2\pi\sigma\sqrt{v}}, \qquad \lambda = \frac{\mu_\infty^Q}{\sigma^2}, \qquad \gamma = \frac{\sqrt{\left(\mu_\infty^Q\right)^2 + \frac{\sigma^2}{v}}}{\sigma^2}.$$

- **Generalized hyperbolic (GH):** the characteristic exponent is

$$\psi_Q(x) = i\mu\xi + \frac{\nu}{2}\log\frac{\alpha^2-\beta^2}{\alpha^2-(\beta+i\xi)^2} + \log\frac{\mathbf{K}_\nu\left(\delta\sqrt{\alpha^2-(\beta+i\xi)^2}\right)}{\mathbf{K}_\nu\left(\delta\sqrt{\alpha^2-\beta^2}\right)},$$

where $\mathbf{K}$ is the modified Bessel function of the second kind and the drift coefficient $\mu = \mu_\infty^Q = E^Q[X_1]$ is determined by the martingale condition $\psi_Q(-i) = r$:

$$\mu_\infty^Q = r - \frac{\nu}{2}\log\frac{\alpha^2-\beta^2}{\alpha^2-(1+\beta)^2} + \log\frac{\mathbf{K}_\nu\left(\delta\sqrt{\alpha^2-(1+\beta)^2}\right)}{\mathbf{K}_\nu\left(\delta\sqrt{\alpha^2-\beta^2}\right)}.$$

The parameters $\alpha, \beta, \delta, \nu$ satisfy the conditions (13.90) and $\alpha > |1+\beta|$.

13.5.2 Incompleteness and choice of an EMM

As it was explained in Section 10.2, option pricing in the absence of arbitrage can be reduced to calculating the expected values of the discounted payoff with respect to an EMM Q which is also called a risk-neutral measure, because the expected returns of the risky asset under Q equals the risk-free interest rate. We saw in Section 10.2.6 that uniqueness of the EMM amounts to the fact that any contingent claim can be priced in a unique way, that is, every contingent claim can be replicated by a dynamic trading strategy. While the Black-Scholes model is complete, typically a generic Lévy market is incomplete. The incompleteness of Lévy models (with the exception of the Brownian motion and of the Poisson process, which however makes no sense economically) can be proved in view of the "predictable representation property" (see Nualart and Schoutens [268]) which is possessed only by few processes.

The incompleteness property means that there exist infinitely many martingale measures compatible with the no-arbitrage requirement. Indeed the no-arbitrage assumption provides an interval of arbitrage-free prices: Eberlein and Jacod [112], Bellamy and Jeanblanc [38] showed, in the case of a pure-jump model and of jump-diffusions respectively, that the option values computed under different EMMs span the entire trivial no-arbitrage interval. This implies that if a trader charges a (super-hedging) price to option buyers in order to eliminate all risks, then this price would be forbiddingly high, that is, super-hedging is not a realistic strategy. A more realistic approach is to charge a reasonable price for partial hedging the risks and bearing some residual risk. Consequently, one is faced with the problem of selecting an EMM which is "optimal" in some sense. We will not enter into the details of such a challenging question and we will present only the popular choices for an EMM which is based on the Esscher transform. There are many papers devoted to the guiding criterion in this choice. One economic criterion is to maximize the utility function describing the traders' preferences and is also related to the minimization of relative entropy. One argument is to choose the EMM which is closest to the historic measure P in an entropy sense, that is, it minimizes

$$\int \log\left(\frac{dQ}{dP}\right) dQ.$$

The minimal entropy martingale measure for Lévy models is studied in Chan [71], Fujiwara and Miyahara [145], Hubalek and Sgarra [174], where it is related to the Esscher transform method. We conclude with the remark that however it is not clear what kind of measure transform the market chooses in reality.

In practice, in the framework of option pricing, the parameter fitting (calibration) for EMM from a given class seems to be a more reasonable procedure, that is, financial intermediaries first decide a specific model for the underlying price process, then they calibrate the parameters of the model to market data (e.g., prices of vanilla options) to determine the probability measure chosen by the market, and finally use it for the pricing and hedging of other derivatives. The existence of many EMMs leads to a great flexibility of Lévy models for calibrating market prices of options. In other words, market incompleteness "may not seem to be a desirable property at first sight", but "it is a realistic property that these models share with real markets" (Cont and Tankov [76]).

13.5.3 Esscher transform

From the theoretical point of view, a method for the construction of martingale measures starting from the real-world measure P, is based on the Esscher transformation. It has been used in actuarial practice since Esscher's paper [123] and then applied in finance by Gerber and Shiu [155]. Kallsen and Shiryaev [196] introduced the Esscher martingale measure for exponential processes and for linear processes: the Esscher transform of a probability measure P is defined as follows.

Definition 13.65 *Let $(\Omega, P, \mathcal{F})$ be a probability space, X a random variable and $\theta \in \mathbb{R}$. The Esscher transform P^θ is defined by*

$$\frac{dP^\theta}{dP} = \frac{e^{\theta X}}{E\left[e^{\theta X}\right]},$$

provided that the expectation $E\left[e^{\theta X}\right]$ exists.

Notice that the measure P^θ is equivalent to P. Similarly one can define the Esscher transform for a Lévy process $(X_t)_{t\in[0,T]}$. Suppose that $\theta \in \mathbb{R}$ and $E\left[e^{\theta X_T}\right] < \infty$. Let $\ell(z)$ denote the Laplace exponent (cf. (13.84)) of the Lévy process: more precisely, we set

$$\ell(z) = \psi(-iz) = \mu z + \frac{\sigma^2 z^2}{2} + \int_{\mathbb{R}} \left(e^{izx} - 1 - zx\mathbb{1}_{\{|x|<1\}}\right) \nu(dx).$$

Since $\ell(i\xi) = \psi(\xi)$ for all real ξ, then $\ell(z)$ is well defined at least for $z \in \mathbb{C}$ with $\mathrm{Re}(z) = 0$ and $E\left[e^{zX_t}\right] = e^{t\ell(z)}$. By Theorem 13.50, the process

$$Z_t^\theta = e^{\theta X_t - t\ell(\theta)}$$

is a real-valued and strictly positive martingale, while $e^{i\theta X_t - t\psi(\theta)}$ is a complex-valued martingale. That is why it is convenient to work with Laplace exponents in order to define a new probability measure by means of Z^θ. Let us set

$$\frac{dP_T^\theta}{dP} = e^{\theta X_T - T\ell(\theta)}. \tag{13.100}$$

Theorem 13.66 *Let $X = (X_t)_{t\in[0,T]}$ be a Lévy process. Then X is a Lévy process also with respect to P_T^θ in* (13.100) *and its Laplace exponent ℓ_θ under P_T^θ is given by*

$$\ell_\theta(z) = \ell(z+\theta) - \ell(\theta). \tag{13.101}$$

Proof. In view of Bayes's formula (Theorem A.113), for all $0 \le s \le t \le T$ we have

$$\begin{aligned}
E^{P_T^\theta}\left[e^{z(X_t - X_s)} \mid \mathcal{F}_s\right] &= \frac{E^P\left[e^{z(X_t-X_s)} Z_T^\theta \mid \mathcal{F}_s\right]}{E^P\left[Z_T^\theta \mid \mathcal{F}_s\right]} \\
&= \frac{1}{Z_s^\theta} E^P\left[E^P\left[e^{z(X_t-X_s)} Z_T^\theta \mid \mathcal{F}_t\right] \mid \mathcal{F}_s\right] \\
&= \frac{1}{Z_s^\theta} E^P\left[e^{z(X_t-X_s)} Z_t^\theta \mid \mathcal{F}_s\right] \\
&= E^P\left[e^{(z+\theta)(X_t-X_s)-(t-s)\ell(\theta)} \mid \mathcal{F}_s\right] \\
&= e^{(t-s)(\ell(z+\theta)-\ell(\theta))}.
\end{aligned}$$

Then the process $(X_t)_{t\in[0,T]}$ has stationary, independent increments and (13.101) holds. □

Theorem 13.67 *Let (μ, σ^2, ν) be the triplet of a Lévy process X with respect to the measure P. Then the triplet $(\mu_\theta, \sigma_\theta^2, \nu_\theta)$ of X with respect to the measure P_T^θ in* (13.100) *is determined as follows:*

$$\begin{aligned}
\mu_\theta &= \mu + \theta\sigma^2 + \int_{\mathbb{R}} (e^{\theta x} - 1) x \mathbb{1}_{\{|x|<1\}} \nu(dx), \\
\sigma_\theta^2 &= \sigma^2, \\
\nu_\theta(dx) &= e^{\theta x}\nu(dx).
\end{aligned} \tag{13.102}$$

Proof. In view of (13.101) we have:

$$\begin{aligned}
\ell_\theta(z) =&\mu z + \frac{\sigma^2}{2}\left((z+\theta)^2-\theta^2\right) + \int_{\mathbb{R}} \left((e^{zx}-1)e^{\theta x} - zx\mathbb{1}_{\{|x|<1\}}\right)\nu(dx) \\
=& \left(\mu+\theta\sigma^2\right)z + \frac{\sigma^2 z^2}{2} + \int_{\mathbb{R}} \left(e^{zx}-1-z\mathbb{1}_{\{|x|<1\}}\right)e^{\theta x}\nu(dx) \\
&+ z\int_{\mathbb{R}} (e^{\theta x}-1)x\mathbb{1}_{\{|x|<1\}}\nu(dx) \\
=&\mu_\theta z + \frac{\sigma_\theta^2 z^2}{2} + \int_{\mathbb{R}} \left(e^{zx}-1-zx\mathbb{1}_{|x|\leq 1}(x)\right)\nu_\theta(dx).
\end{aligned}$$

□

Now we are interested in a condition making the discounted stock price $\widetilde{S}_t = S_0 e^{X_t - rt}$ a martingale under P_T^θ. We have

$$E^{P_T^\theta}\left[\widetilde{S}_t \mid \mathcal{F}_s\right] = \widetilde{S}_s$$

if and only if

$$e^{(t-s)\ell_\theta(1)} = E^{P_T^\theta}\left[e^{X_t - X_s} \mid \mathcal{F}_s\right] = e^{r(t-s)},$$

or equivalently $\ell_\theta(1) = r$. In view of (13.101), we get the following result.

Theorem 13.68 *Suppose that θ^* is a solution to*

$$\ell(1+\theta^*) - \ell(\theta^*) = r. \tag{13.103}$$

If $E^P\left[e^{\theta^* X_T}\right] < \infty$ (to assure that $P_T^{\theta^*}$ exists) and $E^P\left[e^{(\theta^*+1)X_T}\right] < \infty$ (so that $E^{P_T^{\theta^*}}[S_T]$ exists), then $P_T^{\theta^*}$ is an equivalent martingale measure.

Under the assumptions of Theorem 13.68, $P_T^{\theta^*}$ is called the Esscher martingale transform of the objective measure P. Let us denote $P_T^{\theta^*}$ by Q: in view of (13.101), with obvious notation we have

$$\ell_Q(z) = \ell_P(z+\theta^*) - \ell_P(\theta^*). \tag{13.104}$$

In terms of the characteristic exponent we have that if $\theta^* \in \mathbb{R}$ solves

$$\psi_P(-i\theta^*) - \psi_P(-i\theta^* - i) = r,$$

then the characteristic exponent under the EMM Q is given by

$$\psi_Q(\xi) = \psi_P(\xi - i\theta^*) - \psi_P(-i\theta^*).$$

Note that $\psi_Q(-i) = \psi_P(-i - i\theta^*) - \psi_P(-i\theta^*) = -r$, which is the EMM-condition (13.92).

Example 13.69 (Brownian motion with drift) Let $\ell_P(z) = \mu z + \frac{\sigma^2 z^2}{2}$ be the Laplace exponent under the historic measure P. The solution to (13.103) is

$$\theta^* = \frac{1}{\sigma^2}\left(r - \mu - \frac{\sigma^2}{2}\right),$$

and by (13.102) the characteristic exponent is

$$\psi_Q(\xi) = i\left(r - \frac{\sigma^2}{2}\right)\xi - \frac{\sigma^2\xi^2}{2}.$$

It is clear that in this case the Esscher transform is equivalent to the Girsanov transform. □

Example 13.70 (Shifted Poisson process) Let N_t be a Poisson process with intensity parameter λ and let $X_t = \alpha N_t - \beta t$ with $\alpha, \beta > 0$. Since

$$\ell_P(z) = -\beta z + \lambda(e^{\alpha z} - 1),$$

the solution to (13.103) is $\theta^* = \frac{1}{\alpha}\log\frac{r+\beta}{\lambda(e^{\alpha}-1)}$. Thus

$$\ell_Q(z) = -\beta z + \lambda^*(e^{\alpha z} - 1)$$

with $\lambda^* = \lambda e^{\theta^* y} = \frac{r+\beta}{e^y - 1}$. In terms of the characteristic exponent we have:

$$\psi_Q(z) = -i\beta\xi + \lambda^*\left(e^{i\alpha\xi} - 1\right). \quad □$$

Example 13.71 (NIG process) Let

$$\ell_P(z) = \mu z + \delta\left((\alpha^2 - \beta^2)^{\frac{1}{2}} - (\alpha^2 - (\beta + z)^2)^{\frac{1}{2}}\right)$$

with $-\alpha - \beta \le \operatorname{Re} z \le \alpha - \beta$, $|\beta| < \alpha$. Note that both $\ell_P(\theta)$ and $\ell_P(1+\theta)$ exist if $\alpha \ge \frac{1}{2}$ and $-\alpha - \beta \le \theta \le \alpha - \beta - 1$. Moreover (13.103) has a solution[16]

$$\theta^* = -\frac{1}{2} - \beta - \frac{\mu - r}{2\delta}\sqrt{\frac{4\alpha^2}{1 + \left(\frac{\mu - r}{\delta}\right)^2} - 1}$$

if $|\mu - r| \le \delta\sqrt{4\alpha^2 - 1}$. Then (13.104) yields

$$\ell_Q(z) = \mu z + \delta\left((\alpha^2 - \beta^{*2})^{1/2} - (\alpha^2 - (\beta^* + z)^2)^{1/2}\right)$$

with $\beta^* = -\frac{1}{2} - \frac{\mu - r}{2\delta}\sqrt{\frac{4\alpha^2}{1 + \left(\frac{\mu - r}{\delta}\right)^2} - 1}$. □

[16] Equation (13.103) can be easily solved for the unknown $x = (\beta + \theta^*)(\beta + \theta^* + 1)$ and then θ^* is obtained.

Exercise 13.72 Consider the VG process whose Laplace exponent can be represented as follows

$$\ell_P(z) = \mu z - C \log \frac{GM}{GM + (M-G)z - z^2}$$

with $-G < \operatorname{Re} z < M$. Show that if $G + M > 1$ an Esscher parameter exists and compute $\ell_Q(z)$. □

Exercise 13.73 Consider a Lévy process based on a generalized hyperbolic distribution and such that

$$\ell_P(z) = \mu z + \frac{\nu}{2} \log \frac{\alpha^2 - \beta^2}{\alpha^2 - (\beta + z)^2} + \log \frac{\mathbf{K}_\nu(\delta\sqrt{\alpha^2 - (\beta + z)^2})}{\mathbf{K}_\nu(\delta\sqrt{\alpha^2 - \beta^2})}.$$

Show that an Esscher transform with parameter $\theta^* \in]-\alpha - \beta, \alpha - \beta[$ corresponds to a shift of parameter $\beta \longrightarrow \beta + \theta^*$, where θ^* solves the equation

$$\mu\theta - \frac{\nu}{2} \log \frac{\alpha^2 - (\beta + \theta + 1)^2}{\alpha^2 - (\beta + \theta)^2} + \log \frac{\mathbf{K}_\nu(\delta\sqrt{\alpha^2 - (\beta + \theta + 1)^2})}{\mathbf{K}_\nu(\delta\sqrt{\alpha^2 - (\beta + \theta)^2})} = r. \qquad \square$$

The Esscher transform method can be viewed as a special case of more general methods for constructing EMMs. Let us first characterize all structure preserving EMMs Q under which X remains a Lévy process.

Theorem 13.74 *Let X be a Lévy process with triplet (μ, σ^2, ν) under some probability measure P. Then the following two conditions are equivalent:*

i) there is a probability measure Q, equivalent to P, such that X is a Lévy process with triplet $(\widetilde{\mu}, \widetilde{\sigma}^2, \widetilde{\nu})$ under Q;
ii) all of the following conditions hold:
ii.a) $\widetilde{\nu}(dx) = H(x)\nu(dx)$ for some Borel function $H : \mathbb{R} \longrightarrow \mathbb{R}_{>0}$;
ii.b) $\widetilde{\sigma} = \sigma$;
ii.c) $\widetilde{\mu} = \mu + \int_{|x|<1} x(H(x) - 1)\nu(dx) + \sigma\eta$ for some $\eta \in \mathbb{R}$;
ii.d) $\int_{\mathbb{R}} \left(1 - \sqrt{H(x)}\right)^2 \nu(dx) < \infty$.

Proof. See Sato [297], Theorem 33.1. □

We emphasize that Eberlein and Jacod [112] have shown that if there exists a non-structure preserving EMM, there exists always a structure preserving EMM Q under which X is a Lévy process. Now let us impose to a generic equivalent and structure preserving measure that the discounted stock price $\widetilde{S}_t = e^{X_t - rt}$ is a martingale: in view of Corollary 5.2.2. in Applebaum [11] and of the previous theorem, we obtain the following condition:

$$\mu + \frac{\sigma^2}{2} - r + \sigma\eta + \int_{\mathbb{R}} \left(H(x)(e^x - 1) - x\mathbb{1}_{\{|x|<1\}}\right) \nu(dx) = 0$$

under the condition

$$\int_{|x|\geq 1} H(x)(e^x - 1)\nu(dx) < \infty.$$

Note that there are infinitely many solutions pairs (η, H) of this equation. If (η, H) is a solution and f is ν-integrable, then $(\hat{\eta}, \hat{H})$ with

$$\hat{\eta} = \eta + \frac{1}{\sigma}\int_{\mathbb{R}} f(x)\nu(dx), \qquad \hat{H}(x) = H(x) + \frac{f(x)}{e^x - 1}$$

is a solution as well.

We show here two trivial examples of Lévy processes that admit a unique solution (η, H) and in fact give a complete market.

Example 13.75 (Brownian motion) In this case $\nu \equiv 0$ and $\sigma > 0$. The unique solution is $\eta = \frac{r-\mu}{\sigma} - \frac{\sigma}{2}$. This implies that the new drift and volatility parameters under Q are: $\widetilde{\mu} = r - \frac{\sigma^2}{2}$ and $\widetilde{\sigma} = \sigma$. □

Example 13.76 (Shifted Poisson process) Let $X_t = \alpha N_t + \mu t$ with $\mu < r$ and $\alpha > 0$. For simplicity, we take $\alpha > 1$. In this case $\sigma = 0$ and $\nu = \lambda\delta_\alpha$ where λ is the intensity parameter. Then H is constant

$$H = \frac{r - \mu}{\lambda(e^\alpha - 1)}$$

and the new Lévy measure under Q is $\widetilde{\nu} = \lambda^*\delta_\alpha$ with $\lambda^* = \frac{r-\mu}{e^\alpha - 1}$; moreover $\widetilde{\mu} = \mu$. □

13.5.4 Exotic option pricing

Exotic derivatives have gained an increasing importance as financial instruments. Especially path-dependent options are traded in large volumes in the OTC markets. In this section we give a short survey of the literature on European exotic options under a Lévy market. Exotics of American style and multi-asset options are omitted. We will deal only with Asian options in detail and refer to Agliardi [4], Hubalek [172], Kyprianou, Schoutens and Wilmott [221] for the valuation formulas for other classes of options.

Asian options. In Section 7.6 we studied Asian options in the classical Gaussian framework. Here we give a generalization to the Lévy framework in the case of geometric Asian options. We confine ourselves to forward-start fixed strike Asian options and refer to Fusai and Meucci [146] for the floating-strike case and an approximation method for arithmetic Asian options. We also mention Albrecher [5], Albrecher and Predota [6], Eberlein and Papapantoleon [116].

Let K denote the strike price. At first we consider discrete Asian options, whose payoff depends on a discrete average of the asset price at M monitoring times, $T_1 < ... < T_M$. Then the payoff is

$$\max\{w(G_M - K), 0\}, \qquad G_M = \left(\prod_{j=1}^{M} S_{T_j}\right)^{\frac{1}{M}},$$

where w is the binary indicator ($w = \pm 1$). In terms of $X_t = \log S_t$ the payoff is:

$$w\left(\prod_{j=1,\dots,M} \exp\left(\frac{X_{T_j}}{M}\right) - K\right) \mathbb{1}_{\{\frac{w}{M}\sum_{j=1}^{M} X_{T_j} \geq w \log K\}}.$$

Let us compute the characteristic exponent of

$$Y = \sum_{m=1}^{M} \frac{X_m}{M},$$

where X_m denotes X_{T_m}, for brevity's sake. Since

$$Y = \sum_{m=1}^{M} \frac{X_0}{M} + \sum_{j=1}^{M} (X_j - X_{j-1}) \sum_{m=j}^{M} \frac{1}{M}$$

with $X_0 = \log S_t$, the independence of the increments of X_t yields:

$$\begin{aligned} E\left[e^{i\xi Y} \mid \log S_t\right] &= e^{i\xi X_0} \prod_{j=1}^{M} E\left[\exp\left(i\xi (X_j - X_{j-1}) \frac{M-j+1}{M}\right) \mid \log S_t\right] \\ &= \exp\left(i\xi \log S_t + \sum_{j=1}^{M} (T_j - T_{j-1}) \psi\left(\frac{M-j+1}{M}\xi\right)\right), \end{aligned}$$

where ψ is the characteristic exponent of X. Now we assume that the Lévy measure satisfies:

$$\int_{-\infty}^{-1} e^{\lambda_+ x} \nu(dx) + \int_{1}^{+\infty} e^{-\lambda_- x} \nu(dx) < \infty$$

for some constants $\lambda_- < 0 < \lambda_+$. Then the current value of an option whose payoff is

$$\mathbb{1}_{\left\{\frac{w}{M}\sum_{j=1}^{M} X_{T_j} \geq w \log(K)\right\}}$$

is given by the formula (cf. Proposition 15.6)

$$\frac{w e^{-r(T_M - t)}}{\pi} \left(\frac{S_t}{K}\right)^{\alpha w} \int_0^{+\infty} \frac{F(\xi)}{w\alpha - i\xi} d\xi$$

where

$$F(\xi) := \exp\left(-i\xi \log \frac{S_t}{K} + \sum_{j=1}^{M} (T_j - T_{j-1})\psi\left(-\frac{M-j+1}{M}(\xi + i\alpha w)\right)\right)$$

and $T_0 = t$, $\alpha > 0$ and $\frac{M-j+1}{M} w\alpha \in]-\lambda_+, -\lambda_-[$ for $j = 1, ..., M$. More generally, the current value of an option whose payoff is

$$\exp\left(\sum_{m=1}^{M} \gamma_m X_m\right) \mathbb{1}_{\left\{\frac{w}{M} \sum_{j=1}^{M} X_{T_j} \geq w \log K\right\}}$$

is given by:

$$\frac{we^{-r(T_M-t)}}{\pi} S_t^{\sum_{m=1}^{M} \gamma_m} (\frac{S_t}{K})^{\alpha w} \int_0^{+\infty} e^{-i\xi \log \frac{S_t}{K}} \frac{F(\xi)}{w\alpha - i\xi} d\xi$$

where

$$F(\xi) := \exp\left(\sum_{j=1}^{M} (T_j - T_{j-1})\psi\left(-\frac{M-j+1}{M}(\xi + i\alpha w) - i\sum_{m=j}^{M} \gamma_m\right)\right)$$

and $\frac{M-j+1}{M} w\alpha + \gamma_m \in]-\lambda_+, -\lambda_-[$ for $j = 1, ..., M$. Thus the current value of the (discretely monitored) fixed strike geometric Asian option is:

$$\frac{e^{-r(T_M-t)}}{\pi} S_t^{\alpha w} K^{1-\alpha w} \int_0^{+\infty} \frac{F(\xi)}{(i\xi - w\alpha)(i\xi - w\alpha + 1)} d\xi$$

where

$$F(\xi) = \exp\left(-i\xi \log \frac{S_t}{K} + \sum_{j=1}^{M} (T_j - T_{j-1})\psi\left(-\frac{M-j}{M}(\xi + i\alpha w)\right)\right),$$

with $T_0 = t$, $\alpha \in]1, -\lambda_-[$ if $w = 1$, and $\alpha \in]0, \lambda_+[$ if $w = -1$. The pricing formula for the continuous-time monitoring case, where the geometric average is

$$\exp\left(\frac{1}{T - T'} \int_{T'}^{T} \log S_t dt\right),$$

follows from the discrete pricing formula just letting $M \to \infty$. In particular, for the continuous case, one has:

$$\frac{Ke^{-r(T-t)}}{\pi} S_t^{\alpha w} K^{1-\alpha w} \int_0^{+\infty} \frac{F(\xi)}{(i\xi - w\alpha)(i\xi - w\alpha + 1)} d\xi,$$

where

$$F(\xi) = \exp\left(-i\xi \log \frac{S_t}{K} + \int_0^1 \psi((\xi + i\alpha w)(y - 1))dy\right).$$

Exercise 13.77 Compute the price of a geometric Asian option in the Gaussian framework, as a special case. □

Barrier options. The right of the holder of a barrier option is conditioned on the underlying asset crossing (or not crossing) a prespecified barrier before the expiry date. There exists eight barrier options types, depending on the barrier being above or below the initial value (up or down), on the barrier knocking in (activating) or out (extinguishing), and on the Call/Put attribute. Barrier options in a Lévy framework are studied in Boyarchenko and Levendorskii [56], Nguyen and Yor [265]. Discretely monitored barrier options were priced by Heynen and Kat [167] in the traditional Gaussian framework and in Feng and Linetsky [130] in Lévy process models. A survey on discrete barrier options is found in Kou [218].

Lookback options. The payoff of a lookback option depends on the extremal price of the underlying asset over the life of the option. Floating strike lookback Call and Put options have payoffs of the form

$$S_T - \min_{0\le t\le T} S_t \qquad \text{and} \qquad \max_{0\le t\le T} S_t - S_T$$

respectively, while fixed strike lookback Call and Put options have payoffs

$$\left(\max_{0\le t\le T} S_t - K\right) \qquad \text{and} \qquad \left(K - \min_{0\le t\le T} S_t\right)^+$$

respectively, where K denotes the strike price. The valuation of lookback options is a hard problem, because the distribution of the supremum or infimum of a Lévy process is not known explicitly. Explicit formulas are known in the Gaussian case. The Lévy case is studied in Nguyen and Yor [265]. Discrete lookback options in a Lévy framework are considered in Agliardi [4] and Hubalek [172]. The Gaussian case was studied in Heynen and Kat [166].

Compound options. Compound options are options on options. An extension of Geske's formula (see Geske [156]) to a more general Lévy environment is obtained in Agliardi [3]: see also [4] for the N-fold compound options.

Chooser options. These options give their holder the right to decide at a prespecified date before the maturity whether they would like the option to be a Call or a Put option. A valuation formula in a Lévy framework is obtained in [4].

13.5.5 Beyond Lévy processes

We sketch some directions of development of the theory of Lévy processes in finance towards further generalizations. A crucial property of Lévy processes is the assumption of stationary increments. For financial markets, it is not clear whether this condition is valid. Some recent empirical analysis tends to reject such an assumption. For example, Bassler, McCauley and Gunaratne [34] find evidence of non-stationary increments in return for the Euro-Dollar exchange rate through an analysis of intraday increments. Moreover, time inhomogeneous jump-diffusions (possibly depending also on the state variable) seem to provide a good fit for some financial markets. Therefore some extension of Lévy modeling relaxes the stationarity of increments, just assuming *additive processes*, i.e. càdlàg stochastic processes with independent increments and stochastic continuity. Nevertheless, as observed by Cont and Tankov [76], "additive processes have the same drawback as local volatility models: they require time dependent parameters to accommodate the observed term structure of implied volatility. More importantly, when the risk-neutral dynamics of the log-price is given by a Lévy process or an additive process, the implied volatility surface follows a deterministic evolution: there is no *Vega* risk in such models."

An alternative stream of research considers some Feller processes which are natural generalizations of Lévy processes and employs pseudo-differential methods to derive approximate valuation formulas. Such methods are the natural extension to the Lévy framework of the Black-Scholes models with variable coefficients, which are used to reproduce the smiles. A presentation of the theory of Feller process can be found in Jacob [183]. We refer to Barndorff-Nielsen and Levendorskii [27], Boyarchenko and Levendorskii [56] for a description of financial applications of Feller processes.

Another direction of research is followed by the stochastic volatility models which allow to reproduce a realistic implied volatility term structure. In a more general framework, solutions to stochastic differential equations with jumps can be considered: indeed, combining jumps (which are needed for the short-term behaviour) and stochastic volatility (which works especially for long-term smiles) allows for a better calibration of the implied volatility surface. To this end, in Chapter 14 we introduce the basics of stochastic calculus for jumps processes and in Section 14.3 we examine some stochastic volatility model with jumps.

14

Stochastic calculus for jump processes

In this chapter we introduce the basics of stochastic calculus for jump processes. We follow the approaches proposed by Protter [287] for the general theory of stochastic integration and by Applebaum [11] for the presentation of Lévy-type stochastic integrals. We extend to this framework, the analysis performed in the previous chapters for continuous processes: in particular, we prove Itô formula and a Feynman-Kač type representation theorem for solutions to SDEs with jumps. For simplicity, most statements are given in the one-dimensional case. Then we show how to derive the integro-differential equation for a quite general exponential model driven by the solution of a SDE with jumps: these results open the way for the use of deterministic and probabilistic numerical methods, such as finite difference schemes (see, for instance, Cyganowski, Grüne and Kloeden [82]), Galerkin schemes (see, for instance, Platen and Bruti-Liberati [281]) and Monte Carlo methods (see, for instance, Glasserman [158]). In the last part of the chapter, we examine some stochastic volatility models with jumps: in particular, we present the Bates and the Barndorff-Nielsen and Shephard models.

14.1 Stochastic integrals

On a filtered probability space $(\Omega, \mathcal{F}, P, (\mathcal{F}_t))$ satisfying the usual hypotheses, we consider a stochastic process $S = (S_t)_{t\in[0,T]}$ representing a financial asset. Hereafter we assume that S is *càdlàg* (i.e. right continuous with finite left limits), the typical example being a Lévy processes. As in Paragraph 3.4, we are naturally interested in the study of sums of the type

$$\sum_{k=1}^{N} e_k \left(S_{T_k} - S_{T_{k-1}}\right)$$

that represent the gain of the self-financing strategy which consists in holding the quantity e_k of the asset during the k-th period $]T_{k-1}, T_k]$. Typically S has

Pascucci A.: PDE and Martingale Methods in Option Pricing

unbounded variation, such as in the case of Brownian motion: for this reason, in Chapter 4 we developed a theory of stochastic integration based on the key idea, due to Itô, of restricting the class of integrands to the *progressively measurable* processes, that is to the strategies that cannot see into the future increments.

We now remark that if the integrator S is discontinuous, then we need to further restrict the class of integrands for which the stochastic integral makes sense both from the mathematical and financial point of view. To give reasons for this last claim, we analyze some simple example.

Example 14.1 Let $T_1 > 0$ and consider the *deterministic* function

$$t \mapsto S_t = \mathbb{1}_{[T_1,\infty[}(t).$$

Then, as in Example 3.68, for any *continuous* function u, the Riemann-Stieltjes integral is well-defined and

$$\int_0^t u_s dS_s = \begin{cases} 0 & \text{if } t < T_1, \\ u_{T_1} & \text{if } t \geq T_1. \end{cases} \tag{14.1}$$

□

Example 14.2 The integral with respect to S in Example 14.1 can be extended to discontinuous functions in the Lebesgue-Stieltjes sense. Let us briefly recall that if g is an increasing function and we set

$$\mu_g(]a,b[) = g(b-) - g(a+), \qquad a, b \in \mathbb{R},\ a < b,$$

then μ_g can be extended to a measure on $\mathscr{B}(\mathbb{R})$ called *the Lebesgue-Stieltjes measure induced by* g: for instance, in the case $g(t) = t$ we find the Lebesgue measure; in the case of $g = S$ in Example 14.1, we find the Dirac delta centered at T_1, that is $\mu_S = \delta_{T_1}$. Hence, for any deterministic function u, the Lebesgue-Stieltjes integral is given by

$$\int_{[0,t]} u_s dS_s = \int_{[0,t]} u_s \delta_{T_1}(ds) = \begin{cases} 0 & \text{if } t < T_1, \\ u_{T_1} & \text{if } t \geq T_1, \end{cases}$$

and, if u is continuous, it coincides with the Riemann-Stieltjes integral in (14.1). □

Example 14.3 We now examine the convergence of Riemann-Stieltjes sums in case u is discontinuous and S as in Example 14.1. We consider $t \geq T_1$ and a sequence of partitions $\varsigma_n = (t_0^n, \dots, t_{N_n}^n)$ of $[0,t]$ such that $T_1 \in \varsigma_n$ for any n (say, $T_1 = t_{k_n}^n$) and $|\varsigma_n| \to 0$ as n goes to infinity. Let

$$\sum_{k=1}^{N} u_{\tau_k^n} \left(S_{t_k^n} - S_{t_{k-1}^n} \right) = u_{\tau_{k_n}^n}$$

be the Riemann-Stieltjes sum where as usual $\tau_k^n \in [t_{k-1}^n, t_k^n]$ for $k = 1, \dots, N_n$ and in particular $\tau_{k_n}^n \in [t_{k_n-1}^n, T_1]$. Passing to the limit as $n \to \infty$, the above Riemann-Stieltjes sum converges to u_{T_1-} if $\tau_{k_n}^n \in [t_{k_n-1}^n, T_1[$ and to u_{T_1} if $\tau_{k_n}^n = T_1$. Thus the Riemann-Stieltjes integral is well-defined (and independent on the choice of the points τ_k^n) if and only if u is *left-continuous*[1]. Moreover, in that case it coincides with the Lebesgue-Stieltjes integral. □

Example 14.4 Let $S_t = \lambda t - N_t$ where N is a Poisson process with intensity λ. Then S is a martingale because $-S$ is a compensated Poisson process: intuitively, S is a fair investment giving zero gain in average, because the deterministic increase λt is compensated by the sudden falls $-N_t$ which occur randomly and represent the source of risk. We denote by T_n the jump times of S and consider the strategy

$$u_t = \mathbb{1}_{[0,T_1]}(t) \tag{14.2}$$

which consists in buying (at zero price) the asset at $t = 0$ and selling it at the time of the first jump. Note that u is left-continuous and therefore, as we have seen before, the integral of u with respect to S is well-defined: in the Riemann-Stieltjes sense (that is, by considering the integral as a limit of Riemann-Stieltjes sums) we have

$$\begin{aligned} G_t := \int_0^t u_s dS_s &= S_{T_1 \wedge t} - S_0 \\ &= \lambda(t \wedge T_1) - N_{T_1 \wedge t} \\ &= \lambda(t \wedge T_1) - \mathbb{1}_{\{t \geq T_1\}}. \end{aligned}$$

We obtain the same value by integrating in the Lebesgue-Stieltjes sense (that is, by integrating with respect to the Lebesgue-Stieltjes measure induced by S):

$$\begin{aligned} \int_{[0,t]} u_s dS_s &= \lambda \int_{[0,t]} u_s ds - \int_{[0,t]} u_s dN_s \\ &= \lambda(t \wedge T_1) - \sum_{n \geq 1} u_{T_n} \mathbb{1}_{\{t \geq T_n\}} \\ &= \lambda(t \wedge T_1) - \mathbb{1}_{\{t \geq T_1\}}, \end{aligned} \tag{14.3}$$

where the last equality follows from the fact that $u_{T_1} = 1$ because u is left-continuous. The common value G represents the gain of the strategy u: precisely, the first term in (14.3) represents the gain due to the deterministic positive drift λt, while the second term represents the losses due to negative

[1] We remark that in Example 3.67 we showed that the *continuity* if the integrand is necessary for the Riemann-Stieltjes integral to be well-defined with respect to a generic BV function S: however, in that pathological example the function S was not càdlàg.

jumps caused by the Poisson process. Losses compensate gains so there are no arbitrage opportunities: mathematically, this is expressed by the fact that G *is a martingale.* This can be seen more easily by using the Riemann-Stieltjes definition of integral:

$$E\left[G_t - G_s \mid \mathcal{F}_s\right] = E\left[S_{T_1 \wedge t} - S_{T_1 \wedge s} \mid \mathcal{F}_s\right] = 0, \qquad s \le t,$$

as a consequence of the optional sampling theorem and the fact that S is a martingale. □

Example 14.5 Let S be as in Example 14.4. We consider the strategy

$$u_t = \mathbb{1}_{[0,T_1[}(t) \tag{14.4}$$

which consists in buying (at zero price) the asset at $t = 0$ and selling it "right before" it crashes (see Protter [287], p. 65, Cont and Tankov [76], Example 8.1). Note that u is *càdlàg* (right-continuous) instead of *càglàd* (left-continuous) as in (14.2). Now the Riemann-Stieltjes of u with respect to S is not well-defined as explained in Example 14.3; however we may try to use instead the Lebesgue-Stieltjes definition and we get

$$\begin{aligned} G_t := \int_{[0,t]} u_s dS_s &= \lambda \int_{[0,t]} u_s ds - \int_{[0,t]} u_s dN_s \\ &= \lambda(t \wedge T_1) - \sum_{n \ge 1} u_{T_n} \mathbb{1}_{\{t \ge T_n\}} \\ &= \lambda(t \wedge T_1), \end{aligned} \tag{14.5}$$

where the last equality follows from the fact that $u_{T_1} = 0$ because u is right-continuous. Now the gain G_t is strictly positive for any $t > 0$: this means that *u is an arbitrage strategy and G is not a martingale anymore.* It is clear that such a strategy is not acceptable from the financial point of view; also mathematically we lose the fundamental property that the stochastic integral with respect to a martingale is again a martingale.

Intuitively, in order to exploit the strategy in (14.4) we would need to know when the price is going to fall *just before* it happens: in other terms, assuming that u is right-continuous amounts to say that we are able to *predict* the jumps of the Poisson process so to avoid the risk of crashes. Notice also that this situation is peculiar of financial models with jumps: indeed this example is based on the fact that the process S is discontinuous. □

14.1.1 Predictable processes

We consider a process of the form

$$u_t = \sum_{k=1}^{N} e_k \mathbb{1}_{]T_{k-1},T_k]}(t), \qquad t \in [0,T], \tag{14.6}$$

where $0 = T_1 < \cdots < T_N = T$ and, for $1 \le k \le N$, e_k is a *bounded* random variable that is $\mathcal{F}_{T_{k-1}}$-measurable. In order to treat more conveniently the case of jump processes, where jumps occur at random times, it seems natural to assume that T_k, $1 \le k \le N$, are *stopping times* and not simply deterministic times as we did in the Brownian case, cf. Section 4.2. Then we say that $\varsigma = (T_1, \dots, T_N)$ is a *random partition* of the interval $[0, T]$ and that u in (14.6) is a *simple predictable process.*

We remark explicitly that the process in (14.6) is *càglàd* by definition, i.e. it is *left-continuous with finite right limits.* The importance of this simple fact could not be appreciated in the framework of continuous processes. Nevertheless in Examples 14.1-14.5, we showed that this is really a crucial point when dealing with jump processes: if we assume that the asset S is a *càdlàg* (right-continuous) process, then the simple trading strategies should be *càglàd* (left-continuous) processes as in (14.6) to rule out arbitrage opportunities.

In order to describe more generally the hypotheses we are going to assume, we give the following:

Definition 14.6 *We let $\mathbb{L}$ denote the space of càglàd (left-continuous) adapted processes and $\mathbb{D}$ denote the space of càdlàg (right-continuous) adapted processes. The predictable σ-algebra $\mathcal{P}$ is the smallest σ-algebra on $[0, T] \times \Omega$ making all processes in $\mathbb{L}$ measurable. The optional σ-algebra $\mathcal{O}$ is the smallest σ-algebra on $[0, T] \times \Omega$ making all processes in $\mathbb{D}$ measurable.*

Let us consider a stochastic process X as a mapping $(t, \omega) \mapsto X_t(\omega)$ from $[0, T] \times \Omega$ to $\mathbb{R}$. We say that X is *predictable* if it is $\mathcal{P}$-measurable; we say that X is *optional* if it is $\mathcal{O}$-measurable. Clearly, by definition, every process in $\mathbb{L}$ is predictable: in particular, the simple predictable processes of the form (14.6) are predictable. For example, $u_t = X_{t-}$ where X is a compound Poisson process is simple predictable. However, not all predictable processes are left-continuous. Analogously, every process in $\mathbb{D}$ is optional but an optional process is not necessarily right-continuous.

Now we recall that the Brownian integral was defined for *progressively measurable* integrands, that is for processes X such that the mapping $(s, \omega) \mapsto X_s(\omega)$ of $[0, t] \times \Omega$ into $\mathbb{R}$, is $\mathscr{B}([0, t]) \otimes \mathcal{F}_t$-measurable for any $t \le T$. In general, one has the following relationships (see, for instance, Chapter III in Protter [287]):

$$\mathcal{P} \subseteq \mathcal{O} \subseteq \mathcal{A}$$

where $\mathcal{A}$ denotes the σ-algebra on $[0, T] \times \Omega$ generated by all progressively measurable processes. Thus predictable processes form a particular subclass of the family of all progressively measurable processes.

We remark explicitly that

$$\mathbb{L} \subset \mathbb{L}^2_{\mathrm{loc}}$$

where (cf. Definition 4.33) $\mathbb{L}^2_{\mathrm{loc}}$ denotes the family of processes $(u_t)_{t\in[0,T]}$ that are progressively measurable and such that

$$\int_0^T u_t^2\, dt < \infty \quad \text{a.s.}$$

Definition 14.7 *Let S be càdlàg adapted, $S \in \mathbb{D}$, and let $u \in \mathbb{L}$ be a simple predictable process of the form (14.6). The stochastic integral of u with respect to S is defined as*[2]

$$X_t = \int_0^t u_s dS_s := \sum_{k=1}^{N} e_k \left(S^t_{T_k} - S^t_{T_{k-1}} \right), \qquad t \in [0,T]. \tag{14.7}$$

We also use the "differential" notation

$$dX_t = u_t dS_t.$$

The stochastic integral of a simple predictable process has three remarkable properties which follows directly from the definition. Let X be as in (14.7), then we have:

i) X is càdlàg adapted, i.e. $X \in \mathbb{D}$. In particular X is an integrator and the *associativity property* holds, i.e. for any simple predictable process v we have
$$\int_0^t v_s dX_s = \int_0^t u_s v_s dS_s.$$
In differential terms, this can be expressed as follows:
$$dY_t = v_t dX_t \text{ and } dX_t = u_t dS_t \quad \Longrightarrow \quad dY_t = u_t v_t dS_t; \tag{14.8}$$
ii) if S is a martingale then X is also a martingale;
iii) the jump process $\Delta X_t = X_t - X_{t-}$ is indistinguishable from $u_t \Delta S_t$.

In particular, ii) generalizes property (5) of Theorem 4.5 for the Brownian integral and also the proof is essentially analogous: for completeness, we repeat it here. First of all, X is integrable because u is bounded and S is integrable. Moreover, in order to show that $E[X_T \mid \mathcal{F}_t] = X_t$ it is sufficient to prove that

$$E\left[e_k \left(S_{T_k} - S_{T_{k-1}} \right) \mid \mathcal{F}_t \right] = e_k \left(S_{T_k \wedge t} - S_{T_{k-1} \wedge t} \right),$$

for each k. Now we have

$$E\left[e_k \left(S_{T_k} - S_{T_{k-1}} \right) \mid \mathcal{F}_t \right] = E_1 + E_2 + E_3$$

[2] We recall the notation $S^t_T = S_{T\wedge t}$.

where[3]

$$\begin{aligned}E_1 &= E\left[\mathbb{1}_{\{t>T_k\}} e_k \left(S_{T_k} - S_{T_{k-1}}\right) \mid \mathcal{F}_t\right]\\ &= \mathbb{1}_{\{t>T_k\}} e_k \left(S_{T_k} - S_{T_{k-1}}\right),\\ E_2 &= E\left[\mathbb{1}_{\{T_{k-1}<t\le T_k\}} e_k \left(S_{T_k} - S_{T_{k-1}}\right) \mid \mathcal{F}_t\right]\\ &= \mathbb{1}_{\{T_{k-1}<t\le T_k\}} e_k E\left[S_{T_k} - S_{T_{k-1}} \mid \mathcal{F}_t\right]\\ &= \mathbb{1}_{\{T_{k-1}<t\le T_k\}} e_k \left(S_t - S_{T_{k-1}}\right),\end{aligned}$$

and

$$\begin{aligned}E_3 &= E\left[\mathbb{1}_{\{t\le T_{k-1}\}} e_k \left(S_{T_k} - S_{T_{k-1}}\right) \mid \mathcal{F}_t\right]\\ &= \mathbb{1}_{\{t\le T_{k-1}\}} E\left[E\left[e_k \left(S_{T_k} - S_{T_{k-1}}\right) \mid \mathcal{F}_{T_{k-1}}\right] \mid \mathcal{F}_t\right]\\ &= \mathbb{1}_{\{t\le T_{k-1}\}} E\left[e_k E\left[\left(S_{T_k} - S_{T_{k-1}}\right) \mid \mathcal{F}_{T_{k-1}}\right] \mid \mathcal{F}_t\right] = 0.\end{aligned}$$

Therefore we have

$$\begin{aligned}E\left[e_k \left(S_{T_k} - S_{T_{k-1}}\right) \mid \mathcal{F}_t\right] =& \mathbb{1}_{\{t>T_k\}} e_k \left(S_{T_k} - S_{T_{k-1}}\right)\\ &+ \mathbb{1}_{\{T_{k-1}<t\le T_k\}} e_k \left(S_t - S_{T_{k-1}}\right)\\ =& e_k \left(S_{T_k\wedge t} - S_{T_{k-1}\wedge t}\right).\end{aligned}$$

Example 14.8 Let N be a Poisson process with jump times $(T_n)_{n\ge1}$ (we also set $T_0 = 0$). Then

$$N_t = n \quad \text{for} \quad t \in [T_n, T_{n+1}[, \ n \ge 0,$$

and $N \in \mathbb{D}$. We already noted that

$$N_{t-} = n \quad \text{for} \quad t \in]T_n, T_{n+1}], \ n \ge 0,$$

is simple predictable. If we set $N_T^t = N_{T\wedge t}$ as usual, then we have

$$\int_0^t N_{s-} dN_s = \sum_{k\ge1} (k-1)\left(N_{T_k}^t - N_{T_{k-1}}^t\right) =$$

(for $t \in [T_{n-1}, T_n[$)

$$= \sum_{k=1}^{n-1} (k-1)\left(N_{T_k} - N_{T_{k-1}}\right) = \sum_{k=1}^{n-1} (k-1) = \frac{(n-1)(n-2)}{2}.$$

Therefore we have

$$\int_0^t N_{s-} dN_s = \frac{N_t(N_t - 1)}{2}, \qquad t \ge 0. \tag{14.9}$$

□

[3] Note that $\mathbb{1}_{\{t>T_k\}}$, $\mathbb{1}_{\{T_{k-1}<t\le T_k\}}$ and $\mathbb{1}_{\{t\le T_{k-1}\}}$ are $\mathcal{F}_t$-measurable because T_k and T_{k-1} are stopping times.

14.1.2 Semimartingales

In Section 14.1.1 we have defined the stochastic integral

$$\int_0^T u_s dS_s$$

for u simple predictable and $S \in \mathbb{D}$. In order to extend the definition of integral to the class $\mathbb{L}$, we need two ingredients:

- the density (under some suitable norm) of the simple predictable processes in the space $\mathbb{L}$;
- some continuity property of the stochastic integral: roughly speaking, if the sequence (u^n) converges to u, then the stochastic integral of u^n must approximate the stochastic integral of u. In order to get this continuity property, we have to restrict the class of integrators. Actually, the "good integrators" will be exactly the processes in $\mathbb{D}$ which ensure the continuity of the integral: those processes will be called *semimartingales.*

Definition 14.9 *A process $S = (S_t)_{t\in[0,T]}$ in $\mathbb{D}$ is called a semimartingale if, for any sequence (u^n) of simple predictable processes such that*

$$\lim_{n\to\infty} \sup_{[0,T]\times\Omega} |u^n| = 0,$$

we have

$$\lim_{n\to\infty} \int_0^T u_t^n dS_t = 0 \qquad \textit{in probability.}$$

The following examples show that many familiar classes of processes are semimartingales.

Example 14.10 If a process $S \in \mathbb{D}$ has bounded variation a.s. then it is a semimartingale. Indeed, for any simple predictable process u, we have

$$\left| \int_0^T u_t dS_t \right| \le V_{[0,T]}(S) \sup_{[0,T]\times\Omega} |u|,$$

where $V_{[0,T]}(S)$ denotes the first variation of S over $[0,T]$ which, by assumption, is finite a.s. The thesis follows from the fact that a.s. convergence implies convergence in probability (cf. Theorem A.136). In particular, *the Poisson process is a semimartingale.* □

Example 14.11 Every square integrable martingale $S \in \mathbb{D}$ is a semimartingale. Indeed, for any simple predictable process u, we have

$$E\left[\left(\int_0^T u_t dS_t\right)^2\right] = E\left[\left(\sum_{k=1}^N e_k \left(S_{T_k} - S_{T_{k-1}}\right)\right)^2\right] =$$

(by the orthogonality of the increments of L^2-martingales, cf. (4.31))

$$
\begin{aligned}
&= E\left[\sum_{k=1}^{N} e_k^2 \left(S_{T_k} - S_{T_{k-1}}\right)^2\right] \\
&\le E\left[\sum_{k=1}^{N} \left(S_{T_k} - S_{T_{k-1}}\right)^2\right] \sup_{[0,T]\times\Omega} u^2 \\
&= E\left[\sum_{k=1}^{N} \left(S_{T_k}^2 - S_{T_{k-1}}^2\right)\right] \sup_{[0,T]\times\Omega} u^2 \\
&= E\left[S_{T_N}^2 - S_0^2\right] \sup_{[0,T]\times\Omega} u^2 \le
\end{aligned}
$$

(by Doob's inequality)

$$
\le 8E\left[S_T^2\right] \sup_{[0,T]\times\Omega} u^2.
$$

The thesis follows since L^2-convergence implies convergence in probability. In particular, *a Brownian motion is a semimartingale.* □

Example 14.12 *A Lévy process is a semimartingale.* Indeed, by the decomposition in Corollary 13.38, every Lévy process is the sum of a càdlàg L^2-martingale with a BV process. Since the set of semimartingales forms a vector space, the thesis is a consequence of the two examples above. □

More generally, it can be proved, but this is a deep and difficult result (cf. Theorem III-1 in Protter [287]), that a process $X \in \mathbb{D}$ is a semimartingale if and only there exist processes M and Z, with $M_0 = Z_0 = 0$, such that

$$
X_t = X_0 + M_t + Z_t
$$

where M is a local martingale and Z has bounded variation.

The continuity property of Definition 14.9 involves the convergence of random variables, namely $X_t^n = \int_0^t u_s^n dS_s$ with $t = T$ fixed. As in Section 4.3, in order to define properly the stochastic integral *as a process* $t \mapsto X_t$ for $t \in [0,T]$, it is necessary to show that the integral is also continuous as an operator that maps processes into processes. To be more specific, we introduce the notion of *uniform convergence in probability.*

Definition 14.13 *We say that a sequence of processes* $(u_t^n)_{t\in[0,T]}$ *converges to a process u uniformly in probability if*

$$
\lim_{n\to\infty} \sup_{t\in[0,T]} |u_t^n - u_t| = 0 \qquad \textit{in probability.} \tag{14.10}
$$

We are now in position to extend the class of integrands from simple predictable processes to $\mathbb{L}$ by continuity with respect to the uniform convergence in probability.

Theorem 14.14 *The space of simple predictable processes is dense in $\mathbb{L}$ under the uniform convergence in probability: for any $u \in \mathbb{L}$ there exists a sequence (u^n) of simple predictable processes such that (14.10) holds. The stochastic integral with respect to a semimartingale S is continuous under the uniform convergence in probability: if (u^n) is a sequence of simple predictable processes such that*

$$\lim_{n\to\infty} \sup_{t\in[0,T]} |u_t^n| = 0 \qquad \textit{in probability,}$$

then

$$\lim_{n\to\infty} \sup_{t\in[0,T]} \left| \int_0^t u_s^n dS_s \right| = 0 \qquad \textit{in probability.}$$

For any semimartingale S, the continuous linear mapping from $\mathbb{L}$ to $\mathbb{D}$ obtained as the extension of the integral of Definition 14.7 is called stochastic integral: for any $u \in \mathbb{L}$, we denote it by

$$X_t = \int_0^t u_s dS_s, \qquad t \in [0,T],$$

and also write

$$dX_t = u_t dS_t.$$

Proof. See Protter [287], Theorems II-10 and II-11. □

By limit arguments, the main properties of the stochastic integral of simple processes extend to $\mathbb{L}$. Let $u \in \mathbb{L}$, S be a semimartingale and

$$X_t = \int_0^t u_s dS_s.$$

Then we have

i) X is a semimartingale and the *associativity property* holds, i.e. for any $v \in \mathbb{L}$ we have
$$\int_0^t v_s dX_s = \int_0^t u_s v_s dS_s;$$
ii) if S is a square integrable martingale and $u \in \mathbb{L}$ is bounded, then X is also a square integrable martingale. More generally, if S is a local martingale and $u \in \mathbb{L}$ then X is also a local martingale;
iii) the jump process $\Delta X_t = X_t - X_{t-}$ is indistinguishable from $u_t \Delta S_t$.

We recall Example 14.4 where we showed in a very particular case that the notions of Riemann-Stieltjes and Lebesgue-Stieltjes integral coincide if the integrand is left-continuous. Now we state a remarkable theorem which claims that a similar result holds for the stochastic integral of a process u in $\mathbb{L}$ or in $\mathbb{D}$ (in this case, *by taking its left-continuous version u_{t-}*). In the following statement $\varsigma_n = \left(T_1^n, \dots, T_{N_n}^n\right)$ denotes a sequence of random[4] partitions of the interval $[0, T]$ such that

$$\lim_{n\to\infty} |\varsigma_n| = 0 \quad \text{a.s., where} \quad |\varsigma_n| := \max_{1\le k\le N_n} \left|T_k^n - T_{k-1}^n\right|. \tag{14.11}$$

Theorem 14.15 *Let S be a semimartingale and u be a process in $\mathbb{D}$ or in $\mathbb{L}$. Then the Riemann-Stieltjes sum*[5]

$$\sum_{k=1}^{N_n} u_{T_{k-1}^n} \left(S_{T_k^n}^t - S_{T_{k-1}^n}^t\right)$$

converges uniformly in probability as $n \to \infty$ to the stochastic integral

$$\int_0^t u_{s-} dS_s.$$

Proof. See Theorem II-21 in Protter [287]. □

Remark 14.16 Stochastic integration of $\mathbb{L}$ processes is sufficient to prove Itô formula, Girsanov-Meyer theorem and other important results in mathematical finance. On the other hand, the general theory allows to define the stochastic integral of any *predictable processes* (not necessarily left-continuous): this further non-trivial extension is required in the study of more advanced topics such as martingale representation theorems and local times. □

14.1.3 Integrals with respect to jump measures

Let S be a Lévy process with jump measure J and Lévy measure ν. In Section 13.3.3 (cf. (13.42)) we defined the integral

$$M_t = \int_0^t \int_{\mathbb{R}^d} f(s, x)\widetilde{J}(ds, dx)$$

of a deterministic function f with respect to the compensated random measure $\widetilde{J}$: that type of integrals played a crucial role in the Lévy-Itô decomposition and in particular in the representation of small jumps of a Lévy process.

In this section, we extend the definition to the case of *stochastic functions*. This generalization allows to get a clearer expression and a deeper comprehension of the stochastic integral

$$\int_0^t u_s dS_s$$

when the integrator S is a Lévy process.

---

4 $(T_k^n)_{n,k\in\mathbb{N}}$ are stopping times.

5 As usual $S_T^t = S_{T\wedge t}$.

As usual, we introduce the definition gradually, starting from "simple" stochastic functions. By analogy with simple predictable processes, we say that a function

$$\varphi : [0,T] \times \mathbb{R} \times \Omega \longrightarrow \mathbb{R}$$

is a *simple predictable function with respect to* J if it is of the form

$$\varphi(t,x) = \sum_{k=1}^{N} \sum_{i=1}^{m} e_{ki} \mathbb{1}_{]T_{k-1},T_k] \times H_i}(t,x)$$

where $0 = T_1 < \cdots < T_N = T$ are stopping times, $(e_{ki})_{k=1,\dots,N}$ are *bounded* $\mathcal{F}_{T_{k-1}}$-measurable random variables and $(H_i)_{i=1,\dots,m}$ are disjoint Borel subsets of $\mathbb{R}$ such that $J([0,T] \times H_i) < \infty$. Then we define the stochastic integral

$$\begin{aligned}\int_0^T \int_{\mathbb{R}} \varphi(t,x) J(dt,dx) &:= \sum_{k=1}^{N} \sum_{i=1}^{m} e_{ki} J\left(]T_{k-1}, T_k] \times H_i\right) \\ &:= \sum_{k=1}^{N} \sum_{i=1}^{m} e_{ki} \left(J_{T_k}(H_i) - J_{T_{k-1}}(H_i) \right)\end{aligned}$$

where $J_t(H) = J([0,t] \times H)$ is the process defined in Lemma 13.33. More generally, we define the integral as a stochastic process by setting

$$t \mapsto X_t := \int_0^t \int_{\mathbb{R}} \varphi(s,x) J(ds,dx) = \sum_{k=1}^{N} \sum_{i=1}^{m} e_{ki} \left(J_{T_k \wedge t}(H_i) - J_{T_{k-1} \wedge t}(H_i) \right).$$

By construction, $(X_t)_{t \in [0,T]}$ is a càdlàg adapted process; next we show that its compensated version is a L^2-martingale: the following result extends Theorem 13.34. As in Lemma 13.33, we set

$$\widetilde{J}_t(H) = J_t(H) - t\nu(H).$$

Proposition 14.17 *For any simple predictable function* φ*, the stochastic integral*

$$\begin{aligned}M_t &= \int_0^t \int_{\mathbb{R}} \varphi(s,x) \widetilde{J}(ds,dx) \\ &= \sum_{k=1}^{N} \sum_{i=1}^{m} e_{ki} \left(\widetilde{J}_{T_k \wedge t}(H_i) - \widetilde{J}_{T_{k-1} \wedge t}(H_i) \right)\end{aligned}$$

is a L^2*-martingale and it verifies the isometry*

$$\operatorname{var}(M_t) = E\left[M_t^2\right] = E\left[\int_0^t \int_{\mathbb{R}} \varphi^2(t,x) \nu(dx) ds\right]. \qquad (14.12)$$

Proof. By Lemma 13.33, the processes $S_t^i := \widetilde{J}_t(H_i)$, $i = 1, \dots, m$, are independent martingales. Moreover M can be rewritten as a stochastic integral of the type introduced in Section 14.1.1, as follows

$$M_t = \sum_{i=1}^{m} \int_0^t e_s^i dS_s^i,$$

where

$$e_t^i = \sum_{k=1}^{N} e_{ki} \mathbb{1}_{]T_{k-1},T_k]}(t), \qquad i = 1, \dots, m,$$

are simple predictable processes: in particular, M is a martingale with null expectation.

Finally, we prove (14.12) for simplicity only in the case $t = T$: since $(S^i)_{i=1,\dots,m}$ are independent, we have

$$\begin{aligned}\operatorname{var}(M_T) &= \sum_{k=1}^{N} \sum_{i=1}^{m} E\left[e_{ki}^2 \left(S_{T_k}^i - S_{T_{k-1}}^i\right)^2\right] \\ &= \sum_{k=1}^{N} \sum_{i=1}^{m} E\left[e_{ki}^2 E\left[\left(S_{T_k}^i - S_{T_{k-1}}^i\right)^2 \mid \mathcal{F}_{T_{k-1}}\right]\right] =\end{aligned}$$

(by (13.44))

$$= \sum_{k=1}^{N} \sum_{i=1}^{m} E\left[e_{ki}^2 (T_k - T_{k-1}) \nu(H_i)\right]$$

which yields (14.12). □

By means of the isometry (14.12), the class of integrands for

$$M_t = \int_0^t \int_{\mathbb{R}} \varphi(s,x) \widetilde{J}(ds, dx) \tag{14.13}$$

can be enlarged by a procedure which is analogous to that adopted in the Brownian case and based on the Itô isometry. To be more specific, we first extend the notion of predictability (see Definition 14.6).

Definition 14.18 *We denote by $\hat{\mathbb{L}}$ the class of the stochastic functions*

$$\varphi : [0,T] \times \mathbb{R} \times \Omega \longrightarrow \mathbb{R}$$

such that

i) for each $t \in [0,T]$, the mapping $(x,\omega) \mapsto \varphi(t,x,\omega)$ is $\mathscr{B} \otimes \mathcal{F}_t$-measurable;
ii) for each $(x,\omega) \in \mathbb{R} \times \Omega$, the mapping $t \mapsto \varphi(t,x,\omega)$ is càglàd (left-continuous with finite right-limits).

Let ν be the Lévy measure of a Lévy process S: we denote by

- $\mathbb{L}^2_\nu$ *the class of stochastic functions* $\varphi \in \hat{\mathbb{L}}$ *such that*

$$E\left[\int_0^T \int_{\mathbb{R}} \varphi^2(t,x)\nu(dx)dt\right] < \infty; \tag{14.14}$$

- $\mathbb{L}^2_{\nu,\mathrm{loc}}$ *the class of stochastic functions* $\varphi \in \hat{\mathbb{L}}$ *such that*

$$\int_0^T \int_{\mathbb{R}} \varphi^2(t,x)\nu(dx)dt < \infty \qquad \text{a.s.} \tag{14.15}$$

It turns out that $\mathbb{L}^2_\nu$, equipped with the norm (14.14), is a Banach space and the simple predictable functions are dense in $\mathbb{L}^2_\nu$ (see, for instance, Lemmas 4.1.3 and 4.1.4 in Applebaum [11]). Therefore the integral in (14.13) can be extended to $\mathbb{L}^2_\nu$ by means of the isometry (14.12) and we have

Proposition 14.19 *For any $\varphi \in \mathbb{L}^2_\nu$, the process M in (14.13) is a square-integrable martingale such that*

$$E[M_t] = 0, \qquad \mathrm{var}(M_t) = E\left[\int_0^t \int_{\mathbb{R}} \varphi^2(t,x)\nu(dx)ds\right].$$

Proof. The thesis follows from Proposition 14.17 by a limit argument: for full details see, for instance, Theorem 4.2.3 in Applebaum [11]. □

The integral can be further extended to the class $\mathbb{L}^2_{\nu,\mathrm{loc}}$ and in this case we have

Proposition 14.20 *For any $\varphi \in \mathbb{L}^2_{\nu,\mathrm{loc}}$, the integral process M in (14.13) is a local martingale.*

Proof. See, for instance, Theorem 4.2.12 in Applebaum [11]. □

Notation 14.21 *We use the differential notation and we write indifferently*

$$M_t = \int_0^t \int_{\mathbb{R}^d} \varphi(s,x)\widetilde{J}(ds,dx) \qquad \text{or} \qquad dM_t = \int_{\mathbb{R}^d} \varphi(t,x)\widetilde{J}(dt,dx).$$

Remark 14.22 By definition of jump measure, for any stochastic function φ and $R > 0$, we have

$$\int_0^t \int_{|y|\geq R} \varphi(s,y)J(ds,dy) = \sum_{0<s\leq t} \varphi(s,\Delta S_s)\mathbb{1}_{\{|\Delta S_s|\geq R\}} \tag{14.16}$$

where the sum is finite because it has a.s. only a finite number of terms. More generally, as in Theorem 13.34, if

$$\int_0^t \int_{|x|\leq\varepsilon} |\varphi(s,x)|\nu(dx)ds < \infty \qquad \text{a.s.}$$

for some $\varepsilon > 0$, then there exists a.s. the limit, as $R \to 0^+$, of the integral in (14.16) and the series on the right hand side of (14.16) converges absolutely a.s. so that

$$\int_0^t \int_{\mathbb{R}^d} \varphi(s,x) J(ds,dx) = \sum_{\substack{0<s\le t \\ \Delta X_s \neq 0}} \varphi(s, \Delta X_s) < \infty \qquad \text{a.s.} \tag{14.17}$$

□

Remark 14.23 The "extended predictable σ-algebra" $\hat{\mathcal{P}}$ is the σ-algebra on $[0,T] \times \mathbb{R} \times \Omega$ generated by all the stochastic functions in $\hat{\mathbb{L}}$. We say that a stochastic function φ is predictable if it is $\hat{\mathcal{P}}$-measurable: note that a predictable stochastic function φ is not necessarily left-continuous. By analogy with the stochastic integral (cf. Remark 14.16), the notion of integral

$$\int_0^t \int_{\mathbb{R}^d} \varphi(s,x) \widetilde{J}(ds,dx)$$

can be extended to the class of predictable stochastic functions φ verifying (14.15). We refer to Applebaum [11] for details. □

14.1.4 Lévy-type stochastic integrals

In this section we obtain an explicit representation of the stochastic integral of a process in $\mathbb{L}$ with respect to a Lévy process. We recall that $\mathbb{L}$ denotes the class of càglàd adapted processes (cf. Definition 14.6).

Proposition 14.24 *Let S be a one dimensional Lévy process with Lévy-Itô decomposition*

$$S_t = \mu_R t + \sigma W_t + S_t^R + M_t^R \tag{14.18}$$

where W is a standard Brownian motion and

$$S_t^R = \int_0^t \int_{|y|\ge R} y J(ds,dy), \qquad M_t^R = \int_0^t \int_{|y|<R} y \widetilde{J}(ds,dy).$$

Then for any $u \in \mathbb{L}$ we have

$$\int_0^t u_s dS_s = A_t + M_t \tag{14.19}$$

where

$$A_t = \mu_R \int_0^t u_s ds + \int_0^t \int_{|y|\ge R} u_s y J(ds,dy) \tag{14.20}$$

and

$$M_t = \sigma \int_0^t u_s dW_s + \int_0^t \int_{|y|<R} u_s y \widetilde{J}(ds,dy). \tag{14.21}$$

The process A has bounded variation and M is a local martingale. Further, if

$$E\left[\int_0^T u_t^2 dt\right] < \infty$$

then M is a square integrable martingale with null expectation.

Remark 14.25 Proposition 14.24 extends the associativity property (14.8) of the stochastic integral. The above result can be written in the convenient shorthand form: if $u \in \mathbb{L}$ and

$$dS_t = \mu_R dt + \sigma dW_t + \int_{|y|\geq R} yJ(dt, dy) + \int_{|y|<R} y\widetilde{J}(dt, dy),$$

then

$$u_t dS_t = \mu_R u_t dt + \sigma u_t dW_t + \int_{|y|\geq R} u_t yJ(dt, dy) + \int_{|y|<R} u_t y\widetilde{J}(dt, dy). \quad (14.22)$$

□

Proof. Any $u \in \mathbb{L}$ is progressively measurable and such that

$$\int_0^T u_t^2 dt < \infty \qquad \text{a.s.}$$

or in other terms $u \in \mathbb{L}^2_{\text{loc}}$ (cf. Definition 4.1) so that the Brownian integral in (14.21) is well-defined and it is a local martingale. Moreover,

$$\varphi(t, y) := u_t y \mathbb{1}_{\{|y|<R\}} \in \mathbb{L}^2_{\nu,\text{loc}} \quad (14.23)$$

and therefore also the second integral in (14.21) is well-defined and, by Proposition 14.20, it is a local martingale. On the other hand, by (14.16) we have

$$\int_0^t \int_{|y|\geq R} u_s yJ(ds, dy) = \sum_{0<s\leq t} u_s \Delta S_s \mathbb{1}_{\{|\Delta S_s|\geq R\}}$$

where the sum has a.s. only a finite number of terms: then it is clear that A in (14.20) has bounded variation.

Now by the Lévy-Itô decomposition (14.18) and by linearity, we have

$$\begin{aligned}\int_0^t u_s dS_s =& \mu_R \int_0^t u_s ds + \sigma \int_0^t u_s dW_s \\ &+ \int_0^t u_s dS_s^R + \int_0^t u_s dM_s^R.\end{aligned}$$

Moreover, if u is a simple predictable process of the form (14.6), we have

$$\begin{aligned}\int_0^t u_s dM_s^R &= \sum_{k=1}^N e_k \left(M_{T_k \wedge t}^R - M_{T_{k-1} \wedge t}^R \right) \\ &= \sum_{k=1}^N e_k \int_{T_{k-1}\wedge t}^{T_k \wedge t} \int_{|y|<R} y \widetilde{J}(ds, dy) \\ &= \int_0^t \int_{|y|<R} u_s y \widetilde{J}(ds, dy),\end{aligned}$$

and analogously

$$\int_0^t u_s dS_s^R = \int_0^t \int_{|y|\geq R} u_s y J(ds, dy).$$

Then the proof can be completed by a limit argument. □

In some particular cases, it is possible to put $R = 0$ or $R = \infty$ in (14.20)-(14.21) (cf. Section 13.3.3):

- if the jump part of S has bounded variation, that is if

  $$\int_{|x|\leq 1} |x| \nu(dx) < \infty,$$

  then S admits the Lévy-Itô decomposition (cf. (13.55))

  $$S_t = \mu_0 t + B_t + \int_0^t \int_{\mathbb{R}^d} x J(ds, dx),$$

  and formula (14.19) holds with

  $$\begin{aligned}A_t &= \mu_0 \int_0^t u_s ds + \int_0^t \int_{\mathbb{R}} u_s y J(ds, dy) \\ M_t &= \sigma \int_0^t u_s dW_s,\end{aligned}$$

  where, as in (13.54),

  $$\mu_0 = \mu_R - \int_{|x|\leq R} x \nu(dx).$$

  In this case, the contribution of the jump part of S is completely included in the BV part of the integral. For instance, this is the case of a compound Poisson process or a α-stable process with $\alpha \in]0, 1[$;

- if S is integrable, that is

$$\int_{|y|\geq 1} |y|\nu(dy) < \infty, \tag{14.24}$$

then formula (14.19) holds with

$$A_t = \mu_\infty \int_0^t u_s ds$$
$$M_t = \sigma \int_0^t u_s dW_s + \int_0^t \int_{\mathbb{R}} u_s y \widetilde{J}(ds, dy),$$

where $\mu_\infty = E[S_1]$. We remark explicitly that the second integral in the expression of M is well defined because it can be rewritten as the sum $I_1 + I_2 + I_3$ where

$$I_1 = \int_0^t \int_{|y|<1} u_s y \widetilde{J}(ds, dy)$$

is well defined by (14.23),

$$I_2 = \int_0^t \int_{|y|\geq 1} u_s y J(ds, dy) = \sum_{0<s\leq t} u_s \Delta S_s \mathbb{1}_{\{|\Delta S_s|\geq 1\}}$$

is a sum with a finite number of terms a.s. and

$$I_3 = -\int_0^t \int_{|y|\geq 1} u_s y \nu(dy) ds$$

is finite a.s. by the integrability assumption (14.24).

14.2 Stochastic differentials

14.2.1 Itô formula for discontinuous functions

In Section 3.4.2 we defined the Riemann-Stieltjes integral

$$\int_0^t u_s dX_s, \tag{14.25}$$

for $u \in C([0,t])$ and $s \mapsto X_s$ deterministic function with bounded variation. We also proved the standard Itô formula for continuous BV functions (cf. Theorem 3.70): if $f \in C^1(\mathbb{R})$ and $X \in \mathrm{BV} \cap C([0,t])$ then

$$f(X_t) - f(X_0) = \int_0^t f'(X_s) dX_s. \tag{14.26}$$

The main result of this section is the following generalization of formula (14.26) to discontinuous BV functions.

Theorem 14.26 (Deterministic Itô formula) *Let $f \in C^1(\mathbb{R})$ and X be a càdlàg (deterministic) function with bounded variation. Then we have*

$$\begin{aligned} f(X_t) - f(X_0) = & \int_0^t f'(X_{s-})dX_s \\ & + \sum_{0<s\le t} \left(f(X_s) - f(X_{s-}) - f'(X_{s-})\Delta X_s\right). \end{aligned} \tag{14.27}$$

Proof. We show that the series in (14.27) is convergent: first of all, it has a countable number of terms because $X \in \mathrm{BV}$. For the same reason, $\|X\|_\infty = \sup_{t\in[0,T]} |X_t| < \infty$ and therefore we have

$$\begin{aligned} \sum_{0<s\le t} |f(X_s) - f(X_{s-}) - f'(X_{s-})\Delta X_s| & \le 2 \sup_{|y|\le\|X\|_\infty} |f'(y)| \sum_{0<s\le t} |\Delta X_s| \\ & \le 2 \sup_{|y|\le\|X\|_\infty} |f'(y)|\, V_{[0,t]}(X) < \infty. \end{aligned} \tag{14.28}$$

Next we consider the case when X has only a discontinuity at the point t: hence, if we set $\hat{X}_s = X_{s-}$ and $\hat{X}_0 = X_0$, then we have $\hat{X}_s = X_s$ for $s < t$ and $\hat{X}_t = X_t - \Delta X_t$ with $\Delta X_t = X_t - X_{t-} \neq 0$. Since $\hat{X} \in \mathrm{BV} \cap C([0,t])$, by the standard Itô formula (14.26), we have

$$f(\hat{X}_t) - f(\hat{X}_0) = \int_0^t f'(\hat{X}_s)d\hat{X}_s =$$

(as in Example 14.1)

$$\begin{aligned} & = \int_0^t f'(\hat{X}_s)dX_s - f'(\hat{X}_t)\Delta X_t \\ & = \int_0^t f'(X_{s-})dX_s - f'(X_{t-})\Delta X_t, \end{aligned}$$

and therefore we have

$$f(X_t) - f(X_0) = \int_0^t f'(X_{s-})dX_s + f(X_t) - f(X_{t-}) - f'(X_{t-})\Delta X_t.$$

Now the general case can be proved by using a limit argument combined with estimate (14.28). □

14.2.2 Quadratic variation

In this section we generalize the notion of quadratic variation process given in Section 4.3.3. The quadratic variation plays a crucial role in the Itô formula for semimartingales which will be presented in Section 14.2.3.

Definition 14.27 *The quadratic variation process of a semimartingale* $S = (S_t)_{t\in[0,T]}$ *is defined by*

$$\langle S\rangle_t = S_t^2 - 2\int_0^t S_{s-}dS_s, \qquad t\in[0,T]. \tag{14.29}$$

The next theorem gives some elementary properties of $\langle S\rangle$ and shows that Definition 14.27 extends the classical notion of quadratic variation of deterministic functions given in Definition 3.72.

Theorem 14.28 *If* S *is a semimartingale, then* $\langle S\rangle$ *is a càdlàg, adapted and increasing process such that* $\langle S\rangle_0 = S_0^2$ *and*[6]

$$\Delta\langle S\rangle_t = (\Delta S_t)^2, \qquad t\in[0,T]. \tag{14.30}$$

Moreover, if $\varsigma_n = \left(T_1^n,\dots,T_{N_n}^n\right)$ *is a sequence of random partitions of the interval* $[0,T]$ *verifying (14.11), then we have*

$$\langle S\rangle_t = S_0^2 + \lim_{n\to\infty}\sum_{k=1}^{N_n}\left(S^t_{T_k^n} - S^t_{T_{k-1}^n}\right)^2 \tag{14.31}$$

uniformly in probability (cf. Definition 14.13).

Proof. We use repeatedly the elementary equality

$$(b-a)^2 = b^2 - a^2 - 2a(b-a), \qquad a,b\in\mathbb{R}. \tag{14.32}$$

It is clear that, by definition of stochastic integral, $\langle S\rangle \in \mathbb{D}$. Moreover, by (14.32) we have

$$(\Delta S_t)^2 = (S_t - S_{t-})^2 = S_t^2 - S_{t-}^2 - 2S_{t-}(S_t - S_{t-}) = \Delta S_t^2 - 2S_{t-}\Delta S_t,$$

and (14.30) follows from the fact that

$$\Delta\int_0^t S_{s-}dS_s = S_{t-}\Delta S_t.$$

Again by (14.32), we have

$$\begin{aligned}\sum_{k=1}^{N_n}\left(S_{T_k^n\wedge t} - S_{T_{k-1}^n\wedge t}\right)^2 =& \sum_{k=1}^{N_n}\left(S^2_{T_k^n\wedge t} - S^2_{T_{k-1}^n\wedge t}\right)\\ &- 2\sum_{k=1}^{N_n} S_{T_{k-1}^n\wedge t}\left(S_{T_k^n\wedge t} - S_{T_{k-1}^n\wedge t}\right)\\ =& S_t^2 - S_0^2 - 2\sum_{k=1}^{N_n} S_{T_{k-1}^n}\left(S_{T_k^n\wedge t} - S_{T_{k-1}^n\wedge t}\right)\end{aligned}$$

[6] As usual, for any càdlàg process X, we set $\Delta X_t = X_t - X_{t-}$.

and (14.31) follows from Theorem 14.15. Finally, $\langle S\rangle$ is an increasing process because by (14.31) it is the limit of increasing processes: in particular, note that $\langle S\rangle$ *is a semimartingale with bounded variation* (cf. Example 14.10). □

Remark 14.29 Let S be a semimartingale. Since $\langle S\rangle$ is an increasing process, it has at most a countable number of positive jumps and, by (14.30), we have

$$\sum_{0<s\le t} (\Delta S_s)^2 = \sum_{0<s\le t} \Delta\langle S\rangle_s \le \langle S\rangle_t \tag{14.33}$$

where the last inequality follows from the fact that $\langle S\rangle$ is increasing. The process

$$\langle S\rangle_t^c := \langle S\rangle_t - S_0^2 - \sum_{0<s\le t} (\Delta S_s)^2$$

is called *continuous part of* $\langle S\rangle$. Note that $\langle S\rangle_0^c = 0$. Moreover, by (14.30) the process $\langle S\rangle$ is continuous, i.e. $\langle S\rangle = \langle S\rangle^c$, if and only if S is continuous. □

Example 14.30 By (14.31) and Theorem 3.74, for a real Brownian motion W we have

$$\langle W\rangle_t = \langle W\rangle_t^c = t.$$

By (14.31) and Proposition 4.24, the quadratic variation of a Brownian integral

$$X_t = \int_0^t u_s dW_s,$$

with $u \in \mathbb{L}^2_{\mathrm{loc}}$, is given by

$$\langle X\rangle_t = \langle X\rangle_t^c = \int_0^t u_s^2 ds.$$ □

Example 14.31 If N is a Poisson process, then by definition of quadratic variation and Example 14.8, we have

$$\langle N\rangle_t = N_t.$$

Moreover

$$\langle N\rangle_t^c = N_t - \sum_{0<s\le t} (\Delta N_s)^2 = 0.$$ □

Example 14.32 If S is a *continuous* semimartingale with bounded variation then $\langle S\rangle_t = S_0^2$, $t \in [0,T]$. The proof is the same as in the deterministic case (cf. Proposition 3.73) and uses the characterization (14.31) of the quadratic variation process. In particular the quadratic variation of $S_t = \mu t$, $\mu \in \mathbb{R}$, is $\langle S\rangle_t = 0$.

More generally, if S is càdlàg, adapted and has bounded variation then $\langle S\rangle_t^c = 0$ (see Theorem II-26 in Protter [287]), that is

$$\langle S\rangle_t = S_0^2 + \sum_{0<s\le t} (\Delta S_s)^2 .$$ □

Example 14.33 If S is a one-dimensional Lévy process with characteristic triplet (μ_1, σ, ν) then

$$\langle S\rangle_t = \sigma^2 t + \sum_{0<s\leq t} (\Delta S_s)^2 = \sigma^2 t + \int_{\mathbb{R}} x^2 J(ds, dx), \tag{14.34}$$

where J_t denotes the jump measure of S (cf. (13.59)). Moreover

$$\langle S\rangle_t^c = \sigma^2 t. \tag{14.35}$$

As already remarked in Section 13.3.3, the quadratic variation of a Lévy process is always well-defined, even if the variance may be infinite. □

14.2.3 Itô formula for semimartingales

Theorem 14.34 (Itô formula) *Let $f = f(t,x) \in C^{1,2}(\mathbb{R}^2)$ and let X be a semimartingale. Then $f(t, X_t)$ is a semimartingale and we have*

$$\begin{aligned} f(t, X_t) - f(0, X_0) = & \int_0^t \partial_s f(s, X_{s-})\, ds + \int_0^t \partial_x f(s, X_{s-})\, dX_s \\ & + \frac{1}{2}\int_0^t \partial_{xx} f(s, X_{s-})\, d\langle X\rangle_s^c \\ & + \sum_{0<s\leq t} \left(f(s, X_s) - f(s, X_{s-}) - \partial_x f(s, X_{s-})\, \Delta X_s \right) \end{aligned} \tag{14.36}$$

or in differential form

$$\begin{aligned} df(t, X_t) = & \partial_t f(t, X_{t-})\, dt + \partial_x f(t, X_{t-})\, dX_t + \frac{1}{2}\partial_{xx} f(t, X_{t-})\, d\langle X\rangle_t^c \\ & + \left(f(t, X_t) - f(t, X_{t-}) - \partial_x f(t, X_{t-})\, \Delta X_t \right). \end{aligned} \tag{14.37}$$

Sketch of the proof. We only consider the case $f = f(x)$. We first remark that

$$\int_0^t f''(X_{s-})\, d\langle X\rangle_s = \int_0^t f''(X_{s-})\, d\langle X\rangle_s^c + \sum_{0<s\leq t} f''(X_{s-})\, (\Delta X_s)^2$$

where the last series is convergent by (14.33). Therefore the Itô formula can be rewritten as follows

$$\begin{aligned} f(X_t) - f(X_0) = & \int_0^t f'(X_{s-})\, dX_s + \frac{1}{2}\int_0^t f''(X_{s-})\, d\langle X\rangle_s \\ & + \sum_{0<s\leq t} \left(f(X_s) - f(X_{s-}) - f'(X_{s-})\, \Delta X_s - \frac{1}{2} f''(X_{s-})\, (\Delta X_s)^2 \right). \end{aligned} \tag{14.38}$$

In this form it can be obtained by writing $f(X_t)$ as the sum of its increments and then using a second order Taylor expansion to get a Riemann-Stieltjes sum. With respect to the proof in the continuous case (cf. Theorem 5.5), we have to control the influence of jumps, especially when infinitely many of them occur. To this end, given $\varepsilon > 0$, we split the set of jump times of X in $[0,t]$ as the disjoint union $A_\varepsilon \cup B_\varepsilon$ where A_ε is *finite* and

$$\sum_{s\in B_\varepsilon} (\varDelta X_s)^2 \le \varepsilon.$$

This is possible because of (14.33). Next, we consider a sequence of random partitions $\varsigma_n = (T_1^n, \dots, T_{N_n}^n)$ of the interval $[0,t]$, such that $|\varsigma_n| \to 0$ as $n \to \infty$. We have

$$\begin{aligned} f(X_t) - f(X_0) &= \sum_{k=1}^{N_n} \left(f(X_{T_k^n}) - f(X_{T_{k-1}^n}) \right) \\ &= \sum_{k=1}^{N_n} f'(X_{T_{k-1}^n})(X_{T_k^n} - X_{T_{k-1}^n}) + \frac{1}{2}\sum_{k=1}^{N_n} f''(X_{T_{k-1}^n})(X_{T_k^n} - X_{T_{k-1}^n})^2 \\ &+ \sum_{A_e \cap]T_{k-1}^n, T_k^n] \ne \emptyset} \left(f(X_{T_k^n}) - f(X_{T_{k-1}^n}) - f'(X_{T_{k-1}^n})(X_{T_k^n} - X_{T_{k-1}^n}) \right. \\ &\left. - \frac{1}{2} f''(X_{T_{k-1}^n})(X_{T_k^n} - X_{T_{k-1}^n})^2 \right) + R_t(n,\varepsilon), \end{aligned}$$

where $R_t(n,\varepsilon)$ is a remainder term. By Theorem 14.15, the first two sums on the right hand side above converge to the following expressions as $n \to \infty$:

$$\sum_{k=1}^{N_n} f'(X_{T_{k-1}^n})(X_{T_k^n} - X_{T_{k-1}^n}) \longrightarrow \int_0^t f'(X_{s-})dX_s$$

$$\frac{1}{2}\sum_{k=1}^{N_n} f''(X_{T_{k-1}^n})(X_{T_k^n} - X_{T_{k-1}^n})^2 \longrightarrow \frac{1}{2}\int_0^t f''(X_{s-})d\langle X\rangle_s.$$

The third sum converges to

$$\sum_{s\in A_\varepsilon} \left(f(X_s) - f(X_{s-}) - f'(X_{s-})\varDelta X_s - \frac{1}{2} f''(X_{s-})(\varDelta X_s)^2 \right).$$

Finally, the proof can be completed by showing that

$$\limsup_{n\to\infty} R_t(n,\varepsilon) \le r(\varepsilon)\langle X\rangle_t$$

where $r(\varepsilon) \to 0$ as $\varepsilon \to 0^+$. For further details we refer, for instance, to Protter [287], Theorem II-32. □

Example 14.35 We apply the Itô formula with $f(x) = x^2$ and $X_t = N_t$, Poisson process. We get

$$\begin{aligned} N_t^2 &= 2\int_0^t N_{s-} dN_s + \sum_{0<s\le t} \left(N_s^2 - N_{s-}^2 - 2N_{s-}\Delta N_s\right) \\ &= 2\int_0^t N_{s-} dN_s + \sum_{k=1}^{N_t} \left(k^2 - (k-1)^2 - 2(k-1)\right) \\ &= 2\int_0^t N_{s-} dN_s + N_t, \end{aligned}$$

and therefore

$$\int_0^t N_{s-} dN_s = \frac{N_t\,(N_t - 1)}{2}$$

as we already found in Example 14.8 by using directly the definition of stochastic integral of simple processes. □

14.2.4 Itô formula for Lévy processes

Generally speaking, one major reason why the Itô formula is so important in finance is that it allows decomposing any smooth function of a semimartingale (typically representing the price of an asset) in its drift and martingale parts. In this section, by means of the representation of Lévy-type stochastic integrals in Proposition 14.24, we obtain an alternative version of the general Itô formula (14.36) for Lévy processes: since there are several variants of this formula, we try to put more emphasis on the version that is suitable to be used more easily in financial applications.

Lemma 14.36 *Let X be a one dimensional Lévy process with R-triplet (μ_R, σ^2, ν) and $f = f(t,x) \in C^{1,2}([0,T] \times \mathbb{R})$. Then we have*

$$\begin{aligned} df(t,X_t) =& (\mathcal{A}_R + \partial_t) f(t, X_{t-})\,dt + \sigma \partial_x f(t, X_{t-})\,dW_t \\ &+ \int_{|y|<R} \left(f(t, X_{t-} + y) - f(t, X_{t-})\right) \widetilde{J}(dt, dy) \\ &+ \int_{|y|\ge R} \left(f(t, X_{t-} + y) - f(t, X_{t-})\right) J(dt, dy) \end{aligned} \tag{14.39}$$

where

$$\begin{aligned} \mathcal{A}_R f(t,x) =& \mu_R \partial_x f(t,x) + \frac{\sigma^2}{2} \partial_{xx} f(t,x) \\ &+ \int_{|y|<R} \left(f(t, x+y) - f(t,x) - y\partial_x f(t,x)\right) \nu(dy). \end{aligned} \tag{14.40}$$

Proof. First of all we show that each term in (14.39) and (14.40) is well defined. Let us recall that (cf. (13.46))

$$\int_{|y|<1} y^2 \nu(dy) < \infty, \tag{14.41}$$

and let us denote by

$$D(z,R) =]z - R, z + R[$$

the interval centered at z, with radius R. Then the first integral in (14.39) is well defined because the stochastic function

$$\varphi(t,y) := (f(t, y + X_{t-}) - f(t, X_{t-})) \mathbb{1}_{D(0,R)}(y)$$

belongs to $\mathbb{L}^2_{\nu,\mathrm{loc}}$ (cf. Definition 14.18). Indeed we have

$$|\varphi(t,y)| \le |y| \sup_{t\in[0,T]} \sup_{z\in D(X_{t-},R)} |\partial_x f(t,z)|$$

and the claim follows from (14.41). On the other hand, the last integral in (14.39) is equal to a sum with a finite number of terms corresponding to large jumps. Further, we have

$$|f(t, x + y) - f(t,x) - y\partial_x f(t,x)| \mathbb{1}_{\{|y|<R\}} \le y^2 \sup_{[0,T]\times D(x,R)} |\partial_{xx} f|,$$

and therefore, again by (14.41), the last integral in (14.40) is convergent for any $x \in \mathbb{R}$.

Next, by the general Itô formula (14.36), using the expression (14.35) of the continuous part of the quadratic variation of a Lévy process, formula (14.17) and the identity $X_t = X_{t-} + \Delta X_t$, we have

$$\begin{aligned} df(t, X_t) = &\left(\partial_t f + \frac{\sigma^2}{2}\partial_{xx} f\right)(t, X_{t-})\, dt + \partial_x f\,(t, X_{t-})\, dX_t \\ &+ \int_{\mathbb{R}} (f(t, y + X_{t-}) - f(t, X_{t-}) - y\partial_x f(t, X_{t-}))\, J(dt, dy). \end{aligned} \tag{14.42}$$

Now, by Proposition 14.24 (see also (14.22)), we have

$$\begin{aligned} \partial_x f\,(t, X_{t-})\, dX_t = &\mu_R \partial_x f\,(t, X_{t-})\, dt + \partial_x f\,(t, X_{t-})\, \sigma dW_t \\ &+ \int_{|y|<R} \partial_x f\,(t, X_{t-})\, y \widetilde{J}(dt, dy) \\ &+ \int_{|y|\ge R} \partial_x f\,(t, X_{t-})\, y J(dt, dy). \end{aligned} \tag{14.43}$$

Plugging (14.43) into (14.42) and recalling that

$$\widetilde{J}(dt, dy) = J(dt, dy) - \nu(dy)dt,$$

after some cancellation, we get (14.39). □

Under some additional assumption, we can easily obtain an Itô formula where the drift and martingale parts are detached and have an explicit representation.

Theorem 14.37 (Itô formula) *Let X be a one dimensional Lévy process with R-triplet (μ_R, σ^2, ν) and $f = f(t,x) \in C^{1,2}([0,T] \times \mathbb{R})$. If f is bounded then we have*

$$\begin{aligned} df(t,X_t) =& (\mathcal{A}_R + \partial_t) f(t, X_{t-})\, dt + \sigma \partial_x f(t, X_{t-})\, dW_t \\ &+ \int_{\mathbb{R}} (f(t, y + X_{t-}) - f(t, X_{t-}))\, \widetilde{J}(dt, dy) \end{aligned} \tag{14.44}$$

where

$$\begin{aligned} \mathcal{A}_R g(x) =& \mu_R \partial_x g(x) + \frac{\sigma^2}{2} \partial_{xx} g(x) \\ &+ \int_{\mathbb{R}} \left(g(x+y) - g(x) - \partial_x g(x) y \mathbb{1}_{\{|y|<R\}} \right) \nu(dy). \end{aligned} \tag{14.45}$$

The integro-differential operator $\mathcal{A}_R$ in (14.45) *is called the R-characteristic operator of the Lévy process X.*

Proof. Formulas (14.44)-(14.45) follow directly by adding to (14.40), and subtracting from (14.39), the integral term

$$\int_{|y| \geq R} (f(t, y + X_{t-}) - f(t, X_{t-}))\, \nu(dy) dt$$

that is well defined because f is bounded and

$$\int_{|y| \geq R} \nu(dy) < \infty$$

for any $R > 0$, by the general properties of Lévy measures. □

Remark 14.38 The boundedness assumptions in Theorem 14.37 can be weakened and the Itô formula (14.44) holds true under more general hypotheses on f that guarantee that the stochastic function

$$\varphi(t,y) = f(t, y + X_{t-}) - f(t, X_{t-})$$

belongs to $\mathbb{L}^2_{\nu,\text{loc}}$. □

Remark 14.39 Assume that f and $\partial_x f$ are bounded functions: then the term

$$\int_0^t \sigma \partial_x f(s, X_{s-})\, dW_s$$

in (14.44) is a martingale because $\partial_x f(t, X_{t-}) \in \mathbb{L}^2$. Moreover, we have

$$|f(t, X_{t-} + y) - f(t, X_{t-})| \leq \begin{cases} \|\partial_x f\|_\infty |y| & \text{for } |y| \leq 1, \\ 2\|f\|_\infty & \text{for } |y| > 1, \end{cases} \tag{14.46}$$

so that $f(t, X_{t-} + y) - f(t, X_{t-}) \in \mathbb{L}^2_\nu$ and also the term

$$\int_0^t \int_{\mathbb{R}} (f(s, y + X_{s-}) - f(s, X_{s-}))\, \widetilde{J}(ds, dy)$$

is a martingale. □

Example 14.40 (Lévy exponential) Let X be a one dimensional Lévy process with R-triplet (μ_R, σ^2, ν) and characteristic exponent ψ_X. We assume that

$$\int_{|y|\geq 1} e^{2y}\nu(dy) < \infty. \tag{14.47}$$

By Proposition 13.49, condition (14.47) is equivalent to the existence of the second moment of X:

$$E\left[e^{2X_t}\right] = e^{t\psi_X(-2i)}. \tag{14.48}$$

We consider the function $f(t,x) = e^x$ that typically arises in exponential Lévy models. If $\mathcal{A}_R$ is the differential operator in (14.45), we have

$$\begin{aligned}(\mathcal{A}_R + \partial_t) f(t,x) &= \left(\mu_R + \frac{\sigma^2}{2} + \int_{\mathbb{R}} \left(e^y - 1 - y\mathbb{1}_{\{|y|<R\}}\right) \nu(dy)\right) f(t,x) \\ &= \psi_X(-i) f(t,x).\end{aligned}$$

Therefore, if $S_t := e^{X_t}$, by the Itô formula (14.44) we get

$$dS_t = \psi_X(-i) S_{t-}dt + \sigma S_{t-}dW_t + S_{t-}\int_{\mathbb{R}} (e^y - 1)\, \widetilde{J}(dt, dy). \tag{14.49}$$

We remark explicitly that $S \in \mathbb{L}^2$ because

$$E\left[\int_0^T S_t^2 dt\right] = \int_0^T E\left[e^{2X_t}\right] dt = \int_0^T e^{t\psi_X(-2i)} dt$$

which is finite by (14.47)-(14.48): in particular, the Brownian integral in (14.49) is a square-integrable martingale. On the other hand, the stochastic function

$$\varphi(t,y) := (e^y - 1)S_{t-}$$

belongs to $\mathbb{L}^2_\nu$; indeed we have

$$E\left[\int_0^T \varphi^2(t,y)\nu(dy)dt\right] = E\left[\int_0^T S_t^2 dt\right] \int_{\mathbb{R}} (e^y - 1)^2\, \nu(dy)$$

that is finite because $(e^y - 1)^2$ behaves like y^2 close to the origin and like e^{2y} for positive large values of y: thus, by (14.47), $(e^y - 1)^2$ is ν-integrable. By Proposition 14.19 we conclude that also the last term in (14.49) is a square-integrable martingale. □

Under the assumptions of Theorem 14.37, we consider two remarkable cases:

- if the jump part of X has bounded variation, that is if

$$\int_{|y|\le 1} |y|\nu(dy) < \infty, \tag{14.50}$$

then we can let R go to infinity in (14.45): indeed, for any $x \in \mathbb{R}$, the function

$$F_{R,t,x}(y) := f(t, x+y) - f(t,x) - y\partial_x f(t,x)\mathbb{1}_{D(0,R)}(y) \tag{14.51}$$

can be dominated, uniformly in $t \in [0,T]$ and $R \in]0,1[$, by a ν-integrable function as follows:

$$|F_{R,t,x}(y)| \le \begin{cases} 2|y| \sup\limits_{[0,T]\times D(x,1)} |\partial_x f| & \text{for } |y| \le 1, \\ 2\sup|f| & \text{for } |y| > 1. \end{cases}$$

Hence the Itô formula (14.44) holds true with $R = 0$, that is with the following simplified version of the characteristic operator:

$$\mathcal{A}_0 g(x) = \mu_0 \partial_x g(x) + \frac{\sigma^2}{2}\partial_{xx} g(x) + \int_{\mathbb{R}} (g(x+y) - g(x))\,\nu(dy), \tag{14.52}$$

where, as in (13.54),

$$\mu_0 = \mu_1 - \int_{|x|\le 1} x\nu(dx).$$

In particular, if X has bounded variation so that $\sigma = 0$ (cf. Proposition 13.43) then $\mathcal{A}_0$ reduces to a first order integro-differential operator;

- if X is integrable, that is

$$\int_{|y|\ge 1} |y|\nu(dy) < \infty, \tag{14.53}$$

then we can let R go to infinity in (14.45): indeed, for any $x \in \mathbb{R}$, the function $F_{R,t,x}(y)$ in (14.51) can be dominated, uniformly in $t \in [0,T]$ and $R \ge 1$, by a ν-integrable function as follows:

$$|F_{R,t,x}(y)| \le \begin{cases} y^2 \sup\limits_{[0,T]\times D(x,1)} |\partial_{xx} f| & \text{for } |y| \le 1, \\ 2\sup|f| + |y| \sup\limits_{t\in[0,T]} |\partial_x f(t,x)| & \text{for } |y| > 1. \end{cases}$$

Hence the Itô formula (14.44) holds true with $R = \infty$, that is with the following simplified version of the characteristic operator:

$$\begin{aligned} \mathcal{A}_\infty g(x) =& \mu_\infty \partial_x g(x) + \frac{\sigma^2}{2}\partial_{xx} g(x) \\ &+ \int_{\mathbb{R}} (g(x+y) - g(x) - y\partial_x g(x))\,\nu(dy), \end{aligned} \tag{14.54}$$

where $\mu_\infty = E[X_1]$.

14.2.5 SDEs with jumps and Itô formula

In this section we turn to the study of stochastic differential equations with jumps: our presentation follows Applebaum [11], Section 6.2, to which we refer for a complete treatment of the theory.

Equation (14.49) is the simplest and non trivial example of SDE with jumps. Here we consider a Lévy process with Lévy measure ν defined on a filtered probability space $(\Omega, \mathcal{F}, P, (\mathcal{F}_t))$ that satisfies the usual hypotheses: we also denote by W and J the related independent Brownian motion and jump measure respectively. Moreover, we consider $Z \in \mathbb{R}$ and some locally bounded and measurable mappings

$$\begin{aligned}
&\bar{b} = \bar{b}(t,x) : [0,T] \times \mathbb{R} \longrightarrow \mathbb{R}, && \text{(drift coefficient)}\\
&\sigma = \sigma(t,x) : [0,T] \times \mathbb{R} \longrightarrow \mathbb{R}, && \text{(diffusion coefficient)}\\
&\widetilde{a} = \widetilde{a}(t,x,y) : [0,T] \times \mathbb{R} \times \{|y| < 1\} \longrightarrow \mathbb{R}, && \text{(small jumps coefficient)}\\
&\bar{a} = \bar{a}(t,x,y) : [0,T] \times \mathbb{R} \times \{|y| \geq 1\} \longrightarrow \mathbb{R}, && \text{(large jumps coefficient).}
\end{aligned}$$

Definition 14.41 *A solution relative to W and J of the SDE with coefficients $Z, \bar{b}, \sigma, \widetilde{a}, \bar{a}$ is a process $(X_t)_{t\in[0,T]} \in \mathbb{D}$ such that*

i) $\sigma(t, X_{t-}) \in \mathbb{L}^2_{\text{loc}}$;
ii) $\widetilde{a}(t, X_{t-}, y)\mathbb{1}_{\{|y|<1\}} \in \mathbb{L}^2_{\nu,\text{loc}}$;
iii) we have that

$$\begin{aligned}
X_t = Z &+ \int_0^t \bar{b}(s, X_{s-})ds + \int_0^t \sigma(s, X_{s-})dW_s\\
&+ \int_0^t \int_{|y|<1} \widetilde{a}(s, X_{s-}, y)\widetilde{J}(ds, dy)\\
&+ \int_0^t \int_{|y|\geq 1} \bar{a}(s, X_{s-}, y)J(ds, dy), \qquad t \in [0,T].
\end{aligned}$$

In differential form, we write

$$\begin{aligned}
dX_t =& \bar{b}(t, X_{t-})dt + \sigma(t, X_{t-})dW_t\\
&+ \int_{|y|<1} \widetilde{a}(t, X_{t-}, y)\widetilde{J}(dt, dy)\\
&+ \int_{|y|\geq 1} \bar{a}(t, X_{t-}, y)J(dt, dy), \qquad X_0 = Z.
\end{aligned} \tag{14.55}$$

Example 14.42 By the Lévy-Itô decomposition, a Lévy process X with triplet (μ_1, σ^2, ν) can be written as

$$dX_t = \mu_1 dt + \sigma dW_t + \int_{|y|<1} y\widetilde{J}(dt, dy) + \int_{|y|\geq 1} yJ(dt, dy)$$

and therefore it can be considered as a solution of a SDE of type (14.55) where

$$\bar{b}(t,x) = \mu_1, \quad \sigma(t,x) = \sigma, \quad \widetilde{a}(t,x,y) = \bar{a}(t,x,y) = y. \qquad \square$$

Many results for SDEs based on Brownian motion can be extended to the case of SDEs with jumps. Concerning *strong* solutions (i.e. solutions of SDEs where W and J are specified in advance) we have the following classical existence and uniqueness result that can be proved as in Section 9.1 by Gronwall's inequality and a fixed-point technique.

Theorem 14.43 *Assume the following Lipschitz and growth conditions:*

i) for every $n \in \mathbb{N}$ there exists a constant K_n such that

$$\left|\bar{b}(t,x_1) - \bar{b}(t,x_2)\right|^2 + |\sigma(t,x_1) - \sigma(t,x_2)|^2 + \int_{|y|<1} |\widetilde{a}(t,x_1,y) - \widetilde{a}(t,x_2,y)|^2 \nu(dy) \le K_n |x_1 - x_2|^2,$$

for $|x_1|, |x_2| \le n$, $t \in [0,T]$;

ii) there exists a positive constant K such that

$$\bar{b}^2(t,x) + \sigma^2(t,x) + \int_{|y|<1} \widetilde{a}^2(t,x,y)\nu(dy) \le K(1+|x|^2) \qquad x \in \mathbb{R},\ t \in [0,T];$$

iii) the function $(t,x) \mapsto \bar{a}(t,x,y)$ is continuous for any y, $|y| \ge 1$.

Then there exists a pathwise unique solution X to (14.55).

The following Itô formula extends that of Lemma 14.36.

Lemma 14.44 *Under the hypotheses of Theorem 14.43, let X be a strong solution of the SDE* (14.55) *and $f = f(t,x) \in C^{1,2}([0,T] \times \mathbb{R})$. Then we have*

$$\begin{aligned} df(t,X_t) = &(\mathcal{A} + \partial_t) f(t,X_{t-})dt + \sigma(t,X_{t-})\partial_x f(t,X_{t-})\, dW_t \\ &+ \int_{|y|<1} \left(f(t, \widetilde{a}(t,X_{t-},y) + X_{t-}) - f(t,X_{t-})\right) \widetilde{J}(dt,dy) \\ &+ \int_{|y|\ge 1} \left(f(t, \bar{a}(t,X_{t-},y) + X_{t-}) - f(t,X_{t-})\right) J(dt,dy) \end{aligned} \tag{14.56}$$

where $\mathcal{A}$ is the integro-differential operator with variable coefficients

$$\begin{aligned} \mathcal{A}f(t,x) = &\bar{b}(t,x)\partial_x f(t,x) + \frac{\sigma^2(t,x)}{2}\partial_{xx} f(t,x) \\ &+ \int_{|y|<1} \left(f(t,x+\widetilde{a}(t,x,y)) - f(t,x) - \widetilde{a}(t,x,y)\partial_x f(t,x)\right) \nu(dy). \end{aligned} \tag{14.57}$$

As we already mentioned, in the financial applications it is important to separate the drift from the martingale part of the solution of a SDE with jumps. Under suitable integrability conditions[7], this can be easily done by

[7] Which guarantee that the integral

$$\int_{|y|\ge 1} \bar{a}(t,x,y)\nu(dy)$$

converges: for instance, see condition (14.47) in the case of an exponential Lévy process.

rewriting the SDE (14.55) in the more convenient form

$$dX_t = b(t, X_{t-})dt + \sigma(t, X_{t-})dW_t + \int_{\mathbb{R}} a(t, X_{t-}, y)\widetilde{J}(dt, dy), \qquad (14.58)$$

with initial datum $X_0 = Z$, where

$$b(t,x) = \bar{b}(t,x) + \int_{|y|\geq 1} \bar{a}(t,x,y)\nu(dy),$$
$$a(t,x,y) = \widetilde{a}(t,x,y)\mathbb{1}_{\{|y|<1\}} + \bar{a}(t,x,y)\mathbb{1}_{\{|y|\geq 1\}}.$$

For an SDE of the form (14.58), Theorem 14.43 can be reformulated as follows.

Theorem 14.45 *Assume the following Lipschitz and growth conditions:*

i) for every $n \in \mathbb{N}$ there exists a constant K_n such that

$$\begin{aligned}&|b(t,x_1) - b(t,x_2)|^2 + |\sigma(t,x_1) - \sigma(t,x_2)|^2 \\ &+ \int_{\mathbb{R}} |a(t,x_1,y) - a(t,x_2,y)|^2 \nu(dy) \leq K_n|x_1 - x_2|^2,\end{aligned} \qquad (14.59)$$

for $|x_1|, |x_2| \leq n$, $t \in [0,T]$;

ii) there exists a positive constant K such that

$$b^2(t,x) + \sigma^2(t,x) + \int_{\mathbb{R}} a^2(t,x,y)\nu(dy) \leq K(1+|x|^2) \qquad x \in \mathbb{R},\ t \in [0,T]. \qquad (14.60)$$

Then there exists a pathwise unique solution X to (14.58). Moreover, there exists a positive constant C, depending on K, K_n and T only, such that

$$E\left[X_t^2\right] \leq C\left(1 + X_0^2\right), \qquad t \in [0,T]. \qquad (14.61)$$

Example 14.46 The SDE (14.49) for an exponential Lévy process $S_t = e^{X_t}$ is of the form (14.58), that is

$$dS_t = b(t, S_{t-})dt + \sigma(t, S_{t-})dW_t + \int_{\mathbb{R}} a\left(t, S_{t-}, y\right)\widetilde{J}\left(dt, dy\right)$$

where

$$b(t,s) = s\psi_X(-i), \quad \sigma(t,s) = \sigma s, \quad a(t,s,y) = s\left(e^y - 1\right).$$

Under condition (14.47) on the Lévy measure ν, the Lipschitz and growth conditions (14.59)-(14.60) are satisfied. □

Theorem 14.47 (Itô formula for SDEs with jumps) *Under the hypotheses of Theorem 14.45, let X be a solution of the SDE (14.58) and f be a bounded function in $C^{1,2}([0,T]\times\mathbb{R})$. Then we have*

$$df(t,X_t) = (\mathcal{A}+\partial_t)\, f(t,X_{t-})dt + \sigma(t,X_{t-})\partial_x f\,(t,X_{t-})\, dW_t + \int_{\mathbb{R}} \left(f\,(t,X_{t-}+a(t,X_{t-},y)) - f(t,X_{t-})\right) \widetilde{J}(dt,dy) \tag{14.62}$$

where $\mathcal{A}$ is the integro-differential operator with variable coefficients

$$\mathcal{A}f(t,x) = b(t,x)\partial_x f(t,x) + \frac{\sigma^2(t,x)}{2}\partial_{xx} f(t,x) + \int_{\mathbb{R}} \left(f(t,x+a(t,x,y)) - f(t,x) - a(t,x,y)\partial_x f(t,x)\right)\nu(dy). \tag{14.63}$$

Remark 14.48 If f and $\partial_x f$ are bounded functions then the term

$$\int_0^t \sigma(s,X_{s-})\partial_x f\,(s,X_{s-})\, dW_s$$

in (14.62) is a square-integrable martingale: indeed

$$\sigma(t,X_{t-})\partial_x f\,(t,X_{t-}) \in \mathbb{L}^2$$

because, by (14.60) we have

$$|\sigma(t,X_t)\partial_x f\,(t,X_t)|^2 \le \|\partial_x f\|_{L^\infty} K\left(1+X_t^2\right)$$

and the claim follows from estimate (14.61).

On the other hand, consider the stochastic function

$$\varphi(t,y) = f\,(t, a(t,X_{t-},y) + X_{t-}) - f(t,X_{t-}).$$

We have

$$|\varphi(t,y)| \le \begin{cases} \|\partial_x f\|_\infty\, |a(t,X_{t-},y)| & \text{for } |y|<1, \\ 2\|f\|_\infty & \text{for } |y|\ge 1, \end{cases}$$

so that

$$E\left[\int_0^T \int_{\mathbb{R}} \varphi^2(t,y)\nu(dy)dt\right]$$
$$\le 4T\|f\|_\infty^2 + \|\partial_x f\|_\infty^2 E\left[\int_0^T \int_{|y|<1} a^2(t,X_{t-},y)\nu(dy)dt\right] \le$$

(by (14.60))

$$\le 4T\|f\|_\infty^2 + \|\partial_x f\|_\infty^2 K E\left[\int_0^T \left(1+X_{t-}^2\right) dt\right]$$

which is finite by (14.61). Hence $\varphi \in \mathbb{L}^2_\nu$ and also the term

$$\int_0^t \int_{\mathbb{R}} \left(f\left(s, X_{s-} + a(s, X_{s-}, y)\right) - f(s, X_{s-}) \right) \widetilde{J}(ds, dy)$$

is a square-integrable martingale. □

14.2.6 PIDEs and Feynman-Kač representation

As in the diffusive case, the Itô formula establishes a deep connection between SDEs with jumps and partial integro-differential equations (PIDEs). Under the assumptions of the previous section, we consider a SDE with jumps of the form (14.58), that is

$$dX_t = b(t, X_{t-})dt + \sigma(t, X_{t-})dW_t + \int_{\mathbb{R}} a(t, X_{t-}, y)\widetilde{J}(dt, dy),$$

and, for any $t \in [0, T[$ and $x \in \mathbb{R}$, we denote by $(X_s^{t,x})_{s\in[t,T]}$ the corresponding solution with initial condition $X_t^{t,x} = x$. We denote by $\mathcal{A}$ the characteristic operator of X defined in (14.63):

$$\begin{aligned}\mathcal{A}f(t,x) &= b(t,x)\partial_x f(t,x) + \frac{\sigma^2(t,x)}{2}\partial_{xx} f(t,x) \\ &\quad + \int_{\mathbb{R}} \left(f(t, x + a(t,x,y)) - f(t,x) - a(t,x,y)\partial_x f(t,x) \right) \nu(dy).\end{aligned}$$

Notice that $\mathcal{A}$ reduces to an elliptic-parabolic differential operator with variable coefficients when $\nu = 0$.

Definition 14.49 *Let φ and r be bounded continuous functions, $\varphi = \varphi(x) \in C_b(\mathbb{R})$ and $r = r(t,x) \in C_b(\mathbb{R}^2)$. A classical solution of the (backward) Cauchy problem for $\mathcal{A} + \partial_t$ with final datum φ, is a bounded function $f \in C^{1,2}([0,T[\times\mathbb{R}) \cap C([0,T]\times\mathbb{R})$ such that*

$$\begin{cases} \mathcal{A}f(t,x) + \partial_t f(t,x) = r(t,x) f(t,x), & (t,x) \in [0,T[\times\mathbb{R}, \\ f(T,x) = \varphi(x) & x \in \mathbb{R}. \end{cases} \tag{14.64}$$

The following remarkable result is a direct consequence of the Itô formula (14.62).

Theorem 14.50 (Feynman-Kač representation) *If a classical solution f to problem (14.64) exists and is such that $f, \partial_x f \in L^\infty(]0,T[\times\mathbb{R})$ then it has the stochastic representation*

$$f(t,x) = E\left[e^{-\int_t^T r(s, X_s^{t,x})ds} \varphi\left(X_T^{t,x}\right) \right], \qquad t \in [0,T[, \ x \in \mathbb{R}.$$

Proof. For simplicity, we only consider the case $r = 0$. If f is a classical solution to (14.64), then by the Itô formula we have

$$f(T, X_T^{t,x}) - f(t,x) = \int_t^T \sigma(s, X_{s-}^{t,x})\partial_x f\left(s, X_{s-}^{t,x}\right) dW_s$$
$$+ \int_t^T \int_{\mathbb{R}} \left(f\left(s, X_{s-}^{t,x} + a(s, X_{s-}^{t,x}, y)\right) - f(s, X_{s-}^{t,x})\right) \widetilde{J}(ds, dy).$$

By Remark 14.48 and the hypothesis $f, \partial_x f \in L^\infty(]0,T[\times\mathbb{R})$, the right-hand side of the above equation is a square-integrable martingale with null expectation. Thus the thesis follows by taking expectations and using the final condition. □

Theorem 14.50 is a uniqueness result: comparing it with the standard Feynman-Kač representation for diffusions in Theorem 9.45, it is clear that the above result is not optimal and the assumptions on the solution can be refined to determine more explicitly (as in Chapter 6) the uniqueness classes of the PIDE. Existence and uniqueness of classical solutions to PIDEs are discussed in Garroni and Menaldi [148], Bensoussan and Lions [42] for the case $\sigma > 0$ and in Mikulevičius and Pragarauskas [255], [256], Cancelier [65] for the case $\sigma = 0$. Regularity of elliptic ($s > 0$) PIDEs is studied in Garroni and Menaldi [147], [148] and for pure jump processes ($\sigma = 0$) in Eskin [122], Bismut [46], Bichteler and Jacod [45], Cancelier [65]. A Feynman-Kač formula for *backward* stochastic differential equations related to Lévy processes was proved by Nualart and Schoutens [269].

More recently, PIDEs have been studied in the framework of the theory of weak solutions in the viscosity sense: the notion of viscosity solution was generalized to integro-differential equations by Fleming and Soner [132] and Sayah [15] for first order operators and by Alvarez and Tourin [8], Barles Buckdahn and Pardoux [22], Pham [280], Amadori [9] for second order operators.

The Feynman-Kač representation formula allows us to generalize the results of Paragraph 9.4.4 on the transition density of a diffusion. Let us discuss this matter in a heuristic way: by analogy with the classical PDEs theory, we say that the operator $\mathcal{A} + \partial_t$ has a fundamental solution $\Gamma(t,x;T,y)$ if, for every $\varphi \in C_b(\mathbb{R})$, the function

$$f(t,x) = \int_{\mathbb{R}} \varphi(y)\Gamma(t,x;T,y)dy$$

is the classical bounded solution of the Cauchy problem (14.64). If a fundamental solution exists, by the Feynman-Kač formula, we have that

$$E\left[\varphi(X_T^{t,x})\right] = \int_{\mathbb{R}^N} \varphi(y)\Gamma(t,x;T,y)dy.$$

Since φ is arbitrary, we infer that the function

$$y \mapsto \Gamma(t, x; T, y)$$

is the density of the random variable $X_T^{t,x}$, that is Γ is the *transition density* of X. In other words, *if the integro-differential operator $\mathcal{A}+\partial_t$ has a fundamental solution Γ, then Γ is the transition density of X.*

Now let us assume that X is a Lévy process with triplet (μ_1, σ^2, ν) and characteristic operator

$$\begin{aligned} \mathcal{A}g(x) =& \mu_1 \partial_x g(x) + \frac{\sigma^2}{2} \partial_{xx} g(x) \\ &+ \int_{\mathbb{R}} \left(g(x+y) - g(x) - y\partial_x g(x) \mathbb{1}_{\{|y|<1\}} \right) \nu(dy). \end{aligned} \tag{14.65}$$

PIDEs associated to Lévy processes can be considered the natural extension of *constant coefficients* parabolic PDEs, the heat equation being the prototype. Thus we may try to employ the classical procedure, illustrated in Appendix A.3 and based on the Fourier transform, to construct explicitly the fundamental solution. We proceed formally (i.e. without a rigorous justification of the steps) and we apply the Fourier transform operator $\mathcal{F}$, acting only in the spatial variable x, to the Cauchy problem (14.64) for $\mathcal{A}$ as in (14.65). We have

$$\begin{aligned} \mathcal{F}(\mathcal{A}f(t,\cdot))(\xi) &= -i\mu_1 \xi \hat{f}(t,\xi) - \frac{\sigma^2}{2}\xi^2 \hat{f}(t,\xi) \\ &+ \int_{\mathbb{R}} e^{i\xi x} \int_{\mathbb{R}} \left(f(t,x+y) - f(t,x) - y\partial_x f(t,x)\mathbb{1}_{\{|y|<1\}} \right) \nu(dy) dx \\ &= -i\mu_1 \xi \hat{f}(t,\xi) - \frac{\sigma^2}{2}\xi^2 \hat{f}(t,\xi) + \hat{f}(t,\xi) \int_{\mathbb{R}} \left(e^{-iy\xi} - 1 + iy\xi \mathbb{1}_{\{|y|<1\}} \right) \nu(dy) \\ &= \psi_X(-\xi) \hat{f}(t,\xi) \end{aligned}$$

where ψ_X is the characteristic exponent of X appearing in the Lévy-Khintchine representation (cf. (13.60)). Hence, for any $\xi \in \mathbb{R}$, the function $t \mapsto \hat{f}(t,\xi)$ solves the ordinary Cauchy problem with final condition

$$\begin{cases} \partial_t \hat{f}(t,\xi) = -\psi_X(-\xi)\hat{f}(t,\xi) & t \in [0, T[, \\ \hat{f}(T,\xi) = \hat{\varphi}(\xi), \end{cases}$$

and therefore we have

$$\hat{f}(t,\xi) = \hat{\varphi}(\xi) e^{(T-t)\psi_X(-\xi)}.$$

Denoting by $\mathcal{F}^{-1}$ the inverse Fourier transform operator, we get

$$f(t,\cdot) = \mathcal{F}^{-1}\left(\hat{\varphi}(\xi) e^{(T-t)\psi_X(-\xi)} \right) \tag{14.66}$$

$$= \varphi * \mathcal{F}^{-1}\left(e^{(T-t)\psi_X(-\xi)} \right), \tag{14.67}$$

where "$*$" denotes the convolution operation. We conclude that, if the anti-transform in (14.67) is well defined, then the fundamental solution exists and is given by

$$\Gamma(t,x,T,y) = \mathcal{F}^{-1}\left(e^{(T-t)\psi_X(-\xi)}\right)(x-y).$$

Example 14.51 (Cauchy distribution) We consider a α-stable process with $\alpha = 1$: this is a pure jump process with no diffusion component, see Section 13.4.2. By (13.80), the characteristic exponent takes the form

$$\psi(\xi) = i\mu\xi - \sigma|\xi|\left(1 + i\theta\operatorname{sgn}(\xi)\log|\xi|\right),$$

and in the case $\theta = 0$ we get

$$\begin{aligned}\Gamma(t,x,T,y) &= \mathcal{F}^{-1}\left(e^{(T-t)\psi_X(-\xi)}\right)(x-y) \\ &= \frac{(T-t)\sigma}{\pi\left((x-y+(T-t)\mu)^2 + (T-t)^2\sigma^2\right)}\end{aligned}$$

which is the density of the Cauchy distribution. □

Even if the expression of ψ_X is known, in most cases it is not possible to compute explicitly Γ by Fourier inversion. However, formula (14.66) is used[8] in the practical applications because it can be inverted numerically in various efficient ways: numerical methods in option pricing based on Fourier methods are analyzed thoroughly in Chapter 15.

14.2.7 Linear SDEs with jumps

Given a Lévy process Z and $B \in \mathbb{R}$, $B \neq 0$, we consider the one-dimensional SDE with jumps

$$dX_t = -BX_t dt + dZ_t \tag{14.68}$$

which is the non-Gaussian analogue of the linear SDE (9.70). Using the Lévy-Itô decomposition of Z, equation (14.68) can be written in the canonical form (14.55) and the existence of a pathwise unique solution is guaranteed by Theorem 14.43. On the other hand, by the Itô formula it is easy to verify directly that the solution to (14.68) is given explicitly by

$$X_t = e^{-Bt}\left(X_0 + \int_0^t e^{Bs} dZ_s\right). \tag{14.69}$$

The characteristic function of X can be computed explicitly as the following result shows.

[8] Notice that (14.66) is equivalent to the pricing formula (15.7) derived in Chapter 15.

Proposition 14.52 *Let Z be a Lévy process with characteristic exponent ψ. Then the characteristic function of X in (14.69) is equal to*

$$\varphi_{X_t}(\xi) = E\left[e^{i\xi X_t}\right] = \exp\left(i\xi X_0 e^{-Bt} + \int_0^t \psi\left(\xi e^{B(s-t)}\right) ds\right), \qquad \xi \in \mathbb{R}. \tag{14.70}$$

Proof. The thesis is a consequence of the identity

$$E\left[e^{i\int_0^t f(s)dZ_s}\right] = e^{\int_0^t \psi(f(s))ds} \tag{14.71}$$

which holds for any continuous function $f : [0,T] \longrightarrow \mathbb{R}$. Let us prove (14.71): if f is simple, that is

$$f(t) = \sum_{k=1}^{N} f_k \mathbb{1}_{]T_{k-1},T_k]}(t),$$

then we have

$$E\left[e^{i\int_0^t f(s)dZ_s}\right] = \prod_{k=1}^{N} E\left[e^{if_k\left(Z_{T_k}-Z_{T_{k-1}}\right)}\right]$$
$$= \prod_{k=1}^{N} e^{(T_k-T_{k-1})\psi(f_k)} = e^{\int_0^t \psi(f(s))ds}.$$

Then (14.71) follows by approximation. □

Example 14.53 If $Z_t = \mu t + \sigma W_t$ is a Brownian motion with drift and characteristic exponent

$$\psi_Z(\xi) = i\mu\xi - \frac{\sigma^2\xi^2}{2},$$

then, by (14.70), the characteristic function of the solution X in (14.68) is equal to

$$\varphi_{X_t}(\xi) = \exp\left(i\xi\bar{\mu}_t - \frac{\bar{\sigma}_t^2\xi^2}{2}\right),$$

where

$$\bar{\mu}_t = \left(X_0 e^{-Bt} + \frac{\mu\left(1-e^{-Bt}\right)}{B}\right), \qquad \bar{\sigma}_t = \sqrt{\frac{1-e^{-2Bt}}{2B}}.$$

In this case, as we already showed in Section 9.5, X_t is a Gaussian random variable. □

A case of particular interest in finance is when the process Z is a subordinator, i.e. Z is an increasing Lévy process (cf. Section 13.4.4). Indeed, in this case and if $X_0 > 0$, we have that the process X is positive a.s. by definition. Positive solutions of linear SDEs with jumps have been proposed by Barndorff-Nielsen and Shephard [28] as a model for the square volatility process: for more details, we refer to Section 14.3.3.

Example 14.54 If Z is a Gamma subordinator with characteristic exponent

$$\psi_Z(\xi) = -a \log\left(1 - \frac{i\xi}{b}\right),$$

then the characteristic function of X in (14.68) is equal to

$$\varphi_{X_t}(\xi) = \exp\left(ie^{-Bt}X_0\xi + \frac{a}{B}\left(\mathrm{Li}_2\left(\frac{i\xi}{b}\right) - \mathrm{Li}_2\left(\frac{ie^{-Bt}\xi}{b}\right)\right)\right),$$

where $\mathrm{Li}_n(x)$ is the polylogarithm function defined by

$$\mathrm{Li}_n(x) = \sum_{k=1}^{\infty} \frac{x^k}{k^n}.$$

□

14.3 Lévy models with stochastic volatility

14.3.1 Lévy-driven models and pricing PIDEs

In this section, we introduce a generalization of the exponential pricing models studied in Chapter 10. We only consider the one-dimensional case and assume that the price of the underlying asset is of the form

$$S_t = S_0 e^{X_t}, \tag{14.72}$$

where X is the solution of the SDE with jumps

$$dX_t = \bar{b}(t, X_{t-})dt + \bar{\sigma}(t, X_{t-})dW_t + \int_{\mathbb{R}} \bar{a}(t, X_{t-}, y)\widetilde{J}(dt, dy) \tag{14.73}$$

with initial condition $X_0 = 1$. Under the hypotheses of Theorem 14.45, which we assume hereafter, X is uniquely defined and its characteristic operator is given by

$$\begin{aligned}\bar{\mathcal{A}}f(t,x) &= \bar{b}(t,x)\partial_x f(t,x) + \frac{\bar{\sigma}^2(t,x)}{2}\partial_{xx}f(t,x)\\ &\quad + \int_{\mathbb{R}} \left(f(t, x + \bar{a}(t,x,y)) - f(t,x) - \bar{a}(t,x,y)\partial_x f(t,x)\right)\nu(dy).\end{aligned}$$

Concerning the locally non-risky asset, we assume that it is given by

$$B_t = e^{\int_0^t \bar{r}(s, X_s)ds}, \qquad t \in [0, T],$$

where $\bar{r}$ is a (deterministic) bounded and measurable function.

Applying operator $\bar{\mathcal{A}}$ to the exponential function, we get

$$\bar{\mathcal{A}}e^x = e^x \Psi(t,x)$$

where

$$\Psi(t,x) = \bar{b}(t,x) + \frac{\bar{\sigma}^2(t,x)}{2} + \int_{\mathbb{R}} \left(e^{\bar{a}(t,x,y)} - 1 - \bar{a}(t,x,y) \right) \nu(dy). \tag{14.74}$$

Therefore, by the Itô formula (14.62), we have

$$\begin{aligned} dS_t &= \Psi(t, X_{t-}) S_{t-} dt \\ &\quad + \bar{\sigma}(t, X_{t-}) S_{t-} dW_t + S_{t-} \int_{\mathbb{R}} \left(e^{\bar{a}(t,X_{t-},y)} - 1 \right) \widetilde{J}(dt, dy). \end{aligned}$$

In particular, for the discounted price process $\widetilde{S}_t = \frac{S_t}{B_t}$ to be a martingale, the following drift condition must necessarily hold true:

$$\Psi(t,x) = \bar{r}(t,x). \tag{14.75}$$

Thus the dynamics of S under an EMM becomes

$$\begin{aligned} dS_t &= r(t, S_{t-}) S_{t-} dt \\ &\quad + \sigma(t, S_{t-}) S_{t-} dW_t + S_{t-} \int_{\mathbb{R}} \left(e^{a(t,S_{t-},y)} - 1 \right) \widetilde{J}(dt, dy), \end{aligned} \tag{14.76}$$

where

$$r(t,S) = \bar{r}\left(t, g(S)\right), \quad \sigma(t,S) = \bar{\sigma}\left(t, g(S)\right), \quad a(t,S,y) = \bar{a}\left(t, g(S), y\right),$$

and $g(S) = \log \frac{S}{S_0}$.

Example 14.55 In an exponential Lévy model (cf. Example 14.40), the underlying asset is of the form (14.72) where X is a Lévy process satisfying the SDE

$$dX_t = \mu_\infty dt + \sigma dW_t + \int_{\mathbb{R}} y \widetilde{J}(dt, dy)$$

under condition

$$\int_{|y| \geq 1} e^{2y} \nu(dy) < \infty. \tag{14.77}$$

We assume that the interest rate r is constant. In this case, the function Ψ in (14.74) takes the explicit form

$$\Psi(t,x) = \mu_\infty + \frac{\sigma^2}{2} + \int_{\mathbb{R}} \left(e^y - 1 - y \right) \nu(dy) = \psi_X(-i)$$

where ψ_X is the characteristic exponent of X. Hence, the drift condition (14.75) becomes

$$\psi_X(-i) = r. \tag{14.78}$$

More precisely, as in Example 14.40, we have that $\widetilde{S}$ is a martingale if and only if (14.78) is satisfied. Thus, under an EMM the drift parameter μ_∞ is determined by (14.78):

$$\mu_\infty = r - \frac{\sigma^2}{2} - \int_{\mathbb{R}} (e^y - 1 - y)\, \nu(dy).$$

We recall that the drift condition (14.78) was obtained independently in Proposition 13.64 by means of the Lévy-Khintchine formula for X.

As already proved in Example 14.40, by (14.76) the dynamics of S under an EMM is given by

$$dS_t = rS_{t-}dt + \sigma S_{t-}dW_t + S_{t-}\int_{\mathbb{R}} (e^y - 1)\widetilde{J}(dt, dy),$$

that is usually written in the shorthand form

$$\frac{dS_t}{S_{t-}} = rdt + \sigma dW_t + \int_{\mathbb{R}} (e^y - 1)\widetilde{J}(dt, dy). \tag{14.79}$$

□

Now we use the Feynman-Kač Theorem 14.50 to obtain the expression of the risk neutral price of a derivative in terms of the solution of a Cauchy problem for a PIDE. We consider a derivative of the form $H = \varphi(S_T)$, where φ is the deterministic payoff function. Moreover we assume that the risk neutral dynamics of the underlying asset, under the selected EMM, is given by the SDE with jumps (14.76): we denote by $(S_\tau^{t,s})_{\tau\in[t,T]}$ the solution of (14.76) with initial condition $S_t^{t,s} = s$, and by

$$\begin{aligned}\mathcal{A}f(t,s) &= r(t,s)s\partial_s f(t,s) + \frac{\sigma^2(t,s)s^2}{2}\partial_{ss}f(t,s)\\ &+\int_{\mathbb{R}} \left(f(t, se^{a(t,s,y)}) - f(t,s) - s\left(e^{a(t,s,y)} - 1\right)\partial_s f(t,s)\right)\nu(dy)\end{aligned}$$

the characteristic operator of S.

Theorem 14.56 *If the Cauchy problem*

$$\begin{cases}(\mathcal{A} + \partial_t) f(t,s) = r(t,s)f(t,s), & (t,s) \in [0,T[\times\mathbb{R}_{>0},\\ f(T,s) = \varphi(s), & s \in \mathbb{R}_{>0},\end{cases}$$

has a classical solution f (cf. Definition 14.49), then

$$f(t,s) = E\left[e^{-\int_t^T r(\tau, S_\tau^{t,s})d\tau}\varphi\left(S_T^{t,s}\right)\right]$$

is the risk-neutral price of H under the selected EMM.

For the exponential Lévy model in (14.79), the pricing operator is given by

$$\mathcal{A}f(t,s) = rs\partial_s f(t,s) + \frac{\sigma^2 s^2}{2}\partial_{ss} f(t,s) + \int_{\mathbb{R}} \left(f(t,se^y) - f(t,s) - s\left(e^y - 1\right)\partial_s f(t,s)\right)\nu(dy).$$

Apart from the Black&Scholes' type drift and diffusion parts, $\mathcal{A}$ contains a new integral term which depends on the Lévy measure ν. In this particular case, the non-uniqueness of the martingale measure is reflected in the pricing PIDE which is defined in terms of the jump component of the underlying.

Clearly, the models analyzed in this section are the natural extension of standard local volatility models to the framework of jump processes. Stochastic volatility with jumps can be considered as well and in the next sections we examine two remarkable examples, the Bates model and the Barndorf-Nielsen and Shepard model. Essentially, stochastic volatility appears to be needed to explain the variation in strike at longer time, but it is well known that it performs poorly across different maturities, especially at shorter term. Adding jumps to the price and/or the volatility gives a greater flexibility and allows to explain the variation in strike at shorter term which is difficulty captured by standard stochastic volatility models like the Heston and SABR models.

14.3.2 Bates model

The Bates [35] model for currency options combines the Heston stochastic volatility model with the Merton jump-diffusion model. Let use recall that in the Heston model the dynamics of the asset and its variance is given by (15.4), that is

$$\begin{aligned} dS_t &= \mu S_t dt + \sqrt{v_t} S_t dW_t^1, \\ dv_t &= k(v_\infty - v_t)dt + \eta\sqrt{v_t} dW_t^2, \end{aligned}$$

where $W = (W^1, W^2)$ is a correlated Brownian motion with correlation parameter ϱ. By Itô formula, the process $X_t = \log \frac{S_t}{S_0}$ verifies

$$dX_t = \left(\mu - \frac{v_t}{2}\right) dt + \sqrt{v_t} dW_t^1.$$

In the Bates model, the log-price X satisfies the SDE with jumps

$$\begin{aligned} dX_t &= -\frac{v_t}{2} dt + \sqrt{v_t} dW_t^1 + dZ_t \\ &= \left(\mu_\infty - \frac{v_t}{2}\right) dt + \sqrt{v_t} dW_t^1 + \int_{\mathbb{R}} y\widetilde{J}(dt, dy), \end{aligned}$$

where Z is a compound Poisson process with intensity λ and distribution of jumps $\mathcal{N}_{m,\delta^2}$: as usual, $\widetilde{J}$ denotes the compensated jump measure. The processes W and Z are assumed to be independent.

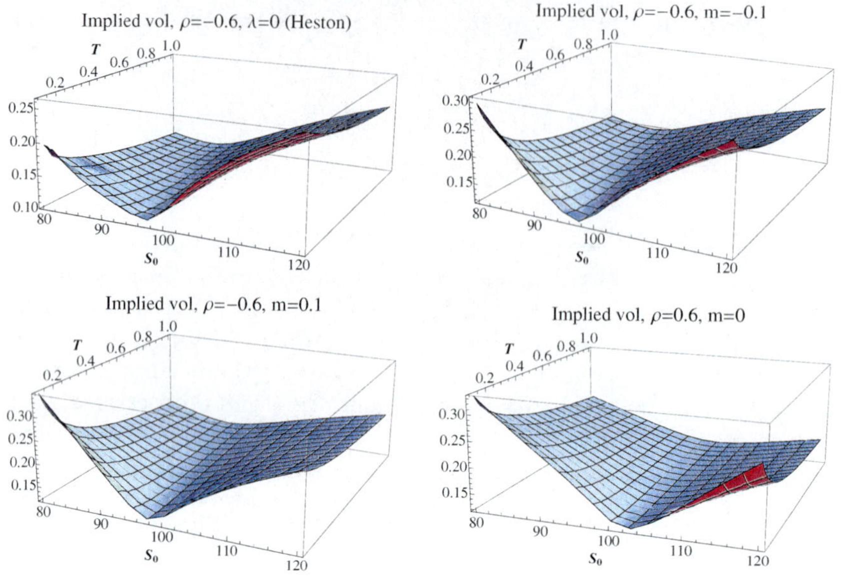

Fig. 14.1. Volatility surfaces in the Bates model for different values of ϱ and m: the other parameters are $v_0 = 0.04$, $v_\infty = 0.02$, $k = 1$, $\eta = 1$, $\lambda = 0.2$, $\delta = 1$

Imposing the drift condition (14.75), we get

$$r = \left(\mu_\infty - \frac{v_t}{2}\right) + \frac{v_t}{2} + \lambda \int_{\mathbb{R}} (e^y - 1 - y) \mathcal{N}_{m,\delta^2}(dy)$$

and therefore the drift coefficient under an EMM is determined by

$$\mu_\infty = r - \lambda\left(e^{m+\frac{\delta^2}{2}} - 1 - m\right).$$

The model has eight parameters: ϱ (correlation), k (speed of mean reversion), v_∞ (long-term mean of the variance v), v_0 (initial variance), η (volatility of the variance), λ (jump intensity), m (mean of the jumps of the log-price), δ (volatility of the jumps of the log-price).

Since jumps are independent from the continuous part, the characteristic function of X can be obtained multiplying the characteristic functions of the Heston and Merton models: more precisely, by (13.96) and (15.24), we have

$$\begin{aligned} \varphi_{X_T}(\xi) = &\exp\left(\frac{v_0}{\eta^2}\left(\frac{1-e^{-D(\xi)T}}{1-G(\xi)e^{-D(\xi)T}}\right)(k - i\rho\eta\xi - D(\xi))\right)\cdot \\ &\cdot\exp\left(\frac{kv_\infty}{\eta^2}\left(T(k - i\rho\eta\xi - D(\xi)) - 2\log\left(\frac{1-G(\xi)e^{-D(\xi)T}}{1-G(\xi)}\right)\right)\right)\cdot \\ &\cdot\exp\left(i\xi T\left(r - \lambda\left(e^{m+\frac{\delta^2}{2}} - 1\right)\right) + T\lambda\left(e^{im\xi - \frac{\xi^2\delta^2}{2}}\right)\right). \end{aligned} \tag{14.80}$$

Since the characteristic function is available in explicit form, European option prices in the Bates model can be computed by using one of the Fourier transform methods of Chapter 15. Moreover the short and long-term smile and skew can be adjusted separately by fine-tuning the stochastic volatility and jump parameters. It turns out that the Bates model is one of the simplest and effective models for reproducing the typical patterns of the implied volatility surfaces. Figure 14.1 depicts the implied volatility surface in the Bates model for typical values of the parameters.

14.3.3 Barndorff-Nielsen and Shephard model

Barndorff-Nielsen and Shephard propose in [29] a stochastic volatility model where the variance is the solution of a linear SDE with jumps of the form (14.68). They assume that the driving Lévy process is a subordinator in order to guarantee that the variance is positive. Specifically, the model is of the form

$$\begin{aligned} S_t &= S_0 e^{X_t}, \\ dX_t &= b_t dt + \sqrt{v_t} dW_t + \varrho dZ_t, \\ dv_t &= -\lambda v_t dt + dZ_t, \end{aligned}$$

where ϱ is a non-positive correlation parameter, $\lambda > 0$, W is a standard Brownian motion and Z is a subordinator with Lévy measure ν and Laplace exponent

$$\ell(\xi) = \int_0^\infty \left(e^{\xi y} - 1\right) \nu(dy).$$

Rewriting Z in terms of its jump measure, we have (cf. (13.58))

$$dZ_t = \mu_\infty dt + \int_0^\infty y \widetilde{J}(dt, dy), \qquad \mu_\infty = E\left[Z_1\right],$$

and

$$dX_t = \left(b_t + \varrho \mu_\infty\right) dt + \sqrt{v_t} dW_t + \varrho \int_0^\infty y \widetilde{J}(dt, dy).$$

Imposing the drift condition (14.75), we get

$$\begin{aligned} r &= b_t + \mu_\infty + \frac{v_t}{2} + \int_0^\infty \left(e^{\varrho y} - 1 - \ y\right) \nu(dy) \\ &= b_t + \frac{v_t}{2} + \int_0^\infty \left(e^{\varrho y} - 1\right) \nu(dy) \\ &= b_t + \frac{v_t}{2} + \ell(\varrho). \end{aligned}$$

Thus the generic risk-neutral dynamics is given by

$$\begin{aligned} dX_t &= \left(r - \frac{v_t}{2} - \ell(\varrho)\right) dt + \sqrt{v_t} dW_t + \varrho dZ_t, \\ dv_t &= -\lambda v_t dt + dZ_t. \end{aligned}$$

Also in this case, it is possible to have the explicit expression of the characteristic function of X, so that Fourier methods for option pricing apply: specifically, we have

$$\varphi_{X_T}(\xi) = \exp\left(i\xi(r - \ell(\varrho))T + \frac{v_0\xi}{2\lambda}(1 - i\xi)\left(1 - e^{-\lambda T}\right)\right) \cdot$$
$$\cdot \exp\left(\int_0^T \ell\left(i\varrho\xi - \frac{\xi}{2\lambda}(i + \xi)\left(1 - e^{-\lambda(T-t)}\right)\right) dt\right).$$

Several experiments on the calibration of stochastic volatility models with jumps in the asset and/or the variance can be found, for instance, in Sepp [305], Schoutens [301], Carr, Geman, Madan, and Yor [68], Zhu [348].

15

Fourier methods

As already explained in the previous chapters, in order to reproduce the real market dynamics it is necessary to introduce more sophisticated models than the Black-Scholes one. These models have to be calibrated to the market in order to approximate the quoted implied volatility surfaces: once this is done, they can give prices to exotic derivatives that are consistent with plain vanilla options.

Now it is worth mentioning that the calibration, as well as other risk management processes, poses severe restrictions on the class of feasible models. Indeed, the numerical procedures involved in a calibration process are very demanding and time-consuming: a typical optimization algorithm requires, at each step of minimization of the error, the computation of several option prices for different strikes and maturities: this is practicable only if closed form solutions or accurate analytical approximations of option prices are available. From this perspective, Fourier inversion is the computational method of choice for a fast calculation of option prices in models with an analytically tractable characteristic function.

The main idea is that, while the probability density of many relevant asset processes is not known explicitly, on the contrary its Fourier transform (i.e. the characteristic function) in some relevant case is available. Therefore, in the Fourier space it is possible to price options analytically: thus, by means of a Fourier inversion algorithm (for instance, the fast Fourier transform, FFT) the original option prices can be computed efficiently.

In this chapter we present two methods for the approximation of option prices by means of Fourier methods. Specifically, we establish pricing formulas based on the *Fourier transform* and on the *Fourier-cosine series expansion*. We analyze the main features of the two approaches and compare their numerical precision and efficiency.

To the best of our knowledge, Fourier transform applications in option pricing where initiated by Heston [165]. Starting from Carr and Madan [69], several authors have proposed extensions and analysis of valuation formulas with Fourier transform methods. The focus of Carr and Madan is on the application of the FFT algorithm to Call options: one of the key points is that the Fourier transform is taken with respect to the log-strike. The approach by Lewis [237] and Raible [288] is similar, except that their transform is taken with respect to log-prices. Other price representations were proposed by Zhu [347], Attari [13] and more recently by Eberlein, Glau and Papapantoleon [111]. Fourier transform valuation formulas for exotic options were considered by Borovkov and Novikov [52]. These methods were also applied for hedging purposes by Hubalek, Kallsen and Krawczyk [173], and in insurance mathematics by Dufresne, Garrido and Morales [105]. The books by Schoutens [301], Cont and Tankov [76], Boyarchenko and Levendorskii[56] also discuss Fourier transform methods in option pricing. We also quote the recent monographs by Zhu [348], Cherubini, Della Lunga, Mulinacci and Rossi [72] and the book on Fourier methods for interest rate derivatives by Bouziane [55]. In his Ph.D. thesis [323] Surkov develops a generic framework based on the Fourier transform for pricing and hedging of various options in equity, commodity, currency, and insurance markets.

The alternative approach, presented in Paragraph 15.3 on Fourier-cosine series expansion, is very recent and was proposed by Fang and Oosterlee [127] who also studied the case of options with the early-exercise feature in [128].

15.1 Characteristic functions and branch cut

Hereafter we consider a rather general setting where *the interest rate r is constant*[1] and there is only one risky asset of the form

$$S_T = S_0 e^{X_T}, \tag{15.1}$$

where $S_0 > 0$ and X_T is a random variable; for convenience, sometimes we also write $S_T = e^{X_T + x_0}$ with $x_0 = \log S_0$. We assume that the characteristic function of X_T, under the selected EMM Q, is well defined as

$$\varphi_{X_T}(\xi) = E^Q\left[e^{i\xi X_T}\right] = \int_{\mathbb{R}} e^{i\xi x} Q^{X_T}(dx), \tag{15.2}$$

where, as usual, Q^{X_T} denotes the distribution of X_T under Q.

Notation 15.1 *In this chapter, we let $\xi \in \mathbb{C}$ in* (15.2) *whenever the integral defining $\varphi_{X_T}(\xi)$ converges: therefore, we might more properly call φ_{X_T} the extended characteristic function of X_T.*

Example 15.2 In the Black-Scholes model, we have

$$S_T = S_0 e^{X_T}, \qquad X_T = \left(r - \frac{\sigma^2}{2}\right) T + \sigma W_T,$$

[1] When the *interest rate is stochastic* such as in the case of interest rate derivatives, one should use the techniques of this chapter applied to the general pricing formula (10.102) of Section 10.4.3.

and

$$\varphi_{X_T}(\xi) = e^{i\left(r-\frac{\sigma^2}{2}\right)T\xi - \frac{\sigma^2\xi^2}{2}T}.$$

□

Fourier methods in option pricing are based on the explicit knowledge of the characteristic function in (15.2). Typically φ_{X_T} is given in terms of the solution of first order differential equations or may have an explicit representation in terms of elementary functions. In any case, the expression of the φ_{X_T} involves functions whose argument is a complex number that must be handled carefully in the implementation.

Indeed, as a preliminary remark, let us recall that, by the definition (15.2), φ_{X_T} is a *continuous* function. As we shall see in a moment, even simple functions such as the logarithm or the square root (that typically occur in some representation of φ_{X_T}) when defined on the complex plane become multi-valued functions. Now, the most common mathematical softwares such as *Matlab* or *Mathematica* automatically select the principal branch of these multi-valued functions: this may generate discontinuities and, as a matter of fact, a wrong representation of the characteristic function[2]. The result is that the Fourier inversion may give completely incorrect option prices.

To be more specific, let us recall that a complex number

$$z = x + iy, \qquad x = \mathrm{Re}(z),\ y = \mathrm{Im}(z),$$

can be represented in polar form as

$$z = |z|e^{i(\theta+2k\pi)}, \qquad k \in \mathbb{Z}, \tag{15.3}$$

where $|z| = \sqrt{x^2+y^2}$ is the modulus of z and $\theta = \mathrm{Arg}(z) \in]-\pi,\pi]$ is the principal argument of z: for $x, y > 0$, we have $\theta = \arctan\frac{y}{x}$. More generally, we recall that the software *Mathematica* has the built-in command $\texttt{ArcTan}[x,y]$ ($\texttt{atan2}(y,x)$ in *Matlab*) that gives the arc-tangent of $\frac{y}{x}$ taking into account which quadrant the point (x,y) is, so that the results are always in the range $]-\pi,\pi]$: thus we have $\mathrm{Arg}(z) = \texttt{ArcTan}[\mathrm{Re}(z),\mathrm{Im}(z)]$.

Now, formally we have

$$\log z = \log\left(|z|e^{i(\theta+2k\pi)}\right) = \log|z| + i\left(\theta + 2k\pi\right), \qquad k \in \mathbb{Z},$$

and similarly

$$z^\alpha = |z|^\alpha e^{i\alpha(\theta+2k\pi)};$$

thus the logarithm and the square root are functions whose value depends on k, that is on the argument of the complex number z in the representation (15.3).

[2] From a theoretical point of view, the characteristic function is correctly represented in terms of the *distinguished complex logarithm* defined in Lemma 13.20.

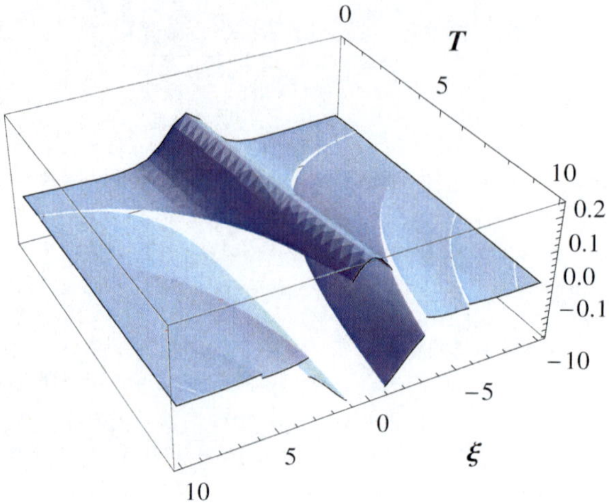

Fig. 15.1. Discontinuities in the integrand of the Fourier pricing formula (15.11) for a Call option in the Heston model with the formulation (15.5) of the characteristic function

As an illustrative example, we consider the Heston model where the asset price S and its variance ν satisfy the system of SDEs

$$\begin{aligned} dS_t &= rS_t dt + \sqrt{\nu_t} S_t dW_t^1, \\ d\nu_t &= k(\nu_\infty - \nu_t)dt + \eta\sqrt{\nu_t} dW_t^2. \end{aligned} \tag{15.4}$$

In the original formulation by Heston [165], also found in the papers by Lee [234], Kahl and Jäckel [192], the characteristic function is given by

$$\begin{aligned} \varphi_{X_T}(\xi) = &\exp\left(i\xi rT + \frac{\nu_0}{\eta^2}\left(\frac{1-e^{-D(\xi)T}}{1-G(\xi)e^{-D(\xi)T}}\right)(k - i\rho\eta\xi - D(\xi))\right)\cdot \\ &\cdot\exp\left(\frac{k\nu_\infty}{\eta^2}\left(T(k - i\rho\eta\xi + D(\xi)) - 2\log\left(\frac{1-\frac{e^{D(\xi)T}}{G(\xi)}}{1-\frac{1}{G(\xi)}}\right)\right)\right) \end{aligned} \tag{15.5}$$

where ν_0 is the initial variance, ϱ is the correlation parameter of the Brownian motions and

$$D(\xi) = \sqrt{(k - i\rho\eta\xi)^2 + (\xi + i)\,\xi\eta^2}, \qquad G(\xi) = \frac{k - i\rho\eta\xi - D(\xi)}{k - i\rho\eta\xi + D(\xi)}. \tag{15.6}$$

Figure 15.1 shows the discontinuities of the integrand appearing in the Fourier representation of the price of a Call option (see (15.11) below) in the Heston model: here the formulation (15.5) of φ_{X_T} has been used and the complex logarithm is automatically restricted to its principal branch by the numerical software, producing the discontinuities. The parameters of the model are taken from the paper of Lord and Kahl [241]:

$$r = 0, \; k = 1.5768, \; \nu_\infty = 0.0398, \; \nu_0 = 0.0175, \; \varrho = -0.5711, \; \eta = 0.5751.$$

For the Heston model, an alternative representation of φ_{X_T} (cf. (15.24)) has been given by Bakshi, Cao and Chen [17] (see also Duffie, Pan and Singleton [104] and Gatheral [150]). Lord and Kahl [241] recently proved that this second formulation, although algebraically equivalent to the previous one, does not produce discontinuities when the main argument of the complex numbers is used. For more details, we refer to Example 15.15 where the Fourier integral method in the Heston model is analyzed.

The problem of the discontinuous representation of characteristic functions was first mentioned by Schöbel and Zhu [299] since similar problems arise in the stochastic volatility model named after these authors. Recently, Lord and Kahl [241] also provided a continuous representation of the characteristic function for the Schöbel-Zhu model.

15.2 Integral pricing formulas

We denote by f the payoff function of a derivative and by

$$H(S_0, T) = e^{-rT} E^Q \left[f(X_T + \log S_0) \right]$$

the corresponding Q-risk-neutral price under the assumptions (15.1)-(15.2) on S. For example, for a Call option with strike K, we set

$$f^{\text{Call}}(x) = (e^x - K)^+, \qquad x \in \mathbb{R},$$

so that the Call price for the maturity T is given by

$$H^{\text{Call}}(S_0, K, T) = e^{-rT} E^Q \left[f^{\text{Call}}(X_T + \log S_0) \right].$$

Analogously, we set

$$f^{\text{Put}}(x) = (K - e^x)^+, \qquad x \in \mathbb{R}.$$

The general idea of Fourier methods in option pricing is rather simple: let us assume, for simplicity, that $r = 0 = \log S_0$; then the option price is formally given by

$$E^Q \left[f(X_T) \right] = \int_{\mathbb{R}} f(x) Q^{X_T}(dx) =$$

(by the Fourier inversion formula in Remark A.66)

$$= \int_{\mathbb{R}} \left(\frac{1}{\pi} \int_0^\infty e^{-ix\xi} \hat{f}(\xi) d\xi \right) Q^{X_T}(dx) =$$

(changing the order of integration)

$$
\begin{aligned}
&= \frac{1}{\pi}\int_0^\infty \hat{f}(\xi)\int_{\mathbb{R}} e^{-ix\xi} Q^{X_T}(dx)d\xi \\
&= \frac{1}{\pi}\int_0^\infty \hat{f}(\xi)\varphi_{X_T}(-\xi)d\xi.
\end{aligned}
\tag{15.7}
$$

Assuming that the functions $\hat{f}$ and φ_{X_T} are known explicitly, the pricing formula (15.7) lends itself to an application of the fast Fourier transform (FFT) or direct numerical integration methods. We recall that a pricing formula equivalent to (15.7) was found in Section 14.2.6, starting from the pricing partial integro-differential equation obtained by means of the Itô formula for Lévy processes.

Actually, the previous arguments are not rigorous and some points need to be fixed. First of all, the classical Fourier transform $\hat{f}$ is defined for $f \in L^1(\mathbb{R})$ and also in the simplest case of a Call or a Put option this fails to be true. Secondly, the classical inversion formula requires that $\hat{f} \in L^1(\mathbb{R})$, but if this is case, f is necessarily continuous and this rules out several interesting payoff functions (e.g. the digital option). Finally, the change the order of integration needs to be rigorously justified. In the following sections we analyze these and other related issues.

15.2.1 Damping method

A classical tool in Fourier analysis for the inversion of non-integrable function, is the damping method (cf., for instance, Dubner and Abate [101]): for a given $\alpha \in \mathbb{R}$, we define the penalized (or damped) function

$$
f_\alpha(x) = e^{-\alpha x} f(x), \qquad x \in \mathbb{R}.
$$

The following example shows the effect of the penalization on the payoff functions of plain vanilla options.

Example 15.3 We have

$$
f_\alpha^{\mathrm{Call}}(x) = e^{-\alpha x}\left(e^x - K\right)^+, \qquad f_\alpha^{\mathrm{Put}}(x) = e^{-\alpha x}\left(K - e^x\right)^+,
$$

and therefore $f_\alpha^{\mathrm{Call}} \in L^1(\mathbb{R})$ for any $\alpha > 1$ and $f_\alpha^{\mathrm{Put}} \in L^1(\mathbb{R})$ for any $\alpha < 0$.□

Since we are going to use the inversion formula, we are also interested in the integrability properties of the Fourier transform of the damped function: the following lemma provides a simple and quite general integrability criterion.

Lemma 15.4 *If $g \in W^{1,2}(\mathbb{R})$, that is $g \in L^2(\mathbb{R})$ and the weak derivative $Dg \in L^2(\mathbb{R})$, then $\hat{g} \in L^1(\mathbb{R})$.*

Proof. It is well known that if $g \in W^{1,2}(\mathbb{R})$ then $g, \hat{g} \in L^2(\mathbb{R})$ and

$$\widehat{Dg}(\xi) = -i\xi\hat{g}(\xi).$$

Then we have

$$\int_{\mathbb{R}} \left(|\hat{g}(\xi)|^2 + \left|\widehat{Dg}(\xi)\right|^2 \right) d\xi = \int_{\mathbb{R}} |\hat{g}(\xi)|^2 \left(1 + |\xi|^2\right) d\xi,$$

and by the Hölder's inequality, we get

$$\begin{aligned}\int_{\mathbb{R}} |\hat{g}(\xi)|\, d\xi &= \int_{\mathbb{R}} |\hat{g}(\xi)| \frac{1+|\xi|}{1+|\xi|} d\xi \\ &\le \left(\int_{\mathbb{R}} |\hat{g}(\xi)|^2 (1+|\xi|)^2 d\xi\right)^{\frac{1}{2}} \left(\int_{\mathbb{R}} \frac{1}{(1+|\xi|)^2}\right)^{\frac{1}{2}} < \infty.\end{aligned}$$

□

Example 15.5 By Lemma 15.4, $f_\alpha^{\mathrm{Call}}, \widehat{f_\alpha^{\mathrm{Call}}} \in L^1(\mathbb{R})$ for any $\alpha > 1$. Indeed $f_\alpha^{\mathrm{Call}} \in W^{1,2}(\mathbb{R})$ because $f_\alpha^{\mathrm{Call}} \in L^2$ and

$$Df_\alpha^{\mathrm{Call}}(x) = \begin{cases} 0 & \text{if } x < \log K, \\ (1-\alpha)e^{(1-\alpha)x} + \alpha K e^{-\alpha x} & \text{if } x > \log K, \end{cases}$$

is a square integrable function for $\alpha > 1$. Analogously, $f_\alpha^{\mathrm{Put}}, \widehat{f_\alpha^{\mathrm{Put}}} \in L^1(\mathbb{R})$ for any $\alpha < 0$, since we have

$$Df_\alpha^{\mathrm{Put}}(x) = \begin{cases} -\alpha K e^{-\alpha x} - (1-\alpha)e^{(1-\alpha)x} & \text{if } x < \log K, \\ 0 & \text{if } x > \log K. \end{cases}$$

□

15.2.2 Pricing formulas

Our first result is a valuation formula for options with continuous payoff function, like Call and Put options.

Proposition 15.6 *Assume that there exists $\alpha \in \mathbb{R}$ such that*

i) $f_\alpha, \hat{f}_\alpha \in L^1(\mathbb{R})$;
ii) $E^Q[S_T^\alpha]$ is finite.

Then the following pricing formula holds:

$$H(S_0, T) = \frac{e^{-rT} S_0^\alpha}{\pi} \int_0^\infty e^{-i\xi \log S_0} \varphi_{X_T}(-(\xi + i\alpha)) \hat{f}(\xi + i\alpha) d\xi. \tag{15.8}$$

Proof. First of all we show that conditions i) and ii) guarantee that the integral in (15.8) converges. Indeed, by i) we have that $\xi \mapsto \hat{f}(\xi + i\alpha)$ is integrable because

$$\hat{f}(\xi + i\alpha) = \int_{\mathbb{R}} e^{i(\xi+i\alpha)x} f(x)dx = \int_{\mathbb{R}} e^{i\xi x} f_\alpha(x)dx = \hat{f}_\alpha(\xi). \tag{15.9}$$

On the other hand, by ii), the function $\xi \mapsto S_0^{-i(\xi+i\alpha)}\varphi_{X_T}(-(\xi + i\alpha))$ is bounded because

$$\left|S_0^{-i(\xi+i\alpha)}\varphi_{X_T}(-(\xi + i\alpha))\right| \le \left|S_0^{-i(\xi+i\alpha)}\right| \int_{\mathbb{R}} \left|e^{-ix(\xi+i\alpha)}\right| Q^{X_T}(dx)$$
$$= S_0^\alpha \int_{\mathbb{R}} e^{\alpha x} Q^{X_T}(dx) = E^Q\left[S_T^\alpha\right] < \infty.$$

For simplicity, we set $x_0 = \log S_0$: then we have

$$H(S_0, T) = e^{-rT} E^Q\left[f(X_T + x_0)\right]$$
$$= e^{-rT} \int_{\mathbb{R}} e^{\alpha(x+x_0)} f_\alpha\left(x + x_0\right) Q^{X_T}(dx) =$$

(by the Fourier inversion formula, since $f_\alpha, \hat{f}_\alpha \in L^1(\mathbb{R})$ by i))

$$= e^{-rT} S_0^\alpha \int_{\mathbb{R}} e^{\alpha x} \left(\frac{1}{\pi}\int_0^\infty e^{-i\xi(x+x_0)} \hat{f}_\alpha(\xi)d\xi\right) Q^{X_T}(dx) =$$

(changing the order of integration, by Fubini's theorem[3])

$$= \frac{e^{-rT} S_0^\alpha}{\pi} \int_0^\infty e^{-i\xi x_0} \left(\int_{\mathbb{R}} e^{-i(\xi+i\alpha)x} Q^{X_T}(dx)\right) \hat{f}_\alpha(\xi)d\xi =$$

(by (15.9))

$$= \frac{e^{-rT} S_0^\alpha}{\pi} \int_0^\infty e^{-i\xi x_0} \varphi_{X_T}(-(\xi + i\alpha))\hat{f}(\xi + i\alpha)d\xi,$$

and this concludes the proof. □

Remark 15.7 Condition i) of Proposition 15.6 is an assumption on the payoff function f and implies that f is a *continuous* function (because $\hat{f}_\alpha \in L^1$). Theorem 15.10 below provides a valuation formula for discontinuous payoff functions.

---

[3] Again by i) and ii), we have

$$\int_{\mathbb{R}} e^{\alpha x} \int_0^\infty \left|e^{-i\xi(x+x_0)} \hat{f}_\alpha(\xi)\right| d\xi Q^{X_T}(dx) = \|\hat{f}_\alpha\|_{L^1} E^Q\left[e^{\alpha X_T}\right] < \infty.$$

Condition *ii)* is an assumption on the Q-distribution of X_T and is equivalent to

$$E^Q\left[S_T^\alpha\right] = S_0^\alpha E^Q\left[e^{\alpha X_T}\right] = S_0^\alpha \int_{\mathbb{R}} e^{\alpha x} Q^{X_T}(dx) < \infty$$

that is, the measure $e^{\alpha x} Q^{X_T}(dx)$ is finite. Also this condition can be weakened in order to deal with particular distributions. We also remark that it is not difficult to generalize Proposition 15.6 to the multi-dimensional case. □

As a corollary of Proposition 15.6, we also give the Fourier formula for the Delta of an option: similar formulas can be easily obtained for all the other Greeks.

Corollary 15.8 (Delta) *Under the assumptions of Proposition 15.6, if in addition one of the functions* $\xi \mapsto (1+|\xi|)\,\varphi_{X_T}(-(\xi+i\alpha))$ *or* $\widehat{Df_\alpha}$ *is integrable, then we have*

$$\begin{aligned}\text{Delta}(S_0,T) :=& \partial_{S_0} H(S_0,T)\\ =& \frac{e^{-rT} S_0^{\alpha-1}}{\pi} \int_0^\infty e^{-i\xi \log S_0} (\alpha - i\xi)\varphi_{X_T}(-(\xi+i\alpha))\hat{f}(\xi+i\alpha)d\xi.\end{aligned} \tag{15.10}$$

Proof. The thesis follows by differentiating formula (15.8) with respect to S_0: the additional assumptions guarantee that we can exchange the integral and differential signs. □

Theorem 15.9 (Call option) *For any* $\alpha > 1$ *such that* $E^Q\left[S_T^\alpha\right]$ *is finite, we have the following pricing formula for a Call option with strike* K *and maturity* T*:*

$$\text{Call}(S_0,K,T) = \frac{e^{-rT} S_0^\alpha K^{1-\alpha}}{\pi} \int_0^\infty e^{-i\xi \log \frac{S_0}{K}} \frac{\varphi_{X_T}(-(\xi+i\alpha))}{(i\xi-\alpha)(i\xi-\alpha+1))} d\xi. \tag{15.11}$$

The same formula for $\alpha < 0$ *gives the price of the Put option*[4]*. Moreover, under the assumptions of Corollary 15.8, the Delta is given by*

$$\text{Delta}^{\text{Call}}(S_0,K,T) = -\frac{e^{-rT} S_0^{\alpha-1} K^{1-\alpha}}{\pi} \int_0^\infty e^{-i\xi \log \frac{S_0}{K}} \frac{\varphi_{X_T}(-(\xi+i\alpha))}{i\xi-\alpha+1} d\xi. \tag{15.12}$$

[4] Under the assumption that $E^Q\left[S_T^\alpha\right]$ is finite.

Proof. We recall Example 15.5 and use the pricing formula (15.8) of Proposition 15.6. To this end, we compute $\widehat{f^{\text{Call}}}(\xi + i\alpha)$: by (15.9), we have

$$\begin{aligned}\widehat{f^{\text{Call}}}(\xi + i\alpha) = \widehat{f^{\text{Call}}_\alpha}(\xi) &= \int_{\log K}^{\infty} e^{(i\xi-\alpha)x}\left(e^x - K\right)dx\\ &= -\frac{K^{i\xi-\alpha+1}}{i\xi-\alpha+1} + K\frac{K^{i\xi-\alpha}}{i\xi-\alpha}\\ &= \frac{K^{1-\alpha}e^{i\xi\log K}}{(i\xi-\alpha)(i\xi-\alpha+1))}.\end{aligned} \tag{15.13}$$

Moreover, a direct computation shows that $\widehat{f^{\text{Call}}_\alpha} = \widehat{f^{\text{Put}}_\alpha}$. Then the thesis follows plugging (15.13) into formulas (15.8) and (15.10). □

Next we state a valuation formula for *discontinuous* payoff functions. The proof is a slight modification of the argument used in Proposition 15.6 and is based on the Dirichlet-Jordan inversion Theorem A.67.

Note that, from the point of view of the numerical implementation, formulas (15.8) and (15.14) are equivalent since the numerical integration is always performed on a bounded domain.

Theorem 15.10 *Assume that there exists $\alpha \in \mathbb{R}$ such that*

i) $f_\alpha \in L^1(\mathbb{R})$;
ii) $E^Q\left[S_T^\alpha\right]$ *is finite;*
iii) the map $x \mapsto E^Q\left[f(X_T + x)\right]$ *is continuous at* $x_0 = \log S_0$ *and has bounded variation (cf. Definition 3.59) in a neighborhood of* x_0.

Then the following pricing formula holds:

$$H(S_0, T) = \frac{e^{-rT}S_0^\alpha}{\pi}\lim_{R\to+\infty}\int_0^R e^{-i\xi\log S_0}\varphi_{X_T}(-(\xi+i\alpha))\hat{f}(\xi+i\alpha)d\xi. \tag{15.14}$$

In the financial literature, several alternative Fourier representation formulas for the price of a Call option have been proposed: Heston [165] first used a general Black-Scholes type formula based on the results of Theorem 10.67. This method is analyzed and broadened by Zhu [348]. The approach of Carr and Madan [69] is closer in spirit to that of (15.11): they first introduced in the area of option pricing the idea of damping functions in order to get the L^1-integrability of the payoff. The focus of Carr and Madan was on the use of the fast Fourier transform to retrieve the option values for a wide range of strikes and maturities in a single run. Similar representations were provided by Raible [288] and Lewis [237] who took the Fourier transform with respect to the log-price instead of the log-strike price as in [69]; moreover the formulas by Lewis are expressed in terms of contour integral in the complex plane.

We close this section by giving a short list of payoff functions of exotic options to which the previous Fourier pricing formulas apply. To shorten the notation, we set $z = \xi + i\alpha$: for any option, we give the payoff function with its extended Fourier transform and the interval of allowed values of the damping parameter α.

- Digital:

$$f(x) = \mathbb{1}_{\{e^x > K\}}, \qquad \hat{f}(z) = -\frac{K^{iz}}{iz}, \qquad \alpha > 0.$$

- Asset or nothing:

$$f(x) = e^x \mathbb{1}_{\{e^x > K\}}, \qquad \hat{f}(z) = -\frac{K^{1+iz}}{1+iz}, \qquad \alpha > 1.$$

- Double digital ($K_1 < K_2$):

$$f(x) = \mathbb{1}_{\{K_1 < e^x < K_2\}}, \qquad \hat{f}(z) = \frac{K_2^{iz} - K_1^{iz}}{iz}, \qquad \alpha \neq 0.$$

- Self-quanto:

$$f(x) = e^x \left(e^x - K\right)^+, \qquad \hat{f}(z) = \frac{K^{2+iz}}{(1+iz)(2+iz)}, \qquad \alpha > 2.$$

- Power:

$$f(x) = \left((e^x - K)^+\right)^2, \qquad \hat{f}(z) = \frac{2K^{2+iz}}{iz(1+iz)(2+iz)}, \qquad \alpha > 2.$$

15.2.3 Implementation

In the representation formula (15.11), the price of a Call option is given in terms of a direct Fourier transform: indeed, (15.11) can be rewritten as

$$\mathrm{Call}(S_0, K, T) = \frac{A_\alpha(S_0, K)}{\pi} \int_0^\infty e^{-i\xi \log M} F_\alpha(\xi) d\xi, \tag{15.15}$$

where $M = \frac{S_0}{K}$ is the moneyness,

$$A_\alpha(S_0, K) = e^{-rT} S_0^\alpha K^{1-\alpha} \quad \text{and} \quad F_\alpha(\xi) = \frac{\varphi_{X_T}(-(\xi + i\alpha))}{(i\xi - \alpha)(i\xi - \alpha + 1)}. \tag{15.16}$$

As discussed by Carr and Madan [69], this representation lends itself to an application of the fast Fourier transform (FFT). Chourdakis [73] also proposed the use of the fractional FFT algorithm. We refer to Cerny [70] and Matsuda [249] for an elementary introduction to FFT methods in option pricing.

More recently several authors (cf., for instance, Kilin [207], Zhu [348], Lord and Kahl [240]) criticized the use of FFT methods for computing option prices. Indeed, a direct integration (DI) can also be used to evaluate the pricing formula (15.15) and it turns out that in most cases DI outperforms FFT and fractional FFT methods in terms of accuracy and speed (see, for instance, the comparison in [348], Section 4.5).

Remark 15.11 It has to be noticed that the function F_α in (15.16), which contains the characteristic function and is the computationally expensive part, *does not depend on* S_0 *and* K*:* therefore, it is sufficient to evaluate F_α only once to compute the Call prices with different strikes. This simple observation allows the use of cache techniques that speed up considerably the computation of option prices and the related Greeks. Kilin [207] states that using cache technique makes the calibration with the direct integration method at least seven times faster than the calibration with the fractional FFT method. In the case of Call options, "a vector input" of strikes can be dealt with in an efficient way: so, we have a vector of input strikes which gives us a vector of output values. This speeds up the calibration process significantly. □

The following experiments have been performed on an Intel(R) Core(TM)2 CPU 2.40 GHz, using the built-in numerical integration command `NIntegrate` of *Mathematica*® based on an adaptive algorithm. The CPU time to compute one price or a Greek, regardless of the model, is of the order of 0.01 seconds.

Remark 15.12 When pricing Call options, in order to reduce the error in the truncation of the integration domain, it is preferable to compute Put prices and then recover the Calls via the Put-Call parity:

$$H^{\text{Call}}(S_0, K, T) = H^{\text{Put}}(S_0, K, T) - Ke^{-rT} + S_0.$$

This allows the exploit the boundedness of the payoff function of the Put. □

To test the efficiency of the pricing formulas (15.11) and (15.12), we first consider the case of the standard Black-Scholes model in Example 15.2 and compare the Fourier prices with the closed form solution given by the Black-Scholes formula of Corollary 7.15. Figure 15.2 shows the relative errors of Call prices and Deltas, defined as

$$\text{RelErr}^{\text{Call}} = \frac{\text{Call}(S_0, K, T) - \text{CallBS}(S_0, K, T)}{\text{CallBS}(S_0, K, T)},$$

$$\text{RelErr}^{\text{Delta}} = \frac{\text{Delta}(S_0, K, T) - \text{DeltaBS}(S_0, K, T)}{\text{DeltaBS}(S_0, K, T)},$$

where Call and Delta are the Fourier prices computed by (15.11)-(15.12), and CallBS and DeltaBS are the exact values obtained by the Black-Scholes formula; we set the relative error equal to zero whenever the CallBS price is less that one basis point of the underlying price, that is $\text{CallBS}(S_0, K, T) < \frac{S_0}{10^4}$: this is the value under which it is commonly agreed that the option is worthless. Parameters selected for this test are

$$K = 100, \quad S_0 \in [50, 150], \quad r = 5\%, \quad \sigma = 30\%, \quad \alpha = 2; \tag{15.17}$$

moreover the maturity ranges from $T = \frac{1}{52}$ (one week) to $T = 1$ (one year). As the figures show, the relative errors for both Call price and Delta are negligible, since they are of the order of 10^{-8} for any moneyness and maturity.

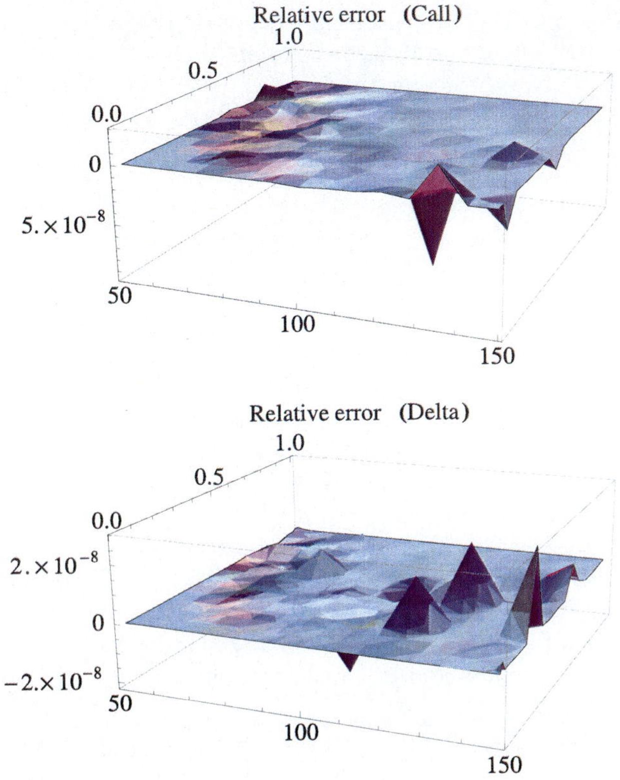

Fig. 15.2. Relative errors for Call price and Delta in the Black-Scholes model, with the parameters as in (15.17)

15.2.4 Choice of the damping parameter

Before considering other more general models, let us briefly discuss the problem of the choice of the damping parameter α. Theoretically, the pricing formula (15.15) is exact for any $\alpha > 1$ such that $E\left[S_T^\alpha\right]$ is finite: on the other hand, it has been recognized by many authors that the integrand in (15.15) can become either strongly peaked when α gets close to 1 or highly oscillatory when α reaches the maximum allowed value. From the numerical point of view, this fact is of extreme importance. For instance, the accuracy of the prices calculated with the FFT or fractional FFT strongly depends on the choice α since the FFT algorithm is heavily affected by the oscillations.

In the financial literature there has not been much work in the study of the optimal value of α. Some "empirical" recommendation are given by Carr and Madan [69], Schoutens, Simons and Tistaert [302]. Lee [234] suggests to choose α so that it minimizes the errors in the approximation of the Fourier integral by the discrete Fourier transform. A more general approach has been proposed by Lord and Kahl [240] who suggest the minimization of the variation (cf.

Definition 3.59) of the integrand in (15.11) as a criterion for selecting α: the idea is to try to control and reduce the oscillations.

In Figure 15.3 we represent the integrand function of formula (15.11) in the Black-Scholes model, that is

$$I(\xi, \alpha, M, T) := e^{-i\xi \log M} \frac{\varphi_{X_T}(-(\xi + i\alpha))}{(i\xi - \alpha)(i\xi - \alpha + 1))}, \tag{15.18}$$

where α is the damping parameter, $M = \frac{S_0}{K}$ is the moneyness, T is the time to maturity and φ_{X_T} is the Gaussian characteristic function of Example 15.2. In the left picture we set $M = 0.8$, $T = 1$ and we represent I as a function of $\xi \geq 0$ and $\alpha > 1$; in the right picture we set $M = 0.8$, $\alpha = 2$ and we represent I as a function of $\xi \geq 0$ and $T \geq 0$. Notice the shape of I that is peaked for α close to 1 and oscillatory for large α. Figure 15.4 shows the behaviour of I for α close to 1 and for large α: in this case, oscillations are much pronounced (of the order of 10^9).

By examining the graph of I, it is possible to select a range of values of α which exclude large oscillations and can be used for the numerical integration of the Fourier pricing formulas (15.11)-(15.12). For instance, in the Black&Scholes model with the specified parameters, $\alpha = 2$ seems a good choice for any moneyness and any maturity, as already confirmed by the results in Figure 15.2.

In general the optimal choice of α depends on the maturity T, the moneyness M and, what is really important, on the parameters of the model. In Figure 15.5 we plot the graph of the integrand I in (15.18) for the Black-Scholes model, with $\alpha = 2$, $T = \frac{1}{12}$ (left) and $T = 10$ (right), as a function of $\xi \in [0, 10]$ and $M \in [0.5, 1.5]$.

Next we examine other popular non-Gaussian models.

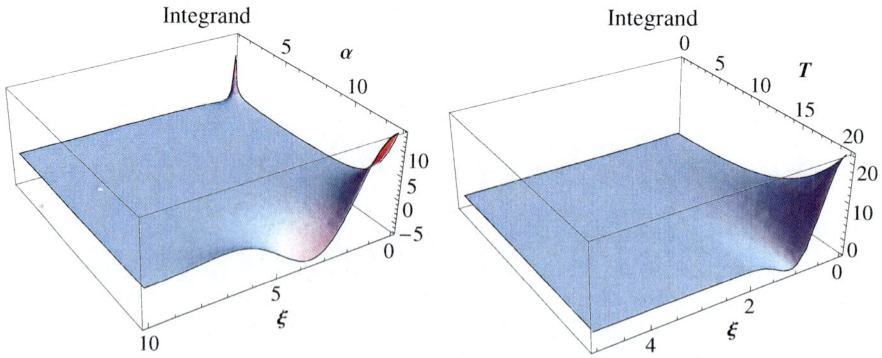

Fig. 15.3. The function $I = I(\xi, \alpha, M, T)$ in (15.18) for the Black-Scholes model with $r = 5\%$ and $\sigma = 30\%$. In the left picture $\xi \in [0, 10]$, $\alpha \in [1.1, 10]$, $M = 0.8$, $T = 1$; in the right picture $\xi \in [0, 10]$, $T \in [0, 20]$, $M = 0.8$ and $\alpha = 2$

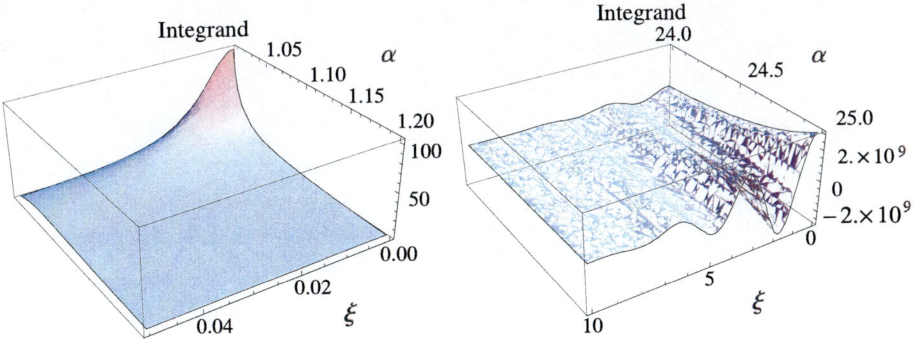

Fig. 15.4. The function $(\xi, \alpha) \mapsto I(\xi, \alpha, M, T)$ in (15.18) for the Black-Scholes model with $r = 5\%$, $\sigma = 30\%$, $T = 1$ and $M = 0.8$. In the left picture $\xi \in [0, 0.05]$, $\alpha \in [1.01, 1.2]$; in the right picture $\xi \in [0, 10]$, $\alpha \in [24, 25]$

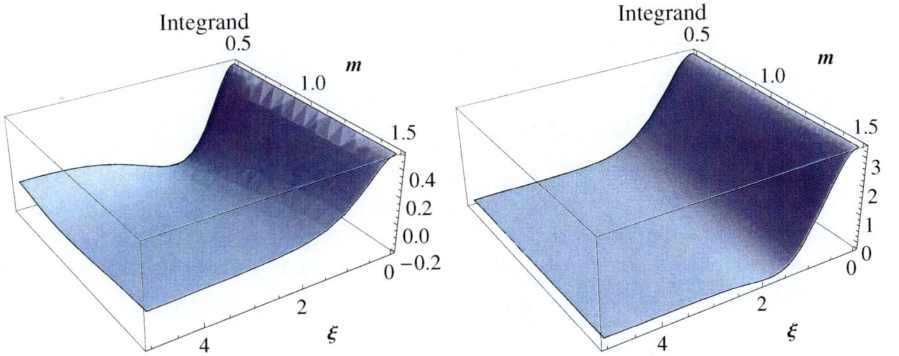

Fig. 15.5. The function $(\xi, M) \mapsto I(\xi, \alpha, M, T)$ in (15.18) for the Black-Scholes model with $r = 5\%$, $\sigma = 30\%$, $\alpha = 2$, $T = \frac{1}{12}$ (left) and $T = 10$ (right)

Example 15.13 (Merton model) In the Merton jump-diffusion model (cf. Example 13.51), the asset price is of the form (15.1) where

$$X_t = \mu t + \sigma W_t + \sum_{n=1}^{N_t} Z_n$$

is the sum of a Brownian motion with drift and a compound Poisson process. Assuming that the Z_n are normally distributed $Z_n \sim \mathcal{N}_{m,\delta^2}$, the drift coefficient under an EMM takes the form

$$\mu = r - \frac{\sigma^2}{2} + \lambda \left(1 - e^{m + \frac{\delta^2}{2}}\right),$$

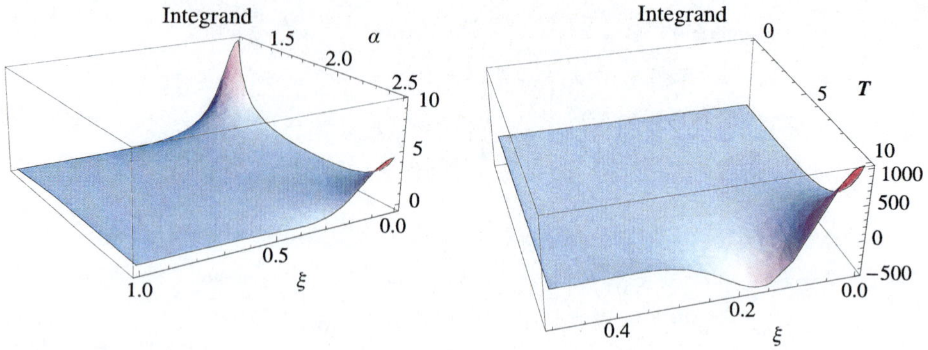

Fig. 15.6. The function $I = I(\xi, \alpha, M, T)$ in (15.18) for the Merton model with the parameters as in (15.20). In the left picture $\xi \in [0, 10]$, $\alpha \in [1.1, 2.5]$, $M = 0.8$, $T = 1$; in the right picture $\xi \in [0, 0.5]$, $T \in [0, 10]$, $M = 0.8$ and $\alpha = 2$

and the characteristic function in (15.2) is given by

$$\varphi_{X_T}(\xi) = \exp\left(i\mu T\xi - \frac{\sigma^2\xi^2}{2}T + \lambda T\left(e^{im\xi - \frac{\delta^2\xi^2}{2}} - 1\right)\right). \qquad (15.19)$$

Thus, we have four parameters: the diffusion volatility σ, the jump intensity λ, the mean jump size m and the standard deviation of jump size δ.

Figure 15.6 shows the graph the Fourier integrand $I = I(\xi, \alpha, M, T)$ in (15.18). In the left picture, we set the moneyness $M = \frac{S_0}{K} = 0.8$, the maturity $T = 1$ and we consider I as a function of $\xi \in [0, 1]$ and $\alpha \in [1.1, 2.5]$: the values of the other parameters are

$$r = 5\%, \qquad \sigma = 15\%, \qquad \lambda = m = 10\%, \qquad \delta = 1. \qquad (15.20)$$

This suggests that α in a neighborhood of 2 can be a good choice of the damping parameter. Indeed, as the right picture in Figure 15.6 shows, for $\alpha = 2$ the integrand I is quite stable for all the maturities $T \in [0, 10]$.

To confirm our result, we analyze the percentage differences of Call prices obtained by using different values of α. More precisely, we denote by $C_\alpha(T)$ the price of a Call option with maturity T, computed by numerically integrating formula (15.15) with $\alpha > 1$: Figure 15.7 shows the graph of the percentage differences

$$\alpha \mapsto \frac{C_\alpha(T) - C_2(T)}{C_2(T)}, \qquad \alpha \in [1.1, 2.6], \qquad (15.21)$$

for $T = \frac{1}{52}, \frac{1}{12}, 1, 5$ and moneyness $M = 0.8$. In all cases, the percentage differences are almost negligible. Other experiments that we do not report here, show that similar results are also valid for different moneyness and for a wide range of values of the parameters of the model. This last fact is particularly important when considering the calibration of the model: indeed, in this case the Fourier pricing formulas are used for different sets of parameters and therefore the stability with respect to the damping parameter is crucial.

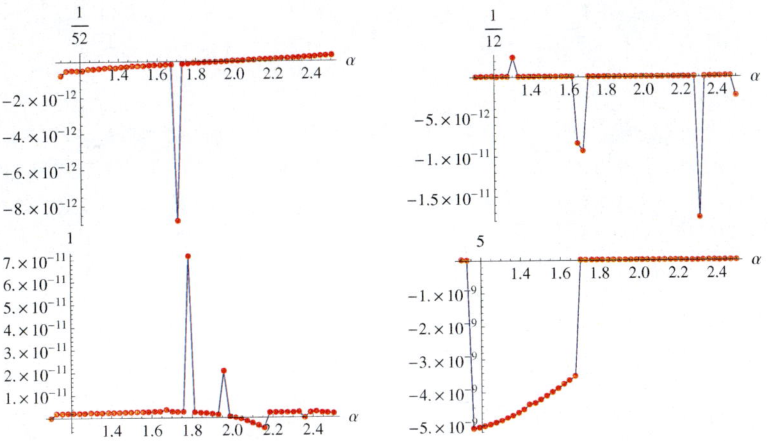

Fig. 15.7. Percentage differences of prices computed with $\alpha \in [1.1, 2.5]$, of Call options with moneyness $M = 0.8$ and maturities $T = \frac{1}{52}, \frac{1}{12}, 1, 5$ in the Merton model

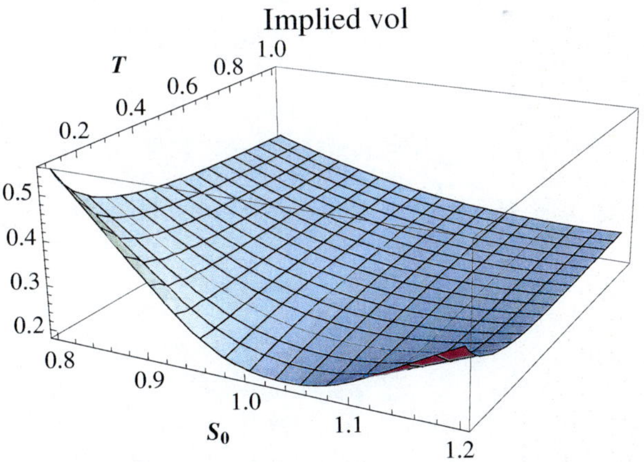

Fig. 15.8. Implied volatility surface in the Merton model with the parameters as in (15.20)

Finally, Figure 15.8 depicts the implied volatility surface computed in the Merton model with the parameters as in (15.20). □

Example 15.14 (VG model) We examine the Variance-Gamma (VG) model which belongs to the class of models with infinite activity Lévy processes (see Example 13.61). The log-price process is defined by a Brownian motion with drift μ and volatility σ, subordinated (i.e. time-changed) by a Gamma process with unit mean and variance ν. The (risk-neutral) characte-

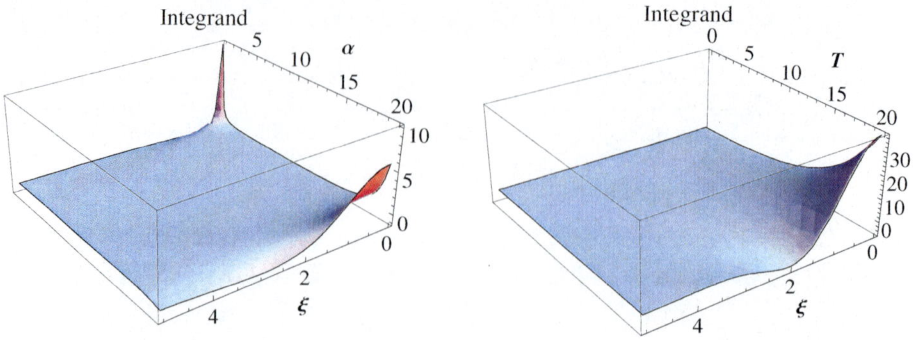

Fig. 15.9. The integrand $I = I(\xi, \alpha, M, T)$ in (15.18) for the VG model with the parameters as in (15.23). In the left picture $\xi \in [0, 5]$, $\alpha \in [1.1, 20]$, $M = 0.8$, $T = 1$; in the right picture $\xi \in [0, 5]$, $T \in [0, 20]$, $M = 0.8$ and $\alpha = 2$

ristic function in (15.2) takes the form

$$\varphi_{X_T}(\xi) = e^{imT\xi} \left(\frac{1}{1 - i\xi\mu\nu + \frac{1}{2}\nu\xi^2\sigma^2} \right)^{\frac{T}{\nu}} \tag{15.22}$$

where

$$m = r + \frac{1}{\nu} \log \left(1 - \mu\nu - \frac{\sigma^2\nu}{2} \right).$$

In our experiment we set

$$\mu = 4\%, \qquad \sigma = 12\%, \qquad \nu = 20\%, \qquad r = 10\%. \tag{15.23}$$

As for the other models, we first represent the Fourier integrand $I = I(\xi, \alpha, M, T)$ in (15.18). In the left picture in Figure 15.9, we consider I as a function of $\xi \in [0, 1]$ and $\alpha \in [1.1, 2.5]$, with the moneyness $M = 0.8$ and the maturity $T = 1$. Also in this case, for α close to 2 the integrand I is stable for long maturities (cf. the right picture in Figure 15.9). However, for small maturities I oscillates and the numerical efficiency of the Fourier inversion is limited by cancellation errors: indeed, since the integrand I assumes positive and negative values, the integral of the absolute value of I may be much larger than the integral of I (which gives the option price). As an illustration, in Figure 15.10 we depict the graph of $\xi \mapsto I(\xi, \alpha, M, T)$ for $T = \frac{1}{52}$, $M = 0.8$ and $\alpha = 2, 5, 10, 20$.

It turns out that for short maturities, the outcome of the Fourier pricing formulas is sensitive to the choice of the parameter α. Indeed, only for maturities greater than one month we obtain results which are stable with respect to the choice of α, for different moneyness and values of the parameters of the VG model. More precisely, in Figure 15.11 we report the percentage differences

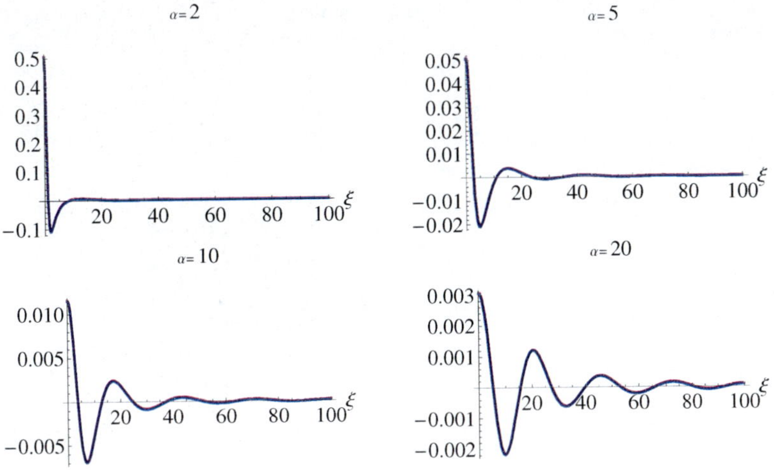

Fig. 15.10. The function $\xi \mapsto I(\xi, \alpha, M, T)$ in the VG model with the parameters as in (15.23), $T = \frac{1}{52}$, $M = 0.8$ and $\alpha = 2, 5, 10, 20$

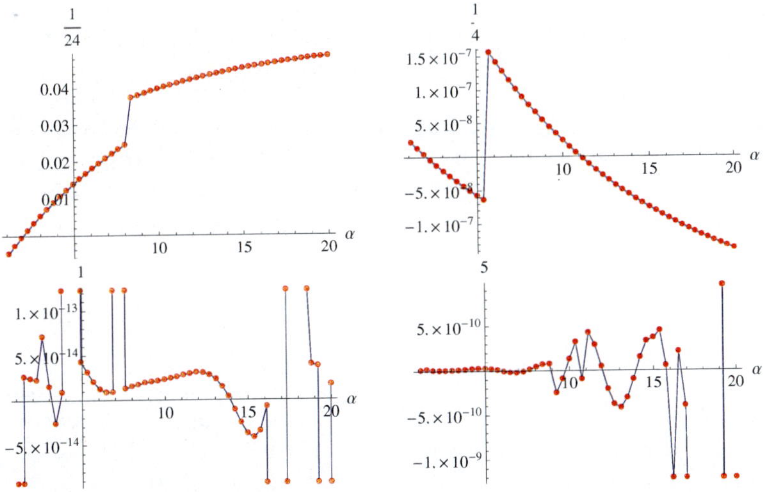

Fig. 15.11. Percentage differences (cf. (15.21)) of prices in the VG model, computed with $\alpha \in [1.1, 20]$, of Call options with moneyness $M = 0.9$ and maturities $T = \frac{1}{24}, \frac{2}{12}, 1, 5$

(defined as in (15.21)) of Call prices computed with moneyness $M = 0.9$ and maturities $T = \frac{1}{24}, \frac{2}{12}, 1, 5$. The same results are obtained for the percentage differences of the values of the Delta. Note that, for the maturity of two weeks (i.e. $T = \frac{1}{24}$), the maximum percentage difference is of the order of 4% for α ranging in $[1.1, 20]$. On the other hand, for maturities greater than one month, the result are quite stable and independent on the choice of+α.

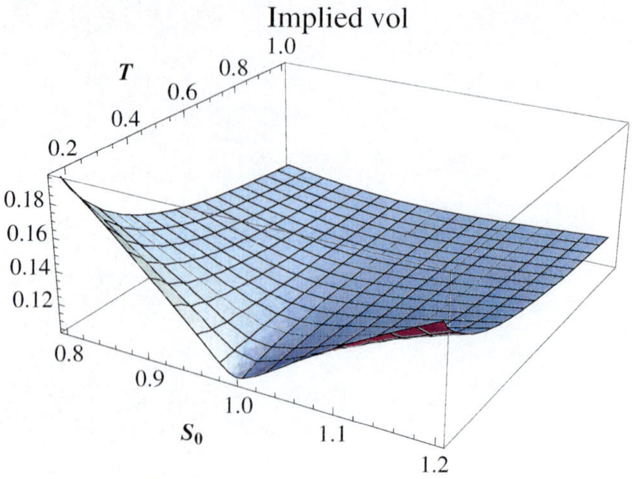

Fig. 15.12. Implied volatility surface in the VG model with the parameters as in (15.23)

Finally, Figure 15.12 depicts the implied volatility surface computed in the VG model with the parameters as in (15.23). □

Example 15.15 (Heston model) We consider another example where the option prices are sensitive to the choice of the damping parameter α. In the Heston stochastic volatility model (cf. Example 10.33) the risk-neutral dynamics of the asset and its variance is given by (15.4), that is

$$\begin{aligned} dS_t &= rS_t dt + \sqrt{\nu_t} S_t dW_t^1, \\ d\nu_t &= k(\nu_\infty - \nu_t)dt + \eta\sqrt{\nu_t} dW_t^2, \end{aligned}$$

where r is the short rate, k is the speed of mean reversion, ν_∞ is the long-term mean of ν and η is the volatility of the variance. Moreover W is a two-dimensional correlated Brownian motion with

$$d\langle W^1, W^2\rangle_t = \varrho dt.$$

To avoid complex discontinuities (cf. Paragraph 15.1), we use the risk-neutral characteristic function given by Bakshi, Cao and Chen [17], which takes the form

$$\begin{aligned} \varphi_{X_T}(\xi) = &\exp\left(i\xi rT + \frac{\nu_0}{\eta^2}\left(\frac{1-e^{-D(\xi)T}}{1-G(\xi)e^{-D(\xi)T}}\right)(k - i\rho\eta\xi - D(\xi))\right)\cdot \\ &\cdot \exp\left(\frac{k\nu_\infty}{\eta^2}\left(T(k - i\rho\eta\xi - D(\xi)) - 2\log\left(\frac{1-G(\xi)e^{-D(\xi)T}}{1-G(\xi)}\right)\right)\right), \end{aligned} \tag{15.24}$$

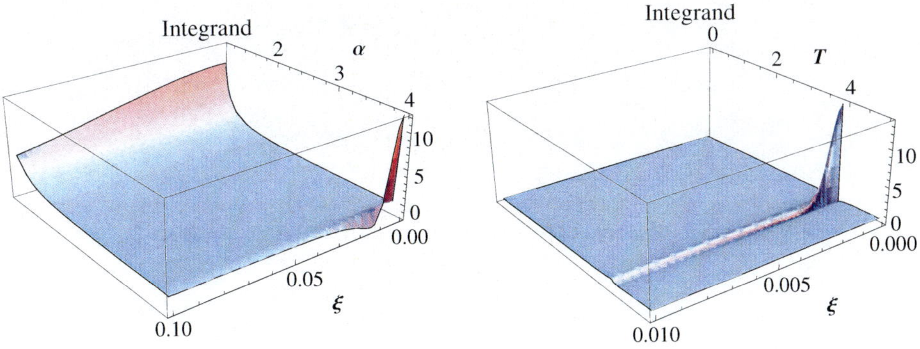

Fig. 15.13. The function $I(\xi, \alpha, M, T)$ in (15.18) for the Heston model with the parameters as in (15.26). In the left picture $\xi \in [0, 0.1]$, $\alpha \in [1.1, 4]$, $M = 0.8$, $T = 1$; in the right picture $\xi \in [0, 0.01]$, $T \in [0, 10]$, $M = 0.8$ and $\alpha = 2$

where ν_0 is the initial variance and D, G are the functions in (15.6), that is

$$D(\xi) = \sqrt{(k - i\rho\eta\xi)^2 + (\xi + i)\,\xi\eta^2}, \qquad G(\xi) = \frac{k - i\rho\eta\xi - D(\xi)}{k - i\rho\eta\xi + D(\xi)}. \tag{15.25}$$

We consider the following values of the parameters:

$$r = 5\%, \quad k = 0.5, \quad \nu_0 = 0.04, \quad \nu_\infty = 0.02, \quad \eta = 1.5, \quad \varrho = -0.4. \tag{15.26}$$

As in the previous examples, we represent the integrand function I in (15.18) as a function of ξ and α, with moneyness $M = 0.8$ and maturity $T = 1$ (left picture in Figure 15.13), and as a function of ξ and T, with moneyness $M = 0.8$ and $\alpha = 2$ (right picture in Figure 15.13). We see that, when the maturity T equals 1, a choice of $\alpha \in [1.1, 3.5]$ is acceptable for a wide range of the parameters of the Heston model (see also Figure 15.14). However, as the maturity increases, we need to choose α closer to 1.1 in order to avoid singularities: the right picture in Figure 15.13 shows the oscillations of I for $\alpha = 2$, $T = 4$ and the parameters as in (15.26).

Finally, in Figure 15.14 we report the percentage differences

$$\alpha \mapsto \frac{C_\alpha(T) - C_{1.5}(T)}{C_{1.5}(T)}$$

of Call prices computed with moneyness $M = 0.8$ and maturities $T = \frac{1}{24}, \frac{1}{12}, 1, 5$. Similar results are obtained for the percentage differences of the values of the Delta and for different values of the parameters. For options with moneyness greater than 0.9, the percentage differences are almost negligible for any maturity. Figure 15.15 depicts the implied volatility surface computed in the Heston model with the parameters as in (15.26).

Recently, a *time-dependent* Heston model was considered by Benhamou, Gobet and Miri in [40] who derived an accurate analytical formula for the price of vanilla options. □

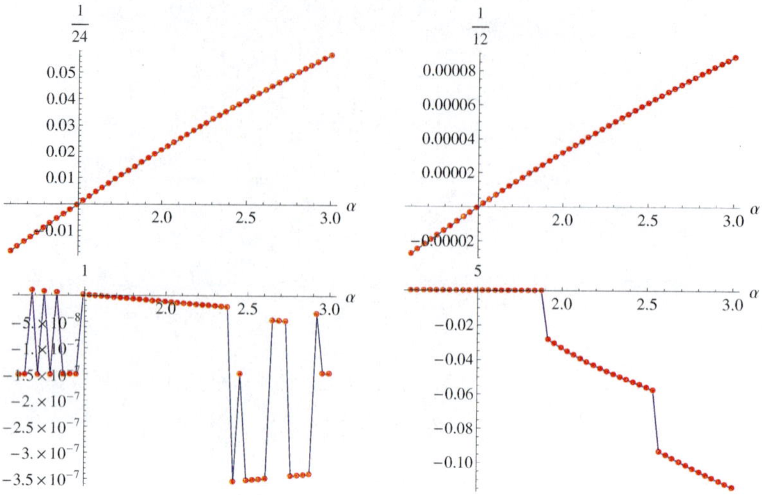

Fig. 15.14. Percentage differences of prices in the Heston model, computed with $\alpha \in [1.1, 3]$, of Call options with moneyness $M = 0.8$ and maturities $T = \frac{1}{24}, \frac{1}{12}, 1, 5$

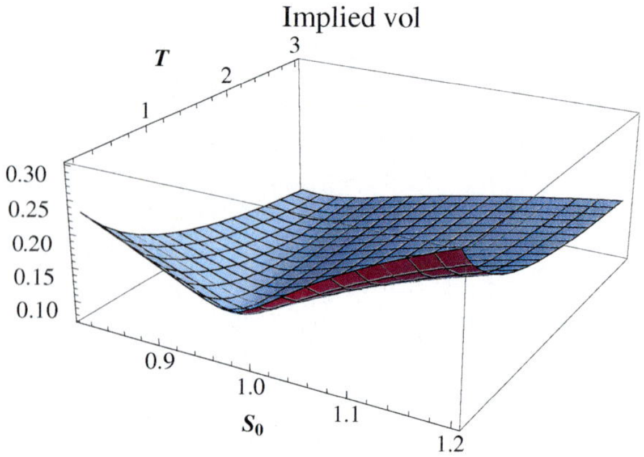

Fig. 15.15. Implied volatility surface in the Heston model with the parameters as in (15.26)

15.3 Fourier-cosine series expansions

An interesting pricing method based on Fourier series expansions was recently proposed by Fang and Oosterlee [127]. In order to present this alternative approach, we first recall some classical result about Fourier series. Let

$$f : [-\pi, \pi] \longrightarrow \mathbb{R}$$

be an integrable function: the Fourier series of f is defined by

$$S_f(x) = \frac{a_0}{2} + \sum_{k=1}^{\infty} \left(a_k \cos(kx) + b_k \sin(kx)\right), \qquad x \in \mathbb{R},$$

where

$$\begin{aligned} a_k &= \frac{1}{\pi}\int_{-\pi}^{\pi} f(\xi)\cos(k\xi)d\xi, \qquad & k \geq 0, \\ b_k &= \frac{1}{\pi}\int_{-\pi}^{\pi} f(\xi)\sin(k\xi)d\xi, \qquad & k \geq 1. \end{aligned}$$

Notice that if f is an even function, then $b_k = 0$ for any $k \geq 1$. A classical result on the convergence of Fourier series which encompasses all the standard situations in option pricing, is the following[5]:

Theorem 15.16 (Jordan) *If f has bounded variation in $[-\pi, \pi]$, then its Fourier series converges for any x and its sum equals*[6]

$$\frac{f(x+) + f(x-)}{2}.$$

Clearly, if f has bounded variation and it is continuous at x_0, then $S_f(x_0) = f(x_0)$: for simplicity, in the sequel we shall always assume that the Fourier series of f converges and its sum equals f.

Through a simple change of variables, we may also consider functions supported on any other finite interval. Indeed, if

$$f : [a, b] \longrightarrow \mathbb{R},$$

then we consider the change of variables $\theta = k\pi\frac{x-a}{b-a}$: more precisely, we set

$$g(\theta) = f\left(\frac{b-a}{\pi}\theta + a\right), \qquad \theta \in [0, \pi],$$

and by symmetry we extend g to $[-\pi, \pi]$ so that it is an even function. Then we get the following *Fourier-cosine series expansion* of f:

$$f(x) = \frac{a_0}{2} + \sum_{k=1}^{N} a_k \cos\left(k\pi\frac{x-a}{b-a}\right), \tag{15.27}$$

where

$$a_k = \frac{2}{b-a}\int_a^b f(\xi)\cos\left(k\pi\frac{\xi-a}{b-a}\right)d\xi, \qquad k \geq 0.$$

It is known that the Fourier-cosine expansion of f in x equals the Chebyshev series expansion of $f\left(\cos^{-1}(t)\right)$ in t.

Next we split the problem of the Fourier approximation of option prices in two steps:

---

[5] For the proof see, for instance, Boyd [57].

[6] See also Remark 3.65.

First step. We consider $f \in L^1(\mathbb{R})$ and assume that we know explicitly its Fourier transform $\hat{f}$. We choose an interval $[a, b]$ and approximate the coefficients of the Fourier-cosine expansion of f as follows:

$$\begin{aligned} a_k &= \frac{2}{b-a}\int_a^b f(\xi)\mathrm{Re}\left(e^{ik\pi\frac{\xi-a}{b-a}}\right)d\xi \\ &= \frac{2}{b-a}\mathrm{Re}\left(e^{-ik\pi\frac{a}{b-a}}\int_a^b f(\xi)e^{ik\pi\frac{\xi}{b-a}}d\xi\right) \\ &\approx \frac{2}{b-a}\mathrm{Re}\left(e^{-ik\pi\frac{a}{b-a}}\hat{f}\left(\frac{k\pi}{b-a}\right)\right) =: A_k, \end{aligned} \tag{15.28}$$

where we used the approximation

$$\hat{f}(\xi) = \int_{\mathbb{R}} e^{ix\xi}f(x)dx \approx \int_a^b e^{ix\xi}f(x)dx.$$

By truncating the series summation in (15.27) to a suitably large index N, we get the following Fourier-cosine type approximation:

$$f(x) \approx \frac{A_0}{2} + \sum_{k=1}^{N} A_k \cos\left(k\pi\frac{x-a}{b-a}\right), \qquad x \in [a, b], \tag{15.29}$$

with A_k as in (15.28).

Second step. We consider a pricing model where there is only one risky asset S which, under the selected EMM Q, takes the form

$$S_T = e^{X_T + x_0}.$$

Here $x_0 = \log S_0$, $S_0 > 0$, and X_T is a random variable whose characteristic function φ_{X_T} (cf. (15.2)) is known. As usual, we denote by $H(S_0, T)$ the price of a European option with maturity T and payoff function f: for instance, $f(x) = (e^x - K)^+$ for a Call option with strike K.
Then, we assume that X_T has a density function Φ_{X_T} and we have

$$\begin{aligned} H(S_0, T) &= e^{-rT}E^Q\left[f(X_T + x_0)\right] \\ &= e^{-rT}\int_{\mathbb{R}} f(x + x_0)\Phi_{X_T}(x)dx \approx \end{aligned}$$

(we truncate the infinite integration range)

$$\approx e^{-rT}\int_a^b f(x + x_0)\Phi_{X_T}(x)dx \approx$$

(we use the Fourier-cosine approximation (15.29) of Φ_{X_T}, with A_k as in (15.31) below)

$$
\begin{aligned}
&\approx e^{-rT}\frac{A_0}{2}\int_a^b f(x+x_0)dx \\
&\quad + e^{-rT}\sum_{k=1}^{N} A_k \int_a^b f(x+x_0)\cos\left(k\pi\frac{x-a}{b-a}\right)dx.
\end{aligned}
$$

Hence we have the following approximation of the option price

$$
H(S_0,T) \approx e^{-rT}\left(\frac{A_0 B_0(S_0)}{2} + \sum_{k=1}^{N} A_k B_k(S_0)\right), \tag{15.30}
$$

where

$$
A_k = \frac{2}{b-a}\mathrm{Re}\left(e^{-ik\pi\frac{a}{b-a}}\varphi_{X_T}\left(\frac{k\pi}{b-a}\right)\right), \tag{15.31}
$$

$$
B_k(S_0) = \int_a^b f(x+\log S_0)\cos\left(k\pi\frac{x-a}{b-a}\right)dx. \tag{15.32}
$$

Note that, up to a multiplicative factor, B_k is the coefficient of the Fourier-cosine expansion of f that, in several interesting cases, can be computed explicitly.

It is worth mentioning that the specification of the pricing model only enters in the A-coefficients that depend on the characteristic function of X_T, but are independent on S_0 and on the payoff function f. Therefore, options with different payoff functions (e.g. Call options with different moneyness) can be computed simultaneously. In particular, as described in Remark 15.11 (see also [127]), we can give "a vector input" of strikes to obtain a vector of output values: from the practical point of view, this speeds up the calibration process significantly.

An error analysis of the Fourier-cosine approximation, based on classical results from Fourier analysis, is performed in [127]: here we confine ourselves to an illustration of the method by some examples. Firstly, we provide the coefficients of the Fourier-cosine expansions of the prices and the Deltas of Call and Put options. More generally, the Fourier coefficients of the Greeks or other exotic payoff functions can be easily obtained by using a symbolic computation software.

Example 15.17 (Call price and Delta) For a Call option we have

$$
B_0^{\mathrm{Call}}(S_0) = \int_a^b \left(e^{x+\log S_0} - K\right)^+ dx =
$$

(assuming $a < -\log\frac{S_0}{K}$)

$$= \int_{-\log\frac{S_0}{K}}^{b} \left(e^{x+\log S_0} - K\right) dx = e^b S_0 - K\left(1 + b + \log\frac{S_0}{K}\right),$$

and, for $k \geq 1$,

$$B_k^{\text{Call}}(S_0) = \int_a^b \left(e^{x+\log S_0} - K\right)^+ \cos\left(k\pi\frac{x-a}{b-a}\right) dx =$$

(assuming $a < -\log\frac{S_0}{K}$)

$$= \int_{-\log\frac{S_0}{K}}^{b} \left(e^{x+\log S_0} - K\right) \cos\left(k\pi\frac{x-a}{b-a}\right) dx$$
$$= \frac{(b-a)^2\left((-1)^k e^b k\pi S_0 - kK\pi\cos\gamma + (b-a)K\sin\gamma\right)}{(b-a)^2 k\pi + k^3\pi^3},$$

with

$$\gamma = \frac{k\pi\left(a + \log\frac{S_0}{K}\right)}{a-b}. \tag{15.33}$$

The coefficients of the Delta of a Call option can be obtained by differentiating with respect to S_0:

$$B_0^{\text{Delta}}(S_0) = \frac{d}{dS_0} B_0^{\text{Call}}(S_0) = e^b - \frac{K}{S_0},$$

and, for $k \geq 1$,

$$B_k^{\text{Delta}}(S_0) = \frac{d}{dS_0} B_k^{\text{Call}}(S_0)$$
$$= \frac{(a-b)\left((-1)^k(a-b)e^b S_0 + (b-a)K\cos\gamma + kK\pi\sin\gamma\right)}{\left((b-a)^2 + k^2\pi^2\right) S_0}$$

with γ as in (15.33). □

Example 15.18 (Put price and Delta) For a Put option we have

$$B_0^{\text{Put}}(S_0) = e^a S_0 - K\left(1 + a + \log\frac{S_0}{K}\right)$$

and, for $k \geq 1$,

$$B_k^{\text{Put}}(S_0) = \frac{(b-a)^2\left(e^a S_0 k\pi - kK\pi\cos\gamma + (b-a)K\sin\gamma\right)}{(b-a)^2 k\pi + k^3\pi^3},$$

with γ as in (15.33). The coefficients of the Delta of a Put option are given by:

$$B_0^{\text{Delta-Put}}(S_0) = e^a - \frac{K}{S_0},$$

and, for $k \geq 1$,

$$B_k^{\text{Delta-Put}}(S_0) = \frac{(a-b)K\left((a-b)e^a S_0 + (b-a)K\cos\gamma + kK\pi\sin\gamma\right)}{K\left((b-a)^2 + k^2\pi^2\right)S_0}.$$

□

15.3.1 Implementation

The implementation of the Fourier-cosine method is straightforward, even if some further indication is needed for the choice of the truncation range $[a, b]$. A natural choice for a, b has been proposed in [127] based on the observation that the shape of the density functions can be estimated using the cumulants of the process. We recall (cf. Appendix A.4) that the n-th cumulant c_n of X_T is defined by

$$c_n = \frac{1}{i^n}\frac{d^n}{d\xi^n}g(\xi)|_{\xi=0}, \qquad g(\xi) = \log\varphi_{X_T}(\xi) = \log E^Q\left[e^{i\xi X_T}\right].$$

For instance, in the Black-Scholes model of Example 15.2 where

$$\varphi_{X_T}(\xi) = e^{i\left(r-\frac{\sigma^2}{2}\right)T\xi - \frac{\sigma^2\xi^2}{2}T},$$

we have

$$c_1 = \left(r - \frac{\sigma^2}{2}\right)T, \qquad c_2 = \sigma^2 T, \tag{15.34}$$

that are the mean and the variance of X_T. Though the expression of the cumulants can be very lengthy in some cases (cf. for instance the Heston model), generally it can be obtained explicitly using a symbolic computation software.

Coming back to the choice of a and b, Fang and Oosterlee [127] propose the following:

$$a = c_1 - L\sqrt{c_2 + \sqrt{c_4}}, \qquad b = c_1 + L\sqrt{c_2 + \sqrt{c_4}}, \tag{15.35}$$

with $L = 10$. The cumulant c_4 gives a contribution in controlling the sharp peaks and fat tails that the density of some models (typically, associated to Lévy processes) exhibits. The choice of $L = 10$ seems appropriate for some models even if this parameter can be adjusted to improve the precision: in general, when increasing the value of L, a larger value of N must be chosen to get the same accuracy, as the following examples show. In some cases we simply put

$$a = c_1 - L\sqrt{c_2}, \qquad b = c_1 + L\sqrt{c_2} \tag{15.36}$$

since this seems to be sufficient to get fairly good results.

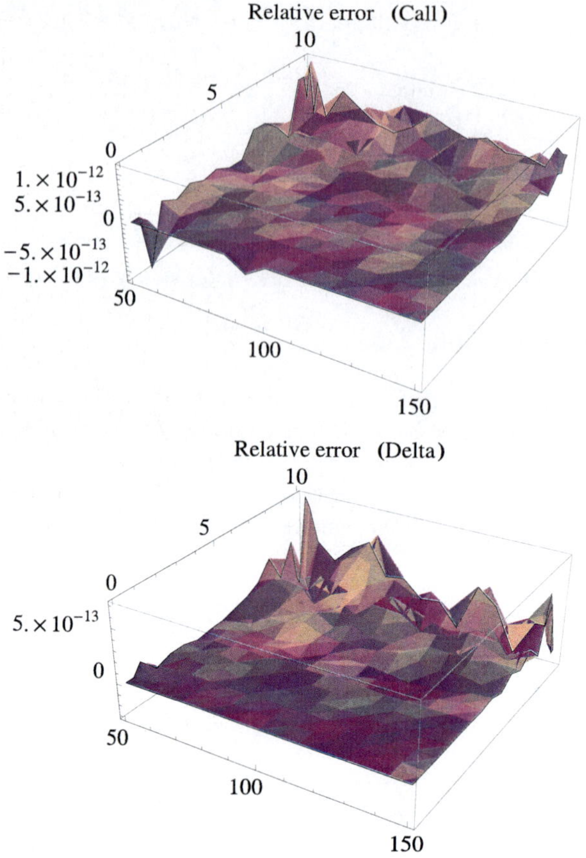

Fig. 15.16. Relative errors for Call price and Delta in the Black-Scholes model, with the parameters $K = 100$, $S_0 \in [50, 150]$, $r = 5\%$ and $\sigma = 30\%$. Moreover, a, b are as in (15.36)-(15.34) and $N = 50$

Remark 15.19 In order to reduce the sensitivity of the method regarding the choice of the parameter L, it is often preferable to price Call options via the Put-Call parity formula as in Remark 15.12. □

In the first test, we consider the Black-Scholes model and compute the relative errors defined as in Section 15.2.3, that is:

$$\text{RelErr}^{\text{Call}} = \frac{\text{Call}(S_0, K, T) - \text{CallBS}(S_0, K, T)}{\text{CallBS}(S_0, K, T)},$$

$$\text{RelErr}^{\text{Delta}} = \frac{\text{Delta}(S_0, K, T) - \text{DeltaBS}(S_0, K, T)}{\text{DeltaBS}(S_0, K, T)},$$

where Call and Delta are the prices computed by the Fourier-cosine approximation with a, b as in (15.36) and $N = 50$. CallBS and DeltaBS denote

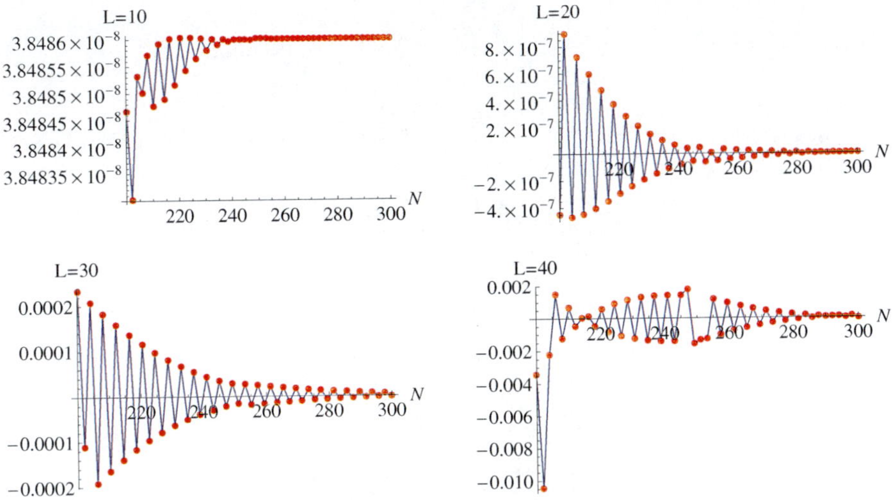

Fig. 15.17. Percentage difference of Fourier-cosine approximation of the Call price in the Heston model: $S_0 = K = 100$, $T = 1$, $L = 10, 20, 30, 40$ and $N \in [200, 300]$. The reference value RV = 5.785155434

the exact values obtained by the Black-Scholes formula. The parameters are $r = 5\%$, $\sigma = 30\%$, $K = 100$ and we consider $S_0 \in [50, 150]$ and maturities from one week, $T = \frac{1}{52}$, to ten years, $T = 10$. In this case, the Fourier-cosine expansion reaches a higher level of accuracy with respect to the Fourier integral approximation, while using significantly less CPU time.

A detailed analysis and several tests to evaluate the efficiency and accuracy of the Fourier-cosine expansion in comparison with other Fourier integral approximations have been performed in [127]: the results are basically in accord with other tests we performed on the variety of models considered in Chapter 10.5 and show that the Fourier-cosine expansion is very fast and robust, also under extreme conditions. Fang and Oosterlee [127] claim that in some cases (eg. the Heston model) the Fourier-cosine method appears to be approximately a factor 20 faster than the FFT method for the same level of accuracy. We also quote the interesting paper [342] where the use of graphics processing units with the Fourier-cosine method is studied.

Let us recall that the method has been proposed in literature only for one-dimensional problems (i.e. one underlying asset): however the generalization to high dimensional option pricing problems seems at hand.

Next we briefly discuss the dependence of the Fourier-cosine expansion on choice of the parameters L in (15.35) and N in the approximation formula (15.30). We consider three models (Heston, VG and CGMY) with the values of the parameters taken by [127].

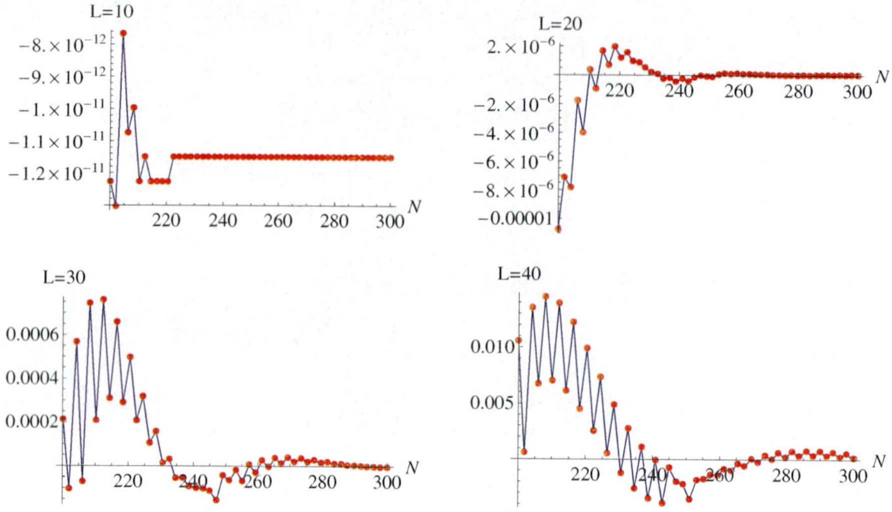

Fig. 15.18. Percentage difference of Fourier-cosine approximation of the Call price in the Heston model: $S_0 = 90$, $K = 100$, $T = \frac{1}{12}$, $L = 10, 20, 30, 40$ and $N \in [200, 300]$. The reference value $\mathrm{RV} = 1.156269518 \times 10^{-3}$

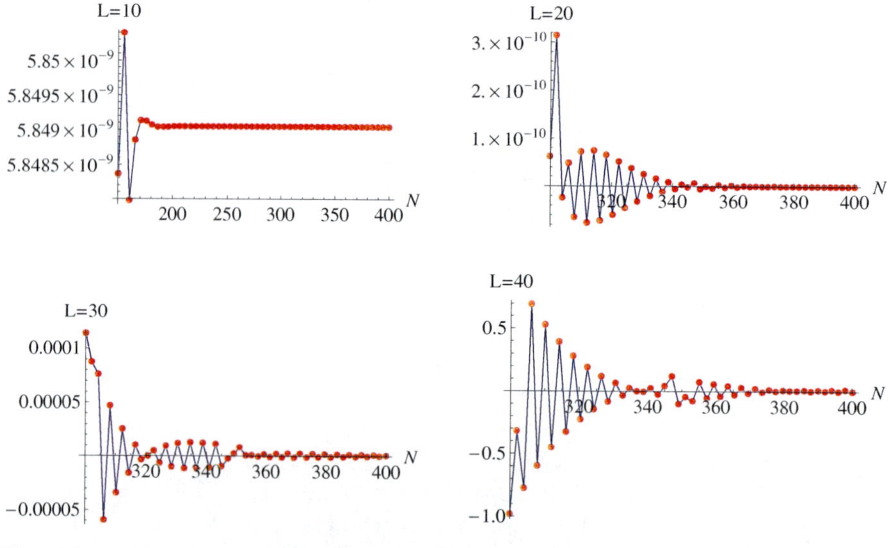

Fig. 15.19. Percentage difference of Fourier-cosine approximation of the Call price in the Heston model: $S_0 = 130$, $K = 100$, $T = 5$, $L = 10, 20, 30, 40$ and $N \in [300, 400]$. The reference value $\mathrm{RV} = 38.14326939$

Example 15.20 (Heston) We consider the Heston model (cf. Example 15.15) with

$$r = 0, \; k = 1.5768, \; \nu_0 = 0.0175, \; \nu_\infty = 0.0398, \; \eta = 0.5751, \; \varrho = -0.5711.$$

We analyze the Fourier-cosine approximation with a, b as in (15.36) and different choices of L and N. The first two cumulants are given by

$$\begin{aligned} c_1 =& e^{-kT} \frac{-u + u_0 + e^{kT}(u - u_0 - Tuk + 2kT\mu)}{2k}, \\ c_2 =& \frac{e^{-2kT}}{8k^3} (u(8e^{kT}k^2 \left(1 + e^{kT}(Tk-1)\right) \\ &+ \eta^2 \left(1 + e^{kT}\left(4 + 4kT + e^{kT}(-5 + 2kT)\right)\right) \\ &- 8e^{kT}\eta k \left(2 + kT + e^{kT}(-2 + kT)\right)\rho) \\ &+ 2u_0(-\eta^2 + e^{2kT}\left(\eta^2 + 4k^2 - 4\eta k \rho\right) \\ &+ 2e^{kT}k(-2k + 2\eta\rho - T\eta(\eta - 2k\rho)))). \end{aligned}$$

We consider the percentage difference, defined as

$$\frac{\text{Call}_{L,N} - \text{RV}}{\text{RV}} \tag{15.37}$$

where RV is the reference value and $\text{Call}_{L,N}$ is the Fourier-cosine approximation of the Call price with integral truncation range L and series truncation index N.

In particular, we compute the price of a Call with $S_0 = K = 100$ and maturity $T = 1$: for $L = 10, 20, 30, 40$ and $N = 1000$, we obtain the same price that is our reference value RV $= 5.785155434$. Figure 15.17 shows the percentage difference (15.37) between RV and the prices obtained by the Fourier-cosine approximation for $N \in [200, 300]$. We see that, as already mentioned, if we truncate the integration domain with a larger interval, then we need to increase the value of N to maintain the accuracy of the approximation. Note that for $L = 10$, the difference is negligible but not identically zero, due to the integration range truncation error: on the contrary, in the other three cases the difference tends to zero as N increases.

We repeat the experiment in other two cases, for short and long maturities. Figure 15.18 presents the percentage differences when $S_0 = 90$, $K = 100$ and $T = \frac{1}{12}$: the reference value is $1.156269518 \times 10^{-3}$ which is obtained with $N = 1000$ and is independent on L.

Finally, we consider the case $S_0 = 130$, $K = 100$ and $T = 5$: the reference value is 38.14326939 which is obtained with $N = 10000$: the percentage difference is represented in Figure 15.19. □

Example 15.21 (Variance-Gamma) We consider the VG model (cf. Example 15.14) with parameters

$$r = 0.1, \qquad \sigma = 0.12, \qquad \mu = -0.14, \qquad \nu = 0.2. \tag{15.38}$$

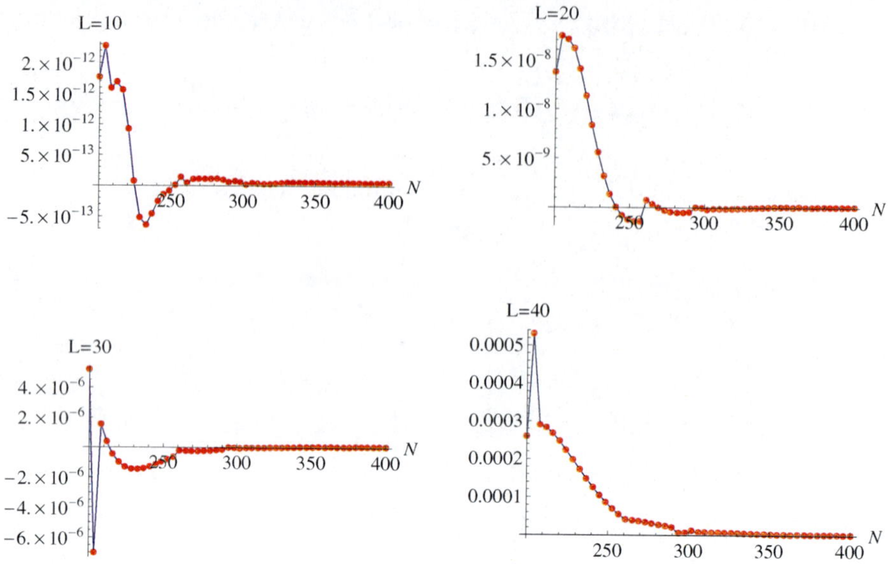

Fig. 15.20. Percentage difference of Fourier-cosine approximation of the Call price in the VG model with parameters as in (15.38). Moreover, $S_0 = 130$, $K = 100$, $T = 5$, $L = 10, 20, 30, 40$ and $N \in [300, 400]$. The reference value RV $= 4.424162989$

We assume a, b of the form (15.35) for typical values of L: the cumulants are given by (see (13.85))

$$\begin{aligned} c_1 &= T\left(r + \mu + \frac{\log\left(1 - \frac{1}{2}\nu\left(2\mu + \sigma^2\right)\right)}{\nu}\right), \\ c_2 &= T\left(\mu^2\nu + \sigma^2\right), \\ c_4 &= 3T\nu\left(2\mu^4\nu^2 + 4\mu^2\nu\sigma^2 + \sigma^4\right). \end{aligned} \tag{15.39}$$

In Figure 15.20 we plot the percentage differences among Fourier-cosine approximations of the Call prices for $L = 10, 20, 30, 40$ and N ranging from 200 to 400. Here $S_0 = 100$, $K = 90$ and $T = 1$: the reference value, computed with $N = 1000$, is 19.099354724. □

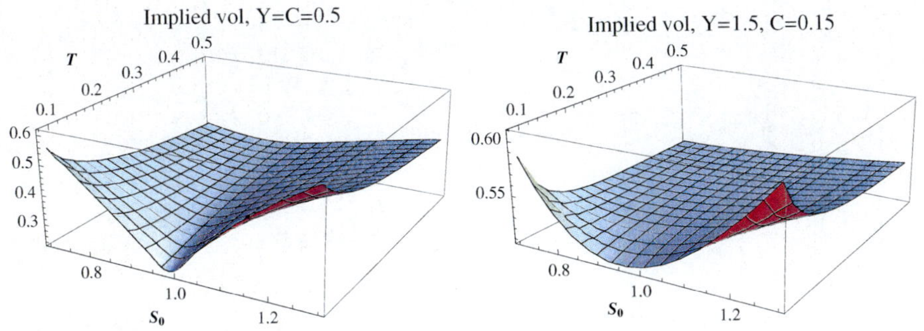

Fig. 15.21. Implied volatility surface in the CGMY model with the parameters as in (15.40) and $Y = C = 0.5$ (left) and $Y = 1.5$, $C = 0.15$ (right)

Example 15.22 (CGMY) The CGMY model is based on a particular tempered stable process (cf. Section 13.4.3) and encompasses the VG and Black-Scholes models: as usual, the underlying asset is in the form $S_T = S_0 e^{X_T}$ and by (13.98)-(13.99) the risk-neutral characteristic exponent given by

$$\psi(\xi) = i\mu_\infty^Q \xi + C\Big((M - i\xi)^Y - M^Y + (G + i\xi)^Y - G^Y + i\xi Y \left(M^{Y-1} - G^{Y-1}\right)\Big)\Gamma(-Y),$$

$$\mu_\infty^Q = r + C\Gamma(-Y)\Big(G^Y - (1+G)^Y + M^Y - (M-1)^Y + Y\left(G^{-1+Y} - M^{-1+Y}\right)\Big),$$

where Γ is the Euler Gamma function. In general, the parameters satisfy $C > 0$, $G > 0$, $M > 1$ and $Y < 2$. For $\sigma = Y = 0$ we get the VG model.

In the following, we set

$$r = 5\%, \quad M = 5, \quad G = 2, \tag{15.40}$$

and we consider different values of Y and C. Figure 15.15 shows the implied volatility surface computed in the CGMY model with the parameters as in (15.40) and $Y = C = 0.5$ (left picture), $Y = 1.5$, $C = 0.15$ (right picture), obtained by the Fourier-cosine expansion with $L = 10$ and $N = 200$.

As in the previous examples, we examine the effect of the choice of the truncation parameters L and N on the quality of the approximation. We

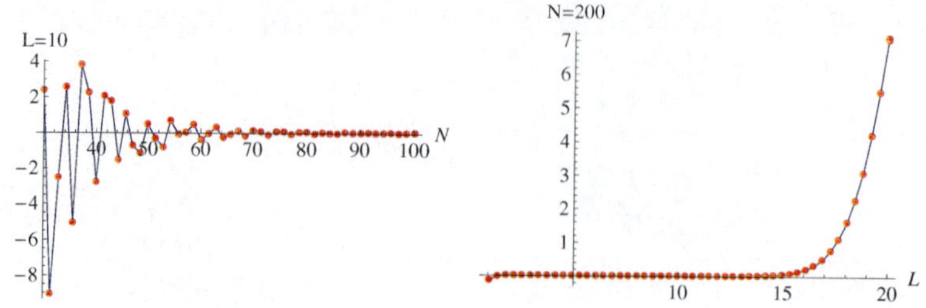

Fig. 15.22. Percentage differences in the Fourier-cosine approximation of the CGMY-Call price as a function of $N \in [30, 100]$ with fixed $L = 10$ (left) and as a function of $L \in [1, 20]$ with fixed $N = 200$ (right). The parameters are as in (15.40), $Y = C = 0.5$, $S_0 = 80$, $K = 100$ and $T = 1$

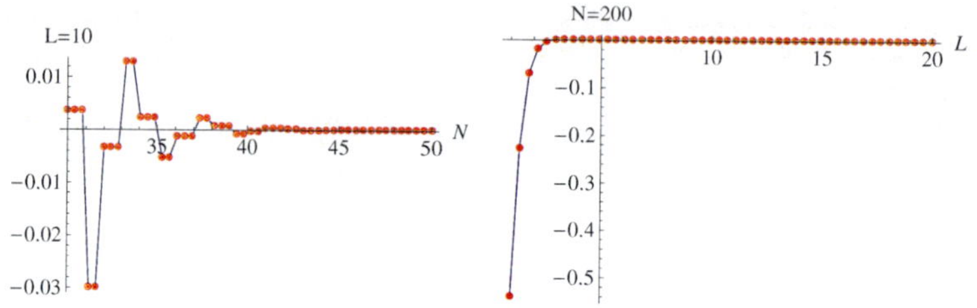

Fig. 15.23. Same as in Figure 15.22 but with $Y = 1.5$ and $C = 0.15$

assume a, b of the form (15.35) where the cumulants are given by

$$
\begin{aligned}
c_1 =& T\Big(r - \frac{\sigma^2}{2} + C\Gamma(-Y)\Big(-(1+G)^Y \\
&+ G^{-1+Y}(G+Y) - (-1+M)^Y + M^{-1+Y}(-Y+M)\Big)\Big), \\
c_2 =& CT(Y-1)Y\Gamma(-Y)\left(G^{Y-2} + M^{Y-2}\right), \\
c_4 =& CT\left(G^{-4+Y} + M^{-4+Y}\right)\Gamma(4-Y).
\end{aligned}
$$

In Figure 15.22 we plot the percentage differences (15.37) among Fourier-cosine approximations of the Call price as a function of $N \in [30, 100]$ with fixed $L = 10$ (left picture) and as a function of $L \in [1, 20]$ with fixed $N = 200$ (right picture). The parameters are as in (15.40), $Y = C = 0.5$, $S_0 = 80$, $K = 100$ and $T = 1$: the reference value, computed with $L = 10$ and $N = 10000$, is 6.112730405.

In Figure 15.23 we repeat the same experiment in the case $Y = 1.5$ and $C = 0.15$: the reference value, computed with $L = 10$ and $N = 10000$, is 11.82598063. For $L \in [5, 15]$, the Fourier-cosine method converges quickly and gives accurate results with a relatively small value of N. As noted in [127], the convergence rate for $Y = 1.5$ is faster than for $Y = 0.5$ since fat-tailed densities can often be well represented by cosine basis functions. In particular

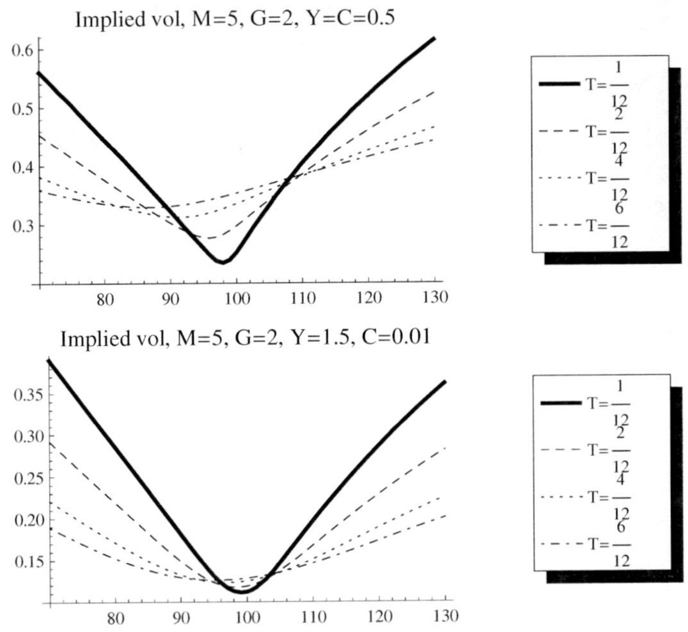

Fig. 15.24. Implied volatility in the CGMY model for different maturities

the method seems to be robust with values of the parameters (for instance, Y close to 2) for which others numerical methods, like partial integro-differential equation methods, have reported to have some difficulty (see [7] and [337]).

In some further experiments, we examine how the implied volatility varies for different choices of the parameters. Figure 15.24 shows the implied volatility in the CGMY model for different maturities ($T = \frac{1}{12}, \frac{2}{12}, \frac{4}{14}, \frac{6}{12}$). Figure 15.25 shows the implied volatility in the CGMY model for different values of the parameter $M = 2, 5, 10, 20$. Figure 15.26 shows the implied volatility in the CGMY model for different values of the parameter $Y = 0.1, 0.3, 0.5, 0.7, 1.1, 1.3, 1.5, 1.7$. □

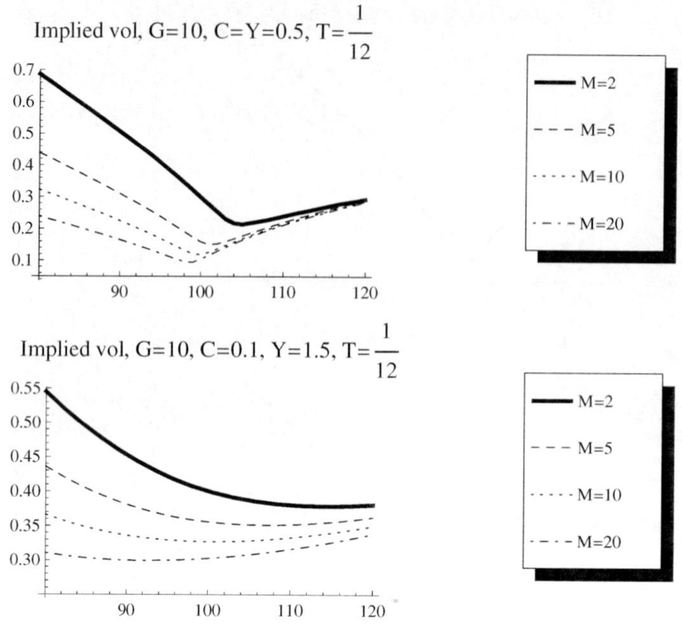

Fig. 15.25. Implied volatility in the CGMY model for different values of M

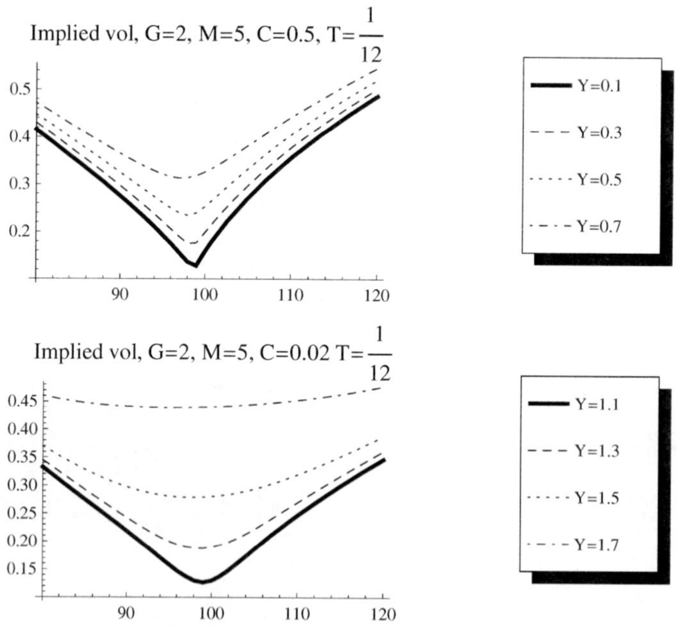

Fig. 15.26. Implied volatility in the CGMY model for different values of Y

16

Elements of Malliavin calculus

This chapter offers a brief introduction to Malliavin calculus and its applications to mathematical finance, in particular the computation of the Greeks by the Monte Carlo method. As we have seen in Section 12.4.2, the simplest way to compute sensitivities by the Monte Carlo method consists in approximating the derivatives by incremental ratios obtained by simulating the payoffs corresponding to close values of the underlying asset. If the payoff function is *not regular* (for example, in the case of a digital option with strike K and payoff function $\mathbb{1}_{[K,+\infty[}$) this technique is not efficient since the incremental ratio has typically a very large variance. In Section 12.4.2 we have seen that the problem can be solved by integrating by parts and differentiating the density function of the underlying asset, provided it is sufficiently regular: if the underlying asset follows a geometric Brownian motion, this is possible since the explicit expression of the density is known.

In a more general setting, the Malliavin calculus allows obtaining explicit integration-by-parts formulas even if the density of the underlying asset is not known and so it provides an effective tool to approximate the Greeks numerically (see, for example, the experiments in [137] where different methods of approximating the Greeks are compared).

The applications of Malliavin calculus to mathematical finance are relatively recent: Malliavin's results [244] initially attracted great interest in view of the proof and extension of Hörmander's hypoellipticity theorem [170] (cf. Section 9.5.2). From a theoretical point of view, a remarkable financial application is the Clark-Ocone formula [270], proved in Paragraph 16.2.1, that improves the martingale representation theorem and allows expressing the hedging strategy of an option in terms of the stochastic derivative of its price.

We also recall that Malliavin calculus was recently used to approximate numerically the price of American options by the Monte Carlo method: see, for instance, Fournié, Lasry, Lebuchoux, Lions and Touzi [138], Fournié, Lasry, Lebuchoux and Lions [137], Kohatsu-Higa and Pettersson [212], Bouchard, Ekeland and Touzi [53], Bally, Caramellino and Zanette [20].

In this chapter we give some basic ideas of Malliavin calculus by analyzing several examples of the applications to the computation of the Greeks. We confine ourselves to the one-dimensional case, choosing simplicity instead of generality; furthermore, some proofs will only be sketched for the sake of brevity. For an organic presentation of the theory, we refer to the monographs by Nualart [267], Shigekawa [308], Sanz-Solé [296], Bell [37], Da Prato [83], Di Nunno, Oksendal and Proske [96]. We mention also some more coincise presentations, mainly application-oriented, that are available on the web: Kohatsu-Higa and Montero [211], Friz [144], Bally [19], Oksendal [272] and Zhang [343].

16.1 Stochastic derivative

In this paragraph we introduce the concept of stochastic (or Malliavin) derivative: the idea is to define the notion of differentiability within the family of random variables that are equal to (or can be approximated by) functions of independent increments of Brownian motion. Under suitable assumptions, we see that this family is wide enough to contain the solution of stochastic differential equations.

Unfortunately the notations that are necessary to introduce Malliavin calculus are a bit burdensome: at the beginning courage must not be lost and a little patience is needed to get acquainted with the notation. On first reading we advise the reader not to dwell too much on the details.

Let us consider a real Brownian motion W on the probability space $(\Omega, \mathcal{F}, P)$, endowed with the Brownian filtration $\mathcal{F}^W = \left(\mathcal{F}_t^W\right)_{t\in[0,T]}$. For the sake of simplicity, since this is not really restrictive, we suppose that $T = 1$ and, for $n \in \mathbb{N}$, let

$$t_n^k := \frac{k}{2^n}, \qquad k = 0, \dots, 2^n$$

be the $(k+1)$-th element of the n-th order dyadic partition of the interval $[0,T]$. Let

$$I_n^k :=]t_n^{k-1}, t_n^k], \qquad \Delta_n^k := W_{t_n^k} - W_{t_n^{k-1}},$$

be the k-th interval of the partition and the k-th increment of the Brownian motion, for $k = 1, \dots, 2^n$, respectively. Furthermore, we denote by

$$\Delta_n := \left(\Delta_n^1, \dots, \Delta_n^{2^n}\right)$$

the $\mathbb{R}^{2^n}$-vector of the n-th order Brownian increments and by C_{pol}^∞ the family of smooth functions that, together with their derivatives of any order, have at most polynomial growth.

Definition 16.1 *Given $n \in \mathbb{N}$, the family of simple n-th order functionals is defined by*

$$\mathcal{S}_n := \{\varphi(\Delta_n) \mid \varphi \in C_{\text{pol}}^\infty(\mathbb{R}^{2^n}; \mathbb{R})\}.$$

We denote by

$$x_n = (x_n^1, \dots, x_n^{2^n}) \tag{16.1}$$

the point in $\mathbb{R}^{2^n}$. It is apparent that $W_T = \varphi(\Delta_n) \in \mathcal{S}_n$ for every $n \in \mathbb{N}$ with $\varphi(x_n^1, \dots, x_n^{2^n}) = x_n^1 + \dots + x_n^{2^n}$.

We also remark that

$$\mathcal{S}_n \subseteq \mathcal{S}_{n+1}, \qquad n \in \mathbb{N},$$

and we define

$$\mathcal{S} := \bigcup_{n \in \mathbb{N}} \mathcal{S}_n,$$

the family of simple functionals. By the growth assumption on φ, $\mathcal{S}$ is a subspace of $L^p(\Omega, \mathcal{F}_T^W)$ for every $p \geq 1$. Further, $\mathcal{S}$ is dense[1] in $L^p(\Omega, \mathcal{F}_T^W)$. We introduce now a very handy notation, that will be often used:

Notation 16.2 *For every $t \in]0, T]$, let $k_n(t)$ be the only element $k \in \{1, \dots, 2^n\}$ such that $t \in I_n^k$.*

Definition 16.3 *For every $X = \varphi(\Delta_n) \in \mathcal{S}$, the stochastic derivative of X at time t is defined by*

$$D_t X := \frac{\partial \varphi}{\partial x_n^{k_n(t)}}(\Delta_n).$$

Remark 16.4 Definition 16.3 is well-posed i.e. *it is independent of n*: indeed it is not difficult to see that, if we have for $n, m \in \mathbb{N}$

$$X = \varphi_n(\Delta_n) = \varphi_m(\Delta_m) \in \mathcal{S},$$

with $\varphi_n, \varphi_m \in C^\infty_{\text{pol}}$, then, for every $t \leq T$, we have

$$\frac{\partial \varphi_n}{\partial x_n^{k_n(t)}}(\Delta_n) = \frac{\partial \varphi_m}{\partial x_m^{k_m(t)}}(\Delta_m).$$

□

Now we endow $\mathcal{S}$ with the norm

$$\begin{aligned} \|X\|_{1,2} :=& E\left[X^2\right]^{\frac{1}{2}} + E\left[\int_0^T (D_s X)^2 ds\right]^{\frac{1}{2}} \\ =& \|X\|_{L^2(\Omega)} + \|DX\|_{L^2([0,T]\times\Omega)}. \end{aligned}$$

Definition 16.5 *The space $\mathbb{D}^{1,2}$ of the Malliavin-differentiable random variables is the closure of $\mathcal{S}$ with respect to the norm $\|\cdot\|_{1,2}$.*

In other terms, $X \in \mathbb{D}^{1,2}$ if and only if there exists a sequence (X_n) in $\mathcal{S}$ such that

[1] Since we are considering the Brownian filtration!

i) $X = \lim_{n\to\infty} X_n$ in $L^2(\Omega)$;
ii) the limit $\lim_{n\to\infty} DX_n$ exists in $L^2([0,T] \times \Omega)$.

In this case it seems natural to define the Malliavin derivative of X as

$$DX := \lim_{n\to\infty} DX_n, \qquad L^2([0,T] \times \Omega).$$

This definition is *well-posed* in view of the following:

Lemma 16.6 *Let (X_n) be a sequence in $\mathcal{S}$ such that*

i) $\lim_{n\to\infty} X_n = 0$ *in* $L^2(\Omega)$;
ii) there exists $U := \lim_{n\to\infty} DX_n$ *in* $L^2([0,T] \times \Omega)$.

Then $U = 0$ *a.e.*[2]

Remark 16.7 The proof of Lemma 16.6 is not obvious since the differentiation operator D is linear but *not bounded*, i.e.

$$\sup_{X\in\mathcal{S}} \frac{\|DX\|_{L^2}}{\|X\|_{L^2}} = +\infty.$$

Indeed it is quite simple to find an example of a sequence (X_n) bounded in $L^2(\Omega)$ and such that (DX_n) is not bounded in $L^2([0,T] \times \Omega)$: for fixed $\bar{n} \in \mathbb{N}$, it suffices to consider $X_n = \varphi_n(\Delta_{\bar{n}})$ with (φ_n) converging in $L^2(\mathbb{R}^{2^{\bar{n}}})$ to a suitable non-regular function. □

We defer the proof of Lemma 16.6 to Paragraph 16.2 and now we analyze some fundamental examples.

16.1.1 Examples

Example 16.8 For fixed t, let us prove that $W_t \in \mathbb{D}^{1,2}$ and[3]

$$D_s W_t = \mathbb{1}_{[0,t]}(s). \tag{16.2}$$

Indeed, recalling Notation 16.2, we consider the sequence

$$X_n = \sum_{k=1}^{k_n(t)} \Delta_n^k, \qquad n \in \mathbb{N}.$$

We have $X_n = W_{t_n^{k_n(t)}} \in \mathcal{S}_n$ and so

$$D_s X_n = \begin{cases} 1 & \text{if } s \le t_n^{k_n(t)}, \\ 0 & \text{if } s > t_n^{k_n(t)}, \end{cases}$$

i.e. $D_s X_n = \mathbb{1}_{[0,t_n^{k_n(t)}]}$. Then (16.2) follows from the fact that

---

[2] In $\mathscr{B} \otimes \mathcal{F}_T^W$.

[3] The stochastic derivative is defined as an L^2-limit, up to sets with null Lebesgue measure: thus, $D_s W_t$ is also equal to $\mathbb{1}_{]0,t[}(s)$ or to $\mathbb{1}_{[0,t]}(s)$.

i) $\lim_{n\to\infty} W_{t_n^{k_n(t)}} = W_t$ in $L^2(\Omega)$;
ii) $\lim_{n\to\infty} \mathbb{1}_{[0,t_n^{k_n(t)}]} = \mathbb{1}_{(0,t)}$ in $L^2([0,T]\times\Omega)$.

□

Remark 16.9 If $X \in \mathbb{D}^{1,2}$ is $\mathcal{F}_t^W$-measurable, then

$$D_s X = 0, \qquad s > t.$$

Indeed, up to approximation, it suffices to consider the case $X = \varphi(\Delta_n) \in \mathcal{S}_n$ for some n: if X is $\mathcal{F}_t^W$-measurable, then it is independent[4] from Δ_n^k for $k > k_n(t)$. Therefore, for fixed $s > t$,

$$\frac{\partial \varphi}{\partial x_n^{k_n(s)}}(\Delta_n) = 0,$$

at least if n is large enough, in such a way that t and s belong to disjoint intervals of the n-th order dyadic partition. □

Example 16.10 Let $u \in L^2(0,T)$ be a (deterministic) function and

$$X = \int_0^t u(r)dW_r.$$

Then $X \in \mathbb{D}^{1,2}$ and

$$D_s X = \begin{cases} u(s) & \text{for } s \le t, \\ 0 & \text{for } s > t. \end{cases}$$

Indeed the sequence defined by

$$X_n = \sum_{k=1}^{k_n(t)} u(t_n^{k-1})\Delta_n^k$$

is such that

$$D_s X_n = \varphi(t_n^{k_n(s)})$$

if $s \le t_n^{k_n(t)}$ and $D_s X_n = 0$ for $s > t_n^{k_n(t)}$. Further, X_n and $D_s X_n$ approximate X and $u(s)\mathbb{1}_{[0,t]}(s)$ in $L^2(\Omega)$ and $L^2([0,T]\times\Omega)$ respectively. □

---

4 Recalling Remark A.43, since $t \in]t_n^{k_n(t)-1}, t_n^{k_n(t)}]$ we have:

i) if $t < t_n^{k_n(t)}$, then X is a function of $\Delta_n^1, \dots, \Delta_n^{k_n(t)-1}$ only;
ii) if $t = t_n^{k_n(t)}$, then X is a function of $\Delta_n^1, \dots, \Delta_n^{k_n(t)}$ only.

16.1.2 Chain rule

If $X, Y \in \mathbb{D}^{1,2}$, then the product XY in general is not square integrable and so it does not belong to $\mathbb{D}^{1,2}$. For this reason, sometimes it is worthwhile to use, instead of $\mathbb{D}^{1,2}$ the slightly smaller space (but closed under products):

$$\mathbb{D}^{1,\infty} = \bigcap_{p \geq 2} \mathbb{D}^{1,p}$$

where $\mathbb{D}^{1,p}$ is the closure of $\mathcal{S}$ with respect to the norm

$$\|X\|_{1,p} = \|X\|_{L^p(\Omega)} + \|DX\|_{L^p([0,T]\times\Omega)}.$$

We observe that $X \in \mathbb{D}^{1,p}$ if and only if there exists a sequence (X_n) in $\mathcal{S}$ such that

i) $X = \lim_{n\to\infty} X_n$ in $L^p(\Omega)$;
ii) the limit $\lim_{n\to\infty} DX_n$ exists in $L^p([0,T]\times\Omega)$.

If $p \leq q$, by Hölder's inequality we get

$$\|\cdot\|_{L^p([0,T]\times\Omega)} \leq T^{\frac{q-p}{pq}} \|\cdot\|_{L^p([0,T]\times\Omega)},$$

and so

$$\mathbb{D}^{1,p} \supseteq \mathbb{D}^{1,q}.$$

In particular, for every $X \in \mathbb{D}^{1,p}$, with $p \geq 2$, and an approximating sequence (X_n) in L^p, we have

$$\lim_{n\to\infty} DX_n = DX, \qquad \text{in } L^2([0,T]\times\Omega).$$

Example 16.11 By using the approximating sequence in Example 16.8, it is immediate to verify that $W_t \in \mathbb{D}^{1,\infty}$ for every t. □

Proposition 16.12 (Chain rule) *Let*[5] $\varphi \in C^\infty_{\text{pol}}(\mathbb{R})$. *Then:*

i) if $X \in \mathbb{D}^{1,\infty}$, *then* $\varphi(X) \in \mathbb{D}^{1,\infty}$ *and*

$$D\varphi(X) = \varphi'(X)DX; \tag{16.3}$$

ii) if $X \in \mathbb{D}^{1,2}$ *and* φ, φ' *are bounded, then* $\varphi(X) \in \mathbb{D}^{1,2}$ *and* (16.3) *holds.*

Further, if $\varphi \in C^\infty_{\text{pol}}(\mathbb{R}^N)$ *and* $X_1, \dots, X_N \in \mathbb{D}^{1,\infty}$, *then* $\varphi(X_1, \dots, X_N) \in \mathbb{D}^{1,\infty}$ *and we have*

$$D\varphi(X_1, \dots, X_N) = \sum_{i=1}^N \partial_{x_i}\varphi(X_1, \dots, X_N) DX_i.$$

[5] Actually it suffices that $\varphi \in C^1$ and that both φ and its first-order derivative have at most polynomial growth.

Proof. We prove only *ii)* since the other parts can be proved essentially in an analogous way. If $X \in \mathcal{S}$, $\varphi \in C^1$ and both φ and its first-order derivative are bounded, then $\varphi(X) \in \mathcal{S}$ and the claim is obvious.

If $X \in \mathbb{D}^{1,2}$, then there exists a sequence (X_n) in $\mathcal{S}$ converging to X in $L^2(\Omega)$ and such that (DX_n) converges to DX in $L^2([0,T] \times \Omega)$. Then, by the dominated convergence theorem, $\varphi(X_n)$ tends to $\varphi(X)$ in $L^2(\Omega)$. Further, $D\varphi(X_n) = \varphi'(X_n)DX_n$ and

$$\|\varphi'(X_n)DX_n - \varphi'(X)DX\|_{L^2} \leq I_1 + I_2,$$

where

$$I_1 = \|(\varphi'(X_n) - \varphi'(X))DX\|_{L^2} \xrightarrow[n\to\infty]{} 0$$

by the dominated convergence theorem and

$$I_1 = \|\varphi'(X_n)(DX - DX_n)\|_{L^2} \xrightarrow[n\to\infty]{} 0$$

since (DX_n) converges to DX and φ' is bounded. □

Example 16.13 By the chain rule, $(W_t)^2 \in \mathbb{D}^{1,\infty}$ and

$$D_s W_t^2 = 2W_t \mathbb{1}_{[0,t]}(s).$$

□

Example 16.14 Let $u \in \mathbb{L}^2$ such that $u_t \in \mathbb{D}^{1,2}$ for every t. Then

$$X := \int_0^t u_r dW_r \in \mathbb{D}^{1,2}$$

and for $s \leq t$

$$D_s \int_0^t u_r dW_r = u_s + \int_s^t D_s u_r dW_r.$$

Indeed, for fixed t, we consider the sequence defined by

$$X_n := \sum_{k=1}^{k_n(t)} u_{t_n^{k-1}} \Delta_n^k, \qquad n \in \mathbb{N},$$

approximating X in $L^2(\Omega)$. Then $X_n \in \mathbb{D}^{1,2}$ and, by the chain rule, we get

$$D_s X_n = u_{t_n^{k_n(s)-1}} + \sum_{k=1}^{k_n(t)} D_s u_{t_n^{k-1}} \Delta_n^k =$$

(since u is adapted and so, by Remark 16.9, $D_s u_{t_n^k} = 0$ if $s > t_n^k$)

$$= u_{t_n^{k_n(s)-1}} + \sum_{k=k_n(s)+1}^{k_n(t)} D_s u_{t_n^{k-1}} \Delta_n^k \xrightarrow[n\to\infty]{} u_s + \int_s^t D_s u_r dW_r$$

in $L^2([0,T] \times \Omega)$. □

Example 16.15 If $u \in \mathbb{D}^{1,2}$ for every t, then we have

$$D_s \int_0^t u_r dr = \int_s^t D_s u_r dr. \qquad \square$$

Example 16.16 Let us consider the solution (X_t) of the SDE

$$X_t = x + \int_0^t b(r, X_r)dr + \int_0^t \sigma(r, X_r)dW_r, \tag{16.4}$$

with $x \in \mathbb{R}$ and the coefficients $b, \sigma \in C_b^1$. Then $X_t \in \mathbb{D}^{1,2}$ for every t and we have

$$D_s X_t = \sigma(s, X_s) + \int_s^t \partial_x b(r, X_r) D_s X_r dr + \int_s^t \partial_x \sigma(r, X_r) D_s X_r dW_r. \tag{16.5}$$

We do not go into the details of the proof of the first claim. The idea is to use an approximation argument based on the Euler scheme (cf. Paragraph 12.2): more precisely, the claim follows from the fact that (X_t) is the limit of the sequence of piecewise constant processes defined by

$$X_t^n = X_{t_n^{k-1}}^n \mathbb{1}_{I_n^k}(t), \qquad t \in [0, T],$$

with $X_{t_n^k}^n$ defined recursively by

$$X_{t_n^k}^n = X_{t_n^{k-1}}^n + b(t_n^{k-1}, X_{t_n^{k-1}}^n)\frac{1}{2^n} + \sigma(t_n^{k-1}, X_{t_n^{k-1}}^n)\Delta_n^k,$$

for $k = 1, \dots, 2^n$. Once we have proved that $X_t \in \mathbb{D}^{1,2}$, (16.5) is an immediate consequence of Examples 16.14, 16.15 and of the chain rule. $\square$

Now we use the classical method of variation of constants to get an explicit expression of $D_s X_t$. Under the assumptions of Example 16.16, we consider the process

$$Y_t = \partial_x X_t, \tag{16.6}$$

solution of the SDE

$$Y_t = 1 + \int_0^t \partial_x b(r, X_r) Y_r dr + \int_0^t \partial_x \sigma(r, X_r) Y_r dW_r. \tag{16.7}$$

Lemma 16.17 *Let Y be as in (16.7) and Z be solution of the SDE*

$$Z_t = 1 + \int_0^t ((\partial_x \sigma)^2 - \partial_x b)(r, X_r) Z_r dr - \int_0^t \partial_x \sigma(r, X_r) Z_r dW_r. \tag{16.8}$$

Then $Y_t Z_t = 1$ for every t.

Proof. We have $Y_0 Z_0 = 1$ and, omitting the arguments, by the Itô formula we have

$$\begin{aligned} d(Y_t Z_t) =& Y_t dZ_t + Z_t dY_t + d\langle Y, Z\rangle_t \\ =& Y_t Z_t \big(\left((\partial_x \sigma)^2 - (\partial_x b) \right) dt - \partial_x \sigma dW_t \\ & + \partial_x b dt + \partial_x \sigma dW_t - (\partial_x \sigma)^2 dt \big) = 0, \end{aligned}$$

and the claim follows by the uniqueness of the representation for an Itô process, Proposition 5.3. □

Proposition 16.18 *Let X, Y, Z be the solutions of the SDEs* (16.4), (16.7) *and* (16.8), *respectively. Then*

$$D_s X_t = Y_t Z_s \sigma(s, X_s). \tag{16.9}$$

Proof. We recall that, for fixed s, the process $D_s X_t$ verifies the SDE (16.5) over $[s, T]$ and we prove that $A_t := Y_t Z_s \sigma(s, X_s)$ verifies the same equation: the claim will then follow from the uniqueness results for SDE.

By (16.7) we have

$$Y_t = Y_s + \int_s^t \partial_x b(r, X_r) Y_r dr + \int_s^t \partial_x \sigma(r, X_r) Y_r dW_r;$$

multiplying by $Z_s \sigma(s, X_s)$ and using Lemma 16.17

$$\begin{aligned} \underbrace{Y_t Z_s \sigma(s, X_s)}_{=A_t} = \underbrace{Y_s Z_s}_{=1} \sigma(s, X_s) &+ \int_s^t \partial_x b(r, X_r) \underbrace{Y_r Z_s \sigma(s, X_s)}_{=A_r} dr \\ &+ \int_s^t \partial_x \sigma(r, X_r) \underbrace{Y_r Z_s \sigma(s, X_s)}_{=A_r} dW_r, \end{aligned}$$

whence the claim. □

Remark 16.19 The concept of stochastic derivative and the results that we proved up to now can be extended to the multi-dimensional case without major difficulties, but for the heavy notation. If $W = (W^1, \dots, W^d)$ is a d-dimensional Brownian motion and we denote the derivative with respect to the i-th component of W by D^i, then we can prove that, for $s \leq t$

$$D_s^i W_t^j = \delta_{ij}$$

where δ_{ij} is Kronecker's delta. More generally, if X is a random variable depending only on the increments of W^j, then $D^i X = 0$ for $i \neq j$. Further, for $u \in \mathbb{L}^2$

$$D_s^i \int_0^t u_r dW_r = u_s^i + \int_s^t D_s^i u_r dW_r.$$

□

16.2 Duality

In this paragraph we introduce the adjoint operator of the Malliavin derivative and we prove a duality result that is the core tool to demonstrate the stochastic integration-by-parts formula.

Definition 16.20 *For fixed $n \in \mathbb{N}$, the family $\mathcal{P}_n$ of the n-th order simple processes consists of the processes U of the form*

$$U_t = \sum_{k=1}^{2^n} \varphi_k(\Delta_n) \mathbb{1}_{I_n^k}(t), \tag{16.10}$$

with $\varphi_k \in C^\infty_{\mathrm{pol}}(\mathbb{R}^{2^n}; \mathbb{R})$ for $k = 1, \dots, 2^n$.

Using Notation 16.2, formula (16.10) can be rewritten more simply as

$$U_t = \varphi_{k_n(t)}(\Delta_n).$$

We observe that

$$\mathcal{P}_n \subseteq \mathcal{P}_{n+1}, \qquad n \in \mathbb{N},$$

and we define

$$\mathcal{P} := \bigcup_{n \in \mathbb{N}} \mathcal{P}_n$$

the family of simple functionals. It is apparent that

$$D : \mathcal{S} \longrightarrow \mathcal{P}$$

i.e. $DX \in \mathcal{P}$ for $X \in \mathcal{S}$. By the growth assumption on the functions φ_k in (16.10), $\mathcal{P}$ is a subspace of $L^p([0,T] \times \Omega)$ for every $p \geq 1$ and furthermore $\mathcal{P}$ is dense in $L^p([0,T] \times \Omega, \mathscr{B} \otimes \mathcal{F}_T^W)$.

Now we recall notation (16.1) and we define the adjoint operator of D.

Definition 16.21 *Given a simple process $U \in \mathcal{P}$ of the form (16.10), we set*

$$D^*U = \sum_{k=1}^{2^n} \left(\varphi_k(\Delta_n)\Delta_n^k - \partial_{x_n^k} \varphi_k(\Delta_n) \frac{1}{2^n} \right). \tag{16.11}$$

D^*U is called Skorohod integral [313] of U: in the sequel we also write

$$D^*U = \int_0^T U_t \diamond dW_t. \tag{16.12}$$

We observe that Definition (16.11) is well-posed since *it does not depend on n*. Further, we note that, differently from the Itô stochastic integral, for the Skorohod integral we do not require the process U to be adapted. For this reason D^* is also called *anticipative stochastic integral.*

Remark 16.22 If U is adapted, then φ_k in (16.10) is $\mathcal{F}^W_{t_n^{k-1}}$-measurable and so, by Remark 16.9, $\partial_{x_n^k}\varphi_k = 0$. Consequently we have

$$\int_0^T U_t \diamond dW_t = \sum_{k=1}^{2^n} \varphi_k(\Delta_n)\Delta_n^k = \int_0^T U_t dW_t.$$

In other terms, *for an adapted stochastic process, the Skorohod integral coincides with the Itô integral.* □

A central result in Malliavin calculus is the following:

Theorem 16.23 (Duality relation) *For every $X \in \mathcal{S}$ and $U \in \mathcal{P}$ we have*

$$E\left[\int_0^T (D_t X) U_t dt\right] = E\left[X \int_0^T U_t \diamond dW_t\right]. \tag{16.13}$$

Remark 16.24 (16.13) can be written equivalently in the form

$$\langle DX, U\rangle_{L^2([0,T]\times\Omega)} = \langle X, D^* U\rangle_{L^2(\Omega)}$$

that justifies calling the Skorohod integral the adjoint operator of D.

Proof. Let U be in the form (16.10) and let $X = \varphi_0(\Delta_m)$ with $\varphi \in C^\infty_{\text{pol}}(\mathbb{R}^{2^m};\mathbb{R})$: evidently it is not restrictive to assume $m = n$. We put $\delta = \frac{1}{2^n}$ and for every $j \in \{1,\dots,2^n\}$ and $k \in \{0,\dots,2^n\}$,

$$\varphi_k^{(j)}(x) = \varphi_k(\Delta_n^1, \dots, \Delta_n^{j-1}, x, \Delta_n^{j+1}, \dots, \Delta_n^{2^n}), \qquad x \in \mathbb{R}.$$

Then we have

$$E\left[\int_0^T (D_t X) U_t dt\right] = \delta E\left[\sum_{k=1}^{2^n} \partial_{x_n^k}\varphi_0(\Delta_n)\varphi_k(\Delta_n)\right] =$$

(since the Brownian increments are independent and identically distributed, $\Delta_n^k \sim \mathcal{N}_{0,\delta}$)

$$= \delta \sum_{k=1}^{2^n} E\left[\int_{\mathbb{R}} \left(\frac{d}{dx}\varphi_0^{(k)}(x)\right)\varphi_k^{(k)}(x)\frac{e^{-\frac{x^2}{2\delta}}}{\sqrt{2\pi\delta}}dx\right] =$$

(integrating by parts)

$$= \delta \sum_{k=1}^{2^n} E\left[\int_{\mathbb{R}} \varphi_0^{(k)}(x)\left(\frac{x}{\delta}\varphi_k^{(k)}(x) - \frac{d}{dx}\varphi_k^{(k)}(x)\right)\frac{e^{-\frac{x^2}{2\delta}}}{\sqrt{2\pi\delta}}dx\right] =$$

$$= E\left[\varphi_0(\Delta_n)\sum_{k=1}^{2^n}\left(\varphi_k(\Delta_n)\Delta_n^k - \partial_{x_n^k}\varphi_k(\Delta_n)\delta\right)\right],$$

and this, in view of the definition of the Skorohod integral, concludes the proof. □

As a consequence of the duality relation, we prove Lemma 16.6.

Proof (of Lemma 16.6). Let (X_n) be a sequence in $\mathcal{S}$ such that

i) $\lim_{n\to\infty} X_n = 0$ in $L^2(\Omega)$;
ii) there exists $U := \lim_{n\to\infty} DX_n$ in $L^2([0,T]\times\Omega)$.

To prove that $U = 0$, we consider $V \in \mathcal{P}$: we have, by *ii)*,

$$E\left[\int_0^T U_t V_t dt\right] = \lim_{n\to\infty} E\left[\int_0^T (D_t X_n) V_t dt\right] =$$

(by the duality relation and then by *i)*)

$$= \lim_{n\to\infty} E\left[X_n \int_0^T V_t \diamond dW_t\right] = 0.$$

The claim follows from the density of $\mathcal{P}$ in $L^2([0,T]\times\Omega, \mathscr{B}\otimes\mathcal{F}_T^W)$. □

Remark 16.25 In an analogous way we prove that, if (U^n) is a sequence in $\mathcal{P}$ such that

i) $\lim_{n\to\infty} U^n = 0$ in $L^2([0,T]\times\Omega)$;
ii) there exists $X := \lim_{n\to\infty} D^* U^n$ in $L^2(\Omega)$,

then $X = 0$ a.s. Then, if $p \geq 2$ and U is such that there exists a sequence (U^n) in $\mathcal{P}$ such that

i) $U = \lim_{n\to\infty} U^n$ in $L^p([0,T]\times\Omega)$;
ii) the limit $\lim_{n\to\infty} D^* U^n$ exists in $L^p(\Omega)$,

we say that U *is p-th order Skorohod-integrable* and the following definition of Skorohod integral is well-posed:

$$D^* U = \int_0^T U_t \diamond dW_t := \lim_{n\to\infty} D^* U^n, \qquad \text{in } L^2(\Omega).$$

Further, the following duality relation

$$E\left[\int_0^T (D_t X) U_t dt\right] = E\left[X \int_0^T U_t \diamond dW_t\right]$$

holds, for every $X \in \mathbb{D}^{1,2}$ and U which is Skorohod-integrable of order two.□

16.2.1 Clark-Ocone formula

The martingale representation theorem asserts that, for every $X \in L^2(\Omega, \mathcal{F}_T^W)$, there exists $u \in \mathbb{L}^2$ such that

$$X = E[X] + \int_0^T u_s dW_s. \tag{16.14}$$

If X is Malliavin differentiable, using Example 16.14 we are able to obtain the expression of u: indeed, formally[6] we have

$$D_t X = u_t + \int_t^T D_t u_s dW_s$$

and so, taking conditional expectation, we can conclude that

$$E\left[D_t X \mid \mathcal{F}_t^W\right] = u_t. \tag{16.15}$$

(16.14)-(16.15) are known as Clark-Ocone formula. Now we proceed to prove it rigorously.

Theorem 16.26 (Clark-Ocone formula) *If* $X \in \mathbb{D}^{1,2}$, *then*

$$X = E[X] + \int_0^T E\left[D_t X \mid \mathcal{F}_t^W\right] dW_t.$$

Proof. It is not restrictive to suppose $E[X] = 0$. For every simple adapted process $U \in \mathcal{P}$ we have, by the duality relation of Theorem 16.23,

$$E[XD^*U] = E\left[\int_0^T (D_t X) U_t dt\right] =$$

(since U is adapted)

$$= E\left[\int_0^T E\left[D_t X \mid \mathcal{F}_t^W\right] U_t dt\right].$$

On the other hand, the Skorohod integral of the adapted process U coincides with the Itô integral and by (16.14) we get

$$E[XD^*U] = E\left[\int_0^T u_t dW_t \int_0^T U_t dW_t\right] =$$

(by Itô isometry)

$$= E\left[\int_0^T u_t U_t dt\right].$$

The claim follows by density, since U is arbitrary. □

Remark 16.27 As an interesting consequence of the Clark-Ocone formula we have that, if $X \in \mathbb{D}^{1,2}$ and $DX = 0$, then X is a.s. constant. □

---

[6] Assuming that $u_t \in \mathbb{D}^{1,2}$ for every t.

Now we dwell on the financial interpretation of the Clark-Ocone formula: we suppose that $X \in L^2(\Omega, \mathcal{F}_T^W)$ is the payoff of a European option on an asset S. We assume that the dynamics of the discounted price under the EMM is given by

$$d\widetilde{S}_t = \sigma_t \widetilde{S}_t dW_t.$$

Then, if (α, β) is a replicating strategy for the option, we have (cf. (10.57))

$$\widetilde{X} = E\left[\widetilde{X}\right] + \int_0^T \alpha_t d\widetilde{S}_t = E\left[\widetilde{X}\right] + \int_0^T \alpha_t \sigma_t \widetilde{S}_t dW_t.$$

On the other hand, by the Clark-Ocone formula we get

$$\widetilde{X} = E\left[\widetilde{X}\right] + \int_0^T E\left[D_t \widetilde{X} \mid \mathcal{F}_t^W\right] dW_t,$$

and so we obtain the expression of the replicating strategy:

$$\alpha_t = \frac{E\left[D_t \widetilde{X} \mid \mathcal{F}_t^W\right]}{\sigma_t \widetilde{S}_t}, \qquad t \in [0, T].$$

16.2.2 Integration by parts and computation of the Greeks

In this section we prove a stochastic integration-by-parts formula and by means of some remarkable examples, we illustrate its application to the computation of the Greeks by the Monte Carlo method. As we have already said in the introduction, the techniques based on Malliavin calculus can be effective also when *poor regularity properties* are assumed on the payoff function F, i.e. just where the direct application of the Monte Carlo method gives unsatisfactory results, even if the underlying asset follows a simple geometric Brownian motion.

The stochastic integration by parts allows removing the derivative of the payoff function, thus improving the numerical approximation: more precisely, let us suppose that we want to determine $\partial_\alpha E[F(S_T)Y]$ where S_T denotes the final price of the underlying asset depending on a parameter α (e.g. α is S_0 in the case of the Delta, α is the volatility in the case of the Vega) and Y is some random variable (e.g. a discount factor). The idea is to try to express $\partial_\alpha F(S_T)Y$ in the form

$$\int_0^T D_s F(S_T) Y U_s ds,$$

for some adapted integrable process U. By using the duality relation, formally we obtain

$$\partial_\alpha E[F(S_T)Y] = E[F(S_T) D^*(YU)],$$

that, as we shall see in the following examples, can be used to get a good numerical approximation.

In this section we want to show how to apply a technique, rather than dwelling on the mathematical details, so the presentation will be somewhat informal, starting already from the next statement.

Theorem 16.28 (Stochastic integration by parts) *Let $F \in C_b^1$ and let $X \in \mathbb{D}^{1,2}$. Then the following integration by parts holds:*

$$E\left[F'(X)Y\right] = E\left[F(X)\int_0^T \frac{u_t Y}{\int_0^T u_s D_s X ds} \diamond dW_t\right], \tag{16.16}$$

for every random variable Y and for every stochastic process u for which (16.16) is well-defined.

Sketch of the proof. By the chain rule we have

$$D_t F(X) = F'(X) D_t X;$$

multiplying by $u_t Y$ and integrating from 0 to T we get

$$\int_0^T u_t Y D_t F(X) dt = F'(X) Y \int_0^T u_t D_t X dt,$$

whence, provided that

$$\frac{1}{\int_0^T u_t D_t X dt}$$

has good integrability properties, we have

$$F'(X)Y = \int_0^T D_t F(X) \frac{u_t Y}{\int_0^T u_s D_s X ds} dt,$$

and, taking the mean

$$E\left[F'(X)Y\right] = E\left[\int_0^T D_t F(X) \frac{u_t Y}{\int_0^T u_s D_s X ds} dt\right] =$$

(by the duality relation)

$$= E\left[F(X)\int_0^T \frac{u_t Y}{\int_0^T u_s D_s X ds} \diamond dW_t\right].$$

□

Remark 16.29 The regularity assumptions on the function F can be greatly weakened: by using a standard regularization procedure, it is possible to prove the validity of the integration-by-parts formula for weakly differentiable (or even differentiable in a distributional sense) functions.

The process u in (16.16) can be often chosen in a suitable way in order to simplify the expression of the integral on the right-hand side (cf. Examples 16.36 and 16.37).

If $u = 1$ and $Y = \partial_\alpha X$, (16.16) becomes

$$E\left[\partial_\alpha F(X)\right] = E\left[F(X)\int_0^T \frac{\partial_\alpha X}{\int_0^T D_s X ds} \diamond dW_t\right]. \tag{16.17}$$

□

In the following Examples 16.30, 16.33 and 16.34, we consider the Black-Scholes dynamics for the underlying asset of an option under the EMM and we apply the integration-by-parts formula with $X = S_T$ where

$$S_T = x\exp\left(\sigma W_T + \left(r - \frac{\sigma^2}{2}\right)T\right). \tag{16.18}$$

Example 16.30 (Delta) We observe that $D_s S_T = \sigma S_T$ and $\partial_x S_T = \frac{S_T}{x}$. Then, by (16.17) we have the following expression for the Black-Scholes Delta

$$\begin{aligned}
\Delta &= e^{-rT}\partial_x E\left[F(S_T)\right] \\
&= e^{-rT} E\left[F(S_T)\int_0^T \frac{\partial_x S_T}{\int_0^T D_s S_T ds} \diamond dW_t\right] \\
&= e^{-rT} E\left[F(S_T)\int_0^T \frac{1}{\sigma T x} dW_t\right] \\
&= \frac{e^{-rT}}{\sigma T x} E\left[F(S_T) W_T\right].
\end{aligned} \tag{16.19}$$

□

We know that in general it is not allowed to "take out" a random variable from an Itô integral (cf. Section 4.3.2): let us see now how this can be made in the case of the anticipative stochastic integral.

Proposition 16.31 *Let $X \in \mathbb{D}^{1,2}$ and let U be a second-order Skorohod-integrable process. Then*

$$\int_0^T X U_t \diamond dW_t = X\int_0^T U_t \diamond dW_t - \int_0^T (D_t X) U_t dt. \tag{16.20}$$

Proof. For every $Y \in \mathcal{S}$, by the duality relation, we have

$$E\left[Y D^*(XU)\right] = E\left[\int_0^T (D_t Y) X U_t dt\right] =$$

(by the chain rule)

$$= E\left[\int_0^T \left(D_t(YX) - Y D_t X\right) U_t dt\right] =$$

(by the duality relation)

$$= E\left[Y\left(XD^*U - \int_0^T D_t X U_t dt\right)\right],$$

and the claim follows by density. □

Formula (16.20) is crucial for the computation of Skorohod integrals. The typical case is when U is adapted: then (16.20) becomes

$$\int_0^T XU_t \diamond dW_t = X\int_0^T U_t dW_t - \int_0^T (D_t X) U_t dt,$$

and so it is possible to express the Skorohod integral as the sum of an Itô integral and of a Lebesgue integral.

Example 16.32 By a direct application of (16.20), we have

$$\int_0^T W_T \diamond dW_t = W_T^2 - T.$$

□

Example 16.33 (Vega) Let us compute the Vega of a European option with payoff function F in the Black-Scholes model: we first notice that

$$\partial_\sigma S_T = (W_T - 2\sigma T)S_T, \qquad D_s S_T \sigma S_T.$$

Then

$$\mathcal{V} = e^{-rT}\partial_\sigma E\left[F(S_T)\right] =$$

(by the integration-by-parts formula (16.17))

$$= e^{-rT} E\left[F(S_T)\int_0^T \frac{W_T - \sigma T}{\sigma T} \diamond dW_t\right] =$$

(by (16.20))

$$= e^{-rT} E\left[F(S_T)\left(\frac{W_T - \sigma T}{\sigma T} W_T - \frac{1}{\sigma}\right)\right].$$

□

Example 16.34 (Gamma) We compute the Gamma of a European option with payoff function F in the Black-Scholes model:

$$\Gamma = e^{-rT}\partial_{xx} E\left[F(S_T)\right] =$$

(by Example 16.30)

$$= \frac{e^{-rT}}{\sigma T} E\left[\partial_x\left(\frac{F(S_T)}{x}\right) W_T\right] = -\frac{e^{-rT}}{\sigma T x^2} E\left[F(S_T)W_T\right] + \frac{e^{-rT}}{\sigma T x} J,$$

where

$$J = E\left[\partial_x F(S_T)W_T\right] = E\left[F'(S_T)\partial_x S_T W_T\right] =$$

(applying (16.16) with $u = 1$ and $Y = (\partial_x S_T)\, W_T = \frac{S_T W_T}{x}$)

$$= E\left[F(S_T)\int_0^T \frac{W_T}{\sigma T x} \diamond dW_T\right] =$$

(by (16.20))

$$= \frac{1}{\sigma T x} E\left[F(S_T)(W_T^2 - T)\right].$$

In conclusion

$$\Gamma = \frac{e^{-rT}}{\sigma T x^2} E\left[F(S_T)\left(\frac{W_T^2 - T}{\sigma T} - W_T\right)\right].$$

□

16.2.3 Examples

Example 16.35 We give the expression of the Delta of an arithmetic Asian option with Black-Scholes dynamics (16.18) for the underlying asset. We denote the average by

$$X = \frac{1}{T}\int_0^T S_t dt$$

and we observe that $\partial_x X = \frac{X}{x}$ and

$$\int_0^T D_s X ds = \int_0^T \int_0^T D_s S_t dt ds = \sigma \int_0^T \int_0^t S_t ds dt = \sigma \int_0^T t S_t dt. \quad (16.21)$$

Then we have

$$\Delta = e^{-rT}\partial_x E\left[F(X)\right] = \frac{e^{-rT}}{x} E\left[F'(X)X\right] =$$

(by (16.17) and (16.21))

$$= \frac{e^{-rT}}{\sigma x} E\left[F(X)\int_0^T \frac{\int_0^T S_s ds}{\int_0^T s S_s ds} \diamond dW_t\right].$$

Now formula (16.20) can be used to compute the anticipative integral: some calculation leads to the following formula (cf., for example, [211]):

$$\Delta = \frac{e^{-rT}}{x} E\left[F(X)\left(\frac{1}{I_1}\left(\frac{W_T}{\sigma} + \frac{I_2}{I_1}\right) - 1\right)\right],$$

where

$$I_j = \frac{\int_0^T t^j S_t dt}{\int_0^T S_t dt}, \qquad j = 1, 2.$$

□

Example 16.36 (Bismut-Elworthy formula) We extend Example 16.30 to the case of a model with local volatility

$$S_t = x + \int_0^t b(s, S_s) ds + \int_0^t \sigma(s, S_s) dW_s.$$

Under suitable assumptions on the coefficients, we prove the following Bismut-Elworthy formula:

$$E\left[\partial_x F(S_T) G\right] = \frac{1}{T} E\left[F(S_T)\left(G \int_0^T \frac{\partial_x S_t}{\sigma(t, S_t)} dW_t - \int_0^T D_t G \frac{\partial_x S_t}{\sigma(t, S_t)} dt\right)\right], \tag{16.22}$$

for every $G \in \mathbb{D}^{1,\infty}$.

We recall that, by Proposition 16.18, we have

$$D_s S_T = Y_T Z_s \sigma(s, S_s), \tag{16.23}$$

since

$$Y_t := \partial_x S_t =: Z_t^{-1}.$$

Let us apply (16.16) after choosing

$$X = S_T, \qquad Y = G Y_T, \qquad u_t = \frac{Y_t}{\sigma(t, S_t)},$$

to get

$$\begin{aligned} E\left[\partial_x F(S_T) G\right] &= E\left[F'(S_T) Y_T G\right] \\ &= E\left[F(S_T) \int_0^T \frac{G Y_T Y_t}{\sigma(t, S_t)} \frac{1}{\int_0^T D_s S_T \frac{Y_s}{\sigma(s, S_s)} ds} \diamond dW_t\right] \end{aligned}$$

(by (16.23))

$$= E\left[F(S_T) \int_0^T \frac{G Y_t}{\sigma(t, S_t)} \diamond dW_t\right]$$

and (16.22) follows from Proposition 16.31, since $\frac{Y_t}{\sigma(t, S_t)}$ is adapted. □

Example 16.37 In this example, taken from [19], we consider the Heston model

$$\begin{cases} dS_t = \sqrt{\nu_t} S_t dB_t^1, \\ d\nu_t = k(\bar{\nu} - \nu_t)dt + \eta\sqrt{\nu_t}dB_t^2, \end{cases}$$

where (B^1, B^2) is a correlated Brownian motion

$$B_t^1 = \sqrt{1-\varrho^2}W_t^1 + \varrho W_t^2, \qquad B_t^2 = W_t^2,$$

with W a standard 2-dimensional Brownian motion and $\varrho \in]-1,1[$. We want to compute the sensitivity of the price of an option with payoff F with respect to the correlation parameter ϱ.

First of all we observe that

$$S_T = S_0 \exp\left(\sqrt{1-\varrho^2}\int_0^T \sqrt{\nu_t}dW_t^1 + \varrho\int_0^T \sqrt{\nu_t}dW_t^2 - \frac{1}{2}\int_0^T \nu_t dt\right),$$

and so

$$\partial_\varrho S_T = S_T G, \qquad G := -\frac{\varrho}{\sqrt{1-\varrho^2}}\int_0^T \sqrt{\nu_t}dW_t^1 + \int_0^T \sqrt{\nu_t}dW_t^2. \tag{16.24}$$

Further, if we denote by D^1 the Malliavin derivative relative to the Brownian motion W^1, by Remark 16.19, we get $D_s^1\nu_t = 0$ and

$$D_s^1 S_T = S_T\sqrt{1-\varrho^2}\sqrt{\nu_s}. \tag{16.25}$$

Then

$$\partial_\varrho E[F(S_T)] = E[F'(S_T)\partial_\varrho S_T] =$$

(by integrating by parts and choosing $X = S_T$, $Y = \partial_\varrho S_T$ and[7] $u_t = \frac{1}{\sqrt{\nu_t}}$ in (16.16))

$$= E\left[F(S_T)\int_0^T \frac{\partial_\varrho S_T}{\sqrt{\nu_t}\int_0^T \frac{D_s^1 S_T}{\sqrt{\nu_s}}ds} \diamond dW_t^1\right] =$$

(by (16.24) and (16.25))

$$= \frac{1}{T\sqrt{1-\varrho^2}}E\left[F(S_T)\int_0^T \frac{G}{\sqrt{\nu_t}} \diamond dW_t^1\right] =$$

(by Proposition 16.31 and since ν is adapted)

$$= \frac{1}{T\sqrt{1-\varrho^2}}E\left[F(S_T)\left(G\int_0^T \frac{1}{\sqrt{\nu_t}}dW_t^1 - \int_0^T \frac{D_t^1 G}{\sqrt{\nu_t}}dt\right)\right] =$$

(since $D_t^1 G = -\varrho\sqrt{\frac{\nu_t}{1-\varrho^2}}$)

$$= \frac{1}{T\sqrt{1-\varrho^2}} E\left[F(S_T)\left(G\int_0^T \frac{1}{\sqrt{\nu_t}} dW_t^1 + \frac{\varrho T}{\sqrt{1-\varrho^2}}\right)\right]. \qquad \square$$

Appendix: a primer in probability and parabolic PDEs

In this Appendix we gather the rudiments of probability theory and we show the connections to parabolic differential equations with constant coefficients. The goal is to collect some background material, assuming the knowledge of standard differential and integral calculus in one or more variables. Some of the classical results are presented without proof and references to the literature are provided.

A.1 Probability spaces

Let Ω be a non-empty set. A *σ-algebra* $\mathcal{F}$ is a collection of subsets of Ω such that:

i) $\Omega \in \mathcal{F}$;
ii) if $F \in \mathcal{F}$ then[1] $F^c := (\Omega \setminus F) \in \mathcal{F}$;
iii) for every sequence $(F_n)_{n\in\mathbb{N}}$ of elements of $\mathcal{F}$, $\bigcup_{n=1}^{\infty} F_n \in \mathcal{F}$.

Let $\mathscr{M}$ be a collection of subsets of Ω. The smallest σ-algebra containing $\mathscr{M}$ is denoted by

$$\sigma(\mathscr{M}) := \bigcap_{\substack{\mathcal{F}\ \sigma-algebra\\ \mathcal{F}\supseteq\mathscr{M}}} \mathcal{F}.$$

We say that $\sigma(\mathscr{M})$ is the *σ-algebra generated by $\mathscr{M}$*. Note that the intersection of σ-algebras is still a σ-algebra.

Example A.1 The Borel σ-algebra $\mathscr{B}(\mathbb{R}^N)$ is the σ-algebra generated by the Euclidean topology of $\mathbb{R}^N$, i.e.

$$\mathscr{B}(\mathbb{R}^N) = \sigma(\{A \mid A \text{ open set in } \mathbb{R}^N\}).$$

[1] When we write $A := B$ we mean that A is equal by definition to B.

Where no confusion can arise, we simply write $\mathscr{B} = \mathscr{B}(\mathbb{R}^N)$. If H is a Borel subset of $\mathbb{R}^N$, we set also $\mathscr{B}(H) = \{H \cap B \mid B \in \mathscr{B}\}$. Note that

$$\mathscr{B}(\mathbb{R}) = \sigma(\mathcal{I}) = \sigma(\mathcal{J}),$$

where $\mathcal{I} = \{\,]a, b[\, \mid a, b \in \mathbb{Q},\ a < b\}$ and $\mathcal{J} = \{\,]-\infty, b] \mid b \in \mathbb{Q}\}$. □

An element of a σ-algebra $\mathcal{F}$ is called a *measurable set*. A *measure* P on $\mathcal{F}$ is a map

$$P : \mathcal{F} \to \mathbb{R}_{\geq 0}$$

such that:

i) $P(\emptyset) = 0$;
ii) P is countably additive that is, for every sequence $(F_n)_{n\in\mathbb{N}}$ of pairwise disjoint elements of $\mathcal{F}$, we have

$$P\Big(\bigcup_{n\geq 1} F_n\Big) = \sum_{n\geq 1} P(F_n).$$

If $P(\Omega) < \infty$, we say that P is a *finite* measure. Further, if

iii) $P(\Omega) = 1$,

then we say that P is a *probability measure*.

A *measure space* is a triple $(\Omega, \mathcal{F}, P)$ with $\mathcal{F}$ a σ-algebra on Ω and P a measure on $\mathcal{F}$. If P is a probability measure then $(\Omega, \mathcal{F}, P)$ is called *probability space* and the set Ω is called *sample space*. A function $f : \Omega \longrightarrow \mathbb{R}^N$ is $\mathcal{F}$-*measurable* (or, simply, measurable) if $f^{-1}(H) \in \mathcal{F}$ for any $H \in \mathscr{B}$.

We can think of every element ω of Ω as the result of an experiment or the state of a phenomenon: for example, the spatial position of a particle or the price of a stock. An element E of $\mathcal{F}$ is also called *event* and $P(E)$ is called *the probability of the event* E. To fix the ideas, if $\Omega = \mathbb{R}_{>0} :=]0, +\infty[$ is the sample space representing the possible prices of a risky asset, then $P(]a, b[)$ represents the probability that the price is greater than a and smaller than b. We say that $E \in \mathcal{F}$ is a *negligible* event (*certain*) if $P(E) = 0$ ($P(E) = 1$). We denote by $\mathcal{N}_P$ the collection of the P-negligible events.

It is not restrictive[2] to assume that P is *complete*, that is for any $A \subseteq E$ with $E \in \mathcal{N}_P$, we have $A \in \mathcal{N}_P$.

---

[2] If P is not complete, we may consider

$$\mathcal{N} = \{A \in \mathcal{F} \mid A \subseteq E \text{ for some } E \in \mathcal{N}_P\},$$

and extend P on $\bar{\mathcal{F}} = \sigma(\mathcal{F} \cup \mathcal{N})$ in a trivial way by using the fact that

$$\bar{\mathcal{F}} = \{B \subseteq \Omega \mid B = E \cup A \text{ with } E \in \mathcal{F},\ A \in \mathcal{N}\}.$$

A.1.1 Dynkin's theorems

Dynkin's theorems are quite technical results that are indeed useful or even essential tools for our analysis. Typically they allow proving the validity of some property for a wide family of measurable sets (or functions), provided that the property is verified for the elements of a particular sub-family: for example, the open intervals in the case of Borel sets, or the characteristic functions of intervals in the case of measurable functions. Dynkin's theorems are based on the following convergence property of measures of monotone sequence of measurable sets.

Lemma A.2 *Let $(\Omega, \mathcal{F}, P)$ be a measurable space and $(A_n)_{n\in\mathbb{N}}$ an increasing*[3] *sequence in $\mathcal{F}$. Then*

$$P\Big(\bigcup_{n\geq 1} A_n\Big) = \lim_{n\to\infty} P(A_n). \tag{A.1}$$

Proof. Since P is countably additive, we have

$$\begin{aligned}
\lim_{n\to\infty} P(A_n) &= \lim_{n\to\infty} P\Big(A_1 + \bigcup_{k=1}^{n-1} (A_{k+1} \setminus A_k)\Big)\\
&= \lim_{n\to\infty} \Big(P(A_1) + \sum_{k=1}^{n-1} P(A_{k+1} \setminus A_k)\Big)\\
&= P(A_1) + \sum_{k=1}^{\infty} P(A_{k+1} \setminus A_k) = P\Big(\bigcup_{n\geq 1} A_n\Big).
\end{aligned}$$

□

Definition A.3 *A family $\mathcal{M}$ of subsets of Ω is called monotone if*

i) $\Omega \in \mathcal{M}$;
ii) if $A, B \in \mathcal{M}$ with $A \subseteq B$ then $B \setminus A \in \mathcal{M}$;
iii) given an increasing sequence $(A_n)_{n\in\mathbb{N}}$ of elements of $\mathcal{M}$ we have that $\bigcup_{n=1}^{\infty} A_n \in \mathcal{M}$.

The only differences between the definitions of σ-algebra and monotone family are in properties *ii)* and *iii)*: in particular, a monotone family is "stable" only under countable unions of *increasing* sequences of measurable sets. Clearly any σ-algebra is monotone. It is also clear that any σ-algebra $\mathcal{F}$ is $\cap$-*stable*, that is the intersections of elements of $\mathcal{F}$ still belongs to $\mathcal{F}$. Conversely, we have:

[3] $(A_n)_{n\in\mathbb{N}}$ is an increasing sequence if $A_n \subseteq A_{n+1}$, for every $n \in \mathbb{N}$.

Lemma A.4 *Every monotone and $\cap$-stable family is a σ-algebra.*

Proof. Clearly if $\mathcal{M}$ is monotone, then it verifies the first two conditions of the definition of σ-algebra. It remains to show that the countable union of elements of $\mathcal{M}$ belongs to $\mathcal{M}$. First of all we observe that, if A and B belong to $\mathcal{M}$, then their union belongs to $\mathcal{M}$ as well: in fact it suffices to note that

$$A \cup B = (A^c \cap B^c)^c.$$

Now, if (A_n) is a sequence of elements of $\mathcal{M}$, we set

$$B_n = \bigcup_{k=1}^{n} A_k.$$

In view of what we have seen earlier, we have that (B_n) is an increasing sequence of elements of $\mathcal{M}$. So, by the third condition in the definition of monotone family, we get

$$\bigcup_{n=1}^{+\infty} A_n = \bigcup_{n=1}^{+\infty} B_n \in \mathcal{M}.$$

□

Theorem A.5 (Dynkin's first theorem) *Let $\mathcal{M}(\mathcal{A})$ be the monotone family generated*[4] *by $\mathcal{A}$, where $\mathcal{A}$ is a family of subsets of Ω. If $\mathcal{A}$ is $\cap$-stable, then*

$$\mathcal{M}(\mathcal{A}) = \sigma(\mathcal{A}). \tag{A.2}$$

Proof. In view of the previous lemma, it suffices to prove that $\mathcal{M}(\mathcal{A})$, denoted by $\mathcal{M}$ for the sake of brevity, is $\cap$-stable. Then it will follow that $\mathcal{M}$ is a σ-algebra and so $\sigma(\mathcal{A}) \subseteq \mathcal{M}$. On the other hand, since every σ-algebra is a monotone family, we have that $\mathcal{M} \subseteq \sigma(\mathcal{A})$ hence (A.2).

We set

$$\mathcal{M}_1 = \{A \in \mathcal{M} \mid A \cap I \in \mathcal{M}, \forall I \in \mathcal{A}\}.$$

We prove that $\mathcal{M}_1$ is a monotone family: since $\mathcal{A} \subseteq \mathcal{M}_1$, it will follow that $\mathcal{M} \subseteq \mathcal{M}_1$ and consequently $\mathcal{M} = \mathcal{M}_1$. We have:

i) $\Omega \in \mathcal{M}_1$;
ii) for any $A, B \in \mathcal{M}_1$ with $A \subseteq B$, we have

$$(B \setminus A) \cap I = (B \cap I) \setminus (A \cap I) \in \mathcal{M}, \qquad I \in \mathcal{A},$$

so that $B \setminus A \in \mathcal{M}_1$;

---

[4] $\mathcal{M}(\mathcal{A})$ is the smallest monotone family containing $\mathcal{A}$.

iii) let (A_n) be an increasing sequence in $\mathcal{M}_1$ and let us denote by A the union of the A_n. Then we have

$$A \cap I = \bigcup_{n \geq 1} (A_n \cap I) \in \mathcal{M}, \qquad I \in \mathcal{A},$$

so that $A \in \mathcal{M}_1$.

This proves that $\mathcal{M} = \mathcal{M}_1$. Now we put

$$\mathcal{M}_2 = \{A \in \mathcal{M} \mid A \cap I \in \mathcal{M}, \forall I \in \mathcal{M}\}.$$

In view of what we have shown above, we have that $\mathcal{A} \subseteq \mathcal{M}_2$. Moreover, following the lines above, we can prove that $\mathcal{M}_2$ is a monotone family: it follows that $\mathcal{M} \subseteq \mathcal{M}_2$ and so $\mathcal{M}$ is $\cap$-stable. □

As a consequence we prove the following useful uniqueness result:

Proposition A.6 *Let $\mathcal{A}$ be a $\cap$-stable family of subsets of Ω. Let P, Q be measures defined on $\sigma(\mathcal{A})$ such that $P(\Omega) = Q(\Omega)$ and*

$$P(M) = Q(M), \qquad M \in \mathcal{A}.$$

Then $P = Q$.

Proof. We set

$$\mathscr{M} = \{M \in \sigma(\mathcal{A}) \mid P(M) = Q(M)\},$$

and we show that $\mathscr{M}$ is a monotone family. Indeed $P(\Omega) = Q(\Omega)$ by assumption, i.e. $\Omega \in \mathscr{M}$, and for any $E, F \in \mathcal{M}$ with $E \subseteq F$, we have

$$P(F \setminus E) = P(F) - P(E) = Q(F) - Q(E) = Q(F \setminus E),$$

so that $(F \setminus E) \in \mathscr{M}$. Finally, if (M_n) is an increasing sequence in $\mathscr{M}$ and M denotes the union of (M_n), then by Lemma A.2

$$P(M) = \lim_{n \to \infty} P(M_n) = \lim_{n \to \infty} Q(M_n) = Q(M),$$

and so $M \in \mathscr{M}$. Since $\mathscr{M}$ is a monotone family containing $\mathcal{A}$ and contained (by construction) in $\sigma(\mathcal{A})$, from Theorem A.5 it follows immediately that $\mathscr{M} = \sigma(\mathcal{A})$. □

Example A.7 The families $\mathcal{I}$ and $\mathcal{J}$ in Example A.1 are $\cap$-stable and generate $\mathscr{B}$. As a consequence of Proposition A.6, in order to prove that two probability measures P, Q on $\mathscr{B}$ are equal it is enough to verify that

$$P(]a, b[) = Q(]a, b[), \qquad a, b \in \mathbb{Q},\ a < b,$$

or that

$$P(] - \infty, b]) = Q(] - \infty, b]), \qquad b \in \mathbb{Q}.$$

An analogous result holds in higher dimension. □

Definition A.8 (Monotone family of functions) *Let $\mathcal{H}$ be a family of bounded functions on Ω with real values. We say that $\mathcal{H}$ is a monotone family of functions if*

i) $\mathcal{H}$ is a (real) linear space;
ii) $\mathcal{H}$ contains the constant function equal to 1;
iii) if (f_n) is an increasing sequence of non-negative functions in $\mathcal{H}$, whose pointwise limit is a bounded function f, then $f \in \mathcal{H}$.

Theorem A.9 (Dynkin's second theorem) *Let $\mathcal{H}$ be a monotone family of functions. If $\mathcal{H}$ contains the indicator functions of the elements of a $\cap$-stable family $\mathcal{A}$, then it contains also every bounded $\sigma(\mathcal{A})$-measurable[5] function.*

Proof. Let us denote by $\mathbb{1}_A$ the indicator function of the set A, defined by

$$\mathbb{1}_A(x) = \begin{cases} 1, & x \in A, \\ 0, & x \notin A. \end{cases}$$

First of all we prove that

$$\mathbb{1}_A \in \mathcal{H}, \qquad \forall A \in \sigma(\mathcal{A}). \tag{A.3}$$

To this end, we set

$$\mathcal{M} = \{A \in \sigma(\mathcal{A}) \mid \mathbb{1}_A \in \mathcal{H}\}.$$

Then $\mathcal{A} \subseteq \mathcal{M}$ by assumption. Moreover $\mathcal{M}$ is a monotone family: indeed we have

i) $\Omega \in \mathscr{M}$ since $1 = \mathbb{1}_\Omega \in \mathcal{H}$;
ii) if $A, B \in \mathscr{M}$ with $A \subseteq B$, then the $B \setminus A \in \mathscr{M}$ because $\mathbb{1}_{B\setminus A} = (\mathbb{1}_B - \mathbb{1}_A) \in \mathcal{H}$;
iii) if $(A_n)_{n\in\mathbb{N}}$ is an increasing sequence in $\mathscr{M}$ then $(\mathbb{1}_{A_n})_{n\in\mathbb{N}}$ is an increasing sequence of non-negative functions converging to the bounded function $\mathbb{1}_A$ where A is the union of (A_n). Then $\mathbb{1}_A \in \mathcal{H}$ so that $A \in \mathscr{M}$.

In view of Theorem A.5 we have $\mathcal{M} = \sigma(\mathcal{A})$ and this proves (A.3). The proof can be concluded now by using the standard results of pointwise approximation of measurable functions: in particular it is well known that, if f is a non-negative bounded and $\sigma(\mathcal{A})$-measurable function there exists an increasing sequence (f_n) of simple non-negative and $\sigma(\mathcal{A})$-measurable functions (and so in $\mathcal{H}$) converging to f. So by iii) in Definition A.8 we have that $f \in \mathcal{H}$. Finally if f, $\sigma(\mathcal{A})$-measurable and bounded, takes both positive and negative values it suffices to decompose it into the difference of its positive and negative parts. □

[5] f is called $\sigma(\mathcal{A})$-measurable if $f^{-1}(H) \in \sigma(\mathcal{A})$ for any $H \in \mathscr{B}$.

As an example of application of the previous theorem we give the following:

Corollary A.10 *Let X, Y be random variables on $(\Omega, \mathcal{F})$. Then X is $\sigma(Y)$-measurable if and only if there exists a $\mathscr{B}$-measurable function f such that $X = f(Y)$.*

Proof. It suffices to consider the case X is bounded, otherwise one can consider the composition of X with a bounded measurable function (e.g. Arctangent). Further, it suffices to prove that, if X is $\sigma(Y)$-measurable, then $X = f(Y)$ for some bounded and $\mathscr{B}$-measurable function f, since the converse is obvious.

We use Dynkin's second theorem and we set

$$\mathcal{H} = \{f(Y) \mid f \text{ bounded and } \mathscr{B}\text{-measurable}\}.$$

Then $\mathcal{H}$ is a monotone family of functions, in fact it is apparent that $\mathcal{H}$ is a linear space containing the constant functions. Further, if $(f_n(Y))_{n\in\mathbb{N}}$ is a monotone increasing sequence of non-negative functions in $\mathcal{H}$ and such that

$$f_n(Y) \le C$$

for some constant C, then, if we put $f = \sup_{n\in\mathbb{N}} f_n$ we have that f is bounded, $\mathscr{B}$-measurable and

$$f_n(Y) \uparrow f(Y) \qquad \text{as } n \to \infty.$$

To conclude, we show that $\mathcal{H}$ contains the characteristic functions of elements of $\sigma(Y)$. If $F \in \sigma(Y) = Y^{-1}(\mathscr{B})$, then there exists $H \in \mathscr{B}$ such that $F = Y^{-1}(H)$ and so

$$\mathbb{1}_F = \mathbb{1}_H(Y)$$

whence we infer that $\mathbb{1}_F \in \mathcal{H}$ for every $F \in \sigma(Y)$. □

A.1.2 Distributions

Probability measures defined on the Euclidean space play an essential role.

Definition A.11 *A probability measure on $(\mathbb{R}^N, \mathscr{B})$ is called distribution.*

The next result is a direct consequence of the well-known properties of the Lebesgue integral: it shows how simple it is to construct a distribution by means of Lebesgue measure.

Proposition A.12 *Let $g : \mathbb{R}^N \longrightarrow \mathbb{R}_{\ge 0}$ be a non-negative $\mathscr{B}$-measurable function such that*

$$\int_{\mathbb{R}^N} g(x)dx = 1.$$

Then P defined by

$$P(H) = \int_H g(x)dx, \qquad H \in \mathscr{B}, \tag{A.4}$$

is a distribution. We say that g is the density of P with respect to Lebesgue measure.

Example A.13 (Uniform distribution) Let $a, b \in \mathbb{R}$ with $a < b$: the distribution with density

$$g(x) = \frac{1}{b-a} \mathbb{1}_{[a,b]}(x), \qquad x \in \mathbb{R}, \tag{A.5}$$

is called uniform distribution on $[a, b]$ and denoted by $\text{Unif}_{[a,b]}$. In what follows we denote the Lebesgue measure of the Borel set H indifferently by $|H|$ or $m(H)$. Then we have

$$\text{Unif}_{[a,b]}(H) = \frac{1}{b-a} |H \cap [a, b]|, \qquad H \in \mathscr{B}.$$

Intuitively, $\text{Unif}_{[a,b]}$ assigns the probability that a "particle" (or the price of an asset) lies in $[a, b]$ uniformly over $[a, b]$: on the contrary, it is impossible that the particle lies outside of $[a, b]$. □

For a distribution P of the form (A.4) we necessarily have

$$|H| = 0 \quad \Longrightarrow \quad P(H) = 0. \tag{A.6}$$

When (A.6) holds, we say that P is *absolutely continuous with respect to Lebesgue measure.* Not all the distributions are of the form (A.4), i.e. not all the distributions have a density with respect to Lebesgue measure as the following example shows.

Example A.14 (Dirac's delta) Let $x_0 \in \mathbb{R}^N$: the Dirac distribution concentrated at x_0 is defined by

$$\delta_{x_0}(H) = \mathbb{1}_H(x_0) = \begin{cases} 1, & x_0 \in H, \\ 0, & x_0 \notin H, \end{cases}$$

for $H \in \mathscr{B}$. Intuitively this distribution represents the certainty of "locating the particle" at the point x_0. This distribution does not have a density with respect to m, since it is not null on the event $\{x_0\}$ that has null Lebesgue measure, so (A.6) is not satisfied for $P = \delta_{x_0}$. □

We consider now other examples of distributions defined by specifying their density with respect to Lebesgue measure.

Example A.15 (Exponential distribution) For any $\lambda > 0$, the distribution with density

$$g_\lambda(t) = \lambda e^{-\lambda t} \mathbb{1}_{]0,+\infty[}(t), \qquad t \in \mathbb{R},$$

is called exponential distribution with parameter λ and denoted by Exp_λ. □

Example A.16 (Cauchy distribution) The distribution with density

$$g(x) = \frac{1}{\pi} \frac{1}{1+x^2}, \qquad x \in \mathbb{R},$$

is called Cauchy distribution.

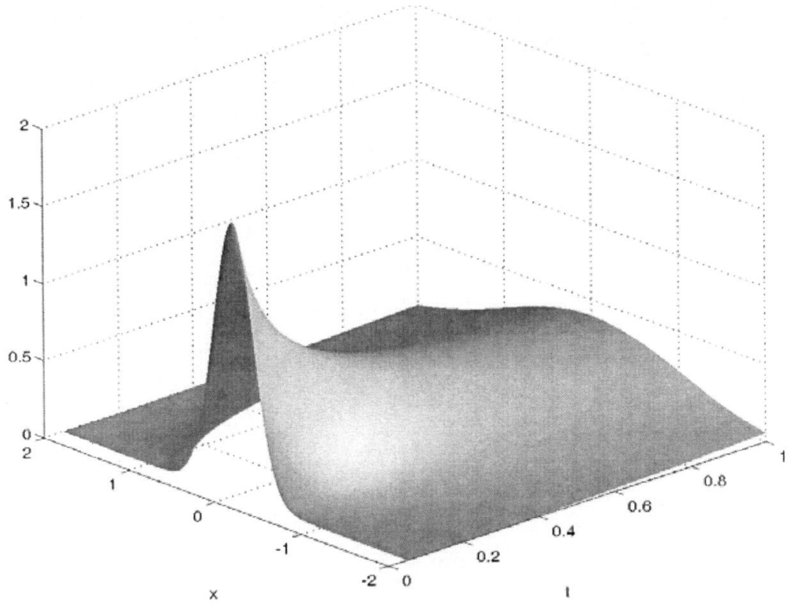

Fig. A.1. Graph of the Gaussian density $\Gamma(t,x)$

□

Example A.17 (Real normal distribution) We set

$$\Gamma(t,x) = \frac{1}{\sqrt{2\pi t}} \exp\left(-\frac{x^2}{2t}\right), \qquad x \in \mathbb{R},\ t > 0. \tag{A.7}$$

For any $\mu \in \mathbb{R}$ and $\sigma > 0$, the distribution $\mathcal{N}_{\mu,\sigma^2}$ with density

$$g(x) = \Gamma(\sigma^2, x - \mu)$$

is called real normal or Gaussian distribution with parameters μ, σ. Then we have

$$\begin{aligned} \mathcal{N}_{\mu,\sigma^2}(H) &= \int_H \Gamma(\sigma^2, x-\mu)dx \\ &= \frac{1}{\sigma\sqrt{2\pi}} \int_H \exp\left(-\frac{1}{2}\left(\frac{x-\mu}{\sigma}\right)^2\right) dx, \qquad H \in \mathscr{B}. \end{aligned}$$

We also extend the definition to $\sigma = 0$ by setting $\mathcal{N}_{\mu,0} = \delta_\mu$. □

Let us remark explicitly that all the functions in the previous examples are densities, i.e. they are $\mathscr{B}$-measurable, non-negative and their integral over $\mathbb{R}$ equals one.

A.1.3 Random variables

A *random variable* (briefly r.v.) on the probability space $(\Omega, \mathcal{F}, P)$, is a measurable function X from Ω with values in $\mathbb{R}^N$, i.e. a function

$$X : \Omega \to \mathbb{R}^N \quad \text{such that} \quad X^{-1}(H) \in \mathcal{F}, \quad H \in \mathscr{B}.$$

Notation A.18 *We denote by $m\mathscr{B}$ (and $m\mathscr{B}_b$) the collection of functions on $\mathbb{R}^N$ with real values that are $\mathscr{B}$-measurable (and bounded and $\mathscr{B}$-measurable).*

Let X be a random variable: we define the map

$$P^X : \mathscr{B} \to [0, 1]$$

by putting

$$P^X(H) = P(X^{-1}(H)), \qquad H \in \mathscr{B}.$$

Then P^X is a distribution that is called *distribution (or law) of* X and we write

$$X \sim P^X.$$

Since

$$X^{-1}(H) = \{\omega \in \Omega \mid X(\omega) \in H\},$$

hereafter we write more intuitively $P(X \in H)$ to denote $P(X^{-1}(H))$. Therefore

$$P^X(H) = P(X \in H)$$

is the probability that the r.v. X belongs to the Borel set H.

Example A.19 In the classical example of dice rolling, we set

$$\Omega = \{(m, n) \in \mathbb{N} \times \mathbb{N} \mid 1 \le m, n \le 6\},$$

$\mathcal{F} = \mathscr{P}(\Omega)$ and we define the measure P by $P(\{(m, n)\}) = \frac{1}{36}$ for every $(m, n) \in \Omega$. We consider the r.v. $X(m, n) = m + n$: then we have

$$P^X(\{7\}) = P(X = 7) = P(X^{-1}(\{7\})) = \frac{6}{36},$$

since there are 6 combinations of throws by which we can obtain 7 among all the possible 36 throws. Analogously we have

$$P(3 \le X < 6) = P\left(X^{-1}([3, 6[)\right) = \frac{2 + 3 + 4}{36} = \frac{1}{4}. \qquad \square$$

Note that different random variables X, Y (possibly defined on different probability spaces as well) can have the same distribution: in that case we write

$$X \overset{d}{=} Y.$$

For instance a r.v. X defined on the probability space $(\Omega, \mathcal{F}, P)$ has the same distribution P^X of the identity r.v. *id*, $id(y) \equiv y$, defined on $(\mathbb{R}, \mathscr{B}, P^X)$. Moreover, if $A, B \in \mathcal{F}$ have the same probability, $P(A) = P(B)$, then $\mathbb{1}_A \overset{d}{=} \mathbb{1}_B$. As a matter of fact, *many financial models are based on the knowledge of the distribution of a r.v. X rather than of its explicit expression and the probability space on which it is defined.*

Remark A.20 Let

$$X : (\Omega, \mathcal{F}) \longrightarrow (\widetilde{\Omega}, \widetilde{\mathcal{F}}),$$

and suppose that $\widetilde{\mathcal{F}} = \sigma(\mathscr{M})$ where $\mathscr{M}$ is (any) collection of subsets of $\widetilde{\Omega}$. We observe that, if

$$X^{-1}(\mathscr{M}) \subseteq \mathcal{F},$$

then X is measurable, i.e. $X^{-1}(\widetilde{\mathcal{F}}) \subseteq \mathcal{F}$. A particularly interesting case is $\widetilde{\Omega} = \mathbb{R}$ and $\mathscr{M} = \{[a, b] \mid a < b\}$.

Indeed

$$\mathcal{G} = \{F \in \widetilde{\mathcal{F}} \mid X^{-1}(F) \in \mathcal{F}\}$$

is a σ-algebra, since

$$X^{-1}(F)^c = X^{-1}(F^c) \quad \text{and} \quad X^{-1}\left(\bigcup_{n \geq 1} F_n\right) = \bigcup_{n \geq 1} X^{-1}(F_n).$$

Further, $\mathscr{M}$ is included in $\mathcal{G}$ and consequently also $\widetilde{\mathcal{F}} = \sigma(\mathscr{M}) \subseteq \mathcal{G}$, i.e. X is measurable. □

Let X be a r.v. on $(\Omega, \mathcal{F}, P)$ with values in $\mathbb{R}^N$. The *distribution function* of X is the function

$$\Phi^X : \mathbb{R}^N \to [0, 1]$$

defined by

$$\Phi^X(y) := P(X \leq y), \qquad y \in \mathbb{R}^N,$$

where $X \leq y$ means that $X_i \leq y_i$ for every $i = 1, \dots, N$.

If X is a real r.v. then the distribution function Φ^X is (not necessarily strictly) monotone increasing and right-continuous. In particular all distribution functions are *càdlàg*[6] functions. Furthermore, we have

$$\lim_{x \to -\infty} \Phi^X(x) = 0, \qquad \lim_{x \to +\infty} \Phi^X(x) = 1.$$

Remark A.21 By Proposition A.6 the distribution function Φ^X determines the distribution P^X uniquely. In the case $N = 1$, if P^X has a density f then

$$\Phi^X(y) = \int_{-\infty}^{y} f(x) dx.$$

[6] Right continuous with left limits, from the French "continue à droite, limitée à gauche".

In particular, if f is continuous at x_0, then Φ^X is differentiable at x_0 and we have

$$\frac{d}{dy}\Phi^X(x_0) = f(x_0). \tag{A.8}$$

More generally, if f is merely integrable, (A.8) holds in a weak sense (cf. Proposition A.162). □

A.1.4 Integration

One of the fundamental concepts associated to a r.v. X is the *mean or expectation*: intuitively it corresponds to an average of the values of X, weighted by the probability P. To make all this rigorous, we have to define the integral of X over the space $(\Omega, \mathcal{F}, P)$:

$$\int_\Omega X dP. \tag{A.9}$$

The construction of the integral in (A.9) is analogous to that of the Lebesgue integral over $\mathbb{R}^N$. Here we only give a succinct outline of the main ideas:

First step. We start by defining the integral of simple random variables. We say that a r.v. $X : \Omega \to \mathbb{R}$ is simple if $X(\Omega)$ has finite cardinality, i.e.

$$X(\Omega) = \{\alpha_1, \dots, \alpha_n\}.$$

In this case, if we set $A_k = X^{-1}(\alpha_k) \in \mathcal{F}$ for $k = 1, \dots, n$, we have

$$X = \sum_{k=1}^{n} \alpha_k \mathbb{1}_{A_k} \tag{A.10}$$

that is, X is a linear combination of indicator functions. In order to have

$$\int_\Omega \mathbb{1}_A dP = P(A), \qquad A \in \mathcal{F},$$

and to make the integral a linear functional, it is natural to define

$$\int_\Omega X dP := \sum_{k=1}^{n} \alpha_k P(A_k) = \sum_{k=1}^{n} \alpha_k P(X = \alpha_k). \tag{A.11}$$

This definition is similar to that of the Riemann integral: as a matter of fact, the concept of simple random variables is analogous to that of piecewise constant functions in $\mathbb{R}$. It is also remarkable that by definition we have

$$\begin{aligned}\int_\Omega f(X) dP &= \sum_{k=1}^{n} f(\alpha_k) P(X = \alpha_k) \\ &= \sum_{k=1}^{n} f(\alpha_k) P^X(f = f(\alpha_k)) = \int_{\mathbb{R}} f dP^X,\end{aligned} \tag{A.12}$$

for any measurable function $f : \mathbb{R} \to \mathbb{R}$.

Second step. For any non-negative real r.v. X we set

$$\int_\Omega X dP = \sup\left\{\int_\Omega Y dP \mid Y \text{ simple r.v., } 0 \le Y \le X\right\}. \tag{A.13}$$

Clearly, the definition (A.13) coincides with (A.11) for simple non-negative random variables, but in general $\int_\Omega X dP \le +\infty$, this meaning that the integral of X may not converge.

Third step. For any real r.v. X, we set

$$X^+ = \max\{0, X\} \qquad \text{and} \qquad X^- = \max\{0, -X\}.$$

Then X^+ and X^- are non-negative random variables and we have $X = X^+ - X^-$. If at least one of the two integrals $\int_\Omega X^+ dP$ and $\int_\Omega X^- dP$ (defined in the second step) are finite, we say that X is *P-semi-integrable* and we define

$$\int_\Omega X dP = \int_\Omega X^+ dP - \int_\Omega X^- dP.$$

In general $\int_\Omega X dP$ can be finite or infinite ($\pm\infty$). If both $\int_\Omega X^+ dP$ and $\int_\Omega X^- dP$ are finite, we say that X is *P-integrable* and we write $X \in L^1(\Omega, P)$: in this case

$$\int_\Omega |X| dP = \int_\Omega X^+ dP + \int_\Omega X^- dP < \infty.$$

Fourth step. Finally, if $X : \Omega \to \mathbb{R}^N$ is a r.v. and $X = (X_1, \dots, X_N)$, we set

$$\int_\Omega X dP = \left(\int_\Omega X_1 dP, \dots, \int_\Omega X_N dP\right).$$

With this definition of the integral, *all the main results of Lebesgue integration theory over* $\mathbb{R}^N$ *hold true:* in particular the fundamental Beppo Levi theorem, Fatou lemma and Lebesgue's dominated convergence theorem.

Notation A.22 *To write explicitly the variable of integration, sometimes we use the notation*

$$\int_\Omega X dP = \int_\Omega X(\omega) P(d\omega).$$

In particular, denoting as usual by m the Lebesgue measure, we write indifferently

$$\int_{\mathbb{R}^N} f\, dm = \int_{\mathbb{R}^N} f(x) m(dx) = \int_{\mathbb{R}^N} f(x) dx.$$

For any $p \ge 1$, we denote by $L^p = L^p(\Omega, \mathcal{F}, P)$ the space of real $\mathcal{F}$-measurable and p-th order P-integrable functions, i.e. such that

$$\|X\|_p := \left(\int_\Omega |X|^p dP\right)^{\frac{1}{p}} < \infty.$$

We note that $\|\cdot\|_p$ is a seminorm[7] in L^p.

[7] In particular $\|X\|_p = 0$ if and only if $X = 0$ a.s.

Theorem A.23 *Let $X : \Omega \to \mathbb{R}^N$ be a r.v. on the probability space $(\Omega, \mathcal{F}, P)$ and $f : \mathbb{R}^N \to \mathbb{R}^n$ a measurable function. Then*

$$f \circ X \in L^1(\Omega, P) \iff f \in L^1(\mathbb{R}^N, P^X)$$

and in this case we have

$$\int_\Omega f(X)dP = \int_{\mathbb{R}^N} f\, dP^X. \tag{A.14}$$

Proof. We only consider the case $N = n = 1$. We set

$$\mathcal{H} = \{f \in m\mathscr{B}_b \mid f \text{ satisfies (A.14)}\}.$$

Then $\mathcal{H}$ contains the indicator functions $f = \mathbb{1}_H$ with $H \in \mathscr{B}$: indeed we have

$$\begin{aligned}\int_\Omega \mathbb{1}_H(X)dP &= \int_{X^{-1}(H)} dP = P(X \in H)\\ &= P^X(H) = \int_H dP^X = \int_{\mathbb{R}} \mathbb{1}_H dP^X.\end{aligned}$$

Moreover, by Beppo Levi theorem, $\mathcal{H}$ is a monotone family (cf. Definition A.8). Then by Dynkin's Theorem A.9, (A.14) holds for any $f \in m\mathscr{B}_b$. Again by Beppo Levi theorem we pass on to measurable and non-negative f. Eventually we decompose f into its negative and positive parts and, again by linearity, we conclude the proof. □

Remark A.24 Let us suppose that the distribution of X is absolutely continuous with respect to Lebesgue measure and so P^X is of the form (A.4) with density g. Then by applying Dynkin's Theorem A.9 as above we can show that, for $f \in L^1(\mathbb{R}^N, P^X)$, we have

$$\int_\Omega f(X)dP = \int_{\mathbb{R}^N} f(x)g(x)dx. \tag{A.15}$$

For example, if X has exponential distribution Exp_λ, then

$$\int_\Omega f(X)dP = \lambda \int_0^{+\infty} f(x)e^{-\lambda x}dx.$$

□

A.1.5 Mean and variance

Let $X : \Omega \to \mathbb{R}$ be an integrable r.v.: the *mean* (or expectation) and the *variance* of X are defined by

$$E[X] = \int_\Omega X dP \quad \text{and} \quad \text{var}(X) = E\left[(X - E[X])^2\right],$$

respectively. The variance estimates how much X differs in average from its expectation. The covariance of two real random variables X, Y is defined by

$$\mathrm{cov}(X,Y) = E\left[(X - E[X])(Y - E[Y])\right], \tag{A.16}$$

provided that $(X - E[X])(Y - E[Y])$ is integrable. We point out that

$$\mathrm{var}(X) = E\left[X^2\right] - E[X]^2, \tag{A.17}$$

and

$$\mathrm{var}(X+Y) = \mathrm{var}(X) + \mathrm{var}(Y) + 2\mathrm{cov}(X,Y). \tag{A.18}$$

If $X = (X_1, \dots, X_N)$ is a r.v. with values in $\mathbb{R}^N$, the *covariance matrix* is defined by

$$\mathrm{Cov}(X) = \left(\mathrm{cov}(X_i, X_j)\right)_{i,j=1,\dots,N}$$

and, in matrix form,

$$\mathrm{Cov}(X) = E\left[(X - E[X])(X - E[X])^*\right].$$

Remark A.25 If X is a real r.v. and $\alpha, \beta \in \mathbb{R}$, then by linearity we have

$$E[\alpha X + \beta] = \alpha E[X] + \beta, \qquad \mathrm{var}(\alpha X + \beta) = \alpha^2 \mathrm{var}(X).$$

More generally, if X is a r.v. in $\mathbb{R}^N$, α is a $(d \times N)$-matrix and $\beta \in \mathbb{R}^d$, then

$$E[\alpha X + \beta] = \alpha E[X] + \beta, \qquad \mathrm{Cov}(\alpha X + \beta) = \alpha \, \mathrm{Cov}(X) \, \alpha^*. \tag{A.19}$$

□

Next we use Theorem A.23 in order to compute mean, variance and distribution function of the distributions examined previously.

Example A.26 (Uniform distribution) Let U be a r.v. with uniform distribution over $[a, b]$:

$$U \sim \mathrm{Unif}_{[a,b]}.$$

The mean of U is

$$E[U] = \int_\Omega U dP = \int_{\mathbb{R}} y P^U(dy) = \int_{\mathbb{R}} \frac{y}{b-a} \mathbb{1}_{[a,b]}(y) dy = \frac{a+b}{2}.$$

Further, we have

$$\mathrm{var}(U) = \int_\Omega (U - E[U])^2 dP = \int_{\mathbb{R}} \left(y - \frac{a+b}{2}\right)^2 P^U(dy) = \frac{(a-b)^2}{12}.$$

Moreover

$$\Phi^U(x) = P(U \le x) = \begin{cases} 0 & x < a, \\ \frac{x-a}{b-a} & a \le x < b, \\ 1 & x \ge b, \end{cases}$$

and in particular

$$\Phi^U(x) = x, \qquad x \in [0,1], \tag{A.20}$$

if $U \sim \mathrm{Unif}_{[0,1]}$.

□

Example A.27 (Dirac distribution) If $X \sim \delta_{x_0}$, then

$$E[X] = \int_{\mathbb{R}} y\delta_{x_0}(dy) = \int_{\{x_0\}} y\delta_{x_0}(dy) = x_0\, \delta_{x_0}(\{x_0\}) = x_0,$$

$$\mathrm{var}(X) = \int_{\mathbb{R}} (y - x_0)^2 \delta_{x_0}(dy) = 0,$$

$$P(X \le x) = \begin{cases} 0 & x < x_0, \\ 1 & x \ge x_0. \end{cases}$$

□

Remark A.28 (Inverse transform method) Given a r.v. X, we consider the inverse $\left(\Phi^X\right)^{-1}$ of the distribution function of X. Clearly $\left(\Phi^X\right)^{-1}$ is well defined if Φ^X is strictly increasing; on the other hand, if $\Phi^X(x) = y$ for more that one value of x, we may set

$$\left(\Phi^X\right)^{-1}(y) = \inf\{x \mid \Phi^X(x) = y\}.$$

Note that, if Φ^X is constant over $[a, b]$, then

$$0 = \Phi^X(b) - \Phi^X(a) = P\left(a < X \le b\right),$$

that is X assumes the values in $[a, b]$ with null probability. We show that

$$X \overset{d}{=} \left(\Phi^X\right)^{-1}(U), \qquad U \sim \mathrm{Unif}_{[0,1]}, \tag{A.21}$$

that is, X and $\left(\Phi^X\right)^{-1}(U)$ have the same distribution: in particular, *the problem of simulating the r.v. X can be reduced to the simulation of a standard uniformly distributed variable.*

In order to prove (A.21), by Remark A.21 it suffices to prove that X and $\left(\Phi^X\right)^{-1}(U)$ have the same distribution function:

$$P\left(\left(\Phi^X\right)^{-1}(U) \le x\right) = P\left(U \le \Phi^X(x)\right) = \Phi^X(x),$$

by (A.20).

□

Example A.29 (Exponential distribution) If $T \sim \mathrm{Exp}_\lambda$ then

$$E[T] = \int_0^{+\infty} t\lambda e^{-\lambda t} dt = \frac{1}{\lambda}, \qquad \mathrm{var}(T) = \frac{1}{\lambda^2}.$$

Moreover we have

$$\Phi^T(t) = P(T \le t) = 1 - e^{-\lambda t} \quad \text{and} \quad P(T > t) = e^{-\lambda t}, \qquad t \ge 0.$$

The r.v. T can be interpreted as the random time at which a particular event occurs. The following important property of "absence of memory" holds:

$$P(T > t + s \mid T > t) = \frac{e^{-\lambda(t+s)}}{e^{\lambda t}} = P(T > s), \qquad t, s \ge 0,$$

that is, the distribution of $T - t$, knowing that $T > t$, is the same as the distribution of T itself.

The distribution function Φ^T is invertible and its inverse is given by

$$\left(\Phi^T\right)^{-1}(y) = -\frac{1}{\lambda}\log(1-y), \qquad y \in [0,1[.$$

Then, by the inverse transform method of Remark A.28, we have

$$T \overset{d}{=} -\frac{1}{\lambda}\log U, \qquad U \sim \text{Unif}_{[0,1]}, \tag{A.22}$$

because U and $1-U$ have the same distribution. This result is useful to simulate random variables with exponential distribution arising in models with jumps. □

Example A.30 (Cauchy distribution) Since the function $g(y) = y$ is not integrable with respect to the Cauchy distribution, the mean of a r.v. with Cauchy distribution is not defined. □

Example A.31 (Normal distribution) If $X \sim \mathcal{N}_{\mu,\sigma^2}$, then

$$E[X] = \int_{\mathbb{R}} y\mathcal{N}_{\mu,\sigma^2}(dy) = \int_{\mathbb{R}} \frac{y}{\sigma\sqrt{2\pi}} \exp\left(-\frac{(y-\mu)^2}{2\sigma^2}\right) dy =$$

(with the change of variables $z = \frac{y-\mu}{\sigma\sqrt{2}}$)

$$= \frac{1}{\sqrt{\pi}}\int_{\mathbb{R}} ze^{-z^2}dz + \frac{\mu}{\sqrt{\pi}}\int_{\mathbb{R}} e^{-z^2}dy = \mu.$$

Further,

$$\text{var}(X) = \int_{\Omega}(X-\mu)^2 dP = \int_{\mathbb{R}} \frac{(y-\mu)^2}{\sigma\sqrt{2\pi}} \exp\left(-\frac{(y-\mu)^2}{2\sigma^2}\right) dy =$$

(with the change of variables $z = \frac{y-\mu}{\sigma\sqrt{2}}$)

$$= \sigma^2 \int_{\mathbb{R}} \frac{2z^2}{\sqrt{\pi}} e^{-z^2} dz = \sigma^2.$$

As an exercise, the reader should verify the last equality by integrating by parts. □

Remark A.32 Given $X \sim \mathcal{N}_{\mu,\,\sigma^2}$ and $\alpha, \beta \in \mathbb{R}$, we have

$$(\alpha X + \beta) \sim \mathcal{N}_{\alpha\mu+\beta,\,\alpha^2\sigma^2}.$$

Indeed, for $\alpha = 0$ the result is obvious, and if $\alpha \neq 0$, then, by Theorem A.23, for every $H \in \mathscr{B}$ we have

$$P((\alpha X + \beta) \in H) = \int_{\Omega} \mathbb{1}_H(\alpha X + \beta)dP = \int_{\mathbb{R}} \frac{\mathbb{1}_H(\alpha y + \beta)}{\sigma\sqrt{2\pi}} e^{-\frac{(y-\mu)^2}{2\sigma^2}} dy =$$

(with the change of variables $z = \alpha y + \beta$)

$$= \int_H \frac{1}{\sigma\alpha\sqrt{2\pi}} e^{-\frac{(z-\alpha\mu-\beta)^2}{2\alpha^2\sigma^2}} dz = \mathcal{N}_{\alpha\mu+\beta,\alpha^2\sigma^2}(H).$$

In particular

$$\frac{X-\mu}{\sigma} \sim \mathcal{N}_{0,1} \tag{A.23}$$

where $\mathcal{N}_{0,1}$ is called *standard normal distribution*. Further,

$$P(X \le y) = P\left(\frac{X-\mu}{\sigma} \le \frac{y-\mu}{\sigma}\right) = \Phi\left(\frac{y-\mu}{\sigma}\right) \tag{A.24}$$

where

$$\Phi(x) = \frac{1}{\sqrt{2\pi}} \int_{-\infty}^{x} e^{-\frac{y^2}{2}} dy, \tag{A.25}$$

is called *standard normal distribution function*. It is easy to verify the following useful property of Φ:

$$\Phi(-x) = 1 - \Phi(x), \qquad x \in \mathbb{R}. \tag{A.26}$$

□

Remark A.33 (Change of density) Let Z be a real random variable with density f and $F \in C^1(\mathbb{R})$ a strictly increasing function. Then the random variable $X = F(Z)$ has density

$$f(G(x))G'(x), \qquad x \in F(\mathbb{R}),$$

where $G = F^{-1}$ is the inverse function of F. Indeed, for every $\varphi \in m\mathscr{B}_b$, we have

$$E[\varphi(Y)] = E[\varphi(F(Z))] = \int_{\mathbb{R}} \varphi(F(z))f(z)dz =$$

(with the change of variables $z = G(x)$)

$$= \int_{F(\mathbb{R})} \varphi(x)f(G(x))G'(x)dx.$$

□

Example A.34 (Log-normal distribution) If $X = e^Z$ with $Z \sim \mathcal{N}_{\mu,\sigma^2}$, then we say that X has a log-normal distribution. If $W \sim \mathcal{N}_{0,1}$ then we have

$$E\left[e^{\sigma W}\right] = \frac{1}{\sqrt{2\pi}} \int_{\mathbb{R}} e^{\sigma x - \frac{x^2}{2}} dx = \frac{e^{\frac{\sigma^2}{2}}}{\sqrt{2\pi}} \int_{\mathbb{R}} e^{-\frac{(x-\sigma)^2}{2}} dx = e^{\frac{\sigma^2}{2}},$$

and so

$$E[X] = e^{\mu + \frac{\sigma^2}{2}}, \tag{A.27}$$

and

$$\mathrm{var}(X) = E\left[X^2\right] - E\left[X\right]^2 = e^{2\mu+\sigma^2}\left(e^{\sigma^2} - 1\right). \tag{A.28}$$

Using Remark A.33, we derive the expression of the log-normal density: in this case $F(z) = e^z$ and $G(x) = F^{-1}(x) = \log x$, $x \in F(\mathbb{R}) = \mathbb{R}_{>0}$. Therefore the density of X is

$$\frac{1}{x\sqrt{2\pi\sigma^2}} \exp\left(-\frac{(\mu - \log x)^2}{2\sigma^2}\right), \qquad x > 0.$$

We also refer to Section 5.1.4. □

Example A.35 (Chi-square distribution) Let $X \sim \mathcal{N}_{0,1}$. The chi-square distribution is the distribution of the r.v. X^2. Clearly, for $y \leq 0$, we have

$$P(X^2 \leq 0) = 0;$$

and for $y > 0$ we have

$$\begin{aligned} P(X^2 \leq y) &= P\left(-\sqrt{y} \leq X \leq \sqrt{y}\right) \\ &= \frac{1}{\sqrt{2\pi}} \int_{-\sqrt{y}}^{\sqrt{y}} e^{-\frac{x^2}{2}} dx = \frac{1}{\sqrt{2\pi}} \int_0^{\sqrt{y}} 2e^{-\frac{x^2}{2}} dx = \end{aligned}$$

(with the change of variables $\xi = x^2$)

$$= \frac{1}{\sqrt{2\pi}} \int_0^y \frac{e^{-\frac{\xi}{2}}}{\sqrt{\xi}} d\xi.$$

In conclusion, recalling (A.8), the chi-square density is given by

$$f(y) = \begin{cases} 0, & y \leq 0, \\ \frac{1}{\sqrt{2\pi y}} e^{-\frac{y}{2}}, & y > 0. \end{cases}$$

If Y has a chi-square distribution, then

$$E\left[Y\right] = E\left[X^2\right] = 1$$

and

$$\mathrm{var}(Y) = E\left[Y^2\right] - E\left[Y\right]^2 = E\left[X^4\right] - 1 = 2.$$

□

Exercise A.36 Consider a r.v. X having as a distribution a linear combination of Dirac's deltas:

$$X \sim p\delta_u + (1-p)\delta_d,$$

where $p \in]0,1[$ and $u, d \in \mathbb{R}$, $d < u$. Therefore X can assume only two values: u with probability p and d with probability $1-p$. By (A.11), we have

$$E\left[X\right] = pu + (1-p)d.$$

Prove that

$$\operatorname{var}(X) = (u-d)^2 p(1-p) = (u - E[X])(E[X] - d), \tag{A.29}$$

and

$$E\left[X^2\right] = (u+d)E[X] - ud. \tag{A.30}$$

A.1.6 σ-algebras and information

Given a r.v. X on the probability space $(\Omega, \mathcal{F}, P)$, we denote by $\sigma(X)$ the *σ-algebra generated by X*, i.e. the σ-algebra generated by the inverse images under X of the Borel sets:

$$\sigma(X) = \sigma\left(\{X^{-1}(H) \mid H \in \mathscr{B}\}\right).$$

Clearly $\sigma(X) \subseteq \mathcal{F}$ and we have

$$\sigma(X) = X^{-1}(\mathscr{B}) = \{X^{-1}(H) \mid H \in \mathscr{B}\}.$$

In many applications and especially in mathematical finance, σ-algebras are routinely used to represent the concept of information. To clear up this statement, that might sound a little obscure, let us consider the following simple example.

Example A.37 We aim to study the probability that, rolling a die, the outcome is an even or an odd number. Thus we let $\Omega = \{n \in \mathbb{N} \mid 1 \le n \le 6\}$, $\mathcal{F}$ the collection of all subsets of Ω and

$$X(n) = \begin{cases} 1, & \text{if } n \text{ is even,} \\ -1, & \text{if } n \text{ is odd.} \end{cases}$$

We have

$$\sigma(X) = \{\emptyset, \Omega, \{2,4,6\}, \{1,3,5\}\}$$

which is strictly contained in $\mathcal{F}$: in order to study the phenomenon described by X, it is necessary to know the probability of the events in $\sigma(X)$. In this sense $\sigma(X)$ contains the information on X. □

Moreover, let X, Y be two random variables on $(\Omega, \mathcal{F})$: to fix the ideas, we can think of X and Y as the price of two risky assets. The condition "X is $\sigma(Y)$-measurable" is often understood as "*X depends on the information on Y*" (or simply on Y). From a mathematical point of view, this is justified by Corollary A.10, stating that X is $\sigma(Y)$-measurable if and only if X is function of Y or, more precisely, there exists a $\mathscr{B}$-measurable function f such that $X = f(Y)$. More generally, if $\mathcal{G}$ is a σ-algebra and X is $\mathcal{G}$-measurable, then we say that *X depends on the information contained in $\mathcal{G}$*.

Example A.38 If X is measurable with respect to the trivial σ-algebra $\mathcal{F} = \{\emptyset, \Omega\}$, then X is constant. Indeed, for a fixed $\bar{\omega} \in \Omega$ we put $a = X(\bar{\omega})$. Then $X^{-1}(\{a\}) \neq \emptyset$ but, by assumption, $X^{-1}(\{a\}) \in \mathcal{F}$ and so $X^{-1}(\{a\}) = \Omega$ i.e. $X(\omega) = a$ for every $\omega \in \Omega$.

More generally, if X is measurable with respect to the σ-algebra $\sigma(\mathcal{N})$ that contains only negligible and certain events, then X is almost surely constant. Proving this fact in detail is a little bit more tricky: since

$$\Omega = \bigcup_{n \geq 1} X^{-1}([-n, n])$$

and X is $\sigma(\mathcal{N})$-measurable by assumption, there exists $\bar{n} \in \mathbb{N}$ such that $P(X^{-1}([-\bar{n}, \bar{n}])) = 1$. Now we can construct two sequences a_n, b_n such that

$$-\bar{n} \leq a_n \leq a_{n+1} < b_{n+1} \leq b_n \leq \bar{n}, \qquad n \in \mathbb{N},$$

and

$$\lim_{n \to \infty} a_n = \lim_{n \to \infty} b_n =: \ell$$

with $P(A_n) = 1$ for every $n \in \mathbb{N}$ where $A_n := X^{-1}([a_n, b_n])$. Finally $P(A) = 1$ where

$$A = X^{-1}(\{\ell\}) = \bigcap_{n \geq 1} A_n$$

and this proves the claim. □

A.1.7 Independence

Given a probability space $(\Omega, \mathcal{F}, P)$ and a non-negligible event B, the conditional probability $P(\cdot \mid B)$ given B is the probability measure on $(\Omega, \mathcal{F})$ defined by

$$P(A|B) = \frac{P(A \cap B)}{P(B)}, \qquad A \in \mathcal{F}.$$

Intuitively the conditional probability $P(A|B)$ represents the probability that the event A occurs, if B has occurred as well. It is easy to verify that $P(\cdot \mid B)$ is a probability measure on $(\Omega, \mathcal{F})$.

Definition A.39 *We say that two events $A, B \in \mathcal{F}$ are independent if:*

$$P(A \cap B) = P(A)P(B). \tag{A.31}$$

If $P(B) > 0$, (A.31) is equivalent to $P(A|B) = P(A)$, i.e. the probability of the event A is independent of the fact that B has or has not occurred. We observe that the property of independence depends on the probability measure that we consider: in other words, two events may be independent under a measure but not independent under another one. As an exercise, prove that, if two events A, B are independent, then their complements A^c, B^c are independent as well. Further, also A^c and B are independent.

Definition A.40 *We say that the two collections $\mathcal{G}, \mathcal{H}$ of events in Ω are independent if*

$$P(A \cap B) = P(A)P(B), \qquad A \in \mathcal{G},\ B \in \mathcal{H}.$$

We say that two random variables X, Y on $(\Omega, \mathcal{F}, P)$ are independent if the correspondent σ-algebras $\sigma(X)$ and $\sigma(Y)$ are independent.

We observe that it is not possible to establish whether two random variables are independent knowing only their distribution. In practice, to verify that two σ-algebras or two random variables are independent, one can use the following result that is a useful consequence of Dynkin's first Theorem A.5.

Lemma A.41 *Consider the σ-algebras $\mathcal{G} = \sigma(\mathcal{I})$ and $\mathcal{H} = \sigma(\mathcal{J})$ generated by the collections of events $\mathcal{I}, \mathcal{J}$ and suppose that $\mathcal{I}, \mathcal{J}$ are $\cap$-stable. Then $\mathcal{G}$ and $\mathcal{H}$ are independent if and only if $\mathcal{I}$ and $\mathcal{J}$ are independent.*

Proof. Let us suppose that $\mathcal{I}, \mathcal{J}$ are independent. For fixed $I \in \mathcal{I}$, the measures

$$H \mapsto P(I \cap H), \qquad H \mapsto P(I)P(H),$$

are equal for $H \in \mathcal{J}$, they satisfy the assumptions of Proposition A.6 and so they coincide on $\mathcal{H} = \sigma(\mathcal{J})$. Since I is arbitrary, we have thus proved that

$$P(I \cap H) = P(I)P(H), \qquad I \in \mathcal{I},\ H \in \mathcal{H}.$$

Now, for a fixed $H \in \mathcal{H}$, we apply Proposition A.6 again, in order to prove that the measures

$$G \mapsto P(G \cap H), \qquad G \mapsto P(G)P(H), \qquad G \in \mathcal{G},$$

coincide and this concludes the proof. □

As a consequence of Lemma A.41, we have that two real random variables X, Y on $(\Omega, \mathcal{F}, P)$ are independent if and only if

$$P(X \le x,\ Y \le y) = P(X \le x)P(Y \le y), \qquad x, y \in \mathbb{R}.$$

Indeed the families $\mathcal{I} = \{X^{-1}(]-\infty, x]) \mid x \in \mathbb{R}\}$ and $\mathcal{J} = \{Y^{-1}(]-\infty, y]) \mid y \in \mathbb{R}\}$ are $\cap$-stable and moreover $\sigma(X) = \sigma(\mathcal{I})$ and $\sigma(Y) = \sigma(\mathcal{J})$.

The next exercise will be useful later on, so we strongly suggest that the reader does it now.

Exercise A.42 Let X, Y be independent random variables. Prove that:

i) if Z is a $\sigma(Y)$-measurable r.v., then X and Z are independent;

ii) if f, g are real $\mathscr{B}$-measurable functions, then the random variables $f(X)$ and $g(Y)$ are independent.

Remark A.43 If X, Y are independent random variables and X is $\sigma(Y)$-measurable, then X is a.s. constant. Indeed

$$P(A \cap B) = P(A)P(B), \qquad A \in \sigma(X),\ B \in \sigma(Y),$$

and since $\sigma(X) \subseteq \sigma(Y)$, then we have

$$P(A) = P(A)^2, \qquad A \in \sigma(X).$$

This is only possible if $\sigma(X) \subseteq \sigma(\mathcal{N})$ and therefore the claim follows from Exercise A.38. □

We now prove an important property of independent random variables: the expectation of the product of independent random variables is equal to the product of their expectations.

Theorem A.44 *If $X, Y \in L^1(\Omega, P)$ are real independent random variables, then*

$$XY \in L^1(\Omega, P), \qquad E[XY] = E[X]\, E[Y].$$

Proof. By an argument analogous to that used in the proof of Theorem A.23, it is enough to prove the statement in the case of indicator functions: $X = \mathbb{1}_E$, $Y = \mathbb{1}_F$ with $E, F \in \mathcal{F}$. By assumption, X, Y are independent, and so E, F are independent as well (please, check this!). Then we have

$$\int_\Omega XY dP = \int_{E\cap F} dP = P(E \cap F) = P(E)P(F) = E[X]\, E[Y]. \qquad \square$$

Exercise A.45 Give an example of random variables $X, Y \in L^1(\Omega, P)$ such that $XY \notin L^1(\Omega, P)$.

As a consequence of the previous theorem, if X, Y are independent, then

$$\operatorname{cov}(X, Y) = 0.$$

In particular, recalling (A.18), we have

$$\operatorname{var}(X + Y) = \operatorname{var}(X) + \operatorname{var}(Y).$$

Note that the converse is not generally true: if two random variables X, Y are such that $\operatorname{cov}(X, Y) = 0$, then they are not necessarily independent.

The concept of independence can be extended to the case of N random variables in a natural way. We say that the collections of events $\mathcal{H}_1, \dots, \mathcal{H}_N$ are independent if

$$P(H_{n_1} \cap \dots \cap H_{n_k}) = P(H_{n_1}) \cdots P(H_{n_k}),$$

for any choice of $H_{n_i} \in \mathcal{H}_{n_i}$ with different indexes $1 \le n_1, \dots, n_k \le N$. We say that the random variables $X_1, \dots, X_N$ are independent if their σ-

algebras $\sigma(X_1), \dots, \sigma(X_N)$ are independent. For example, three events E, F, G are independent if

i) they are pairwise independent;
ii) $P(E \cap F \cap G) = P(E)P(F)P(G)$.

In particular we note that E, F, G can be pairwise independent without being necessarily independent. The next result generalizes Theorem A.44.

Theorem A.46 *If $X_1, \dots, X_N \in L^1(\Omega, P)$ are real independent random variables then*

$$X_1 \cdots X_N \in L^1(\Omega, P) \quad \text{and} \quad E[X_1 \cdots X_N] = E[X_1] \cdots E[X_N].$$

Consequently we have

$$\operatorname{var}(X_1 + \dots + X_N) = \operatorname{var}(X_1) + \dots + \operatorname{var}(X_N).$$

A.1.8 Product measure and joint distribution

Consider two random variables

$$X : \Omega \longrightarrow \mathbb{R}^N, \qquad Y : \Omega \longrightarrow \mathbb{R}^M,$$

on the space $(\Omega, \mathcal{F})$. In this section we examine the relation among the distributions of X, Y and of the r.v.

$$(X, Y) : \Omega \longrightarrow \mathbb{R}^N \times \mathbb{R}^M.$$

The distribution of (X, Y) is usually called *joint distribution* of X and Y; conversely, the distributions of X and Y are called *marginal distributions* of (X, Y). As a remarkable example, we shall prove in Proposition A.94 that the marginal distributions of a multi-normal r.v. are multi-normal.

In order to treat the topic with a higher degree of generality, we recall the definition and some basic properties of the product of two measures. The results in this section can be extended to the case of more than two measures in a natural fashion.

Definition A.47 *Given two finite-measure spaces*

$$(\Omega_1, \mathcal{G}_1, \mu_1) \quad \text{and} \quad (\Omega_2, \mathcal{G}_2, \mu_2),$$

we define

$$\mathcal{G} = \mathcal{G}_1 \otimes \mathcal{G}_2 := \sigma(\{H \times K \mid H \in \mathcal{G}_1, \ K \in \mathcal{G}_2\}),$$

the product σ-algebra of $\mathcal{G}_1$ and $\mathcal{G}_2$.

Clearly $\mathcal{G}$ is a σ-algebra on the Cartesian product $\Omega := \Omega_1 \times \Omega_2$.

Exercise A.48 Prove that $\mathscr{B}(\mathbb{R}^2) = \mathscr{B}(\mathbb{R}) \otimes \mathscr{B}(\mathbb{R})$.

The following theorem contains the definition of the product measure of μ_1 and μ_2.

Theorem A.49 *There exists a unique probability measure μ on $\mathcal{G}$ such that*

$$\mu(H \times K) = \mu_1(H)\mu_2(K), \qquad H \in \mathcal{G}_1, \ K \in \mathcal{G}_2. \tag{A.32}$$

μ is called product measure of μ_1 and μ_2 and we write $\mu = \mu_1 \otimes \mu_2$.

For the existence, one can see, for example, Chapter 8 in Williams [339]. The uniqueness follows from Proposition A.6 and from the fact that the collection $\{H \times K \mid H \in \mathcal{G}_1, \ K \in \mathcal{G}_2\}$ is $\cap$-stable and generates $\mathcal{G}$.

Now we state the classical Fubini-Tonelli theorem.

Theorem A.50 (Fubini-Tonelli theorem) *Let*

$$f = f(\omega_1, \omega_2) : \Omega_1 \times \Omega_2 \longrightarrow \mathbb{R}^N$$

be a $\mathcal{G}$-measurable function. If $f \geq 0$ or if $f \in L^1(O, \mu)$ then we have

i) for every $\omega_1 \in \Omega_1$ the function $\omega_2 \mapsto f(\omega_1, \omega_2)$ is $\mathcal{G}_2$-measurable and the function $\omega_1 \mapsto \int_{\Omega_2} f(\omega_1, \omega_2)\mu_2(d\omega_2)$ is $\mathcal{G}_1$-measurable (and an analogous result holds if we exchange ω_1 with ω_2);

ii)

$$\begin{aligned}\int_\Omega f d\mu &= \int_{\Omega_1} \left(\int_{\Omega_2} f(\omega_1, \omega_2)\mu_2(d\omega_2) \right) \mu_1(d\omega_1) \\ &= \int_{\Omega_2} \left(\int_{\Omega_1} f(\omega_1, \omega_2)\mu_1(d\omega_1) \right) \mu_2(d\omega_2).\end{aligned}$$

Remark A.51 Theorems A.49 and A.50 remain true more generally if the measures are σ-finite. We recall that a measure μ on $(\Omega, \mathcal{F})$ is *σ-finite* if there exists a sequence $(\Omega_n)_{n\in\mathbb{N}}$ in $\mathcal{F}$ such that

$$\Omega = \bigcup_{n\in\mathbb{N}} \Omega_n \quad \text{and} \quad \mu(\Omega_n) < \infty, \quad n \in \mathbb{N}.$$

For example, the Lebesgue measure on $\mathbb{R}^N$ is σ-finite but it is not finite. □

Corollary A.52 *Suppose that the joint distribution $P^{(X,Y)}$ of the random variables X, Y has a density $f_{(X,Y)}$. Then*

$$f_X(z) := \int_{\mathbb{R}^N} f_{(X,Y)}(z, \zeta) d\zeta \qquad \text{and} \qquad f_Y(\zeta) := \int_{\mathbb{R}^M} f_{(X,Y)}(z, \zeta) dz$$

are the densities of the distributions of X and Y, respectively.

Proof. The thesis is a consequence of Theorem A.50. □

The following proposition solves the important problem of reconstructing the joint distribution of *independent* random variables from the marginal distributions.

Proposition A.53 *The following statements are equivalent*[8]*:*

i) X *and* Y *are independent on* $(\Omega, \mathcal{F}, P)$*;*
ii) $P^{(X,Y)} = P^X \otimes P^Y$*;*
iii) $\Phi^{(X,Y)} = \Phi^X \Phi^Y$.

Further, if X, Y *have a joint density* $f_{(X,Y)}$ *then i) is equivalent to*

iv) $f_{(X,Y)} = f_X f_Y$.

Proof. We only prove that i) implies ii): the rest is left as an exercise. For every $H, K \in \mathscr{B}$ we have

$$P^{(X,Y)}(H \times K) = P((X,Y) \in H \times K) = P(X^{-1}(H) \cap Y^{-1}(K)) =$$

(since X, Y are independent random variables)

$$= P(X \in H)P(Y \in K) = P^X(H)P^Y(K).$$

The claim follows then from the uniqueness of the product measure. □

As an application of the previous proposition, we have the following result on the density of the sum of two independent random variables.

Corollary A.54 *Let* $X, Y : \Omega \longrightarrow \mathbb{R}$ *be two random variables with joint density* $f_{(X,Y)}$*. Then the r.v.* $X + Y$ *has density*

$$f_{X+Y}(z) = \int_{\mathbb{R}} f_{(X,Y)}(\zeta, z - \zeta)d\zeta.$$

In particular, if X *and* Y *are independent, then*

$$f_{X+Y}(z) = (f_X * f_Y)(z) := \int_{\mathbb{R}} f_X(\zeta) f_Y(z - \zeta)d\zeta. \tag{A.33}$$

Proof. For every $z \in \mathbb{R}$, we have

$$P(X + Y \le z) = \iint_{\{x+y\le z\}} f_{(X,Y)}(x,y)dxdy =$$

(by the change of variables $\zeta = x + y$ and by Fubini's theorem)

$$= \int_{-\infty}^{z} \left(\int_{\mathbb{R}} f_{(X,Y)}(x, x - \zeta)dx \right) d\zeta.$$

Since the family $\{]-\infty, z] \mid z \in \mathbb{R}\}$ is $\cap$-stable and generates $\mathscr{B}$, by Dynkin's Theorem A.5 we have the claim. Moreover (A.33) follows from Proposition A.53. □

[8] We recall that P^X and Φ^X denote the distribution and the distribution function of the r.v. X, respectively.

Exercise A.55 Determine the density of the sum of two independent normal random variables.

A.1.9 Markov inequality

The following result is sometimes useful to study the integrability properties of a random variable.

Proposition A.56 *Let X be a r.v. and let $f \in C^1(\mathbb{R}_{\geq 0})$ such that $f' \geq 0$ or $f' \in L^1(\mathbb{R}_{\geq 0}, P^{|X|})$. Then*

$$E\left[f(|X|)\right] = f(0) + \int_0^{+\infty} f'(\lambda)P(|X| \geq \lambda)d\lambda. \tag{A.34}$$

Proof. We have

$$\begin{aligned} E\left[f(|X|)\right] &= \int_0^{+\infty} f(y)P^{|X|}(dy) = \\ &= \int_0^{+\infty} \left(f(0) + \int_0^y f'(\lambda)d\lambda\right) P^{|X|}(dy) = \end{aligned}$$

(reversing the order of integration, by Fubini's Theorem A.50)

$$\begin{aligned} &= f(0) + \int_0^{+\infty} f'(\lambda) \int_\lambda^{+\infty} P^{|X|}(dy)d\lambda = \\ &= f(0) + \int_0^{+\infty} f'(\lambda)P(|X| \geq \lambda)d\lambda. \end{aligned}$$

□

Example A.57 If $f(x) = x^p$, $p \geq 1$, by (A.34) we have

$$E\left[|X|^p\right] = p\int_0^{+\infty} \lambda^{p-1} P\left(|X| \geq \lambda\right) d\lambda.$$

Consequently, to prove the p-th order integrability of X it is enough to have an estimate of $P\left(|X| \geq \lambda\right)$, at least for sufficiently large λ. □

The next classical Markov inequality gives an estimate in the opposite direction.

Proposition A.58 (Markov inequality) *Let X be a random variable and let $\lambda \in \mathbb{R}_{>0}$, $1 \leq p < +\infty$. Then*

$$P(|X| \geq \lambda) \leq \frac{E\left[|X|^p\right]}{\lambda^p}. \tag{A.35}$$

In particular, if X is a integrable real r.v., we have

$$P(|X - E\left[X\right]| \geq \lambda) \leq \frac{\operatorname{var}(X)}{\lambda^2}. \tag{A.36}$$

Proof. We prove (A.35) only. We have

$$E\left[|X|^p\right] \geq \int\limits_{\{|X|\geq\lambda\}} |X|^p dP \geq \lambda^p P(|X| \geq |\lambda|). \qquad \square$$

In the same order of ideas we also prove the following useful "identity principle".

Proposition A.59 *Let X be a r.v. with strictly positive density on $H \in \mathscr{B}$. If $g \in m\mathscr{B}$ is such that $g(X) = 0$ a.s. ($g(X) \geq 0$ a.s.) then $g = 0$ ($g \geq 0$) almost everywhere (with respect to Lebesgue measure) on H. In particular if g is continuous, then $g = 0$ ($g \geq 0$) on H.*

Proof. We set

$$H_n = \left\{x \in H \mid |g(x)| \geq \frac{1}{n}\right\}, \qquad n \in \mathbb{N},$$

and we denote the density of X by f. Assuming that $g(X) = 0$ a.s., we have

$$0 = E\left[|g(X)|\right] \geq \frac{1}{n}P(X \in H_n) = \frac{1}{n}\int_{H_n} f(x)dx$$

and so, since by assumption f is strictly positive, H_n must have null Lebesgue measure. We conclude by observing that

$$\{g \neq 0\} = \bigcup_{n\in\mathbb{N}} H_n.$$

If $g(X) \geq 0$ a.s., we proceed in an analogous way, by considering the sequence of sets $H_n = \{g < -\frac{1}{n}\}$, $n \in \mathbb{N}$. $\square$

A.2 Fourier transform

The Fourier transform of a function $f \in L^1(\mathbb{R}^N)$ is defined as follows:

$$\hat{f} : \mathbb{R}^N \longrightarrow \mathbb{C}, \qquad \hat{f}(\xi) := \int_{\mathbb{R}^N} e^{i\langle\xi,x\rangle} f(x)dx, \qquad \xi \in \mathbb{R}^N. \tag{A.37}$$

Sometimes the notation $\mathcal{F}(f) = \hat{f}$ will be used.

Example A.60 We compute the Fourier transform of $f(x) = e^{-\lambda x^2}$, $x \in \mathbb{R}$, where λ is a positive constant. We note that

$$ix\xi - \lambda x^2 = -\lambda\left(x - \frac{i\xi}{2\lambda}\right)^2 - \frac{\xi^2}{4\lambda},$$

so that we have

$$\hat{f}(\xi) = e^{-\frac{\xi^2}{4\lambda}} \int_{\mathbb{R}} e^{-\lambda\left(x-\frac{i\xi}{2\lambda}\right)^2} dx.$$

Formally, by the change of variable $y = x - \frac{i\xi}{2\lambda}$, we get

$$\hat{f}(\xi) = e^{-\frac{\xi^2}{4\lambda}} \int_{\mathbb{R}} e^{-\lambda y^2} dy = \sqrt{\frac{\pi}{\lambda}} e^{-\frac{\xi^2}{4\lambda}}, \qquad \text{(A.38)}$$

where we used the fact that $\int_{\mathbb{R}} e^{-y^2} dy = \sqrt{\pi}$. Actually, we cannot directly perform the above change of variable since y would be a complex variable. However the argument can be made rigorous by means of Cauchy's residue theorem: for more details see, for instance, Rudin [293]. □

Given a finite measure μ on $(\mathbb{R}^N, \mathscr{B})$, we define the Fourier transform of μ as follows:

$$\mathcal{F}(\mu) = \hat{\mu} : \mathbb{R}^N \longrightarrow \mathbb{C}, \qquad \hat{\mu}(\xi) := \int_{\mathbb{R}^N} e^{i\langle \xi, x\rangle} \mu(dx), \qquad \xi \in \mathbb{R}^N. \qquad \text{(A.39)}$$

Example A.61 If δ_{x_0} is the Dirac delta centered at $x_0 \in \mathbb{R}^N$, then

$$\hat{\delta}_{x_0}(\xi) = \int_{\mathbb{R}} e^{i\langle \xi, x\rangle} \delta_{x_0}(dx) = e^{i\langle x_0, \xi\rangle}, \qquad \xi \in \mathbb{R}^N.$$

□

Example A.62 Consider the normal distribution $\mathcal{N}_{\mu,\sigma^2}$ where now $\mu \in \mathbb{R}$ and $\sigma > 0$. Then we have

$$\mathcal{F}\left(\mathcal{N}_{\mu,\sigma^2}\right)(\xi) = \frac{1}{\sqrt{2\pi\sigma^2}} \int_{\mathbb{R}} e^{ix\xi - \frac{(x-\mu)^2}{2\sigma^2}} dx =$$

(by the change of variable $y = x - \mu$)

$$= \frac{e^{i\mu\xi}}{\sqrt{2\pi\sigma^2}} \int_{\mathbb{R}} e^{ix\xi - \frac{y^2}{2\sigma^2}} dy =$$

(by (A.38) with $\lambda = \frac{1}{2\sigma^2}$)

$$= e^{i\mu\xi - \frac{(\sigma\xi)^2}{2}}.$$

Note that, for $\sigma = 0$, we find the Fourier transform of the Dirac delta centered at μ, in accord with Example A.61. □

As a straightforward consequence of the definition of Fourier transform, we have the following:

Proposition A.63 *If $f \in L^1(\mathbb{R}^N)$ and μ is a finite measure, then*

i) $|\hat{f}(\xi)| \le \|f\|_{L^1(\mathbb{R}^N)}$ *and* $|\hat{\mu}(\xi)| \le \mu(\mathbb{R}^N)$*;*
ii) $\hat{f}, \hat{\mu} \in C(\mathbb{R}^N)$*;*
iii) $\lim\limits_{|\xi|\to+\infty} \hat{f}(\xi) = \lim\limits_{|\xi|\to+\infty} \hat{\mu}(\xi) = 0.$

We remark that in general the Fourier transform of an integrable function (or of a finite measure) is not integrable: for example, consider the Fourier transform of the indicator function of the interval $[-1, 1]$.

We recall that the operation of convolution of two functions $f, g \in L^1(\mathbb{R}^N)$ is defined by

$$(f * g)(x) = \int_{\mathbb{R}^N} f(x-y)g(y)dy, \qquad x \in \mathbb{R}^N, \tag{A.40}$$

and $f * g \in L^1(\mathbb{R}^N)$. Indeed

$$\|f * g\|_{L^1(\mathbb{R}^N)} \le \int_{\mathbb{R}^N}\int_{\mathbb{R}^N} |f(x-y)g(y)|dydx =$$

(changing the order of integration)

$$= \int_{\mathbb{R}^N}\int_{\mathbb{R}^N} |f(x-y)g(y)|dxdy = \|f\|_{L^1(\mathbb{R}^N)}\|g\|_{L^1(\mathbb{R}^N)}.$$

The following theorem sums up some other remarkable property of the Fourier transform.

Theorem A.64 *Let $f, g \in L^1(\mathbb{R}^N)$. Then*

i) $\mathcal{F}(f * g) = \mathcal{F}(f)\mathcal{F}(g)$*;*
ii) if $\partial_{x_k} f \in L^1(\mathbb{R}^N)$ *then*

$$\mathcal{F}(\partial_{x_k} f)(\xi) = -i\xi_k \mathcal{F}(f)(\xi); \tag{A.41}$$

iii) if $(x_k f) \in L^1(\mathbb{R}^N)$*, then* $\partial_{\xi_k}\hat{f}$ *exists and*

$$\partial_{\xi_k}\hat{f}(\xi) = i\mathcal{F}(x_k f)(\xi). \tag{A.42}$$

Proof. i) We have

$$\begin{aligned}\hat{f}(\xi)\hat{g}(\xi) &= \int_{\mathbb{R}^N} e^{i\langle\xi,w\rangle} f(w)dw \int_{\mathbb{R}^N} e^{i\langle\xi,y\rangle} g(y)dy \\ &= \int_{\mathbb{R}^N}\int_{\mathbb{R}^N} e^{i\langle\xi,w+y\rangle} f(w)g(y)dwdy =\end{aligned}$$

(by the change of variable $x = y + w$)

$$= \int_{\mathbb{R}^N} e^{i\langle\xi,x\rangle}\int_{\mathbb{R}^N} f(x-y)g(y)dydx.$$

ii) For the sake of simplicity, we prove (A.41) only if $N = 1$:

$$\mathcal{F}(f')(\xi) = \int_{\mathbb{R}} e^{ix\xi} f'(x)dx =$$

(integrating by parts[9])

$$= -\int_{\mathbb{R}} \frac{d}{dx} e^{ix\xi} f(x)dx = -i\xi\mathcal{F}(f)(\xi).$$

iii) Again in the case $N = 1$, we consider the incremental ratio

$$R(\xi, \delta) = \frac{\hat{f}(\xi + \delta) - \hat{f}(\xi)}{\delta}, \qquad \delta \in \mathbb{R} \setminus \{0\}.$$

We have to prove that

$$\lim_{\delta \to 0} R(\xi, \delta) = \mathcal{F}(ixf)(\xi).$$

We observe that

$$R(\xi, \delta) = \int_{\mathbb{R}} e^{ix\xi} \frac{e^{ix\delta} - 1}{\delta} f(x)dx$$

and that, by the mean-value theorem, we have

$$\left| e^{ix\xi} \frac{e^{ix\delta} - 1}{\delta} f(x) \right| \le |xf(x)| \in L^1(\mathbb{R})$$

by assumption. We can apply now the dominated convergence theorem and since $\lim_{\delta \to 0} \frac{e^{ix\delta}-1}{\delta} = ix$, we get

$$\lim_{\delta \to 0} R(\xi, \delta) = \int_{\mathbb{R}} e^{ix\xi} ixf(x)dx$$

whence the claim. □

We state without proof a classical result on the inversion of the Fourier transform.

Theorem A.65 *Let $f \in L^1(\mathbb{R}^N)$ and let μ be a finite measure. If $\hat{f} \in L^1(\mathbb{R}^N)$, then*

$$f(x) = \frac{1}{(2\pi)^N} \int_{\mathbb{R}^N} e^{-i\langle x, \xi \rangle} \hat{f}(\xi)d\xi. \tag{A.43}$$

Analogously if $\hat{\mu} \in L^1(\mathbb{R}^N)$, then μ has density $g \in C(\mathbb{R}^N)$ and

$$g(x) = \frac{1}{(2\pi)^N} \int_{\mathbb{R}^N} e^{-i\langle x, \xi \rangle} \hat{\mu}(\xi)d\xi. \tag{A.44}$$

---

[9] This step can be justified approximating f by functions belonging to C_0^∞.

Remark A.66 If $f, \hat{f} \in L^1(\mathbb{R})$ and f is a real function, then[10] $\hat{f}(-\xi) = \overline{\hat{f}(\xi)}$ and consequently $\xi \mapsto e^{-i\langle x,\xi\rangle}\hat{f}(\xi)$ is an even function. Then we have

$$f(x) = \frac{1}{2\pi}\int_{\mathbb{R}} e^{-ix\xi}\hat{f}(\xi)d\xi = \frac{1}{\pi}\int_0^{\infty} e^{-ix\xi}\hat{f}(\xi)d\xi. \qquad \square$$

Eventually, we state a result on the Fourier inversion of discontinuous functions in one dimension (for the proof and a multi-dimensional extension see Taylor [326]).

Theorem A.67 (Dirichlet-Jordan) *If $f \in L^1(\mathbb{R})$ has bounded variation (cf. Definition 3.59 and Notation 3.49) in a neighborhood of $x \in \mathbb{R}$, then we have*

$$\lim_{R\to+\infty} \frac{1}{2\pi}\int_{-R}^{R} e^{-ix\xi}\hat{f}(\xi)d\xi = \frac{f(x+) + f(x-)}{2}.$$

A.3 Parabolic equations with constant coefficients

We denote by (t, x) the point in $\mathbb{R} \times \mathbb{R}^N$ and we consider the following partial differential equation

$$Lu := \frac{1}{2}\sum_{j,k=1}^{N} c_{jk}\partial_{x_j x_k}u + \sum_{j=1}^{N} b_j\partial_{x_j}u - au - \partial_t u = 0. \tag{A.45}$$

Hereafter we always assume that $\mathcal{C} = (c_{jk})$ is a *symmetric and positive definite* $N \times N$ matrix (we write $\mathcal{C} > 0$), that is

$$\langle \mathcal{C}x, x\rangle > 0, \qquad x \in \mathbb{R}^N \setminus \{0\}. \tag{A.46}$$

Moreover $b = (b_1, \dots, b_N) \in \mathbb{R}^N$ and $a \in \mathbb{R}$. The factor $\frac{1}{2}$ in the second-order part of L appears only in order to obtain an expression consistent with the probabilistic notations (in particular in connection with the multi-normal distribution, cf. Paragraph A.4). Since in this section we assume that c_{jk}, b_j, a are real and constant, we say that (A.45) is a partial differential equation of parabolic type with constant coefficients. The prototype of this kind of equations is the heat equation, corresponding to the case in which $\mathcal{C}$ is the identity matrix, b and a are null:

$$\frac{1}{2}\triangle u - \partial_t u = 0, \tag{A.47}$$

where

$$\triangle = \sum_{j=1}^{N} \partial_{x_j x_j}$$

[10] Here $\bar{z}$ denotes the conjugate of the complex number z.

is the Laplace differential operator. The heat equation is well known in physics since it models the diffusion process of heat in a physical body.

We consider the classical *Cauchy problem* for the operator L in (A.45)

$$\begin{cases} Lu = 0, & \text{in }]0, +\infty[\times \mathbb{R}^N, \\ u(0, x) = \varphi(x), & x \in \mathbb{R}^N, \end{cases} \tag{A.48}$$

where φ is a given continuous bounded function on $\mathbb{R}^N$, $\varphi \in C_b(\mathbb{R}^N)$, that is called *initial datum* of the problem.

Notation A.68 *We denote by $C^{1,2}$ the class of functions with continuous second-order derivatives in the x variables and continuous first-order derivative in the t variable.*

A *classical solution* of the Cauchy problem is a function

$$u \in C^{1,2}(\mathbb{R}_{>0} \times \mathbb{R}^N) \cap C(\mathbb{R}_{\geq 0} \times \mathbb{R}^N)$$

satisfying (A.48). If L is the heat operator, $u(t, x)$ represents the temperature, at time t and at the point x, of a physical body whose initial temperature at initial time $t = 0$ is equal to φ.

Definition A.69 *Fundamental solution of L is a function $\Gamma(t, x)$, defined on $]0, +\infty[\times \mathbb{R}^N$, such that, for every $\varphi \in C_b(\mathbb{R}^N)$, the function defined by*

$$u(t, x) = \begin{cases} \int_{\mathbb{R}^N} \Gamma(t, x - y)\varphi(y)dy, & t > 0, \ x \in \mathbb{R}^N, \\ \varphi(x), & t = 0, \ x \in \mathbb{R}^N, \end{cases} \tag{A.49}$$

is a classical solution of the Cauchy problem (A.48).

We present now a standard method, based on the use of the Fourier transform, to construct a fundamental solution of L.

A.3.1 A special case

We consider first the case in which the coefficients b, a are null. As we shall see, in general we can go back to this case by a suitable change of variables.

We proceed formally (i.e. without a rigorous justification of the steps) to obtain a solution formula that will be verified a posteriori. By applying the Fourier transform *only in the x variables* to equation (A.45) and by using (A.41), we get:

$$\mathcal{F}\left(\frac{1}{2} \sum_{j,k=1}^{N} c_{jk} \partial_{x_j x_k} u(t, x) - \partial_t u(t, x) \right)(\xi)$$
$$= -\frac{1}{2} \sum_{j,k=1}^{N} c_{jk} \xi_j \xi_k \hat{u}(t, \xi) - \partial_t \hat{u}(t, \xi) = 0,$$

or, in other terms,

$$\partial_t \hat{u}(t,\xi) = -\frac{1}{2}\langle \mathcal{C}\xi,\xi\rangle \hat{u}(t,\xi), \tag{A.50}$$

to which we associate the initial condition

$$\hat{u}(0,\xi) = \hat{\varphi}(\xi), \qquad \xi \in \mathbb{R}^N. \tag{A.51}$$

The ordinary Cauchy problem (A.50)-(A.51) has solution

$$\hat{u}(t,\xi) = \hat{\varphi}(\xi) e^{-\frac{t}{2}\langle \mathcal{C}\xi,\xi\rangle}.$$

Therefore, by using i) in Theorem A.64, we get[11]:

$$u(t,x) = \mathcal{F}^{-1}\left(\hat{\varphi}(\xi) e^{-\frac{t}{2}\langle \mathcal{C}\xi,\xi\rangle}\right) = \left(\mathcal{F}^{-1}\left(e^{-\frac{t}{2}\langle \mathcal{C}\xi,\xi\rangle}\right) * \varphi\right)(x), \tag{A.52}$$

where "$*$" denotes the convolution operation in (A.40). Now we use the following lemma whose proof is postponed to the end of the section.

Lemma A.70 *Consider the function*

$$\Gamma(t,x) = \frac{1}{\sqrt{(2\pi t)^N \det \mathcal{C}}} \exp\left(-\frac{1}{2t}\langle \mathcal{C}^{-1}x, x\rangle\right), \tag{A.53}$$

defined for $x \in \mathbb{R}^N$ *and* $t > 0$. *We have*

$$\mathcal{F}(\Gamma(t,\cdot))(\xi) = e^{-\frac{t}{2}\langle \mathcal{C}\xi,\xi\rangle}. \tag{A.54}$$

In Lemma A.70, the assumption (A.46), i.e. $\mathcal{C} > 0$, plays a crucial role. We also remark explicitly that, for $N = 1$ and $\mathcal{C} = 1$, Γ is the density of the normal distribution in (A.7). By (A.54), (A.52) becomes

$$\begin{aligned} u(t,x) &= \int_{\mathbb{R}^N} \Gamma(t, x-y)\varphi(y)dy \\ &= \frac{1}{\sqrt{(2\pi t)^N \det \mathcal{C}}} \int_{\mathbb{R}^N} \exp\left(-\frac{1}{2t}\langle \mathcal{C}^{-1}(x-y), (x-y)\rangle\right) \varphi(y)dy, \end{aligned} \tag{A.55}$$

for $x \in \mathbb{R}^N$ and $t > 0$.

Remark A.71 Until now we have assumed $\varphi \in C_b$, but in the actual applications we need to consider more general initial data, possibly unbounded. As a matter of fact, the convergence of the integral in (A.55) requires much less than the boundedness of φ: it is enough to impose a suitable growth condition on φ at infinity. Indeed, for every fixed $(t,x) \in \mathbb{R}_{>0} \times \mathbb{R}^N$, we have

$$\exp\left(-\frac{1}{2t}\langle \mathcal{C}^{-1}(x-y), (x-y)\rangle\right) \le e^{-c|x-y|^2}, \qquad y \in \mathbb{R}^N,$$

[11] Here we write formally $u = \mathcal{F}^{-1}(v)$ to indicate that $v = \mathcal{F}(u)$.

where $c = \frac{\lambda}{2t} > 0$ and λ is the smallest eigenvalue of $\mathcal{C}^{-1}$. Therefore it is enough[12] to assume that there exist some positive constants c_1, c_2, γ with $\gamma < 2$ such that

$$|\varphi(y)| \leq c_1 e^{c_2|y|^\gamma}, \qquad y \in \mathbb{R}^N, \tag{A.57}$$

to guarantee that the integral in (A.55) converges for every $t > 0$ and $x \in \mathbb{R}^N$. □

Next we aim at proving rigorously that (A.55) gives a representation formula for the classical solution of the Cauchy problem (A.48).

Theorem A.72 *If φ is continuous and verifies condition* (A.57), *then the function u defined by* (A.55) *for $t > 0$ and by $u(0, \cdot) = \varphi$, is a classical solution of the Cauchy problem* (A.48). *In particular Γ in* (A.53) *is a fundamental solution of L in* (A.45) *with $a = 0$ and $b = 0$.*

Proof. For the sake of simplicity we only consider the case of the heat operator. First of all, we verify that the function

$$\Gamma(t, x) = \frac{1}{(2\pi t)^{\frac{N}{2}}} \exp\left(-\frac{|x|^2}{2t}\right), \qquad x \in \mathbb{R}^N, \ t > 0,$$

is a solution of the heat equation: for $k = 1, \dots, N$, we have

$$\begin{aligned}
\partial_{x_k} \Gamma(t, x) &= -\frac{x_k}{t} \Gamma(t, x), \\
\partial_{x_k x_k} \Gamma(t, x) &= \left(\frac{x_k^2}{t^2} - \frac{1}{t}\right) \Gamma(t, x), \\
\partial_t \Gamma(t, x) &= \frac{1}{2}\left(\frac{|x|^2}{t^2} - \frac{N}{t}\right) \Gamma(t, x),
\end{aligned} \tag{A.58}$$

and so it follows immediately that $\frac{1}{2}\triangle\Gamma(t, x) = \partial_t \Gamma(t, x)$.

In order to prove that u in (A.55) is solution of the heat equation, it is enough to use a standard result on differentiation under the integral sign and verify that

$$\left(\frac{1}{2}\triangle - \partial_t\right) u(t, x) = \int_{\mathbb{R}^N} \left(\frac{1}{2}\triangle - \partial_t\right) \Gamma(t, x - y)\varphi(y) dy = 0, \tag{A.59}$$

for every $x \in \mathbb{R}^N$ and $t > 0$. Now, for a fixed $\bar{x} \in \mathbb{R}^N$ and $t, \delta > 0$, we have

$$\partial_{x_k} \Gamma(t, x - y)\varphi(y) = -\left(\frac{x_k - y_k}{t} \Gamma(t, x - y) e^{\frac{|y|^2}{\delta}}\right)\left(\varphi(y) e^{-\frac{|y|^2}{\delta}}\right)$$

---

[12] Actually we can assume the existence of two positive constants c_1, c_2 such that

$$|\varphi(y)| \leq c_1 e^{c_2|y|^2}, \qquad y \in \mathbb{R}^N, \tag{A.56}$$

to make sure that the integral in (A.55) is finite, at least for $t < \frac{\lambda}{2c_2}$.

where $\varphi(y)e^{-\frac{|y|^2}{\delta}}$ is integrable over $\mathbb{R}^N$ by condition (A.57). Then assuming that $\delta > 4t$ and x belongs to a bounded neighborhood of the point $\bar{x}$, the function

$$y \mapsto \frac{x_k - y_k}{t} \Gamma(t, x-y) e^{\frac{|y|^2}{\delta}}$$

is bounded. Hence the dominated convergence theorem guarantees that

$$\partial_{x_k} u(t,\bar{x}) = \int_{\mathbb{R}^N} \partial_{x_k} \Gamma(t, \bar{x} - y)\varphi(y)dy.$$

Analogously we show that

$$\partial_{x_k x_k} u(t,x) = \int_{\mathbb{R}^N} \partial_{x_k x_k} \Gamma(t, x - y)\varphi(y)dy,$$
$$\partial_t u(t,x) = \int_{\mathbb{R}^N} \partial_t \Gamma(t, x - y)\varphi(y)dy,$$

for every $x \in \mathbb{R}^N$ and $t > 0$. This concludes the proof of (A.59).

It remains to prove that the function u is continuous up to $t = 0$: more precisely, we show that, for every fixed $x_0 \in \mathbb{R}^N$, we have

$$\lim_{(t,x)\to(0,x_0)} u(t,x) = \varphi(x_0).$$

Since

$$\int_{\mathbb{R}^N} \Gamma(t, x-y)dy = 1, \qquad x \in \mathbb{R}^N,\ t > 0,$$

we have

$$|u(t,x) - \varphi(x_0)| \le \frac{1}{(2\pi t)^{\frac{N}{2}}} \int_{\mathbb{R}^N} \exp\left(-\frac{|x-y|^2}{2t}\right) |\varphi(y) - \varphi(x_0)|dy =$$

(by the change of variables $\eta = \frac{x-y}{\sqrt{2t}}$)

$$= \frac{1}{\pi^{\frac{N}{2}}} \int_{\mathbb{R}^N} e^{-|\eta|^2} |\varphi(x - \eta\sqrt{2t}) - \varphi(x_0)|d\eta. \tag{A.60}$$

By condition (A.57), for every (t,x) in a neighborhood of $(0,x_0)$, we have

$$e^{-|\eta|^2} |\varphi(x - \eta\sqrt{2t}) - \varphi(x_0)| \le c e^{-\frac{|\eta|^2}{2}}, \qquad \eta \in \mathbb{R}^N,$$

for some positive constant c. Then the claim follows by taking the limit in (A.60) as $(t,x) \to (0,x_0)$ and applying the dominated convergence theorem. □

Example A.73 We consider the Cauchy problem in $\mathbb{R}^2$

$$\begin{cases} \frac{1}{2}\partial_{xx} u(t,x) - \partial_t u(t,x) = 0, & (t,x) \in \mathbb{R}_{>0} \times \mathbb{R}, \\ u(0,x) = e^x & x \in \mathbb{R}. \end{cases}$$

By formula (A.55) we have

$$u(t,x) = \frac{1}{\sqrt{2\pi t}} \int_{\mathbb{R}} e^{-\frac{(x-y)^2}{2t}+y} dy =$$

(by the change of variables $\eta = \frac{x-y}{\sqrt{2t}}$)

$$= \frac{e^{x+\frac{t}{2}}}{\sqrt{\pi}} \int_{\mathbb{R}} e^{-\left(\eta+\frac{t}{2}\right)^2} d\eta = e^{x+\frac{t}{2}}.$$

□

Exercise A.74 Determine the solution of the Cauchy problem

$$\begin{cases} \frac{1}{2}\partial_{xx}u(t,x) - \partial_t u(t,x) = 0, & (t,x) \in \mathbb{R}_{>0} \times \mathbb{R}, \\ u(0,x) = (e^x - 1)^+, & x \in \mathbb{R}, \end{cases}$$

where $\varphi^+ = \max\{0,\varphi\}$ is the positive part of the function φ.

We close this section by giving the:

Proof (of Lemma A.70). In the case $N = 1$, the thesis is a consequence of Example A.60. In general we use the following trick: we set

$$\widetilde{\Gamma}(t,\xi) = \mathcal{F}(\Gamma(t,\cdot))(\xi)$$

with Γ as in (A.53). Denoting by $\nabla_\xi = (\partial_{\xi_1}, \dots, \partial_{\xi_N})$ the gradient in $\mathbb{R}^N$, by property (A.42) we have

$$\nabla_\xi \widetilde{\Gamma}(t,\xi) = i\mathcal{F}\left(x\Gamma(t,x)\right)(\xi) = -it\mathcal{F}\left(-\mathcal{C}\,\frac{\mathcal{C}^{-1}x}{t}\Gamma(t,x)\right)(\xi) =$$

(since $\nabla_x \langle \mathcal{C}^{-1}x, x\rangle = 2\mathcal{C}^{-1}x$)

$$= -it\mathcal{F}\left(\mathcal{C}\nabla_x \Gamma(t,x)\right)(\xi) =$$

(by property (A.41))

$$= -t\,\mathcal{C}\xi\,\widetilde{\Gamma}(t,\xi).$$

In conclusion, for every positive t, $\widetilde{\Gamma}(t,\cdot)$ is solution of the Cauchy problem

$$\begin{cases} \nabla_\xi \widetilde{\Gamma}(t,\xi) = -t\,\mathcal{C}\xi\,\widetilde{\Gamma}(t,\xi), \\ \widetilde{\Gamma}(t,0) = \int_{\mathbb{R}^N} \Gamma(t,x)dx = 1, \end{cases}$$

and consequently, because of the uniqueness of the solution, we get the claim:

$$\widetilde{\Gamma}(t,\xi) = e^{-\frac{t}{2}\langle \mathcal{C}\xi,\xi\rangle}.$$

□

A.3.2 General case

Now we consider the operator L in its most general form

$$Lu = \frac{1}{2}\sum_{j,k=1}^{N} c_{jk}\partial_{x_j x_k} u + \sum_{j=1}^{N} b_j \partial_{x_j} u - au - \partial_t u.$$

We prove that by a simple substitution we can go back to the previous case: indeed by setting

$$u_0(t,x) = e^{at} u(t, x - tb),$$

we have

$$\begin{aligned}
\partial_t u_0(t,x) &= e^{at}(au(t,x-tb) + (\partial_t u)(t,x-tb) - \langle b, (\nabla u)(t,x-tb)\rangle), \\
\partial_{x_j} u_0(t,x) &= e^{at}\left(\partial_{x_j} u\right)(t,x-tb), \\
\partial_{x_j x_k} u_0(t,x) &= e^{at}\left(\partial_{x_j x_k} u\right)(t,x-tb), \qquad j,k=1,\dots,N.
\end{aligned}$$

Therefore, if we set

$$L_0 = \frac{1}{2}\sum_{j,k=1}^{N} c_{jk}\partial_{x_j x_k} - \partial_t$$

we get

$$\begin{aligned}
L_0 u_0(t,x) = & e^{at}\big((L_0 u)(t,x-tb) + \langle b, (\nabla u)(t,x-tb)\rangle - au(t,x-tb) \\
& - (\partial_t u)(t,x-tb)\big) \\
= & e^{at}(Lu)(t,x-tb).
\end{aligned}$$

In conclusion we obtain that the fundamental solution of L is given by

$$\Gamma(t,x) = e^{-at}\Gamma_0(t,x-tb), \qquad (t,x) \in \mathbb{R}_{>0} \times \mathbb{R}^N,$$

where Γ_0 is the fundamental solution of L_0. More explicitly

$$\Gamma(t,x) = \frac{1}{\sqrt{(2\pi t)^N \det C}} \exp\left(-\frac{1}{2t}\langle C^{-1}(x-tb), x-tb\rangle - at\right). \qquad \text{(A.61)}$$

Next we note that, for every $s \in \mathbb{R}$ and $\varphi \in C_b$, the function

$$u(t,x) = \int_{\mathbb{R}^N} \Gamma(t-s, x-y)\varphi(y)dy, \qquad t > s,\ x \in \mathbb{R}^N,$$

is a classical solution of the Cauchy problem

$$\begin{cases} Lu = 0, & \text{in }]s,+\infty[\times\mathbb{R}^N, \\ u(s,x) = \varphi(x), & x \in \mathbb{R}^N. \end{cases} \qquad \text{(A.62)}$$

This justifies the following:

Definition A.75 *The function*

$$\Gamma(t,x;s,y) = \Gamma(t-s, x-y), \qquad x,y \in \mathbb{R}^N,\ t>s,$$

is called fundamental solution of L with pole in (s,y) and computed at (t,x).

The explicit expression of Γ is

$$\Gamma(t,x;s,y) = \frac{1}{\sqrt{(2\pi(t-s))^N \det \mathcal{C}}}\cdot$$
$$\cdot \exp\left(-\frac{1}{2(t-s)}\langle \mathcal{C}^{-1}(x-y-(t-s)b), x-y-(t-s)b\rangle - a(t-s)\right).$$

A.3.3 Locally integrable initial datum

Formula

$$u(t,x) = \int_{\mathbb{R}^N} \Gamma(t,x;s,y)\varphi(y)dy, \qquad t>s,\ x\in\mathbb{R}^N, \tag{A.63}$$

defines a solution of the Cauchy problem (A.62) even under weaker regularity conditions on the initial datum: let us assume that $\varphi \in L^1_{\text{loc}}(\mathbb{R}^N)$ and there exist some positive constants c, R, β with $\gamma < 2$ such that

$$|\varphi(x)| \le c e^{c|x|^\gamma}, \tag{A.64}$$

for almost all $x \in \mathbb{R}^N$ with $|x| \ge R$. This extension can be useful for example if we want to price digital options, where the initial datum is a discontinuous function of the form

$$\varphi(x) = \begin{cases} 1, & x \ge 0, \\ 0, & x < 0. \end{cases}$$

Definition A.76 *We say that $u \in C^{1,2}(]s,+\infty[\times\mathbb{R}^N)$ assumes the initial datum φ in the sense of L^1_{loc} if, for every compact set K in $\mathbb{R}^N$, we have that*

$$\lim_{t\to s^+} \|u(t,\cdot)-\varphi\|_{L^1(K)} = \lim_{t\to s^+} \int_K |u(t,x)-\varphi(x)|dx = 0.$$

We confine ourselves to stating the following existence result: the proof can be found, for example, in DiBenedetto [97], p. 240.

Theorem A.77 *If $\varphi \in L^1_{\text{loc}}(\mathbb{R}^N)$ satisfies condition* (A.64), *then the function u in* (A.63) *is a classical solution of the equation $Lu = 0$ in $]s,+\infty[\times\mathbb{R}^N$ and assumes the initial datum φ in the sense of L^1_{loc}.*

Remark A.78 Convergence in the sense of L^1_{loc} implies pointwise convergence

$$\lim_{t\to s^+} u(t,x) = \varphi(x),$$

for almost all $x \in \mathbb{R}^N$. Nevertheless, if the initial datum is assumed in the sense of pointwise convergence, this is not sufficient to guarantee the uniqueness of the solution of the Cauchy problem. Consider for instance the function

$$u(t,x) = \frac{x}{t^{\frac{3}{2}}} e^{-\frac{x^2}{2t}}, \qquad (t,x) \in]0,+\infty[\times\mathbb{R}.$$

Then u satisfies the heat equation and for every $x \in \mathbb{R}$ we have

$$\lim_{t\to 0^+} u(t,x) = 0.$$

Moreover u is strictly positive and verifies (A.64) with $\gamma = 2$. This example does not contradict the uniqueness results of Chapter 6 since u is *not* continuous on $\mathbb{R}_{\geq 0} \times \mathbb{R}$ and does not take the initial datum in the classical sense.

Note also that the fundamental solution of the heat equation $\Gamma(t,x)$ in (A.7) tends to zero as $t \to 0^+$ in $\mathbb{R}^N \setminus \{0\}$. However, for every $R > 0$, we have

$$\int_{|x|<R} \Gamma(t,x)dx = \pi^{-\frac{N}{2}} \int_{|y|<\frac{R}{\sqrt{2t}}} e^{-|y|^2} dy \xrightarrow[t\to 0^+]{} 1;$$

therefore Γ does not satisfies the null initial condition in the sense of L^1_{loc}. □

A.3.4 Non-homogeneous Cauchy problem

We consider the non-homogeneous Cauchy problem

$$\begin{cases} Lu = f, & \text{in }]0,T[\times\mathbb{R}^N, \\ u(0,\cdot) = \varphi, & \text{on } \mathbb{R}^N, \end{cases} \tag{A.65}$$

where $\varphi \in L^1_{\mathrm{loc}}(\mathbb{R}^N)$ satisfies condition (A.64) and f is continuous and verifies the growth condition

$$|f(t,x)| \leq c e^{c|x|^\gamma}, \qquad (t,x) \in]0,T[\times\mathbb{R}^N, \tag{A.66}$$

where c, γ are positive constants and $\gamma < 2$. Furthermore, we assume that f is locally Hölder continuous in x, uniformly with respect to t, i.e. for every compact set K in $\mathbb{R}^N$ we have that

$$|f(t,x) - f(t,y)| \leq c_K |x-y|^\beta, \qquad t \in]0,T[,\ x,y \in K,$$

with β, c_K positive constants. Then we have the following:

Theorem A.79 *The function defined on* $]0,T[\times\mathbb{R}^N$ *by*

$$u(t,x)=\int_{\mathbb{R}^N}\Gamma(t,x;0,y)\varphi(y)dy-\int_0^t\int_{\mathbb{R}^N}\Gamma(t,x;s,y)f(s,y)dyds, \qquad (A.67)$$

belongs to $C^{1,2}(]0,T[\times\mathbb{R}^N)$*, solves the equation* $Lu=f$ *in* $]0,T[\times\mathbb{R}^N$ *and assumes the initial datum* φ *in the sense of* L^1_{loc}*.*

Proof (Outline). If we set

$$F(t,x)=\int_0^t\int_{\mathbb{R}^N}\Gamma(t,x;s,y)f(s,y)dyds,$$

the claim is proved as soon as we have verified that

$$\lim_{(t,x)\to(0,x_0)}F(t,x)=0, \qquad x_0\in\mathbb{R}^N, \qquad (A.68)$$

$$LF(t,x)=-f(t,x), \qquad (t,x)\in]0,T[\times\mathbb{R}^N. \qquad (A.69)$$

The limit (A.68) is straightforward since we have the estimate

$$\left|\int_{\mathbb{R}^N}\Gamma(t,x;s,y)f(s,y)dy\right|\le Ce^{C|x|^2},$$

that can be proved by proceeding as in Remark 6.16. Concerning (A.69), formally we have

$$\begin{aligned}LF(t,x)&=\int_0^t\int_{\mathbb{R}^N}\underbrace{L\Gamma(t,x;s,y)}_{=0}f(s,y)dyds\\&\quad-\int_{\mathbb{R}^N}\underbrace{\Gamma(t,x;t,y)}_{=\delta_x(y)}f(t,y)dy=-f(t,x).\end{aligned}$$

In order to justify the previous steps it is necessary a careful study of some singular integrals in which the second-order derivatives of the fundamental solution appear: the proof is not trivial and is based on the crucial hypothesis of Hölder continuity of f. For instance, we refer to DiBenedetto [97], Chapter V, for all the details. □

A.3.5 Adjoint operator

Let L be the differential operator in (A.45), that is

$$L=\frac{1}{2}\sum_{j,k=1}^N c_{jk}\partial_{x_jx_k}+\sum_{j=1}^N b_j\partial_{x_j}-a-\partial_t, \qquad (t,x)\in\mathbb{R}^{N+1}.$$

For every $u, v \in C^2(\mathbb{R}^{N+1})$ with compact support, integrating by parts we obtain the following relation

$$\int_{\mathbb{R}^{N+1}} uLv = \int_{\mathbb{R}^{N+1}} vL^*u,$$

where

$$L^* = \frac{1}{2}\sum_{j,k=1}^{N} c_{jk}\partial_{x_j x_k} - \sum_{j=1}^{N} b_j \partial_{x_j} - a + \partial_t \tag{A.70}$$

is called *adjoint operator of* L. For example, the adjoint of the heat operator is simply

$$\frac{1}{2}\Delta + \partial_t.$$

Definition A.80 *A fundamental solution of the operator L^* is a function $\Gamma^*(t,x;T,y)$ defined for every $x, y \in \mathbb{R}^N$ and $t < T$, such that, for every $\varphi \in C_b(\mathbb{R}^N)$, the function*

$$v(t,x) = \int_{\mathbb{R}^N} \Gamma^*(t,x;T,y)\varphi(y)dy, \qquad t < T,\ x \in \mathbb{R}^N,$$

is a classical solution of the backward Cauchy problem

$$\begin{cases} L^*v = 0, & \text{in }]-\infty, T[\,\times \mathbb{R}^N, \\ v(T,x) = \varphi(x), & x \in \mathbb{R}^N. \end{cases} \tag{A.71}$$

We note that L^* is a backward operator and the related Cauchy problem (A.71) involves *a final datum*. The following result establishes the duality relation between the fundamental solutions of L and L^*.

Theorem A.81 *For any $x, y \in \mathbb{R}^N$ and $t < T$, we have*

$$\Gamma^*(t,x;T,y) = \Gamma(T,y;t,x). \tag{A.72}$$

Proof. We construct Γ^* by using the technique in the proof of Theorem A.72, based on the Fourier transform. Then we check directly the validity of formula (A.72). □

Remark A.82 By a simple change of variables, the backward problem for the heat operator is equivalent to the correspondent direct problem: indeed u is solution of

$$\begin{cases} \frac{1}{2}\Delta u - \partial_t u = 0, & \text{in }]0, +\infty[\,\times \mathbb{R}^N, \\ u(0,x) = \varphi(x), & x \in \mathbb{R}^N, \end{cases}$$

if and only if $v(t,x) = u(T-t,x)$ is solution of

$$\begin{cases} \frac{1}{2}\Delta v + \partial_t v = 0, & \text{in }]-\infty, T[\,\times \mathbb{R}^N, \\ v(T,x) = \varphi(x), & x \in \mathbb{R}^N. \end{cases}$$

□

Remark A.83 The problem for the heat equation with *final* datum

$$\begin{cases} \frac{1}{2}\Delta u(t,x) - \partial_t u(t,x) = 0, & (t,x) \in]0,T[\times\mathbb{R}^N, \\ u(T,x) = \varphi(x) & x \in \mathbb{R}^N, \end{cases} \tag{A.73}$$

is in general ill-posed and not solvable. More precisely, even if we have a final datum φ which is bounded and smooth, the solution may become irregular for T arbitrarily small. Indeed, it is enough to consider the solution of (A.73) with $\varphi(x) = \Gamma(T-\varepsilon, x)$, $\varepsilon > 0$, where Γ the fundamental solution of the heat operator. This fact corresponds to the physical phenomenon of heat diffusion that is not in general reversible, i.e. it is not possible to derive the initial state from the knowledge of the temperature at final time. □

A.4 Characteristic function and normal distribution

Definition A.84 *The characteristic function of the r.v. X, with values in $\mathbb{R}^N$, is the function*

$$\varphi_X : \mathbb{R}^N \longrightarrow \mathbb{C}$$

defined by

$$\varphi_X(\xi) = E\left[e^{i\langle \xi, X\rangle}\right], \qquad \xi \in \mathbb{R}^N.$$

In other terms, since

$$\varphi_X(\xi) = \int_{\mathbb{R}^N} e^{i\langle \xi, y\rangle} P^X(dy), \qquad \xi \in \mathbb{R}^N,$$

φ_X is simply the Fourier transform of the distribution P^X of X. In particular, if P^X has a density f, then $\varphi_X = \mathcal{F}(f)$.

The distribution of a r.v. is determined by its characteristic function: indeed the following generalized version of the Fourier inversion theorem holds (for the proof we refer, for example, to Chung [74], Chapter 6.2).

Theorem A.85 *Two random variables with the same characteristic function are identically distributed: more precisely, $\varphi_X(\xi) = \varphi_Y(\xi)$ for every $\xi \in \mathbb{R}^N$ if and only if $X \overset{d}{=} Y$. Further, if $\varphi_X \in L^1(\mathbb{R}^N)$ then the distribution of the r.v. X has a density $f \in C(\mathbb{R}^N)$ defined by*

$$f(x) = \frac{1}{(2\pi)^N} \int_{\mathbb{R}^N} e^{-i\langle x, \xi\rangle} \varphi_X(\xi) d\xi. \tag{A.74}$$

As a consequence of Proposition A.63 and of Theorem A.64 (see also Lemma A.142), the following simple properties of the characteristic function hold; their proof is left as a useful exercise.

Lemma A.86 *If X is a r.v. in $\mathbb{R}^N$, then φ_X is a continuous function such that $\varphi_X(0) = 1$ and*

$$|\varphi_X(\xi)| \le 1, \qquad \xi \in \mathbb{R}^N.$$

If X is a real r.v. in $L^p(\Omega, P)$, with $p \in \mathbb{N}$, then φ_X is differentiable p times and

$$\frac{d^p}{d\xi^p} \varphi_X(\xi)|_{\xi=0} = i^p E[X^p]. \tag{A.75}$$

If X is a *real* random variable in $L^p(\Omega, P)$, then

$$m_p(X) = E[X^p] = \frac{1}{i^p} \frac{d^p}{d\xi^p} \varphi_X(\xi)|_{\xi=0} \tag{A.76}$$

is called the *p-th moment of* X, while

$$\bar{m}_p(X) = E[(X - E[X])^p]$$

is called the *p-th centered moment of* X.

Since $\varphi_X(0) = 1$ and φ_X is a continuous function, it is possible to prove that

$$\varphi_X(\xi) = e^{\psi_X(\xi)} \tag{A.77}$$

in a neighborhood of the origin, for some continuous function ψ_X such that $\psi_X(0) = 0$ (cf. Sato [297], Lemma 7.6.). This result is not trivial since φ_X is a complex-valued function: ψ_X is called the *distinguished complex logarithm* of φ_X. The function ψ_X is also called the *cumulant generating function* of X because, by analogy with (A.76), the *cumulants* of X are defined by

$$c_p(X) = \frac{1}{i^p} \frac{d^p}{d\xi^p} \psi_X(\xi)|_{\xi=0}. \tag{A.78}$$

Differentiating (A.77), the p-th cumulant can be expressed as a polynomial function of the moments m_k or $\bar{m}_k$ with $k = 1, \dots, p$: for instance, we have

$$\begin{aligned}
c_1(X) &= m_1(X) = E[X], \\
c_2(X) &= \bar{m}_2(X) = m_2(X) - m_1(X)^2 = \mathrm{var}(X), \\
c_3(X) &= \bar{m}_3(X) = m_3(X) - 3m_2(X)m_1(X) + 2m_1(X)^3, \\
c_4(X) &= \bar{m}_4(X) - 3\bar{m}_2(X).
\end{aligned}$$

For example, if $X \sim \mathcal{N}_{\mu,\sigma^2}$ then, by Example A.62, we have

$$\varphi_X(\xi) = e^{\psi_X(\xi)} \qquad \text{with} \qquad \psi_X(\xi) = i\mu\xi - \frac{(\sigma\xi)^2}{2}.$$

In this case, ψ_X is a second order polynomial and therefore $c_p(X) = 0$ for any $p \ge 3$. Usually, the cumulants $c_p(X)$ with $p \ge 3$ are considered measures

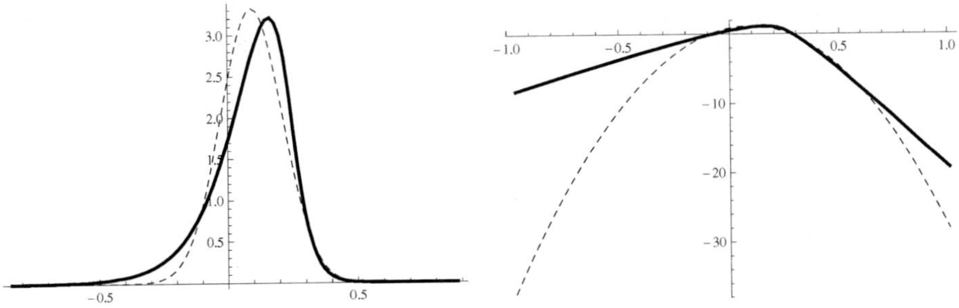

Fig. A.2. Variance-Gamma (thick line) and normal (dashed line) densities (left). Plot of the log-densities (right)

of deviation from normality. In particular, the third and fourth cumulants, scaled by the standard deviation,

$$s(X) = \frac{c_3(X)}{c_2(X)^{\frac{3}{2}}}, \qquad k(X) = \frac{c_4(X)}{c_2(X)^2}, \tag{A.79}$$

are called the *skewness* and *kurtosis* coefficients of X, respectively: if $s(X) > 0$ then X is said to be positively skewed; if $k(X) > 0$ then X is said to be leptokurtic or fat-tailed.

Figure A.2 compares the normal density with the density of X_1 where $(X_t)_{t\geq 0}$ is a Variance-Gamma process with parameters as in Example 15.21: the cumulants $c_1(X_t)$, $c_2(X_t)$ and $c_4(X_t)$ are given in (15.39); we also have

$$c_3(X_t) = t\mu\nu\left(2\mu^2\nu + 3\sigma^2\right).$$

The left picture shows the skewness of X_t. The right picture shows the graphs of the logarithms of the densities: the normal log-density behaves as $x \mapsto -x^2$ as $|x| \to +\infty$, while the VG log-density exhibits skewness and fat tails.

A.4.1 Multi-normal distribution

We say that a r.v. X on $(\Omega, \mathcal{F}, P)$ with values in $\mathbb{R}^N$, is multi-normal if it has a density of the form

$$\frac{1}{\sqrt{(2\pi)^N \det \mathcal{C}}} \exp\left(-\frac{1}{2}\langle \mathcal{C}^{-1}(x-\mu), (x-\mu)\rangle\right), \qquad x \in \mathbb{R}^N, \tag{A.80}$$

where μ is a given vector in $\mathbb{R}^N$ and $\mathcal{C} = (c_{jk})$ is a symmetric and positive definite $(N \times N)$-matrix: in this case we use the notation $X \sim \mathcal{N}_{\mu,\mathcal{C}}$ (see also Remark A.90 below).

As in Example A.31, a direct computation shows that

$$E[X] = \mu, \qquad \text{cov}(X_j, X_k) := E[(X_j - \mu_j)(X_k - \mu_k)] = c_{jk},$$

for $j,k=1,\dots,N$, where $X=(X_1,\dots,X_N)$. So μ is the expectation of X and $\mathcal{C}$ is the $(N\times N)$-dimensional covariance matrix of X:

$$\mathcal{C}=E\left[(X-\mu)(X-\mu)^*\right]. \tag{A.81}$$

Remark A.87 Let φ be a bounded continuous function on $\mathbb{R}^N$. By the representation formula (A.55) and by Theorem A.72, the classical solution of the Cauchy problem

$$\begin{cases}\frac{1}{2}\sum\limits_{j,k=1}^{N} c_{jk}\partial_{x_jx_k}u-\partial_t u=0, & (t,x)\in\,]0,+\infty[\times\mathbb{R}^N,\\ u(0,x)=\varphi(x), & x\in\mathbb{R}^N,\end{cases}$$

has the following probabilistic representation

$$u(t,x)=\int_{\mathbb{R}^N}\varphi(y)\Gamma(t,x-y)dy=E\left[\varphi(X^{t,x})\right]$$

where $X^{t,x}\sim\mathcal{N}_{x,t\mathcal{C}}$. Probabilistic numerical methods (Monte Carlo) for the heat equation are based on representations of this kind. □

Example A.88 If $X\sim\mathcal{N}_{\mu,\sigma^2}$ then, by Example A.62, we have

$$\varphi_X(\xi)=e^{i\mu\xi-\frac{(\sigma\xi)^2}{2}}. \tag{A.82}$$

In particular, if $X\sim\delta_\mu$ then $\varphi_X(\xi)=e^{i\mu\xi}$: notice that, in this case $|\varphi_X(\xi)|=1$ and so φ_X is not integrable[13]. □

Theorem A.89 *The r.v. X is multi-normal, $X\sim\mathcal{N}_{\mu,\mathcal{C}}$, if and only if*

$$\varphi_X(\xi)=\exp\left(i\langle\xi,\mu\rangle-\frac{1}{2}\langle\mathcal{C}\xi,\xi\rangle\right),\qquad \xi\in\mathbb{R}^N. \tag{A.83}$$

Proof. The claim is a direct consequence of Theorem A.85 and can be proved exactly as (A.54). □

Remark A.90 Formula (A.83) gives a characterization of the multi-normal distribution that makes sense also if $\mathcal{C}$ is symmetric and *positive semi-definite*[14]. We can therefore generalize the definition of a multi-normal random variable: we say that $X\sim\mathcal{N}_{\mu,\mathcal{C}}$, with $\mu\in\mathbb{R}^N$ and $\mathcal{C}=(c_{jk})$ symmetric and positive semi-definite, if (A.83) holds. For example, if $\mathcal{C}=0$ then $X\sim\delta_\mu$ since $\varphi_X(\xi)=\mathcal{F}(\delta_\mu)(\xi)=\exp(i\langle\xi,\mu\rangle)$ for $\xi\in\mathbb{R}^N$. □

---

[13] Compare this remark with Theorem A.65.

[14] An $(N\times N)$-dimensional matrix $\mathcal{C}$ is positive semi-definite (we write $\mathcal{C}\geq 0$) if

$$\langle\mathcal{C}x,x\rangle\geq 0,\qquad x\in\mathbb{R}^N.$$

Corollary A.91 *The random variables $X_1, \dots, X_m$ are independent if and only if*

$$\varphi_{(X_1,\dots,X_m)}(\xi_1, \dots, \xi_m) = \varphi_{X_1}(\xi) \cdots \varphi_{X_m}(\xi), \qquad \xi_1, \dots, \xi_m \in \mathbb{R}^N.$$

Proof. We prove the claim only in the case $m = 2$. By Proposition A.53, X, Y are independent if and only if $P^{(X,Y)} = P^X \otimes P^Y$. Now, for $\xi, \eta \in \mathbb{R}^N$, we have

$$\varphi_{(X,Y)}(\xi, \eta) = \iint_{\mathbb{R}^N \times \mathbb{R}^N} e^{i(\langle \xi, x\rangle + \langle \eta, y\rangle)} P^{(X,Y)}(dx\, dy),$$

$$\varphi_X(\xi)\varphi_Y(\eta) = \iint_{\mathbb{R}^N \times \mathbb{R}^N} e^{i(\langle \xi, x\rangle + \langle \eta, y\rangle)} P^X \otimes P^Y(dx\, dy),$$

and so the claim follows from Theorem A.85. □

Exercise A.92 Let X, Y be real independent normally-distributed random variables: $X \sim \mathcal{N}_{\mu, \sigma^2}$ and $Y \sim \mathcal{N}_{\nu, \varrho^2}$. Prove that

$$X + Y \sim \mathcal{N}_{\mu+\nu,\, \sigma^2+\varrho^2}.$$

Solution. Since X, Y are independent, by Lemma A.86, we have

$$\varphi_{X+Y}(\xi) = \varphi_X(\xi)\varphi_Y(\xi) = e^{i(\mu+\nu)\xi - \frac{\xi^2}{2}(\sigma^2+\varrho^2)}$$

by (A.82) and the claim follows from Theorem A.89. □

Remark A.93 We extend Remark A.32: let $X \sim \mathcal{N}_{\mu, \mathcal{C}}$, $\beta \in \mathbb{R}^d$ and let $\alpha = (\alpha_{ij})$ be a generic constant $(d \times N)$-matrix. Then the r.v. $\alpha X + \beta$ is multi-normal with mean $\alpha\mu + \beta$ and covariance matrix $\alpha \mathcal{C} \alpha^*$, where α^* denotes the transpose matrix of α, i.e.:

$$\alpha X + \beta \sim \mathcal{N}_{\alpha\mu+\beta,\, \alpha\mathcal{C}\alpha^*}.$$

As an exercise, verify that $\alpha\mathcal{C}\alpha^*$ is a $(d \times d)$-symmetric and positive semi-definite matrix. □

We give another characterization of multi-normal random variables.

Proposition A.94 *The r.v. X is multi-normal if and only if $\langle \lambda, X\rangle$ is normal for every $\lambda \in \mathbb{R}^N$. More precisely, $X \sim \mathcal{N}_{\mu, \mathcal{C}}$ if and only if*

$$\langle \lambda, X\rangle = \sum_{j=1}^{N} \lambda_j X_j \sim \mathcal{N}_{\langle \lambda, \mu\rangle, \langle \mathcal{C}\lambda, \lambda\rangle}, \qquad \text{(A.84)}$$

for every $\lambda \in \mathbb{R}^N$.

Proof. If $X \sim \mathcal{N}_{\mu,\mathcal{C}}$, then for every $\xi \in \mathbb{R}$ we have

$$\varphi_{\langle \lambda, X\rangle}(\xi) = E\left[e^{i\xi\langle \lambda, X\rangle}\right] = e^{i\xi\langle \lambda,\mu\rangle - \frac{\xi^2}{2}\langle \mathcal{C}\lambda,\lambda\rangle},$$

therefore, by Theorem A.89, $\langle \lambda, X\rangle \sim \mathcal{N}_{\langle \lambda,\mu\rangle,\, \langle \mathcal{C}\lambda,\lambda\rangle}$.

Conversely, if $\langle \lambda, X\rangle \sim \mathcal{N}_{m,\sigma^2}$ then, by (A.82), we have

$$\varphi_X(\lambda) = E\left[e^{i\langle \lambda, X\rangle}\right] = \varphi_{\langle \lambda, X\rangle}(1) = e^{im - \frac{\sigma^2}{2}}$$

where $m = \langle \lambda, E[X]\rangle$ and, putting $\mu_i = E[X_i]$, we have

$$\begin{aligned}\sigma^2 &= E\left[(\langle \lambda, X\rangle - \langle \lambda, E[X]\rangle)^2\right] \\ &= E\left[(\langle \lambda, X - E[X]\rangle)^2\right] = \sum_{i,j=1}^{N} \lambda_i\lambda_j E\left[(X_i - \mu_i)(X_j - \mu_j)\right],\end{aligned}$$

so X has multi-normal distribution, by Theorem A.89. □

A.5 Conditional expectation

In financial applications the price of an asset is generally modeled by a r.v. X and the amount of information available is described by a σ-algebra $\mathcal{G}$: as a consequence, it is natural to introduce the notion of *conditional expectation of X given $\mathcal{G}$*, usually denoted by

$$E[X \mid \mathcal{G}]$$

to describe the best estimate of the price X based on the information $\mathcal{G}$.

A.5.1 Radon-Nikodym theorem

Given any two measures P, Q on $(\Omega, \mathcal{F})$, we say that Q is *P-absolutely continuous on $\mathcal{F}$* if, for every $A \in \mathcal{F}$ such that $P(A) = 0$, we have $Q(A) = 0$. In this case we write $Q \ll P$ or $Q \ll_{\mathcal{F}} P$ if we want to highlight the σ-algebra that we are considering; indeed it is apparent that the notion of absolute continuity depends on the σ-algebra under consideration: if $\mathcal{G} \subseteq \mathcal{F}$ are σ-algebras, then $Q \ll_{\mathcal{G}} P$ does not necessarily imply that $Q \ll_{\mathcal{F}} P$.

If $Q \ll P$ and $P \ll Q$, then we say that the measures P and Q are *equivalent* and we write $P \sim Q$. In case P, Q are probability measures, $Q \ll_{\mathcal{F}} P$ implies that the P-negligible events in $\mathcal{F}$ are also Q-negligible, but the converse may not be true. Obviously, if $Q \ll_{\mathcal{F}} P$, then for every $A \in \mathcal{F}$ such that $P(A) = 1$ we have $Q(A) = 1$, i.e. the certain events for P are certain also for Q, but the converse is not generally true.

Example A.95 We have already seen that, for $\sigma > 0$, $\mathcal{N}_{\mu,\sigma^2}$ is absolutely continuous with respect to Lebesgue measure m in $(\mathbb{R}, \mathscr{B})$. As an exercise prove the converse, that is $m \ll_{\mathscr{B}} \mathcal{N}_{\mu,\sigma^2}$. Note that the distribution δ_{x_0} is not absolutely continuous with respect to Lebesgue measure, since $m(\{x_0\}) = 0$ but $\delta_{x_0}(\{x_0\}) = 1$. □

If P is a distribution of the type (A.4), i.e. with a density with respect to Lebesgue measure m, then $P \ll_{\mathscr{B}} m$. We may wonder whether *all* the measures P such that $P \ll m$ have the form (A.4). The following classical result gives an affirmative answer. For its proof, we refer to Williams [339].

Theorem A.96 (Radon-Nikodym theorem) *Let $(\Omega, \mathcal{F}, P)$ be a finite-measure space. If Q is a finite measure on $(\Omega, \mathcal{F})$ and $Q \ll_{\mathcal{F}} P$, then there exists $L : \Omega \to \mathbb{R}$, $L \geq 0$, such that*

i) L is $\mathcal{F}$-measurable;
ii) L is P-integrable;
iii) $Q(A) = \int_A LdP$ for every $A \in \mathcal{F}$.

Further, L is P-almost surely unique (i.e. if L' verifies the same properties of L, then $P(L = L') = 1$). We say that L is the density of Q with respect to P on $\mathcal{F}$ or also the Radon-Nikodym derivative of Q with respect to P on $\mathcal{F}$ and we write without distinction $L = \frac{dQ}{dP}$ or $dQ = LdP$. In order to emphasize the dependence on $\mathcal{F}$, we also write

$$L = \frac{dQ}{dP} |_{\mathcal{F}} . \tag{A.85}$$

Remark A.97 Let P, Q be probability measures on the space $(\Omega, \mathcal{F})$ with $Q \ll P$ and set $L = \frac{dQ}{dP}$. Using Dynkin's theorem as in the proof of Theorem A.23, we can show that $X \in L^1(\Omega, Q)$ if and only if $XL \in L^1(\Omega, P)$ and in that case

$$E^Q[X] = E^P[XL], \tag{A.86}$$

where E^P and E^Q denote the expectations under the probability measures P and Q respectively. In other words

$$\int_\Omega XdQ = \int_\Omega X \left(\frac{dQ}{dP}\right) dP$$

and this justifies the notation (A.85). □

Theorem A.96 can be extended to the case of P, Q σ-finite, with the exception of the second point. It follows in particular that the distributions with density with respect to Lebesgue measure m are only those that are m-absolutely continuous.

A.5.2 Conditional expectation

The goal of this paragraph is to introduce gradually the rigorous definition of conditional expectation. Those who are familiar with this notion, can skip the paragraph and proceed to Definition A.99 directly.

Given a real integrable r.v. X and an event B with positive probability, we define *the conditional expectation of* X *given the event* B as the mean of X with respect to the measure $P(\cdot \mid B)$ defined in Section A.1.7 and more precisely

$$E[X|B] = \frac{1}{P(B)} \int_B X dP. \tag{A.87}$$

Given $B \in \mathcal{F}$ such that $0 < P(B) < 1$, we denote by $\mathcal{G}$ the σ-algebra generated by B:

$$\mathcal{G} = \{\emptyset, \Omega, B, B^c\}. \tag{A.88}$$

The conditional expectation of X given $\mathcal{G}$, $E[X \mid \mathcal{G}]$, is defined by

$$E[X|\mathcal{G}](\omega) = \begin{cases} E[X|B], & \omega \in B, \\ E[X|B^c], & \omega \in B^c. \end{cases} \tag{A.89}$$

We remark explicitly that $E[X \mid \mathcal{G}]$ *is a random variable.*

Remark A.98 We can prove directly that

i) $E[X|\mathcal{G}]$ is $\mathcal{G}$-measurable;
ii) $\int_G X dP = \int_G E[X| \mathcal{G}] dP$ for every $G \in \mathcal{G}$.

Further, if Y is a r.v. verifying such properties, then $Y = E[X|\mathcal{G}]$ P-a.s., that is properties i) and ii) characterize $E[X|\mathcal{G}]$ P-almost surely. Indeed $G = \{Y > E[X \mid \mathcal{G}]\}$ is a $\mathcal{G}$-measurable event and by ii) we get

$$\int_G (Y - E[X \mid \mathcal{G}]) dP = 0,$$

this implying that $P(G) = 0$. □

We point out that $E[X|\mathcal{G}]$ is $\mathcal{G}$-measurable even if X is not. Intuitively $E[X|\mathcal{G}]$ represents the "expectation of X based on the knowledge of the information in $\mathcal{G}$", i.e. the best approximation of X on the basis of the information in $\mathcal{G}$.

We can generalize the previous definition to the case of a generic σ-algebra in the following way.

Definition A.99 *Let X be a real integrable r.v. on the probability space $(\Omega, \mathcal{F}, P)$ and let $\mathcal{G}$ be a σ-algebra contained in $\mathcal{F}$. Let Y be a r.v. such that*

i) Y is integrable and $\mathcal{G}$-measurable;
ii) $\int_A X dP = \int_A Y dP$ for every $A \in \mathcal{G}$.

Then we say that Y is a version of the conditional expectation (or, more simply, the conditional expectation) of X given $\mathcal{G}$ and we write $Y = E[X \mid \mathcal{G}]$.

In order for Definition A.99 to be well-posed, we have to prove existence and "uniqueness" of the conditional expectation.

Theorem A.100 *Let X be a real integrable r.v. on the probability space $(\Omega, \mathcal{F}, P)$ and let $\mathcal{G}$ be a σ-algebra contained in $\mathcal{F}$. Then there exists a r.v. Y satisfying i),ii) in Definition A.99. Further, those properties characterize Y in the sense that, if Z is another r.v. satisfying i), ii) in Definition A.99, then $Y = Z$ a.s.*

Proof. A simple but indirect[15] proof is based upon the Radon-Nikodym theorem. First of all it is enough to prove the claim in the case that X is a real non-negative r.v. Since $X \in L^1(\Omega, P)$, the equation

$$Q(G) = \int_G X dP, \qquad G \in \mathcal{G},$$

defines a finite measure Q on $\mathcal{G}$. Further, $Q \ll P$ on $\mathcal{G}$ and so by Theorem A.96 there exists a $\mathcal{G}$-measurable r.v. Y such that $Q(G) = \int_G Y dP$ for every $G \in \mathcal{G}$. In order to conclude and prove the uniqueness we can proceed as in Remark A.98. □

Notation A.101 *If $Y = E[X \mid \mathcal{G}]$ and Z is a $\mathcal{G}$-measurable[16] r.v. such that $Z = Y$ a.s. then also $Z = E[X \mid \mathcal{G}]$. So the conditional expectation of X is defined up to a negligible event: the expression $Y = E[X \mid \mathcal{G}]$ does not have to be understood as an equality of random variables, but as a notation to say that Y is a r.v. having the properties i) and ii) of the previous definition. By convention, when we write*

$$E[X \mid \mathcal{G}] = E[Y \mid \mathcal{G}] \quad (\textit{respectively } E[X \mid \mathcal{G}] \le E[Y \mid \mathcal{G}])$$

we mean that, if $A = E[X \mid \mathcal{G}]$ and $B = E[Y \mid \mathcal{G}]$, then

$$A = B \quad \textit{a.s. (respectively } A \le B \quad \textit{a.s.).}$$

Given two random variables X, Y and $F \in \mathcal{F}$, we denote by

$$E[X \mid Y] := E[X \mid \sigma(Y)] \qquad \textit{and} \qquad P(F \mid Y) := E[\mathbb{1}_F \mid Y]$$

the conditional expectation of X and the conditional probability of F, with respect to Y, respectively. Moreover, when $B = \{Y = Y(\omega)\}$ in (A.87), we simply write $E[X \mid Y = Y(\omega)]$ instead of $E[X \mid \{Y = Y(\omega)\}]$. □

[15] We defer a more direct proof to Section A.5.5.
[16] We note that, if $Z = Y$ a.s. and $Y = E[X \mid \mathcal{G}]$, not necessarily Z is $\mathcal{G}$-measurable and so we do not necessarily have $Z = E[X \mid \mathcal{G}]$.

Remark A.102 As a consequence of Dynkin's Theorem A.9, property ii) in Definition A.99 is equivalent to the fact that

$$E[XW] = E[YW], \tag{A.90}$$

for every bounded and $\mathcal{G}$-measurable r.v. W. □

Remark A.103 If Y is $\mathcal{G}$-measurable and

$$\int_A Y dP \geq \int_A X dP, \qquad A \in \mathcal{G}, \tag{A.91}$$

then

$$Y \geq E[X \mid \mathcal{G}].$$

Indeed, if $Z = E[X \mid \mathcal{G}]$ and, by contradiction, $A := \{Y < Z\} \in \mathcal{G}$ is not a negligible event, then we would have

$$\int_A Y dP < \int_A Z dP = \int_A X dP,$$

contradicting assumption (A.91). □

Exercise A.104 Let X be an integrable real r.v. on the probability space $(\Omega, \mathcal{F}, P)$ and let $\mathcal{G}$ be a σ-algebra contained in $\mathcal{F}$. Proceeding as in the proof of Proposition A.6, prove that $Y = E[X \mid \mathcal{G}]$ if and only if:

i) Y is $\mathcal{G}$-measurable;
ii) $\int_A X dP = \int_A Y dP$ for every $A \in \mathcal{A}$ where $\mathcal{A}$ is a $\cap$-stable collection, containing Ω and such that $\mathcal{G} = \sigma(\mathcal{A})$.

A.5.3 Conditional expectation and discrete random variables

In this section we assume that X, Y are *discrete* random variables on $(\Omega, \mathcal{F}, P)$, that is

$$X(\Omega) = \{x_1, \dots, x_n\} \qquad Y(\Omega) = \{y_1, \dots, y_m\}, \tag{A.92}$$

for some $n, m \in \mathbb{N}$.

Proposition A.105 *We have*

$$E[X \mid Y](\omega) = E[X \mid Y = Y(\omega)], \qquad \omega \in \Omega, \tag{A.93}$$

$$E[X \mid Y] = \sum_{k=1}^{n} x_k P(X = x_k \mid Y). \tag{A.94}$$

Proof. We set $Z(\omega) = E[X \mid Y = Y(\omega)]$ for $\omega \in \Omega$. By definition of conditional expectation, (A.93) follows from the fact that:

i) Z is $\sigma(Y)$-measurable, since $Z = z_i := E[X \mid Y = y_i]$ on $\{Y = y_i\}$, $i = 1, \dots, m$, and $Z(\Omega) = \{z_1, \dots, z_m\}$;

ii) for any $i = 1, \dots, m$, we have

$$\int_{\{Y=y_i\}} Z dP = \int_{\{Y=y_i\}} E[X \mid Y = y_i] \, dP$$

(by (A.87))

$$= \int_{\{Y=y_i\}} \left(\frac{1}{P(Y = y_i)} \int_{\{Y=y_i\}} X dP \right) dP$$
$$= \int_{\{Y=y_i\}} X dP,$$

and therefore we also have

$$\int_A Z dP = \int_A X dP, \qquad A \in \sigma(Y).$$

Next we prove (A.94): let $Y(\omega) = y_i$, then we have

$$E[X \mid Y](\omega) = E[X \mid Y = y_i] = \frac{1}{P(Y = y_i)} \int_{\{Y=y_i\}} X dP$$
$$= \frac{1}{P(Y = y_i)} \sum_{k=1}^{n} x_k P((X = x_k) \cap (Y = y_i))$$
$$= \sum_{k=1}^{n} x_k P(X = x_k \mid Y = y_i).$$

□

Proposition A.106 *Let X, Y be discrete random variables such that* (A.92) *holds. Then X, Y are independent if and only if the random variable $P(X = x_k \mid Y)$ is constant (independent on $\omega \in \Omega$) for any $k = 1, \dots, n$.*

Proof. The "only if" part is trivial. Conversely, assume that, for any $k = 1, \dots, n$,

$$p_k := P(X = x_k \mid Y) = P(X = x_k \mid Y = y_i)$$

is constant and does not depend on $i = 1, \dots, m$. Then we have

$$P((X = x_k) \cap (Y = y_i)) = p_k P(Y = y_i) \tag{A.95}$$

and summing for $i = 1, \dots, m$, we get

$$P(X = x_k) = p_k.$$

Inserting this last identity back in (A.95), we deduce

$$P((X = x_k) \cap (Y = y_i)) = P(X = x_k) P(Y = y_i)$$

that proves that X, Y are independent. □

A.5.4 Properties of the conditional expectation

The following properties are a direct consequence of the definition and construction of the conditional expectation. For every $X, Y \in L^1(\Omega, \mathcal{F}, P)$ and $a, b \in \mathbb{R}$ we have:

(1) if X is $\mathcal{G}$-measurable, then $X = E[X|\mathcal{G}]$;
(2) if X and $\mathcal{G}$ are independent (i.e. $\sigma(X)$ and $\mathcal{G}$ are independent), then $E[X] = E[X|\mathcal{G}]$. In particular $E[X] = E[X|\sigma(\mathcal{N})]$;
(3) $E[X] = E[E[X|\mathcal{G}]]$;
(4) [linearity] $aE[X|\mathcal{G}] + bE[Y|\mathcal{G}] = E[aX + bY|\mathcal{G}]$;
(4-b) [linearity] $E^{\lambda P+(1-\lambda)Q}[X|\mathcal{G}] = \lambda E^P[X|\mathcal{G}] + (1-\lambda)E^Q[X|\mathcal{G}]$ for all probability measures P, Q and $\lambda \in [0,1]$;
(5) [monotonicity] if $X \le Y$ a.s., then $E[X|\mathcal{G}] \le E[Y|\mathcal{G}]$.

The following proposition contains other properties of the conditional expectation. Many of them have an analogous counterpart in the properties of the usual integral operator.

Proposition A.107 *Let $X, Y \in L^1(\Omega, \mathcal{F}, P)$ and let $\mathcal{G}, \mathcal{H} \subseteq \mathcal{F}$ be σ-algebras of Ω. Then we have:*

(6) if Y is independent of $\sigma(X, \mathcal{G})$, then $E[XY|\mathcal{G}] = E[X|\mathcal{G}]\,E[Y]$;
(7) if Y is $\mathcal{G}$-measurable and bounded, then $YE[X|\mathcal{G}] = E[XY|\mathcal{G}]$;
(8) if $\mathcal{H} \subseteq \mathcal{G}$, then $E[E[X|\mathcal{G}]\,|\mathcal{H}] = E[X|\mathcal{H}]$;
(9) [Beppo Levi] *if $(X_n)_{n\in\mathbb{N}}$, with $0 \le X_n \in L^1(\Omega, P)$, is a monotone increasing sequence converging pointwise to X a.s. and $Z_n = E[X_n|\mathcal{G}]$, then $\lim_{n\to+\infty} Z_n = E[X|\mathcal{G}]$;*
(10) [Fatou] *let $(X_n)_{n\in\mathbb{N}}$ be a sequence of non-negative random variables in $L^1(\Omega, P)$; then, setting $Z_n = E[X_n|\mathcal{G}]$ and $X = \liminf_{n\to+\infty} X_n$, we have $\liminf_{n\to+\infty} Z_n \ge E[X|\mathcal{G}]$;*
(11) [Dominated convergence] *let $(X_n)_{n\in\mathbb{N}}$ be a sequence converging pointwise to X a.s. and let us suppose that there exists $Y \in L^1(\Omega, P)$ such that $|X_n| \le Y$ a.s. Setting $Z_n = E[X_n|\mathcal{G}]$, then $\lim_{n\to+\infty} Z_n = E[X|\mathcal{G}]$;*
(12) [Jensen's inequality] *if φ is a convex function such that $\varphi(X) \in L^1(\Omega, P)$, then*

$$E[\varphi(X) \mid \mathcal{G}] \ge \varphi\left(E[X \mid \mathcal{G}]\right).$$

Proof. (6) $E[X|\mathcal{G}]\,E[Y]$ is $\mathcal{G}$-measurable and, for every bounded and $\mathcal{G}$-measurable W, we have

$$E[WE[X|\mathcal{G}]\,E[Y]] = E[WE[X|\mathcal{G}]]\,E[Y] =$$

(by Remark A.102)

$$= E[WX]\,E[Y] =$$

(by the independence assumption)

$$= E[WXY],$$

and this proves the claim.

(7) $YE[X|\mathcal{G}]$ is $\mathcal{G}$-measurable by assumption and, for every bounded and $\mathcal{G}$-measurable W, we have

$$E[(WY)E[X|\mathcal{G}]] = E[WYX],$$

by Remark A.102, since WY is $\mathcal{G}$-measurable and bounded.

(8) $E[E[X|\mathcal{G}]|\mathcal{H}]$ is $\mathcal{H}$-measurable and for every bounded and $\mathcal{H}$-measurable (and therefore also $\mathcal{G}$-measurable) r.v. W, we have

$$E[WE[E[X|\mathcal{G}]|\mathcal{H}]] = E[WE[X|\mathcal{G}]] = E[WX].$$

(9) By (5) we have that (Z_n) is an increasing sequence of $\mathcal{G}$-measurable non-negative random variables, so $Z := \sup_{n\in\mathbb{N}} Z_n$ is a $\mathcal{G}$-measurable r.v. Further, for every $G \in \mathcal{G}$, applying twice Beppo Levi's theorem, we have

$$\int_G Z dP = \lim_{n\to\infty} \int_G Z_n dP = \lim_{n\to\infty} \int_G X_n dP = \int_G X dP.$$

(10-11) The proof is analogous to that of (9).

(12) We can proceed as in the proof of the classical Jensen's inequality. We recall that every convex function φ coincides with the upper envelope of the linear functions $\ell \le \varphi$, i.e.

$$\varphi(x) = \sup_{\ell\in\mathcal{L}} \ell(x), \qquad x \in \mathbb{R},$$

where

$$\mathcal{L} = \{\ell : \mathbb{R} \to \mathbb{R} \mid \ell(x) = ax + b,\ \ell \le \varphi\}.$$

Then we have:

$$E[\varphi(X) \mid \mathcal{G}] = E\left[\sup_{\ell\in\mathcal{L}} \ell(X) \mid \mathcal{G}\right] \ge$$

(by (5))

$$\ge \sup_{\ell\in\mathcal{L}} E[\ell(X) \mid \mathcal{G}] =$$

(by (4))

$$= \sup_{\ell\in\mathcal{L}} \ell(E[X \mid \mathcal{G}]) = \varphi([X \mid \mathcal{G}]).$$

□

Lemma A.108 *Let X, Y be random variables on $(\Omega, \mathcal{F}, P)$. Let $\mathcal{G} \subseteq \mathcal{F}$ be a σ-algebra such that*

i) X is independent of $\mathcal{G}$;
ii) Y is $\mathcal{G}$-measurable.

Then, for every $\mathscr{B}$-measurable bounded (or non-negative) function h we have

$$E\left[h(X,Y) \mid \mathcal{G}\right] = g(Y), \qquad \text{where} \quad g(y) = E\left[h(X,y)\right]. \tag{A.96}$$

We write (A.96) *in a more compact form as*

$$E\left[h(X,Y) \mid \mathcal{G}\right] = E\left[h(X,y) \mid \mathcal{G}\right]|_{y=Y}. \tag{A.97}$$

Proof. We have to prove that the r.v. $g(Y)$ is a version of the conditional expectation of $h(X,Y)$. Using the notation P^W to denote the distribution of a given r.v. W, we have

$$g(y) = \int_{\mathbb{R}} h(x,y) P^X(dx).$$

Then, by Fubini's theorem, g is a $\mathscr{B}$-measurable function: consequently, by assumption ii), $g(Y)$ is $\mathcal{G}$-measurable.

Further, given $G \in \mathcal{G}$ and putting $Z = \mathbb{1}_G$, we get

$$\int_G h(X,Y)dP = \int_\Omega h(X,Y)ZdP = \iiint h(x,y)z\, P^{(X,Y,Z)}(d(x,y,z)) =$$

(by the independence assumption i) and Proposition A.53)

$$= \iiint h(x,y)z\, P^X(dx)P^{(Y,Z)}(d(y,z)) =$$

(by Fubini's theorem)

$$= \iint g(y)z\, P^{(Y,Z)}(d(y,z)) = \int_G g(Y)dP. \qquad \square$$

Remark A.109 Under the assumptions of Lemma A.108, by (A.96) we also have

$$E\left[h(X,Y) \mid \mathcal{G}\right] = E\left[h(X,Y) \mid Y\right].$$

Indeed, if $Z = E\left[h(X,Y) \mid \mathcal{G}\right]$, since the function g in (A.96) is $\mathscr{B}$-measurable, we have that Z is $\sigma(Y)$-measurable. Further,

$$\int_G ZdP = \int_G h(X,Y)dP, \qquad G \in \sigma(Y),$$

by definition of Z and since $\sigma(Y) \subseteq \mathcal{G}$. $\square$

We conclude the section with the following useful:

Proposition A.110 *Let X be a r.v. in $\mathbb{R}^N$ and $\mathcal{G} \subseteq \mathcal{F}$ a σ-algebra. Then X and $\mathcal{G}$ are independent if and only if*

$$E\left[e^{i\langle\xi,X\rangle}\right] = E\left[e^{i\langle\xi,X\rangle} \mid \mathcal{G}\right], \qquad \xi \in \mathbb{R}^N. \tag{A.98}$$

Proof. We prove that, if (A.98) holds, then X is independent of every r.v. $Y \in m\mathcal{G}$. For every $\xi, \eta \in \mathbb{R}^N$ we have

$$\varphi_{(X,Y)}(\xi,\eta) = E\left[e^{i(\langle\xi,X\rangle+\langle\eta,Y\rangle)}\right] = E\left[e^{i\langle\eta,Y\rangle} E\left[e^{i\langle\xi,X\rangle} \mid \mathcal{G}\right]\right] =$$

(by assumption)

$$= E\left[e^{i\langle\xi,X\rangle}\right] E\left[e^{i\langle\eta,Y\rangle}\right].$$

The claim follows from Corollary A.91. □

Exercise A.111 Prove that

$$\mathrm{var}(E[X \mid \mathcal{G}]) \le \mathrm{var}(X)$$

i.e. *by conditioning the variance gets smaller.* Furthermore, prove that, if $X_n \to X$ as $n \to \infty$ in $L^1(\Omega, P)$, then

$$\lim_{n\to\infty} E[X_n \mid \mathcal{G}] = E[X \mid \mathcal{G}] \qquad \text{in } L^1(\Omega, P).$$

A.5.5 Conditional expectation in L^2

We consider the space $L^p(\mathcal{F}) := L^p(\Omega, \mathcal{F}, P)$ with $p \ge 1$. If $\mathcal{G}$ is a sub-σ-algebra of $\mathcal{F}$, then $L^p(\mathcal{G})$ is a linear subspace of $L^p(\mathcal{F})$ and for every $X \in L^p(\mathcal{F})$

$$\|E[X \mid \mathcal{G}]\|_p \le \|X\|_p. \tag{A.99}$$

Indeed, by Jensen's inequality with $\varphi(x) = |x|^p$, we have

$$E[|E[X \mid \mathcal{G}]|^p] \le E[E[|X|^p \mid \mathcal{G}]] = E[|X|^p].$$

By (A.99), the conditional expectation $E[\cdot \mid \mathcal{G}]$ is a linear bounded operator from $L^p(\mathcal{F})$ to $L^p(\mathcal{G})$: in other terms, if $\lim\limits_{n\to\infty} X_n = X$ in L^p, i.e. if

$$\lim_{n\to\infty} \|X_n - X\|_p = 0,$$

then we have

$$\lim_{n\to\infty} \|Y_n - Y\|_p = 0,$$

where $Y = E[X \mid \mathcal{G}]$ and $Y_n = E[X_n \mid \mathcal{G}]$. The case $p = 2$ is of particular interest. Let us denote by

$$\langle X, Y\rangle_{L^2} = \int_\Omega XY dP, \qquad X, Y \in L^2,$$

the scalar product[17] in L^2.

Proposition A.112 *For every $X \in L^2(\mathcal{F})$ and $W \in L^2(\mathcal{G})$, we have*

$$\langle X - E[X \mid \mathcal{G}], W\rangle_{L^2} = 0. \tag{A.100}$$

Proof. By (A.90), equation (A.100) holds for every bounded $\mathcal{G}$-measurable W. The claim follows from a standard density argument. □

By (A.100), $X - E[X \mid \mathcal{G}]$ is orthogonal to the subspace $L^2(\mathcal{G})$ and therefore, *the conditional expectation $E[X \mid \mathcal{G}]$ is the projection of X on $L^2(\mathcal{G})$*. Indeed, if $Z = E[X \mid \mathcal{G}]$, we have that $Z \in L^2(\mathcal{G})$ and for every $W \in L^2(\mathcal{G})$

$$\begin{aligned}\|X - W\|_2^2 &= \langle X - Z + Z - W, X - Z + Z - W\rangle_{L^2} \\ &= \|X - Z\|_2^2 + \|Z - W\|_2^2 + 2\underbrace{\langle X - Z, Z - W\rangle_{L^2}}_{=0} \geq\end{aligned}$$

(the last term is null since $X - Z$ is orthogonal to $Z - W \in L^2(\mathcal{G})$)

$$\geq \|X - Z\|_2^2.$$

So $E[X \mid \mathcal{G}]$ minimizes the distance of X from $L^2(\mathcal{G})$ and geometrically it represents the best approximation of X in $L^2(\mathcal{G})$. The characterization of the conditional expectation in terms of projection on the space L^2 can be used in order to give a direct and constructive proof of Theorem A.100 (see, for example, Williams [339]).

A.5.6 Change of measure

On the probability space $(\Omega, \mathcal{F}, P)$ we consider a sub-σ-algebra $\mathcal{G}$ of $\mathcal{F}$ and a probability measure $Q \ll_{\mathcal{F}} P$ (therefore also $Q \ll_{\mathcal{G}} P$). We denote by $L^{\mathcal{F}}$ (resp. $L^{\mathcal{G}}$) the Radon-Nikodym derivative of Q with respect to P on $\mathcal{F}$ (resp. on $\mathcal{G}$). In general $L^{\mathcal{F}} \neq L^{\mathcal{G}}$ since $L^{\mathcal{F}}$ may not be $\mathcal{G}$-measurable. On the other hand, we have

$$L^{\mathcal{G}} = E^P\left[L^{\mathcal{F}} \mid \mathcal{G}\right]. \tag{A.101}$$

Indeed $L^{\mathcal{G}}$ is integrable and $\mathcal{G}$-measurable and we have

$$\int_G L^{\mathcal{G}} dP = Q(G) = \int_G L^{\mathcal{F}} dP, \qquad G \in \mathcal{G},$$

since $\mathcal{G} \subseteq \mathcal{F}$.

[17] Since $\langle X, X\rangle_{L^2} = 0$ if and only if $X = 0$ a.s., $\langle\cdot,\cdot\rangle_{L^2}$ is a scalar product provided that we identify random variables that are a.s. equal.

A result on the change of probability measure for conditional expectations, analogous to formula (A.86), is given by the following:

Theorem A.113 (Bayes' formula) *Let P, Q be probability measures on $(\Omega, \mathcal{F})$ with $Q \ll_{\mathcal{F}} P$. If $X \in L^1(\Omega, Q)$, $\mathcal{G}$ is a sub-σ-algebra of $\mathcal{F}$ and we set $L = \frac{dQ}{dP}\,|_{\mathcal{F}}$, then we have*

$$E^Q[X \mid \mathcal{G}] = \frac{E^P[XL \mid \mathcal{G}]}{E^P[L \mid \mathcal{G}]}.$$

Proof. We put $V = E^Q[X \mid \mathcal{G}]$ and $W = E^P[L \mid \mathcal{G}]$. We have to prove that

i) $Q(W > 0) = 1$;
ii) $VW = E^P[XL \mid \mathcal{G}]$.

Concerning i), since $\{W = 0\} \in \mathcal{G}$, we have

$$Q(W = 0) = \int_{\{W=0\}} L dP = \int_{\{W=0\}} W dP = 0.$$

Concerning ii), VW is obviously $\mathcal{G}$-measurable and for every $G \in \mathcal{G}$ we have

$$\begin{aligned} \int_G VW dP &= \int_G E^P[VL \mid \mathcal{G}]\, dP = \int_G VL dP \\ &= \int_G E^Q[X \mid \mathcal{G}]\, dQ = \int_G X dQ = \int_G XL dP. \end{aligned}$$ □

A.6 Stochastic processes in discrete time

We recall the notation $\mathbb{N}_0 = \mathbb{N} \cup \{0\}$ for the set of non-negative integer numbers.

Definition A.114 *A discrete stochastic process in $\mathbb{R}^N$ is a collection $X = (X_n)_{n \in \mathbb{N}_0}$ of random variables defined on a probability space $(\Omega, \mathcal{F}, P)$ with values in $\mathbb{R}^N$:*

$$X_n : \Omega \longrightarrow \mathbb{R}^N, \qquad n \in \mathbb{N}_0.$$

The collection of σ-algebras $(\mathcal{F}_n^X)_{n \in \mathbb{N}_0}$, defined by

$$\mathcal{F}_n^X = \sigma(X_k,\ 0 \le k \le n),$$

is called natural filtration for X. In general, a filtration on the probability space $(\Omega, \mathcal{F}, P)$ is an increasing collection $(\mathcal{F}_n)_{n \in \mathbb{N}_0}$ (i.e. $\mathcal{F}_n \subseteq \mathcal{F}_{n+1}$ for every n) of sub-σ-algebras of $\mathcal{F}$. We say that the process X is adapted to the filtration $(\mathcal{F}_n)$ if X_n is $\mathcal{F}_n$-measurable, or equivalently $\mathcal{F}_n^X \subseteq \mathcal{F}_n$, for every $n \in \mathbb{N}_0$. We say that X is an integrable process if $X_n \in L^1(\Omega, P)$ for every $n \in \mathbb{N}_0$.

In many applications, stochastic processes are used to describe the evolution in time of a random phenomenon and the index "n" represents the time variable. Since $n \in \mathbb{N}_0$, we often refer to X in the preceding definition as a "stochastic process in discrete time".

To fix the ideas, we can think of X_n as the price of a risky asset at time n. Intuitively $\mathcal{F}_n^X$ represents the information available on the asset X at time n and the filtration $\mathcal{F}^X := (\mathcal{F}_n^X)$ represents the "increasing flow" of information.

In probability theory and in financial applications, the following classes of processes play a central role.

Definition A.115 *Let $M = (M_n)_{n\in\mathbb{N}_0}$ be an integrable adapted stochastic process on the filtered probability space $(\Omega, \mathcal{F}, P, \mathcal{F}_n)$. We say that M is*

- *a discrete martingale (or, simply, a martingale) if*

$$M_n = E\left[M_{n+1} \mid \mathcal{F}_n\right], \qquad n \in \mathbb{N}_0;$$

- *a super-martingale if*

$$M_n \geq E\left[M_{n+1} \mid \mathcal{F}_n\right], \qquad n \in \mathbb{N}_0;$$

- *a sub-martingale if*

$$M_n \leq E\left[M_{n+1} \mid \mathcal{F}_n\right], \qquad n \in \mathbb{N}_0.$$

It is apparent that the martingale property depends on the filtration and on the probability P that are considered. By the linearity of the conditional expectation, martingales form a linear space. Further, linear combinations with non-negative coefficients of super-martingales (sub-martingales) are super-martingales (sub-martingales).

If M is a martingale and $0 \leq k < n$, we have

$$E\left[M_n \mid \mathcal{F}_k\right] = E\left[E\left[M_n \mid \mathcal{F}_{n-1}\right] \mid \mathcal{F}_k\right] = E\left[M_{n-1} \mid \mathcal{F}_k\right] = \dots = M_k, \quad \text{(A.102)}$$

as a consequence of property (8) of conditional expectation. Further, for every n, we have

$$E\left[M_n\right] = E\left[E\left[M_n \mid \mathcal{F}_0\right]\right] = E\left[M_0\right], \qquad \text{(A.103)}$$

hence it follows that *the expectation of a martingale is constant.* Analogously a super-martingale is a stochastic process that "decreases in mean" and a sub-martingale is a stochastic process that "increases in mean".

Remark A.116 If M is a martingale and φ is a *convex* function on $\mathbb{R}$ such that $\varphi(M)$ is integrable, then $\varphi(M)$ is a sub-martingale. Indeed

$$E\left[\varphi(M_{n+1}) \mid \mathcal{F}_n\right] \geq$$

(by Jensen's inequality)

$$\geq \varphi(E\left[M_{n+1} \mid \mathcal{F}_n\right]) = \varphi(M_n).$$

Further, if M is a sub-martingale and φ is a convex and *increasing* function on $\mathbb{R}$ such that $\varphi(M)$ is integrable, then $\varphi(M)$ is a sub-martingale. As remarkable cases, if M is a martingale, then $|M|$ and M^2 are sub-martingales. We point out that the fact that M is a sub-martingale is not enough to conclude that also $|M|$ and M^2 are sub-martingales since the functions $x \mapsto |x|$ and $x \mapsto x^2$ are convex but not increasing. □

Example A.117 Let X be an integrable random variable on the filtered probability space $(\Omega, \mathcal{F}, P, \mathcal{F}_n)$. Then, the process M defined by

$$M_n = E[X \mid \mathcal{F}_n]$$

is a martingale: indeed M is clearly adapted, integrable and we have

$$E[M_{n+1} \mid \mathcal{F}_n] = E[E[X \mid \mathcal{F}_{n+1}] \mid \mathcal{F}_n] = E[X \mid \mathcal{F}_n] = M_n.$$ □

A.6.1 Doob's decomposition

Definition A.118 *We say that a stochastic process A, on a filtered probability space $(\Omega, \mathcal{F}, P, \mathcal{F}_n)$, is predictable if A_n is $\mathcal{F}_{n-1}$-measurable for every $n \geq 1$.*

The following result, which is crucial in the study of American options, sheds some light on the structure of adapted stochastic processes.

Theorem A.119 (Doob's decomposition theorem) *Every adapted and integrable stochastic process X can be decomposed uniquely*[18] *in the sum*

$$X = M + A \tag{A.104}$$

where M is a martingale such that $M_0 = X_0$ and A is a predictable stochastic process such that $A_0 = 0$. Further, X is a super-martingale (sub-martingale) if and only if A is decreasing[19] *(increasing).*

Proof. We define the processes M and A recursively by putting

$$\begin{cases} M_0 = X_0, \\ M_n = M_{n-1} + X_n - E[X_n \mid \mathcal{F}_{n-1}], \qquad n \geq 1, \end{cases} \tag{A.105}$$

and

$$\begin{cases} A_0 = 0, \\ A_n = A_{n-1} - (X_{n-1} - E[X_n \mid \mathcal{F}_{n-1}]), \qquad n \geq 1. \end{cases} \tag{A.106}$$

[18] Up to a negligible event.

[19] A process A is decreasing if $A_n \geq A_{n+1}$ a.s. for every n. A stochastic process A is increasing if $-A$ is decreasing.

More explicitly we have

$$M_n = X_n + \sum_{k=1}^{n} \left(X_{k-1} - E\left[X_k \mid \mathcal{F}_{k-1}\right] \right), \tag{A.107}$$

and

$$A_n = -\sum_{k=1}^{n} \left(X_{k-1} - E\left[X_k \mid \mathcal{F}_{k-1}\right] \right). \tag{A.108}$$

Then it is easy to verify that M is a martingale, A is predictable and (A.104) holds.

Concerning the uniqueness of the decomposition, if (A.104) holds then we have also

$$X_n - X_{n-1} = M_n - M_{n-1} + A_n - A_{n-1},$$

and taking the conditional expectation (under the assumption that M is a martingale and A is predictable), we have

$$E\left[X_n \mid \mathcal{F}_{n-1}\right] - X_{n-1} = A_n - A_{n-1},$$

hence M, A must be defined by (A.105) and (A.106), respectively.

Finally, by (A.106) it is apparent that $A_n \geq A_{n-1}$ a.s. if and only if X is a super-martingale. □

Definition A.120 *Let α and M be two discrete processes. The transform of M by α is the stochastic process $(G_n(\alpha, M))_{n \geq 1}$ defined as*

$$G_n(\alpha, M) = \sum_{k=1}^{n} \alpha_k (M_k - M_{k-1}), \qquad n \in \mathbb{N}. \tag{A.109}$$

The process $G(\alpha, M)$ is the discrete counterpart of the stochastic integral introduced in Chapter 4.

Proposition A.121 *If M is a martingale and α is bounded and predictable, then $G(\alpha, M)$ is a martingale with null expectation. Conversely, if for every predictable and bounded process α, we have*

$$E\left[G_n(\alpha, M)\right] = 0, \qquad n \geq 1, \tag{A.110}$$

then M is a martingale. Moreover if $\alpha \geq 0$ and M is a super-martingale (resp. sub-martingale) then also $G(\alpha, M)$ is a super-martingale (resp. sub-martingale).

Proof. Clearly $G(\alpha, M)$ is an adapted and integrable process. Further, for every n, we have

$$G_{n+1}(\alpha, M) = G_n(\alpha, M) + \alpha_{n+1}(M_{n+1} - M_n),$$

and so

$$E[G_{n+1}(\alpha, M) \mid \mathcal{F}_n] = G_n(\alpha, M) + E[\alpha_{n+1}(M_{n+1} - M_n) \mid \mathcal{F}_n] =$$

(since α is predictable)

$$= G_n(\alpha, M) + \alpha_{n+1} E[M_{n+1} - M_n \mid \mathcal{F}_n] =$$

(since M is a martingale)

$$= G_n(\alpha, M).$$

So $G(\alpha, M)$ is a martingale and

$$\begin{aligned} E[G_n(\alpha, M)] &= E[G_1(\alpha, M)] \\ &= E[\alpha_1(M_1 - M_0)] = E[\alpha_1 E[M_1 - M_0 \mid \mathcal{F}_0]] = 0. \end{aligned}$$

Conversely, we have to prove that $M_{n-1} = E[M_n \mid \mathcal{F}_{n-1}]$: since M is adapted, it suffices to prove that

$$E[M_n \mathbb{1}_A] = E[M_{n-1} \mathbb{1}_A], \qquad A \in \mathcal{F}_{n-1}.$$

Then, for a fixed $A \in \mathcal{F}_{n-1}$, we set

$$\alpha_k = \begin{cases} \mathbb{1}_A, & k = n, \\ 0, & k \neq n. \end{cases}$$

Since the process α is predictable, the claim follows by applying (A.110). □

A.6.2 Stopping times

Definition A.122 *In a filtered space $(\Omega, \mathcal{F}, P, \mathcal{F}_n)$, a random variable*

$$\nu : \Omega \longrightarrow \mathbb{N}_0 \cup \{\infty\}$$

such that

$$\{\nu = n\} \in \mathcal{F}_n, \qquad n \in \mathbb{N}_0, \tag{A.111}$$

is called stopping time.

Intuitively, we can think of a stopping time as the moment in which we make a decision about a random phenomenon (e.g. the decision of exercising an American option). By assuming condition (A.111), we require that the decision depends only on the information available at the moment.

Example A.123 (Exit time) Let X be an adapted process and $H \in \mathscr{B}$. We define the exit time of X from H as

$$\nu(\omega) = \min\{n \mid X_n(\omega) \notin H\}, \qquad \omega \in \Omega,$$

or $\nu(\omega) = \infty$ if $X_n(\omega) \in H$ for any n. Then ν is a stopping time since

$$\{\nu = n\} = \{X_0 \in H\} \cap \{X_1 \in H\} \cap \cdots \cap \{X_{n-1} \in H\} \cap \{X_n \notin H\} \in \mathcal{F}_n.$$

□

Remark A.124 Condition (A.111) has the following simple consequences:

$$\{\nu \le n\} = \bigcup_{k=0}^{n} \{\nu = k\} \in \mathcal{F}_n, \tag{A.112}$$

$$\{\nu \ge n\} = \{\nu > n-1\} = \{\nu \le n-1\}^c \in \mathcal{F}_{n-1}. \tag{A.113}$$

□

Let X and ν be a stochastic process and a stopping time, respectively: the *stopped process* of X by ν is defined by

$$X_n^\nu(\omega) = X_{n\wedge\nu(\omega)}(\omega), \qquad \omega \in \Omega, \tag{A.114}$$

where

$$a \wedge b = \min\{a, b\}.$$

In what follows we write $X_{\nu\wedge n}$ as well instead of X_n^ν. The next lemma gives some simple properties of stopped processes.

Lemma A.125 *Suppose that X is a stochastic process and ν is a stopping time, then:*

i) if X is adapted then also X^ν is adapted;
ii) if X is a martingale then also X^ν is a martingale;
iii) if X is a super-martingale (sub-martingale) then also X^ν is a super-martingale (sub-martingale).

Proof. We have

$$X_{\nu\wedge n} = X_0 + \sum_{k=1}^{n} (X_k - X_{k-1}) \mathbb{1}_{\{k\le\nu\}}, \qquad n \ge 1, \tag{A.115}$$

or in other terms, recalling Definition A.120, X^ν is the transform of X by the process $\alpha = (\alpha_n) := \left(\mathbb{1}_{\{\nu\ge n\}}\right)$. Since α is predictable by (A.113), then X^ν is adapted if X is adapted and the thesis follows by Proposition A.121. □

Given a process X and an a.s. finite stopping time ν, up to a negligible event A, we define the random variable X_ν by setting

$$X_\nu(\omega) = X_{\nu(\omega)}(\omega), \qquad \omega \in \Omega \setminus A.$$

The following result extends the martingale property to random times.

Corollary A.126 (Optional sampling) *If X is a martingale and ν is a stopping time such that $n \le \nu \le N$ a.s. then we have*

$$E\left[X_\nu \mid \mathcal{F}_n\right] = X_n.$$

In particular we also have

$$E\left[X_\nu\right] = E\left[X_0\right]. \tag{A.116}$$

Proof. We have

$$X_n = X_n^\nu =$$

(by Lemma A.125-ii))

$$= E\left[X_N^\nu \mid \mathcal{F}_n\right] = E\left[X_\nu \mid \mathcal{F}_n\right].$$

□

Remark A.127 Identity (A.116) remains valid under the assumption that ν is a.s. finite and X is a uniformly integrable martingale (cf. Definition A.148). Indeed, by Lemma A.125 we have

$$E\left[X_{\nu\wedge n}\right] = E\left[X_0\right], \qquad n \in \mathbb{N},$$

and the claim follows by taking the limit as $n \to \infty$, by Theorem A.149. □

We denote by

$$\mathcal{F}_\nu := \{F \in \mathcal{F} \mid F \cap \{\nu \le n\} \in \mathcal{F}_n \text{ for every } n \in \mathbb{N}\} \tag{A.117}$$

the σ-algebra associated to the stopping time ν. Definition (A.117) is consistent with the standard notation $\mathcal{F}_n$ used for filtrations: in other terms, if ν is the constant stopping time equal to $k \in \mathbb{N}$, then $\mathcal{F}_\nu = \mathcal{F}_k$. Indeed $F \in \mathcal{F}_\nu$ if and only if $F \in \mathcal{F}$ and

$$F \cap \{k \le n\} \in \mathcal{F}_n, \qquad n \in \mathbb{N},$$

i.e. if and only if

$$F \in \mathcal{F}_n, \qquad n \ge k,$$

that is, $F \in \mathcal{F}_k$.

We also remark that, for every $k \in \mathbb{N}$ and stopping time ν, we have

$$\{\nu = k\} \in \mathcal{F}_\nu,$$

since the event

$$\{\nu = k\} \cap \{\nu \le n\} = \begin{cases} \{\nu = k\} & \text{for } k \le n, \\ \emptyset & \text{for } k > n, \end{cases}$$

belongs to $\mathcal{F}_n$ for every $n \in \mathbb{N}$.

Lemma A.128 *If X is an adapted stochastic process and ν is an a.s. finite stopping time, then X_ν is $\mathcal{F}_\nu$-measurable.*

Proof. Since

$$X_\nu = \sum_{n=0}^{\infty} X_n \mathbb{1}_{\{\nu = n\}},$$

it is enough to prove that $X_n \mathbb{1}_{\{\nu=n\}}$ is $\mathcal{F}_\nu$-measurable for every $n \in \mathbb{N}_0$, i.e.

$$\{X_n \mathbb{1}_{\{\nu=n\}} \in H\} \in \mathcal{F}_\nu, \qquad H \in \mathscr{B},\ n \in \mathbb{N}_0.$$

If $H \in \mathscr{B}$ and $0 \notin H$ we have

$$A := \{X_n \mathbb{1}_{\{\nu=n\}} \in H\} = \{X_n \in H\} \cap \{\nu = n\},$$

and so $A \cap \{\nu \le k\} \in \mathcal{F}_k$ for every k, since

$$A \cap \{\nu \le k\} = \begin{cases} A & \text{if } n \le k, \\ \emptyset & \text{if } n > k. \end{cases}$$

On the other hand, if $H = \{0\}$ we have

$$B := \{X_n \mathbb{1}_{\{\nu=n\}} = 0\} = \bigcup_{i \ne n} \{\nu = i\} \cup (\{X_n = 0\} \cap \{\nu = n\}).$$

So $B \in \mathcal{F}_\nu$ since

$$B \cap \{\nu \le k\} = \underbrace{\bigcup_{i \ne n,\ i \le k} \{\nu = i\}}_{\in \mathcal{F}_k} \cup \underbrace{(\{X_n = 0\} \cap \{\nu = n\} \cap \{\nu \le k\})}_{\in \mathcal{F}_k}$$

for every k. □

The following result extends Corollary A.126.

Theorem A.129 (Doob's optional sampling theorem) *Let ν_1, ν_2 be stopping times such that*

$$\nu_1 \le \nu_2 \le N \quad a.s.$$

for some $N \in \mathbb{N}$. If X is a super-martingale then

$$X_{\nu_1} \ge E\left[X_{\nu_2} \mid \mathcal{F}_{\nu_1}\right]. \tag{A.118}$$

Consequently, if X is a martingale then

$$X_{\nu_1} = E\left[X_{\nu_2} \mid \mathcal{F}_{\nu_1}\right].$$

Proof. We first remark that the random variables X_{ν_1}, X_{ν_2} are integrable: indeed

$$|X_{\nu_i}| \le \sum_{k=0}^{N} |X_k|, \qquad i = 1, 2.$$

In order to prove (A.118), we use Remark A.103: since, by Lemma A.128, X_{ν_1} is $\mathcal{F}_{\nu_1}$-measurable, we have to prove that

$$\int_A X_{\nu_1} dP \ge \int_A X_{\nu_2} dP, \qquad A \in \mathcal{F}_{\nu_1}. \tag{A.119}$$

Firstly, we consider the case of constant ν_2, i.e. $\nu_2 = N$. If $A \in \mathcal{F}_{\nu_1}$ we have $A \cap \{\nu_1 = n\} \in \mathcal{F}_n$ and so

$$\int_{A\cap\{\nu_1=n\}} X_{\nu_1} dP = \int_{A\cap\{\nu_1=n\}} X_n dP \geq \int_{A\cap\{\nu_1=n\}} E\left[X_N \mid \mathcal{F}_n\right] dP = \int_{A\cap\{\nu_1=n\}} X_N dP.$$

It follows that

$$\int_A X_{\nu_1} dP = \sum_{n=0}^{N} \int_{A\cap\{\nu_1=n\}} X_{\nu_1} dP \geq \sum_{n=0}^{N} \int_{A\cap\{\nu_1=n\}} X_N dP = \int_A X_N dP. \tag{A.120}$$

Now we consider the general case in which $\nu_2 \leq N$ a.s. By Lemma A.125, X^{ν_2} is a super-martingale and so by applying (A.120) we get

$$\int_A X_{\nu_1} dP = \int_A X_{\nu_1}^{\nu_2} dP \geq \int_A X_N^{\nu_2} dP = \int_A X_{\nu_2} dP.$$ □

A.6.3 Doob's maximal inequality

Among the many noteworthy results of martingale theory, we prove the following Doob's inequality that plays a crucial role in the construction of the stochastic integral.

Theorem A.130 (Doob's maximal inequality) *Let M be a martingale on the filtered space $(\Omega, \mathcal{F}, P, \mathcal{F}_n)$. For every $N \in \mathbb{N}$ and $p \in \mathbb{R}$, $p > 1$, we have*

$$E\left[\max_{0\leq n\leq N} |M_n|^p\right] \leq q^p E\left[|M_N|^p\right], \tag{A.121}$$

where $q = \frac{p}{p-1}$ is the conjugate exponent of p.

Remark A.131 The reverse inequality for (A.121) is trivial:

$$E\left[|M_N|^p\right] \leq \max_{0\leq n\leq N} E\left[|M_n|^p\right] \leq E\left[\max_{0\leq n\leq N} |M_n|^p\right]. \tag{A.122}$$

Indeed, for $k \leq N$ we have

$$|M_k|^p \leq \max_{0\leq n\leq N} |M_n|^p$$

so that

$$E\left[|M_k|^p\right] \leq E\left[\max_{0\leq n\leq N} |M_n|^p\right]$$

and (A.122) follows. □

Proof (of Theorem A.130). We prove a slightly more general statement: if X is a non-negative sub-martingale then

$$E\left[\max_{0\le n\le N} X_n^p\right] \le q^p E\left[X_N^p\right], \tag{A.123}$$

for every $N \in \mathbb{N}$ and $p > 1$. Estimate (A.121) is an immediate consequence of (A.123), applied to the non-negative sub-martingale $X = |M|$.

For fixed $N \in \mathbb{N}$ and $\lambda > 0$, we set

$$\nu_\lambda(\omega) = \min\{n \le N \mid X_n(\omega) \ge \lambda\}$$

and $\nu_\lambda(\omega) = N + 1$ if that set is empty. By Example A.123, ν_λ is a stopping time. Moreover setting

$$W_{n,\lambda} = \mathbb{1}_{\{\nu_\lambda = n\}}, \qquad n = 0, \dots, N,$$

we have obviously

$$\lambda W_{n,\lambda} \le X_n W_{n,\lambda},$$

and, taking expectation,

$$\lambda E\left[W_{n,\lambda}\right] \le E\left[X_n W_{n,\lambda}\right] \le$$

(since, by assumption, X is a sub-martingale)

$$\le E\left[E\left[X_N \mid \mathcal{F}_n\right] W_{n,\lambda}\right] =$$

(by Remark A.102, since $W_{n,\lambda}$ is $\mathcal{F}_n$-measurable and bounded)

$$= E\left[X_N W_{n,\lambda}\right]. \tag{A.124}$$

Now we set $Y = \max_{0\le n\le N} X_n$ and we observe that

$$\mathbb{1}_{\{Y\ge\lambda\}} = \sum_{n=0}^{N} W_{n,\lambda}.$$

Taking expectations and using the estimate (A.124), we get

$$\lambda P(Y \ge \lambda) = \lambda \sum_{n=0}^{N} E\left[W_{n,\lambda}\right] \le E\left[X_N \mathbb{1}_{\{Y\ge\lambda\}}\right]. \tag{A.125}$$

Further, by Example A.57, for every $p > 0$

$$E\left[Y^p\right] = p\int_0^{+\infty} \lambda^{p-1} P\left(Y \ge \lambda\right) d\lambda \le$$

(by (A.125))

$$\leq pE\left[X_N \int_0^{+\infty} \lambda^{p-2} \mathbb{1}_{\{Y \geq \lambda\}} d\lambda\right] = \frac{p}{p-1} E\left[X_N Y^{p-1}\right] \leq$$

(by Hölder's inequality[20])

$$\leq \frac{p}{p-1} E\left[X_N^p\right]^{\frac{1}{p}} E\left[Y^p\right]^{1-\frac{1}{p}},$$

and this concludes the proof. □

We state another classical result according to which, under very general assumptions, a martingale (M_n) converges almost surely as $n \to \infty$. For the proof, we refer to [339].

Theorem A.132 *Let $(X_n)_{n \in \mathbb{N}}$ be a super-martingale such that*

$$\sup_n E\left[X_n^-\right] < +\infty \tag{A.126}$$

where $X^- = \max\{0, -X\}$. Then there exists the limit

$$\lim_{n \to \infty} X_n < \infty \qquad a.s.$$

We point out that, if $(M_n)_{n \in \mathbb{N}}$ is a bounded martingale in L^p for some $p > 1$, i.e. it is such that

$$\lambda := \sup_{n \in \mathbb{N}} E\left[|M_n|^p\right] < \infty,$$

then by Theorem A.130, we have

$$E\left[\sup_{n \in \mathbb{N}} |M_n|^p\right] \leq q^p \lambda,$$

and condition (A.126) is satisfied. Therefore (M_n) converges a.s. as $n \to \infty$ to a random variable M such that $|M| \leq \sup_n |M_n|$. Further, since

$$|M_n - M|^p \leq 2^{p-1}\left(|M_n|^p + |M|^p\right) \leq 2^p \sup_n |M_n|^p,$$

by the dominated convergence theorem we have

$$\lim_{n \to \infty} E\left[|M_n - M|^p\right] = 0.$$

---

[20] For every conjugate exponents $p, q \geq 1$, we have

$$E\left[|XY|\right] \leq E\left[|X|^p\right]^{\frac{1}{p}} E\left[|Y|^q\right]^{\frac{1}{q}}.$$

Therefore we have proved the following:

Corollary A.133 *Every L^p-bounded martingale $(M_n)_{n\in\mathbb{N}}$, with $p > 1$, converges a.s. and in L^p-norm.*

We also prove another useful consequence of the previous result.

Corollary A.134 *Let $(\Omega, \mathcal{F}, P, \mathcal{F}_n)$ be a filtered space and $X \in L^p(\Omega, P)$, with $p > 1$. Then we have*

$$\lim_{n\to\infty} E[X \mid \mathcal{F}_n] = E[X \mid \mathcal{F}_\infty], \qquad \text{in } L^p,$$

where $\mathcal{F}_\infty$ denotes the σ-algebra generated by $(\mathcal{F}_n)_{n\in\mathbb{N}}$.

Proof. The position

$$X_n = E[X \mid \mathcal{F}_n], \qquad n \in \mathbb{N},$$

defines a bounded martingale in L^p with $p > 1$ and so there exists the limit

$$M := \lim_{n\to\infty} X_n, \qquad \text{in } L^p.$$

Then it is enough to prove that

$$M = E[X \mid \mathcal{F}_\infty]. \tag{A.127}$$

We set

$$M_n = E[M \mid \mathcal{F}_n], \qquad n \in \mathbb{N},$$

and we observe that

$$E[|X_n - M_n|] = E[|X_n - E[M \mid \mathcal{F}_n]|] \le E[|X_n - M|] \xrightarrow[n\to\infty]{} 0.$$

Now we fix $\bar{n} \in \mathbb{N}$: for every $F \in \mathcal{F}_{\bar{n}}$ and $n \ge \bar{n}$ we have

$$\int_F (X - M)dP = \int_F E[X - M \mid \mathcal{F}_n]\, dP = \int_F (X_n - M_n)dP \xrightarrow[n\to\infty]{} 0.$$

We infer that

$$\int_F M dP = \int_F X dP, \qquad F \in \mathcal{F}_\infty,$$

and since $M \in m\mathcal{F}_\infty$ we obtain (A.127). □

Remark A.135 By using the notion of uniform integrability, (cf. Section A.7.2), it is possible to extend the convergence result in Corollary A.134 also to the case $p = 1$. □

A.7 Convergence of random variables

We recall the main notions of convergence of random variables. We consider a sequence $(X_n)_{n\in\mathbb{N}}$ of random variables on a probability space $(\Omega, \mathcal{F}, P)$:

i) (X_n) converges *almost surely* to X if

$$P\left(\lim_{n\to\infty} X_n = X\right) = 1,$$

i.e., if the event

$$\{\omega \mid \lim_{n\to\infty} X_n(\omega) = X(\omega)\}$$

is certain. In this case we write

$$X_n \xrightarrow{\text{a.s.}} X;$$

ii) (X_n) converges *in probability* to X if

$$\lim_{n\to\infty} P\left(|X_n - X| > \varepsilon\right) = 0$$

for every $\varepsilon > 0$. In this case we write

$$X_n \xrightarrow{P} X;$$

iii) (X_n) converges *in L^p* to X if

$$\lim_{n\to\infty} E\left[|X_n - X|^p\right] = 0.$$

In this case we write

$$X_n \xrightarrow{L^p} X.$$

The following result sums up the relations among different types of convergence.

Theorem A.136 *The following implications hold:*

i) if $X_n \xrightarrow{a.s.} X$ then $X_n \xrightarrow{P} X$;
ii) if $X_n \xrightarrow{L^p} X$ then $X_n \xrightarrow{P} X$;
iii) if $X_n \xrightarrow{P} X$ then there exists a subsequence (X_{k_n}) such that $X_{k_n} \xrightarrow{a.s.} X$.

In general no other relations hold.

Now we consider a sequence (X_n) of random variables and we denote the distribution of X_n by μ_{X_n}. We recall that, by definition, a sequence of distributions (μ_n) converges to the distribution μ if

$$\lim_{n\to\infty} \int_{\mathbb{R}^N} \varphi d\mu_n = \int_{\mathbb{R}^N} \varphi d\mu, \qquad \varphi \in C_b(\mathbb{R}^N),$$

where $C_b(\mathbb{R}^N)$ denotes the family of continuous and bounded functions.

Definition A.137 *A sequence of random variables* (X_n) *converges in distribution (or in law) to a random variable* X *if the corresponding sequence of distributions* (μ_n) *converges to* μ_X*. In this case we write*

$$X_n \xrightarrow{d} X.$$

Remark A.138 The convergence in distribution of random variables is defined only in terms of the convergence of the respective distributions and consequently it is not necessary that the random variables are defined on the same probability space. If we denote by $(\Omega_n, P_n, \mathcal{F}_n)$ the probability space on which the random variable X_n is defined and $(\Omega, P, \mathcal{F})$ the probability space on which X is defined, then $X_n \xrightarrow{d} X$ if and only if

$$\lim_{n\to\infty} E^{P_n}\left[\varphi(X_n)\right] = E^P\left[\varphi(X)\right], \qquad \varphi \in C_b(\mathbb{R}^N), \tag{A.128}$$

where E^{P_n} and E^P denote the expected values computed under the respective measures. □

The following proposition states that convergence in distribution is the weakest form of convergence.

Proposition A.139 *If* $X_n \xrightarrow{P} X$ *then* $X_n \xrightarrow{d} X$.

Exercise A.140 Let (X_n) be a sequence of random variables on the space $(\Omega, \mathcal{F}, P)$ converging in distribution to a constant r.v. X, i.e. $X_n \xrightarrow{d} X$. Prove that $X_n \xrightarrow{P} X$.

Hint. By contradiction, suppose that (X_n) does not converge to X in probability: then, given $\varepsilon > 0$, there exist $\delta > 0$ and a subsequence (X_{k_n}) such that

$$P(|X_{k_n} - X| > \varepsilon) \geq \delta, \qquad n \in \mathbb{N}.$$

By assumption

$$\lim_{n\to\infty} E\left[\varphi(X_n)\right] = \varphi(X), \qquad \varphi \in C_b.$$

In particular, consider a positive and monotone increasing $\varphi \in C_b$: then, for n large enough, we have

$$\begin{aligned}\varphi(X) + 1 \geq E\left[\varphi(X_{k_n})\right] &\geq \int_{\{|X_{k_n} - X| > \varepsilon\}} \varphi(X_{k_n}) dP \\ &\geq \varphi(X - \varepsilon) P(|X_{k_n} - X| > \varepsilon) \geq \delta\varphi(X - \varepsilon),\end{aligned}$$

and this is absurd, since φ is arbitrary. □

A.7.1 Characteristic function and convergence of variables

Let us denote by

$$\varphi_X(\xi) = E\left[e^{i\langle \xi, X\rangle}\right], \qquad \xi \in \mathbb{R}^N,$$

the characteristic function of the random variable X. We state without proof[21] the following important result that establishes that convergence in distribution of a sequence of random variables is equivalent to pointwise convergence of the corresponding characteristic functions.

Theorem A.141 (Lévy's theorem) *Let (X_n) and (φ_{X_n}) be a sequence of random variables in $\mathbb{R}^N$ and the corresponding sequence of characteristic function, respectively:*

i) if $X_n \xrightarrow{d} X$ where X is some random variable, then

$$\lim_{n\to\infty} \varphi_{X_n}(\xi) = \varphi_X(\xi), \qquad \xi \in \mathbb{R}^N;$$

ii) if $\lim_{n\to\infty} \varphi_{X_n}(\xi)$ exists for every $\xi \in \mathbb{R}^N$ and the function φ defined by

$$\varphi(\xi) = \lim_{n\to\infty} \varphi_{X_n}(\xi), \qquad \xi \in \mathbb{R}^N,$$

is continuous at the origin, then φ is the characteristic function of a random variable X such that $X_n \xrightarrow{d} X$.

The previous theorem can be used to prove some well-known results. We first give a preliminary

Lemma A.142 *Let X be a real random variable such that $E[|X|^p] < \infty$ for some $p \in \mathbb{N}$. Then the following asymptotic expansion holds:*

$$\varphi_X(\xi) = \sum_{k=0}^{p} \frac{(i\xi)^k}{k!} E\left[X^k\right] + \mathrm{o}(\xi^p), \qquad \text{as } \xi \to 0. \tag{A.129}$$

For the definition of the symbol $\mathrm{o}(\cdot)$, see the note on p. 52.

Proof. The thesis is a direct consequence of Lemma A.86, nevertheless we give an alternative proof. By using the Taylor series expansion about $\xi = 0$ with the Lagrange form of the remainder, we have

$$e^{i\xi X} = \sum_{k=0}^{p-1} \frac{(i\xi)^k}{k!} X^k + \frac{(i\xi X)^p}{p!} e^{i\theta\xi X} =$$

(where θ is a random variable such that $|\theta| \le 1$)

$$= \sum_{k=0}^{p-1} \frac{(i\xi)^k}{k!} X^k + \frac{(i\xi)^p}{p!} (X^p + W_p(\xi)),$$

setting $W_p(\xi) = X^p(e^{i\theta\xi X} - 1)$. Taking expectations, we get

$$\varphi_X(\xi) - \sum_{k=0}^{p} \frac{(i\xi)^k}{k!} E\left[X^k\right] = \frac{(i\xi)^p}{p!} E\left[W_p\right] = \mathrm{o}(\xi^p) \qquad \text{as } \xi \to 0,$$

[21] See, for example, Shiryaev [309], Chapter III-3, or Williams [339], Chapter 18.

since

$$\lim_{\xi\to 0} E\left[W_p(\xi)\right] = 0. \tag{A.130}$$

Then (A.130) follows from Lebesgue's dominated convergence theorem: in fact

$$W_p(\xi) = X^p(e^{i\theta\xi X} - 1) \to 0 \qquad \text{as } \xi \to 0$$

and

$$|W_p(\xi)| \le 2|X|^p \in L^1(\Omega, P)$$

by assumption. □

Theorem A.143 (Law of large numbers) *Let (X_n) be a sequence of i.i.d. integrable random variables. Let $\mu = E[X_1]$ and*

$$M_n = \frac{X_1 + \cdots + X_n}{n};$$

then we have

$$M_n \xrightarrow{P} \mu.$$

Proof. We use Lévy's theorem and we consider the sequence of characteristic functions

$$\varphi_{M_n}(\eta) = E\left[e^{i\eta M_n}\right] =$$

(since the random variables X_n are i.i.d.)

$$= \left(E\left[e^{i\frac{\eta X_1}{n}}\right]\right)^n =$$

(by Lemma A.142, in which we take $p = 1$ and $\xi = \frac{\eta}{n}$)

$$= \left(1 + \frac{i\eta\mu}{n} + \mathrm{o}\left(\frac{1}{n}\right)\right)^n \longrightarrow e^{i\eta\mu},$$

as $n \to \infty$. Since $\eta \mapsto e^{i\eta\mu}$ is the Fourier transform of the Dirac's delta δ_μ concentrated at μ, Lévy's theorem implies that $M_n \xrightarrow{d} \mu$. The claim follows from Exercise A.140. □

Remark A.144 If we further assume that $X_1 \in L^2(\Omega)$, we can give a direct and elementary proof of the law of large numbers based on Markov's inequality, Proposition A.58. For the sake of simplicity, let us consider only the 1-dimensional case and we set $\sigma^2 = \mathrm{var}(X_1)$. We have

$$P(|M_n - \mu| \ge \varepsilon) \le \frac{\mathrm{var}(M_n)}{\varepsilon^2} =$$

(since the random variables are i.i.d.)

$$= \frac{n\text{var}\left(\frac{X_1}{n}\right)}{\varepsilon^2} = \frac{\sigma^2}{n\varepsilon^2}. \tag{A.131}$$

Formula (A.131) gives also an explicit estimate of the speed of convergence: indeed it can be rewritten in the equivalent form

$$P\left(|M_n - \mu| \le \varepsilon\right) \ge 1 - \frac{\sigma^2}{n\varepsilon^2}.$$

Then, for a fixed a probability $p \in]0,1[$, we have

$$P\left(|M_n - \mu| \le \frac{\sigma}{\sqrt{n(1-p)}}\right) \ge p.$$

In other terms, for every n, the difference between M_n and μ is, with probability greater than p, smaller than $\frac{C}{\sqrt{n}}$ where $C = \frac{\sigma}{\sqrt{1-p}}$. □

Remark A.145 We recall that the *strong law of large numbers* establishes that, under the hypotheses of Theorem A.143, the sequence converges in a stronger sense:

$$\lim_{n\to\infty} M_n = \mu$$

almost surely and in L^1-norm. □

We have seen that, if X_1 has finite variance, then $M_n - \mu$ tends to zero as $n \to \infty$ with rate of speed equal to $\frac{1}{\sqrt{n}}$. Now one may wonder whether the limit

$$\lim_{n\to\infty} \sqrt{n}\left(M_n - \mu\right)$$

exists and, if it does, what its value is. The answer is given by the following:

Theorem A.146 (Central limit theorem) *Let (X_n) be a sequence of real i.i.d. random variables with $\sigma^2 = var(X_1) < \infty$. As usual we put*

$$M_n = \frac{X_1 + \cdots + X_n}{n}, \qquad \mu = E\left[X_1\right],$$

and we consider the sequence defined by

$$G_n = \sqrt{n}\left(\frac{M_n - \mu}{\sigma}\right), \qquad n \in \mathbb{N}.$$

Then

$$G_n \xrightarrow{d} Z, \qquad Z \sim \mathcal{N}_{0,1}.$$

In particular, for every $x \in \mathbb{R}$,

$$\lim_{n\to\infty} P(G_n \le x) = \Phi(x),$$

where Φ is the standard normal distribution function in (A.25).

Proof. We use Lévy's theorem and we study the convergence of the sequence of characteristic functions

$$\varphi_{G_n}(\eta) = E\left[e^{i\eta G_n}\right] =$$

(since the random variables G_n are i.i.d.)

$$= \left(E\left[e^{i\eta\frac{X_1-\mu}{\sigma\sqrt{n}}}\right]\right)^n =$$

(applying Lemma A.142 with $\xi = \frac{\eta}{\sqrt{n}}$ and $p = 2$)

$$= \left(1 - \frac{\eta^2}{2n} + \mathrm{o}\left(\frac{1}{n}\right)\right) \longrightarrow e^{-\frac{\eta^2}{2}}, \qquad \text{as } n \to \infty,$$

for $\eta \in \mathbb{R}$. Since $e^{-\frac{\eta^2}{2}}$ is the characteristic function of a standard normal random variable, the claim follows from Theorem A.141. □

Remark A.147 The N-dimensional version of the previous theorem states that, if (X_n) is a sequence of i.i.d. random variables in $\mathbb{R}^N$ with finite covariance matrix $\mathcal{C}$, then

$$\sqrt{n}\,(M_n - \mu) \xrightarrow{d} Z$$

with Z multi-normal random variable, $Z \sim \mathcal{N}_{0,\mathcal{C}}$. □

A.7.2 Uniform integrability

We introduce the concept of uniformly integrable family of random variables. This notion allows characterizing L^1-convergence and it is a natural tool for the study of convergence of martingale sequences.

Definition A.148 *A family $\mathcal{X}$ of integrable random variables on a space $(\Omega, \mathcal{F}, P)$ is called uniformly integrable if*

$$\lim_{R\to+\infty} \sup_{X\in\mathcal{X}} \int_{\{|X|\geq R\}} |X| dP = 0.$$

A family consisting in one random variable only, $X \in L^1(\Omega, P)$, is uniformly integrable since

$$\int_{\{|X|\geq R\}} |X| dP \geq RP(|X| \geq R)$$

whence

$$P(|X| \geq R) \leq \frac{\|X\|_1}{R} \xrightarrow[R\to+\infty]{} 0,$$

and, by the dominated convergence theorem,

$$\int_{\{|X|\geq R\}} |X| dP \xrightarrow[R\to+\infty]{} 0.$$

Analogously every family $\mathcal{X}$ of random variables for which there exists $Z \in L^1(\Omega, P)$ such that $|X| \leq Z$, $X \in \mathcal{X}$, is uniformly integrable.

The following noteworthy result extends Lebesgue's dominated convergence theorem.

Theorem A.149 *Let (X_n) be a sequence of random variables in $L^1(\Omega, P)$, a.s. converging pointwise to a random variable X. Then (X_n) converges in L^1-norm to X if and only if it is uniformly integrable.*

For the proof of Theorem A.149 and of the following proposition see, for example, Shiryaev [309].

Proposition A.150 *A family of integrable random variables $\mathcal{X}$ is uniformly integrable if and only if there exists an increasing, convex and positive function*

$$g : \mathbb{R}_{>0} \longrightarrow \mathbb{R}$$

such that

$$\lim_{x\to\infty} \frac{g(x)}{x} = +\infty, \qquad \textit{and} \qquad \sup_{X\in\mathcal{X}} E\left[g(|X|)\right] < \infty.$$

From Proposition A.150 it follows that, in particular, every bounded family[22] in L^p, for some $p > 1$, is uniformly integrable. On the other hand, it is quite easy to construct a sequence of random variables with L^1-norm equal to one, not converging in L^1.

The following result has important applications in martingale theory.

Corollary A.151 *Let $X \in L^1(\Omega, P)$. The family consisting of*

$$\mathcal{X} = \{E\left[X \mid \mathcal{G}\right] \mid \mathcal{G} \textit{ sub-}\sigma\textit{-algebra of } \mathcal{F}\}$$

is uniformly integrable.

Proof. The family consisting only on X is uniformly integrable and therefore there exists a function g for which the properties in Proposition A.150 hold. Then, by Jensen's inequality, we get

$$E\left[g(|E\left[X \mid \mathcal{G}\right]|)\right] \leq E\left[E\left[g(|X|) \mid \mathcal{G}\right]\right] = E\left[g(|X|)\right] < \infty,$$

and the claim follows from Proposition A.150. □

[22] The family $\mathcal{X}$ is bounded in L^p if

$$\sup_{X\in\mathcal{X}} E\left[|X|^p\right] < \infty.$$

A.8 Topologies and σ-algebras

In this paragraph we recall some essential results on topological spaces, dealing in particular with the case of spaces with countable basis and, as a remarkable example, the space of continuous functions on a compact interval.

Definition A.152 *Let Ω be a non-empty set. A topology on Ω is a family $\mathscr{T}$ of subsets of Ω with the following properties:*

i) $\emptyset, \Omega \in \mathscr{T}$;
ii) $\mathscr{T}$ is closed[23] under unions (not necessarily countable unions);
iii) $\mathscr{T}$ is closed under finite intersections.

We say that the pair $(\Omega, \mathscr{T})$ is a topological space: the elements of $\mathscr{T}$ are called open sets.

Given a family $\mathscr{M}$ of subsets of Ω, the intersection of all the topologies that contain $\mathscr{M}$ is a topology that is called topology generated by $\mathscr{M}$ and denoted by $\mathscr{T}(\mathscr{M})$.

Example A.153 Let (Ω, d) be a metric space: we recall that the function

$$d : \Omega \times \Omega \longrightarrow \mathbb{R}_{\geq 0}$$

is a metric (or distance) if it verifies the following properties for every $x, y, z \in \Omega$:

i) $d(x, y) \geq 0$ and $d(x, y) = 0$ if and only if $x = y$;
ii) $d(x, y) = d(y, x)$;
iii) $d(x, y) \leq d(x, z) + d(z, y)$.

If $\mathscr{M}$ is the family of open balls

$$\mathscr{M} = \{D(x, r) \mid x \in \Omega,\ r > 0\},$$

where

$$D(x, r) = \{y \in \Omega \mid d(x, y) < r\},$$

then

$$\mathscr{T}_d := \mathscr{T}(\mathscr{M})$$

is called *topology generated by the distance d*. We consider two remarkable examples:

1) if $\Omega = \mathbb{R}^N$ and $d(x, y) = |x - y|$ is the Euclidean distance, then $\mathscr{T}_d$ is the Euclidean topology;
2) if

$$\Omega = C([a, b]; \mathbb{R}^N) = \{w : [a, b] \longrightarrow \mathbb{R}^N \mid w \text{ continuous}\},$$

[23] The union of elements of $\mathscr{T}$ belongs to $\mathscr{T}$.

the uniform topology on Ω is the topology generated by the uniform distance

$$d(w_1, w_2) = \max_{t \in [a,b]} |w_1(t) - w_2(t)|.$$

The ball in the metric d is defined as

$$D(w_0, r) = \{w \in \Omega \mid |w(t) - w_0(t)| < r,\ t \in [a, b]\}.$$

□

Topological spaces in which the union of open sets can be expressed as countable unions are of utmost importance.

Definition A.154 *A topological space $(\Omega, \mathscr{T})$ has a countable basis if there exists a countable family $\mathcal{A}$ such that $\mathscr{T} = \mathscr{T}(\mathcal{A})$.*

Theorem A.155 *If $(\Omega, \mathscr{T})$ has a countable basis, then every cover of Ω by open sets admits a finite or countable subcover.*

Proof. Let $\{U_i\}$ be a cover of Ω by open sets, i.e. a family of open sets whose union is Ω and let $\mathcal{A} = \{A_n\}_{n \in \mathbb{N}}$ be a countable basis for $\mathscr{T}$. Then every $\omega \in \Omega$ belongs to an open set U_i of the cover and there exists $A_n(\omega) \in \mathcal{A}$ such that $\omega \in A_n(\omega) \subseteq U_i$. The family $\{A_n(\omega)\}_{\omega \in \Omega}$ is finite or countable, so by choosing for every $A_n(\omega)$ an open set U_i containing it, we get a countable or finite subcover. □

If (Ω, d) is a metric space, it is very simple to verify the existence of a countable basis for $\mathscr{T}_d$. First of all, we say that a subset A is *dense in Ω* if, for every $x \in \Omega$ and $n \in \mathbb{N}$, there exists $y_n \in A$ such that $d(x, y_n) \le \frac{1}{n}$, i.e.

$$x = \lim_{n \to \infty} y_n.$$

We say that (Ω, d) is *separable* if there exists a countable dense subset A in Ω. In this case

$$\mathcal{A} = \{D\,(x, 1/n) \mid x \in A,\ n \in \mathbb{N}\} \tag{A.132}$$

is a countable basis for $\mathscr{T}_d$ and we have the following:

Theorem A.156 *A metric space has a countable basis if and only if it is separable.*

Example A.157 The space $C([a, b]; \mathbb{R}^N)$ is separable since the polynomials with rational coefficients are a countable dense subset (by Weierstrass' theorem[24]). Let $\mathscr{B}$ denote the Borel σ-algebra i.e. the σ-algebra generated by the uniform topology: then, since the uniform topology has a countable basis, we have

$$\mathscr{B} = \sigma\left(\{D\,(w, 1/n) \mid w \text{ rational polynomial}, n \in \mathbb{N}\}\right).$$

□

[24] See, for example, Chapter 4 in Folland [133].

A.9 Generalized derivatives

In this paragraph we briefly recall the concepts of weak and distributional derivative. These extensions of the concept of derivative appear naturally in arbitrage-pricing theory. For example, in order to hedge an option, we have to study the derivatives of the pricing function: even in the simplest case of a European option with strike K, the payoff $(x-K)^+$ is a continuous function, but not differentiable in the classical sense at the point $x = K$. For an in-depth analysis of the material in this paragraph we refer, for example, to Brezis [62], Folland [133] and Adams [2].

A.9.1 Weak derivatives in $\mathbb{R}$

Let $I =]a, b[\subseteq \mathbb{R}$ be an open, not necessarily bounded, interval. We denote by $L^1_{\text{loc}} = L^1_{\text{loc}}(I)$ the space of locally integrable functions u over I, i.e. the space of the measurable functions u such that

$$\int_H |u(x)|dx < \infty$$

for every compact subset H of I. The space $C_0^\infty(I)$ of functions on I with compact support[25] and with continuous derivatives of every order, is usually called test-function space.

We say that $u \in L^1_{\text{loc}}(I)$ is weakly differentiable if there exists a function $h \in L^1_{\text{loc}}(I)$, called weak derivative of u, such that

$$\int_I u(x)\varphi'(x)dx = -\int_I h(x)\varphi(x)dx, \qquad \text{(A.133)}$$

for every $\varphi \in C_0^\infty(I)$. In other words the integration-by-parts formula must hold for every test function.

Lemma A.158 *If $u \in L^1_{\text{loc}}(I)$ and*

$$\int_I u\varphi = 0, \qquad \varphi \in C_0^\infty(I),$$

then $u = 0$ a.e.

In view of the previous lemma, h is defined by (A.133) up to a set whose Lebesgue measure is null: therefore, by identifying functions that are equal a.e., we write[26] $h = Du$ to denote that h is the function in L^1_{loc} verifying (A.133). Let us point out that the definition of classical derivative is given pointwise, whilst the notion of weak derivative is "global".

[25] The support of a continuous function u is the closure of the set $\{x \mid u(x) \neq 0\}$ and it is denoted by the symbol $\text{supp}(u)$.

[26] We denote the weak derivative by Du in order to distinguish it from the classical derivative u'.

Example A.159 We consider

$$u(x) = (x - K)^+, \qquad x \in \mathbb{R}. \tag{A.134}$$

For any $\varphi \in C_0^\infty$, we have

$$\int_{\mathbb{R}} u(x)\varphi'(x)dx = \int_K^{+\infty} (x - K)\varphi'(x)dx =$$

(by the standard integration-by-parts formula)

$$= \left[(x - K)\varphi(x)\right]_{x=K}^{+\infty} - \int_K^{+\infty} \varphi(x)dx =$$

(since φ has compact support)

$$= \int_K^{+\infty} \varphi(x)dx.$$

Thus we conclude that the function

$$h(x) = \begin{cases} 1 & x > K, \\ 0 & x < K, \end{cases} \tag{A.135}$$

is the weak derivative of u. □

Notation A.160 *We denote by $W_{\mathrm{loc}}^{1,1}(I)$ the space of weakly differentiable functions over I. For any $k \in \mathbb{N}$ and $p \in [1, +\infty]$, we denote by $W^{k,p}(I)$ the space of functions $u \in L^p(I)$ that admit weak derivatives up to order k in $L^p(I)$. The spaces $W^{k,p}$ are called Sobolev spaces.*

As a consequence of the classical integration-by-parts formula, if a function is continuously differentiable in the classical sense, then it is also differentiable in the weak sense and the two notions of derivative coincide: more precisely, $C^1 \subset W_{\mathrm{loc}}^{1,1}$ and $u' = Du$ for every $u \in C^1$.

A remarkable class of weakly differentiable functions is that of locally Lipschitz continuous functions: we recall that u is locally Lipschitz continuous on I, and we write $u \in \mathrm{Lip}_{\mathrm{loc}}(I)$ if, for every compact subset H of I, there exists a constant l_H such that

$$|u(x) - u(y)| \le l_H|x - y|, \qquad x, y \in H. \tag{A.136}$$

If the estimate (A.136) holds for a constant l that is independent of H, then we say that u is globally Lipschitz continuous on I, or simply Lipschitz continuous, and we write $u \in \mathrm{Lip}(I)$. For example, the function $u(x) = (x - K)^+$ is Lipschitz continuous on $\mathbb{R}$ with constant $l = 1$. By the mean value theorem, every function C^1 with bounded derivative is Lipschitz continuous.

Proposition A.161 *If $u \in \mathrm{Lip}_{\mathrm{loc}}$, then u is differentiable in the classical sense almost everywhere. Moreover $u \in W^{1,1}_{\mathrm{loc}}$ and $u' = Du$.*

Proof. The first part of the claim is a classical result: we refer, for instance, to Chapter VI in Fomin-Kolmogorov [215]. The second part is a simple consequence of the dominated convergence theorem: in fact, for every test function φ, we have

$$\int_{\mathbb{R}} u(x)\varphi'(x)dx = \lim_{\delta\to 0}\int_{\mathbb{R}} u(x)\frac{\varphi(x+\delta)-\varphi(x)}{\delta}dx$$
$$= \lim_{\delta\to 0}\int_{\mathbb{R}} \frac{u(x-\delta)-u(x)}{\delta}\varphi(x)dx =$$

(by the dominated convergence theorem, using the fact that u is differentiable almost everywhere and that, by (A.136), the incremental ratio is locally bounded)

$$= -\int_{\mathbb{R}} u'(x)\varphi(x)dx.$$

□

Now we state some classical results of differential calculus that can be extended to the case of weakly differentiable functions. The next proposition generalizes the fundamental theorem of integral calculus.

Proposition A.162 *Let $h \in L^1_{\mathrm{loc}}(I)$ and $x_0 \in I$. The function*[27]

$$u(x) := \int_{x_0}^{x} h(y)dy, \qquad x \in I,$$

belongs to $W^{1,1}_{\mathrm{loc}} \cap C(I)$ and $h = Du$.

The next proposition essentially states that a function u in $W^{1,1}_{\mathrm{loc}}$ is a "primitive" of its weak derivative.

Proposition A.163 *Every $u \in W^{1,1}_{\mathrm{loc}}(I)$ is equal almost everywhere to a continuous function*[28]*: if $u \in W^{1,1}_{\mathrm{loc}} \cap C(I)$ we have*

$$u(x) = u(x_0) + \int_{x_0}^{x} Du(y)dy, \qquad x, x_0 \in I.$$

---

[27] By convention we assume that

$$\int_{x_0}^{x} h(y)dy = -\int_{x}^{x_0} h(y)dy,$$

for $x < x_0$.

[28] If u is continuous almost everywhere, then it may not be equal almost everywhere to a continuous function: one can think of the weak derivative of the function $(x-K)^+$. Furthermore, if u is equal almost everywhere to a continuous function it may not be continuous almost everywhere: one can think of the Dirichlet function that is equal to 0 on $\mathbb{Q}$ and to 1 on $\mathbb{R}\setminus\mathbb{Q}$.

In particular, if $Du = 0$ then u is constant.

By the previous result, in functional analysis it is customary to identify any element in $W^{1,1}_{\text{loc}}$ with its continuous representative.

Now we state an extension of the integration-by-parts formula.

Proposition A.164 *Let $u, v \in W^{1,1}_{\text{loc}} \cap C(I)$. The $uv \in W^{1,1}_{\text{loc}}(I)$ and we have*

$$\int_{x_0}^{x} uDv = u(x)v(x) - u(x_0)v(x_0) - \int_{x_0}^{x} vDu, \qquad x, x_0 \in I.$$

Moreover, if $f \in C^1(\mathbb{R})$, then $f(u) \in W^{1,1}_{\text{loc}}(I)$ and $Df(u) = f'(u)Du$.

A.9.2 Sobolev spaces and embedding theorems

Let O be an open set in $\mathbb{R}^N$ and $1 \le p \le \infty$.

Definition A.165 *The Sobolev space $W^{1,p}(O)$ is the space of functions $u \in L^p(O)$ for which there exist $h_1, \dots, h_N \in L^p(O)$ such that*

$$\int_O u\partial_{x_i}\varphi = -\int_O h_i\varphi, \qquad \varphi \in C_0^\infty(O),\ i = 1, \dots, N.$$

By Lemma A.158, the functions $h_1, \dots, h_N$ are uniquely determined a.e.: thus we set $D_i u := h_i$ for $i = 1, \dots, N$ and we say that $Du = (h_1, \dots, h_N)$ is the gradient of u.

The space $W^{1,p}(O)$, endowed with the norm

$$\|u\|_{W^{1,p}} := \|u\|_{L^p} + \|Du\|_{L^p},$$

is a Banach space. If $u \in C^1 \cap L^p$ and $\partial_{x_i} u \in L^p$ for every $i = 1, \dots, N$, then $u \in W^{1,p}$ and $\partial_{x_i} u = D_i u$. The higher-order Sobolev spaces can be defined recursively as follows.

Definition A.166 *For any $k \in \mathbb{N}$, $k \ge 2$, we set*

$$W^{k,p}(O) = \{u \in W^{k-1,p}(O) \mid Du \in W^{k-1,p}(O)\}.$$

To give a more explicit representation of the space $W^{k,p}$, we introduce the following:

Notation A.167 *Given a multi-index $\alpha = (\alpha_1, \dots, \alpha_N) \in \mathbb{N}_0^N$, we put*

$$\partial_x^\alpha = \partial_{x_1}^{\alpha_1} \dots \partial_{x_N}^{\alpha_N}$$

and we say that the number

$$|\alpha| = \sum_{i=1}^{N} \alpha_i$$

is the weight of α.

Then $u \in W^{k,p}(O)$ if and only if, for every multi-index α, with $|\alpha| \le k$, there exists a function $h_\alpha \in L^p(O)$ such that

$$\int_O u \partial_x^\alpha \varphi = (-1)^{|\alpha|} \int_O h_\alpha \varphi, \qquad \varphi \in C_0^\infty(O).$$

In that case we write $h_\alpha = D^\alpha u$. The space $W^{k,p}$, endowed with the norm

$$\|u\|_{W^{k,p}} := \sum_{0 \le |\alpha| \le k} \|D^\alpha u\|_{L^p},$$

is a Banach space. We state now the fundamental:

Theorem A.168 (Sobolev-Morrey embedding theorem) *There exists a constant C, depending on p and N only, such that, for every $u \in W^{1,p}(\mathbb{R}^N)$ we have:*

i) if $1 \le p < N$,

$$\|u\|_{L^q(\mathbb{R}^N)} \le C\|u\|_{W^{1,p}(\mathbb{R}^N)}, \qquad p \le q \le \frac{pN}{N-p};$$

ii) if $p > N$, then $u \in L^\infty(\mathbb{R}^N)$ and

$$|u(x) - u(y)| \le C\|Du\|_{L^p(\mathbb{R}^N)}|x-y|^\delta \qquad \text{for almost all } x, y \in \mathbb{R}^N,$$

with $\delta = 1 - \frac{N}{p}$.

Further, if $p = N$, then

$$\|u\|_{L^q(\mathbb{R}^N)} \le c_q\|u\|_{W^{1,N}(\mathbb{R}^N)}, \qquad q \in [p, \infty[,$$

with c_q constant depending on p, q, N only and such that $c_q \to +\infty$ as $q \to \infty$.

A.9.3 Distributions

The following example is preliminary to the concept of distribution or generalized function.

Example A.169 The function u in (A.134) does not admit a second-order derivative in the weak sense. Indeed if there existed $g = Dh$ for h in (A.135), then we would have

$$\int_{\mathbb{R}} g\varphi = -\int_{\mathbb{R}} h\varphi' = -\int_K^{+\infty} \varphi' = \varphi(K), \qquad \varphi \in C_0^\infty(\mathbb{R}). \tag{A.137}$$

In particular

$$\int_{\mathbb{R}} g\varphi = 0, \qquad \varphi \in C_0^\infty(\mathbb{R} \setminus \{K\}),$$

whence, in view of Lemma A.158, $g = 0$ almost everywhere, contradicting (A.137). □

As we have just shown, the derivative, even in the weak sense, of a function may not always exist if we think of it as a function in the classical sense. In distribution theory the concept of function is extended by interpreting every $u \in L^1_{\text{loc}}$ as a functional[29] associating the integral of $u\varphi$ to $\varphi \in C_0^\infty$, rather than, as usual, a map associating $u(x)$ to the number x. For the sake of simplicity, in this section we consider only the 1-dimensional case.

If u is infinitely differentiable (in the weak sense), then

$$\int D^{(k)}u\varphi = (-1)^k \int u\varphi^{(k)}, \qquad \varphi \in C_0^\infty,\ k \in \mathbb{N}. \tag{A.138}$$

On the other hand, regardless of the fact that u is differentiable, the right-hand side of (A.138) defines a linear functional on C_0^∞ to which we can attribute the meaning of "k-th order derivative of u". This consideration is made precise in the following:

Definition A.170 *A distribution Λ on a non-empty open interval I of $\mathbb{R}$ is a linear functional*

$$\Lambda : C_0^\infty(I) \longrightarrow \mathbb{R}$$

such that, for every compact set $H \subset I$, there exist a positive constant M and $m \in \mathbb{N}$ such that

$$|\Lambda(\varphi)| \le M\|\varphi\|_m, \tag{A.139}$$

for every $\varphi \in C_0^\infty(I)$ with support contained in H, where

$$\|\varphi\|_m = \sum_{k=0}^{m} \max_H |\varphi^{(k)}|. \tag{A.140}$$

The space of distributions on I is denoted by $\mathcal{D}'(I)$ and generally the following notation is used:

$$\langle \Lambda, \varphi \rangle := \Lambda(\varphi).$$

Remark A.171 Inequality (A.139) expresses the continuity property of Λ with respect to a suitable topology over $C_0^\infty(I)$: we refer to Rudin [294], Part II, Chapter 6 for further details. Here we observe that, if (φ_n) is a sequence in $C_0^\infty(I)$ with support included in a compact set H for every $n \in \mathbb{N}$, then by (A.139)

$$\lim_{n\to\infty} \|\varphi_n - \varphi\|_m = 0$$

implies that

$$\lim_{n\to\infty} \langle \Lambda, \varphi_n \rangle = \langle \Lambda, \varphi \rangle.$$

□

---

[29] In general the term "function" is used to denote a map between numerical sets (e.g., a function from $\mathbb{R}^N$ to $\mathbb{R}$), whilst the term "functional" is used to denote a map between function spaces.

Every locally integrable function defines a distribution in a natural way: indeed, given $u \in L^1_{\text{loc}}(I)$, we set

$$\langle \Lambda_u, \varphi \rangle = \int_I u(x)\varphi(x)dx, \qquad \varphi \in C_0^\infty(I).$$

Obviously Λ_u is a linear functional and for every compact set H in I, we have

$$|\langle \Lambda_u, \varphi \rangle| \le \|\varphi\|_0 \int_H |u(x)|dx, \qquad \varphi \in C_0^\infty(H).$$

For this reason, it is common practice to identify u with Λ_u and then write $L^1_{\text{loc}} \subset \mathcal{D}'$.

Analogously, if μ is a probability measure on $(\mathbb{R}, \mathscr{B})$ then, setting

$$\langle \Lambda_\mu, \varphi \rangle = \int_{\mathbb{R}} u(x)\mu(dx), \qquad \varphi \in C_0^\infty(\mathbb{R}),$$

we have that Λ_μ is a linear functional and

$$|\langle \Lambda_\mu, \varphi \rangle| \le \|\varphi\|_0 \, \mu(H), \qquad \varphi \in C_0^\infty(H).$$

In other words, Λ_μ is a distribution that is usually identified with μ.

At this point we remark that Definition A.170 is in accordance with Definition A.11 and generalizes the notion of distribution given in Section A.1.2. More generally, it is apparent that every measure on $(\mathbb{R}, \mathscr{B})$, such that $\mu(H) < \infty$ for every compact set H in $\mathbb{R}$, is a distribution.

Definition A.172 *If $\Lambda \in \mathcal{D}'(I)$ and $k \in \mathbb{N}$, the k-th order derivative of Λ is defined by*

$$\langle D^{(k)}\Lambda, \varphi \rangle := (-1)^k \langle \Lambda, \varphi^{(k)} \rangle, \qquad \varphi \in C_0^\infty(I).$$

We note that $D^{(k)}\Lambda \in \mathcal{D}'(I)$, indeed $D^{(k)}\Lambda$ is a linear functional and, for a fixed a compact set H, for every $\varphi \in C_0^\infty(H)$, we have

$$|\langle D^{(k)}\Lambda, \varphi \rangle| = |\langle \Lambda, \varphi^{(k)} \rangle| \le$$

(since $\text{supp}(\varphi^{(k)}) \subseteq \text{supp}(\varphi)$)

$$\le M\|\varphi^{(k)}\|_m \le M\|\varphi\|_{m+k}.$$

Therefore a distribution admits the derivatives of all orders that are themselves distributions and that, in general, are not functions in the classical sense.

Going back to Example A.169, the function $x \mapsto (x-K)^+$ admits first-order weak derivative but not the second-order one. Nevertheless, by (A.137) the second-order distributional derivative is defined by

$$\int_{\mathbb{R}} \varphi(x) D^{(2)}(x-K)^+ dx = \varphi(K) = \delta_K(\varphi), \qquad \varphi \in C_0^\infty(\mathbb{R}).$$

In other terms $D^{(2)}(x-K)^+$ coincides with Dirac's Delta concentrated at K: this is the typical example of a distribution that is not a function in the classical sense.

Now we introduce the notion of translation and convolution in a distributional setting. If φ is a function defined on $\mathbb{R}$ and $x \in \mathbb{R}$, we put

$$T_x\varphi(y) = \varphi(y-x), \qquad \check{\varphi}(y) = \varphi(-y), \qquad y \in \mathbb{R}. \tag{A.141}$$

We note that

$$\int_{\mathbb{R}} \psi T_x \varphi = \int_{\mathbb{R}} \psi(y)\varphi(y-x)dy = \int_{\mathbb{R}} \psi(y+x)\varphi(y)dy = \int_{\mathbb{R}} (T_{-x}\psi)\,\varphi. \tag{A.142}$$

Moreover

$$(T_x\check{\varphi})\,(y) = \check{\varphi}(y-x) = \varphi(x-y),$$

and so

$$(\psi * \varphi)(x) = \int_{\mathbb{R}} \psi(y)\,(T_x\check{\varphi})\,(y)dy. \tag{A.143}$$

By analogy we give the following:

Definition A.173 *Let $\Lambda \in \mathcal{D}'(\mathbb{R})$. The translation $T_x\Lambda$ is the distribution in $\mathcal{D}'(\mathbb{R})$ defined by*

$$\langle T_x\Lambda, \varphi\rangle = \langle \Lambda, T_{-x}\varphi\rangle, \qquad \varphi \in C_0^\infty(\mathbb{R}). \tag{A.144}$$

The convolution of Λ with $\varphi \in C_0^\infty(\mathbb{R})$ is the function defined by

$$(\Lambda * \varphi)\,(x) = \langle \Lambda, T_x\check{\varphi}\rangle, \qquad x \in \mathbb{R}. \tag{A.145}$$

We emphasize that the convolution in (A.145) *is a function.* Further, if Λ is a locally integrable function, definitions (A.144) and (A.145) are in line with (A.142) and (A.143), respectively.

Theorem A.174 *If $\Lambda \in \mathcal{D}'(\mathbb{R})$ and $\varphi, \psi \in C_0^\infty(\mathbb{R})$, then*

i) $\Lambda * \varphi \in C^\infty(\mathbb{R})$ and, for every $k \in \mathbb{N}$, we have

$$(\Lambda * \varphi)^{(k)} = (D^{(k)}\Lambda) * \varphi = \Lambda * (\varphi^{(k)}); \tag{A.146}$$

ii) if Λ has compact support[30] *then $\Lambda * \varphi \in C_0^\infty(\mathbb{R})$;*
iii) $(\Lambda * \varphi) * \psi = \Lambda * (\varphi * \psi)$.

[30] We recall the definition of the support of a distribution: we say that $\Lambda \in \mathcal{D}'(I)$ is null on an open set O in I if $\langle \Lambda, \varphi\rangle = 0$ for every $\varphi \in C_0^\infty(O)$. If W is the union of all the open sets on which Λ is null, then by definition

$$\mathrm{supp}(\Lambda) = I \setminus W.$$

Preliminarily we prove the following:

Lemma A.175 *For every $\Lambda \in \mathcal{D}'(\mathbb{R})$ and $\varphi \in C_0^\infty(\mathbb{R})$ we have*

$$T_x(\Lambda * \varphi) = (T_x\Lambda) * \varphi = \Lambda * (T_x\varphi), \qquad x \in \mathbb{R}. \tag{A.147}$$

Proof. The claim follows from the following equalities:

$$\begin{aligned}
T_x(\Lambda * \varphi)(y) &= (\Lambda * \varphi)(y - x) = \langle \Lambda, T_{y-x}\check{\varphi}\rangle, \\
((T_x\Lambda) * \varphi)(y) &= \langle T_x\Lambda, T_y\check{\varphi}\rangle = \langle \Lambda, T_{-x}T_y\check{\varphi}\rangle = \langle \Lambda, T_{y-x}\check{\varphi}\rangle, \\
(\Lambda * (T_x\varphi))(y) &= \langle \Lambda, T_y(T_x\varphi)\check{}\rangle = \langle \Lambda, T_yT_{-x}\check{\varphi}\rangle = \langle \Lambda, T_{y-x}\check{\varphi}\rangle.
\end{aligned}$$

□

Proof (of Theorem A.174).
i) We set

$$\delta_h = \frac{T_0 - T_h}{h}, \qquad h \neq 0.$$

For every $\varphi \in C_0^\infty(\mathbb{R})$ and $m \in \mathbb{N}$, we have

$$\lim_{h\to 0} \|\delta_h\varphi - \varphi'\|_m = 0,$$

with $\|\cdot\|_m$ defined in (A.140); so

$$\lim_{h\to 0} \|T_x((\delta_h\varphi)\check{}) - T_x((\varphi')\check{})\|_m = 0.$$

Now, by Lemma A.175 we have

$$\delta_h(\Lambda * \varphi)(x) = (\Lambda * (\delta_h\varphi))(x) =$$

(by the definition of convolution)

$$= \langle \Lambda, T_x((\delta_h\varphi)\check{})\rangle(x).$$

Taking the limit as h goes to zero and recalling Remark A.171, we get

$$\frac{d}{dx}(\Lambda * \varphi)(x) = (\Lambda * \varphi')(x).$$

On the other hand, applying Λ to both sides of the following equality

$$T_x((\varphi')\check{}) = -(T_x\check{\varphi})',$$

we get

$$(\Lambda * \varphi')(x) = -\langle \Lambda, (T_x\check{\varphi})'\rangle =$$

(by definition of derivative of Λ)

$$= (D\Lambda * \varphi)(x).$$

An inductive procedure concludes the proof of the first point.

ii) We let

$$K = \text{supp}(\Lambda), \qquad H = \text{supp}(\varphi)$$

and we observe that

$$\text{supp}(T_x \check{\varphi}) = x - H.$$

Then it suffices to remark that, by definition,

$$(\Lambda * \varphi)(x) = \langle \Lambda, T_{-x}\check{\varphi} \rangle = 0$$

if $K \cap (x - H) = \emptyset$, i.e. if

$$x \notin (\text{supp}(\Lambda) + \text{supp}(\varphi)). \tag{A.148}$$

iii) If $\Lambda = \Lambda_u$ with $u \in L^1_{\text{loc}}$, the claim follows immediately by Fubini's theorem. For the general case, we refer to Rudin [294], Theorem 6.30. □

A.9.4 Mollifiers

Every distribution can be approximated by smooth functions using the so-called *mollifiers*. More precisely, let us consider the function

$$\varrho(x) = \begin{cases} c \exp\left(-\frac{1}{1-x^2}\right) & \text{if } |x| < 1, \\ 0 & \text{if } |x| \geq 1, \end{cases}$$

where c is a constant chosen in a way that

$$\int_{\mathbb{R}} \varrho(x) dx = 1.$$

Moreover, we set

$$\varrho_n(x) = n\varrho(nx), \qquad n \in \mathbb{N}.$$

The sequence (ϱ_n) has the typical properties of the so-called *approximations of identity*: in particular, for every $n \in \mathbb{N}$ we have

i) $\varrho_n \in C_0^\infty(\mathbb{R})$;
ii) $\varrho_n(x) = 0$ when $|x| \geq \frac{1}{n}$;
iii) $\int_{\mathbb{R}} \varrho_n(x) dx = 1$.

The functions ϱ_n are also called Friedrichs' mollifiers [142]: if $\Lambda \in \mathcal{D}'(\mathbb{R})$, the convolution

$$\Lambda_n(x) = (\Lambda * \varrho_n)(x), \qquad x \in \mathbb{R},$$

is called *regularization or mollification of* Λ. The following theorem sums up the basic properties of mollification: in the statement Λ_n, u_n denote the mollifications of Λ and u, respectively.

Theorem A.176 *If $\Lambda \in \mathcal{D}'$, then*

i) $\Lambda_n \in C^\infty$;
ii) we have

$$\operatorname{supp}(\Lambda_n) \subseteq \{x \mid \operatorname{dist}(x, \operatorname{supp}(\Lambda)) \le 1/n\}, \tag{A.149}$$

and in particular, if Λ has compact support, then $\Lambda_n \in C_0^\infty$;
iii) if $u \in C$, then $\|u_n\|_\infty \le \|u\|_\infty$ and (u_n) converges uniformly on compact sets to u;
iv) if $u \in L^p$, with $1 \le p < \infty$, then $\|u_n\|_p \le \|u\|_p$ and (u_n) converges in L^p-norm to u;
v) for every $n, k \in \mathbb{N}$ we have

$$\Lambda_n^{(k)} = \left(D^k \Lambda\right)_n \tag{A.150}$$

and consequently

v-a) if $u \in C^k$, then $u_n^{(k)}$ converges to $u^{(k)}$ uniformly on compact sets;

v-b) if $u \in W^{k,p}$, then $D^k u_n$ converges in L^p-norm to $D^k u$;

vi) Λ_n converges to Λ in the sense of distributions, i.e.

$$\lim_{n\to\infty} \langle \Lambda_n, \varphi \rangle = \langle \Lambda, \varphi \rangle,$$

for every $\varphi \in C_0^\infty$.

Proof. Properties i) and ii) follow directly from Theorem A.174: in particular (A.149) follows from (A.148).

iii) Let $u \in C$ with $\|u\|_\infty < +\infty$: we have

$$|u_n(x)| \le \int_{\mathbb{R}} \varrho_n(x-y)|u(y)|dy \le \|u\|_\infty \int_{\mathbb{R}} \varrho_n(x-y)dy = \|u\|_\infty.$$

Moreover, if x belongs to a compact set, we have

$$\begin{aligned}
|u_n(x) - u(x)| &\le \int_{\mathbb{R}} \varrho_n(x-y)|u(y) - u(x)|dy \\
&\le \max_{|x-y|\le 1/n} |u(y) - u(x)| \int_{\mathbb{R}} \varrho_n(x-y)dy \\
&= \max_{|x-y|\le 1/n} |u(y) - u(x)|.
\end{aligned}$$

iv) If $u \in L^p$ we have

$$|u_n(x)| \le \int_{\mathbb{R}} \varrho_n(x-y)|u(y)|dy \le$$

(by Hölder's inequality, with p, q conjugate exponents)

$$\leq \left(\int_{\mathbb{R}} \varrho_n(x-y)dy\right)^{\frac{1}{q}} \left(\int_{\mathbb{R}} \varrho_n(x-y)|u(y)|^p dy\right)^{\frac{1}{p}}$$
$$= \left(\int_{\mathbb{R}} \varrho_n(x-y)|u(y)|^p dy\right)^{\frac{1}{p}},$$

hence it follows that

$$\|u_n\|_p^p \leq \int_{\mathbb{R}}\int_{\mathbb{R}} \varrho_n(x-y)|u(y)|^p dy dx = \int_{\mathbb{R}} |u(y)|^p dy.$$

Similarly we can prove that

$$\|u_n - u\|_p^p \leq \int_{\mathbb{R}} \left(\int_{\mathbb{R}} \varrho_n(x-y)|u(y)-u(x)|dy\right)^p dx \leq$$

(by Hölder's inequality)

$$\leq \int_{\mathbb{R}}\int_{\mathbb{R}} \varrho_n(x-y)|u(y)-u(x)|^p dy dx = \int_{\mathbb{R}} \varrho_n(z) \int_{\mathbb{R}} |u(x-z)-u(x)|^p dz dx,$$

and the claim follows from Lebesgue's dominated convergence theorem and from the L^p-mean continuity, i.e. from the fact[31] that

$$\lim_{z\to 0} \int_{\mathbb{R}} |u(x-z)-u(x)|^p dz = 0.$$

v) (A.150) follows from (A.146).

vi) We have

$$\langle \Lambda, \check{\varphi}\rangle = (\Lambda * \varphi)(0) =$$

(by *v)-a*)

$$= \lim_{n\to\infty} (\Lambda * (\varrho_n * \varphi))(0) =$$

(by Theorem A.174-iii))

$$= \lim_{n\to\infty} ((\Lambda * \varrho_n) * \varphi)(0) = \lim_{n\to\infty} \langle \Lambda_n, \check{\varphi}\rangle.$$

□

31 The L^p-mean continuity can be easily proved using the density of test functions in L^p: for all the details we refer, for example, to Brezis [62].

A.10 Separation of convex sets

In this section we prove a simple result of separation of convex sets in finite dimension that is used in the proof of the fundamental theorems of asset pricing.

Theorem A.177 *Let $\mathscr{C}$ be a closed convex subset in $\mathbb{R}^N$ that does not contain the origin. Then, there exists $\xi \in \mathscr{C}$ such that*

$$|\xi|^2 \le \langle x, \xi \rangle, \qquad x \in \mathscr{C}.$$

Proof. Since $\mathscr{C}$ is closed, there exists $\xi \in \mathscr{C}$ such that

$$|\xi| \le |x|, \qquad x \in \mathscr{C}.$$

Moreover $\mathscr{C}$ is a convex set and therefore we have

$$\xi + t(x - \xi) \in \mathscr{C}, \qquad t \in [0, 1],$$

and so

$$|\xi|^2 \le |\xi + t(x-\xi)|^2 = |\xi|^2 + 2t\langle \xi, x - \xi \rangle + t^2 |x - \xi|^2.$$

Then for $t > 0$ we get

$$0 \le 2\langle \xi, x - \xi \rangle + t|x - \xi|^2,$$

and taking the limit as $t \to 0^+$, we prove the claim. □

Corollary A.178 *Let $\mathscr{K}$ be a convex compact subset of $\mathbb{R}^N$. Let $\mathscr{V}$ be a linear subspace of $\mathbb{R}^N$ such that $\mathscr{V} \cap \mathscr{K} = \emptyset$. Then there exists $\xi \in \mathbb{R}^N$ such that*

$$\langle \xi, x \rangle = 0, \quad x \in \mathscr{V}, \qquad \text{and} \qquad \langle \xi, x \rangle > 0, \quad x \in \mathscr{K}.$$

Proof. The set

$$\mathscr{K} - \mathscr{V} = \{x - y \mid x \in \mathscr{K},\ y \in \mathscr{V}\}$$

is closed[32], convex and it does not contain the origin. Then, by Theorem A.177 there exists $\xi \in \mathbb{R}^N \setminus \{0\}$ such that

$$|\xi|^2 \le \langle x - y, \xi \rangle, \qquad x \in \mathscr{K},\ y \in \mathscr{V}.$$

Since $\mathscr{V}$ is a linear space, it follows that

$$|\xi|^2 \le \langle x, \xi \rangle - t\langle y, \xi \rangle$$

for every $x \in \mathscr{K}$, $y \in \mathscr{V}$ and $t \in \mathbb{R}$. This is possible only if $\langle y, \xi \rangle = 0$ for every $y \in \mathscr{V}$ and this concludes the proof. □

---

[32] To prove that $\mathscr{K} - \mathscr{V}$ is closed we use the assumption that $\mathscr{K}$ is a compact set: we leave the details of the proof to the reader as an exercise.

References

[1] Achdou, Y., and Pironneau, O. *Computational Methods for Option Pricing.* Society for Industrial and Applied Mathematics, 2005.

[2] Adams, R. A. *Sobolev spaces.* Academic Press, New York-London, 1975. Pure and Applied Mathematics, vol. 65.

[3] Agliardi, R. The quintessential option pricing formula under Lévy processes. *Appl. Math. Lett. 22*, 10 (2009), 1626–1631.

[4] Agliardi, R. A comprehensive method for exotic option pricing. *arXiv: 1001.3308v1 [q-fin.PR] 19jan2010* (2010).

[5] Albrecher, H. The valuation of Asian options for market models of exponential Lévy type. In M. Van Maele et al. (Eds.), *Handelingen Contact forum – 2nd Actuarial and financial Mathematics Day. Royal Flemish Academy of Belgium for Science and Arts, Brussels, Belgium 11-20* (2004).

[6] Albrecher, H., and Predota, M. On Asian option pricing for NIG Lévy processes. *J. Comput. Appl. Math. 172*, 1 (2004), 153–168.

[7] Almendral, A., and Oosterlee, C. W. Accurate evaluation of European and American options under the CGMY process. *SIAM J. Sci. Comput. 29*, 1 (2007), 93–117 (electronic).

[8] Alvarez, O., and Tourin, A. Viscosity solutions of nonlinear integro-differential equations. *Ann. Inst. H. Poincaré Anal. Non Linéaire 13*, 3 (1996), 293–317.

[9] Amadori, A. L. Nonlinear integro-differential evolution problems arising in option pricing: a viscosity solutions approach. *Differential Integral Equations 16*, 7 (2003), 787–811.

[10] Ansel, J.-P., and Stricker, C. Lois de martingale, densités et décomposition de Föllmer-Schweizer. *Ann. Inst. H. Poincaré Probab. Statist. 28*, 3 (1992), 375–392.

[11] Applebaum, D. *Lévy processes and stochastic calculus*, second ed., vol. 116 of *Cambridge Studies in Advanced Mathematics.* Cambridge University Press, Cambridge, 2009.

[12] Aronson, D. G. Bounds for the fundamental solution of a parabolic equation. *Bull. Amer. Math. Soc. 73* (1967), 890–896.

[13] Attari, M. Option pricing using Fourier transforms: A numerically efficient simplification. *Available at SSRN: http://ssrn.com/abstract=520042* (2004), 1–7.

[14] Avellaneda, M., and Laurence, P. *Quantitative modeling of derivative securities.* Chapman & Hall/CRC, Boca Raton, FL, 2000. From theory to practice.

[15] Awatif, S. Équations d'Hamilton-Jacobi du premier ordre avec termes intégro-différentiels. I. Unicité des solutions de viscosité. *Comm. Partial Differential Equations 16*, 6-7 (1991), 1057–1074.

[16] Bachelier, L. Théorie de la spéculation. *Ann. Sci. École Norm. Sup. (3) 17* (1900), 21–86.

[17] Bakshi, G., Cao, C., and Chen, Z. Empirical performance of alternative option pricing models. *Journal of Finance 52*, 5 (December 1997), 2003–49.

[18] Ball, C. A., and Roma, A. Stochastic volatility option pricing. *Journal of Financial and Quantitative Analysis 29*, 04 (1994), 589–607.

[19] Bally, V. An elementary introduction to Malliavin calculus. `http://www.inria.fr/rrrt/rr-4718.html` (2007).

[20] Bally, V., Caramellino, L., and Zanette, A. Pricing and hedging American options by Monte Carlo methods using a Malliavin calculus approach. *Monte Carlo Methods Appl. 11*, 2 (2005), 97–133.

[21] Barles, G. Convergence of numerical schemes for degenerate parabolic equations arising in finance theory. In *Numerical methods in finance.* Cambridge Univ. Press, Cambridge, 1997, pp. 1–21.

[22] Barles, G., Buckdahn, R., and Pardoux, E. Backward stochastic differential equations and integral-partial differential equations. *Stochastics Stochastics Rep. 60*, 1-2 (1997), 57–83.

[23] Barles, G., Burdeau, J., Romano, M., and Samsœn, N. Estimation de la frontière libre des options Américaines au voisinage de l'échéance. *C. R. Acad. Sci. Paris Sér. I Math. 316*, 2 (1993), 171–174.

[24] Barlow, M. T. One-dimensional stochastic differential equations with no strong solution. *J. London Math. Soc. (2) 26*, 2 (1982), 335–347.

[25] Barndorff-Nielsen, O. Normal inverse Gaussian distributions and the modelling of stock returns. *Research Report n. 300, Dep. of Theoretical Statistics, Aarhus Univ.* (1995).

[26] Barndorff-Nielsen, O., and Halgreen, C. Infinite divisibility of the hyperbolic and generalized inverse Gaussian distributions. *Z. Wahrscheinlichkeitstheorie und Verw. Gebiete 38*, 4 (1977), 309–311.

[27] Barndorff-Nielsen, O. E., and Levendorskii, S. Z. Feller processes of normal inverse Gaussian type. *Quant. Finance 1*, 3 (2001), 318–331.

[28] Barndorff-Nielsen, O. E., and Shephard, N. Non-Gaussian Ornstein-Uhlenbeck-based models and some of their uses in financial economics. *Journal Of The Royal Statistical Society Series B 63*, 2 (2001), 167–241.

[29] Barndorff-Nielsen, O. E., and Shephard, N. Econometric analysis of realized volatility and its use in estimating stochastic volatility models. *J. R. Stat. Soc. Ser. B Stat. Methodol. 64*, 2 (2002), 253–280.

[30] Barone-Adesi, G., and Whaley, R. Efficient analytic approximation of American option values. *J. of Finance 42* (1987), 301–320.

[31] Barraquand, J., and Martineau, D. Numerical valuation of high dimensional multivariate American securities. *J. of Financial and Quantitative Analysis 30* (1995), 383–405.

[32] Barraquand, J., and Pudet, T. Pricing of American path-dependent contingent claims. *Math. Finance 6*, 1 (1996), 17–51.

[33] Barucci, E., Polidoro, S., and Vespri, V. Some results on partial differential equations and Asian options. *Math. Models Methods Appl. Sci. 11*, 3 (2001), 475–497.

[34] Bassler, K., McCauley, J., and Gunaratne, G. Nonstationary increments, scaling distributions, and variable diffusion processes in financial markets. *Proceedings of the National Academy of Sciences 104* (2007), 17287–17290.

[35] Bates, D. S. Jumps and stochastic volatility: exchange rate processes implicit in Deutsche mark options. *Review of Financial Studies 9*, 1 (1996), 69–107.

[36] Bayraktar, E., and Xing, H. Analysis of the optimal exercise boundary of American options for jump diffusions. *SIAM J. Math. Anal. 41*, 2 (2009), 825–860.

[37] Bell, D. R. *The Malliavin calculus.* Dover Publications Inc., Mineola, NY, 2006. Reprint of the 1987 edition.

[38] Bellamy, N., and Jeanblanc, M. Incompleteness of markets driven by a mixed diffusion. *Finance Stoch. 4*, 2 (2000), 209–222.

[39] Benhamou, E., Gobet, E., and Miri, M. Smart expansion and fast calibration for jump diffusions. *Finance Stoch. 13*, 4 (2009), 563–589.

[40] Benhamou, E., Gobet, E., and Miri, M. Time dependent Heston model. *SIAM J. Financial Math. 1* (2010), 289–325.

[41] Bensoussan, A. On the theory of option pricing. *Acta Appl. Math. 2*, 2 (1984), 139–158.

[42] Bensoussan, A., and Lions, J.-L. *Applications of variational inequalities in stochastic control*, vol. 12 of *Studies in Mathematics and its Applications.* North-Holland Publishing Co., Amsterdam, 1982. Translated from the French.

[43] Benth, F. E. *Option theory with stochastic analysis.* Universitext. Springer-Verlag, Berlin, 2004. An introduction to mathematical finance, Revised edition of the 2001 Norwegian original.

[44] Bertoin, J. *Lévy processes*, vol. 121 of *Cambridge Tracts in Mathematics.* Cambridge University Press, Cambridge, 1996.

[45] Bichteler, K., and Jacod, J. Calcul de Malliavin pour les diffusions avec sauts: existence d'une densité dans le cas unidimensionnel. In *Seminar*

on probability, XVII, vol. 986 of *Lecture Notes in Math.* Springer, Berlin, 1983, pp. 132–157.

[46] Bismut, J.-M. Calcul des variations stochastique et processus de sauts. *Z. Wahrsch. Verw. Gebiete 63*, 2 (1983), 147–235.

[47] Bjork, T. *Arbitrage theory in continuous time.* Second edition. Oxford University Press, Oxford, 2004.

[48] Black, F., Derman, E., and Toy, W. A one-factor model of interest rates and its applications to treasury bond options. *Financial Analysts Journal, Jan.-Feb.* (1990), 33–39.

[49] Black, F., and Scholes, M. The pricing of options and corporate liabilities. *J. Political Economy 81* (1973), 637–654.

[50] Blanchet, A. On the regularity of the free boundary in the parabolic obstacle problem. Application to American options. *Nonlinear Anal. 65*, 7 (2006), 1362–1378.

[51] Blanchet, A., Dolbeault, J., and Monneau, R. On the one-dimensional parabolic obstacle problem with variable coefficients. In *Elliptic and parabolic problems*, vol. 63 of *Progr. Nonlinear Differential Equations Appl.* Birkhäuser, Basel, 2005, pp. 59–66.

[52] Borovkov, K., and Novikov, A. On a new approach to calculating expectations for option pricing. *J. Appl. Probab. 39*, 4 (2002), 889–895.

[53] Bouchard, B., Ekeland, I., and Touzi, N. On the Malliavin approach to Monte Carlo approximation of conditional expectations. *Finance Stoch. 8*, 1 (2004), 45–71.

[54] Bouleau, N., and Lépingle, D. *Numerical methods for stochastic processes.* Wiley Series in Probability and Mathematical Statistics: Applied Probability and Statistics. John Wiley & Sons Inc., New York, 1994.

[55] Bouziane, M. *Pricing interest-rate derivatives*, vol. 607 of *Lecture Notes in Economics and Mathematical Systems.* Springer-Verlag, Berlin, 2008. A Fourier-transform based approach, With a foreword by Rainer Schöbel.

[56] Boyarchenko, S., and Levendorskii, S. *Non-Gaussian Merton-Black-Scholes theory*, vol. 9 of *Advanced Series on Statistical Science & Applied Probability.* World Scientific Publishing Co. Inc., River Edge, NJ, 2002.

[57] Boyd, J. *Chebyshev and Fourier spectral methods*, second ed. Dover Publications Inc., Mineola, NY, 2001.

[58] Brace, A., Gatarek, D., and Musiela, M. The market model of interest rate dynamics. *Math. Finance 7*, 2 (1997), 127–155.

[59] Breeden, D. T., and Litzenberger, R. H. Prices of state-contingent claims implicit in option prices. *J. Business 51*, 4 (1978), 621–651.

[60] Breiman, L. *Probability*, vol. 7 of *Classics in Applied Mathematics.* Society for Industrial and Applied Mathematics (SIAM), Philadelphia, PA, 1992. Corrected reprint of the 1968 original.

[61] Brennan, M. J., and Schwartz, E. S. The valuation of the American put option. *J. of Finance 32* (1977), 449–462.

[62] Brezis, H. *Analyse fonctionnelle*. Collection Mathématiques Appliquées pour la Maîtrise. [Collection of Applied Mathematics for the Master's Degree]. Masson, Paris, 1983. Théorie et applications. [Theory and applications].

[63] Brigo, D., and Mercurio, F. *Interest rate models—theory and practice*, second ed. Springer Finance. Springer-Verlag, Berlin, 2006. With smile, inflation and credit.

[64] Broadie, M., and Detemple, J. The valuation of American options on multiple assets. *Math. Finance 7*, 3 (1997), 241–286.

[65] Cancelier, C. Problèmes aux limites pseudo-différentiels donnant lieu au principe du maximum. *Comm. Partial Differential Equations 11*, 15 (1986), 1677–1726.

[66] Carr, P., and Faguet, D. Valuing finite lived options as perpetual. *Morgan Stanley working paper* (1996).

[67] Carr, P., Geman, H., Madan, D. B., and Yor, M. The fine structure of asset returns: an empirical investigations. *Journal of Business 75* (2002), 305–332.

[68] Carr, P., Geman, H., Madan, D. B., and Yor, M. Stochastic volatility for Lévy processes. *Math. Finance 13*, 3 (2003), 345–382.

[69] Carr, P., and Madan, D. Option valuation using the fast Fourier transform. *J. Comput. Finance 2(4)* (1999), 61–73.

[70] Cerny, A. Introduction to Fast Fourier Transform in Finance. *SSRN eLibrary* (2006).

[71] Chan, T. Pricing contingent claims on stocks driven by Lévy processes. *Ann. Appl. Probab. 9*, 2 (1999), 504–528.

[72] Cherubini, U., Della Lunga, G., Mulinacci, S., and Rossi, P. *Fourier Transform Methods in Finance*. John Wiley & Sons Inc., 2010.

[73] Chourdakis, K. Option Pricing Using the Fractional FFT. *J. Comput. Finance 8(2)* (2005), 1–18.

[74] Chung, K. L. *A course in probability theory*, third ed. Academic Press Inc., San Diego, CA, 2001.

[75] Cont, R. Model uncertainty and its impact on the pricing of derivative instruments. *Math. Finance 16*, 3 (2006), 519–547.

[76] Cont, R., and Tankov, P. *Financial modelling with jump processes*. Chapman & Hall/CRC Financial Mathematics Series. Chapman & Hall/CRC, Boca Raton, FL, 2004.

[77] Corielli, F., Foschi, P., and Pascucci, A. Parametrix approximation of diffusion transition densities. *To appear in SIAM J. Financial Math.* (2011).

[78] Cox, J., Ross, S., and Rubinstein, M. Option pricing: a simplified approach. *J. Financial Econ. 7*, 229–264 (1979).

[79] Cox, J. C. Notes on option pricing I: constant elasticity of variance diffusion. *Working paper, Stanford University, Stanford CA* (1975).

[80] Cox, J. C., Ingersoll, Jr., J. E., and Ross, S. A. A theory of the term structure of interest rates. *Econometrica 53*, 2 (1985), 385–407.

[81] Crank, J., and Nicolson, P. A practical method for numerical evaluation of solutions of partial differential equations of the heat-conduction type. *Proc. Cambridge Philos. Soc. 43* (1947), 50–67.

[82] Cyganowski, S., Grüne, L., and Kloeden, P. E. Maple for jump-diffusion stochastic differential equations in finance. In S. S. Nielsen (Ed.), *Programming Languages and Systems in Computational Economics and Finance.* Kluwer Academic Publishers. Dordrecht, 2002, pp. 441–460.

[83] Da Prato, G. *Introduction to stochastic analysis and Malliavin calculus*, second ed., vol. 7 of *Appunti. Scuola Normale Superiore di Pisa (Nuova Serie) [Lecture Notes. Scuola Normale Superiore di Pisa (New Series)].* Edizioni della Normale, Pisa, 2008.

[84] Dana, R.-A., and Jeanblanc, M. *Financial markets in continuous time.* Springer Finance. Springer-Verlag, Berlin, 2003. Translated from the 1998 French original by Anna Kennedy.

[85] Delbaen, F. Representing martingale measures when asset prices are continuous and bounded. *Math. Finance 2*, 2 (1992), 107–130.

[86] Delbaen, F., and Schachermayer, W. A general version of the fundamental theorem of asset pricing. *Math. Ann. 300*, 3 (1994), 463–520.

[87] Delbaen, F., and Schachermayer, W. The existence of absolutely continuous local martingale measures. *Ann. Appl. Probab. 5*, 4 (1995), 926–945.

[88] Delbaen, F., and Schachermayer, W. The no-arbitrage property under a change of numéraire. *Stochastics Stochastics Rep. 53*, 3-4 (1995), 213–226.

[89] Delbaen, F., and Schachermayer, W. The fundamental theorem of asset pricing for unbounded stochastic processes. *Math. Ann. 312*, 2 (1998), 215–250.

[90] Delbaen, F., and Schachermayer, W. Non-arbitrage and the fundamental theorem of asset pricing: summary of main results. In *Introduction to mathematical finance (San Diego, CA, 1997)*, vol. 57 of *Proc. Sympos. Appl. Math.* Amer. Math. Soc., Providence, RI, 1999, pp. 49–58.

[91] Delbaen, F., and Schachermayer, W. *The mathematics of arbitrage.* Springer Finance. Springer-Verlag, Berlin, 2006.

[92] Delbaen, F., and Shirakawa, H. A note on option pricing for the constant elasticity of variance model. *Asia-Pacific Financial Markets 9* (2002), 85–99. 10.1023/A:1022269617674.

[93] Dempster, M. A. H., and Hutton, J. P. Pricing American stock options by linear programming. *Math. Finance 9*, 3 (1999), 229–254.

[94] Di Francesco, M., and Pascucci, A. On a class of degenerate parabolic equations of Kolmogorov type. *AMRX Appl. Math. Res. Express 3* (2005), 77–116.

[95] Di Francesco, M., Pascucci, A., and Polidoro, S. The obstacle problem for a class of hypoelliptic ultraparabolic equations. *Proc. R. Soc. Lond. Ser. A Math. Phys. Eng. Sci. 464*, 2089 (2008), 155–176.

[96] Di Nunno, G., Oksendal, B., and Proske, F. *Malliavin calculus for Lévy processes with applications to finance.* Universitext. Springer-Verlag, Berlin, 2009.

[97] DiBenedetto, E. *Partial differential equations.* Birkhäuser Boston Inc., Boston, MA, 1995.

[98] Doeblin, W. Sur l'équation de Kolmogoroff. *C. R. Acad. Sci. Paris 210* (1940), 365–367.

[99] Doeblin, W. Sur l'équation de Kolmogoroff. *C. R. Acad. Sci. Paris Sér. I Math. 331*, Special Issue (2000), 1059–1128. Avec notes de lecture du pli cacheté par Bernard Bru. [With notes on reading the sealed folder by Bernard Bru], Sur l'équation de Kolmogoroff, par W. Doeblin.

[100] Dothan, M. U. *Prices in financial markets.* The Clarendon Press Oxford University Press, New York, 1990.

[101] Dubner, H., and Abate, J. Numerical inversion of Laplace transforms by relating them to the finite Fourier cosine transform. *J. Assoc. Comput. Mach. 15* (1968), 115–123.

[102] Duffie, D. *Dynamic asset pricing theory. 3rd ed.* Princeton, NJ: Princeton University Press. xix, 465 p., 2001.

[103] Duffie, D., Filipović, D., and Schachermayer, W. Affine processes and applications in finance. *Ann. Appl. Probab. 13*, 3 (2003), 984–1053.

[104] Duffie, D., Pan, J., and Singleton, K. Transform analysis and asset pricing for affine jump-diffusions. *Econometrica 68*, 6 (2000), 1343–1376.

[105] Dufresne, D., Garrido, J., and Morales, M. Fourier inversion formulas in option pricing and insurance. *Methodol. Comput. Appl. Probab. 11*, 3 (2009), 359–383.

[106] Dumas, B., Fleming, J., and Whaley, R. E. Implied volatility functions: empirical tests. *J. Finance 53* (1998), 2059–2106.

[107] Dupire, B. Pricing and hedging with smiles. In *Mathematics of derivative securities (Cambridge, 1995)*, vol. 15 of *Publ. Newton Inst.* Cambridge Univ. Press, Cambridge, 1997, pp. 103–111.

[108] Durrett, R. *Brownian motion and martingales in analysis.* Ed. Wadsworth Advance Book & Softer, 1984.

[109] Durrett, R. *Probability: theory and examples*, second ed. Duxbury Press, Belmont, CA, 1996.

[110] Eberlein, E. Application of generalized hyperbolic Lévy motions to finance. In *Lévy processes.* Birkhäuser Boston, Boston, MA, 2001, pp. 319–336.

[111] Eberlein, E., Glau, K., and Papapantoleon, A. Analysis of Fourier transform valuation formulas and applications. Quantitative Finance Papers 0809.3405, arXiv.org, 2010.

[112] Eberlein, E., and Jacod, J. On the range of option prices. *Finance and Stochastics 1* (1997), 131–140.

[113] Eberlein, E., and Keller, U. Hyperbolic distributions in finance. *Bernoulli 1*, 3 (1995), 281–299.

[114] Eberlein, E., Keller, U., and Prause, K. New insights into smile, mispricing and value at risk: the hyperbolic model. *Journal of Business 71* (1998), 371–406.

[115] Eberlein, E., Keller, U., and Prause, K. Pricing and hedging in exponential Lévy models: review of recent results. *To appear in Paris-Princeton Lecture Notes in Mathematical Finance* (2010), 371–406.

[116] Eberlein, E., and Papapantoleon, A. Equivalence of floating and fixed strike Asian and lookback options. *Stochastic Process. Appl. 115*, 1 (2005), 31–40.

[117] Eberlein, E., and Prause, K. The generalized hyperbolic model: financial derivatives and risk measures. In *Mathematical finance—Bachelier Congress, 2000 (Paris)*, Springer Finance. Springer, Berlin, 2002, pp. 245–267.

[118] Einstein, A. On the movement of small particles suspended in a stationary liquid demanded by molecular kinetic theory of heat. *Ann. Phys. 17* (1905).

[119] Ekström, E. Bounds for perpetual American option prices in a jump diffusion model. *J. Appl. Probab. 43*, 3 (2006), 867–873.

[120] Elliott, R. J., and Kopp, P. E. *Mathematics of financial markets*, second ed. Springer Finance. Springer-Verlag, New York, 2005.

[121] Epps, T. W. *Pricing derivative securities.* World Scientific, Singapore, 2000.

[122] Eskin, G. I. *Boundary value problems for elliptic pseudodifferential equations*, vol. 52 of *Translations of Mathematical Monographs.* American Mathematical Society, Providence, R.I., 1981. Translated from the Russian by S. Smith.

[123] Esscher, F. On the probability function in the collective theory of risk. *Scandinavian Actuarial Journal 15* (1932), 175–195.

[124] Evans, L. C. *Partial differential equations.* Graduate studies in Mathematics, Volume 19, American Mathematical Society, 1997.

[125] Fama, E. Mandelbrot and the stable Paretian hypothesis. *Journal of Business 36* (1963), 420–429.

[126] Fama, E. The behaviour of stock market prices. *Journal of Business 38* (1965), 34–105.

[127] Fang, F., and Oosterlee, C. W. A novel pricing method for European options based on Fourier-cosine series expansions. *SIAM J. Sci. Comput. 31*, 2 (2008/09), 826–848.

[128] Fang, F., and Oosterlee, C. W. Pricing early-exercise and discrete barrier options by Fourier-cosine series expansions. *Numer. Math. 114*, 1 (2009), 27–62.

[129] Feller, W. Two singular diffusion problems. *Ann. of Math. (2) 54* (1951), 173–182.

[130] Feng, L., and Linetsky, V. Pricing discretely monitored barrier options and defaultable bonds in Lévy process models: a fast Hilbert transform approach. *Math. Finance 18*, 3 (2008), 337–384.

[131] Filipović, D. *Term-Structure Models.* Springer Finance. Springer-Verlag, Berlin, 2009. A grauduate course.

[132] Fleming, W. H., and Soner, H. M. *Controlled Markov processes and viscosity solutions*, second ed., vol. 25 of *Stochastic Modelling and Applied Probability.* Springer, New York, 2006.

[133] Folland, G. B. *Real analysis*, second ed. Pure and Applied Mathematics (New York). John Wiley & Sons Inc., New York, 1999. Modern techniques and their applications. A Wiley-Interscience Publication.

[134] Föllmer, H., and Schied, A. *Stochastic Finance. An Introduction in Discrete Time*, second ed. Gruyter Studies in Mathematics 27, Walter de Gruyter, Berlin, New York, 2002.

[135] Foschi, P., and Pascucci, A. Path dependent volatility. *Decis. Econ. Finance 31*, 1 (2008), 13–32.

[136] Foschi, P., and Pascucci, A. Calibration of a path-dependent volatility model: empirical tests. *Comput. Statist. Data Anal. 53* (2009), 2219–2235.

[137] Fournié, E., Lasry, J.-M., Lebuchoux, J., and Lions, P.-L. Applications of Malliavin calculus to Monte-Carlo methods in finance. II. *Finance Stoch. 5*, 2 (2001), 201–236.

[138] Fournié, E., Lasry, J.-M., Lebuchoux, J., Lions, P.-L., and Touzi, N. Applications of Malliavin calculus to Monte Carlo methods in finance. *Finance Stoch. 3*, 4 (1999), 391–412.

[139] Friedman, A. *Partial differential equations of parabolic type.* Prentice-Hall Inc., Englewood Cliffs, N.J., 1964.

[140] Friedman, A. Parabolic variational inequalities in one space dimension and smoothness of the free boundary. *J. Functional Analysis 18* (1975), 151–176.

[141] Friedman, A. *Variational principles and free-boundary problems*, second ed. Robert E. Krieger Publishing Co. Inc., Malabar, FL, 1988.

[142] Friedrics, K. O. The identity of weak and strong extensions of differential operators. *Trans. AMS 55* (1944), 132–151.

[143] Frittelli, M., and Lakner, P. Almost sure characterization of martingales. *Stochastics Stochastics Rep. 49*, 3-4 (1994), 181–190.

[144] Friz, P. K. An introduction to Malliavin Calculus. `http://www.statslab.cam.ac.uk/~peter` (2005).

[145] Fujiwara, T., and Miyahara, Y. The minimal entropy martingale measures for geometric Lévy processes. *Finance Stoch. 7*, 4 (2003), 509–531.

[146] Fusai, G., and Meucci, A. Pricing discretely monitored Asian options under Lévy processes. *J. Banking Finance 32* (2008), 2076–2088.

[147] Garroni, M. G., and Menaldi, J.-L. *Green functions for second order parabolic integro-differential problems*, vol. 275 of *Pitman Research Notes in Mathematics Series.* Longman Scientific & Technical, Harlow, 1992.

[148] Garroni, M. G., and Menaldi, J. L. *Second order elliptic integro-differential problems*, vol. 430 of *Chapman & Hall/CRC Research Notes in Mathematics.* Chapman & Hall/CRC, Boca Raton, FL, 2002.

[149] Gatarek, D., and Świech, A. Optimal stopping in Hilbert spaces and pricing of American options. *Math. Methods Oper. Res. 50*, 1 (1999), 135–147.

[150] Gatheral, J. *The volatility surface, a practitioner's guide.* Wiley Finance, 2006.

[151] Gatheral, J., Hsu, E. P., Laurence, P., Ouyang, C., and Wang, T.-H. Asymptotics of implied volatility in local volatility models. *To appear in Math. Finance* (2010).

[152] Geman, H. The importance of the forward neutral probability measure in a stochastic approach to interest rates. *Working paper, ESSEC* (1989).

[153] Geman, H. Pure jump Lévy processes for asset price modelling. *Journal of Banking and Finance 26* (2002), 1297–1316.

[154] Geman, H., El Karoui, N., and Rochet, J.-C. Changes of numéraire, changes of probability measure and option pricing. *J. Appl. Probab. 32*, 2 (1995), 443–458.

[155] Gerber, H. U., and Shiu, E. S. W. Option pricing by Esscher transform. *Transaction of Society of Actuaries 46* (1994), 99–140.

[156] Geske, R. The valuation of compound options. *Journal of Financial Economics 7* (1979), 63–81.

[157] Gilbarg, D., and Trudinger, N. S. *Elliptic partial differential equations of second order.* Classics in Mathematics. Springer-Verlag, Berlin, 2001. Reprint of the 1998 edition.

[158] Glasserman, P. *Monte Carlo methods in financial engineering*, vol. 53 of *Applications of Mathematics (New York).* Springer-Verlag, New York, 2004. Stochastic Modelling and Applied Probability.

[159] Hagan, P., Kumar, D., Lesniewski, A., and Woodward, D. Managing smile risk. *Wilmott September* (2002), 84–108.

[160] Hagan, P., and Woodward, D. Equivalent Black volatilities. *Appl. Math. Finance 6*, 6 (1999), 147–159.

[161] Hall, C. A., and Porsching, T. A. *Numerical analysis of partial differential equations.* Prentice Hall Inc., Englewood Cliffs, NJ, 1990.

[162] Han, H., and Wu, X. A fast numerical method for the Black-Scholes equation of American options. *SIAM J. Numer. Anal. 41*, 6 (2003), 2081–2095 (electronic).

[163] Harrison, J. M., and Kreps, D. M. Martingales and arbitrage in multiperiod securities markets. *J. Econom. Theory 20*, 3 (1979), 381–408.

[164] Harrison, J. M., and Pliska, S. R. Martingales and stochastic integrals in the theory of continuous trading. *Stochastic Process. Appl. 11*, 3 (1981), 215–260.

[165] Heston, S. L. A closed-form solution for options with stochastic volatility with applications to bond and currency options. *Rev. Finan. Stud. 6* (1993), 327–343.

[166] Heynen, R. C., and Kat, H. M. Lookback options with discrete and partial monitoring of the underlying price. *Appl. Math. Finance 2* (1995), 273–284.

[167] Heynen, R. C., and Kat, H. M. Discrete partial barrier options with a moving barrier. *Journal of Financial Engineering 5*, 3 (1996), 199–210.

[168] Ho, T., and Lee, S. Term structure movements and pricing interest rate contingent claims. *J. Finance 41* (1986), 1011–1029.

[169] Hobson, D. G., and Rogers, L. C. G. Complete models with stochastic volatility. *Math. Finance 8*, 1 (1998), 27–48.

[170] Hörmander, L. Hypoelliptic second order differential equations. *Acta Math. 119* (1967), 147–171.

[171] Huang, C.-F., and Litzenberger, R. H. *Foundations for financial economics.* North-Holland Publishing Co., New York, 1988.

[172] Hubalek, F. On Fourier methods for simple, multi-asset, and path-dependent options - accuracy and efficiency. *Working Paper* (2007).

[173] Hubalek, F., Kallsen, J., and Krawczyk, L. Variance-optimal hedging for processes with stationary independent increments. *Ann. Appl. Probab. 16*, 2 (2006), 853–885.

[174] Hubalek, F., and Sgarra, C. Esscher transforms and the minimal entropy martingale measure for exponential Lévy models. *Quant. Finance 6*, 2 (2006), 125–145.

[175] Hull, J., and White, A. The pricing of options on assets with stochastic volatilities. *Journal of Finance 42*, 2 (1987), 281–300.

[176] Hull, J., and White, A. Pricing interest rates derivatives securities. *The review of financial studies 3* (1990), 573–592.

[177] Ikeda, N., and Watanabe, S. *Stochastic differential equations and diffusion processes*, second ed., vol. 24 of *North-Holland Mathematical Library.* North-Holland Publishing Co., Amsterdam, 1989.

[178] Ingersoll, J. E. *Theory of Financial Decision Making.* Blackwell, Oxford, 1987.

[179] Itô, K. On stochastic processes. I. (Infinitely divisible laws of probability). *Jap. J. Math. 18* (1942), 261–301.

[180] Itô, K. Stochastic integral. *Proc. Imp. Acad. Tokyo 20* (1944), 519–524.

[181] Ivanov, R. V. On the pricing of American options in exponential Lévy markets. *J. Appl. Probab. 44*, 2 (2007), 409–419.

[182] Jacka, S. Optimal stopping and the American put. *Math. Finance 1*, 2 (1991), 1–14.

[183] Jacob, N. *Pseudo-differential operators and Markov processes*, vol. 94 of *Mathematical Research.* Akademie Verlag, Berlin, 1996.

[184] Jacod, J., and Shiryaev, A. N. *Limit theorems for stochastic processes*, second ed., vol. 288 of *Grundlehren der Mathematischen Wissenschaften [Fundamental Principles of Mathematical Sciences].* Springer-Verlag, Berlin, 2003.

[185] Jaillet, P., Lamberton, D., and Lapeyre, B. Variational inequalities and the pricing of American options. *Acta Appl. Math. 21*, 3 (1990), 263–289.

[186] Jamshidian, F. An exact bond option formula. *J. Finance 44* (1989), 205–209.

[187] Jamshidian, F. Libor and swap market models and measures. *Finance and Stochastics 4* (1997), 293–330.

[188] Jiang, L., and Dai, M. Convergence of binomial tree methods for European/American path-dependent options. *SIAM J. Numer. Anal. 42*, 3 (2004), 1094–1109 (electronic).

[189] Johnson, H., and Shanno, D. Option pricing when the variance is changing. *Journal of Financial and Quantitative Analysis 22*, 02 (June 1987), 143–151.

[190] Jourdain, B., and Martini, C. American prices embedded in European prices. *Ann. Inst. H. Poincaré Anal. Non Linéaire 18*, 1 (2001), 1–17.

[191] Jourdain, B., and Martini, C. Approximation of American put prices by European prices via an embedding method. *Ann. Appl. Probab. 12*, 1 (2002), 196–223.

[192] Kahl, C., and Jackel, P. Not-so-complex logarithms in the Heston model. *Wilmott Magazine September* (2005), 14–103.

[193] Kahl, C., and Jackel, P. Hyp-hyp hooray. *ABN AMRO Financial markets* (2007).

[194] Kallenberg, O. *Random Measures*, fourth ed. Akademie-Verlag and Academic Press, 1976.

[195] Kallenberg, O. *Foundations of modern probability.* Springer-Verlag, New York, 1997.

[196] Kallsen, J., and Shiryaev, A. N. The cumulant process and Esscher's change of measure. *Finance Stoch. 6*, 4 (2002), 397–428.

[197] Kangro, R., and Nicolaides, R. Far field boundary conditions for Black-Scholes equations. *SIAM J. Numer. Anal. 38*, 4 (2000), 1357–1368 (electronic).

[198] Karatzas, I. On the pricing of American options. *Appl. Math. Optim. 17*, 1 (1988), 37–60.

[199] Karatzas, I. Optimization problems in the theory of continuous trading. *SIAM J. Control Optim. 27*, 6 (1989), 1221–1259.

[200] Karatzas, I. *Lectures on the mathematics of finance*, vol. 8 of *CRM Monograph Series.* American Mathematical Society, Providence, RI, 1997.

[201] Karatzas, I., and Shreve, S. E. *Brownian motion and stochastic calculus*, second ed., vol. 113 of *Graduate Texts in Mathematics.* Springer-Verlag, New York, 1991.

[202] Karatzas, I., and Shreve, S. E. *Methods of mathematical finance*, vol. 39 of *Applications of Mathematics (New York).* Springer-Verlag, New York, 1998.

[203] Kato, M. On positive solutions of the heat equation. *Nagoya Math. J. 30* (1967), 203–207.

[204] Keller-Ressel, M. Affine processes: theory and applications to finance. *Ph. D. thesis, TU Vienna* (2008).

[205] Kevorkian, J., and Cole, J. D. *Perturbation methods in applied mathematics*, vol. 34 of *Applied Mathematical Sciences*. Springer-Verlag, New York, 1981.

[206] Kevorkian, J., and Cole, J. D. *Multiple scale and singular perturbation methods*, vol. 114 of *Applied Mathematical Sciences*. Springer-Verlag, New York, 1996.

[207] Kilin, F. Accelerating calibration of the stochastic volatility models. *Available at SSRN: http://ssrn.com/abstract=965248 Frankfurt School of Finance & Management, (CPQF Working Paper Series No. 6)* (2007), 1–19.

[208] Kim, I. J. The analytic valuation of American options. *Rev. Financial Studies 3* (1990), 547–572.

[209] Kinderlehrer, D., and Stampacchia, G. *An introduction to variational inequalities and their applications*, vol. 31 of *Classics in Applied Mathematics*. Society for Industrial and Applied Mathematics (SIAM), Philadelphia, PA, 2000. Reprint of the 1980 original.

[210] Kloeden, P. E., and Platen, E. *Numerical solution of stochastic differential equations*, vol. 23 of *Applications of Mathematics (New York)*. Springer-Verlag, Berlin, 1992.

[211] Kohatsu-Higa, A., and Montero, M. Malliavin calculus in finance. In *Handbook of computational and numerical methods in finance*. Birkhäuser Boston, Boston, MA, 2004, pp. 111–174.

[212] Kohatsu-Higa, A., and Pettersson, R. Variance reduction methods for simulation of densities on Wiener space. *SIAM J. Numer. Anal. 40*, 2 (2002), 431–450 (electronic).

[213] Kolmogorov, A. N. *Selected works. Vol. II*, vol. 26 of *Mathematics and its Applications (Soviet Series)*. Kluwer Academic Publishers Group, Dordrecht, 1992. Probability theory and mathematical statistics, With a preface by P. S. Aleksandrov, Translated from the Russian by G. Lindquist, Translation edited by A. N. Shiryayev [A. N. Shiryaev].

[214] Kolmogorov, A. N. *Selected works of A. N. Kolmogorov. Vol. III*. Kluwer Academic Publishers Group, Dordrecht, 1993. Edited by A. N. Shiryayev.

[215] Kolmogorov, A. N., and Fomin, S. V. *Elementy teorii funktsii i funktsionalnogo analiza*. Izdat. "Nauka", Moscow, 1976. Fourth edition, revised.

[216] Koponen, I. Analytic approach to the problem of convergence of truncated Lévy flights towards the Gaussian stochastic process. *Physical Review 52* (1995), 1197–1199.

[217] Kou, S. A jump-diffusion model for option pricing. *Management Science 48* (2002), 1086–1101.

[218] Kou, S. Discrete barrier and lookback options. In J. R. Birge and V. Linetsky (Eds.), *Handbooks in OR and MS, Elsevier 15* (2008), 343–373.

[219] Kunita, H., and Watanabe, S. On square integrable martingales. *Nagoya Math. J. 30* (1967), 209–245.

[220] Kushner, H. J. *Probability methods for approximations in stochastic control and for elliptic equations.* Academic Press [Harcourt Brace Jovanovich Publishers], New York, 1977. Mathematics in Science and Engineering, vol. 129.

[221] Kyprianou A., Schoutens, W., and Wilmott, P. E. *Exotic Option Pricing and Advanced Lévy Models.* Wiley, Chichester, 2005.

[222] Ladyzhenskaya, O., and Ural'tseva, N. *Linear and quasi-linear equations of parabolic type.* Transl. Math. Monographs 23, American Mathematical Society, Providence, RI, 1968.

[223] Lakner, P. Martingale measures for a class of right-continuous processes. *Math. Finance 3*, 1 (1993), 43–53.

[224] Lamberton, D. Critical price for an American option near maturity. In *Seminar on Stochastic Analysis, Random Fields and Applications (Ascona, 1993)*, vol. 36 of *Progr. Probab.* Birkhäuser, Basel, 1995, pp. 353–358.

[225] Lamberton, D. American options. In David J. Hand and Saul D. Jacka (Eds.), *Statistics in Finance*, Arnold Applications of Statistics Series. London: Arnold. x, 340 p., 1998.

[226] Lamberton, D., and Lapeyre, B. *Introduction to stochastic calculus applied to finance*, second ed. Chapman & Hall/CRC Financial Mathematics Series. Chapman & Hall/CRC, Boca Raton, FL, 2008.

[227] Lamberton, D., and Mikou, M. The critical price for the American put in an exponential Lévy model. *Finance Stoch. 12*, 4 (2008), 561–581.

[228] Lamberton, D., and Pagès, G. Sur l'approximation des réduites. *Ann. Inst. H. Poincaré Probab. Statist. 26*, 2 (1990), 331–355.

[229] Lamberton, D., and Villeneuve, S. Critical price near maturity for an American option on a dividend-paying stock. *Ann. Appl. Probab. 13*, 2 (2003), 800–815.

[230] Lanconelli, E., and Polidoro, S. On a class of hypoelliptic evolution operators. *Rend. Sem. Mat. Univ. Politec. Torino 52*, 1 (1994), 29–63.

[231] Langevin, P. Sur la théorie du mouvement Brownien. *C.R. Acad. Sci. Paris 146* (1908), 530–532.

[232] Laurence, P., and Salsa, S. Regularity of the free boundary of an American option on several assets. *Comm. Pure Appl. Math. 62*, 7 (2009), 969–994.

[233] Lee, E. B., and Markus, L. *Foundations of optimal control theory*, second ed. Robert E. Krieger Publishing Co. Inc., Melbourne, FL, 1986.

[234] Lee, R. Option pricing by transform methods: extensions, unification and error control. *J. Comput. Finance 7(3)* (2004), 51–86.

[235] Levendorskii, S. Z. Pricing of the American put under Lévy processes. *Int. J. Theor. Appl. Finance 7*, 3 (2004), 303–335.

[236] Levi, E. E. Sulle equazioni lineari totalmente ellittiche alle derivate parziali. *Rend. Circ. Mat. Palermo 24* (1907), 275–317.

[237] Lewis, A. A simple option formula for general jump-diffusion and other exponential Lévy processes. Tech. rep., Finance Press, Aug. 2001.

[238] Lieberman, G. *Second order parabolic differential equations.* World Scientific, Singapore, 1996.

[239] Lipton, A. *Mathematical methods for foreign exchange.* World Scientific Publishing Co. Inc., River Edge, NJ, 2001. A financial engineer's approach.

[240] Lord, R., and Kahl, C. Optimal Fourier Inversion in Semi-Analytical Option Pricing. *J. Comput. Finance 10(4)* (2007), 1–30.

[241] Lord, R., and Kahl, C. Complex logarithms in Heston-like models. *SSRN eLibrary* (2008).

[242] MacMillan, L. Analytic approximation for the American put price. *Adv. in Futures and Options Research 1* (1986), 119–139.

[243] Madan, D., and Seneta, E. The variance gamma (VG) model for share market returns. *Journal of Business 63* (1990), 511–524.

[244] Malliavin, P. Stochastic calculus of variation and hypoelliptic operators. Proc. int. Symp. on stochastic differential equations, Kyoto 1976, 195-263 (1978).

[245] Mandelbrot, B. The variation of certain speculative prices. *Journal of Business, 36* (1963), 392–417.

[246] Mantegna, R. N., and Stanley, H. E. Stochastic process with ultraslow convergence to a Gaussian: the truncated Lévy flight. *Phys. Rev. Lett. 73*, 22 (1994), 2946–2949.

[247] Margrabe, W. The value of an option to exchange one asset for another. *J. Finance 33* (1978), 177–186.

[248] Matache, A.-M., Nitsche, P.-A., and Schwab, C. Wavelet Galerkin pricing of American options on Lévy driven assets. *Quant. Finance 5*, 4 (2005), 403–424.

[249] Matsuda, K. Introduction to option pricing with Fourier transform: Option pricing with exponential Lévy models. *Ph.D. thesis, Graduate School and University Center of the City University of New York* (2006).

[250] Merton, R. C. Theory of rational option pricing. *Bell J. Econom. and Management Sci. 4* (1973), 141–183.

[251] Merton, R. C. Option pricing when underlying stock returns are discontinuous. *J. Financ. Econ. 3*, 1-2 (1976), 125–144.

[252] Merton, R. C. *Continuous-time finance.* Cambridge, MA: Blackwell. xix, 732 p., 1999.

[253] Meyer, P. A. *Probability and Potentials.* Blaisdell Publishing Company, Waltham, Mass, 1966.

[254] Meyer, P. A. Un cours sur les intégrales stochastiques. In *Séminaire de Probabilités, X (Seconde partie: Théorie des intégrales stochastiques,*

Univ. Strasbourg, Strasbourg, année universitaire 1974/1975). Springer, Berlin, 1976, pp. 245–400. Lecture Notes in Math., vol. 511.

[255] Mikulevičius, R., and Pragarauskas, H. On the Cauchy problem for certain integro-differential operators in Sobolev and Hölder spaces. *Liet. Mat. Rink. 32*, 2 (1992), 299–331.

[256] Mikulevičius, R., and Pragarauskas, H. On classical solutions of certain nonlinear integro-differential equations. In *Stochastic processes and optimal control (Friedrichroda, 1992)*, vol. 7 of *Stochastics Monogr.* Gordon and Breach, Montreux, 1993, pp. 151–163.

[257] Miltersen, K., Sandmann, K., and Sondermann, D. Closed form solutions for term structure derivatives with log-normal interest rates. *J. Finance 52*, 1 (1997), 409–430.

[258] Mitchell, A. R., and Griffiths, D. F. *The finite difference method in partial differential equations.* John Wiley & Sons Ltd., Chichester, 1980. A Wiley-Interscience Publication.

[259] Mörters, P., and Peres, Y. *Brownian motion*, vol. 30 of *Cambridge Series in Statistical and Probabilistic Mathematics.* Cambridge University Press, Cambridge, 2010.

[260] Mulinacci, S. An approximation of American option prices in a jump-diffusion model. *Stochastic Process. Appl. 62*, 1 (1996), 1–17.

[261] Musiela, M., and Rutkowski, M. *Martingale methods in financial modelling*, second ed., vol. 36 of *Stochastic Modelling and Applied Probability.* Springer-Verlag, Berlin, 2005.

[262] Myneni, R. The pricing of the American option. *Ann. Appl. Probab. 2*, 1 (1992), 1–23.

[263] Nash, J. C. *Compact numerical methods for computers*, second ed. Adam Hilger Ltd., Bristol, 1990. Linear algebra and function minimisation.

[264] Neftci, S. N. *Introduction to the mathematics of financial derivatives. 2nd ed.* Orlando, FL: Academic Press. xxvii, 527 p., 2000.

[265] Nguyen-Ncoc, L., and Yor, M. Lookback and barrier options under general Lévy processes. *in Y. Ait-Sahalia and L. P. Hansen (Eds), Handbook of Financial Econometrics, North-Holland, Amsterdam* (2005).

[266] Novikov, A. A. A certain identity for stochastic integrals. *Teor. Verojatnost. i Primenen. 17* (1972), 761–765.

[267] Nualart, D. *The Malliavin calculus and related topics*, second ed. Probability and its Applications (New York). Springer-Verlag, Berlin, 2006.

[268] Nualart, D., and Schoutens, W. Chaotic and predictable representations for Lévy processes. *Stochastic Process. Appl. 90*, 1 (2000), 109–122.

[269] Nualart, D., and Schoutens, W. Backward stochastic differential equations and Feynman-Kac formula for Lévy processes, with applications in finance. *Bernoulli 7*, 5 (2001), 761–776.

[270] Ocone, D. Malliavin's calculus and stochastic integral representations of functionals of diffusion processes. *Stochastics 12* (1984), 161–185.

[271] Officer, R. R. The distribution of stock returns. *Journal of the American Statistical Association 67* (1972), 807–812.

[272] Oksendal, B. An introduction to Malliavin Calculus with applications to Economics. `http://www.nhh.no/for/dp/1996/wp0396.pdf` (1997).

[273] Oksendal, B., and Reikvam, K. Viscosity solutions of optimal stopping problems. *Stochastics Stochastics Rep. 62*, 3-4 (1998), 285–301.

[274] Oleĭnik, O. A., and Radkevič, E. V. *Second order equations with nonnegative characteristic form.* Plenum Press, New York, 1973.

[275] Paley, R. E. A. C., Wiener, N., and Zygmund, A. Notes on random functions. *Math. Z. 37*, 1 (1933), 647–668.

[276] Pascucci, A. Free boundary and optimal stopping problems for American Asian options. *Finance Stoch. 12*, 1 (2008), 21–41.

[277] Pascucci, A., and Runggaldier, W. J. *Financial mathematics. Theory and problems for multi-periodic models (Finanza matematica. Teoria e problemi per modelli multiperiodali).* Unitext 44. La Matematica per il 3 + 2. Milano: Springer. ix, 264 p., 2009.

[278] Peskir, G., and Shiryaev, A. *Optimal stopping and free-boundary problems.* Lectures in Mathematics ETH Zürich. Birkhäuser Verlag, Basel, 2006.

[279] Pham, H. Optimal stopping, free boundary, and American option in a jump-diffusion model. *Appl. Math. Optim. 35*, 2 (1997), 145–164.

[280] Pham, H. Optimal stopping of controlled jump diffusion processes: a viscosity solution approach. *J. Math. Systems Estim. Control 8*, 1 (1998), 27 pp. (electronic).

[281] Platen, E., and Bruti-Liberati, N. *Numerical solution of stochastic differential equations with jumps in finance.* Stochastic Modelling and Applied Probability 64. Berlin: Springer. xxvi, 856 p., 2010.

[282] Pliska, S. *Introduction to Mathematical Finance: Discrete Time Models.* Blackwell Publisher, Oxford, 1997.

[283] Polidoro, S. On a class of ultraparabolic operators of Kolmogorov-Fokker-Planck type. *Matematiche (Catania) 49*, 1 (1994), 53–105.

[284] Polidoro, S. Uniqueness and representation theorems for solutions of Kolmogorov-Fokker-Planck equations. *Rend. Mat. Appl. (7) 15*, 4 (1995), 535–560.

[285] Polidoro, S. A global lower bound for the fundamental solution of Kolmogorov-Fokker-Planck equations. *Arch. Rational Mech. Anal. 137*, 4 (1997), 321–340.

[286] Press, W. H., Teukolsky, S. A., Vetterling, W. T., and Flannery, B. P. *Numerical recipes in C++.* Cambridge University Press, Cambridge, 2002. The art of scientific computing, Second edition, updated for C++.

[287] Protter, P. E. *Stochastic integration and differential equations*, second ed., vol. 21 of *Applications of Mathematics (New York).* Springer-Verlag, Berlin, 2004. Stochastic Modelling and Applied Probability.

[288] Raible, S. Lévy processes in finance: Theory, numerics, and empirical facts. Tech. rep., PhD thesis, Universität Freiburg, 2000.

[289] Raviart, P.-A., and Thomas, J.-M. *Introduction à l'analyse numérique des équations aux dérivées partielles.* Collection Mathématiques Ap-

pliquées pour la Maîtrise. [Collection of Applied Mathematics for the Master's Degree]. Masson, Paris, 1983.

[290] Renault, E., and Touzi, N. Option hedging and implied volatilities in a stochastic volatility model. *Mathematical Finance 6*, 3 (1996), 279–302.

[291] Rogers, L., and Shi, Z. The value of an Asian option. *J. Appl. Probab. 32*, 4 (1995), 1077–1088.

[292] Rogers, L. C. G., and Veraart, L. A. M. A stochastic volatility alternative to SABR. *J. Appl. Probab. 45*, 4 (2008), 1071–1085.

[293] Rudin, W. *Real and complex analysis*, second ed. McGraw-Hill Book Co., New York, 1974. McGraw-Hill Series in Higher Mathematics.

[294] Rudin, W. *Functional analysis*, second ed. International Series in Pure and Applied Mathematics. McGraw-Hill Inc., New York, 1991.

[295] Samuelson, P. Rational theory of warrant prices. *Indust. Manag. Rev. 6* (1965), 13–31.

[296] Sanz-Solé, M. *Malliavin calculus.* Fundamental Sciences. EPFL Press, Lausanne, 2005. With applications to stochastic partial differential equations.

[297] Sato, K.-i. *Lévy processes and infinitely divisible distributions*, vol. 68 of *Cambridge Studies in Advanced Mathematics.* Cambridge University Press, Cambridge, 1999. Translated from the 1990 Japanese original, Revised by the author.

[298] Savine, A. A theory of volatility. In *Recent developments in mathematical finance (Shanghai, 2001).* World Sci. Publishing, River Edge, NJ, 2002, pp. 151–167.

[299] Schobel, R., and Zhu, J. Stochastic volatility with an Ornstein-Uhlenbeck process: An extension. *European Finance Review 4* (1999), 23–46.

[300] Schoutens, W. The Meixner process in finance. *EURANDOM Report 2001-002* (2001).

[301] Schoutens, W. *Levy Processes in Finance: Pricing Financial Derivatives.* John Wiley & Sons Inc., 2003.

[302] Schoutens, W., Simons, E., and Tistaert, J. A perfect calibration! now what? *Wilmott Magazine March* (2004).

[303] Schroder, M. Computing the constant elasticity of variance option pricing formula. *The Journal of Finance 44*, 1 (1989), pp. 211–219.

[304] Schweizer, M. Martingale densities for general asset prices. *J. Math. Econom. 21*, 4 (1992), 363–378.

[305] Sepp, A. Fourier transform for option pricing under affine jump-diffusions: An overview. *SSRN eLibrary, Paper n.1412333* (2003).

[306] Shahgholian, H. Free boundary regularity close to initial state for parabolic obstacle problem. *Trans. Amer. Math. Soc. 360*, 4 (2008), 2077–2087.

[307] Shaw, W. T. *Modelling financial derivatives with Mathematica.* Cambridge University Press, Cambridge, 1998. Mathematical models and

benchmark algorithms, With 1 CD-ROM (Windows, Macintosh and UNIX).

[308] Shigekawa, I. *Stochastic analysis*, vol. 224 of *Translations of Mathematical Monographs.* American Mathematical Society, Providence, RI, 2004. Translated from the 1998 Japanese original by the author, Iwanami Series in Modern Mathematics.

[309] Shiryaev, A. N. *Probability*, second ed., vol. 95 of *Graduate Texts in Mathematics.* Springer-Verlag, New York, 1996. Translated from the first (1980) Russian edition by R. P. Boas.

[310] Shreve, S. E. *Stochastic calculus for finance. I.* Springer Finance. Springer-Verlag, New York, 2004. The binomial asset pricing model.

[311] Shreve, S. E. *Stochastic calculus for finance. II.* Springer Finance. Springer-Verlag, New York, 2004. Continuous-time models.

[312] Sin, C. A. Complications with stochastic volatility models. *Adv. in Appl. Probab. 30*, 1 (1998), 256–268.

[313] Skorohod, A. V. On a generalization of the stochastic integral. *Teor. Verojatnost. i Primenen. 20*, 2 (1975), 223–238.

[314] Smith, G. D. *Numerical solution of partial differential equations*, third ed. Oxford Applied Mathematics and Computing Science Series. The Clarendon Press Oxford University Press, New York, 1985. Finite difference methods.

[315] Steele, J. M. *Stochastic calculus and financial applications*, vol. 45 of *Applications of Mathematics (New York).* Springer-Verlag, New York, 2001.

[316] Stein, E. M., and Stein, J. C. Stock price distributions with stochastic volatility: An analytic approach. *Review of Financial Studies 4*, 4 (1991), 727–52.

[317] Stricker, C. Arbitrage et lois de martingale. *Ann. Inst. H. Poincaré Probab. Statist. 26*, 3 (1990), 451–460.

[318] Stroock, D. W. *Partial differential equations for probabilists*, vol. 112 of *Cambridge Studies in Advanced Mathematics.* Cambridge University Press, Cambridge, 2008.

[319] Stroock, D. W., and Varadhan, S. R. S. Diffusion processes with continuous coefficients. I. *Comm. Pure Appl. Math. 22* (1969), 345–400.

[320] Stroock, D. W., and Varadhan, S. R. S. Diffusion processes with continuous coefficients. II. *Comm. Pure Appl. Math. 22* (1969), 479–530.

[321] Stroock, D. W., and Varadhan, S. R. S. *Multidimensional diffusion processes*, vol. 233 of *Grundlehren der Mathematischen Wissenschaften [Fundamental Principles of Mathematical Sciences].* Springer-Verlag, Berlin, 1979.

[322] Struwe, M. *Variational methods*, fourth ed., vol. 34 of *Ergebnisse der Mathematik und ihrer Grenzgebiete. 3. Folge. A Series of Modern Surveys in Mathematics [Results in Mathematics and Related Areas. 3rd Series. A Series of Modern Surveys in Mathematics].* Springer-Verlag,

Berlin, 2008. Applications to nonlinear partial differential equations and Hamiltonian systems.

[323] Surkov, V. Option pricing using Fourier space time-stepping framework. *Ph.D. thesis, Graduate Department of Computer Science, University of Toronto* (2009).

[324] Tanaka, H. Note on continuous additive functionals of the 1-dimensional Brownian path. *Z. Wahrscheinlichkeitstheorie und Verw. Gebiete 1* (1962/1963), 251–257.

[325] Tavella, D., and Randall, C. *Pricing Financial Instruments: The Finite Difference Method.* John Wiley & Sons, 2000.

[326] Taylor, M. The Dirichlet-Jordan test and multidimensional extensions. *Proc. Amer. Math. Soc. 129*, 4 (2001), 1031–1035.

[327] Tychonov, A. Théorèmes d'unicité pour l'equation de la chaleur. *Math. Sbornik 42* (1935), 199–216.

[328] Uhlenbeck, G. E., and Ornstein, L. S. On the theory of the Brownian motion. *Physical Review 36* (1930), 823–841.

[329] van der Hoek, J., and Elliott, R. J. *Binomial models in finance.* Springer Finance. Springer-Verlag, New York, 2006.

[330] van Moerbeke, P. On optimal stopping and free boundary problems. *Arch. Rational Mech. Anal. 60*, 2 (1975/76), 101–148.

[331] Varadhan, S. PDE in finance. `http://math.nyu.edu/faculty/varadhan/` (2004).

[332] Vasiček, O. An equilibrium characterization of the term structure. *J. Financial Economics 5* (1977), 177–188.

[333] Veretennikov, A. J. Strong solutions and explicit formulas for solutions of stochastic integral equations. *Mat. Sb. (N.S.) 111(153)*, 3 (1980), 434–452, 480.

[334] Villeneuve, S. Exercise regions of American options on several assets. *Finance Stoch. 3*, 3 (1999), 295–322.

[335] Villeneuve, S., and Zanette, A. Parabolic ADI methods for pricing American options on two stocks. *Math. Oper. Res. 27*, 1 (2002), 121–149.

[336] Wang, A. T. Generalized Ito's formula and additive functionals of Brownian motion. *Z. Wahrscheinlichkeitstheorie und Verw. Gebiete 41*, 2 (1977/78), 153–159.

[337] Wang, R., Wan, J., and Forsyth, P. Robust numerical valuation of European and American options under the CGMY process. *J. Comput. Finance 10* (2007), 31–70.

[338] Widder, D. V. Positive temperatures on a semi-infinite rod. *Trans. Amer. Math. Soc. 75* (1953), 510–525.

[339] Williams, D. *Probability with martingales.* Cambridge Mathematical Textbooks. Cambridge University Press, Cambridge, 1991.

[340] Wilmott, P., Howison, S., and Dewynne, J. *The mathematics of financial derivatives.* Cambridge University Press, Cambridge, 1995. A student introduction.

[341] Zabczyk, J. *Mathematical control theory: an introduction.* Systems & Control: Foundations & Applications. Birkhäuser Boston Inc., Boston, MA, 1992.

[342] Zhang, B., and Oosterlee, C. W. Option pricing with COS method on graphics processing units. *Parallel and Distributed Processing Symposium, International 0* (2009), 1–8.

[343] Zhang, H. The Malliavin Calculus. *http://web.maths.unsw.edu.au/~mathsoc/han_zhang_thesis.pdf* (2004).

[344] Zhang, P. G. *Exotic Options*, second ed. World Scientific, 2001.

[345] Zhang, X. L. Options Américaines et modèles de diffusion avec sauts. *C. R. Acad. Sci. Paris Sér. I Math. 317*, 9 (1993), 857–862.

[346] Zhang, X. L. Numerical analysis of American option pricing in a jump-diffusion model. *Math. Oper. Res. 22*, 3 (1997), 668–690.

[347] Zhu, J. *Modular pricing of options*, vol. 493 of *Lecture Notes in Economics and Mathematical Systems.* Springer-Verlag, Berlin, 2000. An application of Fourier analysis, With a foreword by Rainer Schöbel.

[348] Zhu, J. *Applications of Fourier Transform to Smile Modeling. Theory and Implementation.* Springer Finance. Springer-Verlag, Berlin, 2010.

[349] Zhu, Y.-L., Wu, X., and Chern, I.-L. *Derivative securities and difference methods.* Springer Finance. Springer-Verlag, New York, 2004.

[350] Zvonkin, A. K. A transformation of the phase space of a diffusion process that will remove the drift. *Mat. Sb. (N.S.) 93(135)* (1974), 129–149, 152.

Index